Applied Mathematical Sciences

Volume 34

Applied Mathematical Sciences

1. *John:* Partial Differential Equations, 4th ed.
2. *Sirvoich:* Techniques of Asymptotic Analysis.
3. *Hale:* Theory of Functional Differential Equations, 2nd ed.
4. *Percus:* Combinatorial Methods.
5. *von Mises/Friedrichs:* Fluid Dynamics.
6. *Freiberger/Grenander:* A Short Course in Computational Probability and Statistics.
7. *Pipkin:* Lectures on Viscoelasticity Theory.
8. *Giacoglia:* Perturbation Methods in Non-linear Systems.
9. *Friedrichs:* Spectral Theory of Operators in Hilbert Space.
10. *Stroud:* Numerical Quadrature and Solution of Ordinary Differential Equations.
11. *Wolovich:* Linear Multivariable Systems.
12. *Berkovitz:* Optimal Control Theory.
13. *Bluman/Cole:* Similarity Methods for Differential Equations.
14. *Yoshizawa:* Stability Theory and the Existence of Periodic Solution and Almost Periodic Solutions.
15. *Braun:* Differential Equations and Their Applications, 3rd ed.
16. *Lefschetz:* Applications of Algebraic Topology.
17. *Collatz/Wetterling:* Optimization Problems.
18. *Grenander:* Pattern Synthesis: Lectures in Pattern Theory, Vol. I.
19. *Marsden/McCracken:* Hopf Bifurcation and Its Applications.
20. *Driver:* Ordinary and Delay Differential Equations.
21. *Courant/Friedrichs:* Supersonic Flow and Shock Waves.
22. *Rouche/Habets/Laloy:* Stability Theory by Liapunov's Direct Method.
23. *Lamperti:* Stochastic Processes: A Survey of the Mathematical Theory.
24. *Grenander:* Pattern Analysis: Lectures in Pattern Theory, Vol. II.
25. *Davies:* Integral Transforms and Their Applications, 2nd ed.
26. *Kushner/Clark:* Stochastic Approximation Methods for Constrained and Unconstrained Systems.
27. *de Boor:* A Practical Guide to Splines.
28. *Keilson:* Markov Chain Models—Rarity and Exponentiality.
29. *de Veubeke:* A Course in Elasticity.
30. *Shiatycki:* Geometric Quantization and Quantum Mechanics.
31. *Reid:* Sturmian Theory for Ordinary Differential Equations.
32. *Meis/Markowitz:* Numerical Solution of Partial Differential Equations.
33. *Grenander:* Regular Structures: Lectures in Pattern Theory, Vol. III.
34. *Kevorkian/Cole:* Perturbation Methods in Applied Mathematics.
35. *Carr:* Applications of Centre Manifold Theory.
36. *Bengtsson/Ghil/Källén:* Dynamic Meteorology: Data Assimilation Methods.
37. *Saperstone:* Semidynamical Systems in Infinite Dimensional Spaces.
38. *Lichtenberg/Lieberman:* Regular and Stochastic Motion.
39. *Piccini/Stampacchia/Vidossich:* Ordinary Differential Equations in $\mathbf{R}^n$.
40. *Naylor/Sell:* Linear Operator Theory in Engineering and Science.
41. *Sparrow:* The Lorenz Equations: Bifurcations, Chaos, and Strange Attractors.
42. *Guckenheimer/Holmes:* Nonlinear Oscillations, Dynamical Systems and Bifurcations of Vector Fields.
43. *Ockendon/Taylor:* Inviscid Fluid Flows.
44. *Pazy:* Semigroups of Linear Operations and Applications to Partial Differential Equations.
45. *Glashoff/Gustafson:* Linear Operations and Approximation: An Introduction to the Theoretical Analysis and Numerical Treatment of Semi-Infinite Programs.
46. *Wilcox:* Scattering Theory for Diffraction Gratings.
47. *Hale et al:* An Introduction to Infinite Dimensional Dynamical Systems—Geometric Theory.
48. *Murray:* Asymptotic Analysis.
49. *Ladyzhenskaya:* The Boundary—Value Problems of Mathematical Physics.
50. *Wilcox:* Sound Propagation in Stratified Fluids.
51. *Golubitsky/Schaeffer:* Bifurcation and Groups in Bifurcation Theory, Vol. I.

(continued following index)

J. Kevorkian
J. D. Cole

Perturbation Methods in Applied Mathematics

Springer-Verlag
New York Berlin Heidelberg London
Paris Tokyo Hong Kong Barcelona

J. Kevorkian
Applied Mathematics Program
University of Washington
Seattle, WA 98195
USA

J. D. Cole
Department of Mathematical Sciences
Rensselaer Polytechnic Institute
Troy, NY 12181
USA

Editors

F. John
Courant Institute of Mathematical Sciences
New York University
New York, NY 10012
USA

J. E. Marsden
Department of Mathematics
University of California
Berkeley, CA 94720
USA

L. Sirovich
Division of Applied Mathematics
Brown University
Providence, RI 02912
USA

Mathematics Subject Classification (1980): 34D10, 34E10, 35B20, 35B25

Library of Congress Cataloging in Publication Data

Kevorkian, J
Perturbation methods in applied mathematics.

(Applied mathematical sciences; v. 34)
A revised and updated version of J. D. Cole's
Perturbation methods in applied mathematics, 1968.
Bibliography: p.
Includes index.
1. Differential equations—Numerical solutions.
2. Differential equations—Asymptotic theory.
3. Perturbation (Mathematics) I. Cole, Julian D., joint author. II. Title. III. Series.
QA1.A647 vol. 31 [Q371] 510s [515.3'5] 80-16796

With 79 figures.

9 8 7 6 5 4 3

ISBN 978-1-4419-2812-2

Manufactured in Mexico.

Preface

This book is a revised and updated version, including a substantial portion of new material, of J. D. Cole's text *Perturbation Methods in Applied Mathematics*, Ginn–Blaisdell, 1968. We present the material at a level which assumes some familiarity with the basics of ordinary and partial differential equations. Some of the more advanced ideas are reviewed as needed; therefore this book can serve as a text in either an advanced undergraduate course or a graduate level course on the subject.

The applied mathematician, attempting to understand or solve a physical problem, very often uses a perturbation procedure. In doing this, he usually draws on a backlog of experience gained from the solution of similar examples rather than on some general theory of perturbations. The aim of this book is to survey these perturbation methods, especially in connection with differential equations, in order to illustrate certain general features common to many examples. The basic ideas, however, are also applicable to integral equations, integrodifferential equations, and even to difference equations.

In essence, a perturbation procedure consists of constructing the solution for a problem involving a small parameter ε, either in the differential equation or the boundary conditions or both, when the solution for the limiting case $\varepsilon = 0$ is known. The main mathematical tool used is asymptotic expansion with respect to a suitable asymptotic sequence of functions of ε.

In a regular perturbation problem a straightforward procedure leads to an approximate representation of the solution. The accuracy of this approximation does not depend on the value of the independent variable and gets better for smaller values of ε. We will not discuss this type of problem here as it is well covered in other texts. For example, the problem of calculating the perturbed eigenvalues and eigenfunctions of a self adjoint differential operator is a regular perturbation problem discussed in most texts on differential equations.

Rather, this book concentrates on singular perturbation problems which are very common in physical applications and which require special techniques. Such singular perturbation problems may be divided into two broad categories: layer-type problems and cumulative perturbation problems.

In a layer-type problem the small parameter multiplies a term in the differential equation which becomes large in a thin layer near a boundary (e.g., a boundary-layer) or in the interior (e.g., a shock-layer). Often, but not always, this is the highest derivative in the differential equation and the $\varepsilon = 0$ approximation is therefore governed by a lower order equation which cannot satisfy all the initial or boundary conditions prescribed. In a cumulative perturbation problem the small parameter multiplies a term which never becomes large. However, its cumulative effect becomes important for large values of the independent variable. In some applications both categories occur simultaneously and require the combined use of the two principal techniques we study in this book.

This book is written very much from the point of view of the applied mathematician; much less attention is paid to mathematical rigor than to rooting out the underlying ideas, using all means at our disposal. In particular, physical reasoning is often used as an aid to understanding a problem and to formulating the appropriate approximation procedure.

The first chapter contains some background on asymptotic expansions. The more advanced techniques in asymptotics such as the methods of steepest descents and stationary phase are not covered as there are excellent modern texts including these techniques which, strictly speaking, are not perturbation techniques. In addition, we introduce in this chapter the basic ideas of limit process expansions, matching asymptotic expansions, and general asymptotic expansions.

Chapter 2 gives a deeper exposition of limit process expansions through a sequence of examples for ordinary differential equations. Chapter 3 is devoted to cumulative perturbation problems using the so-called multiple variable expansion procedure. Applications to nonlinear oscillations, flight mechanics and orbital mechanics are discussed in detail followed by a survey of other techniques which can be used for this class of problems.

In Chapter 4 we apply the procedures of the preceding chapters to partial differential equations, presenting numerous physical examples. Finally, the last chapter deals with a typical use of asymptotic expansions, the construction of approximate equations; simplified models such as linearized and transonic aerodynamics, and shallow water theory are derived from more exact equations by means of asymptotic expansions. In this way the full meaning of laws of similitude becomes evident.

The basic ideas used in this book are, as is usual in scientific work, the ideas of many people. In writing the text, no particular attempt has been made to cite the original authors or to have a complete list of references and bibliography. Rather, we have tried to present the "state of the art" in a systematic manner starting from elementary applications and progressing gradually to areas of current research.

For a deeper treatment of the fundamental ideas of layer-type expansions and related problems the reader is referred to the forthcoming book by P. A. Lagerstrom and J. Boa of Caltech.

To a great extent perturbation methods were pioneered by workers in fluid mechanics and these traditional areas are given full coverage. Applications in celestial mechanics, nonlinear oscillations, mathematical biology, wave propagation, and other areas have also been successfully explored since the publication of J. D. Cole's 1968 text. Examples from these more recent areas of application are also covered.

We believe that this book contains a unified account of perturbation theory as it is understood and widely used today.

Fall 1980

J. Kevorkian
J. D. Cole

Contents

Chapter 1

Introduction

1.1 Ordering

We will use the conventional order symbols as a mathematical measure of the relative order of magnitude of various quantities. Although generalizations are straightforward, we need only be concerned with scalar functions of real variables. In the definitions which follow ϕ, ψ, etc. are scalar functions of the variable x (which may be a vector) and the scalar parameter ε. The variable x ranges over some domain D and ε belongs to some interval I.

Large O

Let x be fixed. We say $\phi = O(\psi)$ in I if there exists a $k(x)$ such that $|\phi| \leq k(x)|\psi|$ for all ε in I. Similarly, if ε_0 is a limit point in I we say that $\phi = O(\psi)$ as $\varepsilon \to \varepsilon_0$ if there exists a $k(x)$ and a neighborhood N of ε_0 such that $|\phi| \leq k(x)|\psi|$ for all ε in the intersection of N with I.

We note that if ψ does not vanish in I then the inequality in the above two definitions simply reduces to the statement that ϕ/ψ is bounded.

Small o

Again with x fixed, we say $\phi = o(\psi)$ as $\varepsilon \to \varepsilon_0$ if given any $\delta(x) > 0$, there exists a neighborhood N_δ of ε_0 such that $|\phi| \leq \delta(x)|\psi|$ for all ε in N_δ. Here also the definition simplifies to the statement that $(\phi/\psi) \to 0$ if $\psi \neq 0$ in I. Often, $\phi \ll \psi$ is used as an equivalent notation.

Uniformity

As indicated in the above definitions the quantities k, δ and the neighborhoods N, N_δ will, in general, depend on the value of x. If, however k, δ, N, N_δ

can be found independently of the value of x we say that the order relations hold *uniformly in D*. To illustrate these ideas consider the following examples. In all cases x will be a real variable, the domain D will be the half-open unit interval $0 < x \leq 1$ and I will be the half-open interval $0 < \varepsilon \leq \mu < 1$ with $\varepsilon_0 = 0$.

(i) $$x + \varepsilon = O(1) \quad \text{in } I, \text{ uniformly in } D. \tag{1.1.1}$$

(ii) $$\log(\sin \varepsilon x) = O(\log 2\varepsilon x/\pi) \text{ in } I, \text{ uniformly in } D. \tag{1.1.2}$$

This follows from the fact that $0 < 2z/\pi < \sin z$ for all $0 < z < 1$. Since $0 < \varepsilon x < 1$ always, the inequality in the definition holds with $k = 1$.

(iii) $$\frac{1}{x + \varepsilon} = O(1) \quad \text{in } I. \tag{1.1.3}$$

The statement is true because $1/(x + \varepsilon) < 1/x$ for any given x and all ε in I; thus $k(x) = 1/x$. Now, it is clear that the statement (1.1.3) is not uniformly valid in D because there is no finite constant k for which the required inequality holds for all x in D so long as x is allowed to approach the origin.

For similar reasons the statement $\varepsilon/x(1 - x) = O(\varepsilon)$ in I is not uniformly valid in $0 < x < 1$.

(iv) $$\varepsilon^\alpha = O(\varepsilon^\beta) \quad \text{in } I \text{ for any } \alpha \geq \beta. \tag{1.1.4}$$

This result is trivially true since $\varepsilon^{\alpha-\beta}$ is bounded. In fact, it tends to zero for $\alpha > \beta$, and this is a reminder that the O symbol does not connote equality of order of magnitude but only provides a *one-sided* bound.

(v) $$\sin \frac{x}{\varepsilon} = O(x) \quad \text{as } \varepsilon \to 0. \tag{1.1.5}$$

Here, even though the limit as $\varepsilon \to 0$ of $\sin(x/\varepsilon)$ does not exist for any $x \neq 0$, it is clear that $|\sin x/\varepsilon| \leq 1$ for any x in D. Therefore, (1.1.5) is true with $k(x) = 1/x$ and the statement is not uniformly valid. However, the statement $\sin x/\varepsilon = O(1)$ as $\varepsilon \to 0$ is uniformly valid in D.

(vi) $$\varepsilon^\alpha = o(\varepsilon^\beta) \quad \text{as } \varepsilon \to 0 \text{ if } \alpha > \beta. \tag{1.1.6}$$

(vii) $$\varepsilon^\alpha \log \varepsilon = o(1) \quad \text{as } \varepsilon \to 0 \text{ for any } \alpha > 0. \tag{1.1.7}$$

(viii) $$e^{-x/\varepsilon} = o(\varepsilon^\beta) \quad \text{as } \varepsilon \to 0 \text{ for any } \beta \geq 0 \text{ if } x > 0. \tag{1.1.8}$$

Clearly, the statement (1.1.8) is not uniformly valid, even in the half-open interval $0 < x \leq 1$, and if $x = 0$ the statement is false.

Various operations such as addition, multiplication and integration can be performed with the order relations. In general, differentiation of order relations with respect to ε or x is not permissible. For these and further results the reader may consult Reference 1.1.1.

Reference

1.1.1 A. Erdelyi, *Asymptotic Expansions*, Dover Publications, New York, 1956.

1.2 Asymptotic Sequences and Expansions

Consider a sequence $\{\phi_n(\varepsilon)\}$ $n = 1, 2, \ldots$ of functions of ε. Such a sequence is called an asymptotic sequence if

$$\phi_{n+1}(\varepsilon) = o(\phi_n(\varepsilon)) \quad \text{as } \varepsilon \to \varepsilon_0 \tag{1.2.1}$$

for each $n = 1, 2, \ldots$

If the sequence is infinite and $\phi_{n+1} = o(\phi_n)$ uniformly in n (i.e., the choice of δ and N_δ in the definition given in Section 1.1 does not depend on n) the sequence is said to be uniform in n. Similarly if the ϕ_n also depend on a variable x one can have uniformity with respect to x in some domain D. Some examples of asymptotic sequences are

$$\phi_n(\varepsilon) = (\varepsilon - \varepsilon_0)^n, \quad \text{as } \varepsilon \to \varepsilon_0 \tag{1.2.2}$$

$$\phi_n(\varepsilon) = e^{\varepsilon}\varepsilon^{-\lambda_n}, \quad \text{as } \varepsilon \to \infty, \lambda_{n+1} > \lambda_n \tag{1.2.3}$$

$$\begin{gathered}\phi_0 = \log \varepsilon, \qquad \phi_1 = 1, \qquad \phi_2 = \varepsilon \log \varepsilon, \qquad \phi_3 = \varepsilon \\ \phi_4 = \varepsilon^2 \log^2 \varepsilon, \qquad \phi_5 = \varepsilon^2 \log \varepsilon, \qquad \phi_6 = \varepsilon^2 \ldots, \\ \text{as } \varepsilon \to 0.\end{gathered} \tag{1.2.4}$$

Here again various operations, such as multiplication of two sequences or integration can be used to generate a new sequence. Differentiation with respect to ε may not lead to a new asymptotic sequence. For more details the reader may use Reference 1.1.1.

A sum of terms of the form $\sum_{n=1}^{N} a_n(x)\phi_n(\varepsilon)$ is called an asymptotic expansion of the function $f(x, \varepsilon)$ to N terms (N may be infinite) as $\varepsilon \to \varepsilon_0$ with respect to the sequence $\{\phi_n(\varepsilon)\}$ if

$$f(x, \varepsilon) - \sum_{n=1}^{M} a_n(x)\phi_n(\varepsilon) = o(\phi_M) \quad \text{as } \varepsilon \to \varepsilon_0 \tag{1.2.5}$$

for each $M = 1, 2, \ldots, N$.

If $N = \infty$, the following notation is generally used

$$f(x, \varepsilon) \sim \sum_{n=1}^{\infty} a_n(x)\phi_n(\varepsilon) \quad \text{as } \varepsilon \to \varepsilon_0. \tag{1.2.6}$$

Clearly, an equivalent definition for an asymptotic expansion is that

$$f(x, \varepsilon) - \sum_{n=1}^{M-1} a_n(x)\phi_n(\varepsilon) = O(\phi_M) \quad \text{as } \varepsilon \to \varepsilon_0 \tag{1.2.7}$$

for each $M = 2, \ldots, N$.

An asymptotic expansion is said to be uniformly valid in some domain D in x if the order relations in (1.2.5) or (1.2.7) hold uniformly.

Given a function $f(x, \varepsilon)$ and an asymptotic sequence $\{\phi_n(\varepsilon)\}$, one can uniquely calculate each of the $a_n(x)$ defining the asymptotic expansion of $f(x, \varepsilon)$ by repeated application of the definition (1.2.5). Thus,

$$a_1(x) = \lim_{\varepsilon \to \varepsilon_0} \frac{f(x, \varepsilon)}{\phi_1(\varepsilon)} \tag{1.2.8a}$$

$$a_2(x) = \lim_{\varepsilon \to \varepsilon_0} \frac{f(x, \varepsilon) - a_1(x)\phi_1(\varepsilon)}{\phi_2(\varepsilon)} \tag{1.2.8b}$$

$$a_k(x) = \lim_{\varepsilon \to \varepsilon_0} \frac{f(x, \varepsilon) - \sum_{n=1}^{k-1} a_n(x)\phi_n(\varepsilon)}{\phi_k(\varepsilon)}. \tag{1.2.8c}$$

For example $f(x, \varepsilon) = (x + \varepsilon)^{-1/2}$ has the expansion

$$(x + \varepsilon)^{-1/2} \sim \sum_{n=1}^{\infty} \frac{(-1)^{n-1}}{2^{n-1}(n-1)!} \prod_{k=1}^{n} |2k - 3| \frac{\varepsilon^{n-1}}{x^{(2n-1)/2}} \tag{1.2.9}$$

as $\varepsilon \to 0$, with respect to the sequence $\{\varepsilon^{n-1}\}$. This is also the Taylor series expansion of $(x + \varepsilon)^{-1/2}$ near $\varepsilon = 0$ and is convergent for $\varepsilon < |x|$. Note also that the expansion (1.2.9) is not uniformly valid in any domain in x for which $x = 0$ is a limit point.

A less trivial situation occurs if $f(x, \varepsilon)$ is defined by an integral representation. Consider, for example, the Error function defined by

$$\operatorname{erf} \varepsilon = 1 - \frac{2}{\sqrt{\pi}} \int_{\varepsilon}^{\infty} e^{-t^2}\, dt \tag{1.2.10a}$$

which by setting $t^2 = \tau$ can also be written as

$$\operatorname{erf} \varepsilon = 1 - \frac{1}{\sqrt{\pi}} \int_{\varepsilon^2}^{\infty} e^{-\tau}\tau^{-1/2}\, d\tau. \tag{1.2.10b}$$

We note that after integration by parts once (1.2.10b) becomes

$$\operatorname{erf} \varepsilon = 1 - \frac{1}{\sqrt{\pi}} \left[\frac{e^{-\varepsilon^2}}{\varepsilon} - \frac{1}{2} \int_{\varepsilon^2}^{\infty} e^{-\tau}\tau^{-3/2}\, d\tau \right]$$

and this suggests repeating the process in order to generate an expansion in increasing powers of ε^{-1}. If such an expansion were asymptotic in the limit $\varepsilon \to \infty$, it would be useful for numerical evaluation of erf ε for ε large.

Defining

$$F_n(\varepsilon) = \int_{\varepsilon^2}^{\infty} e^{-\tau}\tau^{-(2n+1)/2}\, d\tau, \qquad n = 0, 1, 2, \ldots \tag{1.2.11}$$

and integrating $F_n(\varepsilon)$ by parts results in the recursion relation

$$F_n(\varepsilon) = \frac{e^{-\varepsilon^2}}{\varepsilon^{2n+1}} - \frac{(2n+1)}{2} F_{n+1}(\varepsilon), \quad n = 0, 1, 2, \ldots \tag{1.2.12}$$

and this can be used to calculate the following *exact* result for F_0

$$F_0(\varepsilon) = e^{-\varepsilon^2}\left[\frac{1}{\varepsilon} - \frac{1}{2\varepsilon^3} + \frac{1.3}{2^2\varepsilon^5} + \cdots + \frac{(-1)^{n-1}1.3.5\ldots(2n-3)}{2^{n-1}\varepsilon^{2n-1}}\right] + (-1)^n\frac{1.3.5\ldots(2n-1)}{2^n}F_n(\varepsilon), \quad n = 1, 2, \ldots. \tag{1.2.13}$$

Thus, (1.2.13) exhibits a formal series in ascending powers of ε^{-1} and an exact expression for the remainder if the series is truncated after n terms. To show that the bracketed expression in (1.2.13) is the asymptotic expansion of F_0, we must verify that (1.2.5) is satisfied, i.e., that

$$F_0(\varepsilon) - e^{-\varepsilon^2}\sum_{n=1}^{M}(-1)^{n-1}\frac{1.3.5\ldots(2n-3)}{2^{n-1}\varepsilon^{2n-1}} = O(\varepsilon^{-(2M-1)}e^{-\varepsilon^2}) \tag{1.2.14}$$

as $\varepsilon \to \infty$.

According to (1.2.13), the above reduces to showing that $S_M(\varepsilon)$ defined by

$$S_M(\varepsilon) = \varepsilon^{2M-1}e^{\varepsilon^2}\frac{(-1)^M 1.3.5\ldots(2M-1)}{2^M}F_M(\varepsilon) \tag{1.2.15}$$

tends to zero as $\varepsilon \to \infty$.

This is easily accomplished once we note that

$$F_M(\varepsilon) \le \frac{1}{\varepsilon^{2M+1}}\int_{\varepsilon^2}^{\infty} e^{-\tau}\,d\tau = \frac{e^{-\varepsilon^2}}{\varepsilon^{2M+1}}. \tag{1.2.16}$$

Therefore,

$$|S_M(\varepsilon)| \le \frac{1.3.5\ldots(2M-1)}{2^M\varepsilon^2} \tag{1.2.17}$$

and hence $S_M = o(1)$ as $\varepsilon \to \infty$.

We note that the asymptotic expansion

$$\operatorname{erf}\varepsilon \sim 1 - \frac{e^{-\varepsilon^2}}{\sqrt{\pi}}\sum_{n=1}^{\infty}\frac{(-1)^{n-1}1.3.5\ldots(2n-3)}{2^{n-1}\varepsilon^{2n-1}} \tag{1.2.18}$$

is *divergent* because the numerical value of the coefficients of ε^{-2n+1} in the series (1.2.18) becomes large as n increases. Actually, (1.2.13) provides an exact expression for the error resulting from using M terms of the expansion (1.2.18) to represent erf ε. It is easily verified that for any fixed ε there is an optimal integer M_0 in the sense that the error is a decreasing function of the number n of terms retained, as long as $n < M_0$. But, if one insists on retaining

$n > M_0$ terms, the error will increase with n. Moreover, M_0 increases with ε and the error of the series with M_0 terms decreases as ε increases. The above features are typical of divergent asymptotic expansions.

The reader may verify that for $\varepsilon = 2$ the series on (1.2.18) gives the best accuracy if 5 terms are used and that the error in this case is only 6.43×10^{-5}, which is remarkable since $\varepsilon = 2$ is not a large number.

Functions defined by integral representations also occur naturally in the solution of linear problems by transform techniques. Various methods have been developed for calculating the asymptotic behavior of such results. A discussion of this topic is beyond the scope of this book. The reader will find an excellent account in Reference 1.2.1.

Problems

1. Calculate the asymptotic behavior as $t \to \infty$ for the initial value problem

$$\frac{d^2y}{dt^2} + y = \frac{1}{t}, \quad \pi \le t < \infty \tag{1.2.19}$$

$$y(\pi) = \frac{dy(\pi)}{dt} = 0 \tag{1.2.20}$$

by two methods.

(a) First, calculate the solution in integral form and use repeated integrations by parts.

(b) Next, observe that

$$y = a \sin t + b \cos t + \sum_{n=1}^{\infty} \frac{C_n}{t^n} \tag{1.2.21}$$

is formally, a general solution for appropriate C_n. Determine the constants a, b, and the C_n and compare your results with those in part (a). Is this asymptotic expansion convergent?

2. Noting that the nonlinear equation

$$\frac{d^2y}{dt^2} - \sin y = -\frac{1}{2} \tag{1.2.22}$$

has the energy integral

$$\frac{1}{2}\left(\frac{dy}{dt}\right)^2 + \cos y + \frac{y}{2} = E = \text{const.} \tag{1.2.23}$$

calculate the first five terms of the asymptotic expansion of the solution of (1.2.22) as $t \to \infty$ for the initial value problem

$$y(0) = 0$$

$$\frac{dy(0)}{dt} = 0.$$

Reference

1.2.1 G. F. Carrier. M. Krook and C. E. Pearson, *Functions of a Complex Variable, Theory and Technique*, McGraw-Hill Book Company, New York, 1966.

1.3 Limit Process Expansions, Matching, General Asymptotic Expansions

Another possible way of defining a function $f(x, \varepsilon)$ is as the solution of a differential equation in which x is the independent variable and ε occurs as a parameter. If one cannot solve this differential equation for arbitrary ε (as, for example, if the differential equation is nonlinear with $\varepsilon \neq 0$) can one calculate the asymptotic expansion of the solution by considering a sequence of simpler differential equations governing each term of this expansion? This is the perturbation idea which will be explored in depth in subsequent chapters. Here, we consider a simple example to introduce some ideas.

The first-order equation

$$\varepsilon \frac{dy}{dx} + y = \frac{\varepsilon[x(\varepsilon - 1) + \varepsilon^2]e^{-x}}{(x + \varepsilon)^2}, \qquad 0 \leq x \leq \infty, 0 < \varepsilon \ll 1 \tag{1.3.1}$$

$$y(0) = 0 \tag{1.3.2}$$

has the exact solution

$$y = f(x, \varepsilon) \equiv e^{-x/\varepsilon} - \frac{\varepsilon e^{-x}}{x + \varepsilon}. \tag{1.3.3}$$

Ignoring temporarily the origin of eq. (1.3.3), we see that $f(x, \varepsilon)$ defines a well behaved function, and it is interesting to consider the asymptotic expansion of this function as $\varepsilon \to 0$. If we fix x to be some positive value and apply the limit process defined by eqs. (1.2.8) with $\phi_n(\varepsilon) = \varepsilon^n$ we find the following expansion for f, called an "outer" expansion

$$\begin{aligned} f &= -\varepsilon \frac{e^{-x}}{x} + \varepsilon^2 \frac{e^{-x}}{x^2} - \varepsilon^3 \frac{e^{-x}}{x^3} + O(\varepsilon^4) \\ &\equiv \sum_{n=0}^{N} \varepsilon^n h_n(x) + O(\varepsilon^{N+1}) \end{aligned} \tag{1.3.4}$$

and the contribution of the $e^{-x/\varepsilon}$ term is smaller than any term in the series in (1.3.4). We shall refer to such a term as a "transcendentally small" term (abbreviated as T.S.T.) in this limit.

Clearly (1.3.4) is not uniformly valid near $x = 0$. In fact, it is singular there, and this expansion is not a good approximation of the function defined by (1.3.3) no matter how small ε is if we allow x also to become small.

It is therefore natural to seek another expansion of (1.3.3) which adequately approximates this function near $x = 0$. Since the combination x/ε occurs in the first term one is led to the change of variables $x^* = x/\varepsilon$

$$y = g(x^*, \varepsilon) \equiv e^{-x^*} - \frac{e^{-\varepsilon x^*}}{x^* + 1}. \tag{1.3.5}$$

With $x^* = x/\varepsilon$ (1.3.5) defines the same function as f. However, the asymptotic expansion of g with x^* *fixed* as $\varepsilon \to 0$ is quite different. It is easy to see that in this limit y has the expansion which will be referred to as the "inner" expansion

$$\begin{aligned} g &= e^{-x^*} - \frac{1}{x^* + 1} + \frac{\varepsilon x^*}{x^* + 1} - \frac{\varepsilon^2 x^{*2}}{2(x^* + 1)} + \frac{\varepsilon^3 x^{*3}}{6(x^* + 1)} + O(\varepsilon^4) \\ &\equiv \sum_{n=0}^{N} \varepsilon^n g_n(x^*) + O(\varepsilon^{N+1}). \end{aligned} \tag{1.3.6}$$

Now, this expansion is accurate for small x. In particular, the condition $y = 0$ at $x = x^* = 0$ is satisfied. However, the result fails to be uniformly valid for x^* large. Thus, the two expansions (1.3.4) and (1.3.6) have mutually exclusive domains of validity. Hence, depending on the magnitude of x compared to ε one expansion or the other should be used.

Several related questions now arise.

(i) Equation (1.3.4) and (1.3.6) give the expansions of the *same* function by different limit processes. Is there another limit process expansion which is contained in both expansions?
(ii) Is it possible to find *one* asymptotic expansion which is uniformly valid for all $x \geq 0$?
(iii) Can one calculate these expansions directly from Equation (1.3.1) without knowing the exact solution?

We will now show that the answer to all these questions is in the affirmative for the present example. Building on the experience gained from this example we will introduce later on in this section the appropriate mathematical framework to define the above ideas.

Let us consider the expansion which would result from (1.3.1) by letting $\varepsilon \to 0$ with $x_\eta = x/\eta(\varepsilon)$ fixed for some $\eta(\varepsilon)$ such that $\varepsilon \ll \eta \ll 1$. Thus, in such an expansion $x \to 0$ in the limit but at a slower rate than in the case leading to (1.3.6). In a sense, the above defines an "intermediate" limit process.

Setting $x = \eta(\varepsilon)x_\eta$, we write (1.3.3) in the form

$$y = l(x_\eta, \eta, \varepsilon) \equiv e^{-\eta x_\eta/\varepsilon} - \frac{\varepsilon e^{-\eta x_\eta}}{\eta x_\eta + \varepsilon}. \tag{1.3.7}$$

Now, applying the limit process $\varepsilon \to 0$ with x_η fixed (1.3.7) has the following expansion, called the "intermediate" expansion

$$l = -\frac{\varepsilon}{\eta x_\eta} + \frac{\varepsilon^2}{\eta^2 x_\eta^2} - \frac{\varepsilon^3}{\eta^3 x_\eta^3} + \varepsilon - \frac{\varepsilon^2}{\eta x_\eta} - \frac{\varepsilon \eta x_\eta}{2} + O\left(\frac{\varepsilon^4}{\eta^4}\right) + O\left(\frac{\varepsilon^3}{\eta^2}\right) + O(\varepsilon^2) + O(\varepsilon\eta^2). \tag{1.3.8}$$

In the above the term $e^{-\eta x_\eta/\varepsilon}$ is transcendentally small and does not appear as long as $\varepsilon|\log \varepsilon| \ll \eta$. Henceforth, we shall always ignore such a term and automatically require that $\varepsilon|\log \varepsilon| \ll \eta$ in our calculations. This restricts somewhat the range of η to $\varepsilon|\log \varepsilon| \ll \eta \ll 1$.

If we now reexpand the outer and inner expansions using the above intermediate limit process, we find that if a sufficient number of terms are included both eventually give (1.3.8). This means that the outer expansion, which was constructed under the assumption $\varepsilon \to 0$ with x fixed $\neq 0$ is actually valid in the extended sense $\varepsilon \to 0$, $x_\eta = x/\eta(\varepsilon)$ fixed for some class of functions $\varepsilon|\log \varepsilon| \ll \eta(\varepsilon)$. Similarly, the inner expansion which was constructed under the assumption $\varepsilon \to 0$, $x^* = x/\varepsilon$ fixed $\neq \infty$ is actually valid in the extended sense $\varepsilon \to 0$, $x_\eta = x/\eta(\varepsilon)$ fixed, for some class of functions $\eta(\varepsilon)$ such that $\eta \ll 1$.

We will demonstrate next that for this example, the extended domains of validity of the inner and outer expansions overlap in the following sense. For each $R = 0, 1, 2, \ldots$ there exist integers P, and Q, and functions $\eta_1(\varepsilon)$ and $\eta_2(\varepsilon)$ with $\eta_1 \ll \eta_2$ such that

$$\lim_{\substack{\varepsilon \to 0 \\ x_\eta \text{ fixed}}} \frac{\left[\sum_{n=0}^{P} h_n(\eta x_\eta)\varepsilon^n - \sum_{n=0}^{Q} g_n(\eta x_\eta/\varepsilon)\varepsilon^n\right]}{\varepsilon^R} = 0 \tag{1.3.9}$$

for all η satisfying $\eta_1 \ll \eta \ll \eta_2$.

Equation (1.3.9) is a matching condition for the inner and outer expansions in their common overlap domain of validity which is defined as the class of functions $\eta(\varepsilon)$ satisfying the condition $\eta_1 \ll \eta \ll \eta_2$.

To demonstrate the result let us first take $R = 0$. Now $h_0 = 0$ and $g_0 = e^{-x^*} - 1/(1 + x^*)$. Assuming that $P = Q = 0$, the question now is whether

$$\lim_{\substack{\varepsilon \to 0 \\ x_\eta \text{ fixed}}} [h_0(\eta x_\eta) - g_0(\eta x_\eta/\varepsilon)] = 0. \tag{1.3.10}$$

Expanding g_0 in terms of x_η gives[1]

$$g_0 = -\frac{\varepsilon}{\eta x_\eta} + \frac{\varepsilon^2}{\eta^2 x_\eta^2} - \frac{\varepsilon^3}{\eta^3 x_\eta^3} + O\left(\frac{\varepsilon^4}{\eta^4}\right) + \text{T.S.T.} \tag{1.3.11}$$

[1] Note that the first two terms in this expansion correspond to the first two terms in the intermediate expansion (1.3.8).

Clearly (1.3.10) is satisfied by any η such that $\varepsilon|\log\varepsilon| \ll \eta$. Thus, the overlap domain for the matching to $O(1)$ is determined by

$$\eta_1 = \varepsilon|\log\varepsilon|, \qquad \eta_2 = 1.$$

To order ε the matching condition (with $P = Q = 1$ and $R = 1$) becomes

$$\lim_{\substack{\varepsilon\to 0\\ x_\eta \text{ fixed}}} \left[\frac{h_0(\eta x_\eta) + \varepsilon h_1(\eta x_\eta) - g_0(\eta x_\eta/\varepsilon) - \varepsilon g_1(\eta x_\eta/\varepsilon)}{\varepsilon}\right] = 0. \qquad (1.3.12)$$

We can use the previous results to calculate

$$\varepsilon h_1 = -\frac{\varepsilon}{\eta x_\eta} + \varepsilon - \frac{\varepsilon\eta x_\eta}{2} + \frac{\varepsilon\eta^2 x_\eta^2}{6} + O(\varepsilon\eta^3) \qquad (1.3.13)$$

$$\varepsilon g_1 = \varepsilon - \frac{\varepsilon^2}{\eta x_\eta} + O\left(\frac{\varepsilon^3}{\eta^2}\right). \qquad (1.3.14)$$

Substituting (1.3.11), (1.3.13) and (1.3.14) into (1.3.12) and cancelling identical terms gives to lowest order

$$\lim_{\substack{\varepsilon\to 0\\ x_\eta \text{ fixed}}} \left[-\frac{\eta x_\eta}{2} - \frac{\varepsilon}{\eta^2 x_\eta^2} + \frac{\varepsilon}{\eta x_\eta} + O\left(\frac{\varepsilon^2}{\eta^3}\right) + O(\eta^2) + O\left(\frac{\varepsilon^2}{\eta^2}\right)\right] = 0. \qquad (1.3.15)$$

All terms automatically vanish as long as $\varepsilon|\log\varepsilon| \ll \eta \ll 1$ except the second term. This will vanish if $\varepsilon/\eta^2 \to 0$, i.e., $\varepsilon^{1/2} \ll \eta$. Since $\varepsilon|\log\varepsilon| \ll \varepsilon^{1/2}$, the overlap domain to $O(\varepsilon)$ is $\varepsilon^{1/2} \ll \eta \ll 1$, i.e., $\eta_1 = \varepsilon^{1/2}$, $\eta_2 = 1$.

To carry out the matching to $O(\varepsilon^2)$, we first attempt to show that

$$\lim_{\substack{\varepsilon\to 0\\ x_\eta \text{ fixed}}} \frac{h_0 + \varepsilon h_1 + \varepsilon^2 h_2 - g_0 - \varepsilon g_1 - \varepsilon^2 g_2}{\varepsilon^2} = 0, \qquad (1.3.16)$$

i.e., we assume that $P = Q = 2$.

We now calculate

$$\varepsilon^2 h_2 = \frac{\varepsilon^2}{\eta^2 x_\eta^2} - \frac{\varepsilon^2}{\eta x_\eta} + \frac{\varepsilon^2}{2} + O(\varepsilon^2\eta) \qquad (1.3.17)$$

$$\varepsilon^2 g_2 = -\frac{\varepsilon\eta x_\eta}{2} + \frac{\varepsilon^2}{2} + O\left(\frac{\varepsilon^3}{\eta}\right). \qquad (1.3.18)$$

After cancelling out all terms which match identically we are left with the requirement

$$\lim_{\substack{\varepsilon\to 0\\ x_\eta \text{ fixed}}} \left[\left(\frac{\eta^2}{\varepsilon}\right)\frac{x_\eta^2}{6} + \frac{\varepsilon}{\eta^3 x_\eta^3} + O\left(\frac{\varepsilon^2}{\eta^4}\right) + O\left(\frac{\eta^3}{\varepsilon}\right) + O\left(\frac{\varepsilon}{\eta^2}\right) + O(\eta) + O\left(\frac{\varepsilon}{\eta}\right)\right] = 0. \qquad (1.3.19)$$

The first term in the above comes from εh_1 while the second comes from g_0. Since the remainders contain the reciprocals of η^2/ε and ε/η^3, it is clear that these terms are singular. Their counter parts must be found in the higher order terms. In fact, it is easy to see that the leading term in $\varepsilon^3 g_3$ will match with $\varepsilon\eta^2 x_\eta^2/6$ and the leading term in $\varepsilon^3 h_3$ will match with $-\varepsilon^3/\eta^3 x_\eta^3$. More precisely,

$$\varepsilon^3 g_3 = \varepsilon\eta^2 \frac{x_\eta^2}{6} + O(\varepsilon^2\eta) \tag{1.3.20}$$

$$\varepsilon^3 h_3 = -\frac{\varepsilon^3}{\eta^3 x_\eta^3} + O\left(\frac{\varepsilon^3}{\eta^2}\right) \tag{1.3.21}$$

Now, the matching condition to $O(\varepsilon^2)$ with $P = Q = 3$ is satisfied, as long as $\eta^3/\varepsilon \to 0$ to eliminate the remainder term in εh_1. We must also require that $(\varepsilon/\eta^2) \to 0$ to eliminate the remainder terms in εg_1 and $\varepsilon^3 h_3$. The above two conditions imply that η must be constrained as follows

$$\varepsilon^{1/2} \ll \eta \ll \varepsilon^{1/3}, \tag{1.3.22}$$

i.e., $\eta_1 = \varepsilon^{1/2}$ and $\eta_2 = \varepsilon^{1/3}$.

This process can be continued indefinitely. We note that the overlap domain shrinks as the order of the matching increases, and this is typical. We have carried out the matching to $O(\varepsilon^2)$ to point out the fact that terms of order ε^3 were needed. In general, the choice of P and Q is not a priori obvious and must be deduced from the matching.

We have adopted the somewhat cumbersome formulation of writing both inner and outer variables in terms of x_η to insure that singularities are taken into account, and perhaps more importantly, to exhibit directly the overlap domain by requiring all remainder terms to vanish to a given order.

One could perform the matching by writing the outer expansion in terms of the inner variable, expanding the result then comparing this with the inner expansion (or vice-versa). However, this procedure which appears more expeditious (and is sometimes used in the literature) does not exhibit the overlap domain nor does it systematically assure the vanishing of all neglected terms.

For example, had we used the above short-cut it would indeed have been difficult at the stage of Equation (1.3.16) to have realized that terms of order ε^3 were needed to carry out the matching.

We note that the intermediate expansion was not needed to perform the matching in this example. Using a sufficient number of terms in each expansion allowed us to exhibit both the matching and the common overlap domain in the $x - \varepsilon$ plane.

Let us examine this idea a little more closely. To any given order of accuracy in ε the inner and outer expansion are constructed in domains as indicated in Figure 1.3.1.

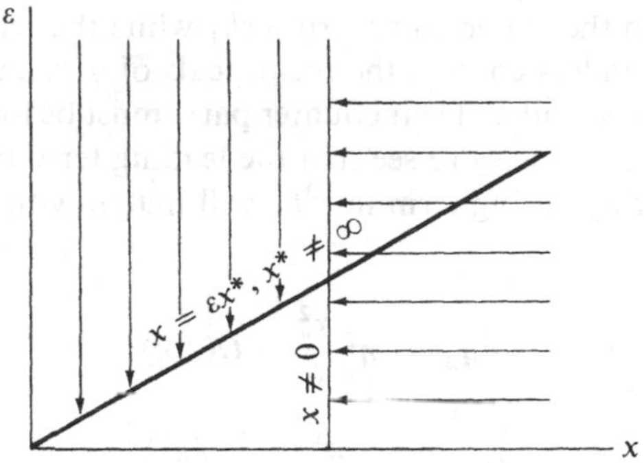

Figure 1.3.1 Assumed Domains of Validity of Inner and Outer Expansions

Thus, in the outer region x is held fixed, not equal to zero as $\varepsilon \to 0$ and the totality of all such paths then covers a domain which in the limit as $\varepsilon \to 0$ tends to a rectangular region away from $x = 0$. Similarly, in the inner limit we let $\varepsilon \to 0$ with x^* fixed and finite. Clearly, the totality of all such paths in the limit defines the triangular region indicated with the x axis excluded.

The requirements $x^* \neq \infty$, $x \neq 0$ thus prevent overlap in the limit $\varepsilon \to 0$. However, the example we considered indicates that the *actual* domains of validity of the inner and outer expansions are larger than assumed and do overlap as indicated in Figure 1.3.2.

As a result of this direct overlap, one need not consider the intermediate expansion at all. Although it is possible to invent pathological examples where this type of direct matching is not possible no matter how many terms are retained, (because $\eta_2 \ll \eta_1$) such examples do not normally occur in physically meaningful problems. Henceforth, we will adopt the point of view of seeking a direct matching of inner and outer expansions and will not compare each individually with an intermediate expansion. The validity of this assumption will be assured once the matching is carried out in a non-empty overlap domain.

We now turn to the question of finding one asymptotic expansion which is uniformly valid for all x. Proceeding heuristically, we assume (since the ingredients of both expansions are needed in their respective regions) that

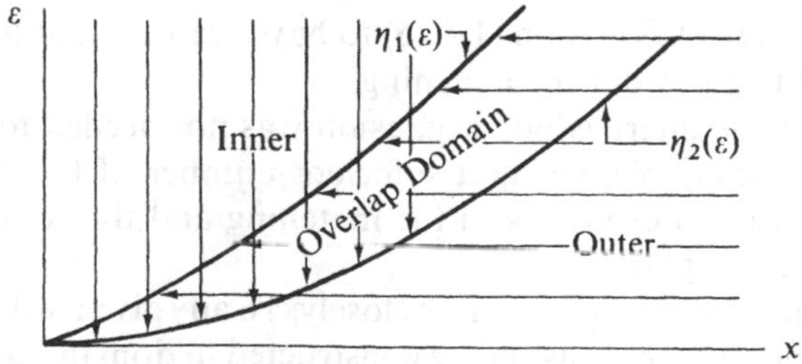

Figure 1.3.2 Actual Domains of Validity of Inner and Outer Expansions

this uniformly valid "composite" expansion consists of the sum of the inner and outer expansions, less those terms which are common to both. The common terms are those that are matched and should not appear twice in the composite expansion. Of course, the final result must be checked to verify that it leads to the inner or outer expansions under the appropriate limit processes.

To $O(1)h_0 = 0$ therefore $g_0(x^*)$ is uniformly valid. This can be verified in two ways, the first directly, by noting that

$$\lim_{\varepsilon \to 0} g(x^*, \varepsilon) - g_0(x^*) = 0 \quad \text{for all } x^* \tag{1.3.23}$$

or indirectly by noting that the outer limit process, when applied to g_0, gives zero. To order ε, we note by comparing Equations (1.3.11), (1.3.13) and (1.3.14) that in the sum $g_0 + \varepsilon(h_1 + g_1)$ the terms $\varepsilon - 1/x^*$ are common to both expansions thus, the composite expansion which is uniformly valid to $O(\varepsilon)$ is

$$\begin{aligned} y = k(x, \varepsilon) &= g_0 + \varepsilon\left(h_1 + g_1 - 1 + \frac{1}{\varepsilon x^*}\right) + O(\varepsilon^2) \\ &= e^{-x^*} - \frac{1}{1 + x^*} + \varepsilon\left[-\frac{e^{-x}}{x} - \frac{1}{1 + x^*} + \frac{1}{\varepsilon x^*}\right] + O(\varepsilon^2). \end{aligned} \tag{1.3.24}$$

To prove that (1.3.24) is indeed uniformly valid we must show that in the limit as $\varepsilon \to 0$, x fixed $\neq 0$ (1.3.24) reduces to εh_1, and that in the limit $\varepsilon \to 0$ x^* fixed it reduces to $g_0 + \varepsilon g_1$.

To show that (1.3.24) leads to the outer expansion in the limit $\varepsilon \to 0$, x fixed $\neq 0$, we rewrite it in the form

$$k = e^{-x/\varepsilon} - \frac{\varepsilon}{\varepsilon + x} + \varepsilon\left[-\frac{e^{-x}}{x} - \frac{\varepsilon}{\varepsilon + x} + \frac{1}{x}\right] + O(\varepsilon^2). \tag{1.3.25}$$

The term $e^{-x/\varepsilon}$ is transcendentally small. Developing $\varepsilon/(\varepsilon + x)$ gives

$$\frac{\varepsilon}{\varepsilon + x} = \frac{\varepsilon}{x(1 + \varepsilon/x)} = \frac{\varepsilon}{x}\left(1 - \frac{\varepsilon}{x} + \cdots\right) = \frac{\varepsilon}{x} - \frac{\varepsilon^2}{x^2} + \cdots \tag{1.3.26}$$

and the leading term in the above cancels the term $1/x$ in the bracketed expression of (1.3.25). Thus, to $O(\varepsilon)$ we have

$$\lim_{\substack{\varepsilon \to 0 \\ x\,\text{fixed}}} \frac{k}{\varepsilon} = -\frac{e^{-x}}{x} = h_1(x). \tag{1.3.27}$$

To show that (1.3.24) gives the inner expansion, we rewrite it in terms of x^* as follows

$$k = e^{-x^*} - \frac{1}{1 + x^*} + \varepsilon\left[\frac{-e^{-\varepsilon x^*} + 1}{\varepsilon x^*} - \frac{1}{1 + x^*}\right] + O(\varepsilon^2). \tag{1.3.28}$$

Developing $1 - e^{-\varepsilon x^*}$ gives $\varepsilon x^* + \cdots$. Therefore

$$\lim_{\substack{\varepsilon \to 0 \\ x^* \text{fixed}}} \frac{k - g_0(x^*)}{\varepsilon} = \frac{x^*}{1 + x^*} = g_1(x^*). \tag{1.3.29}$$

This proves that (1.3.24) is the uniformly valid asymptotic expansion of (1.3.3) to $O(\varepsilon)$. Similar, more tedious arguments can be used to calculate the next term in k.

The important feature to note here is that the uniformly valid expansion is *not* in the form (1.2.5) because the dependence on x and ε *cannot* be factored.

The above ideas lead to the following definitions:

We say that $\sum_{n=1}^{N} f_n(x^*)\mu_n(\varepsilon)$ is a "limit process" expansion as $\varepsilon \to 0$ of $f(x, \varepsilon)$ with respect to the sequence $\{\mu_n\}$ if it can be derived by making the change of variable $x = s(\varepsilon)x^*$ for a specific $s(\varepsilon)$ in f, and then proceeding with usual calculation for the f_n [cf. (1.2.8)] *with x^* fixed in the limit $\varepsilon \to 0$.*

Thus, the outer and inner expansions discussed in the above example are limit process expansions with $s = 1$ and $s = \varepsilon$, respectively. The composite expansion is not a limit process expansion. In fact, it is a general asymptotic expansion in the sense of the following definition

We say that $\sum_{n=1}^{N} k_n(x, \varepsilon)$ is a "general asymptotic expansion" as $\varepsilon \to 0$ of $f(x, \varepsilon)$ to N terms with respect to the sequence $\{\mu_n\}$ if for each $M = 1, 2, \ldots N$

$$\lim_{\varepsilon \to 0} \frac{f(x, \varepsilon) - \sum_{n=1}^{M} k_n(x, \varepsilon)}{\mu_n(\varepsilon)} = 0. \tag{1.3.30}$$

Thus, for the example considered $\mu_1 = 1, \mu_2 = \varepsilon, \ldots$ and

$$k_1 = e^{-x/\varepsilon} - \frac{\varepsilon}{\varepsilon + x} \tag{1.3.31}$$

$$k_2 = \frac{\varepsilon(1 - e^{-x})}{x} - \frac{\varepsilon^2}{\varepsilon + x} \tag{1.3.32}$$

etc.

We now turn to the very important question of whether the outer and inner expansions can be derived from the differential equation (1.3.1) without knowledge of the exact solution (1.3.3). This is crucial, if the above ideas are to be exploited for problems where an exact solution is not available. Assume an outer expansion of the form

$$y = \mu_1(\varepsilon)h_1(x) + \mu_2(\varepsilon)h_2(x) + \mu_3(\varepsilon)h_3(x) + \cdots \tag{1.3.33}$$

with μ_1, μ_2, etc. unknown. Substituting (1.3.33) into (1.3.1) and developing the outer expansion of the right hand side of (1.3.1) gives ($' = d/dx$)

$$\begin{aligned} &\varepsilon\mu_1 h_1' + \varepsilon\mu_2 h_2' + \varepsilon\mu_3 h_3' + \mu_1 h_1 + \mu_2 h_2 + \mu_3 h_3 + O(\mu_4) \\ &= -\varepsilon\frac{e^{-x}}{x} + \varepsilon^2 \frac{e^{-x}}{x^2}(2 + x) - \varepsilon^3 \frac{e^{-x}(3 + x)}{x^3} + O(\varepsilon^4). \end{aligned} \tag{1.3.34}$$

The lowest order term on the left hand side is $\mu_1 h_1$, while the leading term on the right hand side is $-\varepsilon e^{-x}/x$. Therefore, we must choose $\mu_1 = \varepsilon$ and find $h_1 = -e^{-x}/x$.

In the next approximation (with $\mu_1 = \varepsilon$) the equation

$$\varepsilon^2 h_1' + \mu_2 h_2 = \varepsilon^2 \frac{e^{-x}}{x^2}(2 + x) \tag{1.3.35}$$

contains the lowest order terms.

Thus, unless we set $\mu_2 = \varepsilon^2$ we have a contradiction $[h_1' \neq (e^{-x}/x^2) \times (2 + x)]$. This fixes $\mu_2 = \varepsilon^2$ and (1.3.35) defines h_2 to be

$$h_2 = \frac{e^{-x}}{x^2}(2 + x) - h_1' = \frac{e^{-x}}{x^2}. \tag{1.3.36}$$

Similarly, consistency of the terms of order ε^3 requires that $\mu_3 = \varepsilon^3$ and one then calculates

$$h_3 = -\frac{e^{-x}}{x^3} \tag{1.3.37}$$

and the process can be continued indefinitely. We see for this example that applying the outer limit to the differential equation does indeed reproduce the outer expansion. Since we are dealing with a first-order equation the calculation of the outer expansion does not involve the solution of differential equations.

Consider next the inner limit applied to the differential equation (1.3.1). Writing (1.3.1) and (1.3.2) in terms of x^* and expanding the right hand side in the limit process $\varepsilon \to 0$, x^* fixed gives

$$\frac{dy}{dx^*} + y = -\frac{x^*}{(1 + x^*)^2} + \varepsilon\frac{(x^{*2} + x^* + 1)}{(1 + x^*)^2} - \varepsilon^2\frac{(x^{*3} + 2x^{*2} + 2x^*)}{2(1 + x^*)^2} + O(\varepsilon^3) \tag{1.3.38a}$$

$$y(0) = 0. \tag{1.3.38b}$$

We assume an inner expansion of the form

$$y = \nu_0(\varepsilon)g_0(x^*) + \nu_1(\varepsilon)g_1(x^*) + \nu_2(\varepsilon)g_2(x^*) + \cdots \tag{1.3.39}$$

The boundary condition (1.3.38) then implies that $g_n(0) = 0$, $n = 0, 1, \ldots$. Substituting (1.3.39) into (1.3.38) gives the first approximation

$$\frac{dg_0}{dx^*} + g_0 = \begin{cases} 0 & \text{if } \nu_0 \gg 1, \\ \dfrac{-x^*}{(1 + x^*)^2} & \text{if } \nu_0 = 1. \end{cases} \tag{1.3.40}$$

Since the solution of (1.3.40) subject to $g_0(0) = 0$ is identically zero, the first nonvanishing term in (1.3.39) will correspond to the choice $v_0 = 1$ in which case the solution is

$$g_0(x^*) = e^{-x^*} - \frac{1}{1 + x^*}. \tag{1.3.41}$$

Repeating this argument to higher orders, we see that unless we choose $v_1 = \varepsilon$, $v_2 = \varepsilon^2$ etc. the corresponding $g_i(x^*)$ will be identically zero. Setting $v_i = \varepsilon^i$ then provides a set of nonhomogeneous equations which can be solved and which reproduce the inner expansion (1.3.6).

In the next chapter we will build on the above ideas and consider less trivial examples where considering all limit process expansions and matching will provide the uniformly valid general asymptotic expansion for cases where the differential equation is hard (or impossible) to solve exactly if $\varepsilon \neq 0$.

It can be noted here that the ideas introduced in order to construct and match asymptotic expansions depended in no way on the linearity of the particular example. Hence one can expect that these methods can be applied to nonlinear cases also.

Chapter 2

Limit Process Expansions Applied to Ordinary Differential Equations

In this chapter a series of simple examples are considered, some model and some physical, in order to demonstrate the application of various techniques concerning limit process expansions. In general we expect analytic dependence of the exact solution on the small parameter ε, but one of the main tasks in the various problems is to discover the nature of this dependence by working with suitable approximate differential equations. Another problem is to systematize as much as possible the procedures for discovering these expansions.

The main unifying features of problems having two or more limit process expansions is that certain terms in the governing differential equation will change their orders of magnitude depending on the domain in x. Often, (but not in all cases) the highest derivative in the differential equation will be multiplied by the small parameter ε, and this term will be small everywhere except near special points, e.g., boundary points.

In physical problems, ε is considered dimensionless and is found by expressing the entire problem in suitable dimensionless coordinates. Physical problems have another advantage from the point of view of perturbation procedures: very often the general nature of the solution is known and this simplifies the task of finding the appropriate limit process expansions.

2.1 Linear Oscillator: Regular Perturbation

As a first example, consider a case for which the exact solution is easily found: the response of a linear spring-mass-damping system initially at rest to an

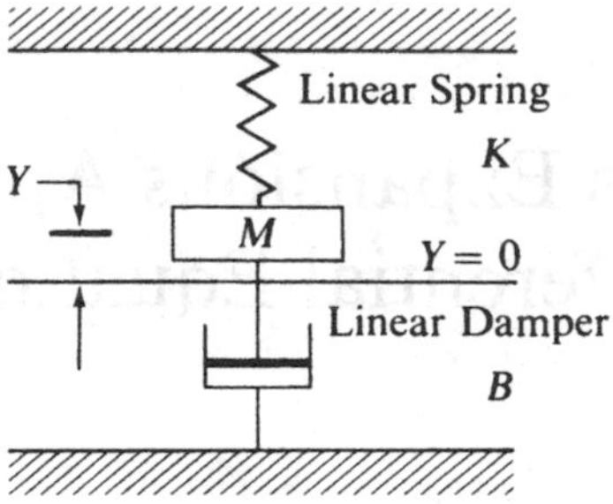

Figure 2.1.1 Spring-Mass-Damping System

impulse I_0 (see Figure 2.1.1). The equation and initial conditions are

$$M\frac{d^2Y}{dT^2} + B\frac{dY}{dT} + KY = I_0\delta(T) \tag{2.1.1a}$$

$$Y(0^-) = \frac{dY(0^-)}{dT} = 0, \tag{2.1.1b}$$

where δ is the Dirac delta function.

Problem (2.1.1) can be replaced by an equivalent one, (2.1.2) by considering an impulse-momentum balance across $T = 0$ or by integrating Equation (2.1.1) from $T = 0^-$ to $T = 0^+$:

$$M\frac{d^2Y}{dT^2} + B\frac{dY}{dT} + KY = 0, \qquad T > 0 \tag{2.1.2a}$$

$$Y(0^+) = 0 \tag{2.1.2b}$$

$$\frac{dY(0^+)}{dT} = \frac{I_0}{M}. \tag{2.1.2c}$$

The solution defined by this problem is the fundamental solution of this linear equation.

The "regular" perturbation is concerned with an approximation for small damping coefficient B. For small B, we expect the motion to be a slightly damped oscillation close to the free simple harmonic oscillation of the system —the solution of Equation (2.1.2) with $B = 0$. For the introduction of dimensionless coordinates, a suitable time scale is $\sqrt{M/K}$, the reciprocal of the natural frequency of free undamped motion, since this scale remains in the limit $B \to 0$. The length scale A, a measure of the amplitude, can be chosen arbitrarily and this choice will not affect the resulting dimensionless differential equation since it is linear. Actually, we will choose A in a form convenient for normalizing the initial velocity.

Setting

$$t^* = \frac{T}{(M/K)^{1/2}}, \qquad y = \frac{Y}{A} \tag{2.1.3}$$

we have

$$\frac{d^2y}{dt^{*2}} + 2\varepsilon^* \frac{dy}{dt^*} + y = 0, \tag{2.1.4}$$

where

$$\varepsilon^* = \frac{B}{2(MK)^{1/2}}.$$

In these variables $y(0^+) = 0$, $dy(0^+)/dt^* = 1$ if we set $A = I_0/(MK)^{1/2}$.

It is seen that the solution of the problem only involves the one parameter ε^*. Small damping corresponds to ε^* small so that an expansion with the limit process $\varepsilon^* \to 0$ is considered.

For this example it is evident that nontrivial correction terms can be obtained only if y is expanded in powers of ε^* as follows

$$y(t^*, \varepsilon^*) = g_0(t^*) + \varepsilon^* g_1(t^*) + \cdots. \tag{2.1.5}$$

By substituting the expansion (2.1.5) into Equation (2.1.4) and the initial conditions or, equivalently, by repeatedly applying the limit process $\varepsilon^* \to 0$, t^* fixed, we obtain the following sequence of linear problems to solve in order

$$\frac{d^2g_0}{dt^{*2}} + g_0 = 0; \qquad g_0(0) = 0, \qquad \frac{dg_0(0)}{dt^*} = 1 \tag{2.1.6a}$$

$$\frac{d^2g_1}{dt^{*2}} + g_1 = -2\frac{dg_0}{dt^*}; \qquad g_1(0) = \frac{dg_1(0)}{dt^*} = 0 \tag{2.1.6b}$$

$$\frac{d^2g_i}{dt^{*2}} + g_i = -2\frac{dg_{i-1}}{dt^*}; \qquad g_i(0) = \frac{dg_i(0)}{dt^*} = 0, \qquad i = 1, 2, \ldots. \tag{2.1.6c}$$

The equation for g_0 is that of the free undamped motion; the correction g_1 is computed by the damping acting on the velocity of the free motion g_0, and so on. The solutions are

$$g_0 = \sin t^*, \tag{2.1.7a}$$

$$g_1 = -t^* \sin t^*, \tag{2.1.7b}$$

etc. Thus, as far as the solution has been carried the result is

$$y = \sin t^* - \varepsilon t^* \sin t^* + O(\varepsilon^2 t^{*2}). \tag{2.1.8}$$

In this approximation, the amplitude decays linearly with time. If we consider the exact solution

$$y(t^*, \varepsilon^*) = \frac{e^{-\varepsilon^* t^*}}{\sqrt{1 - \varepsilon^{*2}}} \sin(\sqrt{1 - \varepsilon^{*2}}\, t^*) \tag{2.1.9}$$

we see that the amplitude $1 - \varepsilon^* t^*$ is the expansion of $e^{-\varepsilon^* t^*}$ to this order. Clearly, this expansion is only valid for t^* *fixed and finite*. Thus, this regular perturbation is identical to an inner limit process expansion as discussed in

Section 1.3. (This also explains the choice of notation for the time). For any finite time interval $0 < t^* < t_0^*$, ε^* can be chosen sufficiently small so that Equation (2.1.8) is a good approximation {uniformly valid to $O(\varepsilon)$} of the exact result of (2.1.9).

Because in this limit $e^{-\varepsilon^* t^*}$ and $\sin\sqrt{1-\varepsilon^{*2}}t^*$ are approximated by power series in $\varepsilon^* t^*$ and $\varepsilon^{*2} t^*$, the results are not uniformly valid over the entire time interval. We shall consider the problem of improving this approximation as $t^* \to \infty$ in chapter 3, but first, we consider a perturbation problem for Equation (2.1.1) which is singular at $T = 0$.

2.2 Linear Oscillator: Singular Problem

The singular problem is connected with approximations of Equation (2.1.1) for small values of the mass M. The difficulty near $T = 0$ arises from the fact that the limit equation with $M = 0$ is first order, so that both initial conditions, Eqs. (2.1.2b) and (2.1.2c), cannot be satisfied. The loss of an initial or boundary condition in a problem leads, in general, to the occurrence of a boundary layer.

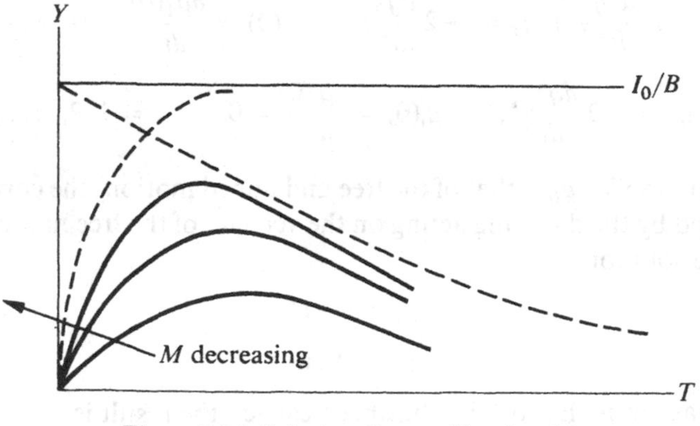

Figure 2.2.1 Solution Curves, Varying M

First, a discussion of this problem is given based on physical reasoning. The general nature of the solution for small values of M is sketched in Figure 2.2.1 with each solid curve corresponding to a fixed M. After a short time interval, it can be expected that the motion of a system is described by the limit form of Equation (2.1.1) with $M = 0$. Thus, we have

$$B\frac{dY}{dT} + KY = I_0\delta(T). \tag{2.2.1}$$

The initial condition in velocity is lost, and the effect of the impulse is to jump the initial displacement from $Y(0^-) = 0$ to

$$Y(0+) = \frac{I_0}{B}. \tag{2.2.2}$$

The solution is

$$Y = \frac{I_0}{B} e^{-KT/B}. \tag{2.2.3}$$

The solution demonstrates the exponential decay after the short initial interval in which the displacement increased infinitely rapidly from 0 to I_0/B. In order to describe the motion during the initial instants, we remark that inertia is certainly dominant at $T = 0$ (impulse-momentum balance) and that due to the large initial velocity, damping is important immediately but the spring is not, since deflection must be achieved before the spring is effective. Thus, in the initial instants, Equation (2.1.1) can be approximated by

$$M\frac{d^2Y}{dT^2} + B\frac{dY}{dT} = I_0\delta(t), \qquad Y(0-) = 0, \qquad \frac{dY}{dT}(0-) = 0 \tag{2.2.4}$$

with the solution

$$Y(T) = \frac{I_0}{B}\{1 - e^{-BT/M}\}. \tag{2.2.5}$$

This solution shows the approach of the deflection in a very short time ($M \to 0$) to the starting value for the decay solution of Equation (2.2.3). The curves are shown dashed in Figure 2.2.1 and give an overall picture of the motion.

Following our physical considerations, we aim to construct suitable asymptotic expansions for expressing these physical ideas and to show how to join these expansions. The method uses expansions valid after a short time (away from the initial point) and expansions valid near the initial point.

For the expansion valid away from the initial point, we find that natural variables are those based on a time scale for decay (B/K) and on an amplitude linear in I_0. Let

$$t = \frac{K}{B}T, \qquad y = B\frac{Y}{I_0}$$

so that Equation (2.1.2) reads

$$\varepsilon\frac{d^2y}{dt^2} + \frac{dy}{dt} + y = 0, \tag{2.2.6}$$

where $\varepsilon = MK/B^2$, with initial conditions

$$y(0) = 0, \qquad \frac{dy}{dt}(0) = \frac{1}{\varepsilon}. \tag{2.2.7}$$

The expansion valid away from the initial point (outer expansion) is associated with the limit

$$\varepsilon \to 0, \qquad t \text{ fixed (outer limit)}. \tag{2.2.8}$$

An asymptotic expansion in the form

$$y(t, \varepsilon) = \nu_1(\varepsilon)h_1(t) + \nu_2(\varepsilon)h_2(t) + \cdots, \tag{2.2.9}$$

is assumed.

The equations resulting from repeated application of the outer limit to Equation (2.2.6) or equating terms of the same order of magnitude when Equation (2.2.9) is substituted in Equation (2.2.6) are

$$\frac{dh_1}{dt} + h_1 = 0, \tag{2.2.10}$$

$$\frac{dh_2}{dt} + h_2 = \begin{cases} -\dfrac{d^2h_1}{dt^2} & \text{if } \dfrac{\nu_1\varepsilon}{\nu_2} = 1, \\[2ex] 0 & \text{if } \dfrac{\nu_1\varepsilon}{\nu_2} \to 0. \end{cases} \tag{2.2.11}$$

The initial conditions for this set of equations, as well as the orders of the various $\nu_i(\varepsilon)$, are unknown and have to be found by matching with an expansion valid near $t = 0$ (inner expansion), which takes account of the initial conditions.

The solutions for the outer expansion are

$$h_1 = A_1 e^{-t} \tag{2.2.12}$$

$$h_2 = A_2 e^{-t} - A_1 t e^{-t}. \tag{2.2.13}$$

The term $-A_1 te^{-t}$ above would be missing if it turned out that $\nu_1\varepsilon/\nu_2 \to 0$.

Various limits can be considered in which a representative time t approaches the origin in the $\varepsilon - t$ space at varying rates. That is, limits and associated asymptotic expansions can be considered for which t_η is fixed, where

$$t_\eta = \frac{t}{\eta(\varepsilon)}, \qquad \eta(\varepsilon) \to 0 \quad \text{as } \varepsilon \to 0.$$

In terms of this variable, Equation (2.2.6) is

$$\frac{\varepsilon}{\eta^2}\frac{d^2y}{dt_\eta^2} + \frac{1}{\eta}\frac{dy}{dt_\eta} + y = 0. \tag{2.2.14}$$

Three cases evidently arise, yielding different limit equations, which would be satisfied by the dominant term of corresponding asymptotic expansions.

Case I

Inner-inner limit $\eta \ll \varepsilon$ or $\varepsilon/\eta \to \infty$,

$$\frac{d^2y}{dt_\eta^2} = 0. \tag{2.2.15}$$

This is not a distinguished limit as Equation (2.2.15) holds for a class of functions of different orders. In this limit the governing equation is of second order so that the initial conditions can be accounted for. However, the use of an inner-inner limit is unnecessary for this problem since the expansion associated with it is contained in Case II. The expansion associated with this limit is valid only in a very small time interval, $t \leq k\eta(\varepsilon)$, around $t = 0$, the inertia-dominated regime.

Case II

Initial-layer (boundary-layer) limit, $\eta = \varepsilon$

$$\frac{d^2y}{dt_\eta^2} + \frac{dy}{dt_\eta} = 0. \tag{2.2.16}$$

This is called a *distinguished* limit since $\eta(\varepsilon)$ cannot belong to a class of functions with different orders but is definitely $O(\varepsilon)$ as $t \to 0$. The corresponding expansion will be a limit-process expansion [cf. Chapter 1]. The initial conditions can be satisfied. This limit yields the boundary-layer equation derived previously by physical reasoning. As t_η in Equation (2.2.16) approaches zero, the solutions of Equation (2.2.15) are obtained and, in that sense, Case I is contained in Case II.

Case III

Intermediate limit, $\varepsilon \ll \eta(\varepsilon) \ll 1$, or $\varepsilon/\eta(\varepsilon) \to 0$,

$$\frac{dy}{dt_\eta} = 0. \tag{2.2.17}$$

This also is not a distinquished limit but consists of a class of limits (and expansions) intermediate to the boundary-layer limit and outer limit. Equation (2.2.17) can handle neither the initial conditions nor the expected behavior at infinity. In this example, limits of Case III are also superfluous since they are contained in both those of Case II and the outer limit.

Thus, consider the asymptotic expansion associated with the initial layer limit

$$y(t;\varepsilon) = \mu_1(\varepsilon)g_1(t^*) + \mu_2(\varepsilon)g_2(t^*) + \cdots, \tag{2.2.18}$$

where $t^* = t/\varepsilon$, and with the sequence of approximate equations which result,

$$\frac{d^2 g_1}{dt^{*2}} + \frac{dg_1}{dt^*} = 0 \tag{2.2.19}$$

$$\frac{d^2 g_2}{dt^{*2}} + \frac{dg_2}{dt^*} = \begin{cases} -g_1 & \text{if } \dfrac{\varepsilon\mu_1}{\mu_2} \to 1, \\ 0 & \text{if } \dfrac{\varepsilon\mu_1}{\mu_2} \to 0. \end{cases} \tag{2.2.20}$$

The initial conditions fix $\mu_1(\varepsilon)$ since

$$\frac{dy}{dt} = \frac{\mu_1}{\varepsilon}\frac{dg_1}{dt^*} + \frac{\mu_2}{\varepsilon}\frac{dg_2}{dt^*} + \cdots = \frac{1}{\varepsilon} \quad \text{as } t^* \to 0.$$

Thus, $\mu_1 = 1$, and the initial conditions associated with (2.2.19) and (2.2.20) are

$$g_1(0) = 0, \qquad \frac{dg_1(0)}{dt^*} = 1, \tag{2.2.21}$$

$$g_2(0) = 0, \qquad \frac{dg_2(0)}{dt^*} = 0. \tag{2.2.22}$$

The solution of the initial-layer equation (2.2.19) is thus

$$g_1 = 1 - e^{-t^*}. \tag{2.2.23}$$

A nonzero term is needed for g_2 if corrections to the first term are to be found. Thus, μ_2 must be chosen equal to ε, and Equation (2.2.20) reads

$$\frac{d^2 g_2}{dt^{*2}} + \frac{dg_2}{dt^*} = -(1 - e^{-t^*}).$$

Integrating once and using the initial conditions (2.2.22) gives

$$\frac{dg_2}{dt^*} + g_2 = B_2 - t^* - (e^{-t^*}) = (1 - e^{-t^*}) - t^*.$$

The solution of this is

$$g_2 = (2 - t^*) - (2 + t^*)e^{-t^*}. \tag{2.2.24}$$

The inner expansion can be carried out in this way to any order.

The matching of inner and outer expansions serves to determine both the orders of magnitude of the ν_i and the constants of integration $A_1, A_2, \ldots$ in the outer expansion. Crudely speaking, the idea of matching is that the behavior of the outer expansion as $t \to 0$ and the inner expansion as $t^* \to \infty$ is in agreement. More formally, as discussed in Chapter 1, there is a domain in which both expansions are valid, an overlap domain, and in which these expansions agree. At first, the outer expansion is assumed valid, that is, truly asymptotic to the exact solution for $t \geq k_1 > 0$ (and for as many terms as are considered). Correspondingly, the inner expansion is valid for $0 \leq t^* \leq k_2$ or $0 \leq t \leq k_2 \varepsilon$. If the regions of validity can be extended so that the outer expansion is valid in $t \geq \eta_1(\varepsilon)$ and the inner expansion extended to cover $0 \leq t \leq \eta_2(\varepsilon)$, where $\eta_2 \gg \eta_1$, then the expansions are valid in an overlap domain. If we assume that overlap exists in this case, as already indicated for the first terms, an intermediate limit (corresponding to Case III) must hold in the overlap domain.

Accordingly, we define the intermediate variable t_η

$$t_\eta = \frac{t}{\eta(\varepsilon)}$$

for some, yet to be specified, class of functions $\eta(\varepsilon)$ contained in the class $\varepsilon \ll \eta(\varepsilon) \ll 1$. We will investigate whether the inner and outer expansions when expressed in terms of t_η agree to some order.

More precisely, matching to $O(1)$ requires[1] that we be able to find a class of functions $\eta(\varepsilon)$ such that

$$\lim_{\substack{\varepsilon \to 0 \\ t_\eta \text{ fixed}}} \left\{ \nu_1(\varepsilon) h_1(\eta t_\eta) - g_1\left(\frac{\eta t_\eta}{\varepsilon}\right) \right\} = 0 \tag{2.2.25a}$$

or, using the expressions that we have calculated for h_1 and g_1, this means that

$$\lim_{\substack{\varepsilon \to 0 \\ t_\eta \text{ fixed}}} \{\nu_1(\varepsilon) A_1 e^{-\eta t_\eta} - [1 - e^{-(\eta/\varepsilon)t_\eta}]\} = 0. \tag{2.2.25b}$$

Since $e^{-(\eta/\varepsilon)t_\eta}$ is transcendentally small, Equation (2.2.25b) can only hold if we choose $\nu_1(\varepsilon) = 1$ and $A_1 = 1$, and the overlap domain for the matching to order unity is

$$\varepsilon |\log \varepsilon| \ll \eta(\varepsilon) \ll 1.$$

Assuming that $\nu_2 = \varepsilon \nu_1 = \varepsilon$ [cf. Equation (2.2.11)], matching to order ε will provide the value of A_2. The matching condition, now assuming that the first two terms in each expansion are sufficient, is

$$\lim_{\substack{\varepsilon \to 0 \\ t_\eta \text{ fixed}}} \frac{1}{\varepsilon} \left\{ \nu_1(\varepsilon) h_1(\eta t_\eta) + \nu_2(\varepsilon) h_2(\varepsilon) - g_1\left(\frac{\eta t_\eta}{\varepsilon}\right) - \varepsilon g_2\left(\frac{\eta t_\eta}{\varepsilon}\right) \right\} = 0. \tag{2.2.26}$$

[1] Here we are assuming that the first term in each expansion is all that is needed in the matching to $O(1)$.

If we take account of terms which have already been matched, expand the exponential $e^{-\eta t_\eta}$, and neglect transcendentally small terms, Eq. (2.2.26) reduces to

$$\lim_{\substack{\varepsilon \to 0 \\ t_\eta \text{ fixed}}} \frac{1}{\varepsilon}\{-\eta t_\eta + O(\eta^2) + \varepsilon A_2 + \eta t_\eta - 2\varepsilon + O(\varepsilon^2)\} = 0. \quad (2.2.27)$$

The $O(\eta^2)$ remainder results form the next term in the expansion of $e^{-\eta t_\eta}$ while the $O(\varepsilon^2)$ remainder accounts for the third term, $\varepsilon^2 g_2$, in the inner expansion.

The terms of order η cancel identically, as they should (for otherwise $\eta/\varepsilon \to \infty$ and matching would fail to order ε). The terms of order ε in the numerator will cancel if we set $A_2 = 2$, and the most important remainder term is $O(\eta^2/\varepsilon)$ and will vanish if $\eta(\varepsilon) \ll \varepsilon^{1/2}$. Thus, the overlap domain for the matching to order ε is $\varepsilon|\log \varepsilon| \ll \eta \ll \varepsilon^{1/2}$ and is smaller than that associated with the $O(1)$ matching.

We note that all the assumptions that were made prior to the matching are justified a posteriori: It was necessary to set $\nu_2 = \varepsilon$ in order to obtain $A_2 = 2$. The choice $\varepsilon \ll \nu_2$ would have required that we set $A_2 = 0$ leaving no term to match the constant 2 in g_2. The number of terms needed in each expansion to order 1 and ε was one and two respectively. Failure of the matching would have required that these assumptions be reexamined and modified or abandoned. For example, in some problems a homogeneous solution (corresponding to the choice $\nu_1 \varepsilon \ll \nu_2$) might be needed to carry out the matching, or the number of terms required in one of the expansions might exceed those in the other, etc. Examples of these possibilities will occur as we study progressively more complicated problems.

The results achieved so far are summarized as follows:

$$y(t;\varepsilon) = e^{-t} + \varepsilon[2 - t]e^{-t} + \cdots \text{(outer)}, \quad (2.2.28a)$$

$$y(t;\varepsilon) = [1 - e^{-t^*}] + \varepsilon[(2 - t^*) - (2 + t^*)e^{-t^*}] + \cdots, \; t^* = \frac{t}{\varepsilon} \text{ (inner)}. \quad (2.2.28b)$$

A uniformly valid asymptotic expansion, of the general form of Equation (1.3.30) can be constructed from Equations (2.2.28). The inner and outer expansions have some terms in common, namely those terms that are matched (cancel out in the matching). If the two expansions in Equations (2.2.28) are added together and the common part subtracted, then a uniformly valid ($0 \le t \le t_1$) representation results. It is clear from the occurrence of the term $\varepsilon t e^{-t}$ in the outer expansion that uniform validity[2] does not extend to $t = \infty$. The remedy for this is discussed below. The common part (cp) is, for this case,

$$\text{cp} = 1 + \varepsilon(2 - t^*) + \cdots.$$

[2] Note that even though $te^{-t} \to 0$ as $t \to \infty$ the expansion in Equation (2.2.28a) is not uniformly valid when $t = O(\varepsilon^{-1})$ because $\varepsilon t e^{-t}$ is then $O(e^{-1/\varepsilon})$ which is the same order as the leading term.

In a boundary-layer problems such as this, it is possible and desirable to subtract the common part from the inner expansion, so that the part of the uniform expansion left in inner variables decays exponentially. The resulting expansion is

$$y(t;\varepsilon) = \{e^{-t} - e^{-t^*}\} + \varepsilon\{(2 - t)e^{-t} - (2 + t^*)e^{-t^*}\} + \cdots. \tag{2.2.29}$$

This expansion falls into the class of general asymptotic expansions of the form

$$y(t;\varepsilon) = F_0(t;\varepsilon) + \varepsilon F_1(t;\varepsilon) + \cdots.$$

defined by Equation (1.3.30). For this case, the general expansion takes the form of a composite expansion; each term is composed of an inner part which decays to zero and an outer part:

$$y(t;\varepsilon) = [h_1(t) + g_1^*(t^*)] + \varepsilon[h_2(t) + g_2^*(t^*)] + \cdots, \qquad t^* = \frac{t}{\varepsilon}, \tag{2.2.30}$$

where the g_i^* are defined in Equation (2.2.29).

Such a form as Equation (2.2.30) can be assumed *a priori* in many boundary-layer problems [cf. Section 3.2.6]. The $g_i^*(t/\varepsilon)$ should have the property of correcting the incorrect (in general) boundary condition of the first outer solution h_1 and should decay exponentially from the boundary.

Finally, note that in this example the first term of the uniformly valid approximation gives a good description of the physical phenomenon for small M. In physical variables, we have

$$y \cong \frac{I_0}{B}\{e^{-KT/B} - e^{-BT/M}\}. \tag{2.2.31}$$

The motion shows a rapid rise to peak at $T \cong (M/B)\log(B^2/KM)$ and an eventual decay.

If we are interested in extending the range of uniform validity of the outer expressions to $t = \infty$ it can be done by considering a more general outer limit process. There is a cumulative effect of inertia which slowly shifts the time scale for decay. For example a variable t^+

$$t^+ = t\{1 + a_1\varepsilon + a_2\varepsilon^2 + \cdots\} \tag{2.2.32}$$

can be considered and the associated limit process has $t^+(t, \varepsilon)$ fixed. The constants a_k are to be found so as to enforce uniform validity at infinity. The outer expansion is now of the form

$$y(t;\varepsilon) = h_1^+(t^+) + \varepsilon h_2^+(t^+) + \cdots. \tag{2.2.33}$$

Note that

$$\frac{dy}{dt} = \frac{dh_1^+}{dt^+} + \varepsilon\left\{\frac{dh_2^+}{dt^+} + a_1\frac{dh_1^+}{dt^+}\right\} + \cdots$$

$$\frac{d^2y}{dt^2} = \frac{d^2h_1^+}{dt^{+2}} + \varepsilon\left\{\frac{d^2h_2^+}{dt^{+2}} + 2a_1\frac{d^2h_1^+}{dt^{+2}}\right\} + \cdots.$$

Thus the sequence replacing (2.2.10), (2.2.11) is

$$\frac{dh_1^+}{dt^+} + h_1^+ = 0 \tag{2.2.34}$$

$$\frac{dh_2^+}{dt^+} + h_2^+ = -\frac{d^2h_1^+}{dt^{+2}} - a_1\frac{dh_1^+}{dt^+}. \tag{2.2.35}$$

The term εte^{-t} which caused trouble at infinity arose from the right hand side of (2.2.11) in particular from the presence of e^{-t}, a solution of the homogeneous equation. This term can be eliminated by the choice of a_1. In this example $a_1 = 1$ evidently makes the right hand side of (2.2.35) equal to zero.

The outer expansion is now

$$y(t;\varepsilon) = A_1e^{-t^+} + \varepsilon A_2 e^{-t^+} + \cdots, \qquad t^+ = t(1 + \varepsilon + \cdots). \tag{2.2.36}$$

This expansion is now uniform near infinity. A similar idea for periodic solutions will be discussed in Section 3.1.

2.3 Linear Singular Perturbation Problems with Variable Coefficients

The ideas of the previous section are now applied to some further examples with the aim of showing how variable coefficients in the equation can affect the nature of the expansion.

Consider first a general boundary-value problem for a finite interval, $0 \le x \le 1$, say, for

$$\varepsilon\frac{d^2y}{dx^2} + a(x)\frac{dy}{dx} + b(x)y = 0, \tag{2.3.1}$$

with the boundary conditions

$$y(0) = A, \qquad y(1) = B. \tag{2.3.2}$$

In general, A and B can depend in a regular way on ε, but here it is assumed that they are independent of ε.

The first question to be discussed is the existence of the solution to the boundary value problem for some range of $\varepsilon > 0$ in the neighborhood of $\varepsilon = 0$. A general result is that the solution to the boundary value problem specified by (2.3.1) and (2.3.2) exists and is unique if the corresponding problem with zero boundary conditions ($y(0) = y(1) = 0$) has only the trivial solution $y = 0$. That is, in order to apply simple boundary layer theory in this example we have to be sure that no eigenvalues exist as $\varepsilon \to 0$. If in fact eigenvalues exist as $\varepsilon \to 0$ the structure of the solution is much more complicated although it is still possible to find asymptotic results.

The conditions under which boundary layer theory can be applied are easily deduced from a canonical form of Equation (2.3.1). Under the transformation

$$y(x) = \exp\left[-\frac{1}{2\varepsilon}\int_0^x a(\xi)d\xi\right]w(x) \tag{2.3.3}$$

(2.3.1) takes the form ($' = d/dx$)

$$\varepsilon\frac{d^2w}{dx^2} - \left[\frac{a^2(x)}{4\varepsilon} + \frac{a'(x)}{2} - b(x)\right]w(x) = 0. \tag{2.3.4}$$

It is only necessary to discuss the lowest eigenvalue and corresponding eigenfunction since its existence (or nonexistence) implies the existence (or nonexistence) of the infinite discrete spectrum. The qualitative shape of the eigenfunctions for $y(x)$, $w(x)$ would be the same if we assumed that

$$\int_0^x a(\xi)d\xi < \infty.$$

Then, the fundamental eigenfunctions would necessarily have the qualitative shape shown in Figure 2.3.1. It can be seen from Equation (2.3.4) that $d^2w/dx^2 > 0$ when $w > 0$ if

$$\frac{a^2(x)}{4\varepsilon} + \frac{a'(x)}{2} - b(x) > 0, \qquad 0 \le x \le 1 \tag{2.3.5}$$

and in this case it is not possible to have an eigenfunction of the shape in Figure 2.3.1. Thus, a sufficient condition for the existence of simple boundary layers as $\varepsilon \to 0$ is merely that

$$a(x) \ne 0, \qquad |a'(x)| < \infty, \qquad b(x) < \infty, \qquad 0 \le x \le 1. \tag{2.3.6}$$

Then it always is possible to find sufficiently small ε so that the inequality in (2.3.5) is satisfied. In all the examples that we will consider in this section the inequality in (2.3.5) will hold.

Assume now that $a(x)$, $b(x)$ are such that the solution to the boundary value problem exists for ε sufficiently small. In the limit problem ($\varepsilon = 0$), the equation drops down to first-order and, in general, both boundary conditions cannot be satisfied. A boundary layer can be expected to occur at

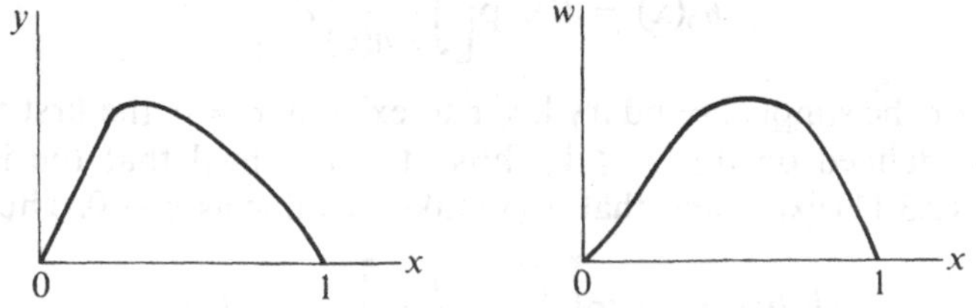

Figure 2.3.1 Shape of Eigenfunctions

either one of the ends (or possibly in the interior). As seen previously, the idea of the boundary layer is that the higher-order terms of Equation (2.3.1) dominate the behavior in the boundary layer. Thus, we have

$$\varepsilon \frac{d^2 y_{BL}}{dx^2} + a(x) \frac{dy_{BL}}{dx} = 0. \tag{2.3.7}$$

Exponential decay (rather than growth) is essential for boundary-layer behavior. Thus, if $a(x) > 0$ in $0 \leq x \leq 1$, the solutions of Equation (2.3.7) can be expected to have exponential decay near $x = 0$ ($\varepsilon > 0$), and the boundary layer occurs there; if $a(x) < 0$, the boundary layer occurs near $x = 1$. The case where $a(x)$ changes sign in the interval $0 \leq x \leq 1$ is evidently more complicated and examples are discussed in Sections 2.3.3 and 2.3.4.

We can make a few remarks about the general form the expansion takes. Assuming that $a(x) > 0$, $0 \leq x \leq 1$, we see that the outer expansion, valid away from $x = 0$, must be of the form of powers in ε, that is,

$$y(x; \varepsilon) = h_0(x) + \varepsilon h_1(x) + \varepsilon^2 h_2(x) + \cdots. \tag{2.3.8}$$

The various h_i all satisfy first-order differential equations and the boundary conditions

$$h_0(1) = B, \qquad h_i(1) = 0, \qquad i = 1, 2, 3, \ldots. \tag{2.3.9}$$

The sequence in powers of ε is necessary to ensure that the various h_i, $i = 1, 2, \ldots$, satisfy nonhomogeneous differential equations. If other orders of ε were used, the corresponding h_j would be identically zero. Thus, a sequence of equations for the outer expansion is obtained:

$$\begin{aligned}
a(x)\frac{dh_0}{dx} + b(x)h_0 &= 0, \\
a(x)\frac{dh_1}{dx} + b(x)h_1 &= -\frac{d^2 h_0}{dx^2}, \\
&\vdots \\
a(x)\frac{dh_i}{dx} + b(x)h_i &= -\frac{d^2 h_{i-1}}{dx^2}, \qquad i = 2, 3, \ldots.
\end{aligned} \tag{2.3.10}$$

The solution of the first equation of (2.3.10), if we take account of the boundary condition, is

$$h_0(x) = B \exp\left[\int_x^1 \frac{b(\xi)}{a(\xi)}\, d\xi\right]. \tag{2.3.11}$$

In order for the simple boundary layer to exist at $x = 0$, the first term $h_0(x)$ should be defined on $0 \leq x \leq 1$; thus, it is assumed that the integral in Equation (2.3.11) exists and that $h_0(x)$ takes a value as $x \to 0$. Thus,

$$h_0(0) = B \exp\left[\int_0^1 \frac{b(\xi)}{a(\xi)}\, d\xi\right] = C \quad \text{(say)}. \tag{2.3.12}$$

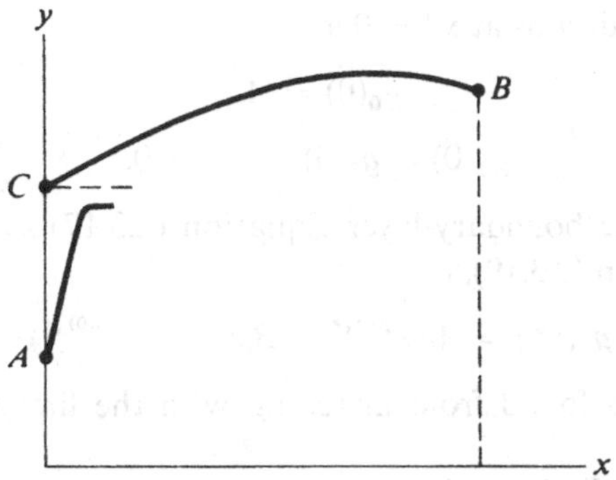

Figure 2.3.2 Boundary Layer at $x = 0$

In general, $h_0(0) \neq A$, so that a boundary layer exists at $x = 0$ (see Figure 2.3.2).

Assuming now that the variable coefficients have regular expansions near $x = 0$,

$$\begin{aligned} a(x) &= a^{(0)} + a^{(1)}x + \cdots, \qquad a^{(0)} > 0, \\ b(x) &= b^{(0)} + b^{(1)}x + \cdots, \end{aligned} \tag{2.3.13}$$

we can construct a boundary-layer expansion. The suitable inner variable is

$$x^* = \frac{x}{\varepsilon} \tag{2.3.14}$$

since the basic equation behaves near the boundary $x = 0$ in a way that is essentially the same as the constant-coefficient equation of the previous Section. The orders in ε of the terms in the asymptotic sequence for the inner expansion are found, strictly, from the condition of matching with the outer expansion. For the functions considered here, a power series in ε is adequate:

$$y(x;\varepsilon) = g_0(x^*) + \varepsilon g_1(x^*) + \cdots. \tag{2.3.15}$$

The coefficients in Equation (2.3.1) are also expressed in terms of the inner coordinate x^*, and thus have the expansion

$$\begin{aligned} a(x) &= a(\varepsilon x^*) = a^{(0)} + \varepsilon a^{(1)}x^* + \cdots, \\ b(x) &= b(\varepsilon x^*) = b^{(0)} + \varepsilon b^{(1)}x^* + \cdots. \end{aligned} \tag{2.3.16}$$

These expansions are useful for the inner-limit process $\varepsilon \to 0$, x^* fixed. Thus, the equations satisfied by the g_i are

$$\frac{d^2 g_0}{dx^{*2}} + a^{(0)} \frac{dg_0}{dx^*} = 0, \tag{2.3.17}$$

$$\frac{d^2 g_1}{dx^{*2}} + a^{(0)} \frac{dg_1}{dx^*} = -b^{(0)} g_0(x^*) - a^{(1)} x^* \frac{dg_0}{dx^*}. \tag{2.3.18}$$

The boundary conditions at $x^* = 0$ are

$$g_0(0) = A, \tag{2.3.19}$$

$$g_1(0) = g_2(0) = \cdots = 0. \tag{2.3.20}$$

The solution of the boundary-layer Equation (2.3.17) satisfying the initial condition, Equation (2.3.19), is

$$g_0(x^*) = Ae^{-a(0)x^*} + B_0(1 - e^{-a(0)x^*}). \tag{2.3.21}$$

The constant B_0 is found from matching with the first term of the outer expansion.

An intermediate limit for this problem is

$$\begin{aligned} &\lim(\varepsilon \to 0,\, x_\eta \text{ fixed}), \qquad x_\eta = \frac{x}{\eta(\varepsilon)}, \\ &\varepsilon \ll \eta \ll 1 \quad \text{or} \quad \frac{\eta}{\varepsilon} \to \infty, \qquad \eta \to 0. \end{aligned} \tag{2.3.22}$$

Exactly as in the case of constant coefficients, the boundary layer $g_0(x^*)$ goes asymptotically as $x^* \to \infty$ to a constant value, the deviation being transcendentally small in the intermediate limit. Thus, we have

$$\begin{aligned} \{h_0 - g_0\} = \Bigg\{ &B \exp\left[\int_{\eta x_\eta}^{1} \frac{b(\xi)}{a(\xi)}\, d\xi\right] - A \exp\left[-a^{(0)} \frac{\eta}{\varepsilon} x_\eta\right] \\ &- B_0\left(1 - \exp\left[-a^{(0)} \frac{\eta}{\varepsilon} x_\eta\right]\right)\Bigg\}. \end{aligned}$$

In the limit, this bracket approaches zero if

$$B_0 = B \exp\left[\int_0^1 \frac{b(\xi)}{a(\xi)}\, d\xi\right] = C. \tag{2.3.23}$$

Expressed simply, the matching is of the type

$$h_0(0) = g_0(\infty). \tag{2.3.24}$$

A uniformly valid approximation to order unity is obtained, as before, by adding h_0 and g_0 and subtracting the common part (Equation 2.3.23). Thus, the uniformly valid $(0 \leq x \leq 1)$ first approximation is

$$\begin{aligned} y_{uv}(x; \varepsilon) &= B \exp\left[\int_x^1 \frac{b(\xi)}{a(\xi)}\, d\xi\right] + \left\{A - B \exp\left[\int_0^1 \frac{b(\xi)}{a(\xi)}\, d\xi\right]\right\} e^{-a(0)x^*} + o(1), \\ x^* &= \frac{x}{\varepsilon}. \end{aligned} \tag{2.3.25}$$

Higher approximations can be carried out. These are illustrated for a special simple case. Also, a composite form of expansion analogous to Equation (2.2.30) can be used in this problem to produce the result of Equation (2.3.25);

the "defect" boundary-layer terms decay exponentially. Four illustrative examples with variable coefficients are now worked out, so that several features connected with the higher-approximation and nonregular behavior of the coefficients can be seen in concrete cases.

2.3.1 Analytic Coefficients

In this subsection we consider the special case

$$\varepsilon \frac{d^2y}{dx^2} + (1 + \alpha x)\frac{dy}{dx} + \alpha y = 0, \qquad 0 \le x \le 1, \alpha = \text{const.} > -1 \quad (2.3.26)$$

with boundary conditions

$$y(0) = 0, \qquad y(1) = 1. \quad (2.3.27)$$

The outer expansion is

$$y(x; \varepsilon) = h_0(x) + \varepsilon h_1(x) + \cdots, \quad (2.3.28)$$

so that

$$(1 + \alpha x)\frac{dh_0}{dx} + \alpha h_0 = 0, \qquad h_0(1) = 1; \quad (2.3.29)$$

$$(1 + \alpha x)\frac{dh_1}{dx} + \alpha h_1 = -\frac{d^2h_0}{dx^2}, \qquad h_1(1) = 0. \quad (2.3.30)$$

The boundary conditions for the outer expansion are taken at the right-hand end of the interval, since the boundary layer occurs at $x = 0$. The solutions of Equations (2.3.29) and (2.3.30) are easily obtained as

$$h_0(x) = \frac{1 + \alpha}{1 + \alpha x}, \quad (2.3.31)$$

$$h_1(x) = -\alpha\left\{\frac{1}{(1 + \alpha)(1 + \alpha x)} - \frac{1 + \alpha}{(1 + \alpha x)^3}\right\}. \quad (2.3.32)$$

Note that $h_0(0) = 1 + \alpha > 0$. The inner expansion valid near the boundary is

$$y(x; \varepsilon) = g_0(x^*) + \varepsilon g_1(x^*) + \cdots, \qquad x^* = \frac{x}{\varepsilon}. \quad (2.3.33)$$

The basic Equation (2.3.26) becomes

$$\frac{1}{\varepsilon}\left\{\frac{d^2g_0}{dx^{*2}} + \varepsilon\frac{d^2g_1}{dx^{*2}} + \cdots\right\} + \frac{1}{\varepsilon}\{1 + \alpha\varepsilon x^*\}\left\{\frac{dg_0}{dx^*} + \varepsilon\frac{dg_1}{dx^*} + \cdots\right\}$$
$$+ \alpha g_0 + \varepsilon\alpha g_1 + \cdots = 0, \quad (2.3.34)$$

so that we have

$$\frac{d^2 g_0}{dx^{*2}} + \frac{dg_0}{dx^*} = 0, \qquad g_0(0) = 0; \tag{2.3.35}$$

$$\frac{d^2 g_1}{dx^{*2}} + \frac{dg_1}{dx^*} = -\alpha x^* \frac{dg_0}{dx^*} - \alpha g_0, \qquad g_1(0) = 0. \tag{2.3.36}$$

The constants of integration for the boundary-layer solutions are found from the boundary condition at $x^* = 0$ and by matching with the outer expansion. It is easily found that

$$g_0(x^*) = B_0(1 - e^{-x^*}). \tag{2.3.37}$$

The intermediate limit for matching is just that of the general example of Equation (2.3.22). Thus, we have

$$\lim_{\substack{\varepsilon \to 0 \\ x_\eta \text{ fixed}}} \{h_0(x) + \cdots - B_0(1 - e^{-x^*})\} = 0$$

or

$$(1 + \alpha) - B_0 = 0. \tag{2.3.38}$$

The boundary layer in this example rises from $x = 0$ to the value of $h_0(0) = 1 + \alpha$; $h_0(0)$ may be any positive value. Using this value of B_0, we find the solution for g_1:

$$g_1(x^*) = C_1 + B_1 e^{-x^*} - \alpha(\alpha + 1)\left\{(x^* - 1) - \frac{x^{*2}}{2} e^{-x^*}\right\}. \tag{2.3.39}$$

and the boundary condition at $x^* = 0$ requires

$$B_1 = -C_1 - \alpha(\alpha + 1).$$

The matching condition to order ε is ($x = \eta x_\eta$, $x^* = (\eta/\varepsilon)x_\eta$)

$$\lim_{\substack{\varepsilon \to 0 \\ x_\eta \text{ fixed}}} \frac{[h_0(x) + \varepsilon h_1(x)] + \cdots - [g_0(x^*) + \varepsilon g_1(x^*) \cdots]}{\varepsilon} = 0.$$

If we neglect the transcendentally small terms in g_0, g_1, the matching condition is

$$\lim_{\substack{\varepsilon \to 0 \\ x_\eta \text{ fixed}}} \frac{1}{\varepsilon}\left\{\frac{1 + \alpha}{1 + \alpha\eta x_\eta} - \varepsilon\alpha\left[\frac{1}{(1 + \alpha)(1 + \alpha\eta x_\eta)} - \frac{1 + \alpha}{(1 + \alpha\eta x_\eta)^3}\right] + O(\varepsilon^2)\right.$$
$$\left. - (1 + \alpha) - \varepsilon\left[C_1 - \alpha(\alpha + 1)\left(\frac{\eta x_\eta}{\varepsilon} - 1\right)\right]\right\} = 0 \tag{2.3.40}$$

or

$$\lim_{\substack{\varepsilon \to 0 \\ x_\eta \text{ fixed}}} \frac{1}{\varepsilon}\left\{(1+\alpha)(1-\alpha\eta x_\eta) + O(\eta^2) - \varepsilon\alpha\left[\frac{1}{1+\alpha} - (1+\alpha) + O(\eta)\right]\right.$$

$$\left. + O(\varepsilon^2) - (1+\alpha) - \varepsilon\left[C_1 - \alpha(\alpha+1)\left(\frac{\eta x_\eta}{\varepsilon} - 1\right)\right]\right\} = 0.$$

The terms $O(\eta)$ are matched, so that choosing

$$C_1 + \alpha(\alpha+1) = -\alpha\left[\frac{1}{1+\alpha} - (1+\alpha)\right]$$

or

$$C_1 = -\frac{\alpha}{1+\alpha}$$

completes the matching.

The intermediate class of limits is now restricted to $\eta^2/\varepsilon \to 0$, $\eta \to 0$, or $\eta \ll \sqrt{\varepsilon}$. Strictly speaking in order to ensure that the exponentially decaying terms are really transcendentally small it is necessary to restrict the range of $\eta(\varepsilon)$ a little beyond $\varepsilon \ll \eta$. In fact, we require η to belong to the class

$$\varepsilon|\log \varepsilon| \ll \eta(\varepsilon) \ll \sqrt{\varepsilon}.$$

The common part (cp) contained in both inner and outer expansions which matches in this overlap region is, thus,

$$\text{cp} = (1+\alpha) - \varepsilon\left(\frac{\alpha}{1+\alpha} + \alpha(1+\alpha)[x^* - 1]\right), \tag{2.3.41}$$

written in the inner variable x^*. Adding the first two terms of inner and outer expansions and subtracting the common part yields the first two terms of a uniformly valid expansion. The inner expansion contributes only transcendentally small terms away from the boundary:

$$y(x;\varepsilon) = (1+\alpha)\left\{\frac{1}{1+\alpha x} - e^{-x^*}\right\} - \varepsilon\left\{\frac{\alpha}{1+\alpha}\left[\frac{1}{1+\alpha x} - e^{-x^*}\right]\right.$$

$$\left. - \alpha(1+\alpha)\left[\frac{1}{(1+\alpha x)^3} - e^{-x^*}\right] - \alpha(\alpha+1)\frac{x^{*2}}{2}e^{-x^*}\right\}$$

$$+ O(\varepsilon^2), \qquad 0 \le x \le 1. \tag{2.3.42}$$

The uniformly valid expansion again has the form of a composite expression,

$$y(x;\varepsilon) = \sum_{k=0} \varepsilon^k\left\{h_k(x) + f_k\left(\frac{x}{\varepsilon}\right)\right\}. \tag{2.3.43}$$

This form could have been taken as the starting point with the requirements that $h_0 + f_0$ satisfy both boundary conditions, $(h_i + f_i, i \neq 0)$ satisfy zero boundary conditions, and all f_i are transcendentally small away from the boundary.

2.3.2 Non-analytic Coefficients

This example is chosen to illustrate the effect of coefficients which are not analytic at the boundary point in modifying the form of the expansion. The basic assumption about the coefficients is that the first term of the outer expansion exists so that $h_0(x)$ is defined on $0 \le x \le 1$ (cf. Equation 2.3.8)

$$\varepsilon \frac{d^2y}{dx^2} + \sqrt{x}\frac{dy}{dx} - y = 0, \qquad 0 \le x \le 1 \tag{2.3.44}$$

$$y(0) = 0, \qquad y(1) = e^2. \tag{2.3.45}$$

In order to see if the solution to the boundary value problem exists we check the coefficient of $w(x)$ in (2.3.4). With

$$a(x) = \sqrt{x}, \qquad b(x) = -1$$

we have

$$\frac{a^2(x)}{4\varepsilon} + \frac{a'(x)}{2} - b(x) = \frac{x}{4\varepsilon} + \frac{1}{4\sqrt{x}} + 1 > 0.$$

Thus the solution to this particular problem exists for all $\varepsilon > 0$.

The outer expansion ($\varepsilon \to 0$, x fixed $\neq 0$) is

$$y(x, \varepsilon) = h_0(x) + \varepsilon h_1(x) + O(\varepsilon^2)$$

with the following equations and boundary conditions at $x = 1$:

$$\sqrt{x}\frac{dh_0}{dx} - h_0 = 0, \qquad h_0(1) = e^2 \tag{2.3.46}$$

$$\sqrt{x}\frac{dh_1}{dx} - h_1 = -\frac{d^2h_0}{dx^2}, \qquad h_1(1) = 0. \tag{2.3.47}$$

The solutions are

$$h_0(x) = e^{2\sqrt{x}} \tag{2.3.48}$$

$$h_1(x) = e^{2\sqrt{x}}\left[-\frac{1}{2x} + \frac{2}{\sqrt{x}} - \frac{3}{2}\right] \tag{2.3.49}$$

$$\varepsilon^2 h_2 = O\left(\frac{\varepsilon^2}{x^{5/2}}\right) \qquad \text{as } x \to 0. \tag{2.3.50}$$

We see that h_0 does not satisfy the boundary condition at $x = 0$, and that $h_1, h_2 \ldots$ have singularities at $x = 0$. Clearly, an inner expansion near $x = 0$ is needed to complete the description of the solution.

In order to construct the inner or boundary-layer expansion, a suitable boundary-layer coordinate must be found, so that derivative terms in Equation (2.3.44) dominate and are of the same order. With $\delta(\varepsilon)$ as an unknown function of ε ($\delta(\varepsilon) \to 0$ as $\varepsilon \to 0$) we let

$$x^* = \frac{x}{\delta(\varepsilon)}$$

be a boundary-layer coordinate, and consider the boundary-layer expansion

$$y(x, \varepsilon) = v_0(\varepsilon)g_0(x^*) + v_1(\varepsilon)g_1(x^*) + \cdots = \sum_{j=0} v_j(\varepsilon)g_j(x^*), \quad (2.3.51)$$

where $\{v_j(\varepsilon)\}$ is an asymptotic sequence to be determined. The basic equation (2.3.44) becomes

$$\frac{\varepsilon}{\delta^2(\varepsilon)}\left[v_0\frac{d^2g_0}{dx^{*2}} + v_1\frac{dg_1^2}{dx^{*2}} + \cdots\right] + \frac{\sqrt{\delta(\varepsilon)}}{\delta(\varepsilon)}\sqrt{x^*}\left[v_0\frac{dg_0}{dx^*} + v_1\frac{dg_1}{dx^*} + \cdots\right] - v_0g_0 - v_1g_1 - \cdots = 0. \quad (2.3.52)$$

The terms involving g_0 have v_0 as a common factor. The term involving dg_0/dx^* dominates the g_0 term. In order to have the richest equation, we must have the second derivative term also of this leading order. Thus, $\delta(\varepsilon)$ must be chosen so that

$$\frac{\varepsilon}{\delta^2(\varepsilon)} = \frac{1}{\sqrt{\delta(\varepsilon)}}$$

or

$$\delta(\varepsilon) = \varepsilon^{2/3}; \qquad x^* = \frac{x}{\varepsilon^{2/3}}. \quad (2.3.53)$$

The dominant boundary-layer equation is

$$\frac{d^2g_0}{dx^{*2}} + \sqrt{x^*}\frac{dg_0}{dx^*} = 0, \qquad g_0(0) = 0. \quad (2.3.54)$$

The solution, if we take account of the boundary condition at $x^* = 0$ is

$$g_0(x^*) = C_0\int_0^{x^*}\exp\left(-\frac{2}{3}\zeta^{3/2}\right)d\zeta. \quad (2.3.55)$$

As $x^* \to \infty$, the integral defining g_0 approaches a constant

$$k = \int_0^{\infty}\exp\left(-\frac{2}{3}\zeta^{3/2}\right)d\zeta$$

which can be expressed in terms of the Gamma function. The integral in Equation (2.3.55) can be transformed to an incomplete Gamma function. The approach of g_0 to its asymptotic value is exponential. In fact, writing

$$\int_0^{x^*} \exp\left(-\frac{2}{3}\zeta^{3/2}\right)d\zeta = \int_0^{\infty} \exp\left(-\frac{2}{3}\zeta^{3/2}\right)d\zeta - \int_{x^*}^{\infty} \exp\left(-\frac{2}{3}\zeta^{3/2}\right)d\zeta \tag{2.3.56}$$

and repeated integrations by parts of the second term on the right-hand side gives

$$g_0 = C_0\left\{k + \left[-\frac{1}{x^{*1/2}} + \frac{1}{2x^{*2}} - \frac{1}{x^{*7/2}} + O(x^{*-5})\right]\exp\left(-\frac{2}{3}x^{*3/2}\right)\right\}. \tag{2.3.57}$$

The matching to $O(1)$ can now be carried out. Introducing the matching variable

$$x_\eta = \frac{x}{\eta(\varepsilon)}$$

for some class of function $\eta(\varepsilon)$ such that $\varepsilon^{2/3} \ll \eta(\varepsilon) \ll 1$, we see that

$$g_0\left(\frac{\eta x_\eta}{\varepsilon^{2/3}}\right) = kC_0 + \text{T.S.T.}$$

$$h_0(\eta x_\eta) = 1 + O(\sqrt{\eta}).$$

Strictly speaking in order that $e^{-\eta^{3/2}x_\eta^{3/2}/\varepsilon}$ should represent a transcendentally small term it is necessary that $\varepsilon^{-M}e^{-\eta^{3/2}/\varepsilon} \to 0$ as $\varepsilon \to 0$ for any finite positive M. Thus, we have matching to $O(1)$ in the overlap domain

$$\varepsilon^{2/3}|\log \varepsilon|^{2/3} \ll \eta(\varepsilon) \ll 1$$

by setting

$$C_0 = \frac{1}{k}, \qquad k = \int_0^{\infty} e^{-(2/3)\xi^{3/2}}\, d\xi = \left(\frac{2}{3}\right)^{1/3}\Gamma\left(\frac{2}{3}\right) = 1.17\cdots.$$

For matching to higher orders, it is necessary to calculate further terms in the inner expansion which is evidently in the form

$$y(x, \varepsilon) = \sum_{i=0} \varepsilon^{i/3} g_i(x^*). \tag{2.3.58}$$

Each g_i satisfies

$$\frac{d^2 g_i}{dx^{*2}} + \sqrt{x^*}\,\frac{dg_i}{dx^*} = g_{i-1}, \qquad i = 1, 2, \ldots$$

or

$$\frac{d}{dx^*}\left[e^{(2/3)x^{*3/2}}\frac{dg_i}{dx^*}\right] = e^{(2/3)x^{*3/2}}g_{i-1}, \qquad i = 1, 2, \ldots, g_i(0) = 0. \tag{2.3.59}$$

Using Equation (2.3.59) one can calculate each g_i by quadrature. Only the asymptotic behavior of the g_i as $x^* \to \infty$ is needed for the matching. This is easily found by using (2.3.59) rather than attempting to expand the exact solution by repeated integrations by parts. For example, we have

$$\frac{d^2 g_1}{dx^{*2}} + \sqrt{x^*}\frac{dg_1}{dx^*} = 1 + \text{T.S.T.} \tag{2.3.60}$$

Clearly, as $x^* \to \infty$, dg_1/dx^* dominates over d^2g_1/dx^{*2}. Therefore, the first approximation must be given by $\sqrt{x^*}\, dg_1/dx^* \approx 1$ which implies that $g_1 \sim 2x^{*1/2}$ as $x^* \to \infty$ [and we confirm indeed that $d^2g_1/dx^{*2} = O(x^{*-3/2})$] and therefore $d^2g_1/dx^{*2} = o(\sqrt{x^*}\, dg_1/dx^*)$ as $x^* \to \infty$.

We are thus led to seek an expansion for g_1 in the form

$$g_1 = 2x^{*1/2} + K_1 + \sum_{n=1}^{\infty} \frac{C_{n/2}}{x^{*n/2}} \quad \text{as } x^* \to \infty. \tag{2.3.61}$$

Substituting the series (2.3.61) into (2.3.60) then defines all the $C_{n/2}$. The constant K_1, which is obviously the value of $g_1 - 2x^{*1/2}$ as $x^* \to \infty$, can only be determined by the matching since g_1 is not uniformly valid for $x^* \to \infty$. At any rate, g_1 must involve an arbitrary constant since we have only imposed one boundary condition on Equation (2.3.60). Thus K_1 can be related to the arbitrary constant in g_1.

Using this procedure we calculate the following expansions for the g_i as $x^* \to \infty$

$$g_0 = 1 + \text{T.S.T.} \tag{2.3.62}$$

$$g_1 = 2x^{*1/2} + K_1 - \frac{1}{2x^*} + O(x^{*-5/2}) \tag{2.3.63}$$

$$g_2 = 2x^* + 2K_1 x^{*1/2} + K_2 + \frac{1}{x^{*1/2}} - \frac{K_1}{2x^*} + O(x^{*-3/2}) \tag{2.3.64}$$

$$g_3 = \tfrac{4}{3}x^{*3/2} + 2K_1 x^* + 2K_2 x^{*1/2} + K_3 + \frac{K_1}{x^{*1/2}} - \frac{K_2}{2x^*} + O(x^{*-3/2}), \tag{2.3.65}$$

where K_2, K_3 etc. are arbitrary constants to be determined by matching.

If we attempt to match the inner and outer expansions to some order $\gamma(\varepsilon) \ll 1$, we must show that

$$\lim_{\substack{\varepsilon \to 0 \\ x_\eta \text{ fixed}}} \frac{h_0(\eta x_\eta) + \varepsilon h_1(\eta x_\eta) - \sum_{j=0}^{3} \varepsilon^{j/3} g_j(\eta x_\eta/\varepsilon^{2/3})}{\gamma(\varepsilon)} = 0 \tag{2.3.66}$$

for some domain in $\eta(\varepsilon)$.

Expanding $h_0, h_1, g_0, \ldots, g_3$ in terms of the intermediate variable x_η gives

$$h_0 = 1 + 2\eta^{1/2}x_\eta^{1/2} + 2\eta x_\eta + \tfrac{4}{3}\eta^{3/2}x_\eta^{3/2} + \tfrac{2}{3}\eta^2 x_\eta^2 + \tfrac{4}{15}\eta^{5/2}x_\eta^{5/2} + O(\eta^3) \tag{2.3.67}$$

$$h_1 = -\frac{1}{2\eta x_\eta} + \frac{1}{\eta^{1/2}x_\eta^{1/2}} + \frac{3}{2} + O(\eta^{1/2}) \tag{2.3.68}$$

$$g_0 = 1 + \text{T.S.T.} \tag{2.3.69}$$

$$g_1 = \frac{2\eta^{1/2}x_\eta^{1/2}}{\varepsilon^{1/3}} + K_1 - \frac{\varepsilon^{2/3}}{2\eta x_\eta} + O(\varepsilon^{5/3}\eta^{-5/2}) \tag{2.3.70}$$

$$g_2 = \frac{2\eta x_\eta}{\varepsilon^{2/3}} + 2K_1\frac{\eta^{1/2}x_\eta^{1/2}}{\varepsilon^{1/3}} + K_2 + \frac{\varepsilon^{1/3}}{\eta^{1/2}x_\eta^{1/2}} - \frac{K_1\varepsilon^{2/3}}{2\eta x_\eta} + O(\varepsilon\eta^{-3/2}) \tag{2.3.71}$$

$$g_3 = \frac{4}{3}\frac{\eta^{3/2}x_\eta^{3/2}}{\varepsilon} + 2K_1\frac{\eta x_\eta}{\varepsilon^{2/3}} + 2K_2\frac{\eta^{1/2}x_\eta^{1/2}}{\varepsilon^{1/3}} + K_3 + \frac{K_1\varepsilon^{1/3}}{\eta^{1/2}x_\eta^{1/2}} - \frac{K_2}{2}\frac{\varepsilon^{2/3}}{\eta x_\eta} + O(\varepsilon\eta^{-3/2}). \tag{2.3.72}$$

We see that the first four terms in the expansion of h_0 respectively match with the leading terms of g_0, g_1, g_2 and g_3. Moreover, the first three terms in h_1 match with corresponding terms in g_1, g_2 and g_3 if we set $K_3 = 3/2$. Now, let us consider an asymptotic sequence of functions $\gamma(\varepsilon)$. Eventually, we must proceed in the matching to functions $\gamma(\varepsilon) \ll \varepsilon^{1/3}$. At this point the term $\varepsilon^{1/3}K_1/\gamma(\varepsilon)$ arising from g_1 will become singular. It is obvious that this term has no counterpart in the outer expansion. Therefore, we must set $K_1 = 0$. A similar argument requires that $K_2 = 0$ because with $\gamma \ll \varepsilon^{2/3}$ the term $\varepsilon^{2/3}K_2/\gamma$ arising in g_2 becomes singular.

This completes the determination of K_1, K_2 and K_3. However, the matching can only be carried out with $\gamma = \varepsilon^{2/3}$ but not with $\gamma = \varepsilon$, without further terms. This can be seen from the following argument.

With $\gamma = \varepsilon^{2/3}$, the leading unmatched term in h_0 is $(2/3)\eta^2 x_\eta^2$. Thus, in Equation (2.3.66) this term becomes $O(\eta^2\varepsilon^{-2/3})$. It will vanish if we choose $\eta \ll \varepsilon^{1/3}$. The leading unmatched term in the inner expansion arises from g_1 and is $O(\varepsilon^{5/3}\eta^{-5/2})$; its contribution in Equation (2.3.66) is $O(\varepsilon^{4/3}\eta^{-5/2})$. Therefore, it will vanish if $\varepsilon^{8/15} \ll \eta$, and this establishes the nonempty overlap domain

$$\varepsilon^{8/15} \ll \eta \ll \varepsilon^{1/3} \tag{2.3.73}$$

for the matching to $O(\varepsilon^{2/3})$.

It is natural to attempt the matching to $O(\varepsilon)$ since both inner and outer expansions have been carried to $O(\varepsilon)$. This fails unless we include g_4 and g_5 in Equation (2.3.66).

To see this we note that with $\gamma = \varepsilon$ the vanishing of the term $(2/3)\eta^2 x_\eta^2$ in h_0 requires that $\eta \ll \varepsilon^{1/2}$. The term of order $\varepsilon^{5/3}\eta^{-5/2}$ is obviously a term proportional to $\varepsilon^2/x^{5/2}$ and must therefore match with a corresponding term in h_2 (cf. Equation (2.3.66)). We certainly do not need to go that far, and must eliminate this term from Equation (2.3.66). But, in order to do so $\varepsilon/\eta^{5/2}$ must tend to zero, i.e., $\varepsilon^{2/5} \ll \eta$ which contradicts the requirement $\eta \ll \varepsilon^{1/2}$.

We observe that each term in the expansion of h_0 was matched with the leading term of successive g_i's. Thus, we expect the two unmatched terms in h_0 to correspond to the leading terms in g_4 and g_5.

That this is indeed the case can be easily verified once we calculate the following expansions of g_4 and g_5 for x^* large

$$\begin{aligned} g_4 &= \tfrac{2}{3}x^{*2} + \tfrac{4}{3}K_1 x^{*3/2} + 2K_2 x^* + O(x^{*1/2}) \\ g_5 &= \tfrac{4}{15}x^{*5/2} + \tfrac{2}{3}K_1 x^{*2} + O(x^{*3/2}) \end{aligned} \tag{2.3.74}$$

Now that the first six terms in h_0 have been matched the remainder in Equation (2.3.66) arising from h_0 is $O(\eta^3\varepsilon^{-1})$ and will vanish if $\eta \ll \varepsilon^{1/3}$ again. This together with the condition $\varepsilon^{2/5} \ll \eta$ determines the nonempty overlap domain

$$\varepsilon^{2/5} \ll \eta \ll \varepsilon^{1/3}.$$

The uniformly valid composite expansion to $O(\varepsilon)$ is then obtained in the form

$$\begin{aligned} y = h_0 + \varepsilon h_1 + \sum_{j=0}^{5} \varepsilon^{j/3} g_j(x^*) - \Bigg[1 + \varepsilon^{1/3}\left(2x^{*1/2} - \frac{1}{2x^*}\right) \\ + \varepsilon^{2/3}\left(2x^* + \frac{1}{x^{*1/2}}\right) + \varepsilon(\tfrac{4}{3}x^{*3/2} + \tfrac{3}{2}) + \varepsilon^{4/3}(\tfrac{2}{3}x^{*2}) \\ + \varepsilon^{5/3}(\tfrac{4}{15}x^{*5/2})\Bigg] + O(\varepsilon^2). \end{aligned} \tag{2.3.75}$$

2.3.3 Interior Layer

This example and the following one are designed to illustrate some of the complications that may arise if $a(x)$ in Equation (2.3.1) changes sign in the domain of interest. We first consider the case where $a'(x)$ is positive throughout the interval with $a(0) = 0$, and choose an equation for which the exact solution can be explicitly calculated. We consider

$$\varepsilon \frac{d^2y}{dx^2} + x\frac{dy}{dx} - y = 0, \qquad -1 \le x \le 1, 0 < \varepsilon \ll 1 \tag{2.3.76}$$

$$y(-1) = 1; \qquad y(1) = 2. \tag{2.3.77}$$

Noting that $y = Cx$ where C is a constant, is one solution[3] of (2.3.76) we can calculate the other linearly independent solution in the form $y = x \int \times (e^{-x^2/2\varepsilon}/x^2)dx$. Integrating this latter solution by parts gives a more convenient representation, and we have the general solution in the form

$$y = C_1 x + C_2\left(e^{-x^2/2\varepsilon} + \frac{x}{\varepsilon}\int_{-1}^{x} e^{-s^2/2\varepsilon}\,ds\right). \tag{2.3.78}$$

Imposing the boundary conditions (2.3.77) gives

$$C_1 = -1 + \frac{3e^{-1/2\varepsilon}}{2e^{-1/2\varepsilon} + I(\varepsilon)/\varepsilon} \tag{2.3.79}$$

$$C_2 = \frac{3}{2e^{-1/2\varepsilon} + I(\varepsilon)/\varepsilon}, \tag{2.3.80}$$

where

$$I(\varepsilon) = \int_{-1}^{1} e^{-s^2/2\varepsilon}\,ds = 2\int_{0}^{\infty} e^{-s^2/2\varepsilon}\,ds - 2\int_{1}^{\infty} e^{-s^2/2\varepsilon}\,ds. \tag{2.3.81}$$

Integration by parts on the last integral shows that:

$$I(\varepsilon) \sim \sqrt{2\pi\varepsilon} + \text{T.S.T.} \tag{2.3.82}$$

Thus,

$$C_1 = -1 + \text{T.S.T.} \tag{2.3.83}$$

and

$$C_2 = 3\sqrt{\frac{\varepsilon}{2\pi}} + \text{T.S.T.} \tag{2.3.84}$$

We now examine the solution (2.3.78) in various regions of the interval $-1 \leq x \leq 1$. With the change of variable $s = \sqrt{\varepsilon}\sigma$ we can now write the solution

$$y = -x + \frac{3x}{\sqrt{2\pi}}\int_{-1/\sqrt{\varepsilon}}^{x/\sqrt{\varepsilon}} e^{-\sigma^2/2}\,d\sigma + \text{T.S.T.} \quad \text{for } x \neq 0 \tag{2.3.85a}$$

$$y = 3\sqrt{\frac{\varepsilon}{2\pi}} + \text{T.S.T.} \quad \text{for } x = 0. \tag{2.3.85b}$$

Therefore, if $x < 0$, the result in (2.3.85) reduces as $\varepsilon \to 0$ to

$$y \sim -x - \text{T.S.T.} \tag{2.3.86}$$

and if $x > 0$, we have

$$y \sim 2x + \text{T.S.T.} \tag{2.3.87}$$

[3] For this example $y = Cx$ is also the outer solution *to all orders*. The second solution is found by assuming the form $y = xw(x; \varepsilon)$, a standard method.

The exact solution thus shows that there are no boundary layers at either end point, and two branches of the outer solution, each satisfying the appropriate boundary condition, hold on each half of the interval $-1 \leq x \leq 1$. Clearly, if $x = O(\sqrt{\varepsilon})$ the above approximations are not valid and we have to examine equation (2.3.78) more carefully. We note that equation (2.3.76) has a distinguished limit with $\varepsilon \to 0$, $x^* = x/\sqrt{\varepsilon}$ fixed, and that in this limit the exact differential equation

$$\frac{d^2y}{dx^{*2}} + x^* \frac{dy}{dx^*} - y = 0 \tag{2.3.88}$$

must be solved. This distinguished limit follows directly from the solution (2.3.78) written in terms of x^* as follows

$$y = -\varepsilon^{1/2}x^* + \frac{3\varepsilon^{1/2}}{\sqrt{2\pi}}\left[e^{-x^{*2}/2} + x^* \int_{-1/\sqrt{\varepsilon}}^{x^*} e^{-\sigma^2/2}\, d\sigma\right] + \text{T.S.T.}$$

Now, letting $\varepsilon \to 0$ with x^* fixed, we obtain

$$y \sim \varepsilon^{1/2}\left[-x^* + \frac{3}{\sqrt{2\pi}}\left(e^{-x^{*2}/2} + x^* \int_{-\infty}^{x^*} e^{-\sigma^2/2}\, d\sigma\right)\right]. \tag{2.3.89}$$

Equation (2.3.89) describes a "corner-layer" solution valid near $x = 0$. We verify that this corner layer matches to the left with $y = -x$ and with $y = 2x$ to the right. Thus, the uniformly valid solution to $O(1)$ is, in fact, the corner-layer solution given by (2.3.89) and the result is sketched in Figure 2.3.3 below.

The reader can easily verify that the above results also follow systematically by seeking inner, interior and outer solutions to Equation (2.3.76). In particular, the fact that no boundary layers are possible at the end points $x = \pm 1$ follows from the observation that the corresponding boundary layer equations give exponential growth in the interior of the $-1 < x < 1$ interval. Thus, one must use the two outer expansions (2.3.86) and (2.3.87). Then, it remains to smooth the resulting corner at $x = 0$ and a solution of the form of Equation (2.3.89) with two arbitrary constants can be calculated for

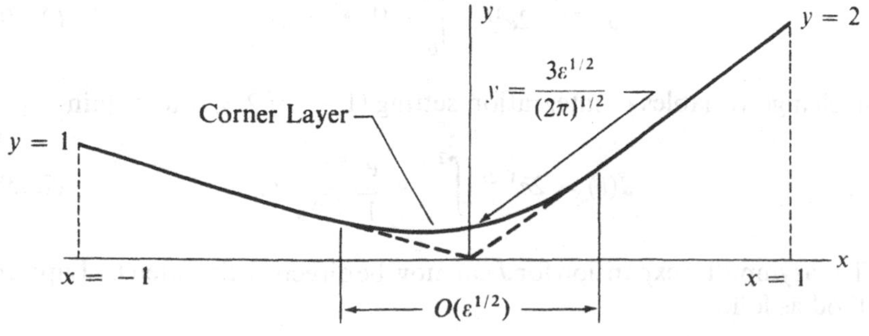

Figure 2.3.3 Corner Layer at $x = 0$

Equations (2.3.88). Matching to the left and right then determines these two constants.

2.3.4 Two Boundary Layers

This example illustrates an interesting collection of difficulties and is one where the techniques of matching of inner and outer expansions *fails* to yield a unique result. We simply change the signs of the second and third terms in Equation (2.3.76) and consider the boundary value problem

$$\varepsilon y'' - xy' + y = 0, \qquad -1 \le x \le 1; 0 < \varepsilon \ll 1 \tag{2.3.90}$$

$$y(-1) = 1; \qquad y(1) = 2. \tag{2.3.91}$$

Again, an exact solution can be explicitly found since $y = Cx$ solves Equation (2.3.90). We write the exact solution in the form

$$y = C_1 x + C_2\left[e^{x^2/2\varepsilon} - \frac{x}{\varepsilon}\int_{-1}^{x} e^{s^2/2\varepsilon}\, ds\right]. \tag{2.3.92}$$

Imposing the boundary conditions gives

$$C_1 = -1 + \frac{3e^{1/2\varepsilon}}{2e^{1/2\varepsilon} - J(\varepsilon)/\varepsilon} \tag{2.3.93}$$

and

$$C_2 = \frac{3}{2e^{1/2\varepsilon} - J(\varepsilon)/\varepsilon}, \tag{2.3.94}$$

where

$$J(\varepsilon) = \int_{-1}^{1} e^{s^2/2\varepsilon}\, ds. \tag{2.3.95}$$

To calculate $J(\varepsilon)$, we write Equation (2.3.95) in the form

$$J(\varepsilon) = 2e^{1/2\varepsilon}\int_{0}^{1} e^{-(1-s^2)/2\varepsilon}\, ds \tag{2.3.96}$$

then change variables of integration setting $(1 - s^2)/2 = u$ to obtain

$$J(\varepsilon) = 2e^{1/2\varepsilon}\int_{0}^{1/2} \frac{e^{-u/\varepsilon}}{\sqrt{1 - 2u}}\, du. \tag{2.3.97}$$

The asymptotic expansion for J can now be directly calculated by Laplace's method as follows

$$J(\varepsilon) = 2\varepsilon e^{1/2\varepsilon}[1 + \varepsilon + 3\varepsilon^2 + O(\varepsilon^3)]. \tag{2.3.98}$$

The asymptotic expansions for C_1 and C_2 thus become

$$C_1 = -\frac{3}{2\varepsilon} + \frac{7}{2} + O(\varepsilon) \tag{2.3.99}$$

$$C_2 = e^{-1/2\varepsilon}\left[-\frac{3}{2\varepsilon} + \frac{9}{2} + O(\varepsilon)\right]. \tag{2.3.100}$$

We will now show that the solution consists of a boundary layer at $x = -1$

$$y \sim -\tfrac{1}{2} + \tfrac{3}{2}e^{-(x+1)/\varepsilon} \tag{2.3.101}$$

a boundary layer at $x = +1$

$$y \sim \tfrac{1}{2} + \tfrac{3}{2}e^{(x-1)/\varepsilon} \tag{2.3.102}$$

and an outer solution

$$y \sim \frac{x}{2}. \tag{2.3.103}$$

The interesting feature of this outer solution valid at all interior points $|x| \neq 1$ is the fact that it does not satisfy either *boundary condition.*

Consider first the behavior of the exact solution near $x = -1$. We introduce the boundary layer variable $x^* = (x+1)/\varepsilon$ and write the solution (2.3.92) in the form

$$y = C_1(-1 + \varepsilon x^*) + C_2 e^{1/2\varepsilon}\left[e^{-x^* + \varepsilon x^{*2}/2} - \left(x^* - \frac{1}{\varepsilon}\right)\int_{-1}^{-1+\varepsilon x^*} e^{-(1-s^2)/2\varepsilon}\, ds\right]. \tag{2.3.104}$$

The integral appearing in equation (2.3.104) can be decomposed as follows

$$\begin{aligned}\int_{-1}^{-1+\varepsilon x^*} e^{-(1-s^2)/2\varepsilon}\, ds &= \int_{-1}^{0} e^{-(1-s^2)/2\varepsilon}\, ds + \int_{0}^{-1+\varepsilon x^*} e^{-(1-s^2)/2\varepsilon}\, ds \\ &= \frac{J(\varepsilon)}{2} e^{-1/2\varepsilon} + \int_{0}^{-1+\varepsilon x^*} e^{-(1-s^2)/2\varepsilon}\, ds.\end{aligned} \tag{2.3.105}$$

To calculate the asymptotic behavior of the integral appearing on the right-hand side of Equation (2.3.105) we first make the change of variable $s + 1 - \varepsilon x^* = u$ to obtain

$$K(x^*, \varepsilon) = -\int_{0}^{-1+\varepsilon x^*} e^{-(1-s^2)/2\varepsilon}\, ds = e^{-x^* + \varepsilon x^{*2}/2}\int_{0}^{1-\varepsilon x^*} e^{-(2u-u^2)/2\varepsilon + ux^*}\, du.$$

Then we set $(2u - u^2)/2 = \sigma$ and develop the integrand near $\sigma = 0$, which is the point giving the dominant contribution. We find

$$K(x^*\varepsilon) = e^{-x^* + \varepsilon x^{*2}/2} \int_0^{1/2 + O(\varepsilon)} e^{-\sigma/\varepsilon}[1 + (x^* + 1)\sigma + O(\sigma^2)]d\sigma. \quad (2.3.106)$$

The result in Equation (2.3.106) can now be directly evaluated by Laplace's method

$$K(x^*, \varepsilon) = \varepsilon e^{-x^* + \varepsilon x^{*2}/2}[1 + (x^* + 1)\varepsilon + O(\varepsilon^2)].$$

Substituting for K, C_1 and C_2 into Equation (2.3.104) then shows that in the limit $\varepsilon \to 0$, x^* fixed we have

$$y = -\tfrac{1}{2} + \tfrac{3}{2}e^{-x^*} + O(\varepsilon). \quad (2.3.107a)$$

A similar calculation near $x = 1$ with $x^{**} = (x - 1)/\varepsilon$, leads to

$$y = \tfrac{1}{2} + \tfrac{3}{2}e^{x^{**}} + O(\varepsilon). \quad (2.3.107b)$$

Thus, the boundary layer solution at the left end tends to $y = -\frac{1}{2}$ and the boundary layer at the right end tends to $y = +\frac{1}{2}$.

To determine the asymptotic limit of the exact solution away from the ends we use the expression for y given by Equation (2.3.92) and write the integral appearing there as follows

$$\int_{-1}^{x} e^{s^2/2\varepsilon}\, ds = \frac{J(\varepsilon)}{2} + e^{x^2/2\varepsilon} \int_0^x e^{-(x^2 - s^2)/2\varepsilon}\, ds.$$

Using the same type of arguments discussed earlier we find

$$y = C_1 x + C_2\left\{e^{x^2/2\varepsilon} - xe^{1/2\varepsilon}[1 + \varepsilon + O(\varepsilon^2)] - 2\left[1 + \frac{\varepsilon}{x^2} + O(x^2)\right]e^{x^2/2\varepsilon}\right\}. \quad (2.3.108)$$

Since $C_2 = O(e^{-1/2\varepsilon}/\varepsilon)$, the only contribution to y in the limit $\varepsilon \to 0$, $|x|$ fixed $\neq 1$ comes from the second term, as all the other terms multiplying C_2 become transcendentally small, and we find

$$\begin{aligned} y &= x\{C_1 - C_2 e^{1/2\varepsilon}[1 + \varepsilon + O(\varepsilon^2)]\} + \text{T.S.T.} \\ &= \frac{x}{2} + \text{T.S.T.} \end{aligned} \quad (2.3.109)$$

Thus, the uniformally valid solution has the form sketched in Figure 2.3.4.

Let us now attempt deriving this solution by matching appropriate boundary layer and outer expansions.

Again, the outer expansion, *to all orders*, is $y = Cx$, where C is an unknown constant. If we postulate a boundary layer at $x = -1$, we seek an expansion

$$y(x, \varepsilon) = g_0^{(L)}(x^*) + \varepsilon g_1^{(L)}(x^*) + \cdots, \quad (2.3.110)$$

where $x^* = (x + 1)/\varepsilon$.

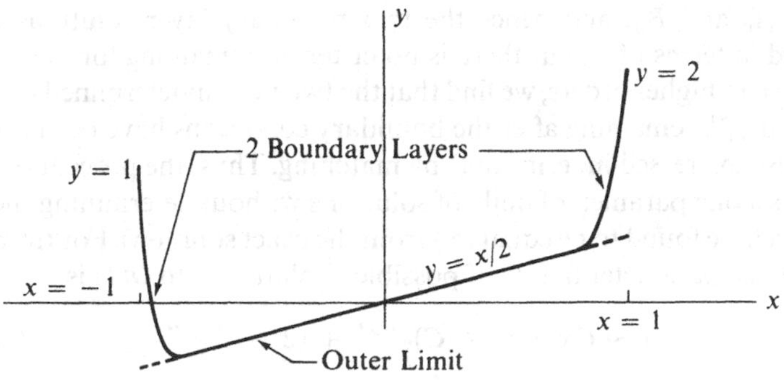

Figure 2.3.4 Two Boundary Layers

The functions $g_0^{(L)}$ and $g_1^{(L)}$ obey

$$\frac{d^2 g_0^{(L)}}{dx^{*2}} + \frac{dg_0^{(L)}}{dx^*} = 0 \tag{2.3.111}$$

$$\frac{d^2 g_1^{(L)}}{dx^{*2}} + \frac{dg_1^{(L)}}{dx^*} = -g_0^{(L)} + x^* \frac{dg_0^{(L)}}{dx^*} \tag{2.3.112}$$

with boundary conditions $g_0^{(L)}(0) = 1$, $g_1^{(L)}(0) = 0$, etc.

The solution of Equation (2.3.111) subject to $g_0^{(L)}(0) = 1$ is

$$g_0^{(L)}(x^*) = A_0 + (1 - A_0)e^{-x^*}, \tag{2.3.113}$$

where A_0 is to be determined by matching.

Similarly, with $x^{**} = (x - 1)/\varepsilon$, we consider a boundary layer near $x = 1$ in the form

$$y(x, \varepsilon) = g_0^{(R)}(x^{**}) + \varepsilon g_1^{(R)}(x^{**}) + \cdots. \tag{2.3.114}$$

Now $g_0^{(R)}$ and $g_1^{(R)}$ obey

$$\frac{d^2 g_0^{(R)}}{dx^{**2}} - \frac{dg_0^{(R)}}{dx^{**}} = 0 \tag{2.3.115}$$

$$\frac{d^2 g_1^{(R)}}{dx^{**2}} - \frac{dg_1^{(R)}}{dx^{**}} = -g_0^{(R)} + x^{**} \frac{dg_0^{(R)}}{dx^{**}} \tag{2.3.116}$$

with boundary conditions $g_0^{(R)}(0) = 2$, $g_1^{(R)}(0) = 0$, etc.

Solving (2.3.115) with $g_0^{(R)} = 2$ gives

$$g_0^{(R)} = B_0 + (2 - B_0)e^{x^{**}}. \tag{2.3.117}$$

The matching of the outer solution Cx with $g_0^{(L)}$ gives

$$-C = A_0$$

and matching Cx with $g_0^{(R)}$ gives

$$C = B_0.$$

Thus A_0 and B_0, and hence the two boundary layer solutions can be expressed in terms of C, but there is no criterion remaining for determining C! If we go to higher orders, we find that the two new undetermined constants in $g_1^{(L)}$ and $g_1^{(R)}$, remaining after the boundary conditions have been imposed, can be also expressed in terms of C by matching. Thus, the procedure simply generates a one parameter family of solutions without determining the value of C (which we found to be equal to $\frac{1}{2}$ from the exact solution). For the present case, the one parameter family of possible "solutions" to $O(1)$ is

$$y \sim Cx + (1 + C)e^{-x^*} + (2 - C)e^{x^{**}}. \tag{2.3.118}$$

The resolution of the difficulty (of not being able to determine C by matching) exhibited in this example is given in Reference 2.3.1 for the general boundary value problem

$$\varepsilon \frac{d^2y}{dx^2} + f(x, \varepsilon)\frac{dy}{dx} + g(x, \varepsilon)y = 0 \tag{2.3.119}$$

$$y(-a, \varepsilon) = \alpha(\varepsilon) \tag{2.3.120a}$$

$$y(b, \varepsilon) = \beta(\varepsilon), \tag{2.3.120b}$$

where f changes sign in the interval $-a \leq x \leq b$.

The basic idea used in Reference 2.3.1 is a consequence of the fact that a solution of Equation (2.3.119) subject to the boundary conditions (2.3.120) must be an *extremal* for the variational principle

$$\delta I = 0; \qquad \delta y(-a, \varepsilon) = \delta y(b, \varepsilon) = 0. \tag{2.3.121}$$

Here I is the integral

$$I = \int_{-a}^{b} L(x, y, y'; \varepsilon)dx \tag{2.3.122}$$

and L is any Lagrangian function for which Equation (2.3.119) is the Euler–Lagrange equation, i.e., Equation (2.3.119) is obtained from

$$\frac{d}{dx}\left(\frac{\partial L}{\partial y'}\right) - \frac{\partial L}{\partial y} = 0. \tag{2.3.123}$$

This is simply Hamilton's principle in dynamics {cf. Reference (2.3.2)} restated for the simple Equation (2.3.119).

Now, assume that we have found a one-parameter family of possible "solutions" of the boundary value problem (2.3.119), (2.3.120). The *correct* solution is the one which satisfies the variational principle (2.3.121), and this is true asymptotically also to any order in ε.

Using the one parameter family of possible "solutions" in Equation (2.3.122) determines I as a function of C and ε. Since all the possible "solutions" were calculated subject to the boundary conditions (2.3.120), the variational principle reduces to the requirement

$$\frac{\partial I}{\partial C}(C, \varepsilon) = 0 \tag{2.3.124}$$

and solving this equation should define the correct value for C.

In Reference (2.3.1) the Lagrangian function associated with Equation (2.3.119) was obtained using a dynamical analogy [cf. Problem 2.3.3]. The Lagrangian function corresponding to a given differential equation of the form (2.3.119) is not unique, and a simpler one can be found by transforming Equation (2.3.119) to the form

$$\frac{d}{dx}\left[p(x)\frac{dy}{dx}\right] + q(x)y = 0. \tag{2.3.125}$$

This is accomplished by multiplying Equation (2.3.119) by the integrating factor

$$a(x, \varepsilon) = \exp\frac{1}{\varepsilon}\int_0^x f(t, \varepsilon)dt \tag{2.3.126}$$

and we find

$$p = \varepsilon a \tag{2.3.127a}$$

$$q = ag. \tag{2.3.127b}$$

Now identifying

$$\frac{\partial L}{\partial y'} = py' \tag{2.3.128a}$$

$$\frac{\partial L}{\partial y} = -qy \tag{2.3.128b}$$

gives the Lagrangian

$$L(x, y, y', \varepsilon) = \tfrac{1}{2}(\varepsilon y'^2 - gy^2)\exp\frac{1}{\varepsilon}\int_0^x f(t, \varepsilon)dt. \tag{2.3.129}$$

It is easily verified that the Euler–Lagrange Equation (2.3.123) that results from the above Lagrangian is, in fact, Equation (2.3.119).

We will now proceed with the calculations leading to Equation (2.3.129) for the simple example of Equation (2.3.90). The reader can find the corresponding results for the general case of Equation (2.3.119) in Reference 2.3.1.

Using Equation (2.3.118) to evaluate y and y', we calculate the following leading term in the expansion of the Lagrangian

$$L = \frac{(1 + C)^2}{2\varepsilon}\exp\left[-\frac{(x + 2)^2}{2\varepsilon}\right] + \frac{(2 - C)^2}{2\varepsilon}\exp\left[-\frac{(x - 2)^2}{2\varepsilon}\right] + O(1). \tag{2.3.130}$$

In the above only the term $\varepsilon y'^2/2$ contributes to $O(\varepsilon^{-1})$ if g is $O(1)$ as it is in our case. The contribution due to the term $-gy^2/2$ is $O(1)$. We also note that to $O(\varepsilon^{-1})$ L and hence I are non-negative.

Using Equation (2.3.130) in Equation (2.3.122) we calculate the leading term of the action I in the form

$$I = \frac{5 - 2C + 2C^2}{\sqrt{2\varepsilon}} \int_{1/\sqrt{2\varepsilon}}^{3/\sqrt{2\varepsilon}} e^{-s^2}\, ds + O(1). \qquad (2.3.131)$$

Now, setting $\partial I/\partial C = 0$ gives $C = \frac{1}{2}$ as we found from the exact solution.

In the general case of Equation (2.3.119), one can also calculate a one parameter family of possible "solutions" to $O(1)$ exhibiting possible boundary layers at both ends. Using the variational principle (2.3.121) fixes the value of the undetermined parameter and thus defines the solution. In particular, depending on the value of C one may have a boundary layer at $x = -a$, $x = b$ or at both points.

Problems

1. Study the solution of

$$\varepsilon y'' + x^{3/2} y' - y = 0;$$
$$\varepsilon y'' + x^{-3/2} y' - y = 0 \qquad 0 \le x \le 1$$
$$y(0) = 0$$
$$y(1) = e^{2/5}$$

to $O(\varepsilon)$ with particular attention to exhibiting the overlap domain in the matching.

2. Calculate the solution of

$$\varepsilon y'' + x^{-1/2} y' - y = 0; \qquad 0 \le x \le 1$$
$$y(0) = 0$$
$$y(1) = e^{2/3}$$

3. As an alternate derivation of a Lagrangian for Equation (2.3.119) introduce the transformation

$$w = y \exp\left[\frac{1}{2\varepsilon} \int_0^x f(t, \varepsilon)\, dt\right]$$

which removes the first derivative term {cf. Equations (2.3.3) and (2.3.4)} and gives the following equation for w

$$\varepsilon w'' + \left(g - \frac{f^2}{4\varepsilon} - \frac{f'}{2}\right) w = 0.$$

Now interpret the above as the equation for an oscillator where ε is the mass, w the displacement, x is the time and the oscillator is acted on by a linear time dependent spring with spring constant k

$$k = g - \frac{f^2}{4\varepsilon} - \frac{f'}{2}.$$

Clearly, the kinetic energy is

$$T = \frac{\varepsilon w'^2}{2}$$

and the potential energy V, defined by

$$\frac{\partial V}{\partial w} = w\left(g - \frac{f^2}{4\varepsilon} - \frac{f'}{2}\right)$$

is

$$V = \frac{w^2}{2}\left(g - \frac{f^2}{4\varepsilon} - \frac{f'}{2}\right).$$

Thus, a Lagrangian for the oscillator equation is

$$L_1 = T - V = \frac{1}{2}\left[\varepsilon w'^2 - w^2\left(g - \frac{f^2}{4\varepsilon} - \frac{f'}{2}\right)\right].$$

Show that the Euler–Lagrange equation resulting from L_1 gives the oscillator equation.

Now transform w to y in L_1 and show that the Lagrangian

$$\tilde{L}(x, y, y'; \varepsilon) = \frac{1}{2}\left[\varepsilon y'^2 + fyy' + \left(\frac{f^2}{2\varepsilon} + \frac{f'}{2} - g\right)y^2\right]\exp\frac{1}{\varepsilon}\int_0^x f(t, \varepsilon)dt$$

which results will also give Equation (2.3.119) as Euler–Lagrange equation. Thus, the Lagrangian corresponding to a given differential equation is not unique.

Note that $\tilde{L}$ is more complicated than the L used in Equation (2.3.129). It must follow that the Euler–Lagrange equation corresponding to $L^* = \tilde{L} - L$ is identically satisfied for any y.

In our case

$$L^* = \left[\frac{fyy'}{2} + \left(\frac{f^2}{4\varepsilon} + \frac{f'}{4}\right)y^2\right]\exp\frac{1}{\varepsilon}\int_0^x f(t, \varepsilon)dt.$$

Verify that

$$\frac{d}{dx}\left[\frac{\partial L^*}{\partial y'}\right] - \frac{\partial L^*}{\partial y} = 0$$

for any y.

Show that in general, for a Lagrangian which is linear in y' of the form

$$L^*(x, y, y'; \varepsilon) = A(x, y; \varepsilon)y' + B(x, y; \varepsilon)$$

the Euler–Lagrange equation is identically satisfied for any y as long as

$$\frac{\partial A}{\partial x} = \frac{\partial B}{\partial y}$$

and this is the case for L^* in our problem.

4. Calculate the uniformly valid solution to $O(1)$ for the following boundary-value problems with $0 < \varepsilon \ll 1$.

   (a) $\varepsilon y'' + (x - \frac{1}{2})y' + \varepsilon y = 0;\ y(0) = 1,\ y(1) = -1$
   (b) $\varepsilon y'' + (\frac{1}{4} - x^2)y' + 2xy = 0;\ y(-1) = 1,\ y(1) = 2.$

5. For

$$\varepsilon y'' + y' - xy = 0, \qquad 0 \le x \le 1, 0 < \varepsilon \ll 1$$

with

$$y(0) = 0, \qquad y(1) = e^{1/2}$$

construct the first term of a uniformly valid asymptotic expansion. Express the exact solution to the boundary-value problem in terms of Bessel functions of order $\frac{1}{3}$, ($I_{1/3}$, $K_{1/3}$). Using the asymptotic properties of these Bessel functions, verify your result.

References

2.3.1 J. Grassman and B. J. Matkowsky, A variational approach to singularly perturbed boundary value problems for ordinary and partial differential equations with turning points, *S.I.A.M. Journal on Applied Mathematics*, **32**, No. 3, May 1977, pp. 588–597.

2.3.2 H. C. Corben and P. Stehle, *Classical Mechanics*, John Wiley and Sons, Inc., New York, 1960.

2.4 Theorem of Erdelyi, Rigorous Results

In Reference 2.4.1 Erdelyi has stated and proved a theorem for a singular boundary-value problem for a general second-order nonlinear ordinary differential equation. Under certain assumptions, he is able to show rigorously that the solution has the composite form of an outer-solution, a boundary-layer, and uniform-error term. Similar results have been given earlier by other authors (References 2.4.2, 2.4.3) under more restricted conditions. For example, in Reference (2.4.2) asymptotic expansions in terms of the small parameter are used, but in Reference (2.4.1) the proof is based on an integral representation of the solution, leading to a more general theorem. The theorem applies to some of the examples of the preceding Section 2.3 as well as the example of the following Section which provides a test of the sharpness of the theorem. Here the theorem is stated without proof. For details, the reader is referred to the original paper.

The problem considered is denoted by P_ε.

$$P_\varepsilon\colon \varepsilon \frac{d^2y}{dx^2} + F\left(x, y, \frac{dy}{dx}; \varepsilon\right) = 0, \qquad 0 \le x \le 1, \tag{2.4.1}$$

$$y(0) = A(\varepsilon), \tag{2.4.2}$$

$$y(1) = B(\varepsilon), \tag{2.4.3}$$

The situation in mind is one where the boundary layer occurs at $x = 0$ as $\varepsilon \to 0$. Fundamental to the theorem is the limit problem P_0 for $\varepsilon = 0$, in which the boundary condition is satisfied at $x = 1$.

$$P_0: F(x, u(x), u'(x); 0) = 0, \tag{2.4.4}$$

$$u(1) = B(0), \tag{2.4.5}$$

where $' = d/dx$. Here $u(x)$ corresponds to the first term of an outer expansion. In terms of $u(x)$, a certain function $\phi(x)$ and domain D_δ are defined.

Let

$$\phi(x) = \int_0^x F_{y'}(\xi, u(\xi), u'(\xi); 0)d\xi \tag{2.4.6}$$

and

$$D_\delta: \left\{(x, y, y'; \varepsilon), 0 \le x \le 1: |y(x) - u(x)| < \delta, \right.$$

$$\left. |y'(x) - u'(x)| < \delta\left(1 + \frac{e^{-\phi(x)/\varepsilon}}{\varepsilon}\right), 0 < \varepsilon < \varepsilon_0 \right\}. \tag{2.4.7}$$

Under the following assumptions, a theorem can be stated and proved.

(A1) Problem P_0 has a twice continuously differentiable solution $u(x)$ in $0 \le x \le 1$.

(A2) For some $\delta > 0$, F has partial derivatives with respect to y, y' in D_δ and F, and these partial derivatives are continuous functions of (x, y, y') for fixed ε.

(A3) As $\varepsilon \to 0$,

$$F(x, u(x), u'(x); \varepsilon) = O(\varepsilon),$$

$$q(x; \varepsilon) = F_y(x, u(x), u'(x); \varepsilon) = O(1),$$

$$p(x; \varepsilon) = F_{y'}(x, u(x), u'(x); \varepsilon) = \phi'(x) + \varepsilon p_1(x; \varepsilon),$$

where $\phi(x)$ is twice continuously differentiable in $0 \le x \le 1$,

$$\phi(0) = 0, \qquad \phi'(x) > 0 \quad \text{and} \quad p_1(x; \varepsilon) = O(1).$$

Also, we have

$$F_{yy}(x, y, y'; \varepsilon) = O(1), \qquad F_{yy'}(x, y, y'; \varepsilon) = O(1),$$

$$F_{y'y'}(x, y, y'; \varepsilon) = O(\varepsilon).$$

(A4) $B(\varepsilon) - B(0) = O(\varepsilon)$.

(A5) $F_y(x, y, y'; \varepsilon) = O(1)$, $F_{y'}(x, y, y'; \varepsilon) \ge \lambda_0 > 0$ in D_δ.

The statement of the theorem follows.

Under assumptions A1–A4, there exists a $\mu_0 > 0$ independent of ε, such that for $|A(\varepsilon) - u(0)| < \mu_0$, the problem P_ε has a solution

$$y = y(x; \varepsilon),$$

for all ε sufficiently small. The solution $y(x; \varepsilon)$ can be written

$$y(x; \varepsilon) = u(x) + v(x; \varepsilon) + w(x; \varepsilon), \tag{2.4.8}$$

where

$$y^*(x; \varepsilon) = u(x) + v(x; \varepsilon)$$

satisfies the differential equation and the boundary condition at $x = 1$. Furthermore, we have

$$v(x; \varepsilon) = O(\varepsilon), \qquad v'(x; \varepsilon) = O(\varepsilon), \qquad 0 \le x \le 1,$$

and

$$w(x; \varepsilon) = O\left(\exp \frac{-\phi(x)}{\varepsilon}\right), \qquad \varepsilon w'(x; \varepsilon) = O\left(\exp \frac{-\phi(x)}{\varepsilon}\right), \qquad 0 \le x \le 1.$$

Under the additional assumption (A5), the solution $y(x; \varepsilon)$, is unique. The significance of the various assumptions is fairly clear. The conditions on $\phi(x)$ in A3 are such to ensure a boundary layer at $x = 0$, and those on the partial derivatives of F to rule out certain nonlinearities.

The content of the theorem is that for $A(\varepsilon)$ sufficiently close to $u(0)$ (independent of ε), the solution can be expressed in the specified form. Here $u(x)$ corresponds to the first term of an outer expansion, $w(x; \varepsilon)$ is that part of a boundary-layer solution which decays exponentially away from the boundary, and $v(x; \varepsilon)$ are all the remaining correction terms which are uniformly small.

The dependence on ε of the boundary-layer term here agrees with that of the previous Section if the boundary-layer variable x^* is expressed in terms of x. In the example of Section 2.3.2, we see that

$$F(x, y, y'; \varepsilon) = \sqrt{x}y' - y, \qquad F_{y'} = \sqrt{x},$$

so that

$$\phi(x) = \frac{2x^{3/2}}{3}.$$

Thus, according to the theorem, $w(x; \varepsilon) = O(\exp(-\frac{2}{3}(x/\varepsilon)^{3/2}))$. The decaying part of the boundary layer as calculated in Section 2.3.2 goes as

$$(x^*)^{-1/2} \exp(-\tfrac{2}{3}x^{*3/2}), \quad \text{where } x^* = \frac{x}{\varepsilon^{2/3}}$$

so that there is agreement in the exponential factor. The example of the next Section shows that a boundary layer of this type may not occur if $A(\varepsilon)$ is too far away from $u(0)$, so that the result on the existence of μ_0 is sharp. For further discussion, see Reference 2.4.4.

There are numerous other studies where rigorous proofs of existence uniqueness and asymptotic properties are given for problems of the type P_ε

and more general cases. We will mention two such works in connection with the example of Section 2.7.2.

In these proofs, the solution is expressed, as in Equation (2.4.8), in the composite form of an outer limit, a boundary-layer, and an error term. In contrast, there are other studies where the solution of the exact nonlinear problem is compared to an appropriate neighboring linear problem.

For example, we outline next some of the results in Reference 2.4.5 without providing the proofs. Consider the nonlinear problem N_ε defined by

$$\varepsilon y'' + f(x, y, y', \varepsilon)y' = 0, \qquad 0 < x < 1 \tag{2.4.9}$$

subject to the linear boundary conditions

$$y'(0, \varepsilon) - ay(0, \varepsilon) = A \tag{2.4.10}$$

$$y'(1, \varepsilon) + by(1, \varepsilon) = B. \tag{2.4.11}$$

It is assumed that ε is positive and small, and that the constants a, b, A and B are independent of ε and satisfy $A \geq 0; a, b, B > 0$. In addition, the function f satisfies certain conditions as in A2, A3, etc. which we do not list. It is then shown in Reference 2.4.5 that both y and y' are close to the solution of an associated linear problem L_ε, defined by

$$\varepsilon w'' + g(x, \varepsilon)w' = 0, \qquad 0 < x < 1 \tag{2.4.12}$$

subject to the same boundary conditions as given in Equations (2.4.10, 11).

The degree of approximation is established to be

$$\left.\begin{aligned} y(x; \varepsilon) - w(x, \varepsilon) &= O(\varepsilon) \\ y'(x, \varepsilon) - w'(x, \varepsilon) &= O(\varepsilon) \end{aligned}\right. \tag{2.4.13}$$

uniformly on the interval $0 \leq x \leq 1$.

The key to the final result lies in finding the function $g(x, \varepsilon)$, which is to be a solution of the nonlinear integral equation

$$g(x, \varepsilon) = f\left\{x, \frac{B}{b}, \left(A + \frac{aB}{b}\right)\exp\left[-\frac{1}{\varepsilon}\int_0^x g(s, \varepsilon)ds\right], \varepsilon\right\}. \tag{2.4.14}$$

Unfortunately, this is generally difficult to solve and we can make no progress in solving the linear problem of Equation (2.4.12) unless g is known. For this reason, a constructive procedure which makes use of an available approximate solution is more useful.

References

2.4.1 A. Erdelyi, Singular perturbation, *Bull Amer. Math. Soc.* **68** (1962), 420–424.

2.4.2 E. A. Coddington and N. Levinson, A boundary value problem for a nonlinear differential equation with a small parameter, *Proc. Amer. Math. Soc.* **3** (1952), 73–81.

2.4.3 W. Wasow, Singular perturbations of boundary value problems for nonlinear differential equations of the second order. *Comm. Pure and Applied Math.* **9** (1956), 93–113.
2.4.4 A. Erdelyi, On a nonlinear boundary value problem involving a small parameter, *J. Aust. Math. Soc. II*, Part 4 (1962), 425–439.
2.4.5 J. J. Shepherd, Asymptotic solution of a nonlinear singular perturbation problem. *S.I.A.M. J. on Appl. Math.* **35** (1978) 176–186.

2.5 Model Nonlinear Example for Singular Perturbations

In this section, a model nonlinear example is studied that illustrates the following points: (1) A consistent study of boundary layers in the general sense enables the correct limit ($\varepsilon = 0$) solutions to be isolated, and (2) a wide variety of phenomena can occur even in a simple looking nonlinear problem. The example is

$$\varepsilon \frac{d^2y}{dx^2} + y\frac{dy}{dx} - y = 0, \qquad 0 \le x \le 1, \tag{2.5.1}$$

$$y(0) = A, \qquad y(1) = B.$$

Here A and B are not considered dependent on ε. The main problem of interest is the study of the dependence of the solutions on the boundary values A and B. Since the problem is nonlinear, the dependence on boundary conditions is nontrivial and can change the qualitative nature of the solution.

Actually, we can use a symmetry argument to cut in half the range of values of A and B that one need consider. We note that (2.5.1) is invariant under the transformation

$$y \leftrightarrow -y, \qquad x \leftrightarrow (1 - x), \qquad A \leftrightarrow -B, \qquad B \leftrightarrow -A.$$

Therefore, if $y = f(x, A, B)$ is a solution of (2.5.1) subject to the boundary conditions $y(0) = A$, $y(1) = B$, then $y = -f(1 - x, -B, -A)$ is also a solution of (2.5.1), and this solution satisfies the boundary conditions $y(0) = -B, y(1) = -A$.

Thus, the solution corresponding to a given point A, B generates a solution for the "reflected point" $-B, -A$. This reflected solution is obtained by the transformation $y \to -y, x \to 1 - x$, i.e. a reflection about the x axis followed by a reflection about the line $x = \frac{1}{2}$. We need therefore only consider values of A and B on one side of the line $B = -A$, and will take $B \ge -A$. Regions of the A, B plane ($B < -A$) with reflected solutions will be labeled with the subscript R.

The outer limit ($\varepsilon \to 0$, x fixed) satisfies the equation

$$h\frac{dh}{dx} - h = 0. \tag{2.5.2}$$

Two branches appear in the limit solution:

$$h = 0 \tag{2.5.3a}$$

$$h = x + c, \qquad c = \text{const.} \tag{2.5.3b}$$

Only the branch $h = x + c$ has a chance of satisfying an arbitrary end condition. Note that since the outer limit (2.5.3) is linear in x it is an exact solution of (2.5.1) and that if an outer expansion is constructed in outer variables

$$y(x;\varepsilon) = h(x) + \varepsilon h_1(x) + \cdots \tag{2.5.4}$$

then $h_1 = h_2 = \cdots = 0$. In this problem, it is not clear a priori where boundary layers will occur, so that various possibilities must be examined. Two outer solutions h_R and h_L are possible depending on whether the boundary condition at the right or left is satisfied respectively,

$$h_R(x) = x + B - 1 \tag{2.5.5}$$

$$h_L(x) = x + A. \tag{2.5.6}$$

These solutions take the values

$$\begin{aligned} h_R(0) &= B - 1 \\ h_L(1) &= A + 1 \end{aligned} \tag{2.5.7}$$

at the other end of the interval. We see that if $B - 1 = A$ (i.e., $h_R(0) = A$ or $h_L(1) = B$) the outer limit $h_R(x) = h_L(x)$ is an exact solution to the boundary value problem (2.5.1).

If $A \neq B - 1$, we expect a boundary or interior layer to occur somewhere on the interval (0, 1). A study of the possible boundary layers is now made with the aim of determining where these may occur and of what types they may be.

If y is not small, the simplest type of boundary layer can occur over a scale of order ε in x. In such a boundary layer, the derivative terms in Equation (2.5.1) are dominant. The corresponding asymptotic expansion is of the form

$$y(x;\varepsilon) = g(x^*) + \varepsilon g_1(x^*) + \cdots, \tag{2.5.8}$$

where

$$x^* = \frac{x - x_d}{\varepsilon}.$$

Here x_d gives the location of the layer. Since Equation (2.5.1) is autonomous the choice of x_d does not alter the equations governing $g, g_1, \ldots$. We have

$$\frac{d^2g}{dx^{*2}} + g\frac{dg}{dx^*} = 0 \tag{2.5.9}$$

which has a first integral

$$\frac{dg}{dx^*} + \frac{g^2}{2} = C, \; C = \text{const.} \tag{2.5.10}$$

Choosing $C < 0$ in (2.5.10) leads to a result which cannot match with the outer solution. Thus, we set $C = \beta^2/2 > 0$, and write the solution of Equation (2.5.10) as

$$g(x^*) = \beta \tanh \frac{\beta}{2}(x^* + k) \tag{2.5.11}$$

or

$$g(x^*) = \beta \coth \frac{\beta}{2}(x^* + k), \tag{2.5.12}$$

where k is a constant of integration.

A sketch of these solutions is given in Figure 2.5.1 where the abscissa is $x^* + k$. Thus, varying k results in a translation of the curves along the x^* axis.

The tanh solution, Equation (2.5.11), increases to its asymptotic value β as $x^* \to \infty$ and decreases to $-\beta$ as $x^* \to -\infty$. The approach is exponential:

$$g(x^*) \sim \beta\{1 - 2e^{-\beta(x^*+k)} + \cdots\} \quad \text{as } x^* \to \infty. \tag{2.5.13}$$

The coth solution, Equation (2.5.12), decreases from infinity at $x^* = -k$ to its asymptotic value β as $x^* \to \infty$ and increases from $-\infty$ at $x^* = -k$ to its asymptotic value $-\beta$ as $x^* \to -\infty$.

Segments of these solutions can be used as boundary layers that match in a simple way with the outer solutions of Equations (2.5.5), (2.5.6). Evidently, the matching conditions are

$$h_R(x_d) = g(\infty) \quad \text{or} \quad h_L(x_d) = g(-\infty). \tag{2.5.14}$$

This type of simplified matching to order unity follows directly from the consideration of suitable intermediate limits as discussed earlier.

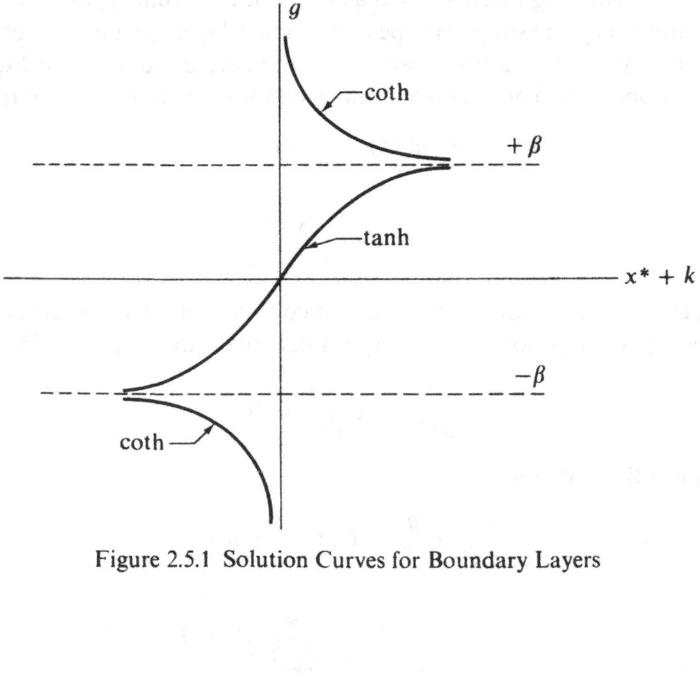

Figure 2.5.1 Solution Curves for Boundary Layers

Before carrying out the discussion of the possibilities for various boundary conditions, notice that another distinguished limit exists for Equation (2.5.1) if y is allowed to be small. The fact that scaling y may give a different distinguished limit is a consequence of the nonlinearity. In fact, if $\bar{y} = y/\sqrt{\varepsilon}$ and $\bar{x} = (x - x_0)/\sqrt{\varepsilon}$, the equation for $\bar{y}$ is free of ε. Considered as a local solution derived from the exact equation by means of the asymptotic expansion

$$y(x; \varepsilon) = \sqrt{\varepsilon} f(\bar{x}) + \varepsilon f_1(\bar{x}) + \cdots, \qquad \bar{x} = \frac{x - x_0}{\sqrt{\varepsilon}} \tag{2.5.15}$$

$f(\bar{x})$ should be an important element in some approximations. This statement is based on the idea that distinguished limits are always significant. The equation for f is the exact Equation (2.5.1)

$$\frac{d^2 f}{d\bar{x}^2} + f\frac{df}{d\bar{x}} - f = 0. \tag{2.5.16}$$

Thus, the local solution (2.5.15) can be calculated only if the boundary conditions appropriate for f simplify the solution of (2.5.16).

Next, consider the range of values of A and B for which solutions can be composed of h and g functions, that is, of outer solutions and boundary layers of order ε in thickness. The situation is represented on the (A, B) diagram of Figure 2.5.2. The line $B = A + 1$ represents solutions with no boundary layer where the exact solution is the outer limit $h_R = h_L = x + A = x + B - 1$. If $A > B - 1$, the outer solution

$$h_R = x + B - 1 \tag{2.5.17}$$

satisfies the right-hand boundary condition and takes a positive value if $B - 1 > 0$.

A boundary layer at $x = 0$ descending to $B - 1$ can then be used to complete the solution. Thus, the triangular domain $A > B - 1 > 0$ consists of a left boundary layer descending by a coth solution to the h_R outer limit. This is abbreviated by (LBL ↓ coth) in the diagram. Such a boundary layer [cf. Figure 2.5.1] matched to h_R is

$$g_L(x^*) = (B - 1)\coth\left(\frac{B - 1}{2}\right)(x^* + k), \qquad x^* = \frac{x}{\varepsilon}. \tag{2.5.18}$$

The value of k is chosen to satisfy the boundary condition at $x^* = 0$.

$$A = (B - 1)\coth\left(\frac{B - 1}{2} k\right), \tag{2.5.19}$$

i.e., $k = [2/(B - 1)]\coth^{-1}[A/(B - 1)]$. Note that the lower boundary of this domain brings us to the limiting case $B = 1$ where the boundary layer solution decays algebraically as $x^* \to \infty$, and we put this case aside temporarily. Corresponding to region I we have region I_R consisting of a right boundary layer ascending by a coth solution for values of A and B such that

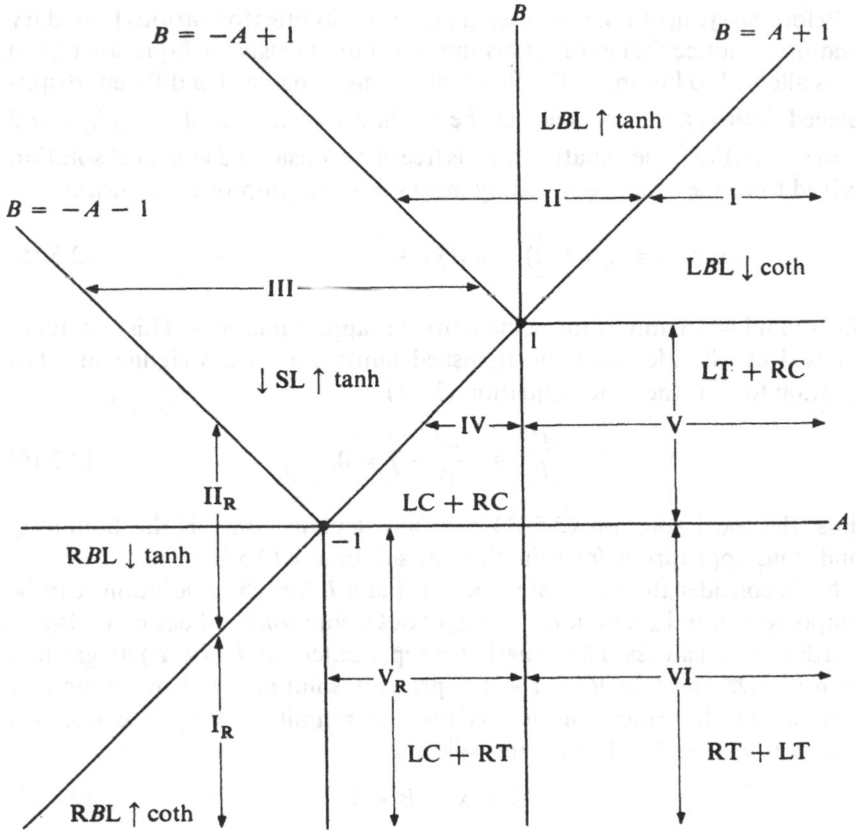

Figure 2.5.2 Possible Solutions in the $A - B$ Plane

$B < A + 1 < 0$. It is easy to verify that the choice of h and g functions is unique here. For example, for region I if we had chosen h_L we would have needed a right boundary layer rising to a positive value and no such boundary layer solution is available.

Next consider the region to the left of the line $B = A + 1$, i.e., $A < B - 1$ but with B still greater than unity. It is still possible to fit in a tanh-type boundary layer at the left end to match h_R, provided that $|A| < B - 1$. The restriction $|A| < B - 1$ insures that the asymptotic value of the tanh solution as $x^* \to \infty$ can be equal to A. Thus, region II is defined by $0 \leq |A| < B - 1$ with a left tanh boundary layer which rises from A to $B - 1$. In region II_R with $|B| < |A + 1|$, $A + 1 < 0$ we have a right tanh boundary layer which descends from B to $A + 1$.

A case between II and II_R has $B > A + 1$, but a tanh boundary layer at the end cannot provide a sufficient rise (or descent) to match the end condition. There is, however, the possibility of using the tanh solution at an interior point x_d and matching both as $x^* \to \infty$ and $x^* \to -\infty$. The boundary layer is so to speak, pushed off the ends and appears in the interior as a shock-layer. This is the case III in Figure 2.5.2 and has $B > A + 1$, $-(B + 1) < A < 1 - B$. The left and right boundary conditions are satisfied by outer solutions $h_L = A + x$, $h_R = x - 1 + B$. The tanh solution of Equation (2.5.11) matches to values $\pm B$ symmetric about $y = 0$ as $x^* \to \pm\infty$. Thus, this solution can serve as a shock layer centered at $x = x_d$, where x_d is defined by the symmetry condition $h_L(x_d) = -h_R(x_d)$, i.e.,

$$A + x_d = -B - x_d + 1$$

which gives

$$x_d = \frac{1 - A - B}{2}. \tag{2.5.20}$$

The inner solution is

$$g(x^*) = \frac{B - A - 1}{2} \tanh \frac{B - A - 1}{4} x^* \tag{2.5.21}$$

$$x^* = \frac{x - (1 - A - B)/2}{\varepsilon}.$$

The possibilities for boundary layers of order ε in thickness are now exhausted but large parts of the (A, B) plane are still inaccessible; for example, $A > 0, B < 0$. A hint at the kind of solutions needed is obtained by considering the special case $A = 0$, $0 < B < 1$. In this example, outer solutions of different branches can be used to satisfy the end conditions

$$h_L(x) = 0, \qquad h_R = x + B - 1 \tag{2.5.22}$$

These solutions intersect in a corner at $x = x_c = 1 - B$. A smooth solution over the full interval can be found if we can exhibit a corner-layer solution centered about $x = x_c$ and matching with h_L and h_R of Equation 2.5.22. Such a corner-layer solution, if it exists, must be contained in the solutions of Equation (2.5.16). The matching conditions for h_L and h_R are such that

$$f(\bar{x}) \to 0 \quad \text{as } \bar{x} \to -\infty; \qquad f(\bar{x}) \to \bar{x} \quad \text{as } \bar{x} \to \infty. \tag{2.5.23}$$

To determine whether such solutions exist, we study the phase plane of Equation (2.5.16),

$$v = \frac{df}{d\bar{x}} \tag{2.5.24}$$

$$\frac{dv}{df} = -\frac{f(v - 1)}{v} \tag{2.5.25}$$

The diagram of the integral curves of Equation (2.5.25) is Figure 2.5.3. Along any path, the direction of increasing $\bar{x}$ is indicated by an arrow, as found from Equation (2.5.24). It is clear that the paths which approach $v = 1 = df/d\bar{x}$ are capable of matching of the type $f(\bar{x}) \to \bar{x}$ as $\bar{x} \to +\infty$.

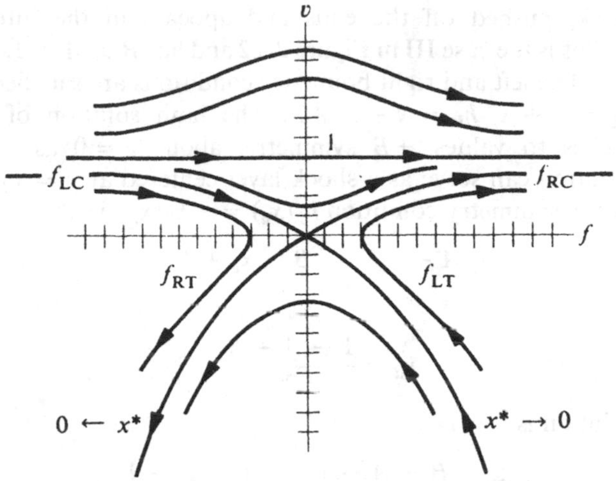

Figure 2.5.3 Phase Plane of Various Solutions

The exceptional path labeled f_{RC}, which starts from the origin, has a chance also to satisfy $f(\bar{x}) \to 0$, $\bar{x} \to -\infty$, since the origin is a singular point. The nature of the singularity is found from the approximate form of Equation (2.5.25) or

$$v^2 - f^2 = \text{const.}$$

Since f_{RC} passes through the origin the constant is equal to zero, so that along f_{RC}

$$v = f + \cdots \quad \text{as } f \to 0.$$

The integration of Equation (2.5.24) with $v = f$ shows that

$$f_{RC} = k_0 e^{\bar{x}} + \cdots \quad \text{as } \bar{x} \to -\infty. \tag{2.5.26}$$

That is, the matching condition as $\bar{x} \to -\infty$ is satisfied with an exponential approach. Here f_{RC} is called a right-corner solution and can be used together with Equation (2.5.22) to complete the solution for $A = 0, 0 < B < 1$. The reflected solution for $B = 0, -1 < A < 0$ will involve the left corner solution labeled f_{LC} in Figure 2.5.3.

The combination of these cases has solutions with both left and right corners and occurs in the triangular region IV of Figure 2.5.2. In this region $B - 1 < A < 0, 0 < B < A + 1$. The outer solution has three pieces

$$h_L(x) = x + A, \qquad 0 \le x \le -A;$$

$$h_m(x) = 0, \qquad -A < x \le 1 - B;$$

$$h_R(x) = x - 1 + B, \qquad 1 - B \le x \le 1.$$

Here f_{LC} provides the match between h_L and h_m, and f_{RC} the match between h_m and h_R.

The other two exceptional paths in the phase plane of Figure 2.5.3 are also necessary to complete the coverage of the (A, B) plane. Consider for example that $B = 0, A > 0$. The outer solution satisfying the right-boundary condition is

$$h_R = 0 \qquad (A > 0, B = 0)$$

$f_{LT}(\bar{x})$ with $\bar{x} = x/\sqrt{\varepsilon}$ can match to this as $\bar{x} \to \infty$ with exponential approach. This can be seen from a discussion of the behavior near $(v = f = 0)$ the same as for f_{RC} [cf. Equation (2.5.25)]. However it is not reasonable to expect to satisfy a boundary condition where y is $O(1)$ with a transition layer where y is $O(\sqrt{\varepsilon})$. Therefore we must try to match this transition layer to a thinner boundary layer around $x = 0$. In order to study this matching we need to know the behavior of f_{LT} as $\bar{x} \to 0$. In this case the behavior of f_{LT} as $\bar{x} \to 0$ can be obtained from the complete integral of Equation (2.5.25) taken along the exceptional path. Equation 2.5.25 can be written

$$\left(1 + \frac{1}{v - 1}\right) dv + f\,df = 0$$

so that the first integral representing paths through the origin is

$$v + \log|1 - v| + \frac{f^2}{2} = 0. \tag{2.5.27}$$

Thus, we see that as v, f approach infinity

$$v \to -\frac{f^2}{2}$$

and

$$d\bar{x} \to -\frac{2\,df}{f^2}$$

or

$$f(\bar{x}) = \frac{2}{\bar{x} + k_0} + \cdots. \tag{2.5.28}$$

It is clear that for matching we must have $f(\bar{x})$ becoming large as $\bar{x} \to 0$ so that y can become $O(1)$. Thus the constant of integration $k_0 = 0$ and

$$f(\bar{x}) = \frac{2}{\bar{x}} + \cdots, \qquad \bar{x} \to 0. \tag{2.5.29}$$

Next we consider an $O(\varepsilon)$ boundary layer at $x = 0$, an inner layer, as in Equation (2.5.9) but in order to have algebraic d cay we choose the constant of integration in Equation (2.5.10) to be zero. Tl us

$$\frac{dg}{dx^*} + \frac{g^2}{2} = 0. \tag{2.5.30}$$

The appropriate solution satisfying the boundary condition $g = A$ at $x^* = 0$ is

$$g(x^*) = \frac{2}{x^* + 2/A}. \tag{2.5.31}$$

Now the $O(1)$ matching can be discussed with the help of an intermediate limit ($\varepsilon \to 0$, x_η fixed) where

$$x_\eta = \frac{x}{\eta(\varepsilon)}, \qquad \varepsilon \ll \eta \ll \sqrt{\varepsilon}.$$

In order to demonstrate that this matching is really valid and that the singularity of $f(\bar{x})$ does not produce too high a singularity in $f_1(\bar{x})$ it is useful to work out both the next term in the expansion of f and the first term of f_1. The occurrence of a log term in the integral of (2.5.27) indicates that a log term appears in the expansion of $f(\bar{x})$ as $\bar{x} \to 0$. By substitution in the basic Equation (2.5.27), the following expansion can be verified

$$f(\bar{x}) = \frac{2}{\bar{x}} + \frac{2}{3}\bar{x} \log \bar{x} + k\bar{x} + O(\bar{x}^2 \log \bar{x}). \tag{2.5.32}$$

Here k is to be regarded as a known constant which could be found, for example, from the complete numerical integration of Equation (2.5.27) for $f(\bar{x})$. The boundary conditions used to specify $f(\bar{x})$ are $f(\bar{x}) \to 0$ as $\bar{x} \to \infty$, $f(\bar{x}) \to 2/\bar{x}$ as $\bar{x} \to 0$ [cf. Equation (2.5.28)].

Next by substituting the expansion (2.5.15) in the original Equation (2.5.1) we find that f_1 satisfies

$$\frac{d^2 f_1}{d\bar{x}^2} + f\frac{df_1}{d\bar{x}} + \left(\frac{df}{d\bar{x}} - 1\right)f_1 = 0. \tag{2.5.33}$$

This linear equation is the variational equation of the original Equation (2.5.1). We can remark that whenever the basic equation is non-linear the equations for the higher approximations are linear. In general these linear equations have variable coefficients which depend on the earlier terms in the expansion. In this particular case it is easy to verify that there is a solution of Equation (2.5.33) such that $f_1(\bar{x}) \to (\text{const.})e^{-\bar{x}}$ as $\bar{x} \to \infty$. Now as $\bar{x} \to 0$ this solution can be expressed as some linear combinations of the two independent solutions. As $\bar{x} \to 0$ Equation (2.5.33) can be approximated by using the asymptotic form of f

$$\frac{d^2 f_1}{d\bar{x}^2} + \left(\frac{2}{\bar{x}} + \cdots\right)\frac{df_1}{d\bar{x}} - \left(\frac{2}{\bar{x}^2} + \cdots\right)f_1 = 0. \tag{2.5.34}$$

By seeking solutions of the form $f_1 \sim \bar{x}^\alpha$ we find the indicial equation

$$\alpha(\alpha - 1) + 2\alpha - 2 = (\alpha + 2)(\alpha - 1) = 0.$$

Thus f_1 has as expansion as $\bar{x} \to 0$

$$f_1 = \frac{k_1}{\bar{x}^2} + \cdots. \tag{2.5.35}$$

The constant k_1 is to be found when matching is carried out to a sufficiently high order.

Next a similar discussion needs to be made for the $g(x^*)$ expansion as $x^* \to \infty$. Because of the occurrence of the log term in the expansion (2.5.32) of $f(\bar{x})$ it turns out that a log term has to be included in the inner expansion. The occurrence of this term would really not be discovered until matching to higher order was attempted. The expansion is

$$y(x; \varepsilon) = g(x^*) + \varepsilon \log \varepsilon g_{11}(x^*) + \varepsilon g_1(x^*) + \cdots \tag{2.5.36}$$

and on substitution in the basic equation we find the following sequence of equations and boundary conditions

$$\frac{d^2 g}{dx^{*2}} + g\frac{dg}{dx^*} = 0; \qquad g(0) = A \tag{2.5.37}$$

$$\frac{d^2 g_{11}}{dx^{*2}} + g\frac{dg_{11}}{dx^*} + \frac{dg}{dx^*} g_{11} = 0; \qquad g_{11}(0) = 0 \tag{2.5.38}$$

$$\frac{d^2 g_1}{dx^{*2}} + g\frac{dg_1}{dx^*} + \frac{dg}{dx^*} g_1 = g; \qquad g_1(0) = 0. \tag{2.5.39}$$

We have already found $g(x^*)$ so that straightforward integration yields the following solutions

$$g(x^*) = \frac{2}{x^* + (2/A)} \tag{2.5.40}$$

$$g_{11}(x^*) = \frac{C_{11}}{3}\left(x^* + \frac{2}{A}\right) - \frac{8}{3A^3}\frac{C_{11}}{(x^* + (2/A))^2} \tag{2.5.41}$$

$$g_1(x^*) = \frac{2}{3}\left(x^* + \frac{2}{A}\right)\log\left(x^* + \frac{2}{A}\right) + C_1\left(x^* + \frac{2}{A}\right) - \frac{(16C_1/A^4) + (32/3A^4)\log(2/A)}{(x^* + (2/A))^3}. \tag{2.5.42}$$

The constants C_{11}, and C_1 would be found in higher order matching.

However there are no arbitrary constants to be found in matching to order unity because both $f(\bar{x})$ and $g(x^*)$ are completely defined; f was defined by conditions of exponential decay at infinity and singular behavior at the origin, g by algebraic decay at infinity and $g(0) = A$. Thus, matching to order unity here is a verification that, in a certain sense, g is contained in f.

In order to match, consider the intermediate limit ($\varepsilon \to 0$, x_η fixed) where

$$x_\eta = \frac{x}{\eta(\varepsilon)}; \qquad \varepsilon \ll \eta \ll \sqrt{\varepsilon} \tag{2.5.43}$$

so that

$$\bar{x} = \frac{x}{\sqrt{\varepsilon}} = \frac{\eta}{\sqrt{\varepsilon}} x_\eta \to 0, \qquad x^* = \frac{x}{\varepsilon} = \frac{\eta}{\varepsilon} x_\eta \to \infty.$$

For the transition layer expansion and the inner expansion to match to $O(1)$ we want

$$\lim_{\substack{\varepsilon \to 0 \\ x_\eta \text{ fixed}}} \left\{ \sqrt{\varepsilon} f\left(\frac{\eta x_\eta}{\sqrt{\varepsilon}}\right) + \varepsilon f_1\left(\frac{\eta x_\eta}{\sqrt{\varepsilon}}\right) + \cdots - g\left(\frac{\eta x_\eta}{\sqrt{\varepsilon}}\right) \right.$$
$$\left. - \varepsilon \log \varepsilon g_{11}\left(\frac{\eta x_\eta}{\varepsilon}\right) - g_1\left(\frac{\eta x_\eta}{\varepsilon}\right) - \cdots \right\} = 0. \tag{2.5.44}$$

Substituting from the various expansions just constructed we find

$$\lim_{\substack{\varepsilon \to 0 \\ x_\eta \text{ fixed}}} \left\{ \sqrt{\varepsilon}\left(\frac{2}{(\eta/\sqrt{\varepsilon})x_\eta} + O\left(\frac{\eta}{\sqrt{\varepsilon}} \log \eta, \frac{\eta}{\sqrt{\varepsilon}} \log \varepsilon\right) + O\left(\frac{\varepsilon^2}{\eta^2}\right)\right) \right.$$
$$\left. - \left(\frac{2}{(\eta x_\eta/\varepsilon) + A} + O(\eta \log \varepsilon, \eta \log \eta)\right) \right\} = 0. \tag{2.5.45}$$

The dominant terms cancel and the other terms all vanish if $\eta(\varepsilon)$ is in the class of $\varepsilon \ll \eta \ll \sqrt{\varepsilon}/\log \varepsilon$. Thus the $O(\sqrt{\varepsilon})$ transition layer can be continued with this inner layer to satisfy an $O(1)$ boundary condition.

In essence our discussion of all the possible solutions is now complete. All solutions with corner layers or transition layers match to an outer solution $h = 0$. $A > 0$, $B < 0$ demands f_{LT}, f_{RT}, $A > 0$, $0 < B < 1$ demands f_{LT}, f_{RC} and the reflection $-1 < A < 0$, $B < 0$ demands f_{LC} and f_{RT}.

Thus the systematic use of boundary-layer theory and matching can successfully cope with the wide variety of problems which can arise in a non-linear case. While it is true that the full equation must be integrated to find $f_{LT}, f_{RT}, f_{LC}, f_{RC}$ the boundary conditions are canonical so that this integration can be done once for all problems.

Problems

1. Study the solutions to order unity for all A and B independent of ε, of the boundary value problem with $y(0) = A$, $y(1) = B$ governed by the following equations:

   (a) $\varepsilon y y'' + y y' = -x$, $0 < \varepsilon \ll 1$.
   (b) $\varepsilon y'' + \frac{1}{2}y^2 y' - y = 0$, $0 < \varepsilon \ll 1$.

2. Solve the following boundary-value problem exactly by differentiating the equation. Are the non-unique solutions which result derivable by perturbations

$$\varepsilon(y'')^2 + (x - \tfrac{1}{2})y' - y = 0, \qquad 0 < \varepsilon \ll 1$$
$$y(0) = y(1) = 1.$$

3. Solve the following nonlinear problem (which is a model of the shock structure for an isothermal shock for large Mach number) exactly

$$y'' + \frac{\varepsilon y'}{y^2} - y' = 0, \qquad -\infty < x < \infty, 0 < \varepsilon \ll 1$$
$$y(-\infty) = 1, \qquad y(\infty) = \varepsilon.$$

Next study the asymptotic behavior of the solution for $x < 0$, and $x \approx 0$ using the differential equation. Verify your results by comparison with the exact solution.

2.6 Relaxation Oscillations of the van der Pol Oscillator

Relaxation oscillations are periodic motions with nearly discontinuous segments. The system typically operates in a fast phase and in a slow phase. In each phase, different physical processes dominate.

From the point of view of perturbation theory, the regions of rapid change in time are boundary layers or shock layers. The problem of perturbation theory is the problem of matching the rapid phases of motion to the slow phases.

A classical example is the relaxation oscillator of van der Pol illustrated in Figure 2.6.1. A linear oscillating circuit with resistance is coupled inductively to a triode which effectively provides a negative resistance for small currents. This negative resistance causes small currents to grow, but the eventual amplitude is limited due to saturation of the triode. An oscillation of definite amplitude and period, depending on the parameters, is produced.

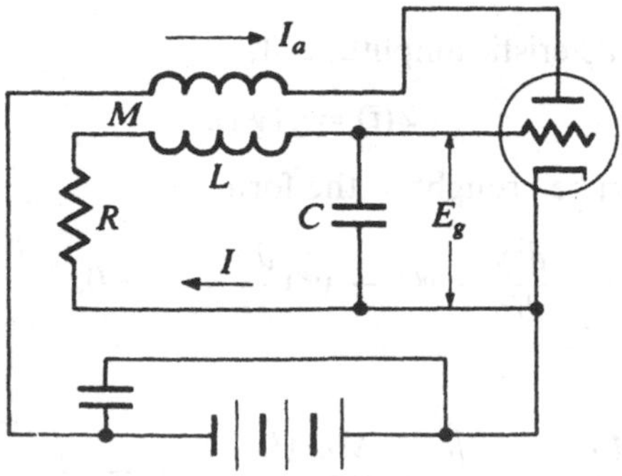

Figure 2.6.1 Circuit for van der Pol Oscillator

This oscillation, or limit cycle, is approached as time increases, independent of the initial conditions. For limiting values of the parameters, the limit cycle has the "jerky" character typical of relaxation oscillations.

The circuit equation corresponding to Figure 2.6.1 is

$$L\frac{dI}{d\tau} + RI + \frac{Q}{C} = M\frac{dI_a}{d\tau}. \tag{2.6.1}$$

Here, τ is the physical time (sec), and the mutual inductance M is positive. The windings are arranged to oppose the voltage drop across L. The grid current is assumed to be negligibly small. The plate current I_a depends mainly on the grid voltage E_g and is given by the characteristic of the vacuum tube. An analytic form which approximates the tube characteristic well for $E_g/E_s < 1$ (E_s = characteristic saturation voltage) is

$$I_a = \sigma\left(E_g - \frac{1}{3}\frac{E_g^3}{E_s^2}\right), \tag{2.6.2}$$

where σ is the tube conductance (mho).

Now, $Q(\tau)$ is the charge on the capacitor, so that

$$E_g(\tau) = \frac{Q}{C}, \tag{2.6.3}$$

$$I = \frac{dQ}{d\tau} = C\frac{dE_g}{d\tau}. \tag{2.6.4}$$

Thus, Equation (2.6.1) can be expressed in terms of the dimensionless grid voltage $V = E_g(\tau)/E_s$ and dimensionless time $\bar{t}$, based on the natural frequency of oscillations of the linear system

$$\frac{d^2V}{d\bar{t}^2} + R\sqrt{\frac{C}{L}}\frac{dV}{d\bar{t}} + V = \frac{M\sigma}{\sqrt{LC}}(1 - V^2)\frac{dV}{d\bar{t}}, \tag{2.6.5}$$

$$\bar{t} = \omega\tau, \qquad \omega = \frac{1}{\sqrt{LC}}.$$

By choosing a characteristic amplitude A,

$$V(\bar{t}) = Ay(\bar{t}),$$

Equation (2.6.5) can be brought to the form

$$\frac{d^2y}{d\bar{t}^2} - \nu(1 - y^2)\frac{dy}{d\bar{t}} + y = 0, \tag{2.6.6}$$

where

$$\nu = \frac{M\sigma}{\sqrt{LC}} - R\sqrt{\frac{C}{L}} = \frac{M\sigma A^2}{\sqrt{LC}}, \qquad A = \sqrt{1 - \frac{RC}{M\sigma}}.$$

This form is suitable for studying weak nonlinear effects; the motion is close to simple harmonic. The period is close to that of the linear oscillator $2\pi/\omega$. However, when ν becomes large, the time scale of the oscillations changes, and the problem assumes a singular form. Let

$$t = \frac{\bar{t}}{\nu} = \frac{\omega}{\nu}\tau. \tag{2.6.7}$$

Then Equation (2.6.6) is

$$\varepsilon\frac{d^2y}{dt^2} - (1 - y^2)\frac{dy}{dt} + y = 0, \qquad \varepsilon = \frac{1}{\nu^2} \ll 1. \tag{2.6.8}$$

We consider now the periodic solution of Equation (2.6.8) which corresponds to the limit cycle, and we study the limit of the limit cycles as $\varepsilon \to 0$, the relaxation oscillation.[4] Equation (2.6.8) with $\varepsilon = 0$ describes the motion over the main part of the cycle. However, the solution is necessarily discontinuous; the limit is singular. We sketch out below the various asymptotic expansions necessary to construct a uniformly valid approximation to the periodic solutions of Equation (2.6.8) as $\varepsilon \to 0$. The outer expansion is associated with the limit process

$$\varepsilon \to 0, \qquad t \text{ fixed}.$$

The equation has been scaled so that the limit solution is $O(1)$. Thus, the expansion has the following form.

Outer Expansion

$$y(t;\varepsilon) = u_0(t) + \varepsilon u_1(t) + \cdots. \tag{2.6.9}$$

The sequence of equations that result is

$$(1 - u_0^2)\frac{du_0}{dt} - u_0 = 0, \tag{2.6.10}$$

$$\frac{d}{dt}[(1 - u_0^2)u_1] - u_1 = \frac{d^2u_0}{dt^2}. \tag{2.6.11}$$

The integral curves for u_0 have the form indicated in Figure 2.6.2. Both branches of the solution are represented by the integral

$$\log|u_0| - \frac{u_0^2}{2} = t + \text{const}. \tag{2.6.12}$$

The periodic solution must be constructed by piecing together segments of the different branches, such as AB and CD. In the periodic case, symmetry

[4] For the physical approximation to be valid, we need $E_g/E_s < 1$, that is $A < 1$ and $\nu \gg 1$.

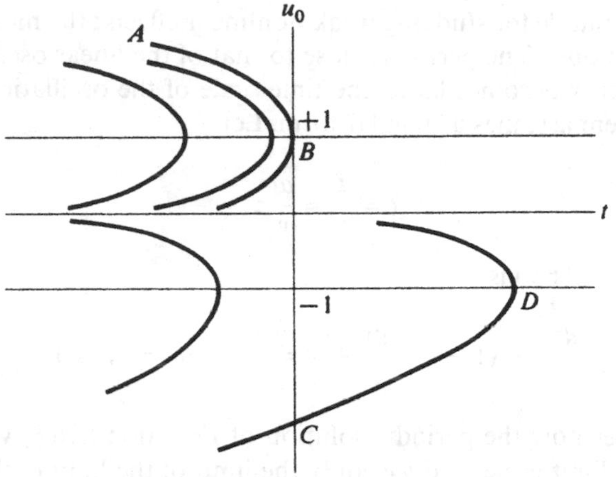

Figure 2.6.2 Outer Solution Curves

requirements demand that CD is a translation and reflection of AB. The branch AB is represented by

$$\log u_0 - \frac{u_0^2 - 1}{2} = t \qquad (AB). \tag{2.6.13}$$

The time origin has been taken at the jumping point B, where $u_0 = 1$. The period of the oscillations, even in the first approximation, is not yet evident. In the uniform approximation, the discontinuous segment, analogous to the shock layer of Section 2.5, is replaced by suitable boundary-layer expansions. Certain details of the solutions u_0, u_1 are necessary for the matching, and these are worked out now. The solutions for u_1 can be obtained by considering $u_1 = u_1(u_0)$ and rewriting Equation (2.6.11) as

$$\frac{-u_0}{u_0^2 - 1}\frac{d}{du_0}[(u_0^2 - 1)u_1] + u_1 = -\frac{u_0}{u_0^2 - 1}\frac{d}{du_0}\left(\frac{u_0}{u_0^2 - 1}\right).$$

Various terms can be combined so that we have

$$\frac{d}{du_0}\left[\frac{u_0^2 - 1}{u_0}u_1\right] = \frac{u_0}{u_0^2 - 1} - \frac{1}{u_0} - \frac{2u_0}{(u_0^2 - 1)^2}.$$

Integration of the last equation yields

$$u_1 = \frac{u_0}{u_0^2 - 1}\left\{A_1 + \frac{1}{u_0^2 - 1} + \frac{1}{2}\log\frac{u_0^2 - 1}{u_0^2}\right\}. \tag{2.6.14}$$

For any given branches AB, and CD with $|u_0| > 1$, u_1 is defined by Equation (2.6.14) with an arbitrary constant of integration A_1.

The behavior of u_0, u_1 as $t \to 0^-$ in branch AB is needed for the matching. From the expansion of Equation (2.6.12) near $t = 0$, we find that

$$u_0 = 1 + \sqrt{-t} + \tfrac{1}{6}(-t) + \tfrac{5}{72}(-t)^{3/2} + O((-t)^{5/2}), \qquad (2.6.15)$$

$$u_1 = \frac{1}{4(-t)} + \frac{1}{4\sqrt{-t}} \log\sqrt{-t} + \frac{1}{4\sqrt{-t}}(2A_1 - \tfrac{1}{3} + \log 2) + \cdots. \qquad (2.6.16)$$

Next, we consider joining the two branches AB and CD with a shock layer. The thickness of the layer is chosen so that the second-derivative terms in Equation (2.6.8) remain, that is, the thickness is $O(\varepsilon)$. Since the time origin for this expansion is not fixed, the limit process associated with the expansion has

$$\left(\varepsilon \to 0, \quad t^* = \frac{t - \delta(\varepsilon)}{\varepsilon} \text{ fixed}\right); \qquad \delta(\varepsilon) \quad \text{is to be found.} \qquad (2.6.17)$$

For matching to u_0 as $t \to 0$, the first term is $O(1)$. The inner expansion is, thus,

$$y(t; \varepsilon) = g_0(t^*) + \beta_1(\varepsilon)g_1(t^*) + \cdots, \qquad (2.6.18)$$

and Equation (2.6.8) becomes

$$\frac{1}{\varepsilon}\left\{\frac{d^2g_0}{dt^{*2}} + \beta_1 \frac{d^2g_1}{dt^{*2}} + \cdots\right\} - \frac{1}{\varepsilon}\{(1 - g_0^2) - 2\beta_1 g_0 g_1 - \cdots\}$$
$$\times \left\{\frac{dg_0}{dt^*} + \beta_1 \frac{dg_1}{dt^*} + \cdots\right\} + g_0 + \beta_1 g_1 + \cdots = 0.$$

Thus, we have

$$\frac{d^2g_0}{dt^{*2}} - (1 - g_0^2)\frac{dg_0}{dt^*} = 0 \qquad (2.6.19)$$

and, if $\beta_1 \gg \varepsilon$,

$$\frac{d^2g_1}{dt^{*2}} - \frac{d}{dt^*}[(1 - g_0^2)g_1] = 0. \qquad (2.6.20)$$

If $\beta_1 = \varepsilon$, the forcing term $-g_0$ appears on the right-hand side of Equation (2.6.20).

The first integral of Equation (2.6.19) is

$$\frac{dg_0}{dt^*} - g_0 + \frac{1}{3}g_0^3 = \text{const.} = k_0. \qquad (2.6.21)$$

In the matching of the inner expansion to the outer expansion, an intermediate class of limits of the form

$$\left(\varepsilon \to 0, \quad t_\eta = \frac{t - \delta(\varepsilon)}{\eta} \text{ fixed}\right), \qquad \frac{\eta}{\varepsilon} \to \infty, \qquad \eta \to 0$$

would be considered, so that

$$t = \eta t_\eta + \delta(\varepsilon) \to (0^-), \qquad (t_\eta < 0), \qquad t^* = \frac{\eta}{\varepsilon} t_\eta \to -\infty.$$

Since $u_0 \to 1$ as $t \to 0^-$, it is clear that if matching is to be possible to the first order, $g_0 \to 1$ as $t^* \to -\infty$. Thus, the constant in Equation (2.6.21) is fixed as $-\frac{2}{3}$:

$$\frac{dg_0}{dt^*} - g_0 + \frac{1}{3} g_0^3 = -\frac{2}{3}$$

or

$$\frac{dg_0}{dt^*} = -\frac{1}{3}(1 - g_0)^2(g_0 + 2). \tag{2.6.22}$$

For matching to the other branch of the outer solution (CD), we expect that $t^* \to +\infty$, and it follows from Equation (2.6.22) that $g_0 \to -2$. Thus, for matching to first order, $u_0 \to -2$ as $t \to 0^+$ (point C). These considerations give the first estimate of the size of the jump (u_0 goes from 1 to -2) and hence, the first approximation to the period. This same result can also be found directly from phase-plane or energy considerations. We now proceed to examine the matching in more detail. It is clear from Equation (2.6.22) that g_0 decays exponentially toward its asymptotic value (Equation 2.6.2) as $t^* \to +\infty$, but decays only algebraically as $t^* \to -\infty$. The form of the curves is indicated in Figure 2.6.3. Some difficulty in matching to higher order can be anticipated as $t^* \to -\infty$ because of the algebraic decay. The integral of Equation (2.6.22) is

$$\frac{1}{3}\log(1 - g_0) - \frac{1}{1 - g_0} - \frac{1}{3}\log(g_0 + 2) = -t^*, \tag{2.6.23}$$

with the constant of integration absorbed in $(-t^*)$. From Equations (2.6.23) or (2.6.22), the expansion as $t^* \to -\infty$ is easily worked out:

$$g_0(t^*) = 1 + \frac{1}{t^*} - \frac{1}{3}\log\frac{(-t^*)}{t^{*2}} + \cdots, \qquad (t^* \to -\infty). \tag{2.6.24}$$

Correspondingly, the integral of Equation (2.6.20) is

$$\frac{dg_1}{dt^*} - (1 - g_0^2)g_1 = \text{const.} = k_1. \tag{2.6.25}$$

The solution of the homogeneous part of Equation (2.6.25) can be written as (cf. Equation 2.6.19)

$$g_1 = h_1 \frac{dg_0}{dt^*}, \qquad h_1 = \text{const.}$$

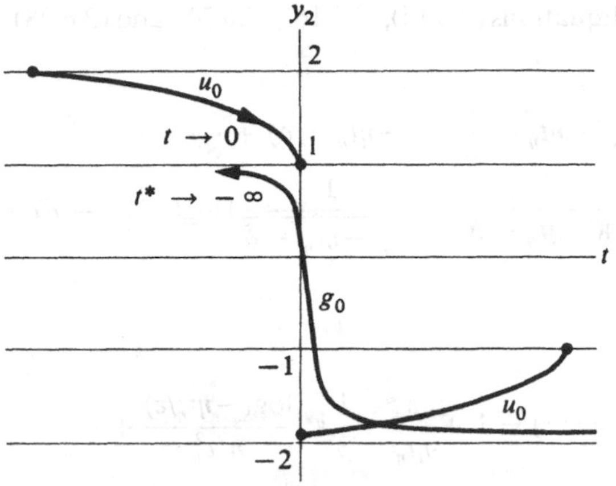

Figure 2.6.3 First-Order Boundary Layer Joining Outer Solutions

so that

$$g_1 = h_1 \frac{dg_0}{dt^*} + g_{1_p}, \tag{2.6.26}$$

where g_{1_p} is a particular solution of Equation (2.6.25). Here h_1, k_1 are the two constants of integration. The behavior of the particular solution as $t^* \to -\infty$ is easily found from the form of Equation (2.6.25) as $t^* \to -\infty$,

$$\frac{dg_{1_p}}{dt^*} - \left(-\frac{2}{t^*} + \frac{2}{3}\frac{\log(-t^*)}{t^{*2}} - \frac{1}{t^{*2}} + \cdots\right) g_{1_p} = k_1,$$

so that

$$g_{1_p} = \frac{k_1}{3}\left\{t^* + \frac{1}{3}\log(-t^*) - \frac{1}{3} + O\left(\frac{\log^2(-t^*)}{t^{*2}}\right)\right\}, \qquad t^* \to -\infty. \tag{2.6.27}$$

Thus, if $k_1 \neq 0$, the particular solution g_{1_p} dominates as $t^* \to -\infty$, and

$$g_1 = \frac{k_1}{3}\left\{t^* + \frac{1}{3}\log(-t^*) - \frac{1}{3} + O\left(\frac{\log^2(-t^*)}{t^{*2}}\right)\right\} + h_1\left\{-\frac{1}{t^{*2}} + O\left(\frac{\log(-t^*)}{t^{*3}}\right)\right\}, \qquad t^* \to -\infty. \tag{2.6.28}$$

Now, consider the attempt at matching the outer and inner expansions by writing both in intermediate variables

$$t = \eta t_\eta + \delta(\varepsilon) \to 0^-, \qquad \frac{\eta}{\varepsilon} \to \infty, \qquad \eta \to 0;$$
$$t^* = \frac{\eta}{\varepsilon} t_\eta \to -\infty, \qquad t_\eta < 0. \tag{2.6.29}$$

We have, from Equations (2.6.15), (2.6.16), (2.6.24), and (2.6.28), the following.

Outer

$$y(t;\varepsilon) = 1 + \sqrt{-\eta t_\eta - \delta} + \tfrac{1}{6}(-\eta t_\eta - \delta) + \cdots$$
$$+ \varepsilon\left\{\frac{1}{4(-\eta t_\eta - \delta)} + \frac{1}{8\sqrt{-\eta t_\eta - \delta}}[\log(-\eta t_\eta - \delta) + \cdots]\right\} + \cdots. \tag{2.6.30}$$

Inner

$$y(t;\varepsilon) = 1 + \frac{\varepsilon}{\eta t_\eta} - \frac{1}{3}\varepsilon^2\frac{\log(-\eta t_\eta/\varepsilon)}{\eta^2 t_\eta^2} + \cdots$$
$$+ \beta_1(\varepsilon)\left\{\frac{k}{3}\left(\frac{\eta t_\eta}{\varepsilon}\right) + \cdots - \frac{h_1\varepsilon^2}{\eta^2 t_\eta^2} + \cdots\right\}. \tag{2.6.31}$$

It is clear that, even with $\delta \ll \eta$, there is no term in the outer expansion capable of matching the term $O(1/\eta t_\eta)$ in the inner expansion. In particular note that $y > 1$ in the outer expansion and $g_0 < 1$ in the inner.

The implication of the failure of matching to higher order is the existence of a distinguished limit and transition expansion between the inner and outer expansions. The transition expansion should match to the outer as $t \to 0^-$ and the inner as $t^* \to -\infty$.

The thickness of the transition layer must be larger than that of the inner layer. Therefore, consider a limit process in which

$$\varepsilon \to 0, \qquad \tilde{t} = \frac{t - \rho(\varepsilon)}{\nu(\varepsilon)} \text{ is fixed}, \qquad 1 \gg \nu(\varepsilon) \gg \varepsilon.$$

Since the first term is already matched, the transition expansion has the form

$$y(t;\varepsilon) = 1 + \sigma_1(\varepsilon)f_1(\tilde{t}) + \sigma_2(\varepsilon)f_2(\tilde{t}) + \cdots. \tag{2.6.32}$$

Thus, Equation (2.6.8) becomes

$$\frac{\varepsilon}{\nu^2}\left\{\sigma_1\frac{d^2f_1}{d\tilde{t}^2} + \sigma_2\frac{d^2f_2}{d\tilde{t}^2} + \cdots\right\} - \frac{1}{\nu}\{-2\sigma_1 f_1 - 2\sigma_2 f_2 - \sigma_1^2 f_1^2 - \cdots\}$$
$$\times\left\{\sigma_1\frac{df_1}{d\tilde{t}} + \sigma_2\frac{df_2}{d\tilde{t}} + \cdots\right\} + 1 + \sigma_1 f_1 + \sigma_2 f_2 + \cdots = 0. \tag{2.6.33}$$

The orders of the terms associated with derivatives of each order are

$$\frac{\varepsilon\sigma_1}{\nu^2} \leftrightarrow \frac{\sigma_1^2}{\nu} \leftrightarrow 1. \tag{2.6.34}$$

The distinguished limit is that in which all these orders are equal. The distinguished equation contains a representative of each of the three basic terms (y'', y', y) in Equation (2.6.8) [in contrast to the inner (y'', y') and outer (y', y) which contain only two]. Thus, we have

$$\nu = \varepsilon^{2/3}, \qquad \sigma_1 = \varepsilon^{1/3}. \tag{2.6.35}$$

If the equation for f_2 is to contain forcing terms, it is necessary that

$$\sigma_2 = \varepsilon^{2/3}. \tag{2.6.36}$$

The resulting expansion and equations are as follows.

Transition Expansion

$$y(t, \varepsilon) = 1 + \varepsilon^{1/3} f_1(\tilde{t}) + \varepsilon^{2/3} f_2(\tilde{t}) + \cdots, \qquad \tilde{t} = \frac{t - \rho(\varepsilon)}{\varepsilon^{2/3}}, \tag{2.6.37}$$

$$\frac{d^2 f_1}{d\tilde{t}^2} + 2f_1 \frac{df_1}{d\tilde{t}} + 1 = 0, \tag{2.6.38}$$

$$\frac{d^2 f_2}{d\tilde{t}^2} + 2\frac{d}{d\tilde{t}}(f_1 f_2) = -f_1^2 \frac{df_1}{d\tilde{t}} - f_1. \tag{2.6.39}$$

Typically, the equation for f_1 is nonlinear and that for f_2, f_3, etc. are linear. If necessary for matching, there is also the possibility of using a term f_{12} of order intermediate to $\varepsilon^{1/3}$ and $\varepsilon^{2/3}$; f_{12} satisfies the homogeneous version of Equation (2.6.39).

Next we study the properties of the solutions of Equations (2.6.38) and (2.6.39) with an eye to matching these with the inner and outer expansions as $t^* \to -\infty$ and $(t \to 0^-, \tilde{t} \to -\infty)$, respectively. The first integral of Equation (2.6.38) is

$$\frac{df_1}{d\tilde{t}} + f_1^2 + \tilde{t} = 0, \tag{2.6.40}$$

where the constant of integration has been absorbed in the time shift $\rho(\varepsilon)$, which is still to be found. Equation (2.6.40) is of Riccati type and can be solved explicitly by means of the transformation

$$f_1 = \frac{V'(\tilde{t})}{V(\tilde{t})} = \frac{d}{d\tilde{t}} \log V. \tag{2.6.41}$$

Here $V(\tilde{t})$ satisfies the Airy equation

$$\frac{d^2 V}{d\tilde{t}^2} + \tilde{t} V = 0, \tag{2.6.42}$$

and, in terms of modified Bessel functions, the solution has the general form

$$V(\tilde{t}) = M\sqrt{-\tilde{t}}\, K_{1/3}\left(\tfrac{2}{3}(-\tilde{t})^{3/2}\right) + N\sqrt{-\tilde{t}}\, I_{1/3}\left(\tfrac{2}{3}(-\tilde{t})^{3/2}\right). \tag{2.6.43}$$

As $\tilde{t} \to -\infty$, these functions behave as

$$K_{1/3}\left(\frac{2}{3}(-\tilde{t})^{3/2}\right) = \frac{1}{2}\sqrt{3\pi}(-\tilde{t})^{-3/4}\exp\left[-\frac{2}{3}(-\tilde{t})^{3/2}\right] \times \left\{1 - \frac{5}{48(-\tilde{t})^{3/2}} + \cdots\right\}, \tag{2.6.44}$$

$$I_{1/3}\left(\frac{2}{3}(-\tilde{t})^{3/2}\right) = \frac{1}{2}\sqrt{\frac{3}{\pi}}(-\tilde{t})^{-3/4}\exp\left[\frac{2}{3}(-\tilde{t})^{3/2}\right]\left\{1 + \frac{5}{48(-\tilde{t})^{3/2}} + \cdots\right\}. \tag{2.6.45}$$

It is clear from Equation (2.6.40) that as $\tilde{t} \to -\infty$, $f_1 = \pm\sqrt{-\tilde{t}} + \cdots$. Further, if matching with the expansion of u_0 as $t \to 0^-$ (Equation 2.6.15) is to be achieved, $f_1 = +\sqrt{-\tilde{t}}$ is the proper behavior. The asymptotic behavior of the $I_{1/3}$ and $K_{1/3}$ functions is such that

$$\log V(\tilde{t}) = \begin{cases} \frac{2}{3}(-\tilde{t})^{3/2} + \cdots & \text{if } N \neq 0, \\ -\frac{2}{3}(-\tilde{t})^{3/2} + \cdots & \text{if } N = 0. \end{cases}$$

Thus, from Equation (2.6.41), f_1 has the proper behavior only if $N = 0$, and Equation (2.6.43) becomes

$$V(\tilde{t}) = \begin{cases} M\sqrt{-\tilde{t}}\,K_{1/3}\left(\frac{2}{3}(-\tilde{t})^{3/2}\right), & \tilde{t} < 0; \\ M\dfrac{\pi}{\sqrt{3}}\sqrt{\tilde{t}}\left\{J_{1/3}\left(\frac{2}{3}\tilde{t}^{3/2}\right) + J^{1/3}\left(\frac{2}{3}\tilde{t}^{3/2}\right)\right\}, & \tilde{t} > 0. \end{cases} \tag{2.6.46}$$

Both constants of integration have now been accounted for in f_1. A sketch of the form of V and f_1 is shown in Figure 2.6.4. Here $\tilde{t}_0$ is the first zero of the Airy function.

We next construct a representation of the solution f_2. The first integral of Equation (2.6.39) is

$$\frac{df_2}{d\tilde{t}} + 2f_1 f_2 = C_2 - \frac{1}{3}f_1^3 - \log V(\tilde{t}), \tag{2.6.47}$$

where f_1, V are defined by Equations (2.6.41) and (2.6.46). Using Equation (2.6.41), we rewrite Equation (2.6.47) as

$$\frac{d}{d\tilde{t}}(V^2 f_2) + \frac{1}{3}\frac{V'^3}{V} + V^2 \log V = C_2 V^2,$$

so that

$$f_2 = \frac{D_2}{V^2} + \frac{C_2}{V^2}\int_{-\infty}^{\tilde{t}} V^2(\lambda)\,d\lambda - \frac{1}{V^2}\int_{-\infty}^{\tilde{t}} V^2 \log V\,d\lambda - \frac{1}{3V^2}\int_{-\infty}^{\tilde{t}} \frac{V'^3}{V}\,d\lambda. \tag{2.6.48}$$

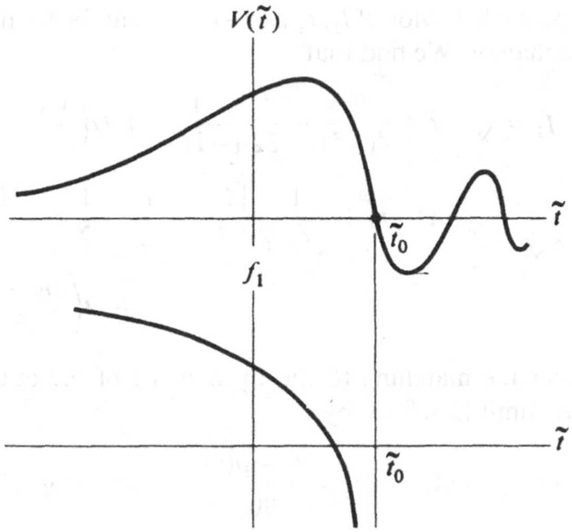

Figure 2.6.4 Transition Layer

Here C_2 and D_2 are the constants of integration. Integrating the last integral by parts, and using Equations (2.6.40) and (2.6.46), we can bring Equation (2.6.48) to the form

$$f_2 = \frac{D_2}{V^2} - \left(C_2 - \frac{1}{3}\log V\right)\frac{df_1}{d\tilde{t}} - \frac{1}{6}f_1^2 - \frac{2}{3V^2}\int_{-\infty}^{\tilde{t}} V^2 \log V \, d\lambda. \quad (2.6.49)$$

Now, as $\tilde{t} \to -\infty$, we have

$$V(\tilde{t}) = \frac{M}{2}\sqrt{3\pi}(-\tilde{t})^{-1/4}\exp[-\tfrac{2}{3}(-\tilde{t})^{3/2}] + \cdots,$$

$$\int_{-\infty}^{\tilde{t}} V^2 \log V \, d\lambda = \frac{M^2}{4}3\pi\int_{-\infty}^{\tilde{t}} \frac{\exp[-\tfrac{4}{3}(-\lambda)^{3/2}]}{\sqrt{-\lambda}}\left(-\frac{2}{3}(-\lambda)^{3/2} + \cdots\right)d\lambda$$

$$= -M^2\pi\sqrt{-\tilde{t}}\exp\left[-\frac{1}{3}(-\tilde{t})^{3/2}\right] + \cdots,$$

$$\frac{1}{V^2}\int_{-\infty}^{\tilde{t}} V^2 \log V \, d\lambda = -\frac{4}{3}(-\tilde{t}) + \cdots,$$

and

$$f_1 = \sqrt{-\tilde{t}} + \cdots, \qquad \frac{df}{d\tilde{t}} = -\frac{1}{2\sqrt{-\tilde{t}}} + \cdots.$$

Hence, to prevent exponential growth as $\tilde{t} \to -\infty$, D_2 is chosen to be zero, and the final form of f_2 is

$$f_2 = -\left(C_2 - \frac{1}{3}\log V\right)\frac{df_1}{d\tilde{t}} - \frac{1}{6}f_1^2 - \frac{2}{3V^2(\tilde{t})}\int_{-\infty}^{\tilde{t}} V^2(\lambda)\log V(\lambda)d\lambda. \quad (2.6.50)$$

Thus, the asymptotic behavior of f_1, f_2 as $\tilde{t} \to -\infty$ can be found from V and its asymptotic behavior. We find that

$$f_1 = \sqrt{-\tilde{t}} + \frac{1}{4(-\tilde{t})} - \frac{5}{32}\frac{1}{(-\tilde{t})^{5/2}} + O\left(\frac{1}{\tilde{t}^4}\right), \tag{2.6.51}$$

$$f_2 = \frac{1}{6}(-\tilde{t}) + \frac{1}{8\sqrt{-\tilde{t}}}\log(-\tilde{t}) + \frac{1}{\sqrt{-\tilde{t}}}\left\{\frac{C_2}{2} - \frac{1}{12} - \frac{1}{2}\log\frac{M\sqrt{3\pi}}{2}\right\} + O\left(\frac{\log(-\tilde{t})}{\tilde{t}^2}\right). \tag{2.6.52}$$

Next, we consider the matching to the branch AB of the outer expansion. The intermediate limit is defined by

$$(\varepsilon \to 0,\ t_\eta \text{ fixed}), \qquad t_\eta = \frac{t - \rho(\varepsilon)}{\eta(\varepsilon)}, \qquad \varepsilon^{2/3} \ll \eta \ll 1,$$

so that we have

$$t = \eta t_\eta + \rho(\varepsilon) \to 0, \qquad \tilde{t} = \frac{\eta}{\varepsilon} t_\eta \to -\infty.$$

The outer expansion written in intermediate variables is as follows from Equations (2.6.15) and (2.6.16).

Outer

$$\begin{aligned} y(t;\varepsilon) = 1 &+ \sqrt{-\eta t_\eta - \rho} + \frac{1}{6}(-\eta t_\eta - \rho) + \cdots \\ &+ \varepsilon\left\{\frac{1}{4(-\eta t_\eta - \rho)} + \frac{1}{8\sqrt{-\eta t_\eta - \rho}}\log(-\eta t_\eta - \rho)\right. \\ &\left.+ \left(\frac{A_1}{2} - \frac{1}{12} + \frac{\log 2}{4}\right)\frac{1}{\sqrt{-\eta t_\eta - \rho}} + \cdots\right\}. \end{aligned} \tag{2.6.53}$$

For $\rho \ll \eta$, we have the following.

Outer

$$\begin{aligned} y(t;\varepsilon) = 1 &+ \sqrt{-\eta t_\eta} - \frac{1}{2}\frac{\rho}{\sqrt{-\eta t_\eta}} + \cdots + \frac{1}{6}(-\eta t_\eta) - \frac{\rho}{6} + \cdots \\ &+ \varepsilon\left\{\frac{1}{4(-\eta t_\eta)} + \frac{\rho}{4(-\eta t_\eta)^2} + \cdots + \frac{1}{8\sqrt{-\eta t_\eta}}\log(-\eta t_\eta)\right. \\ &- \frac{1}{8}\frac{\rho}{(\eta t_\eta)^{3/2}} + \frac{\rho}{16(-\eta t_\eta)^{3/2}}\log(-\eta t_\eta) + \cdots \\ &\left.+ \left(\frac{A_1}{2} - \frac{1}{12} + \frac{\log 2}{4}\right)\frac{1}{\sqrt{-\eta t_\eta}} + \cdots\right\}, \end{aligned} \tag{2.6.54}$$

and the intermediate expansion is as follows from Equations (2.6.51) and (2.6.52).

Transition

$$y(t;\varepsilon) = 1 + \varepsilon^{1/3}\left\{\frac{\sqrt{-\eta t_\eta}}{\varepsilon^{1/3}} + \frac{\varepsilon^{2/3}}{4(-\eta t_\eta)} + \cdots\right\}$$

$$+ \varepsilon^{2/3}\left\{\frac{1}{6}\frac{(-\eta t_\eta)}{\varepsilon^{2/3}} + \frac{1}{8}\frac{\varepsilon^{1/3}}{\sqrt{-\eta t_\eta}}\left(\log(-\eta t_\eta) - \frac{2}{3}\log\varepsilon\right)\right.$$

$$\left. + \frac{\varepsilon^{1/3}}{\sqrt{-\eta t_\eta}}\left(\frac{C_2}{2} - \frac{1}{12} - \frac{1}{2}\log\frac{M\sqrt{3\pi}}{2}\right) + \cdots\right\}. \quad (2.6.55)$$

A comparison of these two expressions shows that all the terms in Equation (2.6.55) can be matched exactly by those in Equation (2.6.54) if the time shift $\rho(\varepsilon)$ and A_1 are chosen so that

$$\rho = \frac{1}{6}\varepsilon\log\varepsilon, \qquad A_1 = C_2 - \log\frac{M\sqrt{3\pi}}{2} - \frac{1}{4}\log 2. \quad (2.6.56)$$

All the terms omitted vanish more rapidly than those matched. Thus, the constants of integration (ρ, C_2) of the first two terms in the transition expansion are found in terms of A_1, from the matching.

The next problem is the matching of the transition expansion to the inner expansion as $t^* \to -\infty$. It is clear from the behavior of f_1 (cf. Figure 2.6.4) that the matching will take place as $\tilde{t} \to \tilde{t}_0$, the first zero of the Airy function. To study the matching, we need to know the behavior of f_1, f_2 as $\tilde{t} \to \tilde{t}_0$. We know that $V(\tilde{t})$ has a simple zero at $\tilde{t} = \tilde{t}_0$; it follows from Equation (2.6.42) that $V(\tilde{t})$ has an expansion near $\tilde{t}_0$ of the form

$$V(\tilde{t}) = M\left\{-K(\tilde{t} - \tilde{t}_0) + \frac{K}{6}\tilde{t}_0(\tilde{t} - \tilde{t}_0)^3 + O((\tilde{t} - \tilde{t}_0)^5)\right\}, \quad (2.6.57)$$

where $K =$ const. Thus, from Equation (2.6.41), as $\tilde{t} \to \tilde{t}_0$, we have

$$f_1(\tilde{t}) = \frac{1}{\tilde{t} - \tilde{t}_0} - \frac{1}{3}\tilde{t}_0(\tilde{t} - \tilde{t}_0) + O((\tilde{t} - \tilde{t}_0)^2), \quad (2.6.58)$$

$$\frac{df_1}{d\tilde{t}}(\tilde{t}) = \frac{1}{\tilde{t} - \tilde{t}_0} - \frac{1}{3}\tilde{t}_0 + O((\tilde{t} - \tilde{t}_0)). \quad (2.6.59)$$

Now, the integral in Equation (2.6.50) approaches a finite value as $\tilde{t} \to \tilde{t}_0$, so that the dominant term in f_2 comes from $\log V\, df_1/d\tilde{t}$, and we have

$$f_2 = -\frac{1}{3}\frac{\log(\tilde{t}_0 - \tilde{t})}{(\tilde{t}_0 - \tilde{t})^2} + O\left(\frac{1}{(\tilde{t}_0 - \tilde{t})^2}\right). \quad (2.6.60)$$

To express the intermediate limit for this case, we first write t^* in terms of $\tilde{t}$,

$$t^* = \frac{t - \delta(\varepsilon)}{\varepsilon} = \frac{\tilde{t} + \frac{1}{6}\varepsilon^{1/3}\log\varepsilon - \varepsilon^{-2/3}\delta(\varepsilon)}{\varepsilon^{1/3}} \tag{2.6.61}$$

and let

$$\delta(\varepsilon) = \varepsilon^{2/3}\{\tilde{t}_0 + \gamma(\varepsilon)\},$$

so that

$$t^* = \frac{\tilde{t} - \tilde{t}_0 - \gamma(\varepsilon) + \frac{1}{6}\varepsilon^{1/3}\log\varepsilon}{\varepsilon^{1/3}} = \frac{\tilde{t} - \tilde{t}_0 - \sigma(\varepsilon)}{\varepsilon^{1/3}}, \tag{2.6.62}$$

where

$$\sigma(\varepsilon) = \gamma(\varepsilon) - \tfrac{1}{6}\varepsilon^{1/3}\log\varepsilon.$$

The intermediate class of limits is, thus,

$$\varepsilon \to 0,\ t_\eta \text{ fixed}, \quad \text{where } t_\eta = \frac{\tilde{t} - \tilde{t}_0 - \sigma(\varepsilon)}{\eta},\ \varepsilon^{1/3} \ll \eta \ll 1, \tag{2.6.63}$$

so that we have

$$t^* = \frac{\eta}{\varepsilon^{1/3}}\, t_\eta \to -\infty, \qquad (t_\eta < 0), \qquad \tilde{t} - \tilde{t}_0 \to -\eta t_\eta + \sigma(\varepsilon).$$

Thus, using Equations (2.6.58) and (2.6.60), we find that the transition expansion is written in intermediate variables.

Transition

$$\begin{aligned} y(t;\varepsilon) = 1 &+ \varepsilon^{1/3}\left\{\frac{1}{\eta t_\eta + \sigma} - \frac{1}{3}\tilde{t}_0(\eta t_\eta + \sigma) + \cdots\right\} \\ &+ \varepsilon^{2/3}\left\{-\frac{\log(-\eta t_\eta - \sigma)}{3(\eta t_\eta + \sigma)^2} + \cdots\right\}. \end{aligned} \tag{2.6.64}$$

For $\sigma \ll \eta$, we have the following.

Transition

$$\begin{aligned} y(t;\varepsilon) = 1 &+ \varepsilon^{1/3}\left\{-\frac{1}{(-\eta t_\eta)} - \frac{\sigma}{(-\eta t_\eta)^2} + \cdots + \frac{1}{3}\tilde{t}_0(-\eta t_\eta) + \cdots\right\} \\ &- \frac{1}{3}\varepsilon^{2/3}\frac{\log(-\eta t_\eta)}{(-\eta t_\eta)^2} + \cdots. \end{aligned} \tag{2.6.65}$$

For the inner expansion, using Equations (2.6.24) and (2.6.28), we find the following.

Inner

$$y(t;\varepsilon) = 1 - \frac{\varepsilon^{1/3}}{(-\eta t_\eta)} - \frac{1}{3}\frac{\varepsilon^{2/3}}{(-\eta t_\eta)^2}\log(-\eta t_\eta) + \frac{1}{9}\frac{\varepsilon^{2/3}\log\varepsilon}{(-\eta t_\eta)^2} + \cdots$$
$$+ \beta_1(\varepsilon)\left\{-\frac{k_1}{3}\frac{(-\eta t_\eta)}{\varepsilon^{1/3}} + \frac{k_1}{3}\log(-\eta t_\eta)\right.$$
$$\left. + \frac{k_1}{9}\log\varepsilon + \cdots - h_1\frac{\varepsilon^{2/3}}{(-\eta t_\eta)^2} + \cdots\right\}. \tag{2.6.66}$$

The terms in Equation (2.6.65) can all be matched to terms in Equation (2.6.66) with the choices

$$\beta_1 = \varepsilon^{2/3}, \qquad k_1 = -\tilde{t}_0, \qquad -\sigma\varepsilon^{1/3} = \tfrac{1}{9}\varepsilon^{2/3}\log\varepsilon. \tag{2.6.67}$$

This is consistent with our previous assumption that $\beta_1 \gg \varepsilon$. It follows that

$$\gamma(\varepsilon) = \tfrac{1}{18}\varepsilon^{1/3}\log\varepsilon, \qquad \delta = \varepsilon^{2/3}\tilde{t}_0 + \tfrac{1}{18}\varepsilon\log\varepsilon. \tag{2.6.68}$$

Thus, the relationship of t^* and t, Equation (2.6.61), is

$$t^* = \frac{t - \varepsilon^{2/3}\tilde{t}_0 - \frac{1}{18}\varepsilon\log\varepsilon}{\varepsilon}. \tag{2.6.69}$$

Lastly, we need to close the cycle by matching the inner expansion as $t^* \to \infty$ to the branch CD of the outer expansion as $t \to 0^+$, $u_0 \to -2$. From Equation (2.6.13), we can write CD as a reflection and translation,

$$\log(-u_0) - \frac{u_0^2 - 1}{2} = t^+, \tag{2.6.70}$$

where $t^+ = t - \frac{1}{2}T(\varepsilon)$ and $T(\varepsilon) =$ period on t scale. The outer expansion now has t^+ fixed.

We can expand Equation (2.6.13) about $u_0 = -2$ to obtain

$$u_0(t^+) = -2 + \tfrac{2}{3}\{t^+ - \log 2 + \tfrac{3}{2}\} + \cdots. \tag{2.6.71}$$

Equation (2.6.14) shows that $u_1 \to$ const. as $u_0 \to -2$.

It also follows from Equation (2.6.22) or (2.6.23) that, as $t^* \to \infty$, we have

$$g_0(t^*) = -2 + O(e^{-3t^*}), \tag{2.6.72}$$

and the particular solution g_{1_p} of Equation (2.6.26) satisfies an equation of the form

$$\frac{dg_{1_p}}{dt^*} + \{3 + O(e^{-3t^*})\}g_{1_p} = k_1 = -\tilde{t}_0. \tag{2.6.73}$$

Thus, we have

$$g_1(t^*) = g_{1_p} = -\tfrac{1}{3}\tilde{t}_0 + \cdots, \qquad t^* \to \infty. \tag{2.6.74}$$

The relationship of t^* and t^+ comes from Equation (2.6.69).

$$t^* = \frac{t^+ + \frac{1}{2}T(\varepsilon) - \varepsilon^{2/3}\tilde{t}_0 - \frac{1}{18}\varepsilon \log \varepsilon}{\varepsilon}, \qquad (2.6.75)$$

so that the intermediate limit for this matching has ($\varepsilon \to 0$, t_η fixed)

$$t_\eta = \frac{t^+ + \frac{1}{2}T(\varepsilon) - \varepsilon^{2/3}\tilde{t}_0 - \frac{1}{18}\varepsilon \log \varepsilon}{\eta(\varepsilon)}, \qquad \varepsilon \ll \eta \ll 1, \qquad (2.7.76)$$

and

$$t^* = \frac{\eta}{\varepsilon} t_\eta \to \infty, \qquad t^+ = -\frac{1}{2}T(\varepsilon) + \varepsilon^{2/3}\tilde{t}_0 + \frac{1}{18}\varepsilon \log \varepsilon + \eta t_\eta.$$

Using Equation (2.6.71) we can write the outer expansion in t_η as follows.

Outer

$$\begin{aligned} y(t; \varepsilon) = -2 + \tfrac{2}{3}\{&-T(\varepsilon) + \varepsilon^{2/3}\tilde{t}_0 + \tfrac{1}{18}\varepsilon \log \varepsilon \\ &+ \eta t_\eta - \log 2 + \tfrac{3}{2}\} + \cdots + O(\varepsilon) \end{aligned} \qquad (2.6.77)$$

and from Equations (2.6.72) and (2.6.74), the inner expansion becomes the following.

Inner

$$y(t; \varepsilon) = -2 + \varepsilon^{2/3}(-\tfrac{1}{3}\tilde{t}_0) + \cdots + \beta_2(\varepsilon)(\ ) + \cdots, \qquad \beta_2 \ll \varepsilon^{2/3}. \qquad (2.6.78)$$

The period $T(\varepsilon)$ is fixed by the requirement that the terms $O(1)$ and $O(\varepsilon^{2/3})$ match. Thus we have

$$T(\varepsilon) = 3 - 2 \log 2 + 3\tilde{t}_0 \varepsilon^{2/3} + O(\varepsilon \log \varepsilon). \qquad (2.6.79)$$

As far as the result above has been carried out, it agrees with that of Dorodnitsyn in Reference 2.6.1. It is not worthwhile to carry out the period to $O(\varepsilon \log \varepsilon)$ without calculating the terms $O(\varepsilon)$ at the same time. The method followed by Dorodnitsyn involved the matching of asymptotic expansions constructed in various regions in the phase plane. Reference 2.6.2 corrects the higher terms in Dorodnitsyn's formula for the period and carries out numerical calculations which show close agreement with the asymptotic formula for the period. It would be possible to find more details of the expansion for the period by carrying out further terms. The basic ideas essential for this procedure have been illustrated in this section.

References

2.6.1 A. A. Dorodnitsyn, Asymptotic Solution of the van der Pol Equations, *Prik. Mat. I Mekh, 11* (1947), 313–328 (in Russian).

2.6.2 M. Urabe, Numerical Study of Periodic Solutions of the van der Pol Equations, in *Proc. Int. Symp. on Non-Linear Differential Equations and Non-Linear Mechanics*, 184–195. Edited by J. P. LaSalle and S. Lefschetz. New York. Academic Press, 1963.

2.7 Singular Boundary Problems

In this section, two problems are discussed in which the expansions in terms of a small parameter are singular, not because of a lowering of the order of the equation in the limit but, rather, because of a difficulty associated with the behavior near the boundary point. Nevertheless, the same method as used in previous sections enables the expansions of the solutions to be found. Different asymptotic expansions valid in different regions are constructed, and the matching of these expansions in an overlap domain enables all unknown constants to be found. Thus, a uniformly valid approximation can be constructed.

2.7.1 Periodic Collision Orbits in the Problem of Two Fixed Force-Centers

Consider the planar motion of a point mass in the gravitational field of two fixed centers of attraction. This is a classical example of an "integrable" dynamical system and is discussed in various texts in dynamics (e.g., Reference 2.7.1).

Using suitable dimensionless variables, the two fixed centers have masses $1 - \mu$ and μ and can be located at $x = 0$ and $x = 1$. The equations of motion will then become

$$\frac{d^2x}{dt^2} = -\frac{(1-\mu)x}{r^3} - \frac{\mu(x-1)}{r_1^3} \tag{2.7.1a}$$

$$\frac{d^2y}{dt^2} = -\frac{(1-\mu)y}{r^3} - \frac{\mu y}{r_1^3}, \tag{2.7.1b}$$

where

$$r = \sqrt{(x^2 + y^2)}, \qquad r_1 = \sqrt{(x-1)^2 + y^2}$$

are the distances from the particle to these two centers as shown in Figure 2.7.1.

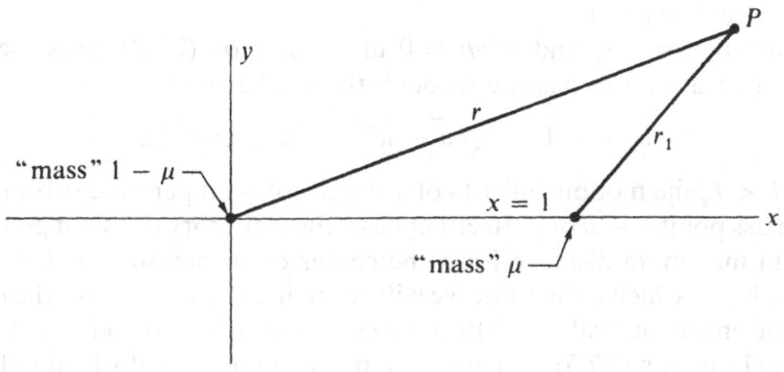

Figure 2.7.1 Planar Motion in the Field of Two Fixed Centers

It is well known that Equations (2.7.1) admit a family of periodic solutions which are confocal ellipses with foci at $x = 0$ and $x = 1$. If the speed at any point on this elliptic orbit is denoted by v, one can show that $v = \sqrt{v_{1-\mu}^2 + v_\mu^2}$ where $(v_{1-\mu})$, (v_μ) is the speed corresponding to the same elliptic orbit but with the point mass $(1 - \mu)$, (μ) as the only center of attraction. In the limit as the eccentricity of this family of orbits tends to unity we have the periodic double collision orbit in the unit interval.

For this limiting case Equations (2.7.1) reduce to

$$\frac{d^2x}{dt^2} = -\frac{(1-\mu)}{x^2} + \frac{\mu}{(x-1)^2}, \qquad 0 \le x \le 1 \tag{2.7.2}$$

and we wish to study the solution for the case $\mu \ll 1$.

This problem was considered in Reference (2.7.2) as a mathematical model for the more realistic problem of motion from earth to moon discussed in Reference (2.7.3).

This work has been extended to the three dimensional case (Reference 2.7.4) and to interplanetary trajectories (Reference 2.7.5). In addition there are numerous other references in the literature which have applied the idea of matching different expressions first discussed in Reference 2.7.2, to fairly complicated problems in dynamics.

For Equation (2.7.2) we have the energy integral

$$\frac{1}{2}\left(\frac{dx}{dt}\right)^2 - \frac{(1-\mu)}{x} - \frac{\mu}{1-x} = h = \text{const.} \tag{2.7.3}$$

Thus, Equation (2.7.3) can be used to express t as a function of x by quadrature. The result involves elliptic integrals and is not very instructive. The qualitative nature of the collision orbits can be determined from the phase-plane path of solutions for various values of h given in Figure 2.7.2. Setting the right-hand side of Equation (2.7.2) equal to zero gives the equilibrium point

$$x_e = \frac{(1-\mu) - \sqrt{\mu - \mu^2}}{1 - 2\mu} = 1 - \mu^{1/2} + O(\mu) \tag{2.7.4}$$

which is a saddle point.

Substituting $x = x_e$ and $dx/dt = 0$ into Equation (2.7.3) gives the value of h for the trajectories passing through the saddle point.

$$h_e = -1 - 2\sqrt{\mu - \mu^2} = -1 + O(\mu^{1/2}). \tag{2.7.5}$$

For $h < h_e$ the motion consists of a single collision periodic orbit relative to the mass point $1 - \mu$ or μ. In either case, the trajectory does not go beyond a certain maximum distance from the center of attraction. For $h > h_e$, for example $h = 0$ which is the value we will adopt in our calculations, the motion spans the entire interval with alternate collisions at $x = 0$ and $x = 1$.

Using Equation (2.7.3), we can calculate t as a function of x by quadrature, and this result is given in Reference 2.7.2. Here we will derive the asymptotic

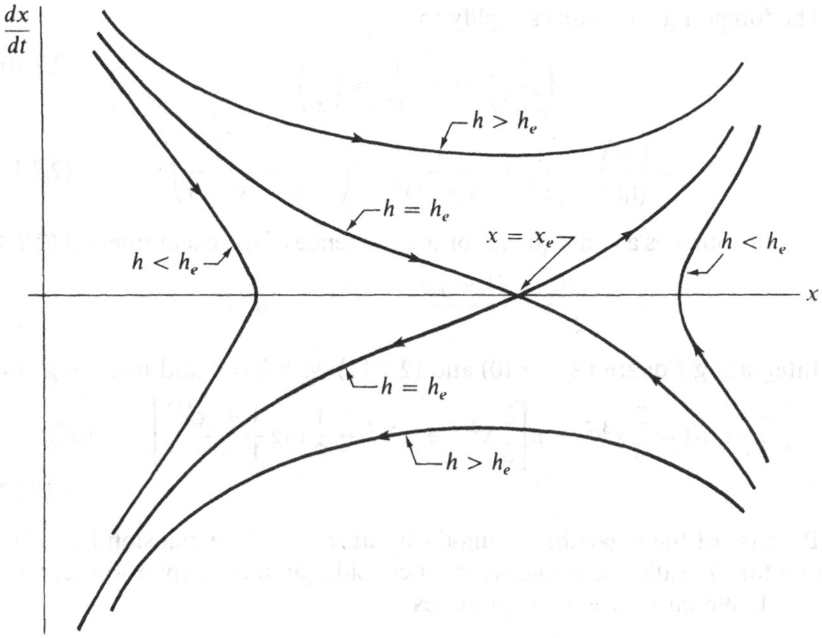

Figure 2.7.2 Phase Plane Trajectories

representation of this solution from Equation (2.7.2) ignoring the existence of the exact integral (2.7.3).

For $\mu \to 0$ Equation (2.7.2) defines a singular perturbation problem because the outer expansion which neglects the term $\mu/(x-1)^2$ to first order is in error in a neighborhood of $x = 1$. In fact, it is clear that regardless of the size of μ, there is a neighborhood of $x = 1$ in which this term is dominant. Since the location of nonuniformity is a priori known it is convenient to let x be the independent variable and regard t as the dependent variable. Denoting derivatives with respect to x by primes, Equation (2.7.2) transforms to

$$-\frac{t''}{t'^3} = -\frac{(1-\mu)}{x^2} + \frac{\mu}{(x-1)^2}. \tag{2.7.6}$$

We now assume an outer expansion

$$t(x, \mu) = t_0(x) + \mu t_1(x) + O(\mu^2) \tag{2.7.7}$$

and derive the following equations for t_0 and t_1.

$$-\frac{t_0''}{t_0'^3} = -\frac{1}{x^2} \tag{2.7.8}$$

$$-\frac{t_1''}{t_0'^3} + \frac{3t_1't_0''}{t_0'^4} = \frac{1}{x^2} + \frac{1}{(x-1)^2}. \tag{2.7.9}$$

The foregoing equations simplify to

$$\left(\frac{1}{2t_0'^2}\right)' = -\frac{1}{x^2} = \left(\frac{1}{x}\right)' \tag{2.7.10}$$

$$-\left(\frac{t_1'}{t_0'^3}\right)' = \frac{1}{x^2} + \frac{1}{(x-1)^2} = \left(-\frac{1}{x} - \frac{1}{x-1}\right)' \tag{2.7.11}$$

which, of course is a consequence of the existence of the exact integral (2.7.3)

$$\frac{1}{2t'^2} - \frac{(1-\mu)}{x} - \frac{\mu}{1-x} = h$$

Integrating Equations (2.7.10) and (2.7.11) with $h = 0$ and $t(0) = 0$ gives

$$\sqrt{2}\,t(x, \mu) = \frac{2}{3}x^{3/2} + \mu\left[\frac{2}{3}x^{3/2} + x^{1/2} - \frac{1}{2}\log\frac{1 + x^{1/2}}{1 - x^{1/2}}\right] + O(\mu^2). \tag{2.7.12}$$

Because of the logarithmic singularity at $x = 1$ the expansion (2.7.12) is not uniformly valid there, and we must consider an inner expansion centered at $x = 1$. We introduce inner variables

$$x_\alpha = \frac{1-x}{\mu^\alpha}; \qquad t_\beta = \frac{t - \tau(\mu)}{\mu^\beta}, \tag{2.7.13}$$

where the constants α and β are to be determined by an analysis of the order of magnitude of the various terms in the differential equation written in inner variables. Here $\tau(\mu)$ is the half-period and will be determined from the matching.

Equation (2.7.2) transforms to

$$-\frac{d^2x_\alpha}{dt_\beta^2} = -\mu^{2\beta-\alpha}\frac{(1-\mu)}{(1-\mu^\alpha x_\alpha)^2} + \frac{\mu^{2\beta-3\alpha+1}}{x_\alpha^2} = \frac{\mu^{2\beta-3\alpha+1}}{x_\alpha^2} + O(\mu^{2\beta-\alpha}). \tag{2.7.14}$$

Since the attraction of the mass at $x = 1$ must be taken into account to first order we must set $2\beta - 3\alpha + 1 = 0$, and this gives a relationship between β and α. For initial conditions which correspond to trajectories spanning the unit interval the velocities of the inner and outer expansions must also match. The velocity dx/dt, calculated from the outer expansion is $O(1)$ in the matching region. According to the inner expansion the velocity $dx/dt = O(\mu^{\alpha-\beta})$. Therefore we must set $\alpha = \beta$ and this gives $\alpha = \beta = 1$ and we have the inner variables

$$x^* = \frac{1-x}{\mu}, \qquad t^* = \frac{t - \tau(\mu)}{\mu}.$$

Since $t = \tau + \mu t^*$, we need only calculate t^* as a function of x^* to $O(1)$ in order to obtain $t(x, \mu)$ to $O(\mu)$. Thus, it suffices to study the solution of the limiting differential equation (x^*, t^* fixed $\mu \to 0$) that ensues from Equation (2.7.2):

$$-\frac{d^2x^*}{dt^{*2}} = \frac{1}{x^{*2}}. \tag{2.7.15a}$$

This has the integral

$$\frac{1}{2}\left(\frac{dx^*}{dt^*}\right)^2 - \frac{1}{x^*} = h^*. \tag{2.7.15b}$$

We calculate h^* by matching the velocities given by the inner and outer limits. According to Equation (2.7.12) (or the energy integral with $h = 0$) $v^2/2 \to 1$ as $x \to 1$ where $v = dx/dt = -dx^*/dt^*$. Equation (2.7.15) gives $v^2/2 \to h^*$ as $x^* \to \infty$. The above simplified matching, which can be justified using intermediate variables gives $h^* = 1$. We can now solve Equation (2.7.15).

Restricting attention to the half-period with positive velocity, we have

$$-\frac{\sqrt{2}}{\mu}\int_t^{\tau} ds = \int_{x^*}^{0}\sqrt{\frac{\xi}{1+\xi}}\,d\xi. \tag{2.7.16}$$

Setting $\tau = \tau_0 + \mu\tau_1 + \cdots$, Equation (2.7.16) integrates to

$$\frac{\sqrt{2}}{\mu}[t - (\tau_0 + \mu\tau_1 + \cdots)] = -\sqrt{x^*(1 + x^*)} + \log(\sqrt{x^*} + \sqrt{1 + x^*}). \tag{2.7.17}$$

If we indicate the outer and inner expansions for $\sqrt{2}\,t(x, \mu)$ by

$$\sqrt{2}\,t(x, \mu) = h_0(x) + \mu h_1(x) + O(\mu^2) \tag{2.7.18}$$

$$\sqrt{2}\,t(x, \mu) = g_0(x^*) + \mu g_1(x^*) + O(\mu^2) \tag{2.7.19}$$

h_0 and h_1 are defined by Equation (2.7.12), and Equation (2.7.17) gives

$$g_0(x^*) = \sqrt{2}\,\tau_0 \tag{2.7.20}$$

$$g_1(x^*) = \sqrt{2}\,\tau_1 - \sqrt{x^*(1 + x^*)} + \log(\sqrt{x^*} + \sqrt{1 + x^*}). \tag{2.7.21}$$

The matching condition to $O(\mu)$ is [with $x_\eta = (1 - x)/\eta(\mu)$]

$$\lim_{\substack{\mu \to 0 \\ x_\eta \text{ fixed}}} \frac{1}{\mu}\left[h_0(1 - \eta x_\eta) + \mu h_1(1 - \eta x_\eta) - g_0\left(\frac{\eta x_\eta}{\mu}\right) - \mu g_1\left(\frac{\eta x_\eta}{\mu}\right)\right] = 0 \tag{2.7.22}$$

for some overlap domain contained in $\mu \ll \eta \ll 1$.

We calculate

$$h_0 + \mu h_1 = \frac{2}{3} - \eta x_\eta + \mu\left[\frac{5}{3} - \frac{1}{2}\log 2 + \frac{1}{2}\log\frac{\eta x_\eta}{2}\right] + O(\eta^2) \quad (2.7.23)$$

$$g_0 + \mu g_1 = \sqrt{2}\tau_0 + \mu\left[\sqrt{2}\tau_1 - \frac{\eta x_\eta}{\mu} - \frac{1}{2} + \log 2 + \frac{1}{2}\log\frac{\eta x_\eta}{\mu}\right] + O\left(\frac{\mu^2}{\eta}\right). \quad (2.7.24)$$

Thus, we must set

$$\sqrt{2}\tau_0 = \tfrac{2}{3} \quad (2.7.25)$$

$$\sqrt{2}\tau_1 = \tfrac{13}{6} - \log 4 + \log \mu^{1/2} \quad (2.7.26)$$

and all singular terms match. The neglected terms will vanish in Equation (2.7.22) as long as $\eta^2/\mu \to 0$ and $\mu/\eta \to 0$ and this determines the overlap domain

$$\mu \ll \eta \ll \mu^{1/2}. \quad (2.7.27)$$

The composite expansion, uniformly valid to order μ on $0 \leq x \leq 1$ with $dx/dt > 0$ is

$$\begin{aligned}\sqrt{2}t &= \frac{2}{3}x^{3/2} + \mu\left[\frac{2}{3}x^{3/2} + x^{1/2} - \frac{1}{2}\log\frac{1 + x^{1/2}}{1 - x^{1/2}}\right] \\ &\quad + \mu[x^* - \sqrt{x^*(1 + x^*)} + \log(\sqrt{x^*} + \sqrt{1 + x^*}) \\ &\quad + \log 2 + \tfrac{1}{2} - \tfrac{1}{2}\log x^*] + O(\mu^2). \end{aligned} \quad (2.7.28)$$

If we denote the solution (2.7.28) for $0 \leq t \leq \tau$ by

$$t = f(x, \mu), \quad (2.7.29a)$$

then the solution for $\tau \leq t \leq 2\tau$ is

$$t = 2\tau - f(x, \mu) \quad (2.7.29b)$$

by symmetry, and periodicity then defines the solution for all times.

2.7.2 A Model Example for the Stokes–Oseen Problem

The mathematical problem of Low Reynolds number flows past an object is outlined in Section 4.3.2. A model example for this flow was proposed and discussed in a seminar given by P. A. Lagerstrom in 1960. An early version of this study is included in the lecture notes of Reference 2.7.6. More recently, rigorous proofs of the validity of the early asymptotic results were given independently, and using different approaches, in References 2.7.7, and 2.7.8. In addition, detailed calculations of the asymptotic solution can be found in Reference 2.7.9.

The model is a singular boundary-value problem in the sense that the form of the expansion comes not from a distinguished limit but from the behavior of the solution near the boundary points.

Consider the equation

$$\frac{d^2u}{dr^2} + \frac{1}{r}\frac{du}{dr} + u\frac{du}{dr} = 0 \tag{2.7.30}$$

with boundary conditions

$$u(\varepsilon) = 0 \tag{2.7.31}$$

$$u(\infty) = 1. \tag{2.7.32}$$

Actually, a slightly more general version is studied in Reference (2.7.7) where Equation (2.7.30) reads

$$\frac{d^2u}{dr^2} + \frac{k-1}{r}\frac{du}{dr} + \alpha u\frac{du}{dr} + \beta\left(\frac{du}{dr}\right)^2 = 0, \tag{2.7.33}$$

where α and β are arbitrary non-negative constants and k is any real number. In Reference (2.7.8) k is taken to be an integer while α is set equal to unity.

Although the above vaguely resembles the radial momentum equation for a viscous imcompressible flow, no physical correspondence is intended as the relation of the model to the Navier–Stokes equations is strictly qualitative.

In this section we will only study the special case $k = 2$, $\alpha = 1$, $\beta = 0$ of Equation (2.7.30) for which it is easy to give a rigorous demonstration of the validity of the asymptotic expansions and matching.

We want the behavior of the solution $u(r; \varepsilon)$ as $\varepsilon \to 0$. The problem also has an analogy with a cylindrically symmetric heat-flow problem having a heat source strength/area proportional to $u(du/dr)$. In these coordinates, as $\varepsilon \to 0$ the size of the cold ($u = 0$) cylinder shrinks to zero. The general shape of the expected solution is shown in Figure 2.7.3. From this, the first

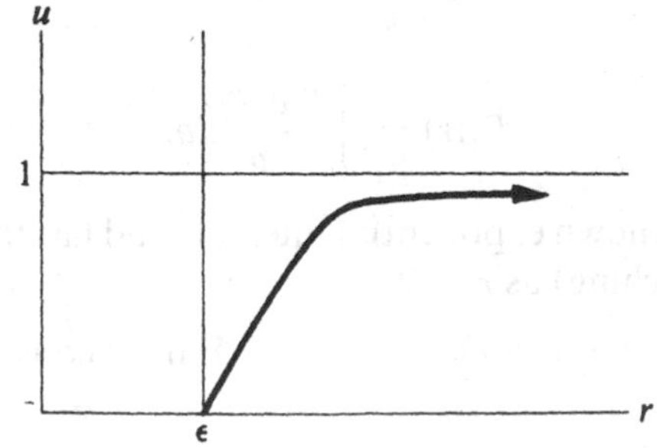

Figure 2.7.3 Stokes–Oseen Model Solution

term of the limiting solution connected with the outer limit ($\varepsilon \to 0$, r fixed) can be intuitively guessed as

$$u \to 1. \tag{2.7.34}$$

That is, the zero-size cold cylinder, does not disturb the temperature field at all. Away from $r = \varepsilon$, one might expect only small perturbations to this solution. Thus, an outer expansion of the form

$$u(r; \varepsilon) = 1 + \mu_1(\varepsilon)h_1(r) + \mu_2(\varepsilon)h_2(r) + \cdots \tag{2.7.35}$$

is assumed, with the idea of satisfying the boundary conditions at infinity and matching to an inner expansion near $r = \varepsilon$. The first term of Equation (2.7.35) is not a good approximation in some neighborhood of $r = \varepsilon$, and the orders in the asymptotic sequence $\mu_i(\varepsilon)$ are not known a priori.

The equations satisfied by h_1 and h_2 are

$$\frac{d^2h_1}{dr^2} + \left(\frac{1}{r} + 1\right)\frac{dh_1}{dr} = 0, \tag{2.7.36}$$

$$\frac{d^2h_2}{dr^2} + \left(\frac{1}{r} + 1\right)\frac{dh_2}{dr} = \begin{cases} 0 & \text{if } \dfrac{\mu_1^2}{\mu_2} \to 0 \\ -h_1\dfrac{dh_1}{dr} & \text{if } \dfrac{\mu_1^2}{\mu_2} = O(1). \end{cases} \tag{2.7.37}$$

The boundary condition at infinity becomes

$$h_1(\infty) = 0, \qquad h_2(\infty) = 0. \tag{2.7.38}$$

An h_2 which is significantly different from h_1 appears only if $\mu_2 = \mu_1^2$, and we can assume that $\mu_2 = \mu_1^2$ with the option of inserting h_1 terms of various orders larger than μ_1^2.

The solutions for h_1 and h_2 satisfying the condition at infinity are easily found. Equation (2.7.36) can be written

$$\frac{d}{dr}\left(re^r\frac{dh_1}{dr}\right) = 0,$$

so that

$$h_1(r) = A_1E_1(r), \tag{2.7.39}$$

where

$$E_1(r) = \int_r^\infty \frac{e^{-\rho}}{\rho}\,d\rho. \tag{2.7.40}$$

Here E_1 is the well-known exponential integral[5] and has the following expansion (useful for matching) as $r \to 0$.

$$E_1(r) = -\log r - \gamma + r + O(r^2), \qquad \gamma = \text{Euler's const.} = 0.577215\ldots. \tag{2.7.41}$$

[5] Sometimes denoted by $-Ei(-r)$.

Similarly, we have

$$\frac{d}{dr}\left(re^r\frac{dh_2}{dr}\right) = A_1^2 E_1(r). \tag{2.7.42}$$

Defining

$$E_n(r) = \int_r^\infty \frac{e^{-\rho}}{\rho^n}\,d\rho, \tag{2.7.43}$$

we can easily show that

$$\int_r^\infty E_n(\rho)d\rho = -rE_n(r) + E_{n-1}(r). \tag{2.7.44}$$

Hence, Equation (2.7.42) becomes

$$\frac{dh_2}{dr} = -A_1^2\frac{e^{-2r}}{r} + A_1^2 e^{-r}E_1(r) - A_2\frac{e^{-r}}{r}$$

and

$$h_2(r) = A_2 E_1(r) + A_1^2\{2E_1(2r) - e^{-r}E_1(r)\}. \tag{2.7.45}$$

Use has been made of the result

$$\int_r^\infty e^{-\rho}E_1(\rho)d\rho = e^{-r}E_1(r) - E_1(2r). \tag{2.7.46}$$

The expansion of h_2(r) as $r \to 0$ is, thus,

$$h_2(r) = -(A_2 + A_1^2)\log r - (A_2 + A_1^2)\gamma - A_1^2 2\log 2 - A_1^2 r\log r$$
$$+ [A_2 + (3-\gamma)A_1^2]r + O(r^2\log r). \tag{2.7.47}$$

Now an inner expansion has to be constructed that can take care of the boundary condition $u = 0$ on $r = \varepsilon$. A suitable inner coordinate is

$$r^* = \frac{r}{\varepsilon},$$

and the limit process has r^* fixed as $\varepsilon \to 0$. The form of this expansion is

$$u(r;\varepsilon) = \nu_0(\varepsilon)g_0(r^*) + \nu_1(\varepsilon)g_1(r^*) + \nu_2 g_2 + \cdots. \tag{2.7.48}$$

Again, choose $\nu_1 = \varepsilon\nu_0^2$ so that the equation for g_1 has a forcing term; other terms similar to g_0 but of order intermediate to ν_0, ν_1 can be inserted in the expansion if necessary. For g_0 we have

$$\frac{d^2g_0}{dr^{*2}} + \frac{1}{r^*}\frac{dg_0}{dr^*} = 0, \qquad g_0(1) = 0, \tag{2.7.49}$$

so that

$$g_0 = B_0\log r^*.$$

Then we have

$$\frac{d^2 g_1}{dr^{*2}} + \frac{1}{r^*}\frac{dg_1}{dr^*} = -g_0 \frac{dg_0}{dr^*} = -B_0^2 \frac{\log r^*}{r^*}, \qquad g_1(1) = 0. \quad (2.7.50)$$

Integration of Equation (2.7.50) yields

$$g_1(r^*) = B_1 \log r^* - B_0^2(r^* \log r^* - 2r^* + 2). \quad (2.7.51)$$

For matching the inner and outer expansions, we can consider an intermediate limit, denoted for brevity by $\lim_\eta$, where

$$\lim_\eta : \varepsilon \to 0, \qquad r_\eta = \frac{r}{\eta} \text{ fixed},$$

where

$$\varepsilon \ll \eta \ll 1 \text{ or } \frac{\eta}{\varepsilon} \to \infty, \qquad \eta \to 0.$$

In this limit $r = \eta r_\eta \to 0$, $r^* = (\eta/\varepsilon) r_\eta \to \infty$. The first-order matching is

$$\lim_\eta \left[1 + \mu_1(\varepsilon) h_1(\eta r_\eta) + \cdots - \nu_0(\varepsilon) g_0\left(\frac{\eta}{\varepsilon} r_\eta\right) \cdots\right] = 0.$$

or

$$\lim_\eta \left[1 + O(\mu_1 \log \eta) + \cdots - \nu_0(\varepsilon) B_0 \log\left(\frac{1}{\varepsilon}\right) + O(\nu_0 \log \eta)\right] = 0. \quad (2.7.52)$$

If we choose

$$\nu_0(\varepsilon) = \frac{1}{\log(1/\varepsilon)}, \qquad B_0 = 1, \quad (2.7.53)$$

the first terms are matched since $\mu_1, \nu_0 \to 0$. Matching to the next order demands that

$$\lim_\eta \frac{1}{\delta_1(\varepsilon)} \left[1 + \mu_1(\varepsilon) h_1(\eta r_\eta) + \mu_1^2(\varepsilon) h_2(\eta r_\eta) + \cdots - \frac{1}{\log(1/\varepsilon)} g_0\left(\frac{\eta}{\varepsilon} r_\eta\right)\right.$$
$$\left. - \frac{\varepsilon}{\log^2(1/\varepsilon)} g_1\left(\frac{\eta}{\varepsilon} r_\eta\right) \cdots\right] = 0 \quad (2.7.54)$$

for some suitable $\delta_1(\varepsilon) \to 0$. Writing out Equation (2.7.54), we have

$$\frac{1}{\delta_1}\left\{1 + \mu_1(\varepsilon)[A_1(-\log \eta r_\eta) - A_1\gamma + A_1 \eta r_\eta + O(\eta^2)]\right.$$
$$+ \mu_1^2(\varepsilon)[-(A_1^2 + A_2)\log \eta r_\eta - (A_1^2 + A_2)\gamma - A_1^2 2 \log 2$$
$$- A_1^2 \eta r_\eta \log \eta r_\eta + (A_2 + (3 - \gamma)A_1^2)\eta r_\eta + \cdots] - \frac{1}{\log(1/\varepsilon)}\left[\log \frac{\eta r_\eta}{\varepsilon}\right]$$
$$\left. - \frac{\varepsilon}{\log^2(1/\varepsilon)}\left[B_1 \log\left(\frac{\eta r_\eta}{\varepsilon}\right) - \frac{\eta r_\eta}{\varepsilon} \log \frac{\eta r_\eta}{\varepsilon} + \frac{2\eta r_\eta}{\varepsilon} - 2\right] \cdots\right\} \to 0. \quad (2.7.55)$$

The terms of order one are already matched. The terms $(\log \eta r_\eta)$ can be matched by choosing

$$\mu_1(\varepsilon) = \frac{1}{\log(1/\varepsilon)}, \qquad A_1 = -1. \tag{2.7.56}$$

A suitable δ_1 has $\mu_1/\delta_1 \to 0$. The next term to be matched (with a suitable function δ_2 replacing δ_1) is $-A_1\gamma\mu_1(\varepsilon)$ in the outer expansion, and we can see that there is no corresponding term in the inner expansion to match this term. Thus, a term

$$\nu^+(\varepsilon)g^+(r^*) = \nu^+(\varepsilon)B^+ \log r^*$$

has to be inserted between g_0, g_1 in the inner expansion with

$$\nu^+(\varepsilon) = \frac{1}{\log^2(1/\varepsilon)}.$$

Thus, omitting the already matched terms, we write Equation (2.7.55) with the intermediate term added in:

$$\frac{1}{\delta_2}\Bigg\{1 + \frac{1}{\log(1/\varepsilon)}[\gamma - \eta r_\eta + O(\eta^2)] + \frac{1}{\log^2(1/\varepsilon)}\Bigg[-(1 + A_2)\log \eta r_\eta$$
$$-(1 + A_2)\gamma - 2\log 2 - \eta r_\eta \log \eta r_\eta\Bigg] + \cdots - \frac{B^+}{\log^2(1/\varepsilon)}\log\frac{\eta r_\eta}{\varepsilon}$$
$$-\frac{\varepsilon}{\log^2(1/\varepsilon)}\left[B_1 \log\frac{\eta r_\eta}{\varepsilon} - \left(\frac{\eta r_\eta}{\varepsilon}\log\frac{\eta r_\eta}{\varepsilon}\right) + \frac{2\eta r_\eta}{\varepsilon} - 2\right]\cdots\Bigg\} \to 0. \tag{2.7.57}$$

The γ term in h_1 is matched with the choice

$$B^+ = \gamma. \tag{2.7.58}$$

This at the same time fixes A_2:

$$A_2 = -\gamma - 1. \tag{2.7.59}$$

This procedure can evidently be continued, with the appropriate insertion of terms of intermediate order in the inner expansion. Summarizing the results, we see that the terms and orders of inner and outer expansions are as shown in the Table below.

The term with g_1 is transcendentally small compared with the g_0 term; it does not enter the matching until further terms such as

$$\left[\frac{(\gamma^2 - 2\log 2)}{\log^2(1/\varepsilon)}\right]$$

are matched by the introduction of an intermediate order $\nu^{++} = 1/[\log^3(1/\varepsilon)]$. This serves to determine A_3 associated with the log term of h_3, etc. The effect of the nonlinearity never appears in the inner expansion, but this effect is in the far field of the outer expansion. The outer expansion contains the inner expansion. In the analogy with viscous flow, the inner or Stokes flow is not adequate for finding the solution and evaluating the skin friction $(du/dr)_0$, but the outer or Oseen flow is.

Order	Term
Outer	
1	1
$\mu_1 = \dfrac{1}{\log(1/\varepsilon)}$	$h_1 = -E_1(r)$
$\mu_2 = \dfrac{1}{\log^2(1/\varepsilon)}$	$h_2 = -(1+\gamma)E_1(r) + E_1(2r) - e^{-r}E_1(r)$
Inner	
$\nu_0 = \dfrac{1}{\log(1/\varepsilon)}$	$g_0 = \log r^*$
$\nu^+ = \dfrac{1}{\log^2(1/\varepsilon)}$	$g^+ = \gamma \log r^*$
$\nu_1 = \dfrac{\varepsilon}{\log^2(1/\varepsilon)}$	$g_1 = B_1 \log r^* - (r^* \log r^* - 2r^* + 2)$

In this example, we can give a proof that our guess of the first term in the outer expansion is really correct. We can write Equation (2.7.30) as

$$\frac{d}{dr}\left(r\frac{du}{dr}\right) + u\left(r\frac{du}{dr}\right) = 0, \qquad u(\varepsilon) = 0,\ u(\infty) = 1. \tag{2.7.60}$$

If we regard Equation (2.7.60) as a linear problem for $r(du/dr)$ and integrate, the problem can be formulated as an integral equation:

$$u(r;\varepsilon) = 1 - \frac{G(r;\varepsilon)}{G(\varepsilon;\varepsilon)}, \tag{2.7.61}$$

where

$$G(r;\varepsilon) = \int_r^\infty \frac{[\exp - \int_\varepsilon^\rho u(\sigma;\varepsilon)d\sigma]}{\rho}\,d\rho.$$

Equation (2.7.60) is invariant under a group of transformations if $ru \sim$ const. and can be reduced to the first-order system

$$\frac{dt}{ds} = \frac{t(1-s)}{t+s}, \qquad \frac{dr}{r} = \frac{dt}{t(1-s)}, \tag{2.7.62}$$

where

$$s = ru, \qquad t = r^2 \frac{du}{dr}.$$

From a study of the integral curves of Equation (2.7.62), we can conclude that the only possible solution of Equation (2.7.60) satisfying the boundary condition has $s \geq 0$, and hence $u \geq 0$. These phase-plane considerations also can be used to prove the existence of a unique solution. It follows from Equation (2.7.61) that

$$0 \leq u \leq 1, \qquad \varepsilon \leq r \leq \infty. \tag{2.7.63}$$

Thus, we have

$$G(\varepsilon; \varepsilon) \geq \int_\varepsilon^\infty \frac{e^{-(\rho-\varepsilon)}}{\rho} d\rho \geq \int_\varepsilon^\infty \frac{e^{-\rho}}{\rho} d\rho \quad \text{or} \quad G(\varepsilon; \varepsilon) \geq E_1(\varepsilon). \tag{2.7.64}$$

Now we can write

$$G(r; \varepsilon) = \int_r^{r_0} \frac{[\exp - \int_\varepsilon^\rho u(\sigma; \varepsilon) d\sigma]}{\rho} d\rho + \int_{r_0}^\infty \frac{[\exp - \int_\varepsilon^\rho u(\sigma; \varepsilon) d\sigma]}{\rho} d\rho.$$

Thus, we have

$$G(r; \varepsilon) \leq \log \frac{r_0}{r} + \frac{1}{r_0 u(r_0; \varepsilon)} \int_{r_0}^\infty u(r; \varepsilon) \exp\left[- \int_\varepsilon^\rho u(\sigma; \varepsilon) d\sigma \right] d\rho \tag{2.7.65}$$

if we use the fact that $u(r)$ is monotonic $(r > 0)$.

Integrating, we have

$$G(r; \varepsilon) \leq \log \frac{r_0}{r} + \frac{1}{r_0 u(r_0; \varepsilon)} \left[\exp - \int_\varepsilon^{r_0} u(\rho; \varepsilon) d\rho \right],$$

$$G(r; \varepsilon) \leq \log \frac{r_0}{r} + \frac{1}{r_0 u(r_0; \varepsilon)}.$$

Now, for any given δ and all $\varepsilon < \delta$, it follows that $u(r_0; \varepsilon) > u(r_0, \delta)$, so that

$$G(r; \varepsilon) \leq \log \frac{r_0}{r} + \frac{1}{r_0 u(r_0; \varepsilon)}, \qquad \varepsilon < \delta. \tag{2.7.66}$$

This, in Equation (2.7.61) the outer limit $\varepsilon \to 0$, r fixed shows that

$$\frac{G(r; \varepsilon)}{G(\varepsilon; \varepsilon)} \leq \frac{1}{E_1(\varepsilon)} \left\{ \log \frac{r_0}{r} + \frac{1}{r_0 u(r_0; \delta)} \right\} \to 0, \qquad \varepsilon \to 0. \tag{2.7.67}$$

The problem of finding the viscous, incompressible flow past a circular cylinder relies on considerations, such as these, although a rigorous proof has not been provided for that case. The considerations given above can actually be extended to demonstrate the overlapping of the two expansions used.

PROBLEMS

1. Consider the initial value problem for Equation (2.7.2)

$$x(0) = 1 - \mu^{1/2}$$

$$\frac{dx}{dt}(0) = \sqrt{2(2 - c^2)}\mu^{1/4}, \qquad 0 \le c^2 \le 2$$

which corresponds to starting near the equilibrium point with a small positive velocity. What are the appropriatc inner variables? Calculate the inner and outer expansions to order μ and match to find the half-period for this casc. Compare your result with the exact expression for the half-period.

2. Consider Equation (2.7.33) for the special case $k = 1$, $\alpha = 1$, $\beta = 0$ and calculate the exact solution. Show that the outer limit, $\varepsilon \to 0$, r fixed, is not $u = 1$. Show also that the incorrect assumption of an outer expansion of the form of Equation (2.3.35) can not be made to match with the inner expansion.

3. Parallel the discussion of Section 2.7.2 for the case $k = 3$, $\alpha = 1$, $\beta = 1$ in Equation (2.7.33).

References

2.7.1 H. C. Corben and P. Stehle, *Classical Mechanics*, John Wiley and Sons, Inc., New York, 1960.

2.7.2 P. A. Lagerstrom and J. Kevorkian, Matched conic approximation to the two fixed force-center problem, *The Astronomical Journal* **68**, No. 2, March 1963, pp. 84–92.

2.7.3 P. A. Lagerstrom and J. Kevorkian, Earth-to-moon trajectories in the restricted three-body problem, *Journal de Mécanique* **2**, No. 2, June 1963, pp. 189–218.

2.7.4 P. A. Lagerstrom and J. Kevorkian, Nonplanar earth-to-moon trajectories, *AIAA Journal* **4**, No. 1, January 1966, pp. 149–152.

2.7.5 J. V. Breakwell and L. M. Perko, Matched asymptotic expansions, patched conics, and the computation of interplanetary trajectories, *Methods in Astrodynamics and Celestial Mechanics* **17**, eds. R. L. Duncombe and V. G. Szebehely, pp. 159–182, 1966.

2.7.6 P. A. Lagerstrom, Méthodes asymptotiques pour l'étude des équations de Navier–Stokes, *Lecture notes*, Institut Henri Poincaré, Paris, 1961. Translated by T. J. Tyson, California Institute of Technology, Pasadena, California, 1965.

2.7.7 D. S. Cohen, A. Fokas, and P. A. Lagerstrom, Proof of some asymptotic results for a model equation for low Reynolds number flow, *S.I.A.M. J. on Appl. Math.* **35** (1978), 187–207.

2.7.8 A. D. MacGillivray, On a model equation of Lagerstrom, *S.I.A.M. J. on Appl. Math.* **34** (1978), 804–812.

2.7.9 W. B. Bush, On the Lagerstrom mathematical model for viscous flow at low Reynolds number, *S.I.A.M. J. on Appl. Math.* **20** (1971), 279–287.

2.8 Higher-Order Example: Beam String

In this section, an elementary example of a higher-order equation is constructed in order to show that the ideas of the previous sections have a natural and general validity.

The engineering theory of an elastic beam with tension which supports a given load distribution leads to the following differential equation, when it is assumed that the deflection W is small (linearized theory):

$$EI\frac{d^4W}{dX^4} - T\frac{d^2W}{dX^2} = P(X), \qquad 0 \le X \le L. \tag{2.8.1}$$

Here

E is the constant modulus of elasticity,
I is the constant moment of inertia of cross section about neutral axis,
T is the constant external tension,
X is the coordinate along beam,
W is the deflection of neutral axis,
$P(X)$ is the external load per length on the beam.

The derivation of an equation of this type from a more exact elastic theory is an example of the kind of approximation that is discussed in Chapter 5 of this book. In the engineering theory of bending, see for example Reference 2.8.1, a model is made for the deformation of the beam under load in which plane cross sections of the beam remain plane under load. The tension and compression forces due to bending which act along the beam are computed by Hooke's law from the stretching of the fibers; the neutral axis is unstressed. Adjacent sections exert a bending moment M on each other proportional to the beam curvature. For small deflections, we have

$$M(X) \doteq -EI\frac{d^2W}{dX^2}. \tag{2.8.2}$$

A vertical shear V at these sections produces a couple to balance the bending moment:

$$V = \frac{dM}{dX}. \tag{2.8.3}$$

The effect of this shear in supporting the external load is expressed in Equation (2.8.1) by the fourth derivative term. The load is carried by the tension in the structure in the usual way in which a string or cable supports an external load. $[T(d^2W)/(dX^2)]$. When the deflection is known, the stresses of interest can be calculated.

Singular perturbation problems arise when the effect of bending rigidity is relatively small in comparison to the tension. In general, when a more complicated model of a physical phenomenon (for example, beam vs. string) is constructed, the order of the differential equations is raised. Correspondingly, the nature of the boundary conditions at the ends is more complicated; due to the higher order of the equations, more conditions are needed. For the string problem, for example, it is sufficient to prescribe the deflection. In a beam problem, the mode of support must also be given. The loss of a boundary condition in passage from the beam-string to the string implies the existence of a boundary layer near the support, a local region in which bending rigidity is important. A similar phenomenon can occur under a region of rapid change of the load, or near a concentrated load. Various types of boundary conditions can be used to represent the end of a beam of which the following are most common and important.

(1) Pin-end: no restoring moment M applied at the end, $d^2W/dX^2 = 0$; deflection prescribed, for example, $W = 0$.
(2) Built-in end: slope at end prescribed, for example, $dW/dX = 0$, deflection prescribed, for example, $W = 0$;
(3) Free end: no bending moment exerted on end $d^2W/dX^2 = 0$, no shear exerted on end $dW/dX = 0$.

Consider now the typical problem for a beam with built-in ends under the distributed load $P(X)$. The problem can be expressed in suitable dimensionless coordinates by measuring lengths in terms of L and using a characteristic load density $\mathscr{P}$, so that

$$P(X) = \mathscr{P}p(x), \qquad 0 \le x \le 1, \tag{2.8.4}$$

where $x = X/L$. The deflection is conveniently measured in terms of that characteristic of the string alone,

$$w(x) = \left(\frac{T}{\mathscr{P}L^2}\right)W(X), \tag{2.8.5}$$

and the resulting dimensionless equation and boundary conditions are

$$\varepsilon\frac{d^4w}{dx^4} - \frac{d^2w}{dx^2} = p(x), \qquad 0 \le x \le 1;$$

$$w(0) = \frac{dw}{dx}(0) = w(1) = \frac{dw}{dx}(1) = 0. \tag{2.8.6}$$

The small parameter ε of the problem is

$$\varepsilon = \frac{EI}{TL^2} \tag{2.8.7}$$

and measures the relative importance of the bending rigidity in comparison to the tension.

Next we construct the inner and outer expansions. For the outer expansion ($x \to 0$, ε fixed), we expect the first term to be independent of ε, and we write

$$w(x; \varepsilon) = h_0(x) + \nu_1(\varepsilon)h_1(x) + \nu_2(\varepsilon)h_2(x) + \cdots. \tag{2.8.8}$$

The corresponding differential equations are

$$-\frac{d^2h_0}{dx^2} = p(x),$$

$$-\frac{d^2h_1}{dx^2} = \begin{cases} -\dfrac{d^4h_0}{dx^4} = p''(x) & \text{if } \dfrac{\nu_1}{\varepsilon} = 1, \\ 0 & \text{if } \dfrac{\nu_1}{\varepsilon} \to 0. \end{cases} \tag{2.8.9}$$

There are no boundary conditions for the h_i, but the constants of integration in the solutions must be obtained by matching with the boundary layers at each end. The solution for $h_0(x)$ is, thus,

$$h_0(x) = B_0 + A_0 x - \int_0^x (x - \lambda)p(\lambda)d\lambda. \tag{2.8.10}$$

For purposes of matching, later it is useful to have the series expansions of $h_0(x)$ near $x = 0$ and $x = 1$.

Near $x = 0$,

$$h_0(x) = h_0(0) + xh_0'(0) + \frac{x^2}{2}h_0''(0) + \cdots$$

$$= B_0 + A_0 x - p(0)\frac{x^2}{2!} - p'(0)\frac{x^3}{3!} + O(x^4). \tag{2.8.11}$$

Near $x = 1$,

$$h_0(x) = B_0 + A_0 - \int_0^1 (1 - \lambda)p(\lambda)d\lambda + \left\{A_0 - \int_0^1 p(\lambda)d\lambda\right\}(x - 1)$$

$$- p(1)\frac{(x-1)^2}{2!} - p'(1)\frac{(x-1)^3}{3!} + O((x-1)^4). \tag{2.8.12}$$

A suitable boundary-layer coordinate x^* is chosen by the requirement that the bending and tension terms are of the same order of magnitude near $x = 0$:

$$x^* = \frac{x}{\sqrt{\varepsilon}}. \tag{2.8.13}$$

The corresponding asymptotic expansion near $x = 0$ is

$$w(x; \varepsilon) = \mu_0(\varepsilon)g_0(x^*) + \mu_1(\varepsilon)g_1(x^*) + \cdots. \tag{2.8.14}$$

Equation (2.8.6) thus becomes

$$\frac{1}{\varepsilon}\left\{\mu_0\frac{d^4g_0}{dx^{*4}} + \mu_1\frac{d^4g_1}{dx^{*4}} + \cdots\right\} - \frac{1}{\varepsilon}\left\{\mu_0\frac{d^2g_1}{dx^{*2}} + \mu_1\frac{d^2g_1}{dx^{*2}} + \cdots\right\}$$

$$= p(0) + \sqrt{\varepsilon}x^*p'(0) + \cdots. \tag{2.8.15}$$

Two possibilities arise: either $\mu_0/\varepsilon \to \infty$ or $\mu_0/\varepsilon = 1$. It can be shown that, in general, the second possibility does not allow matching, and thus only the first is considered here. The effect of the external load, then, does not appear in the first boundary-layer equation but enters at first only through the matching. Of course, it has to be verified that the assumption $\mu_0/\varepsilon \to \infty$ is correct after μ_0 is found from the matching. Thus, we have

$$O\left(\frac{\mu_0}{\varepsilon}\right): \frac{d^4 g_0}{dx^{*2}} - \frac{d^2 g_0}{dx^{*2}} = 0. \tag{2.8.16}$$

Both boundary conditions at $x^* = 0$ are to be satisfied by g_0:

$$\frac{dg_0}{dx^*} = g_0 = 0, \quad \text{at } x^* = 0. \tag{2.8.17}$$

Using the fact that exponential growth (e^{x^*}) cannot match as $x^* \to \infty$ and taking into account the boundary conditions, we obtain a solution with one arbitrary constant,

$$g_0(\tilde{x}) = C_0\{x^* - 1 + e^{-x^*}\}. \tag{2.8.18}$$

An intermediate limit suitable for matching near $x^* = 0$ is given by x_η fixed:

$$x_\eta = \frac{x}{\eta(\varepsilon)}, \qquad \eta \to 0, \frac{\eta}{\sqrt{\varepsilon}} \to \infty, \tag{2.8.19}$$

so that

$$x^* = \frac{\eta}{\sqrt{\varepsilon}} x_\eta \to \infty, \qquad x = \eta x_\eta \to 0. \tag{2.8.20}$$

Matching near $x = 0$ takes the form

$$\lim_{\substack{\varepsilon \to 0 \\ x_\eta \text{ fixed}}} \left\{ h_0(\eta x_\eta) + \nu_1(\varepsilon) h_1(\eta x_\eta) + \cdots - \mu_0(\varepsilon) g_0\left(\frac{\eta}{\sqrt{\varepsilon}} x_\eta\right) - \mu_1(\varepsilon) g_1\left(\frac{\eta x_\eta}{\sqrt{\varepsilon}}\right) \cdots \right\} = 0. \tag{2.8.21}$$

Using the expansion of Equations (2.8.11) and (2.8.18), we find that the first-order matching condition is

$$\lim_{\substack{\varepsilon \to 0 \\ x_\eta \text{ fixed}}} \left\{ B_0 + A_0 \eta x_\eta + \cdots - \mu_0(\varepsilon) C_0\left(\frac{\eta}{\sqrt{\varepsilon}} x_\eta - 1 + \exp\frac{-\eta x_\eta}{\sqrt{\varepsilon}}\right) + \cdots + \nu_1(\varepsilon) h_1(\eta x_\eta) \right\} = 0. \tag{2.8.22}$$

The term linear in x_η dominates g_0, so that matching is only possible if

$$B_0 = 0 \tag{2.8.23}$$

and

$$\mu_0 = \sqrt{\varepsilon}, \qquad A_0 = C_0. \tag{2.8.24}$$

This verifies the fact that $\mu_0/\varepsilon \to \infty$.

Another point can be noticed from Equation (2.8.22). The term $O(1)$ in g_0 can not be matched except by a suitable h_1. That is, we have $v_1(\varepsilon) = \sqrt{\varepsilon}$, and h_1 satisfies the equation of an unloaded string (cf. Equation 2.8.9):

$$\frac{d^2 h_1}{dx^2} = 0. \tag{2.8.25}$$

The solution is

$$h_1 = B_1 + A_1 x. \tag{2.8.26}$$

Thus, considering matching to the next order in Equation (2.8.22), we have

$$B_1 = -C_0. \tag{2.8.27}$$

The final determination of the unknown constants depends on the application of similar considerations at the other end of the beam. Summarizing, for the outer expansion we have thus far

$$w(x;\varepsilon) = h_0(x) + \sqrt{\varepsilon}h_1(x) + \cdots$$

where

$$h_0(x) = C_0 x - \int_0^x (x - \lambda)p(\lambda)d\lambda, \qquad h_1(x) = -C_0 + A_1 x. \tag{2.8.28}$$

The first approximation h_0 satisfies a zero-deflection boundary condition at $x = 0$ as might have been expected from physical consideration. Applying the same reasoning at $x = 1$, we find, for example, that

$$C_0 = \int_0^1 (1 - \lambda)p(\lambda)d\lambda = -M^{(1)}. \tag{2.8.29}$$

Here $M^{(1)}$ represents the total moment of the applied load about $x = 1$. The result of Equation (2.8.29) is now verified by detailed matching, and the unknown constant A_1 is also found.

For the boundary layer near $x = 1$, the coordinate

$$x^+ = \frac{(x-1)}{\sqrt{\varepsilon}} \tag{2.8.30}$$

is used, and the boundary-layer expression is

$$w(x;\varepsilon) = \sqrt{\varepsilon}f_0(x^+) + \cdots. \tag{2.8.31}$$

The equation for f_0 is the same as that for g_0, so that the solution satisfying the boundary condition at $x = 1$, $x^+ = 0$ is

$$f_0(x^+) = D_0\{x^+ + 1 - e^{x^+}\}. \tag{2.8.32}$$

Exponential growth as $x^+ \to -\infty$ is ruled out near $x = 1$.

The intermediate limit near $x^+ = 1$ is defined in terms of x_ξ:

$$x_\xi = \frac{x-1}{\xi(\varepsilon)} < 0, \tag{2.8.33}$$

where

$$x = 1 + \xi x_\xi \to 1, \qquad x^+ = \left(\frac{\xi}{\sqrt{\varepsilon}}\right) x_\xi \to -\infty.$$

The expansion of h_0 near $x = 1$ (Equation 2.8.12) is now

$$h_0(x) = C_0 + M^{(1)} + (C_0 - k)(x - 1) + \cdots, \tag{2.8.34}$$

where

$$k = \int_0^1 p(\lambda)d\lambda = \text{total load on beam.}$$

The matching condition near $x = 1$ is, thus,

$$\lim_{\substack{\xi\to 0\\ x_\xi \text{ fixed}}} \left\{ C_0 + M^{(1)} + (C_0 - k)\xi x_\xi + \cdots + \sqrt{\varepsilon}(-C_0 + A_1 + A_1 \xi x_\xi) + \cdots \right.$$

$$\left. - \sqrt{\varepsilon} D_0 \left(\frac{\xi}{\sqrt{\varepsilon}} x_\xi + 1 - \exp\left(\frac{\xi x_\xi}{\sqrt{\varepsilon}}\right)\right) + \cdots \right\} = 0. \tag{2.8.35}$$

First-order matching shows that, indeed,

$$C_0 = -M^{(1)}$$

and

$$C_0 - k = D_0, \tag{2.8.36}$$

that is,

$$D_0 = \int_0^1 (1-\lambda)p\, d\lambda - \int_0^1 p\, d\lambda = -\int_0^1 \lambda p(\lambda) d\lambda = -M^{(0)}. \tag{2.8.37}$$

Further, the matching of the constant term in f_0 yields

$$A_1 = C_0 + D_0 = -[M^{(1)} + M^{(0)}]. \tag{2.8.38}$$

Thus, finally, the three expansions are fully determined to the orders considered.

$$w(x;\varepsilon) = -\sqrt{\varepsilon} M^{(1)}\{x^* - 1 + e^{-x^*}\} + \cdots \quad \text{near } x = 0,$$

$$w(x;\varepsilon) = -M^{(1)}x - \int_0^x (x-\lambda)p(\lambda)d\lambda$$

$$+ \sqrt{\varepsilon}\{M^{(1)} - (M^{(0)} + M^{(1)})x\} + \cdots \quad \text{away from the ends,}$$

$$w(x;\varepsilon) = -\sqrt{\varepsilon} M^{(0)}\{x^+ + 1 - e^{+x^+}\} + \cdots \quad \text{near } x = 1. \tag{2.8.39}$$

The uniformly valid approximation is constructed, as before, by adding together all three expansions and subtracting the common part, which has canceled out identically in the matching. Thus, we have

$$w_{uv} = -M^{(1)}x - \int_0^x (x-\lambda)p(\lambda)d\lambda + \sqrt{\varepsilon}\{M^{(0)}e^{(x-1)/\sqrt{\varepsilon}} - M^{(1)}e^{-x/\sqrt{\varepsilon}} + M^{(1)} - (M^{(0)} + M^{(1)})x\} + O(\sqrt{\varepsilon}). \tag{2.8.40}$$

The uniformly valid expansion is again recognized as having the form of a composite expansion,

$$\sum_{n=0} \varepsilon^{n/2}\{h_n(x) + G_n(\tilde{x}) + F_n(x^+)\},$$

where the F_n and G_n decay exponentially.

From this expansion, the deflection curve, bending moment, and stresses are easily calculated. For example, the bending moment distribution near $x = 0$, proportioned to d^2w/dx^2, comes only from the boundary-layer term. Near $x = 0$, the moment (and stress) decay exponentially:

$$M = -\frac{EI\mathscr{P}}{T}\frac{d^2w}{dx^2} = +\frac{M^{(1)}}{\sqrt{\varepsilon}}\frac{EI\mathscr{P}}{T}e^{-x/\sqrt{\varepsilon}} = M^{(1)}\sqrt{\frac{EI}{T}}\,L\mathscr{P}\exp\left(-X\sqrt{\frac{T}{EI}}\right). \tag{2.8.41}$$

PROBLEM

1. In Reference 2.8.1 (pp. 277 ff), the deflection theory of suspension bridges is discussed, and the differential equation for the additional cable deffection $w(X)$ (or beam deflection) over that due to dead load is obtained. This equation is of the form studied in this Section,

$$EI\frac{d^4W}{dX^4} - (T + \tau)\frac{d^2W}{dX^2} = P - Q\frac{\tau}{T}, \qquad 0 \le x \le L,$$

where

T = dead load cable tension,
τ = increase in cable tension due to live load,
$Q(X)$ = dead load per length,
$P(X)$ = live load per length.

The increase of the main-cable tension depends on the stretching of the cable and is thus related to W. Assuming a linear elasticity for the cable (and small-cable slopes) we find that

$$\frac{TL_c}{E_cA_c}\tau = \int_0^L W(X)Q(X)dX,$$

where L_c, E_c, A_c are the original cable length, modulus of elasticity, and cross-sectional area, respectively.

According to the usual boundary conditions, the truss (or beam) is considered pin-ended.

Using boundary-layer theory, calculate the deflection $W(X)$ due to a uniform dead load, $Q =$ const. and a concentrated live load, $P = P_0 S(X - L/2)$, at the center. Use either matched or composite expansions. Indicate what kind of problem must be solved to find the additional tension.

Reference

2.8.1 T. von Karman, and M. Biot, *Mathematical Methods in Engineering*, McGraw-Hill Book Co., New York, 1940.

Chapter 3

Multiple-Variable Expansion Procedures

Various physical problems are characterized by the presence of a small disturbance which, because of being active over a long time, has a non-negligible cumulative effect. For example, the effect of a small damping force over many periods of oscillation is to produce a decay in the amplitude of a linear oscillator. A fancier example having the same physical and mathematical features is the motion of a satellite around the earth, where the dominant force is a spherically symmetric gravitational field. If this were the only force acting on the satellite the motion would (for sufficiently low energies) be periodic. The presence of a thin atmosphere, a slightly non-spherical earth, a small moon, a distant sun, etc. all produce small but cumulative effects which after a sufficient number of orbits drastically alter the nature of the motion.

It is the aim of this chapter to discuss one of several possible methods that have been devised in order to provide a systematic way of estimating this cumulative effect. The main effort here, as in the preceding chapter, is the exposition of various aspects of the method by means of a series of examples. In solving a given problem, the aim is to combine appropriate techniques to construct an expansion which is uniformly valid over long time intervals.

A central feature of the method is the nonexistence, for long times, of a limit process expansion of the type used so extensively in the previous chapter. As a result, one is led to representing the solution at the outset in the form of a general asymptotic expansion as discussed in Section 1.3 (cf. Equation 1.3.30). This is in contrast to the situation encountered in Chapter 2 where the general asymptotic expansion arose at the last stage of computations when one combined an inner and outer expansion to define the composite solution. Since limit process expansions are not applicable, successive terms in the solution cannot be calculated by the repeated application of limits, and more importantly, rules must be established for the calculation of these terms. Viewed in this light the method of this chapter becomes a generalization of a method proposed by the astronomer Lindstedt for the calculation of periodic solutions. Thus, it is appropriate to begin this chapter with a brief review of Lindstedt's method although, strictly speaking, this method involves limit process expansions.

3.1 Method of Strained Coordinates (Lindstedt's Method) for Periodic Solutions

In the form we will consider this method, it was discussed in 1892 in Volume II of Poincaré's famous treatise on celestial mechanics, Reference 3.1.1. Although Poincaré gave due credit for the original idea to an obscure reference by Lindstedt in 1882, subsequent authors have generally referred to this as the method of Poincaré. Actually, the idea goes further back to Stokes, Reference 3.1.2, who in 1847 used essentially the same method to calculate periodic solutions for a weakly nonlinear wave propagation problem (cf. Problem 3.1.1). Strictly speaking, one should therefore refer to this as Stokes' method. This has not been the case and many authors have called it the PLK method (P for Poincaré, L for Lighthill who introduced a more general version in 1949, and K for Kuo who applied it to viscous flow problems in 1953). To minimize confusion, we will adhere to Van Dyke's nomenclature of the "method of strained coordinates" and refer the reader to Reference 3.1.3 which contains an extensive discussion of applications in fluid mechanics. Some of these applications deal with partial differential equations (for example, supersonic thin-airfoil theory) and will be discussed further in Chapter 5.

3.1.1 The Weakly Nonlinear Oscillator

Consider the weakly nonlinear oscillator with no damping, modeled in dimensionless variables by

$$\frac{d^2y}{dt^2} + y + \varepsilon y^3 = 0, \qquad 0 < \varepsilon \ll 1 \tag{3.1.1}$$

$$y(0) = 0$$

$$\dot{y}(0) = v > 0. \tag{3.1.2a}$$

We will see presently that for $\varepsilon \geq 0$ all the solutions of (3.1.1) are periodic. Hence, regardless of the initial values of y and dy/dt, the solution will at some later time pass through $y = 0$. Since (3.1.1) is autonomous, there is no loss of generality in choosing the origin of time when $y = 0$.

In a straightforward perturbation procedure, one would assume

$$y(t, \varepsilon) = y_0(t) + \varepsilon y_1(t) + O(\varepsilon^2) \tag{3.1.3}$$

and calculate the following equations and initial conditions for y_0 and y_1,

$$\frac{d^2 y_0}{dt^2} + y_0 = 0; \qquad y_0(0) = 0, \qquad \frac{dy_0(0)}{dt} = v \tag{3.1.4a}$$

$$\frac{d^2 y_1}{dt^2} + y_1 = -y_0^3; \qquad y_1(0) = \frac{dy_1(0)}{dt} = 0. \tag{3.1.4b}$$

Solving (3.1.4a) gives

$$y_0(t) = v \sin t \tag{3.1.5}$$

and when this is used in the right-hand side of (3.1.4b) we find

$$\frac{d^2 y_1}{dt^2} + y_1 = -v^3 \sin^3 t = \frac{v^3}{4} \sin 3t - \frac{3v^3}{4} \sin t. \tag{3.1.6}$$

The solution of (3.1.6) gives

$$y_1 = -\frac{9}{32} v^3 \sin t - \frac{v^3}{32} \sin 3t + \frac{3v^3}{8} t \cos t, \tag{3.1.7}$$

where the last term in (3.1.7) is the response to the term $-(3v^3/4)\sin t$ in (3.1.6).

Since $-(3/4)v^3 \sin t$ is a homogeneous solution of (3.1.6), it gives rise to a resonant response proportional to $t \cos t$. Such a term is referred to as "mixed-secular" in the literature in astronomy. Here mixed denotes the fact that we have the product of an algebraic and a trigonometric term, and secular (which is derived from the French word *siècle* for century) because, generally, in astronomical applications ε is quite small and the effect of a term proportional to $\varepsilon t \cos t$ becomes appreciable only after a very long period, of the order of a century. However, regardless of how small ε is the straightforward perturbation expansion (3.1.3) breaks down when $t = O(1/\varepsilon)$. Thus, (3.1.3) is only valid for $0 \le t \le T$ where T is some finite period not very large compared to unity. This phenomenon was also encountered in Section 2.1 [cf. Equation (2.1.8)].

Actually, the solution of Equation (3.1.1) is bounded. In fact, it happens to be periodic for any v as we will see presently.

We can solve (3.1.1) exactly since it describes a conservative system with the energy integral

$$\frac{1}{2}\left(\frac{dy}{dt}\right)^2 + \frac{1}{2} y^2 + \frac{\varepsilon y^4}{4} = \text{const.} = \frac{v^2}{2} \tag{3.1.8}$$

obtained by multiplying (3.1.1) by dy/dt and noting that the result is integrable.

Since the potential energy $y^2/2 + \varepsilon y^4/4$ is (for $\varepsilon > 0$) a concave function for all y, the solution for any energy level $v^2/2$ describes periodic oscillations in the interval $-y_m \leq y \leq y_m$, where

$$y_m = \left[\frac{-1 + (1 + 2\varepsilon v^2)^{1/2}}{\varepsilon}\right]^{1/2} = v\left[1 - \frac{\varepsilon v^2}{4} + O(\varepsilon^2)\right] \tag{3.1.9}$$

is obtained by solving the quadratic equation which results from (3.1.8) when $dy/dt = 0$.

One can proceed further and calculate the formal solution by integrating (3.1.8) once more as follows

$$t = \pm \int_0^y \frac{ds}{\sqrt{v^2 - s^2 - \varepsilon s^4/2}}, \tag{3.1.10}$$

where the upper sign is to be used when dy/dt is positive and the lower sign when dy/dt is negative.

The above can be expressed as an elliptic integral of the first kind by setting[1]

$$s = -y_m \cos \psi. \tag{3.1.11}$$

We calculate

$$(1 + 2\varepsilon v^2)^{1/4} t = \pm \int_{\pi/2}^{\cos^{-1}(-y/y_m)} \frac{d\psi}{\sqrt{1 - k^2 \sin^2 \psi}}, \tag{3.1.12}$$

where

$$k^2 = \frac{-1 + \sqrt{1 + 2\varepsilon v^2}}{2\sqrt{1 + 2\varepsilon v^2}} = \frac{\varepsilon v^2}{2} + O(\varepsilon^2). \tag{3.1.13}$$

In particular, since the potential energy is an even function of y, one fourth of the period P is the value of t when $y = y_m$, and (3.1.12) gives

$$P(\varepsilon) = \frac{4K(k^2)}{(1 + 2\varepsilon v^2)^{1/4}}, \tag{3.1.14}$$

where $K(k^2)$ is the complete elliptic integral of the first kind defined by

$$K(k^2) = \int_0^{\pi/2} \frac{d\psi}{\sqrt{1 - k^2 \sin^2 \psi}}.$$

Since $k^2 = O(\varepsilon)$, we can use standard tables to derive the following expansion in powers of ε for the period

$$P = 2\pi[1 - \tfrac{3}{8}\varepsilon v^2 + O(\varepsilon^2)]. \tag{3.1.15}$$

We see that with $\varepsilon \neq 0$ the period is amplitude dependent, unlike the situation for the linear case where $P = 2\pi$ for any value of v.

[1] The reader will find an extensive discussion of elliptic functions and the various definitions we use here in Reference 3.1.4.

The result (3.1.15) also follows from (3.1.10) after some algebra if we set the upper limit equal to y_m and expand the resulting definite integral for $P/4$ in powers of ε.

Finally, we can invert (3.1.12) using elliptic functions in the form

$$y = y_m cn[(1 + 2\varepsilon v^2)^{1/4} t + K(k^2), k] \tag{3.1.16}$$

which defines a periodic function of time. A more explicit form of (3.1.16) can be obtained by expressing y as a Fourier series in the form

$$y(t, \varepsilon) = \sum_{n=1}^{\infty} b_n(\varepsilon) \sin \frac{2n\pi t}{P(\varepsilon)}, \tag{3.1.17a}$$

where

$$b_n(\varepsilon) = \frac{4}{P} \int_0^{P/2} y(t, \varepsilon) \sin \frac{2n\pi t}{P} \, dt \tag{3.1.17b}$$

and the coefficients can be calculated in principle using (3.1.16) for y. Either form of the exact solution is cumbersome, particularly if one is only interested in the case $\varepsilon \ll 1$.

We note from (3.1.17a) that the solution actually depends on the variable $t^+ = 2\pi t/P$ instead of t, and that if $\varepsilon \ll 1$, t^+ will have the form

$$t^+ = (1 + \varepsilon\omega_1 + \varepsilon^2\omega_2 + \cdots)t, \tag{3.1.18}$$

where the ω_i are constants independent of ε. Equation (3.1.15) shows, for example, that $\omega_1 = (3/8)v^2$. It is now clear why the limit process expansion (3.1.3) failed. For example, a term of the form

$$\sin \frac{2n\pi t}{P} = \sin n(1 + \varepsilon\omega_1 + \varepsilon^2\omega_2 + \cdots)t$$

occurring in the exact solution would, under the limit process $\varepsilon \to 0$, t fixed, develop into

$$\sin \frac{2n\pi t}{P} = \sin nt + n\varepsilon\omega_1 t \cos nt + O(\varepsilon^2).$$

Thus, the mixed secular terms encountered in solving (3.1.1) in the form (3.1.3) are strictly due to the nonuniform representation of trigonometric functions of the variable t^+.

The remedy is easily discerned; we need a limit process expansion where t^+ is held fixed as $\varepsilon \to 0$.

Thus, to accommodate for trigonometric terms with arguments involving t^+, and to allow the b_n to have expansions in terms of ε, we seek a solution in the form

$$y(t, \varepsilon) = f_0(t^+) + \varepsilon f_1(t^+) + \varepsilon^2 f_2(t^+) + \cdots, \tag{3.1.19}$$

where t^+ is defined by (3.1.18) and the ω_i are unknown constants to be determined by the requirement that the f_i be periodic functions of t^+. This is the essential idea of the method of strained coordinates.

Since (3.1.19) is a limit process expansion, it is convenient to first write (3.1.1) and (3.1.2) in terms of t^+ as follows

$$(1 + \varepsilon\omega_1 + \varepsilon^2\omega_2 + \cdots)^2 \frac{d^2y}{dt^{+2}} + y + \varepsilon y^3 = 0 \tag{3.1.20}$$

$$y(0) = 0 \tag{3.1.21a}$$

$$(1 + \varepsilon\omega_1 + \varepsilon^2\omega_2 + \cdots)\frac{dy(0)}{dt^+} = v. \tag{3.1.21b}$$

Substituting (3.1.19) into the above then gives the following sequence of initial value problems for the f_i:

$$L(f_0) \equiv \frac{d^2f_0}{dt^{+2}} + f_0 = 0; \qquad f_0(0) = 0, \qquad \frac{df_0(0)}{dt^+} = v \tag{3.1.22}$$

$$L(f_1) = -2\omega_1 \frac{d^2f_0}{dt^{+2}} - f_0^3; \qquad f_1(0) = 0, \qquad \frac{df_1(0)}{dt^+} = -\omega_1 v \tag{3.1.23}$$

$$L(f_2) = -(\omega_1^2 + 2\omega_2)f_0 - 2\omega_1 \frac{d^2f_1}{dt^{+2}} - 3f_0^2 f_1;$$

$$f_2(0) = 0; \qquad \frac{df_2(0)}{dt^+} = (\omega_1^2 - \omega_2)v \tag{3.1.24}$$

etc.

We solve (3.1.22) immediately:

$$f_0(t^+) = v \sin t^+. \tag{3.1.25}$$

Using this in (3.1.23) gives

$$L(f_1) = (2\omega_1 v - \tfrac{3}{4}v^3)\sin t^+ + \frac{v^3}{4}\sin 3t^+. \tag{3.1.26}$$

We have seen that a homogeneous solution of $L(f_1) = 0$ such as the first term on the right-hand side of (3.1.26) will have a mixed secular term as a response, which becomes unbounded as $t \to \infty$ and is certainly not periodic. Therefore, to ensure the periodicity of f_1 we must set

$$2\omega_1 v - \tfrac{3}{4}v^3 = 0 \tag{3.1.27}$$

which can only hold (since $v \neq 0$) if

$$\omega_1 = \tfrac{3}{8}v^2. \tag{3.1.28}$$

Thus, we have recovered the earlier result quite efficiently. Moreover, what remains of (3.1.26) can now be solved subject to the appropriate initial conditions in the form

$$f_1(t^+) = -\tfrac{9}{32}v^3 \sin t^+ - \frac{v^3}{32}\sin 3t^+. \tag{3.1.29}$$

The procedure can be continued indefinitely. Removal of homogeneous solutions (terms proportional to $\sin t^+$) from the right hand side of the $L(f_i)$ defines ω_i. A feature typical of weakly nonlinear oscillations will be recognized, namely that higher harmonics of $\sin t^+$ occur to higher orders. In fact, we can deduce for this example that $f_n(t)$ will only involve the $(n + 1)$-functions $\sin t^+, \sin 3t^+, \ldots, \sin(2n + 1)t^+$.

Finally, the foregoing procedure also applies if $\varepsilon < 0$ as long as $|\varepsilon| \ll 1$ because all solutions are periodic for $v = O(1)$.

3.1.2 Rayleigh's Equation

In the preceding example for $\varepsilon > 0$, all the solutions of (3.1.1) were periodic regardless of the value of v. We now consider Rayleigh's equation

$$\frac{d^2y}{dt^2} + y + \varepsilon\left[-\frac{dy}{dt} + \frac{1}{3}\left(\frac{dy}{dt}\right)^3\right] = 0, \qquad 0 < \varepsilon \ll 1 \tag{3.1.30}$$

which has only one periodic solution called a "limit cycle" corresponding to one particular initial value of y when $dy/dt = 0$.

Equation (3.1.30) is related to the van der Pol equation discussed in Section 2.6 through the transformation $w = dy/dt$; differentiating (3.1.30) and setting $dy/dt = w$ gives the van der Pol equation for w.

Although one can rigorously prove the existence of a limit cycle, we will use a heuristic argument on which to base our suspicion for the occurrence of such a solution. Consider (3.1.30) in the phase-plane of y and dy/dt. If $\varepsilon = 0$ the integral curves are circles. For any positive ε, the oscillator is additionally subjected to the "force" $\varepsilon[dy/dt - \frac{1}{3}(dy/dt)^3]$. Now if dy/dt is small, i.e., if the motion starts near the origin of the phase-plane, the term dy/dt is more important than $-\frac{1}{3}(dy/dt)^3$. Hence the net effect of the bracketed term in (3.1.30) is a negative damping, leading to an increase in amplitude. But this cannot go on indefinitely, since eventually the term $-\frac{1}{3}(dy/dt)^3$ would dominate and would produce a decay in the amplitude. Similarly, if the motion were initiated with a large value of v, the tendency is for the amplitude to decrease until a balance is struck between the two opposing forces in the bracketed term. Therefore, we have a strong suspicion that for certain very special initial conditions there exists a closed trajectory in the phase plane, i.e., a periodic solution.

We will use the method of strained coordinates to exhibit this solution. Since the appropriate initial amplitude is unknown we assume that

$$y(0) = a(\varepsilon) = a_0 + \varepsilon a_1 + \varepsilon^2 a_2 \tag{3.1.31a}$$

$$\frac{dy(0)}{dt} = 0, \tag{3.1.31b}$$

where the unknown constants a_i are to be determined. Note, here again, that a periodic solution will always pass through $dy/dt = 0$ and we set the origin of the time scale to be zero when this occurs.

We develop $y(t, \varepsilon)$ in the form

$$y(t, \varepsilon) = f_0(t^+) + \varepsilon f_1(t^+) + \varepsilon^2 f_2(t^+) + \cdots \tag{3.1.32}$$

with

$$t^+ = (1 + \varepsilon\omega_1 + \varepsilon^2\omega_2 + \cdots)t. \tag{3.1.33}$$

The equations and initial conditions governing the f_i now are

$$L(f_0) \equiv \frac{d^2 f_0}{dt^{+2}} + f_0 = 0, \qquad f_0(0) = a_0, \qquad \frac{df_0(0)}{dt^+} = 0; \tag{3.1.34}$$

$$L(f_1) = -2\omega_1 \frac{d^2 f_0}{dt^{+2}} + \frac{df_0}{dt^+} - \frac{1}{3}\left(\frac{df_0}{dt^+}\right)^3, \qquad f_1(0) = a_1, \qquad \frac{df_1(0)}{dt^+} = 0; \tag{3.1.35}$$

$$L(f_2) = -2\omega_1 \frac{d^2 f_1}{dt^{+2}} - (2\omega_2 + \omega_1^2)\frac{d^2 f_0}{dt^{+2}} + \frac{df_1}{dt^+} + \omega_1 \frac{df_0}{dt^+} - \left(\frac{df_0}{dt^+}\right)^2\left[\frac{df_1}{dt^+} + \omega_1 \frac{df_0}{dt^+}\right], \qquad f_2(0) = a_2, \qquad \frac{df_2(0)}{dt^+} = 0. \tag{3.1.36}$$

Clearly

$$f_0(t^+) = a_0 \cos t^+ \tag{3.1.37}$$

and using this result in (3.1.35) gives

$$L(f_1) = 2\omega_1 a_0 \cos t^+ + \left(\frac{a_0^3}{4} - a_0\right)\sin t^+ - \frac{a_0^3}{12}\sin 3t^+. \tag{3.1.38}$$

Periodicity requires that

$$2\omega_1 a_0 = 0 \tag{3.1.39}$$

$$\frac{a_0^3}{4} - a_0 = 0. \tag{3.1.40}$$

We discard the trivial solution $a_0 = 0$, ω_1 arbitrary, and set

$$\omega_1 = 0 \tag{3.1.41}$$

$$a_0 = 2. \tag{3.1.42}$$

This determines y to $O(1)$ with t^+ to $O(\varepsilon)$ and gives a limit cycle amplitude of 2.

Next, we calculate the solution for f_1 from (3.1.38) with only $-\frac{2}{3}\sin 3t^+$ remaining on the right-hand side. This gives

$$f_1(t^+) = -\tfrac{1}{4}\sin t^+ + a_1 \cos t^+ + \tfrac{1}{12}\sin 3t^+. \tag{3.1.43}$$

Using the results calculated so far, we can evaluate the right-hand side of (3.1.36) and find

$$L(f_2) = (4\omega_2 + \tfrac{1}{4})\cos t^+ + 2a_1 \sin t^+ - \tfrac{1}{2}\cos 3t^+ - a_1 \sin 3t^+ + \tfrac{1}{4}\cos 5t^+. \tag{3.1.44}$$

Thus, in order to have f_2 periodic we must set

$$a_1 = 0 \tag{3.1.45a}$$

$$\omega_2 = -\tfrac{1}{16} \tag{3.1.45b}$$

and this procedure can be continued indefinitely. We note that once $f_n(t^+)$ is completely determined, i.e., when a_n is evaluated, we also have evaluated ω_{n+1}.

In this example, the method of strained coordinates defined both the appropriate initial conditions for a periodic solution as well as the corresponding period.

The basic assumption for the applicability of the method is that the exact solution depends to all orders on *one strained coordinate only*. This is certainly true for a periodic solution. The reader is cautioned that for nonperiodic solutions, particularly when applied to partial differential equations, the method might superficially appear to work but could give incorrect results. Examples of this are cited in Reference 3.1.3. Also, it will be pointed out in Chapter 5 that the method fails to higher orders even for the problem of supersonic thin airfoil theory, for which it was first proposed.

Problems

1. Consider the weakly nonlinear wave equation

$$\frac{\partial^2 u}{\partial t^2} - \frac{\partial^2 u}{\partial x^2} + u + \varepsilon u^3 = 0, \qquad 0 < \varepsilon \ll 1.$$

For $\varepsilon = 0$ this equation has the solution

$$u = \sin \theta$$

$$\theta = (kx - \sqrt{k^2 + 1}\,t), \qquad k = \text{arbitrary}$$

corresponding to a periodic wave propagating to the right. Stokes showed that for $0 < \varepsilon \ll 1$ these periodic solutions survive and can be approximated by

$$u(x, t, \varepsilon) = \sin \theta^+ + \varepsilon u_1(\theta^+) + \varepsilon^2 u_2(\theta^+) + \cdots$$

$$\theta^+ = [kx - \sqrt{k^2 + 1}(1 + \varepsilon\omega_1 + \varepsilon^2\omega_2 + \cdots)t].$$

(a) Calculate ω_1 and u_1 and deduce the initial conditions $u(x, 0, \varepsilon)$, $u_t(x, 0, \varepsilon)$ which would produce this solution to $O(\varepsilon)$.

(b) Derive an exact expression for this periodic solution.

2. Calculate the solution of

$$\frac{d^2y}{dt^2} + y + \varepsilon y|y| = 0$$

$$y(0) = 0$$

$$\dot{y}(0) = v$$

to $O(\varepsilon)$. Hint, $\sin t^+ |\sin t^+|$ is an odd periodic function of t^+ and can therefore be developed in a Fourier sine series over the interval $0 \le t^+ \le \pi$.

3. Calculate the limit cycle to $O(\varepsilon)$ for

$$\frac{d^2y}{dt^2} + y + \varepsilon\left[-\frac{dy}{dt} + \frac{dy}{dt}\left|\frac{dy}{dt}\right|\right] = 0.$$

4. Consider Mathieu's equation

$$\frac{d^2y}{dt^2} + [\delta(\varepsilon) + \varepsilon \cos t]y = 0, \qquad 0 < \varepsilon \ll 1.$$

It can be shown that for appropriate values of $\delta(\varepsilon)$ this equation has periodic solutions with period 2π or 4π. Expanding y and δ in the form

$$y(t, \varepsilon) = y_0(t) + \varepsilon y_1(t) + \cdots$$

$$\delta(\varepsilon) = \delta_0 + \varepsilon\delta_1 + \varepsilon^2\delta_2 + \cdots$$

show that a necessary condition for periodic solutions of period 2π or 4π is $\delta_0 = n^2/4$, $n = 0, 1, 2, \ldots$. For the cases $n = 0, 1, 2$ calculate δ_1 and $y_1(t)$.

5. Show that Duffing's equation

$$\frac{d^2y}{dt^2} + y + \varepsilon y^3 = \varepsilon \cos \Omega(\varepsilon)t$$

has periodic solutions with frequency Ω for appropriate initial conditions. Carry out the calculations in detail to $O(\varepsilon)$ for the cases

$$\Omega(\varepsilon) = 1 + \varepsilon\omega_1 + \varepsilon^2\omega_2 + \cdots$$

and

$$\Omega(\varepsilon) = 3 + \varepsilon\mu_1 + \varepsilon^2\mu_2 + \cdots$$

and observe the occurrence of subharmonics (i.e., a response with frequency equal to a fraction of the impressed frequency) in the second case.

6. Consider the periodic function $f(t, \varepsilon)$ having a Fourier series expansion

$$f(t, \varepsilon) = \frac{a_0(\varepsilon)}{2} + \sum_{n=1}^{\infty} [a_n(\varepsilon)\cos n\omega(\varepsilon)t + b_n(\varepsilon)\sin n\omega(\varepsilon)t],$$

where the $a_n(\varepsilon)$, $b_n(\varepsilon)$ and $\omega(\varepsilon)$ can all be expanded asymptotically for $\varepsilon \to 0$ in the form

$$a_n(\varepsilon) \sim \sum_{i=0}^{\infty} a_{ni}\varepsilon^i$$

$$b_n(\varepsilon) \sim \sum_{i=0}^{\infty} b_{ni}\varepsilon^i$$

$$\omega(\varepsilon) \sim 1 + \sum_{i=1}^{\infty} \omega_i\varepsilon^i$$

with the a_{n_i}, b_{n_i} and ω_i constants independent of ε.

Clearly, the asymptotic expansion of $f(t, \varepsilon)$ to any order ε^N by the method of strained coordinates is given by

$$f(t, \varepsilon) = \sum_{i=0}^{N} f_i(t^+)\varepsilon^i + O(\varepsilon^{N+1}),$$

where

$$t^+ = \left[1 + \sum_{j=1}^{N+1} \omega_j\varepsilon^j + O(\varepsilon^{N+2})\right]t$$

$$f_i(t^+) = \frac{a_{0i}}{2} + \sum_{n=1}^{\infty} (a_{ni}\cos nt^+ + b_{ni}\sin nt^+).$$

Show that

$$\frac{f(t, \varepsilon) - \sum_{i=0}^{N} f_i(t^+)\varepsilon^i}{\varepsilon^N} = O(\varepsilon) + O(\varepsilon^2 t) \quad \text{as } \varepsilon \to 0.$$

Thus, the strained coordinate expansion is uniformly valid for $0 \le t \le T(\varepsilon)$, where $T = O(\varepsilon^{-1})$.

References

3.1.1. H. Poincaré. *Les Methodes Nouvelles de la Mécanique Celeste*, Vol. II, Dover, New York, 1957.

3.1.2 G. G. Stokes, On the Theory of Oscillatory Waves, *Cambridge*, *Transactions*, **8**, 1847, pp. 441–473.

3.1.3 M. van Dyke, *Perturbation Methods in Fluid Mechanics*, Annotated Edition, Parabolic Press, Stanford, California, 1975.

3.1.4 P. F. Byrd and M. D. Friedman, *Handbook of Elliptic Integrals for Engineers and Scientists*, 2nd edition, Springer-Verlag, New York, 1971.

3.2 Two-Variable Expansion Procedure, Weakly Nonlinear Autonomous Systems of Second Order

A typical elementary example illustrating the kinds of problems which arise is the effect of a small linear damping on a linear oscillator. This example was discussed previously in Section 2.1 (Equation 2.1.4). Simplifying the notation, we have

$$\frac{d^2y}{dt^2} + 2\varepsilon\frac{dy}{dt} + y = 0 \tag{3.2.1}$$

$$y(0) = 0 \tag{3.2.2a}$$

$$\frac{dy(0)}{dt} = 1, \tag{3.2.2b}$$

where ε is the ratio of the two time scales T_1, T_2

$$\varepsilon = \frac{T_1}{T_2} = \frac{B}{2\sqrt{KM}}. \tag{3.2.3a}$$

Here T_2 is the damping time

$$T_2 = \frac{2M}{B} \tag{3.2.3b}$$

which is long if B is small and T_1 is the oscillation time for $B = 0$

$$T_1 = \sqrt{\frac{M}{K}}, \tag{3.2.3c}$$

which is short compared to T_2.

The physical phenomena described by Equation (3.2.1) occur over these two time scales as can be seen clearly if the exact solution

$$y = \frac{e^{-\varepsilon t}}{\sqrt{1-\varepsilon^2}} \sin\sqrt{1-\varepsilon^2}\,t \tag{3.2.4}$$

is written with dimensional variables Y, and T

$$\frac{Y}{A} = \frac{e^{-T/T_2}}{\sqrt{1-(T_1/T_2)^2}} \sin\sqrt{1-\left(\frac{T_1}{T_2}\right)^2}\left(\frac{T}{T_1}\right). \tag{3.2.5}$$

For $\varepsilon \ll 1$ the "period" of damped oscillations in approximately $2\pi T_1$ and the damping time, which we may define as the time it takes for the damping to have an $O(1)$ effect on the solution, is T_2.

An expansion of the solution of (3.2.1) was derived in Section 2.1 in the form (cf. Equation 2.1.8)

$$y(t, \varepsilon) = \sin t - \varepsilon t \sin t + O(\varepsilon^2 t^2). \tag{3.2.6}$$

This expansion is associated with the limit process $\varepsilon \to 0$, t fixed and is only initially valid due to the presence of the $\varepsilon t \sin t$ term. In this example, the first mixed secular term we encounter is due to the nonuniform representation for large times of the $e^{-\varepsilon t}$ term in the exact solution. Mixed secular terms of the form $\varepsilon^2 t \cos t$ will occur next from the nonuniform representation of the $\sin\sqrt{1-\varepsilon^2}\,t$ term in (3.2.4).

It is also evident that mutually contradictory requirements will arise if we wish to represent both $e^{-\varepsilon t}$ and $\sin\sqrt{1-\varepsilon^2}\,t$ uniformly for long times. In fact, the only uniformly valid representation of $e^{-\varepsilon t}$ for long times is $e^{-\varepsilon t}$ itself; therefore we need the limit process $\varepsilon \to 0$, $\tilde{t} = \varepsilon t$ fixed. However, this limit process *does not exist* for $\sin\sqrt{1-\varepsilon^2}\,t = \sin(\sqrt{1-\varepsilon^2}/\varepsilon)\tilde{t}$ as the argument of the sine tends to infinity. Another way of saying this is that the function defined by (3.2.4) does not have an outer expansion.

On the other hand, as discussed in Section 3.1, we need the limit process expansion $\varepsilon \to 0$, $t^+ = (1 - (\varepsilon^2/2) + \cdots)t$ fixed, i.e., an expansion in terms of the strained coordinate t^+ in order to uniformly represent the trigonometric term. In this case $e^{-\varepsilon t}$ becomes $e^{-\varepsilon t^+(1+\varepsilon^2/2+\cdots)}$ and leads to essentially the same difficulty for long times as the initially valid expansion.

Any general method that is proposed must, thus, be able to cope *simultaneously* with these two difficulties. The essence of the two-variable method is to represent the solution as a general asymptotic expansion where each term can be uniquely expressed as a function of the two time variables

$$t^+ = \left(1 + \sum_{n=2}^{\infty} \varepsilon^n \omega_n\right)t = \text{“fast time”} \tag{3.2.7a}$$

$$\tilde{t} = \varepsilon t = \text{“slow time,”} \tag{3.2.7b}$$

where the ω_n are unknown constants.[2]

More precisely, for problems governed by the weakly nonlinear autonomous equation

$$\frac{d^2y}{dt^2} + y + \varepsilon f\left(y, \frac{dy}{dt}\right) = 0 \tag{3.2.8}$$

we will derive solutions in the form

$$y(t, \varepsilon) \sim \sum_{n=0}^{\infty} F_n(t^+, \tilde{t})\varepsilon^n. \tag{3.2.9}$$

The basic assumption is that each term F_n in the general asymptotic expansion depends on t and ε in such a way that the result can be unambiguously expressed as a function of t^+ and $\tilde{t}$.

In view of the preceding discussion of the roles of the two time scales, the above is a plausible assumption which we do not attempt to justify further. Granted the structure of the general asymptotic expansion in the form (3.2.9), one can deduce a very powerful yet simple consistency argument to calculate each of the F_n.

[2] Note that ω_1 is missing in (3.2.7a). This will be rigorously justified in Section 3.2.4 and hinges on the fact that a trigonometric term with argument t^+ in which $\omega_1 \neq 0$ can always be decomposed into a product of two trigonometric functions, one involving the t^+ of (3.2.7a) and the other involving the $\tilde{t}$ of (3.2.7b).

The argument goes as follows. Since the F_n are unique, and since each F_n is basically a function of t and ε, no term in a given F_n should produce a contribution of a different order when expressed as a function of t and ε. Thus, for example, mixed secular terms of the type $t^+ \sin t^+$, $\tilde{t} \sin t^+$, etc. or even terms tending to zero as $t \to \infty$ of the form

$$\tilde{t}e^{-\tilde{t}} \sin t^+$$

cannot be tolerated because when the linear factors involving t^+ or $\tilde{t}$ are expressed in terms of t and ε they produce contributions which are of a different order. This idea of consistency under relabelling will be further elaborated in the examples.

A general discussion of the two variable expansion procedure is given by the first author in Reference 3.2.1, which is a revision of his earlier work. Many of the examples discussed here appear in that work as do further problems of orbit theory that we will consider in Section 3.4.

For many problems, the method applied here gives the same results as the method of averaging of Krylov and Bogoliubov (Reference 3.2.2). The theoretical basis of the method of averaging is discussed in the books by Bogoliubov and Mitropolsky, Reference 3.2.3 and Mitropolsky, Reference 3.2.4. The relationship between the method of averaging and the two-variable method is discussed in a paper by Morrison (Reference 3.2.5). Morrison shows that the two methods give identical results up to the second approximation for a wide class of problems. The method of averaging is discussed in Section 3.7.1.

In comparing the two methods, the main advantage of the two variable procedure is the simplicity of the formalism. A second advantage is the explicit appearance of the variables which make it possible to apply the ideas of matching to expansions of this type (cf. Sections 3.3.3, 3.5.4, and 3.5.5). A disadvantage of the two variable method is that the proper choice of fast and slow variables is a priori known only for problems of the form (3.2.8) where the perturbation function f is independent of both t and ε. As will be shown in Sections 3.3, 3.4, and 3.5, the proper choice of variable requires careful analysis for more complicated problems.

Struble is another author who works out similar problems by more or less ad hoc arguments about the form of the solution (see Reference 3.2.6). Struble's results also agree exactly with those of the two-variable expansion procedure.

A less general method, applicable only to Hamiltonian differential equations is quite popular with astronomers and is due to von Zeipel (see Reference 3.2.7 and Reference 3.2.8 which gives a more modern discussion and applications). Here again, the two-variable method is far more efficient. This is discussed in Reference 3.2.9 where the two methods are compared for a Hamiltonian system with two degrees of freedom. von Zeipel's method is discussed in Section 3.7.2.

Finally, it should be mentioned that other versions of the two-variable procedure have been proposed by various authors subsequent to the work described in Reference 3.2.1. These are too numerous to list and do not involve any essential new ideas.

3.2.1 Linear Oscillator with Small Damping

We start discussion of the two variable method by studying the damped linear oscillator

$$\frac{d^2y}{dt^2} + 2\varepsilon \frac{dy}{dt} + y = 0 \tag{3.2.10}$$

$$y(0) = 0 \tag{3.2.11a}$$

$$\frac{dy(0)}{dt} = 1. \tag{3.2.11b}$$

We assume that the solution has a general asymptotic expansion of the form

$$y = F_0(t^+, \tilde{t}) + \varepsilon F_1(t^+, \tilde{t}) + \varepsilon^2 F_2(t^+, \tilde{t}) + \cdots \tag{3.2.12}$$

involving the fast and slow times

$$t^+ = (1 + \varepsilon^2\omega_2 + \varepsilon^3\omega_3 + \cdots)t \tag{3.2.13a}$$

$$\tilde{t} = \varepsilon t. \tag{3.2.13b}$$

We then use the chain rule to calculate

$$\frac{dy}{dt} = \frac{\partial F_0}{\partial t^+}(1 + \varepsilon^2\omega_2) + \varepsilon \frac{\partial F_0}{\partial \tilde{t}} + \varepsilon \frac{\partial F_1}{\partial t^+} + \varepsilon^2 \frac{\partial F_1}{\partial \tilde{t}} + \varepsilon^2 \frac{\partial F_2}{\partial t^+} + O(\varepsilon^3) \tag{3.2.14a}$$

$$\frac{d^2y}{dt^2} = \frac{\partial^2 F_0}{\partial t^{+2}}(1 + 2\varepsilon^2\omega_2) + 2\varepsilon \frac{\partial^2 F_0}{\partial t^+ \partial \tilde{t}} + \varepsilon^2 \frac{\partial^2 F_0}{\partial \tilde{t}^2}$$
$$+ \varepsilon \frac{\partial^2 F_1}{\partial t^{+2}} + 2\varepsilon^2 \frac{\partial^2 F_1}{\partial t^+ \partial \tilde{t}} + \varepsilon^2 \frac{\partial^2 F_2}{\partial t^{+2}} + O(\varepsilon^3). \tag{3.2.14b}$$

Thus, the sequence of equations for the F_n are

$$L(F_0) \equiv \frac{\partial^2 F_0}{\partial t^{+2}} + F_0 = 0 \tag{3.2.15a}$$

$$L(F_1) = -2 \frac{\partial^2 F_0}{\partial t^+ \partial \tilde{t}} - 2 \frac{\partial F_0}{\partial t^+} \tag{3.2.15b}$$

$$L(F_2) = -2\omega_2 \frac{\partial^2 F_0}{\partial t^{+2}} - \frac{\partial^2 F_0}{\partial \tilde{t}^2} - 2 \frac{\partial^2 F_1}{\partial t^+ \partial \tilde{t}} - 2 \frac{\partial F_0}{\partial \tilde{t}} - 2 \frac{\partial F_1}{\partial t^+}. \tag{3.2.15c}$$

The first of these is the equation for the free oscillations, while the remainder have the appearance of forced linear oscillations. However, since $F_0 = F_0(t^+, \tilde{t})$, the free linear oscillations which are the solutions to (3.2.15a) have the possibility of being slowly modulated. Thus, we have

$$F_0(t^+, \tilde{t}) = A_0(\tilde{t})\cos t^+ + B_0(\tilde{t})\sin t^+. \tag{3.2.16}$$

According to Equations (3.2.12) and (3.2.14a), the initial conditions $y(0) = 0$, $dy(0)/dt = 1$ become

$$F_0(0, 0) = 0, \qquad \frac{\partial F_0}{\partial t^+}(0, 0) = 1 \tag{3.2.17a}$$

$$F_1(0, 0) = 0, \qquad \frac{\partial F_1}{\partial t^+}(0, 0) = -\frac{\partial F_0}{\partial \tilde{t}}(0, 0) \tag{3.2.17b}$$

$$F_2(0, 0) = 0, \qquad \frac{\partial F_2}{\partial t^+}(0, 0) = -\frac{\partial F_1}{\partial \tilde{t}}(0, 0) - \omega_2 \frac{\partial F_0}{\partial t^+}(0, 0). \tag{3.2.17c}$$

Equation (3.2.17a) yields initial conditions for A_0 and B_0

$$A_0(0) = 0, \qquad B_0(0) = 1. \tag{3.2.18}$$

Nothing more can be found out about $A_0(\tilde{t})$ and $B_0(\tilde{t})$ without considering F_1. This is directly analogous to the situation encountered in Section 3.1 for the method of strained coordinates.

Substituting for F_0 into the right-hand side of Equation (3.2.15b) gives

$$L(F_1) = 2\left[\frac{dA_0}{d\tilde{t}} + A_0\right]\sin t^+ - 2\left[\frac{dB_0}{d\tilde{t}} + B_0\right]\cos t^+. \tag{3.2.19}$$

The bracketed terms on the right-hand side of (3.2.19) are functions of $\tilde{t}$ only. Therefore the particular solutions corresponding to these terms would be functions of $\tilde{t}$ multiplied by the mixed secular terms $t^+ \sin t^+$ or $t^+ \cos t^+$. Such terms cannot be permitted to occur in the solution because they lead to unboundedness with respect to t^+. Alternately, we can use the consistency argument mentioned earlier: A term like $\varepsilon t^+ \sin t^+$ is inconsistent with a unique F_1 since it could equally well be relabelled $\tilde{t} \sin t^+ + O(\varepsilon^3)$ and would become $O(1)$ instead of being $O(\varepsilon)$ as assumed.

Therefore, we must eliminate all homogeneous solutions of $L(F_1) = 0$ and this gives the two first order ordinary differential equations for A_0 and B_0:

$$\frac{dA_0}{d\tilde{t}} + A_0 = 0 \tag{3.2.20a}$$

$$\frac{dB_0}{d\tilde{t}} + B_0 = 0. \tag{3.2.20b}$$

Taking account of the initial conditions, (3.2.18), we find that

$$A_0(\tilde{t}) = 0, \qquad B_0(\tilde{t}) = e^{-\tilde{t}}. \tag{3.2.21}$$

The uniformly valid expansion thus far is

$$y(t, \varepsilon) = e^{-\tilde{t}} \sin t^{+} + \varepsilon\{A_1(\tilde{t})\cos t^{+} + B_1(\tilde{t})\sin t^{+}\} + O(\varepsilon^2). \quad (3.2.22)$$

Comparing this with the exact solution (3.2.4), we see that the first term $e^{-\tilde{t}} \sin t^{+}$ is indeed the correct uniformly valid approximation of the exact result to $O(1)$.

Now, $A_1(\tilde{t})$, $B_1(\tilde{t})$ and the frequency shift ω_2 are to be found from similar considerations applied to (3.2.15c) for F_2. Thus far, we have

$$F_0(t^{+}, \tilde{t}) = e^{-\tilde{t}} \sin t^{+}, \qquad F_1(t^{+}, \tilde{t}) = A_1(\tilde{t})\cos t^{+} + B_1(\tilde{t})\sin t^{+},$$

and

$$L(F_2) = \left[2\left(\frac{dA_1}{d\tilde{t}} + A_1\right) + (2\omega_2 + 1)e^{-\tilde{t}}\right]\sin t^{+} - 2\left[\frac{dB_1}{d\tilde{t}} + B_1\right]\cos t^{+}. \quad (3.2.23)$$

First, repeating the argument that homogeneous solutions of $L(F_2) = 0$ cannot be permitted, we must set the bracketed terms in (3.2.23) equal to zero. Solving the resulting equations for A_1 and B_1 subject to the initial conditions $A_1(0) = B_1(0) = 0$ (which follow from (3.2.17b)) we find

$$A_1(\tilde{t}) = -(\omega_2 + \tfrac{1}{2})\tilde{t}e^{-\tilde{t}} \quad (3.2.24a)$$

$$B_1(\tilde{t}) = 0. \quad (3.2.24b)$$

This means that εF_1 would be proportional to $\varepsilon\tilde{t}e^{-\tilde{t}} \cos t^{+}$. Again, such a term cannot be consistent because it can also be written as $\varepsilon^2 e^{-\tilde{t}} t^{+} \cos t^{+} + O(\varepsilon^3)$ and shift to $O(\varepsilon^2)$ in the expansion. One could also have required that $|F_2/F_1|$ be bounded to disallow such a term. Therefore, we must set

$$\omega_2 = -\tfrac{1}{2}. \quad (3.2.25)$$

All the necessary reasoning has now been explained to carry out the solution to any order and, in fact, to solve a wide variety of weakly nonlinear problems having the form of (3.2.8). For this problem, the result

$$y(t, \varepsilon) = e^{-\tilde{t}} \sin\left[1 - \frac{\varepsilon^2}{2} + O(\varepsilon^3)\right]t + O(\varepsilon^2) \quad (3.2.26)$$

is seen to be the uniformly valid general asymptotic expansion of the exact solution to $O(\varepsilon)$ for times t of order $1/\varepsilon$.

Next we proceed to several examples of weakly nonlinear bounded oscillations.

3.2.2 Oscillator with Small Cubic Damping

In suitable dimensionless variables, an oscillator with cubic damping can be represented by

$$\frac{d^2y}{dt^2} + y + \varepsilon\left(\frac{dy}{dt}\right)^3 = 0, \qquad 0 < \varepsilon \ll 1. \quad (3.2.27)$$

It is sufficient to consider the special initial conditions

$$y(0) = 1 \tag{3.2.28a}$$

$$\frac{dy(0)}{dt} = 0 \tag{3.2.28b}$$

since the problem is autonomous and the solution is oscillatory.

Using the expansion (3.2.12) with t^+ and $\tilde{t}$ as defined before, we calculate

$$L(F_0) = 0 \tag{3.2.29a}$$

$$L(F_1) = -2\frac{\partial^2 F_0}{\partial t^+ \partial \tilde{t}} - \left(\frac{\partial F_0}{\partial t^+}\right)^3 \tag{3.2.29b}$$

$$\begin{aligned} L(F_2) = &-2\frac{\partial^2 F_1}{\partial t^+ \partial \tilde{t}} - 2\omega_2 \frac{\partial^2 F_0}{\partial t^{+2}} - \frac{\partial^2 F_0}{\partial \tilde{t}^2} \\ &- 3\left(\frac{\partial F_0}{\partial t^+}\right)^2\left(\frac{\partial F_1}{\partial t^+}\right) - 3\left(\frac{\partial F_0}{\partial t^+}\right)^2\left(\frac{\partial F_0}{\partial \tilde{t}}\right). \end{aligned} \tag{3.2.29c}$$

The basic solution is again

$$F_0(t^+, \tilde{t}) = A_0(\tilde{t})\cos t^+ + B_0(\tilde{t})\sin t^+. \tag{3.2.30}$$

Now, using the identities

$$\begin{aligned} \sin^3 t &= \tfrac{3}{4}\sin t - \tfrac{1}{4}\sin 3t \\ \sin^2 t\cos t &= \tfrac{1}{4}\cos t - \tfrac{1}{4}\cos 3t \\ \sin t\cos^2 t &= \tfrac{1}{4}\sin t + \tfrac{1}{4}\sin 3t \\ \cos^3 t &= \tfrac{3}{4}\cos t + \tfrac{1}{4}\cos 3t \end{aligned} \tag{3.2.31}$$

we end up with the following equation for F_1

$$\begin{aligned} L(F_1) = &\,2\left[\frac{dA_0}{d\tilde{t}} + \frac{3}{8}A_0(A_0^2 + B_0^2)\right]\sin t^+ - 2\left[\frac{dB_0}{d\tilde{t}} + \frac{3}{8}B_0(A_0^2 + B_0^2)\right]\cos t^+ \\ &- \frac{A_0}{4}(A_0^2 - 3B_0^2)\sin 3t^+ - \frac{B_0}{4}(B_0^2 - 3A_0^2)\cos 3t^+. \end{aligned} \tag{3.2.32}$$

Removal of mixed secular terms in t^+ requires that

$$\frac{dA_0}{d\tilde{t}} + \frac{3}{8}A_0(A_0^2 + B_0^2) = 0 \tag{3.2.33a}$$

$$\frac{dB_0}{d\tilde{t}} + \frac{3}{8}B_0(A_0^2 + B_0^2) = 0 \tag{3.2.33b}$$

which are two coupled nonlinear equations for A_0 and B_0.

The cynical reader might say that we have gained little by replacing a nonlinear second order Equation (3.2.27) by two nonlinear first order Equations (3.2.33) for the slowly varying functions appearing in the first approximation. Actually, the situation is quite a bit better, as Equations (3.2.33) are integrable for any initial values of A_0 and B_0, while (3.2.27) is not integrable at all. We will see in Section 3.2.5 that this is true for an arbitrary problem in the form of Equation (3.2.33).

To see this we merely introduce the slowly varying amplitude $\rho_0(\tilde{t})$ and phase $\phi_0(\tilde{t})$ by

$$A_0 = \rho_0 \cos \phi_0 \tag{3.2.34a}$$

$$B_0 = \rho_0 \sin \phi_0 \tag{3.2.34b}$$

and Equations (3.2.33) transform to[3]

$$\frac{d\rho_0}{d\tilde{t}} + \frac{3}{8}\rho_0^3 = 0 \tag{3.2.35a}$$

$$\frac{d\phi_0}{d\tilde{t}} = 0. \tag{3.2.35b}$$

For our choice of initial conditions $\rho_0(0) = 1$, $\phi_0(0) = 0$ and integrating (3.2.35) gives

$$A_0 = \rho_0 = \frac{1}{\sqrt{1 + 3\tilde{t}/4}} \tag{3.2.36a}$$

$$B_0 = \phi_0 = 0. \tag{3.2.36b}$$

Thus, the uniformly valid solution to $O(1)$ is

$$F_0 = \frac{1}{\sqrt{1 + 3\tilde{t}/4}} \cos[1 + O(\varepsilon^2)]t. \tag{3.2.37}$$

We see that because of the relatively weaker damping than in the linear case the solution now decays algebraically.

We can also solve what remains of (3.2.29b) for F_1 in the form

$$F_1(t^+, \tilde{t}) = A_1(\tilde{t})\cos t^+ + B_1(\tilde{t})\sin t^+ + \frac{\sin 3t^+}{4(3\tilde{t} + 4)^{3/2}}. \tag{3.2.38}$$

It is left as an exercise to the reader to show that

$$A_1(\tilde{t}) = 0, \qquad \omega_2 = 0 \tag{3.2.39a}$$

$$B_1(\tilde{t}) = \frac{1}{(3\tilde{t} + 4)^{1/2}}\left[\frac{3}{8(3\tilde{t} + 4)} + \frac{15}{32}\right]. \tag{3.2.39b}$$

[3] Had we written the Solution (3.2.30) in the equivalent form $F_0 = \rho_0 \cos(t^+ + \phi_0)$, we would have directly obtained Equations (3.2.35).

3.2.3 Rayleigh's Equation

In Section 3.1.2 we studied the periodic solutions of Rayleigh's equation

$$\frac{d^2y}{dt^2} + y + \varepsilon\left[-\frac{dy}{dt} + \frac{1}{3}\left(\frac{dy}{dt}\right)^3\right] = 0. \tag{3.2.40}$$

Here we study the solution for an arbitrary initial displacement a

$$y(0) = a \tag{3.2.41a}$$

$$\frac{dy(0)}{dt} = 0. \tag{3.2.41b}$$

The expansion (3.2.12) is again used to calculate the following equations for F_0, F_1, and F_2.

$$L(F_0) = 0 \tag{3.2.42a}$$

$$L(F_1) = -2\frac{\partial^2 F_0}{\partial t^+ \partial \tilde{t}} + \frac{\partial F_0}{\partial t^+} - \frac{1}{3}\left(\frac{\partial F_0}{\partial t^+}\right)^3 \tag{3.2.42b}$$

$$L(F_2) = -2\omega_2\frac{\partial^2 F_0}{\partial t^{+2}} - 2\frac{\partial^2 F_1}{\partial t^+ \partial \tilde{t}} - \frac{\partial^2 F_0}{\partial \tilde{t}^2} + \frac{\partial F_1}{\partial t^+} - \left(\frac{\partial F_0}{\partial t^+}\right)^2\left[\frac{\partial F_1}{\partial t^+} + \frac{\partial F_0}{\partial \tilde{t}}\right]. \tag{3.2.42c}$$

The solution is similar to the preceding case. We use (3.2.30) for F_0. Now, the linear-damping term and the factor $\frac{1}{3}$ modify (3.2.32) slightly and Equation (3.2.42b) now becomes

$$\begin{aligned} L(F_1) = {} & \left[2\frac{dA_0}{d\tilde{t}} - A_0 + \frac{A_0}{4}(A_0^2 + B_0^2)\right]\sin t^+ \\ & - \left[2\frac{dB_0}{d\tilde{t}} - B_0 + \frac{B_0}{4}(A_0^2 + B_0^2)\right]\cos t^+ \\ & - \frac{A_0}{12}(A_0^2 - 3B_0^2)\sin 3t^+ - \frac{B_0}{12}(B_0^2 - 3A_0^2)\cos 3t^+. \end{aligned} \tag{3.2.43}$$

The slowly varying functions A_0 and B_0, obtained by setting the bracketed terms on the right-hand side of (3.2.43) equal to zero, are

$$A_0(\tilde{t}) = \frac{2\lambda}{\sqrt{1 - ke^{-\tilde{t}}}} \tag{3.2.44a}$$

$$B_0(\tilde{t}) = \frac{2\sqrt{1 - \lambda^2}}{\sqrt{1 - ke^{-\tilde{t}}}}, \tag{3.2.44b}$$

where λ and k are integration constants.

This result clearly shows the approach to the limit cycle of amplitude 2, found in Section 3.1.2, independent now of the initial condition. The approach is exponential over the long time scale. For the initial conditions of (3.2.41) we have $\lambda = 1$, $k = (a^2 - 4)/a^2$, and

$$y(t, \varepsilon) = \frac{2a}{\sqrt{a^2 - (a^2 - 4)e^{-\tilde{t}}}} \cos[1 + O(\varepsilon^2)]t + O(\varepsilon). \qquad (3.2.45)$$

Thus, if the solution starts on the limit cycle ($a = 2$), it remains there.

The reader can verify that proceeding further defines the solution as follows

$$F_1 = B_1(\tilde{t})\sin t^+ + \frac{a^3}{12[a^2 - (a^2 - 4)e^{-\tilde{t}}]} \sin 3t^+ \qquad (3.2.46)$$

$$B_1(\tilde{t}) = \frac{A_0}{8} \log\left(\frac{A_0}{a}\right) + \frac{A_0}{64}(A_0^2 + 5a^2 - 32) \qquad (3.2.47)$$

$$\omega_2 = -\tfrac{1}{16}. \qquad (3.2.48)$$

3.2.4 Scaling

So far, the three examples of the type characterized by Equation (3.2.8) that we have considered have described bounded oscillations. The question arises whether the two variable method also applies to situations where the solution becomes unbounded.

For example, if we let ε be negative in the linear damping case the method does give the correct result. Unfortunately, this linear case is exceptional. For example, if we consider Equation (3.2.27) where we have a small cubic damping, and now let ε be negative, the result given by Equation (3.2.36a) predicts that the amplitude becomes infinite at the finite time $\tilde{t} = \frac{4}{3}$, i.e., $t = -4/3\varepsilon$. This is obviously incorrect as the solution y tends to infinity only as $t \to \infty$. The key here is that in the two variable method we tacitly assume that $L(y) = O[f(y, dy/dt)]$ in the solution domain. This is true for the linear damping case even if $\varepsilon < 0$, but is certainly incorrect if f is nonlinear and leads to an unbounded y.

For the case of a negative cubic damping, since y becomes large, the appropriate problem to be solved can be easily ascertained by rescaling y. Let

$$\tilde{y} = (-\varepsilon)^\alpha y \qquad (3.2.49)$$

for some positive α. Inserting this into Equation (3.2.27) gives

$$(-\varepsilon)^{-\alpha}\frac{d^2\tilde{y}}{dt^2} + (-\varepsilon)^{-\alpha}\tilde{y} - (-\varepsilon)^{1-3\alpha}\left(\frac{d\tilde{y}}{dt}\right)^3 = 0. \qquad (3.2.50)$$

Now with $\varepsilon < 0$ as y gets large, the last term in Eq. (3.2.50) becomes larger and larger until its order of magnitude equals that of the linear terms, i.e., when $\alpha = \frac{1}{2}$. Unfortunately, when y does become large, we must solve the full equation

$$\frac{d^2\tilde{y}}{dt^2} + \tilde{y} - \left(\frac{d\tilde{y}}{dt}\right)^3 = 0. \tag{3.2.51}$$

Thus, the two variable solution, (3.2.36a), for the case $\varepsilon < 0$ is only initially valid. One needs the exact solution of (3.2.51) to describe y for times greater than $O(1)$.

A similar example is the oscillator with a "soft" spring and negative damping

$$\frac{d^2y}{dt^2} + y - \varepsilon\left(\frac{dy}{dt} + y^3\right) = 0, \qquad \varepsilon > 0 \tag{3.2.52}$$

$$y(0) = 1 \tag{3.2.53a}$$

$$\frac{dy(0)}{dt} = 0. \tag{3.2.53b}$$

It is easy to establish the qualitative behavior of the solution by noting that (3.2.52) can be written as

$$\frac{dE}{dt} = \varepsilon\left(\frac{dy}{dt}\right)^2 > 0 \tag{3.2.54}$$

with

$$E = \frac{1}{2}\left[\left(\frac{dy}{dt}\right)^2 + y^2 - \frac{\varepsilon y^4}{2}\right]. \tag{3.2.55}$$

Here E is the energy of the undamped oscillator. Since $dE/dt > 0$, we expect the motion to start as being oscillatory with slowly increasing amplitude. Since after each cycle of oscillation the energy is slightly higher than its initial value, we expect the motion to "climb out" of the potential well of the function $V(y) = \frac{1}{2}y^2 - \varepsilon y^4/4$ and "escape" to either $+\infty$ or $-\infty$ when E exceeds $1/4\varepsilon$. This is sketched in Figure 3.2.1 below for the case when $y \to -\infty$.

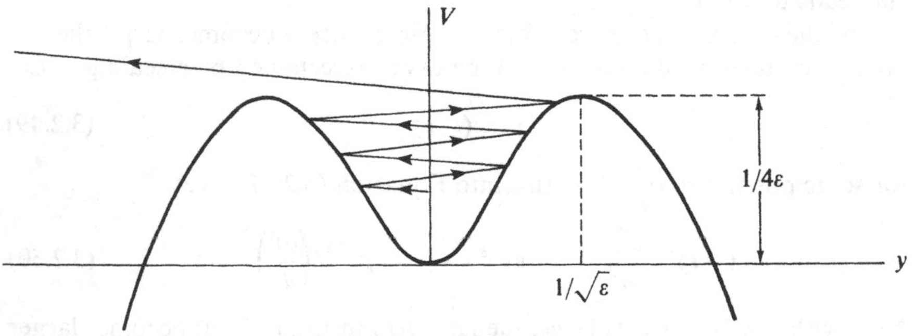

Figure 3.2.1 Escape to $y = -\infty$

Of course, whether y tends to $+\infty$ or $-\infty$ depends very much on the cumulative effect of the negative damping term.

If we attempt to solve (3.2.52) using the two variable method we obtain the nonsensical result that

$$y = F_0(t^+, \tilde{t}) + \varepsilon F_1(t^+, \tilde{t}) + \cdots \tag{3.2.56}$$

with

$$F_0 = \rho_0(\tilde{t})\cos[t + \phi(\tilde{t})] \tag{3.2.57}$$

$$\rho_0 = e^{\tilde{t}/2} \tag{3.2.58}$$

$$\phi_0 = \tfrac{3}{8}(1 - e^{\tilde{t}}) \tag{3.2.59}$$

which predicts that the solution remains oscillatory with a monotonically increasing amplitude and phase.

Here again, the assumption that the perturbation terms remain $O(\varepsilon)$ is incorrect. The correct perturbation problem to be solved is obtained by rescaling y. If we set

$$\tilde{y} = \varepsilon^{\alpha} y$$

with $\alpha > 0$, Equation (3.2.52) transforms to

$$\varepsilon^{-\alpha}\frac{d^2\tilde{y}}{dt^2} + \varepsilon^{-\alpha}\tilde{y} - \varepsilon^{1-\alpha}\frac{d\tilde{y}}{dt} - \varepsilon^{1-3\alpha}\tilde{y}^3 = 0.$$

Clearly, we must set $-\alpha = 1 - 3\alpha$, i.e., $\alpha = \frac{1}{2}$ to obtain the richest equation when $\varepsilon \to 0$. With $\tilde{y}$ so chosen, Equation (3.2.52) becomes

$$\frac{d^2\tilde{y}}{dt^2} + \tilde{y} - \tilde{y}^3 - \varepsilon\frac{d\tilde{y}}{dt} = 0. \tag{3.2.60}$$

Here, the unperturbed problem is the full nonlinear oscillator equation which admits both oscillatory and escape solutions. Thus, the effect of the $-\varepsilon(dy/dt)$ term is to provide a gradual transition from oscillatory motion to escape. Unfortunately, the perturbation Problem (3.2.60) is very difficult to solve. We will consider this type of problem in Section 3.6 for cases where the unperturbed ($\varepsilon = 0$) problem has only periodic solutions, which is not the case here.

It is not only for unbounded oscillations that one must worry about the appropriate scale of the dependent variable. Consider, for example, the problem

$$\frac{d^2y}{dt^2} + y + \varepsilon\left(\frac{dy}{dt}\right)^3 + 3\nu\varepsilon^2\frac{dy}{dt} = 0 \tag{3.2.61}$$

$$y(0) = 1, \qquad \frac{dy}{dt}(0) = 0 \tag{3.2.62}$$

where ν is a positive constant.

Clearly, with $\varepsilon > 0$ the motion is damped since both perturbation terms oppose the motion.

If we assume the usual expansion for y in the form

$$y(t, \varepsilon) = F_0(t^+, \tilde{t}) + \varepsilon F_1(t^+, \tilde{t}) + \cdots \tag{3.2.63}$$

we find that F_0 is identical with the one for Equation (3.2.37) since the term $3\varepsilon^2 v\, dy/dt$ does not affect the determination of F_0. Proceeding further, we find that F_1 is of the form

$$F_1 = A_1(\tilde{t})\cos t^+ + B_1(\tilde{t})\sin t^+ + \frac{1}{4(3\tilde{t} + 4)^{3/2}} \sin 3t^+, \tag{3.2.64}$$

where as before

$$t^+ = [1 + O(\varepsilon^3)]t \tag{3.2.65}$$

$$B_1 = \left[\frac{3}{8(3\tilde{t} + 4)^{3/2}} + \frac{15}{32(3\tilde{t} + 4)^{1/2}}\right]\sin t^+ \tag{3.2.66}$$

and the term multiplied by v now means that A_1 is not zero but given by

$$A_1(\tilde{t}) = -\frac{v}{(3\tilde{t} + 4)^{3/2}}\left(\frac{9}{2}\tilde{t}^2 + 12\tilde{t}\right). \tag{3.2.67}$$

This result is inconsistent. In fact, we see that as $t \to \infty$, $A_1 \to -(9v/2(3)^{3/2})\tilde{t}^{1/2}$ which would say that F_1 becomes unbounded in contradiction to the damped behavior of the solution.

In this example the difficulty is in the fact that assuming that $3v\varepsilon^2(dy/dt) = O[\varepsilon^2(dy/dt)^3]$ (as we do in the usual procedure) becomes incorrect as dy/dt gets smaller. For sufficiently small dy/dt the term $3v\varepsilon^2(dy/dt)$ will be more important than the term $\varepsilon(dy/dt)^3$.

Therefore, we must rescale y as follows

$$y^* = \frac{y}{\varepsilon^\beta} \tag{3.2.68}$$

and inserting this into Equation (3.2.61) shows that $\beta = \frac{1}{2}$ and the correct perturbation problem is defined by

$$\frac{d^2y^*}{dt^2} + y^* + \varepsilon^2\left[\left(\frac{dy^*}{dt}\right)^3 + 3v\frac{dy^*}{dt}\right] = 0 \tag{3.2.69}$$

$$y^*(0) = a = \frac{1}{\varepsilon^{1/2}} \tag{3.2.70a}$$

$$\frac{dy^*(0)}{dt} = 0. \tag{3.2.70b}$$

Now, we can apply the two variable method using

$$t^+ = [1 + O(\varepsilon^4)]t \tag{3.2.71a}$$

$$\tilde{t} = \varepsilon^2 t. \tag{3.2.71b}$$

Assuming an expansion of the form[4]

$$y^*(t, \varepsilon) = F_0(t^+, \tilde{t}) + \varepsilon F_1(t^+, \tilde{t}) + \cdots \tag{3.2.72}$$

we calculate

$$F_0 = \rho_0(\tilde{t})\cos[t^+ + \phi_0(\tilde{t})] \tag{3.2.73a}$$

and

$$\begin{aligned} L(F_1) &= -2\frac{\partial^2 F_0}{\partial t^+ \partial \tilde{t}} - \left(\frac{\partial F_0}{\partial t^+}\right)^3 - 3\nu\frac{\partial F_0}{\partial t^+} \\ &= \left(2\frac{d\rho_0}{d\tilde{t}} + \frac{3}{4}\rho_0^3 + 3\nu\rho_0\right)\sin(t^+ + \phi_0) \\ &\quad + 2\rho_0\frac{d\phi_0}{d\tilde{t}}\cos(t^+ + \phi_0) - \frac{\rho_0^3}{4}\sin 3(t^+ + \phi_0). \end{aligned} \tag{3.2.73b}$$

Eliminating mixed secular terms determines ρ_0 and ϕ_0 as follows

$$\rho_0(\tilde{t}) = \frac{2\nu^{1/2}}{[(1 + 4\nu/a^2)e^{3\nu\tilde{t}} - 1]^{1/2}} \tag{3.2.74a}$$

$$\phi_0(\tilde{t}) = 0 \tag{3.2.74b}$$

and this shows that the amplitude decays with time.

First, we note that

$$\lim_{\nu\to 0} \varepsilon^{1/2}\rho_0 = \frac{2}{\sqrt{4 + 3\tilde{t}}}, \qquad \tilde{t} = \varepsilon t$$

as found earlier. Moreover, if we take the limit as $\varepsilon \to 0$, $\tilde{t} = \varepsilon t$ fixed, the result in (3.2.74a) gives the same nonuniform expansion (3.2.63).

3.2.5 General Theory

Guided by the examples worked out thus far we now consider the two variable expansion for the general second order equation of the form

$$\frac{d^2y}{dt^2} + y + \varepsilon f\left(y, \frac{dy}{dt}\right) = 0 \tag{3.2.75}$$

[4] Strictly speaking, we should also allow the F_i to depend on $\varepsilon^{1/2}$ since this occurs in the initial condition. However, as we will presently see disregarding the order of a leads to no difficulties.

with initial conditions

$$y(0) = a \tag{3.2.76a}$$

$$\frac{dy}{dt}(0) = b. \tag{3.2.76b}$$

Even though there is no essential difficulty in formally working out the general theory for the case where f also involves ε we do not consider this case to avoid the complications discussed in the last example of the preceding section. Also, we restrict attention to perturbation functions for which the solution of (3.2.75) is bounded.

A possible sufficient condition on f for bounded solutions is that $\varepsilon(dy/dt)f$ be positive for all values of y and $\dot{y}$ outside some finite circle R_0 around the origin in the phase plane. To see this we define R by

$$R = \sqrt{y^2 + \left(\frac{dy}{dt}\right)^2}. \tag{3.2.77}$$

Then, using Equation (3.2.75), we calculate

$$\frac{dR}{dt} = \frac{1}{R}\frac{dy}{dt}\left(\frac{d^2y}{dt^2} + y\right) = -\frac{\varepsilon}{R}\frac{dy}{dt}f\left(y, \frac{dy}{dt}\right). \tag{3.2.78}$$

Thus, if $\varepsilon(dy/dt)f(y, dy/dt)$ is positive for all R greater than some R_0, then all the integral curves tend towards R_0 and the solution is bounded.

For example, if $f = (dy/dt)^{2n+1}$, $n = 0, 1, 2, \ldots$ this condition is satisfied for $R_0 = 0$. Unfortunately, this is only a sufficiency condition, and a fairly strong one; therefore not very useful as there are functions f which do not satisfy this condition for any R_0 but which do correspond to bounded solutions. For example, for Rayleigh's equation, we have

$$\varepsilon\frac{dy}{dt}f = -\varepsilon\left(\frac{dy}{dt}\right)^2\left[1 - \frac{1}{3}\left(\frac{dy}{dt}\right)^2\right]$$

and this quantity is only positive in the two half planes $|dy/dt| > \sqrt{3}$. Thus, there is no R_0 which works. Nevertheless, all solutions are bounded!

A more useful sufficiency condition for a bounded solution can be posed in terms of the existence of a suitable Liapunov function. Since a precise definition requires too much of a digression, we refer the reader to Chapter 5 of Reference 3.2.10 for a comprehensive discussion of the stability theory for equations of the type (3.2.75).

With these preliminary remarks in mind, we seek a solution for Equation (3.2.75) in the form

$$y(t, \varepsilon) = F(t^+, \tilde{t}, \varepsilon) = \sum_{n=0}^{N} F_n(t^+, \tilde{t})\varepsilon^n + O(\varepsilon^{N+1}), \tag{3.2.79}$$

where

$$t^+ = \left[\sum_{n=0}^{N} \omega_n \varepsilon^n + O(\varepsilon^{N+1})\right] t, \qquad \omega_0 = 1 \tag{3.2.80}$$

$$\tilde{t} = \varepsilon t. \tag{3.2.81}$$

We then calculate the following expansions for the various terms appearing in (3.2.75).

$$\frac{dy}{dt} = \frac{\partial F_0}{\partial t^+} + \sum_{n=1}^{N}\left(\frac{\partial F_n}{\partial t^+} + \frac{\partial F_{n-1}}{\partial \tilde{t}} + \sum_{k=1}^{n} \omega_k \frac{\partial F_{n-k}}{\partial t^+}\right)\varepsilon^n + O(\varepsilon^{N+1}) \tag{3.2.82}$$

$$\begin{aligned}\frac{d^2y}{dt^2} &= \frac{\partial^2 F_0}{\partial t^{+2}} + \varepsilon\left(\frac{\partial^2 F_1}{\partial t^{+2}} + 2\omega_1 \frac{\partial^2 F_0}{\partial t^{+2}} + 2\frac{\partial^2 F_0}{\partial t^+ \partial \tilde{t}}\right) \\ &\quad + \varepsilon^2 \left\{\sum_{n=0}^{N}\left[\sum_{r=0}^{n+2}\left(\sum_{k=0}^{r} \omega_k \omega_{r-k}\right)\frac{\partial^2 F_{n-r-2}}{\partial t^{+2}}\right.\right. \\ &\quad \left.\left. + 2\sum_{k=0}^{n+1} \omega_k \frac{\partial^2 F_{n-k+1}}{\partial t^+ \partial \tilde{t}} + \frac{\partial^2 F_n}{\partial \tilde{t}^2}\right]\varepsilon^n\right\} + O(\varepsilon^{N+1}).\end{aligned} \tag{3.2.83}$$

Since the arguments of the perturbation function f are expanded in powers of ε, f itself can be expanded in the form

$$f\left(y, \frac{dy}{dt}\right) = f_0 + \varepsilon f_1 + O(\varepsilon^2), \tag{3.2.84}$$

where

$$f_0 = f\left(F_0, \frac{\partial F_0}{\partial t^+}\right) \tag{3.2.85}$$

$$f_1 = F_1 \frac{\partial f}{\partial y}\left(F_0, \frac{\partial F_0}{\partial t^+}\right) + \left(\frac{\partial F_1}{\partial t^+} + \frac{\partial F_0}{\partial \tilde{t}} + \omega_1 \frac{\partial F_0}{\partial t^+}\right)\frac{\partial f}{\partial \dot{y}}\left(F_0, \frac{\partial F_0}{\partial t^+}\right). \tag{3.2.86}$$

Although there is no concise form for the general term in (3.2.84), one can compute these routinely by expanding f in a Taylor series around F_0 and $\partial F_0/\partial t^+$.

Substituting the above into (3.2.75) gives

$$L(F_0) \equiv \frac{\partial^2 F_0}{\partial t^{+2}} + F_0 = 0 \tag{3.2.87a}$$

$$L(F_1) = -2\frac{\partial^2 F_0}{\partial t^+ \partial \tilde{t}} - 2\omega_1 \frac{\partial^2 F_0}{\partial t^{+2}} - f_0 \tag{3.2.87b}$$

$$\begin{aligned}L(F_2) &= -2\frac{\partial^2 F_1}{\partial t^+ \partial \tilde{t}} - (2\omega_2 + \omega_1^2)\frac{\partial^2 F_0}{\partial t^{+2}} - \frac{\partial^2 F_0}{\partial \tilde{t}^2} \\ &\quad - 2\omega_1\left(\frac{\partial^2 F_1}{\partial t^{+2}} + \frac{\partial^2 F_0}{\partial t^+ \partial \tilde{t}}\right) - f_1.\end{aligned} \tag{3.2.87c}$$

The initial conditions, when expanded, give

$$F_0(0, 0) = a; \qquad F_n(0, 0) = 0, \qquad n \neq 0 \tag{3.2.88a}$$

$$\frac{\partial F_0}{\partial t^+}(0, 0) = b; \qquad \frac{\partial F_n}{\partial t^+} = -\left(\frac{\partial F_{n-1}}{\partial \tilde{t}}(0, 0) + \sum_{k=1}^{n} \omega_k \frac{\partial F_{n-k}}{\partial t^+}(0, 0)\right). \tag{3.2.88b}$$

The solution of Equation (3.2.87) was worked out in Reference 3.2.1 explicitly to $O(\varepsilon)$. However, the differential equations governing the slowly varying parameters were only worked out explicitly for the solution to $O(1)$. In Reference (3.2.5), Morrison succeeded in carrying out the details of this calculation explicitly to $O(\varepsilon)$ and we now present his ingenious procedure.

We write the solution of Equation (3.2.87a) in the form

$$F_0 = \alpha_0(\tilde{t})\cos\psi, \tag{3.2.89a}$$

where

$$\psi = t^+ - \phi_0(\tilde{t}) \tag{3.2.89b}$$

and α_0 and ϕ_0 are the slowly varying amplitude and phase of the $O(1)$ solution. The initial conditions imply that

$$\alpha_0(0) = (a^2 + b^2)^{1/2} \tag{3.2.90a}$$

$$\phi_0(0) = \tan^{-1}\frac{b}{a}. \tag{3.2.90b}$$

If we now substitute this result into the right-hand side of Equation (3.2.87b), we find

$$\begin{aligned} L(F_1) = {} & 2\alpha_0' \sin\psi - 2\alpha_0(\phi_0' - \omega_1)\cos\psi \\ & - f(\alpha_0 \cos\psi, -\alpha_0 \sin\psi), \end{aligned} \tag{3.2.91}$$

where a prime denotes $d/d\tilde{t}$.

Now, since the arguments of f depend on $\sin\psi$ and $\cos\psi$, f is a periodic function of ψ with period 2π. If we wish, we could express f in a Fourier series with respect to the variable ψ and the coefficients of this Fourier series *would be functions only of* α_0. This was the approach adopted in Reference 3.2.1, but it proves to be inconvenient in the calculations to higher order. Instead, Morrison in Reference 3.2.5 proposed to use the Fourier series idea only to isolate the first harmonics of periodic functions, as these first harmonics contribute inconsistent mixed secular terms in t^+ in the solution to $O(1)$. The first harmonics of f are

$$\frac{1}{\pi}\int_0^{2\pi} f(\alpha_0 \cos\psi, -\alpha_0 \sin\psi)\sin\psi\, d\psi \equiv 2P_1(\alpha_0) \tag{3:2.92a}$$

$$\frac{1}{\pi}\int_0^{2\pi} f(\alpha_0 \cos\psi, -\alpha_0 \sin\psi)\cos\psi\, d\psi \equiv 2Q_1(\alpha_0) \tag{3.2.92b}$$

and for a given perturbation function f, one can in principle calculate the two functions P_1 and Q_1.

Then, in order to remove terms multiplying $\sin \psi$ and $\cos \psi$ in Equation (3.2.91), we must set

$$\alpha_0' = P_1(\alpha_0) \tag{3.2.93a}$$

$$-\alpha_0(\phi_0' - \omega_1) = Q_1(\alpha_0). \tag{3.2.93b}$$

The above differential equations, subject to the initial conditions given by Equations (3.2.90) define α_0 and ϕ_0. In fact, we can evaluate $\tilde{t}$ in terms of α_0 by quadrature from Equation (3.2.93a) in the form

$$\tilde{t} = \int_{\alpha_0(0)}^{\alpha_0} \frac{ds}{P_1(s)} = K[(\alpha_0, \alpha_0(0)] \tag{3.2.94}$$

which when inverted gives

$$\alpha_0 = I[\tilde{t}, \alpha_0(0)]. \tag{3.2.95}$$

If we now denote $Q_1(\alpha_0)/\alpha_0$ by $J[\tilde{t}, \alpha_0(0)]$, we can also compute ϕ_0 by quadrature in the form

$$\phi_0 = \phi_0(0) + \omega_1 \tilde{t} - \int_0^{\tilde{t}} J[s, \alpha_0(0)]ds. \tag{3.2.96}$$

It is important to note that the system of two first order nonlinear Equations (3.2.93) will *always* uncouple and can be solved by quadrature. [cf. the remarks following Equations (3.2.33)].

Let us now consider the role of the unknown coefficient ω_1 in the solution. If we substitute the result we have for ϕ_0 into Equation (3.2.89a) and write out t^+ and $\tilde{t}$ in terms of ε and t according to the definitions (3.2.80) and (3.2.81) we find

$$F_0 = \alpha_0 \cos\left(t + \varepsilon^2\omega_2 t + \varepsilon^3\omega_3 t + \cdots - \phi_0(0) + \int_0^{\tilde{t}} J\, ds\right). \tag{3.2.97}$$

We see that the $\varepsilon\omega_1 t$ term in t^+ cancels the $\omega_1\tilde{t}$ term in ϕ_0 and *the final result for ψ is independent of ω_1*. In fact, we will see presently that to all orders the solution can be expressed as a function of ψ and $\tilde{t}$. Thus, our earlier claim based on intuition that ω_1 may be set equal to zero [cf. Equation (3.2.7a)] is justified, and henceforth we will adopt this definition of t^+.

The right-hand side of Equation (3.2.91), now free of first harmonics, becomes

$$L(F_1) = 2[P_1(\alpha_0)\sin \psi + Q_1(\alpha_0)\cos \psi] - f_0(\alpha_0, \psi), \tag{3.2.98}$$

where we have used the notation

$$f_0(\alpha_0, \psi) = f(\alpha_0 \cos \psi, -\alpha_0 \sin \psi). \tag{3.2.99}$$

We now seek the general solution of (3.2.98) in the usual way as

$$F_1 = F_1^{(H)} + F_1^{(P)}, \tag{3.2.100}$$

where $F_1^{(H)}$ is the homogeneous solution

$$F_1^{(H)}(t^+, \tilde{t}) = A_1(\tilde{t})\cos\psi + B_1(\tilde{t})\sin\psi \tag{3.2.101a}$$

and $F_1^{(P)}$ is a particular solution assumed in the form

$$F_1^{(P)} = \lambda_1(\alpha_0, \psi)\cos\psi + \alpha_0\mu_1(\alpha_0, \psi)\sin\psi. \tag{3.2.101b}$$

Moreover, since we have already included arbitrary $\cos\psi$, $\sin\psi$ terms in the homogeneous solution, we require these to be absent from $F_1^{(P)}$, i.e., we set

$$\int_0^{2\pi} \lambda_1(\alpha_0, \psi)d\psi = \int_0^{2\pi} \mu_1(\alpha_0, \psi)d\psi = 0. \tag{3.2.102}$$

Using variation of parameters, we find that λ_1 and μ_1 must satisfy

$$\frac{\partial\lambda_1}{\partial\psi} = f_0(\alpha_0, \psi)\sin\psi - P_1(\alpha_0); \tag{3.2.103}$$

$$\alpha_0\frac{\partial\mu_1}{\partial\psi} = -f_0(\alpha_0, \psi)\cos\psi + Q_1(\alpha_0). \tag{3.2.104}$$

Note that the right-hand sides of Equations (3.2.103, 4) are 2π-periodic functions of ψ and according to the definitions of P_1 and Q_1 these periodic functions *have zero average value*. Therefore, the quadrature of Equations (3.2.103, 4) defines λ_1 and μ_1 as periodic functions of ψ. Again, for a given perturbation function f, one can calculate λ_1 and μ_1 explicitly by quadrature so F_1 only involves the two unknown functions of $\tilde{t}$, A_1 and B_1. These will be defined next by requiring F_2 to be consistent.

The right-hand side of Equation (3.2.87c), with $\omega_1 = 0$, is simply

$$L(F_2) = -f_1 - \left(\frac{\partial^2 F_0}{\partial\tilde{t}^2} + 2\frac{\partial^2 F_1}{\partial t^+\partial\tilde{t}}\right) - 2\omega_2\frac{\partial^2 F_0}{\partial t^{+2}}. \tag{3.2.105}$$

We will now isolate the first harmonics of the terms appearing in Equation (3.2.105). Consider first f_1. Using the definition (3.2.86) we need to compute $\partial F_1/\partial t^+$ and $\partial F_0/\partial\tilde{t}$ and these are

$$\frac{\partial F_1}{\partial t^+} = \frac{\partial F_1}{\partial\psi} = [B_1(\tilde{t}) + \alpha_0(\tilde{t})\mu_1(\alpha_0, \psi) - P_1(\alpha_0)]\cos\psi$$
$$- [A_1(\tilde{t}) + \lambda_1(\alpha_0, \psi) - Q_1(\alpha_0)]\sin\psi \tag{3.2.106a}$$

$$\frac{\partial F_0}{\partial\tilde{t}} = \alpha_0'(\tilde{t})\cos\psi + \alpha_0(\tilde{t})\phi_0'(\tilde{t})\sin\psi$$
$$= P_1(\alpha_0)\cos\psi - Q_1(\alpha_0)\sin\psi. \tag{3.2.106b}$$

Therefore,

$$\frac{\partial F_1}{\partial t^+} + \frac{\partial F_0}{\partial \tilde{t}} = (B_1 + \alpha_0 \mu_1)\cos\psi - (A_1 + \lambda_1)\sin\psi. \quad (3.2.107a)$$

Substituting this and the expression

$$F_1 = (A_1 + \lambda_1)\cos\psi + (B_1 + \alpha_0\mu_1)\sin\psi \quad (3.2.107b)$$

into Equation (3.2.86) gives

$$\begin{aligned} f_1 = (A_1 + \lambda_1)&\left[\frac{\partial f}{\partial y}\left(F_0, \frac{\partial F_0}{\partial t^+}\right)\cos\psi - \frac{\partial f}{\partial \dot{y}}\left(F_0, \frac{\partial F_0}{\partial t^+}\right)\sin\psi\right] \\ + (B_1 + \alpha_0\mu_1)&\left[\frac{\partial f}{\partial y}\left(F_0, \frac{\partial F_0}{\partial t^+}\right)\sin\psi + \frac{\partial f}{\partial \dot{y}}\left(F_0, \frac{\partial F_0}{\partial t^+}\right)\cos\psi\right]. \quad (3.2.108) \end{aligned}$$

But, according to the definition (3.2.99) for f_0 we have

$$\frac{\partial f}{\partial y}\left(F_0, \frac{\partial F_0}{\partial t^+}\right)\cos\psi - \frac{\partial f}{\partial \dot{y}}\left(F_0, \frac{\partial F_0}{\partial t^+}\right)\sin\psi = \frac{\partial f_0}{\partial \alpha_0} \quad (3.2.109a)$$

$$\frac{\partial f}{\partial y}\left(F_0, \frac{\partial F_0}{\partial t^+}\right)(-\alpha_0\sin\psi) - \frac{\partial f}{\partial \dot{y}}\left(F_0, \frac{\partial F_0}{\partial t^+}\right)(\alpha_0\cos\psi) = \frac{\partial f_0}{\partial \psi}. \quad (3.2.109b)$$

Therefore,

$$f_1 = (A_1 + \lambda_1)\frac{\partial f_0}{\partial \alpha_0} - \left(\frac{B_1}{\alpha_0} + \mu_1\right)\frac{\partial f_0}{\partial \psi}. \quad (3.2.110)$$

Consider now the expressions for one-half times the coefficients of $\sin\psi$ and $\cos\psi$ in the Fourier expansion for f_1. These are

$$\frac{1}{2\pi}\int_0^{2\pi} f_1 \sin\psi\, d\psi = \frac{1}{2\pi}\int_0^{2\pi}\left[(A_1 + \lambda_1)\frac{\partial f_0}{\partial \alpha_0}\sin\psi - \left(\frac{B_1}{\alpha_0} + \mu_1\right)\frac{\partial f_0}{\partial \psi}\sin\psi\right]d\psi \quad (3.2.111a)$$

and

$$\frac{1}{2\pi}\int_0^{2\pi} f_1 \cos\psi\, d\psi = \frac{1}{2\pi}\int_0^{2\pi}\left[(A_1 + \lambda_1)\frac{\partial f_0}{\partial \alpha_0}\cos\psi - \left(\frac{B_1}{\alpha_0} + \mu_1\right)\frac{\partial f_0}{\partial \psi}\cos\psi\right]d\psi. \quad (3.2.111b)$$

Noting that A_1, B_1 and α_0 are functions only of $\tilde{t}$ and that λ_1, μ_1, $\partial f_0/\partial\alpha_0$ and $\partial f_0/\partial\psi$ are known functions of α_0 and ψ, we write Equations (3.2.111) as

$$\frac{1}{2\pi}\int_0^{2\pi} f_1 \sin\psi\, d\psi = \frac{A_1}{2\pi}\int_0^{2\pi}\frac{\partial f_0}{\partial\alpha_0}\sin\psi\, d\psi - \frac{B_1}{2\pi\alpha_0}\int_0^{2\pi}\frac{\partial f_0}{\partial\psi}\sin\psi\, d\psi + P_2(\alpha_0) \quad (3.2.112a)$$

$$\frac{1}{2\pi}\int_0^{2\pi} f_1 \cos\psi\, d\psi = \frac{A_1}{2\pi}\int_0^{2\pi}\frac{\partial f_0}{\partial\alpha_0}\cos\psi\, d\psi - \frac{B_1}{2\pi\alpha_0}\int_0^{2\pi}\frac{\partial f_0}{\partial\psi}\cos\psi\, d\psi + Q_2(\alpha_0), \quad (3.2.112b)$$

where P_2 and Q_2 are the following known functions of α_0

$$P_2(\alpha_0) = \frac{1}{2\pi}\int_0^{2\pi}\lambda_1(\alpha_0,\psi)\frac{\partial f_0}{\partial\alpha_0}(\alpha_0,\psi)\sin\psi\, d\psi - \frac{1}{2\pi}\int_0^{2\pi}\mu_1(\alpha_0,\psi)\frac{\partial f_0}{\partial\alpha_0}(\alpha_0,\psi)\sin\psi\, d\psi \quad (3.2.113a)$$

$$Q_2(\alpha_0) = \frac{1}{2\pi}\int_0^{2\pi}\lambda_1(\alpha_0,\psi)\frac{\partial f_0}{\partial\alpha_0}(\alpha_0,\psi)\cos\psi\, d\psi - \frac{1}{2\pi}\int_0^{2\pi}\mu_1(\alpha_0,\psi)\frac{\partial f_0}{\partial\psi}(\alpha_0,\psi)\cos\psi\, d\psi. \quad (3.2.113b)$$

We can further simplify Equations (3.2.112) by noting, according to Equations (3.2.92), that

$$\frac{dP_1}{d\alpha_0} = \frac{1}{2\pi}\int_0^{2\pi}\frac{\partial f_0}{\partial\alpha_0}(\alpha_0,\psi)\sin\psi\, d\psi \quad (3.2.114a)$$

and

$$\frac{dQ_1}{d\alpha_0} = \frac{1}{2\pi}\int_0^{2\pi}\frac{\partial f_0}{\partial\alpha_0}(\alpha_0,\psi)\cos\psi\, d\psi. \quad (3.2.114b)$$

Moreover, integration by parts and use of Equations (3.2.92) gives

$$\frac{1}{2\pi}\int_0^{2\pi}\frac{\partial f_0}{\partial\psi}(\alpha_0,\psi)\sin\psi\, d\psi = -Q_1(\alpha_0) \quad (3.2.115a)$$

$$\frac{1}{2\pi}\int_0^{2\pi}\frac{\partial f_0}{\partial\psi}(\alpha_0,\psi)\cos\psi\, d\psi = P_1(\alpha_0). \quad (3.2.115b)$$

Thus, Equations (3.2.112) simplify to

$$\frac{1}{2\pi}\int_0^{2\pi} f_1 \sin\psi \, d\psi = P_2(\alpha_0) + A_1(\tilde{t})\frac{dP_1}{d\alpha_0} + \frac{B_1(\tilde{t})}{\alpha_0} Q_1(\alpha_0) \quad (3.2.116a)$$

$$\frac{1}{2\pi}\int_0^{2\pi} f_1 \cos\psi \, d\psi = Q_2(\alpha_0) + A_1(\tilde{t})\frac{dQ_1}{d\alpha_0} - \frac{B_1(\tilde{t})}{\alpha_0} P_1(\alpha_0). \quad (3.2.116b)$$

Next, we consider the contributions from the two terms $\partial^2 F_0/\partial \tilde{t}^2 + 2\partial^2 F_1/\partial t^+ \partial \tilde{t}$. Denoting

$$g(\psi, \tilde{t}) \equiv \frac{\partial F_0}{\partial \tilde{t}} + 2\frac{\partial F_1}{\partial t^+} \quad (3.2.117)$$

we have

$$\begin{aligned}\frac{\partial^2 F_0}{\partial \tilde{t}^2} + 2\frac{\partial^2 F_1}{\partial t^+ \partial \tilde{t}} &= \frac{\partial}{\partial \tilde{t}}[g(\psi, \tilde{t})] \\ &= \frac{\partial g}{\partial \tilde{t}} + \frac{\partial g}{\partial \psi}\frac{\partial \psi}{\partial \tilde{t}} \\ &= \frac{\partial g}{\partial \tilde{t}} + \frac{Q_1}{\alpha_0}\frac{\partial g}{\partial \psi}.\end{aligned} \quad (3.2.118)$$

Thus, one-half times the coefficients of the first harmonics of $\partial^2 F_0/\partial \tilde{t}^2 + 2\partial^2 F_1/\partial t^+ \partial \tilde{t}$ are given by

$$\frac{1}{2\pi}\int_0^{2\pi} \frac{\partial}{\partial \tilde{t}}[g(\psi, \tilde{t})]\sin\psi \, d\psi = \frac{1}{2\pi}\int_0^{2\pi} \frac{\partial g}{\partial \tilde{t}} \sin\psi \, d\psi + \frac{Q_1}{2\pi\alpha_0}\int_0^{2\pi} \frac{\partial g}{\partial \psi}\sin\psi \, d\psi \quad (3.2.119a)$$

$$\frac{1}{2\pi}\int_0^{2\pi} \frac{\partial}{\partial \tilde{t}}[g(\psi, \tilde{t})]\cos\psi \, d\psi = \frac{1}{2\pi}\int_0^{2\pi} \frac{\partial g}{\partial \tilde{t}} \cos\psi \, d\psi + \frac{Q_1}{2\pi\alpha_0}\int_0^{2\pi} \frac{\partial g}{\partial \psi}\cos\psi \, d\psi. \quad (3.2.119b)$$

We integrate the second terms in Equations (3.2.119) by parts and note that the derivative with respect to $\tilde{t}$ can be moved outside the integrals of the first terms. Thus,

$$\begin{aligned}\frac{1}{2\pi}\int_0^{2\pi} \frac{\partial}{\partial \tilde{t}}[g(\psi, \tilde{t})]\sin\psi \, d\psi &= \frac{1}{2\pi}\frac{d}{d\tilde{t}}\int_0^{2\pi} g(\psi, \tilde{t})\sin\psi \, d\psi \\ &\quad - \frac{Q_1}{2\pi\alpha_0}\int_0^{2\pi} g(\psi, \tilde{t})\cos\psi \, d\psi\end{aligned} \quad (3.2.120a)$$

$$\begin{aligned}\frac{1}{2\pi}\int_0^{2\pi} \frac{\partial}{\partial \tilde{t}}[g(\psi, \tilde{t})]\cos\psi \, d\psi &= \frac{1}{2\pi\alpha_0}\frac{d}{d\tilde{t}}\int_0^{2\pi} g(\psi, \tilde{t})\cos\psi \, d\psi \\ &\quad + \frac{Q_1}{2\pi\alpha_0}\int_0^{2\pi} g(\psi, \tilde{t})\sin\psi \, d\psi.\end{aligned} \quad (3.2.120b)$$

Using Equations (3.2.106), we compute

$$g = 2(B_1 \cos\psi - A_1 \sin\psi) - H(\alpha_0, \psi), \tag{3.2.121a}$$

where H denotes

$$H(\alpha_0, \psi) = (P_1 - 2\alpha_0\mu_1)\cos\psi + (2\lambda_1 - Q_1)\sin\psi. \tag{3.2.121b}$$

We wish to show that H has no contribution in Equations (3.2.120), i.e., that

$$\int_0^{2\pi} H(\alpha_0, \psi)\sin\psi\, d\psi = \int_0^{2\pi} H(\alpha_0, \psi)\cos\psi\, d\psi = 0. \tag{3.2.122}$$

Consider the first part of Equation (3.2.122). Using the definition for H, we have

$$\int_0^{2\pi} H(\alpha_0, \psi)\sin\psi\, d\psi = -\alpha_0 \int_0^{2\pi} \mu_1 \sin 2\psi\, d\psi - \int_0^{2\pi} \lambda_1 \cos 2\psi\, d\psi - \pi Q_1, \tag{3.2.123}$$

where we have used trigonometric identities for $\sin^2\psi$, $\cos^2\psi$ and $\sin\psi\cos\psi$ and we have also noted that λ_1 has a zero average value [cf. Equation (3.2.102)]. If we now use integration by parts and Equations (3.2.103, 4) to evaluate $\partial\mu_1/\partial\psi$ and $\partial\lambda_1/\partial\psi$, the right-hand side of Equation (3.2.123) vanishes identically. A similar calculation confirms the second part of Equation (3.2.122).

Thus, Equations (3.2.120) reduce to

$$\frac{1}{2\pi}\int_0^{2\pi}\left(\frac{\partial^2 F_0}{\partial\tilde{t}^2} + 2\frac{\partial^2 F_1}{\partial t^+ \partial\tilde{t}}\right)\sin\psi\, d\psi = -\left(\frac{dA_1}{d\tilde{t}} + \frac{Q_1 B_1}{\alpha_0}\right) \tag{3.2.124a}$$

$$\frac{1}{2\pi}\int_0^{2\pi}\left(\frac{\partial^2 F_0}{\partial\tilde{t}^2} + 2\frac{\partial^2 F_1}{\partial t^+ \partial\tilde{t}}\right)\cos\psi\, d\psi = \frac{dB_1}{d\tilde{t}} - \frac{Q_1 A_1}{\alpha_0}. \tag{3.2.124b}$$

Finally, we have

$$\frac{\partial^2 F_0}{\partial t^{+2}} = -F_0 = -\alpha_0 \sin\psi. \tag{3.2.125}$$

Collecting the coefficients of the $\sin\psi$ terms on the right-hand side of Equation (3.2.105) and setting these equal to zero gives

$$\frac{dA_1}{d\tilde{t}} - A_1\frac{dP_1}{d\alpha_0} = P_2(\alpha_0). \tag{3.2.126a}$$

Similarly, cancelling the coefficients of the $\cos\psi$ terms gives

$$\frac{dB_1}{d\tilde{t}} - \frac{P_1}{\alpha_0}B_1 = A_1\left(\frac{Q_1}{\alpha_0} - \frac{dQ_1}{d\alpha_0}\right) - Q_2(\alpha_0) + \omega_2\alpha_0. \tag{3.2.126b}$$

These equations, when solved subject to the appropriate initial conditions define $A_1(\tilde{t})$ and $B_1(\tilde{t})$. We determine ω_2, as illustrated by the examples in the preceding section, by requiring B_1 to be consistent.

We note first that these differential equations are explicit in the sense that all the terms appearing can be computed a priori for a given perturbation function f. Furthermore, the equations are linear and uncoupled since one can solve Equation (3.2.126a) first to define $A_1(\tilde{t})$, then use the result in Equation (3.2.126b). The possibility of achieving this remarkable explicit result was not considered in Reference (3.2.1) and is totally Morrison's contribution.

We conclude this section by writing down the explicit solution for y and dy/dt assuming that the quadratures in Equations (3.2.93), (3.2.103), (3.2.104) and (3.2.126) have been carried out:

$$y = \{\alpha_0(\tilde{t}) + \varepsilon[A_1(\tilde{t}) + \lambda_1(\alpha_0, \psi)] + O(\varepsilon^2)\}$$

$$\times \cos\left\{\psi - \varepsilon\left[\mu_1(\alpha_0, \psi) + \frac{B_1(\tilde{t})}{\alpha_0(\tilde{t})}\right] + O(\varepsilon^2)\right\} + O(\varepsilon^2) \quad (3.2.127a)$$

$$\frac{dy}{dt} = -\{\alpha_0 + \varepsilon[A_1 + \lambda_1] + O(\varepsilon^2)\}\sin\left\{\psi - \varepsilon\left[\mu_1 + \frac{B_1}{\alpha_0}\right] + O(\varepsilon^2)\right\} + O(\varepsilon^2)$$

$$(3.2.127b)$$

where

$$\psi = (1 + \varepsilon^2\omega_2 + O(\varepsilon^3))t + \phi_0(\tilde{t}). \quad (3.2.12)$$

In Section 3.7.1, we will return to these results and show that they agree at least to the order carried out with the approximations obtained by the method of averaging.

3.2.6 Applicability of the Two-Variable Method to Boundary-Layer Problems

It is interesting to note that the two variable procedure can also be used to solve problems of boundary layer type as discussed in Chapter 2. Consider the initial value problem for the damped linear oscillator with small mass discussed in Section 2.2. Rewriting Equations (2.2.6) and (2.2.7) in terms of the inner variable $t^* = t/\varepsilon$ gives

$$\frac{d^2y}{dt^{*2}} + \frac{dy}{dt^*} + \varepsilon y = 0 \quad (3.2.128)$$

$$y(0) = 0 \quad (3.2.129a)$$

$$\frac{dy(0)}{dt^*} = 1. \quad (3.2.129b)$$

Thus, in the terminology of this chapter, t^* is the fast variable and $t = \varepsilon t^*$ is the slow variable.

We seek a two variable solution in the form

$$y(t^*, \varepsilon) = F_0(t^*, t) + \varepsilon F_1(t^*, t) + \cdots. \tag{3.2.130}$$

The equations for F_0 and F_1 are easily calculated as

$$\frac{\partial^2 F_0}{\partial t^{*2}} + \frac{\partial F_0}{\partial t^*} \equiv M(F_0) = 0 \tag{3.2.131a}$$

$$M(F_1) = -2\frac{\partial^2 F_0}{\partial t^* \partial t} - \frac{\partial F_0}{\partial t} - F_0. \tag{3.2.131b}$$

Solving (3.2.131a) gives

$$F_0 = A_0(t)e^{-t^*} + B_0(t). \tag{3.2.132}$$

The initial conditions (3.2.129) imply that $A_0(0) + B_0(0) = 0$ and $-A_0(0) = 1$. Thus $B_0(0) = 1$. Substituting (3.2.132) into the right-hand side of (3.2.131a) gives

$$M(F_1) = \left(\frac{dA_0}{dt} - A_0\right)e^{-t^*} - \left(\frac{dB_0}{dt} + B_0\right). \tag{3.2.133}$$

The solution of (3.2.133) is easily calculated as

$$F_1 = A_1(t)e^{-t^*} + B_1(t) - \left(\frac{dA_0}{dt} - A_0\right)t^*e^{-t^*} - \left(\frac{dB_0}{dt} + B_0\right)t^*. \tag{3.2.134}$$

Clearly the terms proportional to $t^*e^{-t^*}$ and t^* are inconsistent because if we relabel these in terms of t we obtain $(t/\varepsilon)e^{-t/\varepsilon}$ and t/ε which would change their order. So, we must set

$$\frac{dA_0}{dt} - A_0 = 0 \tag{3.2.135a}$$

$$\frac{dB_0}{dt} + B_0 = 0. \tag{3.2.135b}$$

Solving these with $A_0(0) = -1$, $B_0(0) = 1$ gives

$$A_0(t) = -e^t \tag{3.2.136a}$$

$$B_0(t) = e^{-t}. \tag{3.2.136b}$$

Thus, the uniformly valid solution to $O(1)$ that we have obtained is

$$F_0 = e^{-t} - e^{-t^*+t}. \tag{3.2.137}$$

Now, in Chapter 2, we calculated the uniformly valid solution to $O(1)$ for this problem by matching an outer and inner limit and obtaining the composite solution [c.f. Equation (2.2.29)]

$$y = e^{-t} - e^{-t^*} + O(\varepsilon) \tag{3.2.138}$$

and there appears to be a discrepancy between the two results. However, this discrepancy is superficial because the difference between (3.2.137) and (3.2.138) is $e^{-t^*}(1 - e^t)$ and is transcendentally small in the outer region. In the inner region it is $O(\varepsilon)$ and is taken into account in the second term in the inner expansion. Actually, if we compare (3.2.137) with the exact solution of (3.2.128),

$$y(t; \varepsilon) = \frac{1}{\sqrt{1 - 4\varepsilon}}\left[\exp(-1 + \sqrt{1 - 4\varepsilon})\frac{t}{2\varepsilon} - \exp(-1 - \sqrt{1 - \varepsilon})\frac{t}{2\varepsilon}\right], \tag{3.2.139}$$

we see that the exponent $(-t^* + t)$ corresponds to the first two terms in the expansion of the second exponential of (3.2.139). Thus, although (3.2.137) and (3.2.138) are asymptotically equivalent to $O(1)$, the former is, in some sense, more accurate. Of course, if we wish to insure uniformity near $t = \infty$, we must again introduce a strained slow variable [cf. (2.2.32)] to adequately represent the behavior of the first exponential in (3.2.139) as $t \to \infty$.

The above ideas also carry over to boundary value problems. For example, see Problem 3.2.5, which is a boundary value problem over the unit interval with a boundary layer near the origin.

In spite of the remarkable efficiency and accuracy of the two variable method for this example one must keep the following disadvantages in mind.

In general, the choice of the fast and slow variables are not given *a priori*, nor does one know the locations of boundary layers, corner layers etc. These questions can be systematically addressed, as we saw in Chapter 2, by constructing and attempting the matching of appropriate limit process expansions. It is only when the structure of the composite expansion is known that one can hope to apply a multiple variable expansion.

Often, limit process expansions correspond to definite physical approximations and are intrinsically important. Constructing the composite solution directly may not be necessary.

At any rate, in this book we will not pursue the question of which method should be used if both are applicable. Whenever a problem does have a solution derivable by matching limit process expansions, we will use this approach. If the problem arises in a form where there is a small cumulative perturbation *but no outer limit*, we will use the multiple variable method.

3.2.7 Forced Motion near Resonance

In this section forced and free motions of a weakly non-linear oscillator are considered. All slow variations are considered to depend on $\tilde{t} = \varepsilon t$. It is shown how a consistent treatment based on two variables can include many previous cases. The general initial value problem will be considered in order to get some idea of how the solution approaches its final state in the cases of forced motion. This kind of analysis can take the place of a stability analysis in showing which "final" states are accessible, that is, stable and which are not.

In variables with physical units the initial value problem for the oscillator can be written

$$M\frac{d^2Y}{dT^2} + B\frac{dY}{dT} + KY + JY^3 = F\cos\Omega T \tag{3.2.140}$$

$$Y(0) = D, \qquad \frac{dY}{dT}(0) = 0 \quad \text{(say)}. \tag{3.2.141}$$

The problem can be expressed in dimensionless variables (y, t). Let $t = \Omega_N T$, $\Omega_N = \sqrt{K/M}$ = natural frequency of free linear oscillations, $y = Y/A$, A = a characteristic amplitude of motion, to be made more precise later. The following parameters then appear

$$\varepsilon = \frac{JA^2}{K} = \frac{\text{non-linear part of spring force}}{\text{linear part of spring force}}$$

$$\varepsilon f = \frac{F}{KA} = \frac{\text{weak driving force}}{\text{spring force}}.$$

Near resonance this force is large enough to cause $O(1)$ displacement, $f = O(1)$.

$\dfrac{\Omega}{\Omega_N} = 1 + \varepsilon\omega$ = driver frequency close to resonance of the linear system, $\omega = O(1)$.

$\varepsilon\beta = \dfrac{B}{\sqrt{KM}}$ = ratio of weak damping to $\frac{1}{2}$ critical linear damping, $\beta = O(1)$.

Thus the problem reads

$$\frac{d^2y}{dt^2} + \varepsilon\beta\frac{dy}{dt} + y + \varepsilon y^3 = \varepsilon f\cos(1 + \varepsilon\omega)t = \varepsilon f\cos(t + \omega\tilde{t}) \tag{3.2.142}$$

$$y(0) = \delta, \qquad \frac{dy}{dt}(0) = 0, \qquad \text{where } \delta = \frac{D}{A}. \tag{3.2.143}$$

The original general problem depends on seven physical constants $(M, B, K, J, D, F, \Omega)$ and by dimensional analysis the dimensionless version should depend on four parameters $(\varepsilon, \beta, f, \omega)$. The extra parameter δ appears because the length A has been introduced for convenience. The perturbation expansion is expressed in terms of ε so that the resulting solution to be studied still depends in general on three parameters and a wide variety of phenomena can occur.

The two-variable expansion has the form

$$y = F_0(t, \tilde{t}) + \varepsilon F_1(t, \tilde{t}) + \cdots. \tag{3.2.144}$$

Shifts in t of the form $t^+ = t(1 + \varepsilon^2\omega_2 + \cdots)$ are not necessary to the order considered here. In the same way as before the sequence of approximating equations is

$$L(F_0) \equiv \frac{\partial^2 F_0}{\partial t^2} + F_0 = 0 \tag{3.2.145}$$

$$F_0(0,0) = \delta, \qquad \frac{\partial F_0}{\partial t}(0,0) = 0 \tag{3.2.146}$$

$$L(F_1) = -F_0^3 + f\cos(t + \omega\tilde{t}) - 2\frac{\partial^2 F_0}{\partial t\,\partial\tilde{t}} - \beta\frac{\partial F_0}{\partial t} \tag{3.2.147}$$

$$F_1(0,0) = 0, \qquad \frac{\partial F_1}{\partial t}(0,0) = -\frac{\partial F_0}{\partial \tilde{t}}(0,0). \tag{3.2.148}$$

Let the solution be represented in terms of a slowly varying amplitude and phase relative to the driver

$$F_0(t,\tilde{t}) = R(\tilde{t})\cos(t + \omega\tilde{t} - \nu(\tilde{t})). \tag{3.2.149}$$

Referring to the initial conditions (3.2.143) we can choose

$$R(0) = \delta, \qquad \nu(0) = 0. \tag{3.2.150}$$

Equations for the slowly varying amplitude and phase are obtained in the usual way from the condition that mixed secular terms do not appear in the solution F_1. The equation for F_1 now reads

$$L(F_1) = -R^3\cos^3(t + \omega\tilde{t} - \nu) + f\cos(t + \omega\tilde{t}) + 2\frac{dR}{d\tilde{t}}\sin(t + \omega\tilde{t} - \nu)$$
$$+ 2\left(\omega - \frac{d\nu}{d\tilde{t}}\right)R\cos(t + \omega\tilde{t} - \nu) + \beta R\sin(t + \omega\tilde{t} - \nu). \tag{3.2.151}$$

The coefficients of $\cos(t + \omega\tilde{t} - \nu)$, $\sin(t + \omega\tilde{t} - \nu)$ must thus both vanish in (3.2.151). To find these equations use $\cos^3 t = \frac{3}{4}\cos t + \frac{1}{4}\cos 3t$ and $\cos(t + \omega\tilde{t}) = \cos(t + \omega\tilde{t} - \nu)\cos\nu - \sin(t + \omega\tilde{t} - \nu)\sin\nu$. Then, the basic system to be studied is

$$\frac{dR}{d\tilde{t}} + \frac{\beta}{2}R = \tfrac{1}{2}f\sin\nu \tag{3.2.152a}$$

$$R\frac{d\nu}{d\tilde{t}} - (\omega - \tfrac{3}{8}R^2)R = \tfrac{1}{2}f\cos\nu. \tag{3.2.152b}$$

Now we will consider a series of special cases to obtain some simple results and to show how some previous results are contained in this formalism.

1. Free Undamped Motion

For this case the force $f = 0$, and the damping $\beta = 0$. The system (3.2.152) reduces to

$$\frac{dR}{d\tilde{t}} = 0, \qquad R(0) = \delta \tag{3.2.153a}$$

$$\frac{dv}{d\tilde{t}} = \omega - \tfrac{3}{8}R^2, \qquad v(0) = 0. \tag{3.2.153b}$$

The solution has $R = \delta = \text{const}$. That is, the amplitude of the motion is preserved and the phase v is

$$v = \omega\tilde{t} - \tfrac{3}{8}\delta^2\tilde{t}. \tag{3.2.154}$$

Returning to the original solution (3.2.144), we find

$$y = \delta\cos(t + \tfrac{3}{8}\delta^2\tilde{t}) + \cdots. \tag{3.2.155}$$

The driver frequency ω drops out, as it must since it is undefined for a free motion problem. The amplitude A in this case can be identified with the initial displacement, $\delta = 1$. Thus $y = \cos(t + \frac{3}{8}\tilde{t})$. This frequency shift is exactly that obtained by the method of strained coordinates in (3.1.28).

2. Free Damped Motion $f = 0$

Then from (3.2.152)

$$\frac{dR}{d\tilde{t}} + \frac{\beta}{2}R = 0, \qquad R(0) = \delta \tag{3.2.156a}$$

$$\frac{dv}{d\tilde{t}} = \omega - \tfrac{3}{8}R^2, \qquad v(0) = 0. \tag{3.2.156b}$$

The solution shows:

$$R(\tilde{t}) = \delta e^{-(\beta/2)\tilde{t}} \tag{3.2.157}$$

so that the decay is exactly the same as in the linear case. Then

$$v(\tilde{t}) = \omega\tilde{t} - \frac{3}{8\beta}\{1 - e^{-\beta\tilde{t}}\}. \tag{3.2.158}$$

The frequency ω again drops out, $\delta = 1$ as above, and

$$y = e^{-(\beta/2)\tilde{t}}\cos\left(t + \frac{3}{8\beta}(1 - e^{-\beta\tilde{t}})\right). \tag{3.2.159}$$

The phase lag goes from zero to $(-(3/8\beta))$ as $\tilde{t} \to \infty$. As $\beta \to 0$ we recover case 1 above.

3. Forced Linear Motion

To achieve this case in the previous framework it is necessary to drop out the term $\frac{3}{8}R^3$ which comes from the spring non-linearity in (3.2.152). Since in the original equation $J = 0$, ε can instead be identified with the weak driving force. That is, let $\delta = 1, A = D, f = 1$ and our basic system for slowly varying amplitude and phase (3.2.152) becomes

$$\frac{dR}{d\tilde{t}} + \frac{\beta}{2}R = \tfrac{1}{2}\sin\nu, \qquad R(0) = 1 \tag{3.2.160a}$$

$$R\frac{d\nu}{d\tilde{t}} - \omega R = \tfrac{1}{2}\cos\nu, \qquad \nu(0) = 0. \tag{3.2.160b}$$

As $\tilde{t} \to \infty$, the slow variations approach a steady state for $\beta > 0$ given by $\beta R = \sin\nu$; $2\omega R = -\cos\nu$ or

$$R(\infty) = \frac{1}{\sqrt{\beta^2 + 4\omega^2}}, \qquad \nu(\infty) = \tan^{-1}\left(-\frac{\beta}{2\omega}\right). \tag{3.2.161}$$

This shows the typical resonance amplification of the forced linear oscillator (see Figure 3.2.2). The phase lag ν, as $\tilde{t} \to \infty$, varies from 0 ($\omega \to -\infty$) to π ($\omega \to \infty$) and is exactly $\pi/2$ on the linear resonance $\omega = 0$. As damping gets relatively smaller, the change in phase takes place in a narrower range of frequency around $\omega = 0$. The approach of the system (3.2.160) to the steady state is most easily expressed in terms of

$$\begin{aligned} A(\tilde{t}) &= R(\tilde{t})\cos\nu(\tilde{t}) \\ B(\tilde{t}) &= R(\tilde{t})\sin\nu(\tilde{t}). \end{aligned} \tag{3.2.162}$$

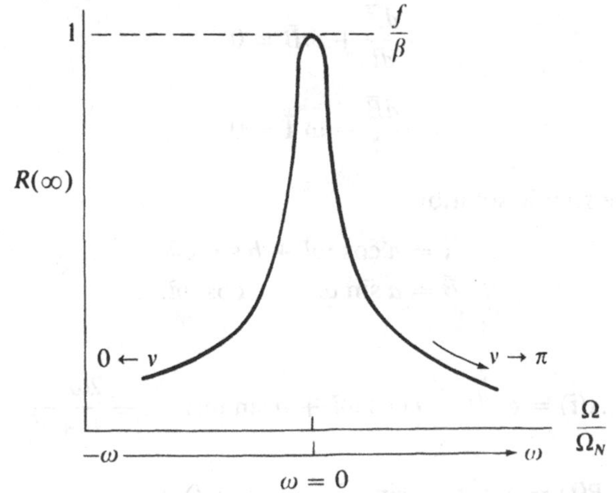

Figure 3.2.2 Asymptotic Form of Linear Resonance Curve ($\tilde{t} \to \infty$)

The original solution (3.2.149) reads

$$F_0(t, \tilde{t}) = A(\tilde{t})\cos(t + \omega\tilde{t}) + B(\tilde{t})\sin(t + \omega\tilde{t}) \tag{3.2.163}$$

in these variables. We easily obtain from (3.2.160) the following system

$$\frac{dA}{d\tilde{t}} + \frac{\beta}{2}A + \omega B = 0 \tag{3.2.164a}$$

$$\frac{dB}{d\tilde{t}} + \frac{\beta}{2}B - \omega A = \tfrac{1}{2} \tag{3.2.164b}$$

for which the steady state corresponding to (3.2.161) is

$$A(\infty) = -\frac{2\omega}{\beta^2 + 4\omega^2}, \qquad B(\infty) = \frac{\beta}{\beta^2 + 4\omega^2}. \tag{3.2.165}$$

The difference from the steady state A^*, B^* where

$$A(\tilde{t}) = A^*(\tilde{t}) - \frac{2\omega}{\beta^2 + 4\omega^2}, \qquad B(\tilde{t}) = B^*(\tilde{t}) + \frac{\beta}{\beta^2 + 4\omega^2}$$

satisfies

$$\frac{dA^*}{d\tilde{t}} + \frac{\beta}{2}A^* + \omega B^* = 0 \tag{3.2.166a}$$

$$\frac{dB^*}{d\tilde{t}} + \frac{\beta}{2}B^* - \omega A^* = 0. \tag{3.2.166b}$$

The form now suggests that exponential damping can be factored out

$$A^* = \bar{A}(\tilde{t})e^{-(\beta/2)\tilde{t}}, \qquad B^* = \bar{B}e^{-(\beta/2)\tilde{t}} \tag{3.2.167}$$

so that

$$\frac{d\bar{A}}{d\tilde{t}} + \omega\bar{B} = 0 \tag{3.2.168a}$$

$$\frac{d\bar{B}}{d\tilde{t}} - \omega\bar{A} = 0 \tag{3.2.168b}$$

which has the simple solution

$$\begin{aligned} \bar{A} &= a\cos\omega\tilde{t} + b\sin\omega\tilde{t} \\ \bar{B} &= a\sin\omega\tilde{t} - b\cos\omega\tilde{t}. \end{aligned} \tag{3.2.169}$$

Thus

$$\begin{aligned} A(\tilde{t}) &= e^{-(\beta/2)\tilde{t}}(a\cos\omega\tilde{t} + b\sin\omega\tilde{t}) - \frac{2\omega}{\beta^2 + 4\omega^2} \\ B(\tilde{t}) &= e^{-(\beta/2)\tilde{t}}(a\sin\omega\tilde{t} - b\cos\omega\tilde{t}) + \frac{\beta}{\beta^2 + 4\omega^2}. \end{aligned} \tag{3.2.170}$$

The initial conditions corresponding to (3.2.160) are $B(0) = 0$, $A(0) = 1$ so that the solution (3.2.109) becomes

$$A(\tilde{t}) = e^{-(\beta/2)\tilde{t}}\left[\left(1 + \frac{2\omega}{\beta^2 + 4\omega^2}\right)\cos\omega\tilde{t} + \frac{\beta}{\beta^2 + 4\omega^2}\sin\omega\tilde{t}\right] - \frac{2\omega}{\beta^2 + 4\omega^2}$$

$$B(\tilde{t}) = e^{-(\beta/2)\tilde{t}}\left[\left(1 + \frac{2\omega}{\beta^2 + 4\omega^2}\right)\sin\omega\tilde{t} - \frac{\beta}{\beta^2 + 4\omega^2}\cos\omega\tilde{t}\right] + \frac{\beta}{\beta^2 + 4\omega^2}. \tag{3.2.171}$$

The approach to the steady state in the form of oscillatory decay from the given initial state is now clear. This type of behavior disappears as the damping $\beta \to 0$

$$\begin{aligned} A(\tilde{t}) &= \left(1 + \frac{1}{2\omega}\right)\cos\omega\tilde{t} - \frac{1}{2\omega} \\ B(\tilde{t}) &= \left(1 + \frac{1}{2\omega}\right)\sin\omega\tilde{t}. \end{aligned} \tag{3.2.172}$$

The solution never reaches the steady state but a slow beating oscillation about the steady state (in $\tilde{t}$) occurs. Finally, if $\beta = 0$ and resonance is approached ($\omega \to 0$) we find

$$\begin{aligned} A(\tilde{t}) &\to 1 \\ B(\tilde{t}) &\to \frac{\tilde{t}}{2}. \end{aligned} \tag{3.2.173}$$

The beat period approaches infinity so that linear growth in $\tilde{t}$ is the ultimate dominating result.

4. Forced Non-linear Motion

According to the basic system (3.2.152) in the general case the slow variation of amplitude and phase can approach a steady state

$$R(\infty) = \rho, \qquad \nu(\infty) = \alpha \tag{3.2.174}$$

given by

$$\begin{aligned} \beta\rho &= f\sin\alpha \\ -2\omega\rho + \tfrac{3}{4}\rho^3 &= f\cos\alpha. \end{aligned} \tag{3.2.175}$$

These steady states are singular points of the system (3.2.152). The nature of the singular points and the structure of the motion can be discussed qualitatively for very small damping by considering at first $\beta = 0$. Then there are two possible branches to discuss which give different relations between frequency ω and amplitude ρ:

Branch (i)

$$\alpha = 0$$

$$\omega(\rho) = \tfrac{3}{8}\rho^2 - \frac{f}{2\rho}. \tag{3.2.176}$$

This branch, in phase with the driver, lies above the free resonance curve $\omega = \frac{3}{8}\rho^2$ (cf. Figure 3.2.3).

Branch (ii)

$$\alpha = \pi$$

$$\omega(\rho) = \tfrac{3}{8}\rho^2 + \frac{f}{2\rho}. \tag{3.2.177}$$

On this branch there is a minimum frequency with corresponding amplitude

$$\omega_m = \frac{3^{4/3}}{2^{7/3}} f^{2/3}, \qquad \rho_m = (\tfrac{2}{3}f)^{1/3}. \tag{3.2.178}$$

This branch lies below the corresponding free branch of the non-linear motion. If the damping β is finite then these two branches are joined and the phase varies continuously.

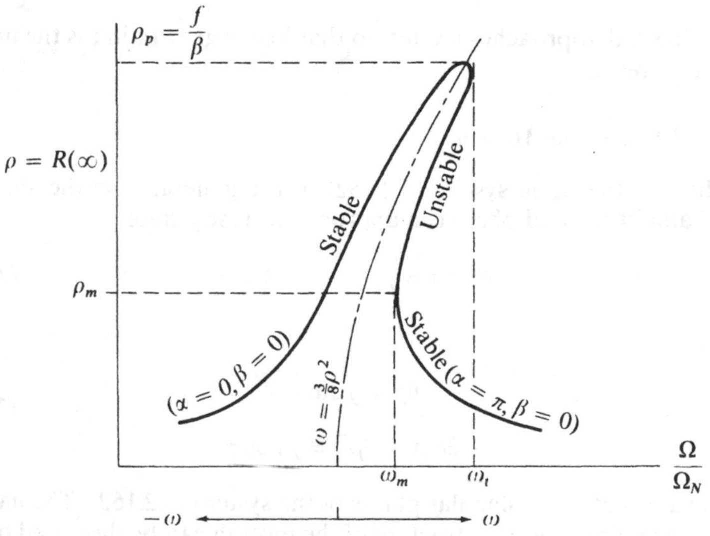

Figure 3.2.3 Non-linear Resonance Diagram f, β Fixed

The complete resonance curve is given by

$$\beta^2\rho^2 + (2\omega\rho - \tfrac{3}{4}\rho^3)^2 = f^2. \tag{3.2.179}$$

Thus it can be shown that the peak amplitude occurs exactly on the free "resonance" curve $\omega = \frac{3}{8}\rho^2$ and has the value

$$\rho_p = \frac{f}{\beta}. \tag{3.2.180}$$

The peak amplitude ρ_p of the non-linear case is thus exactly the same as that of the linear case, if the driver frequency is adjusted properly.

The question of whether all points on branches (i) or (ii) can be reached from an arbitrary initial state can be answered by considering the nature of the singular points. This can be done, at first again, for $\beta = 0$. For branch (ii) $\omega > \omega_m$, $\alpha = \pi$ let

$$\begin{Bmatrix} R(\tilde{t}) = \rho + r(\tilde{t}) \\ v(\tilde{t}) = \alpha + \theta(\tilde{t}) \end{Bmatrix}, \qquad \omega = \tfrac{3}{8}\rho^2 + \frac{f}{2\rho} > \omega_m. \tag{3.2.181}$$

Then linearization of (3.2.152) about the singular point gives

$$\begin{gathered} \frac{dr}{d\tilde{t}} + \frac{f}{2}\theta = 0 \\ \rho\frac{d\theta}{d\tilde{t}} - \omega(\rho)r + \tfrac{9}{8}\rho^2 r = 0. \end{gathered} \tag{3.2.182}$$

For solutions of the form $r = ae^{\lambda\tilde{t}}$, $\theta = be^{\lambda\tilde{t}}$ we find

$$\lambda^2 = \frac{f}{2}\left(\frac{3}{4}\rho - \frac{f}{2\rho^2}\right). \tag{3.2.183}$$

Thus on the upper part of this branch $\rho > \rho_m = (\frac{2}{3}f)^{1/3}$ there are two real roots $\lambda_{1,2}$ of opposite sign. The singularity is a saddle point. Only one exceptional path runs into this saddle point so that this branch is accessible only from a very special set of initial conditions. Therefore, this branch is labeled unstable in Figure 3.2.3. This singular point is analogous to the unstable equilibrium of a pendulum standing vertically. Therefore, the addition of finite damping cannot change the nature of this singular point. But if $\rho < \rho_m$, $\lambda_{1,2}$ are complex conjugates and the singular point is a center with closed paths around $r = \theta = 0$. The motion in this case is beating with a period depending on $\lambda_{1,2}$. But now the addition of damping $\beta > 0$ causes the beats to die out and the steady state is approached; the branch is stable.

Similarly the entire branch (i) is stable and which branch is approached depends, in this view, on the initial conditions. The entire course of all solutions can be studied in the (A, B) phase plane with (R, v) as polar coordinates. This in fact is one of the main advantages of the present approach to the original non-autonomous system. For R large the basic system (3.2.152) shows

$$\frac{dR}{dv} \to \frac{4}{3}\frac{\beta}{R}, \qquad \frac{dR}{d\tilde{t}} \to -\frac{\beta}{2}R \quad \text{or} \quad R = R_0 e^{-(\beta/2)\tilde{t}}. \tag{3.2.184}$$

The motion spirals inward for $\beta > 0$. In accordance with the previous discussion all paths run into one singular point for $\omega < \omega_m$ (cf. Figure 3.2.3), corresponding to the stable branch with frequencies less than $\frac{3}{8}\rho^2$. For $\omega > \omega_t$ all paths run into the singular point corresponding to the stable part of the branch with frequencies $\omega > \frac{3}{8}\rho^2$. For $\omega_m < \omega < \omega_t$ there is a saddle point corresponding to the unstable branch. The paths through the saddle are separatrices which divide the A, B plane into those initial conditions which run into the different singular points corresponding to the stable branches. A detailed picture depends on the numerical values of (f, β, ω).

Problems

1. Carry out the details of the solution to $O(\varepsilon)$ for the examples of Sections 3.2.2 and 3.2.3.

2. Carry out the solution to $O(\varepsilon)$ for equations of the type (3.2.8) with the following perturbation functions f:

   (a) $f = cy^3 + dy/dt$, c = arbitrary constant independent of ε,
   (b) $f = (dy/dt)|dy/dt|^n$, n = positive integer,
   (c) $f = c(d/dt)(y^3) - (dy/dt)^2$, c = positive constant.

3. Comment on the solution of Prob. 3.2.2c when c is negative.

4. Can one use the two variable method to solve

$$\frac{d^2y}{dt^2} - y + \varepsilon y^3 = 0, \qquad \varepsilon > 0,$$

   for which all solutions are periodic, by perturbing about the $\varepsilon = 0$ solution? Show that if the appropriate scaling for y is introduced we no longer have a perturbation problem.

5. Consider the boundary value problem

$$\frac{d^2y}{dx^{*2}} + \frac{dy}{dx^*} - \varepsilon y^2 = 0, \qquad 0 < \varepsilon \ll 1$$

$$y(0) = 0$$

$$y\left(\frac{1}{\varepsilon}\right) = 1.$$

   Clearly this is the same problem as

$$\varepsilon\frac{d^2y}{dx^2} + \frac{dy}{dx} - y^2 = 0$$

$$y(0) = 0, \qquad y(1) = 1, \qquad x = \varepsilon x^*$$

which can be solved by matching an inner and outer expansion. Do this, and compare your result with the solution by the two variable (x^*, x) method. What happens if the sign of the y^2 term is positive?

References

3.2.1 J. Kevorkian, The two variable expansion procedure for the approximate solution of certain nonlinear differential equations. *Lectures in Applied Mathematics* **7**, *Space Mathematics, Part III*, American Mathematical Society, Providence, Rhode Island, 1966, pp. 206–275. This is a revised version of the author's (1961) Ph.D. thesis at Caltech. A brief version appears in the work of J. D. Cole and J. Kevorkian, *Nonlinear Differential Equations and Nonlinear Mechanics*, (LaSalle and Lefschetz, editors), New York, Academic Press, 1963, pp. 113–120.

3.2.2 N. M. Krylov, and N. N. Bogoliubov, *Introduction to Nonlinear Mechanics*, Princeton University Press, Princeton, 1957.

3.2.3 N. N. Bogoliubov, and Y. A. Mitropolsky, *Asymptotic Methods in the Theory of Nonlinear Oscillations*, Hindustan Publishing Co., 1961.

3.2.4 Y. A. Mitropolsky, *Problèmes de la Théorie Asymptotique des Oscillations non Stationnaires*, Gauthier-Villars, Paris, 1966.

3.2.5 J. A. Morrison, Comparison of the Modified Method of Averaging and the Two Variable Expansion Procedure, *S.I.A.M. Review* **8**, 1966, pp. 66–85.

3.2.6 R. A. Struble, *Nonlinear Differential Equations*, McGraw-Hill Book Co., New York, 1962.

3.2.7 H. von Zeipel, Recherche sur le Mouvement des Petites Planètes, *Ark. Astron. Mat. Fys.*, 1916, **11–13**.

3.2.8 D. Brouwer and G. M. Clemence, *Methods of Celestial Mechanics*, Academic Press, New York, 1961.

3.2.9 J. Kevorkian, von Zeipel Method and the Two Variable Expansion Procedure, *Astronomical Journal* **71**, 1966, pp. 878–885.

3.2.10 F. A. Brauer and J. A. Nohel, *The Qualitative Theory of Ordinary Differential Equations*, W. A. Benjamin, Inc., Menlo Park, California, 1969.

3.3 Second Order Equations with Variable Coefficients

In this section we consider the more general class of differential equations

$$\frac{d^2y}{dt^2} + y + \varepsilon f\left(y, \frac{dy}{dt}, t, \varepsilon\right) = 0 \tag{3.3.1}$$

where again if $\varepsilon = 0$ the motion is simple harmonic.

As we saw in Section 3.2.4 the presence of ε in the perturbation term has non-trivial consequences. In general, this necessitated the rescaling of the dependent variable to cast the problem in an appropriate form. If, in addition,

t also occurs in f the situation becomes more complicated. As we will illustrate by means of several examples in this section, we first have to determine the proper choice of fast and slow variables and in some cases we may have to match two multiple scale expansions to arrive at a uniformly valid solution.

3.3.1 Mathieu's Equation

A class of problems involving both bounded and unbounded oscillations to which the two variable procedure can be applied is exemplified by the Mathieu equation:

$$\frac{d^2y}{dt^2} + (\delta + \varepsilon \cos t)y = 0. \tag{3.3.2}$$

The general theory of differential equations with periodic coefficients (Floquet Theory) yields the fundamental result that the (δ, ε) plane is divided into regions of "stability"[5] and instability separated by transition curves (cf. Figure 3.3.1). The general theory shows that along the transition curves, periodic solutions of period 2π and 4π exist (as well as linearly increasing solutions).

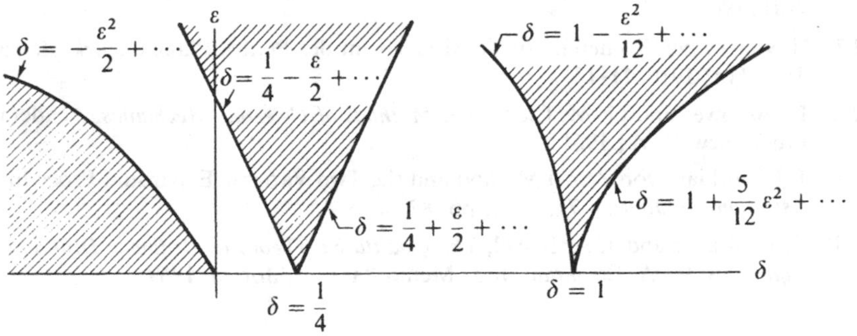

Figure 3.3.1 Mathieu Equation, Shaded Region Unstable

An approximation to the transitional curves for small ε can thus be found by the method of strained coordinates by looking for the $\delta(\varepsilon)$ for which Eq. (3.3.2) possesses periodic solutions with period 2π and 4π (cf. Problem 3.1.4). The general theory also states that the solutions in the unstable region have the form of exponentials times (2π or 4π) periodic functions, and that the transitional curves intersect $\varepsilon = 0$ at the critical points:

$$\delta_c = \frac{n^2}{4}, \qquad n = 0, 1, 2, \ldots . \tag{3.3.3a}$$

[5] Here stability means boundedness as $t \to \infty$.

The interested reader may consult Reference 3.3.1 for proofs of the above statements and further results on the Mathieu equation.

The application of the two variable procedure to Equation (3.3.2) enables one to find not only the transitional curves but also the form of the solutions in neighborhoods of these curves. A uniformly valid approximation on the interval $0 \le t \le T = O(1/\varepsilon)$ is sought for general initial conditions such that the first term in the asymptotic representation shows the transition from stability to unstability.

Separate discussions have to be made for the various critical points, although $n = 1, 2, \ldots$, are very similar. Consider first the neighborhood of $(\delta = \frac{1}{4}, \varepsilon = 0)$. A uniform approximation to Equation (3.3.2) as $\varepsilon \to 0$ is desired. If δ is held fixed $(\delta \neq \frac{1}{4})$, it is clear that only the stable solution consisting of harmonic oscillations will be obtained. Thus, it is necessary to consider $\delta \to \frac{1}{4}$ as $\varepsilon \to 0$ and to represent this limit as

$$\delta = \tfrac{1}{4} + \varepsilon\delta_1 + \varepsilon^2\delta_2 + \cdots. \tag{3.3.3b}$$

Here δ_1, δ_2 are parameters on which the solution depends, parameters indicating a path of approach to δ_c, and, in fact, a particular value of δ_1 corresponds to the transitional curve, $\delta_2 = \delta_2(\delta_1)$, and so on.

One time scale of this problem is associated with the simple harmonic oscillations at the critical points (t-scale); the other is caused by small-scale fluctuations ($\varepsilon \cos t$) in the effective spring force. This small fluctuation can produce a cumulative or long time-scale effect on the solution. It is certainly not clear *a priori* what the proper $\tilde{t}$ is, and the choice of $\tilde{t}$ is to be regarded as part of the problem. It will be seen, however, that there is a connection between the choice of $\tilde{t}$ and the two-variable expansion (Equation 3.2.9).

The problem we consider is

$$n = 1, \qquad \delta_c = \tfrac{1}{4}, \qquad \delta = \tfrac{1}{4} + \varepsilon\delta_1 + \varepsilon^2\delta_2 + \cdots,$$

$$\frac{d^2y}{dt^2} + \{\tfrac{1}{4} + \varepsilon(\delta_1 + \cos t) + \cdots\}y = 0, \tag{3.3.4}$$

$$y(0) = a, \qquad \frac{dy}{dt}(0) = b.$$

Let $\tilde{t} = \varepsilon t$, and try an expansion of the form

$$y(t; \varepsilon) = F_0(t, \tilde{t}) + \varepsilon F_1(t, \tilde{t}) + \cdots. \tag{3.3.5}$$

Thus, the successive equations for F_0 and F_1 are

$$\frac{\partial^2 F_0}{\partial t^2} + \frac{1}{4}F_0 = 0, \tag{3.3.6a}$$

$$\frac{\partial^2 F_1}{\partial t^2} + \frac{1}{4}F_1 = -2\frac{\partial^2 F_0}{\partial t\, \partial \tilde{t}} - (\delta_1 + \cos t)F_0. \tag{3.3.6b}$$

The first solution, as usual, represents modulated oscillations:

$$F_0(t, \tilde{t}) = A_0(\tilde{t})\cos\frac{t}{2} + B_0(\tilde{t})\sin\frac{t}{2}, \tag{3.3.7}$$

where now $A_0(0) = a$, $B_0(0) = 2b$. The equation for F_1 thus reads

$$\frac{\partial^2 F_1}{\partial t^2} + \frac{1}{4}F_1 = -2\left\{-\frac{1}{2}\frac{dA_0}{d\tilde{t}}\sin\frac{t}{2} + \frac{1}{2}\frac{dB_0}{d\tilde{t}}\cos\frac{t}{2}\right\}$$
$$- (\delta_1 + \cos t)\left\{A_0(\tilde{t})\cos\frac{t}{2} + B_0(\tilde{t})\sin\frac{t}{2}\right\}. \tag{3.3.8}$$

Now we use the same argument as before: Unless all resonance producing terms ($\cos t/2$, $\sin t/2$) on the right-hand side are made to vanish, the solution grows on the t scale. Also, εt cannot be permitted to appear in a uniformly valid representation. Thus, using

$$\sin\frac{t}{2}\cos t = \frac{1}{2}\left\{\sin\frac{3t}{2} - \sin\frac{t}{2}\right\}, \qquad \cos\frac{t}{2}\cos t = \frac{1}{2}\left\{\cos\frac{3t}{2} + \cos\frac{t}{2}\right\}, \tag{3.3.9}$$

we find the differential equations for A_0, B_0:

$$\frac{dA_0}{d\tilde{t}} - (\delta_1 - \tfrac{1}{2})B_0 = 0, \tag{3.3.10a}$$

$$\frac{dB_0}{d\tilde{t}} + (\delta_1 + \tfrac{1}{2})A_0 = 0. \tag{3.3.10b}$$

This system has a solution of the form

$$A_0 = a_0 e^{\gamma\tilde{t}}, \qquad B_0 = b_0 e^{\gamma\tilde{t}}, \tag{3.3.11}$$

so that we have

$$\gamma a_0 - (\delta_1 - \tfrac{1}{2})b_0 = 0, \qquad (\delta_1 + \tfrac{1}{2})a_0 + \gamma b_0 = 0. \tag{3.3.12}$$

Thus, the characteristic exponent γ satisfies the equation

$$\gamma^2 + (\delta_1^2 - \tfrac{1}{4}) = 0. \tag{3.3.13}$$

It is immediately seen that real roots $\gamma_{1,2}$ occur for $|\delta_1| < \frac{1}{2}$. These values of δ_1 define, to this order, the region of instability. If, however, $|\delta_1| > \frac{1}{2}$, the roots are purely imaginary, so that the solutions are bounded and have the form of modulated oscillations. The transitional curve is given by $\delta_1 = \pm\frac{1}{2}$, a case of equal zero roots. The solutions of Equations (3.3.10) are easily constructed once $\gamma_{1,2}$ are found. If the initial conditions $A_0(0) = a$, $B_0(0) = 2b$

are accounted for, it is found that

$$A_0(\tilde{t}) = \left(\frac{a}{2} - b\sqrt{\frac{\frac{1}{2} - \delta_1}{\frac{1}{2} + \delta_1}}\right)\exp\left(\sqrt{\frac{1}{4} - \delta_1^2}\,\tilde{t}\right) + \left(\frac{a}{2} + b\sqrt{\frac{\frac{1}{2} - \delta_1}{\frac{1}{2} + \delta_1}}\right)\exp\left(-\sqrt{\frac{1}{4} - \delta_1^2}\,\tilde{t}\right),$$

$$B_0(\tilde{t}) = \left(b - \frac{a}{2}\sqrt{\frac{\frac{1}{2} + \delta_1}{\frac{1}{2} - \delta_1}}\right)\exp\left(\sqrt{\frac{1}{4} - \delta_1^2}\,\tilde{t}\right) + \left(b + \frac{a}{2}\sqrt{\frac{\frac{1}{2} + \delta_1}{\frac{1}{2} - \delta_1}}\right)\exp\left(-\sqrt{\frac{1}{4} - \delta_1^2}\,\tilde{t}\right),$$

$$|\delta_1| < \tfrac{1}{2}. \tag{3.3.14}$$

This form of solution is in precise agreement with the general theory for equations with periodic coefficients. The solution on the transitional boundary which satisfies the same initial conditions is easily obtained from Equation (3.3.14) by the limit $|\delta_1| \to \frac{1}{2}$. Thus, we have

$$A_0(\tilde{t}) = a, \qquad B_0(\tilde{t}) = 2b - a\tilde{t}, \qquad (\delta_1 = \tfrac{1}{2}); \tag{3.3.15}$$

$$A_0(\tilde{t}) = (a - 2b)\tilde{t}, \qquad B_0(\tilde{t}) = 2b, \qquad (\delta_1 = -\tfrac{1}{2}). \tag{3.3.16}$$

The continuation of this solution across the transition curve $|\delta_1| = \frac{1}{2}$ yields the modulated oscillation (cf. Equation (3.3.7)) of the stable region $|\delta_1| > \frac{1}{2}$:

$$\begin{aligned} A_0(\tilde{t}) &= a\cos\sqrt{\delta_1^2 - \frac{1}{4}}\,\tilde{t} \pm 2b\sqrt{\frac{\delta_1 - \frac{1}{2}}{\delta_1 + \frac{1}{2}}}\sin\sqrt{\delta_1^2 - \frac{1}{4}}\,\tilde{t} \\ B_0(\tilde{t}) &= 2b\cos\sqrt{\delta_1^2 - \frac{1}{4}}\,\tilde{t} \mp a\sqrt{\frac{\delta_1 + \frac{1}{2}}{\delta_1 - \frac{1}{2}}}\sin\sqrt{\delta_1^2 - \frac{1}{4}}\,\tilde{t} \end{aligned} \tag{3.3.17}$$

(upper sign $\delta_1 > \frac{1}{2}$, lower sign $\delta_1 < -\frac{1}{2}$).

This approximation, however is not valid down to $\varepsilon = 0$, since $\varepsilon = 0$ corresponds to $|\delta_1| \to \infty$. Validity of the expansion can be expected in a wedge-like region $|\delta_1| < k$, including the transition curve.

Similar considerations apply near the critical point $\delta_c = 1$. However, in this case the slow variable and the associated expansion $\delta(\varepsilon)$ must be chosen (cf. Reference 3.2.1 for details):

$$\tilde{t} = \varepsilon^2 t, \tag{3.3.18}$$

$$\delta(\varepsilon) = 1 + \delta_2\varepsilon^2 + \cdots. \tag{3.3.19}$$

Thus, Equation (3.3.2) can be written as

$$\frac{d^2y}{dt^2} + (1 + \varepsilon\cos t + \varepsilon^2\delta_2 + \cdots)y = 0. \tag{3.3.20}$$

The expansion for y is of the same form as before,

$$y(t;\varepsilon) = F_0(t,\tilde{t}) + \varepsilon F_1(t,\tilde{t}) + \varepsilon^2 F_2(t,\tilde{t}) + \cdots, \tag{3.3.21}$$

but now

$$\frac{dy}{dt} = \frac{\partial F_0}{\partial t} + \varepsilon^2\frac{\partial F_0}{\partial \tilde{t}} + \varepsilon\frac{\partial F_1}{\partial t} + \varepsilon^2\frac{\partial F_2}{\partial \tilde{t}} + \cdots, \tag{3.3.22}$$

$$\frac{d^2y}{dt^2} = \frac{\partial^2 F_0}{\partial t^2} + 2\varepsilon^2\frac{\partial^2 F_0}{\partial t\,\partial \tilde{t}} + \varepsilon\frac{\partial^2 F_1}{\partial t^2} + \varepsilon\frac{\partial^2 F_2}{\partial t^2} + \cdots. \tag{3.3.23}$$

Using these expressions in Equation (3.3.20), we see that the sequence of approximating equations is

$$\frac{\partial^2 F_0}{\partial t^2} + F_0 = 0, \tag{3.3.24}$$

$$\frac{\partial^2 F_1}{\partial t^2} + F_1 = -F_0\cos t, \tag{3.3.25}$$

$$\frac{\partial^2 F_2}{\partial t^2} + F_2 = -2\frac{\partial^2 F_0}{\partial t\,\partial \tilde{t}} - \delta_2 F_0 - F_1\cos t. \tag{3.3.26}$$

The basic solution is a modulation of the oscillations near $(\varepsilon = 0, \delta = 1)$:

$$F_0 = A_0(\tilde{t})\cos t + B_0(\tilde{t})\sin t. \tag{3.3.27}$$

In this case, however, the equation for F_1 does not provide the information necessary to find A_0, B_0. The right-hand side of Equation (3.3.25) is

$$-F_0\cos t = -A_0\cos^2 t - B_0\sin t\cos t = -\frac{A_0}{2}(1+\cos 2t) - \frac{B_0}{2}\sin 2t. \tag{3.3.28}$$

Equation (3.3.28) contains no resonance-producing terms. Equation (3.3.26) must be considered if A_0 and B_0 are to be found. The particular solution of Equation (3.3.25) due to F_0 is F_{1_p},

$$F_{1_p} = \frac{A_0(\tilde{t})}{6}\cos 2t - \frac{A_0(\tilde{t})}{2} + \frac{B_0(\tilde{t})}{6}\sin 2t. \tag{3.3.29}$$

Thus, Equation (3.3.26) becomes

$$\begin{aligned}\frac{\partial^2 F_2}{\partial t^2} + F_2 = {}& -2\left\{-\frac{dA_0}{d\tilde{t}}\sin t + \frac{dB_0}{d\tilde{t}}\cos t\right\} - \delta_2\{A_0\cos t + B_0\sin t\}\\ &- \frac{A_0}{12}\{\cos 3t + \cos t\} + \frac{B_0}{12}\{\sin 3t + \sin t\}\\ &- \{A_1\cos t + B_1\sin t\}\cos t. \end{aligned}\tag{3.3.30}$$

Now, the argument about resonance-producing terms can be applied, and it must be required that

$$2\frac{dA_0}{d\tilde{t}} - (\delta_2 + \tfrac{1}{12})B_0 = 0, \qquad A_0(0) = a; \tag{3.3.31a}$$

$$2\frac{dB_0}{d\tilde{t}} + (\delta_2 - \tfrac{5}{12})A_0 = 0, \qquad B_0(0) = b. \tag{3.3.31b}$$

This pair of equations is of the same form as those found for $\delta_c = \frac{1}{4}$ (Equation (3.3.10)), and the same form of solution (3.3.11) can be used. Now, we have

$$\gamma^2 + \tfrac{1}{4}(\delta_2 - \tfrac{5}{12})(\delta_2 + \tfrac{1}{12}) = 0. \tag{3.3.32}$$

Real roots corresponding to the unstable region occur for $-\frac{1}{12} < \delta_2 < \frac{5}{12}$.

Finally, consider the behavior of the solutions to the Mathieu equation near the stability boundary through the origin, i.e., $\delta = -\frac{1}{2}\varepsilon^2 + \cdots$. The behavior near this boundary is qualitatively different from that to be expected near the other stability boundaries which occur in pairs. Nevertheless, the two-variable method provides an approach to this problem. The expression for the stability boundary again takes the form

$$\delta = \varepsilon\delta_1 + \varepsilon^2\delta_2 + \cdots, \tag{3.3.33}$$

where in fact it is to be shown that $\delta_1 = 0$. The Mathieu equation (Equation (3.3.2)) is

$$\frac{d^2y}{dt^2} + \{\varepsilon(\delta_1 + \cos t) + \varepsilon^2\delta_2 + \cdots\}y = 0. \tag{3.3.34}$$

Consider the general initial conditions

$$y(0) = a, \qquad \frac{dy}{dt}(0) = b$$

as before.

In this case, the two-variable expression is of the form

$$y(t;\varepsilon) = \frac{1}{\varepsilon}F_{-1}(t,\tilde{t}) + F_0(t,\tilde{t}) + \varepsilon F_1(t,\tilde{t}) + \cdots, \tag{3.3.35}$$

where now $\tilde{t} = \varepsilon t$ and, hence,

$$\frac{dy}{dt} = \frac{1}{\varepsilon}\frac{\partial F_{-1}}{\partial t} + \left(\frac{\partial F_0}{\partial t} + \frac{\partial F_{-1}}{\partial \tilde{t}}\right) + \cdots,$$

and the equations of the approximating sequence are

$$\frac{\partial^2 F_{-1}}{\partial t^2} = 0, \tag{3.3.36}$$

$$\frac{\partial^2 F_0}{\partial t^2} = -2\frac{\partial^2 F_{-1}}{\partial t\,\partial\tilde{t}} - (\delta_1 + \cos t)F_{-1} \tag{3.3.37}$$

$$\frac{\partial^2 F_1}{\partial t^2} = -2\frac{\partial^2 F_0}{\partial t\,\partial\tilde{t}} - (\delta_1 + \cos t)F_0 - \frac{\partial^2 F_{-1}}{\partial\tilde{t}^2} - \delta_2 F_{-1}. \tag{3.3.38}$$

The solution of Equation (3.3.36) is

$$F_{-1}(t, \tilde{t}) = A_{-1}(\tilde{t})t + B_{-1}(\tilde{t}), \tag{3.3.39}$$

and all the solutions of homogeneous equations $\partial^2 F_i/\partial t^2 = 0$ are of this form. Assuming boundedness on the fast time scale means that $A_{-1}(\tilde{t}) = 0$. Thus, near the stability boundary we expect solutions to have the form:

$$F_{-1}(t, \tilde{t}) = B_{-1}(\tilde{t}), \tag{3.3.40}$$

of slowly varying functions. This explains why the initial term is $O(1/\varepsilon)$ in order to produce a velocity which is $O(1)$.

Now the equation for F_0, (Equation 3.3.37), reads

$$\frac{\partial^2 F_0}{\partial t^2} = -(\delta_1 + \cos t)B_{-1}(\tilde{t}). \tag{3.3.41}$$

Again, unless $\delta_1 = 0$, two rapidly unbounded terms appear in the solutions of Equation (3.3.41). Thus, we have $\delta_1 = 0$ and

$$F_0(t, \tilde{t}) = B_0(\tilde{t}) + B_{-1}(\tilde{t})\cos t. \tag{3.3.42}$$

The function $B_{-1}(\tilde{t})$, the basic uniformly valid solution, is still unknown, and the equation for F_1 must be considered in this example. Equation (3.3.38) now reads

$$\frac{\partial^2 F_1}{\partial t^2} = 2\frac{dB_{-1}}{d\tilde{t}}\sin t - \{B_0(\tilde{t}) + B_{-1}(\tilde{t})\cos t\}\cos t - \frac{d^2 B_{-1}}{d\tilde{t}^2} - \delta_2 B_{-1}(\tilde{t}). \tag{3.3.43}$$

Using the same argument that terms independent of t must not appear on the right-hand side of Equation (3.3.43) (and using $\cos^2 t = \frac{1}{2}\{\cos 2t + 1\}$), we obtain the required differential equation (in this case *second*-order) for B_{-1},

$$\frac{d^2 B_{-1}}{d\tilde{t}^2} + (\delta_2 + \tfrac{1}{2})B_{-1} = 0. \tag{3.3.44}$$

It is clear, then, that in this case there is only one stability boundary and that unstable solutions exist for $\delta_2 < -\frac{1}{2}$ and bounded solutions exist for $\delta_2 > -\frac{1}{2}$. The original initial conditions are now

$$F_{-1}(0, 0) = 0, \qquad \frac{\partial F_{-1}}{\partial \tilde{t}}(0, 0) = b, \tag{3.3.45}$$

so that the dominant unbounded term depends only on the initial dy/dt. The solution of Equation (3.3.44) is

$$\begin{aligned} B_{-1}(\tilde{t}) &= \frac{b}{\sqrt{\delta_2 + \frac{1}{2}}}\sin\sqrt{\delta_2 + \tfrac{1}{2}}\,\tilde{t}, \qquad \delta_2 > -\tfrac{1}{2} \quad \text{(stable)}, \\ B_{-1}(\tilde{t}) &= \frac{b}{\sqrt{-(\delta_2 + \frac{1}{2})}}\sinh\sqrt{-(\delta_2 + \tfrac{1}{2})}\,\tilde{t}, \qquad \delta_2 < -\tfrac{1}{2} \quad \text{(unstable)}. \end{aligned} \tag{3.3.46}$$

In the limit $\delta_2 = -\frac{1}{2}$, the solution $B_{-1}(\tilde{t}) = b\tilde{t}$, so that $y = bt + \cdots$. The solution actually does grow linearly in t, due to the fact that the average of $\cos t$ is $\frac{1}{2}$.

This expression can be carried to higher orders to represent also the effect of $y(0)$.

3.3.2 Oscillator with Slowly Varying Frequency; Adiabatic Invariance

A natural extension of the two-variable method is to those problems which contain a slowly varying function explicitly. However, the simple example of this Section shows that some thought must be given to the proper choice of variables. The classical example is the motion of a pendulum under slow variations in its length. In this context, of course, "slow" means over a time scale long compared to the natural period. In the version corresponding to small amplitudes, the following equation would apply:

$$\frac{d^2y}{dt^2} + \mu^2(\tilde{t})y = 0, \tag{3.3.47}$$

where $\tilde{t} = \varepsilon t$ is the slow variable, $\mu^2 > 0$, and $\mu = O(1)$. Arbitrary initial conditions can be chosen, for example, as

$$y(0) = a, \qquad \frac{dy}{dt}(0) = b. \tag{3.3.48}$$

An attempt to apply the two-variable procedure directly to Equation (3.3.47) with $\mu(\tilde{t})t$ and $\tilde{t}$ as the fast and slow variables fails as can be quickly verified by the reader. The reason for this failure is in the assumption that μt is the appropriate fast variable for, even though the instantaneous frequency is μ, the phase of the oscillation is not μt, but rather the integral of μ with respect to t since μ is *not* constant.

This can be explicitly demonstrated by transforming the independent variable in Equation (3.3.47) from t to t^+ in such a way that the oscillations will have a constant frequency with respect to t^+. If we let

$$t^+ = f(t; \varepsilon), \qquad f(0, \varepsilon) = 0 \tag{3.3.49}$$

and perform the exact transformation (3.3.49) on Equation (3.3.47) we find

$$\frac{d^2y}{dt^{+2}} + \frac{d^2f/dt^2}{(df/dt)^2}\frac{dy}{dt^+} + \frac{\mu^2(\tilde{t})}{(df/dt)^2}y = 0 \tag{3.3.50}$$

$$y(0) = a, \qquad \frac{dy(0)}{dt^+} = \frac{b}{df(0; \varepsilon)/dt}. \tag{3.3.51}$$

In order for the oscillations to have constant frequency on the t^+ scale we must set df/dt proportional to μ. For convenience we take

$$\frac{df}{dt} = \mu.$$

Now df/dt is the instantaneous frequency on the t scale and the appropriate fast variable t^+ is

$$t^+ = \int_0^t \mu(\varepsilon s)ds. \tag{3.3.52}$$

Since $d^2 f/dt^2 = \varepsilon\, d\mu/d\tilde{t}$, the problem now appears with small, slowly varying damping:

$$\frac{d^2y}{dt^{+2}} + \varepsilon g(\tilde{t}) \frac{dy}{dt^+} + y = 0, \qquad g(\tilde{t}) = \frac{d\mu/d\tilde{t}}{\mu^2} \tag{3.3.53}$$

$$y(0) = a, \qquad \frac{dy(0)}{dt^+} = \frac{b}{\mu(0)}. \tag{3.3.54}$$

Note that the definition (3.3.52) for the fast variable t^+ is a generalization of the usual definition of t^+ [cf. Equation (3.1.18)]. An even more general choice of the form

$$\frac{dt^+}{dt} = \omega_0(\tilde{t}) + \varepsilon\omega_1(\tilde{t}) + \varepsilon^2\omega_2(\tilde{t}) + \cdots$$

will be needed in problems to be discussed in Sections 3.4 and 3.5.

Now to apply the method, let

$$y(t^+; \varepsilon) = F_0(t^+, \tilde{t}) + \varepsilon F_1(t^+, \tilde{t}) + \cdots \tag{3.3.55}$$

and note that

$$\frac{dy}{dt^+} = \frac{\partial F_0}{\partial t^+} + \frac{\partial F_0}{\partial \tilde{t}}\frac{d\tilde{t}}{dt}\frac{dt}{dt^+} + \varepsilon\frac{\partial F_1}{\partial t^+} + \cdots = \frac{\partial F_0}{\partial t^+} + \varepsilon\left[\frac{1}{\mu(\tilde{t})}\frac{\partial F_0}{\partial \tilde{t}} + \frac{\partial F_1}{\partial t^+}\right] + \cdots \tag{3.3.56}$$

and

$$\frac{d^2y}{dt^{+2}} = \frac{\partial^2 F_0}{\partial t^{+2}} + \frac{2\varepsilon}{\mu}\frac{\partial^2 F_0}{\partial \tilde{t}\,\partial t^+} + \varepsilon\frac{\partial^2 F_1}{\partial t^{+2}} + \cdots. \tag{3.3.57}$$

Thus, the sequence of approximating equations is

$$\frac{\partial^2 F_0}{\partial t^{+2}} + F_0 = 0, \tag{3.3.58}$$

$$\frac{\partial^2 F_1}{\partial t^{+2}} + F_1 = -g(\tilde{t})\frac{\partial F_0}{\partial t^+} - \frac{2}{\mu(\tilde{t})}\frac{\partial^2 F_0}{\partial t^+\,\partial \tilde{t}}. \tag{3.3.59}$$

Using

$$F_0(t^+, \tilde{t}) = A_0(\tilde{t})\cos t^+ + B_0(\tilde{t})\sin t^+ \tag{3.3.60}$$

to describe the basic oscillation, we see that the right-hand side of Equation (3.3.59) is

$$\left(A_0 \frac{\mu'(\tilde{t})}{\mu^2(\tilde{t})} + \frac{2}{\mu(\tilde{t})}\frac{dA_0}{d\tilde{t}}\right)\sin t^+ - \left(B_0 \frac{\mu'(\tilde{t})}{\mu^2(\tilde{t})} + \frac{2}{\mu(\tilde{t})}\frac{dB_0}{d\tilde{t}}\right)\cos t^+.$$

These terms must be made to vanish for uniformity of the approximation, so that

$$A_0(\tilde{t}) = A_0(0)\sqrt{\frac{\mu(0)}{\mu(\tilde{t})}}, \qquad B_0(\tilde{t}) = B_0(0)\sqrt{\frac{\mu(0)}{\mu(\tilde{t})}}. \tag{3.3.61}$$

The general solution of the initial value problem is

$$F_0(t^+, \tilde{t}) = \sqrt{\frac{\mu(0)}{\mu(\tilde{t})}}\left\{a \cos t^+ + \frac{b}{\mu(0)} \sin t^+\right\}. \tag{3.3.62}$$

We can use the above explicit solution to verify the well-known result that $(1/2\mu)[(dy/dt)^2 + \mu^2(\tilde{t})y^2]$ is an adiabatic invariant to $O(1)$.

First, we note that there are many possible invariants to $O(1)$, since with $\varepsilon = 0$ Equation (3.3.47) is integrable. For example, if we denote the non-constant instantaneous energy by E:

$$E = \frac{1}{2}\left[\left(\frac{dy}{dt}\right)^2 + \mu^2 y^2\right]$$

then E, $E\mu$, etc. are all invariants to $O(1)$ because dE/dt, $d(E\mu)/dt$, etc. are all $O(\varepsilon)$. We reserve the term adiabatic (i.e., slowly varying) to an invariant whose *derivative is purely oscillatory on the fast scale.* As a consequence of this property, we expect that assuming an adiabatic invariant to be constant will, in some sense, introduce no cumulative errors as $t \to \infty$. A more precise definition of an adiabatic invariant is given in Section 3.7.3. To exhibit the oscillatory nature of $J \equiv E/\mu$ we calculate

$$\frac{dJ}{dt} = \frac{1}{\mu}\frac{dE}{dt} - \frac{\varepsilon}{\mu^2} E \frac{d\mu}{d\tilde{t}} = \frac{\varepsilon}{2}\frac{d\mu}{d\tilde{t}}\left[y^2 - \frac{1}{\mu^2}\left(\frac{dy}{dt}\right)^2\right].$$

Inserting the solution (3.3.62) to $O(1)$ in the above gives

$$\frac{dJ}{dt} = \varepsilon\mu(0)\left[a^2 - \frac{b^2}{\mu(0)^2}\right]\frac{1}{2\mu}\frac{d\mu}{d\tilde{t}} \cos 2t^+ + \varepsilon \frac{ab}{\mu}\frac{d\mu}{d\tilde{t}} \sin 2t^+ + O(\varepsilon^2)$$

which is oscillatory on the scale of t^+ with zero average value. On the other hand an invariant like E, for example, is less nice because

$$\frac{dE}{dt} = \varepsilon\mu \frac{d\mu}{d\tilde{t}} y^2$$

and inserting the solution to $O(1)$ shows that

$$\frac{dE}{dt} = \frac{\varepsilon\mu(0)}{2}\frac{d\mu}{d\tilde{t}}\left\{\left[a^2 + \frac{b^2}{\mu(0)^2}\right] + \frac{2ab}{\mu(0)}\sin 2t^+ + \left[a^2 - \frac{b^2}{\mu(0)^2}\right]\cos 2t^+\right\} + O(\varepsilon^2).$$

Thus, dE/dt has a nonoscillatory component to $O(\varepsilon)$ equal to

$$\frac{\varepsilon\mu(0)}{2}\frac{d\mu}{d\tilde{t}}\left(a^2 + \frac{b^2}{\mu(0)^2}\right)$$

and is not adiabatic.

A discussion based on the work of Reference 3.3.2 of how one can compute an adiabatic invariant to any given order for this problem is given in Section 3.7.3. An equivalent approach is also discussed in Section 3.7.2.

A practical problem leading to Equation (3.3.47) is the motion of a charged particle in a magnetic field almost homogeneous in space and varying slowly in time (cf. Reference 3.3.3). The equations of motion of a particle in the xy-plane, with a magnetic field $B(t)$ in the z-direction, are

$$m\frac{d^2x}{dt^2} = qB\frac{dy}{dt} + qE_x, \qquad m\frac{d^2y}{dt^2} = -qB\frac{dx}{dt} + qE_y, \qquad q = \text{charge}. \tag{3.3.63}$$

The Maxwell equation

$$\text{curl } E = -\frac{\partial B}{\partial t} \tag{3.3.64}$$

has the local solution

$$E_x = -\frac{y}{2}\frac{dB}{dt} + \cdots, \qquad E_y = \frac{x}{2}\frac{dB}{dt} + \cdots, \tag{3.3.65}$$

which can be used in Equation (3.3.63) near the origin. Letting

$$u = x + iy, \tag{3.3.66}$$

we find that the system (Equation (3.3.63)) becomes

$$\frac{d^2u}{dt^2} + i\omega\frac{du}{dt} + \frac{i}{2}\frac{d\omega}{dt}u = 0, \tag{3.3.67}$$

where

$$\omega(t) = \frac{qB(t)}{m}$$

is the cyclotron frequency. For $B = \text{const.}$, the particle motion is a circular orbit about the origin with this frequency. Now, if we introduce the amplitude $\phi(t)$ by

$$u = x + iy = \phi(t)\exp\left[-(i/2)\int_0^t \omega\,dt\right], \tag{3.3.68}$$

$\phi(t)$ satisfies

$$\frac{d^2\phi}{dt^2} + \frac{\omega^2(t)}{4}\phi = 0. \tag{3.3.69}$$

The result (Equation (3.3.62)) shows precisely in what sense the magnetic moment proportional to $\phi_{max}\sqrt{\omega}$ is constant.

3.3.3 Sturm–Liouville Equation; Differential Equation with a Large Parameter

The classical problem of the approximate solution of a differential equation with a large parameter (cf. Equation (3.3.70) below) falls naturally into the discussion of this chapter. The usual asymptotic expansion valid away from turning points turns out not to be a limit process expansion but, rather one of the two-variable type. However, near a turning point, the local behavior dominates, so that a limit-process expansion, valid locally, can be constructed. These two expansions, however, can be matched by the same procedure used for purely limit-process expansions. This extension of our previous ideas should prove useful for many similar problems.

For the consideration of asymptotic distribution of eigenvalues and eigenfunctions and for various other reasons, it is often necessary to obtain the asymptotic behavior of the solutions to the general self-adjoint second-order equation[6]

$$\frac{d}{d\tilde{x}}\left(p(\tilde{x})\frac{dy}{d\tilde{x}}\right) + [\lambda q(\tilde{x}) - r(\tilde{x})]y = 0 \tag{3.3.70}$$

as $\lambda \to \infty$.

A standard method is the transformation (Liouville, Green, 1837) of Equation (3.3.70) to an equation of canonical type by the introduction of

$$y(\tilde{x}) = f(\tilde{x})w(z), \qquad z = g(\tilde{x}). \tag{3.3.71}$$

Over an interval of $\tilde{x}$ where p, q individually have one sign (say positive) Equation (3.3.70) is transformed to

$$\frac{d^2w}{dz^2} + \lambda w = \phi(z)w \tag{3.3.72}$$

[6] The choice of $\tilde{x}$ for independent variable is for consistency of notation and will become evident later on.

by a suitable choice of (f, g). For large λ, the right-hand side of Equation (3.3.72) makes a small contribution that can be estimated by iteration (cf. Reference 3.3.4). Since for $\lambda \to \infty$ the solutions of Equation (3.3.72) have the form of slowly varying oscillations, it is natural to expect the two-variable procedure to apply to this part of the problem. Besides having a certain unity with what has gone before, the two-variable method has the advantage that higher approximations are more easily calculated.

When the original Equation (3.3.70) has a simple turning point ($q(\tilde{x}) = 0$), the extension of the previous method uses a comparison equation,

$$\frac{d^2w}{dz^2} + \lambda z w = \psi(z)w, \tag{3.3.73}$$

which gives the results of the WKBJ ... method. The procedure here is different. A local expansion valid near the turning point is constructed and matched to the expansions valid away from the turning point.

The method used here is similar to that of the previous Section (adiabatic invariants), of a slowly varying frequency. The equation is transformed to the form for an oscillator of constant frequency and small damping. A fast variable is $x = \sqrt{\lambda}\tilde{x}$, and $\tilde{x}$ itself is a slow variable. Thus, Equation (3.3.70) can be written

$$\frac{d^2y}{dx^2} + \varepsilon \frac{dp/d\tilde{x}}{p(\tilde{x})}\frac{dy}{dx} + \left\{\frac{q(\tilde{x})}{p(\tilde{x})} - \varepsilon^2 \frac{r(\tilde{x})}{p(\tilde{x})}\right\} y = 0, \tag{3.3.74}$$

where $\varepsilon = 1/\sqrt{\lambda} \to 0$. Consider, first, Equation (3.3.74) over an interval $(0 < \tilde{x} < 1)$, where $p, q > 0$, and consider also that $\lambda > 0$, as would be typical for an eigenvalue problem. Then, introduce a new fast variable x^+,

$$x^+ = \psi(x), \tag{3.3.75}$$

in order to bring Equation (3.3.74) to the desired form. We have

$$\psi'^2(x)\frac{d^2y}{dx^{+2}} + \psi''(x)\frac{dy}{dx^+} + \varepsilon \frac{dp/d\tilde{x}}{p(\tilde{x})}\psi'(x)\frac{dy}{dx^+} + \left\{\frac{q(\tilde{x})}{p(\tilde{x})} - \varepsilon^2 \frac{r(\tilde{x})}{p(\tilde{x})}\right\} y = 0, \tag{3.3.76}$$

where primes denote derivatives with respect to x. It is clear that $\psi'(x)$ should be chosen so that

$$\psi'(x) = \sqrt{\frac{q(\tilde{x})}{p(\tilde{x})}} = O(1), \tag{3.3.77}$$

and the relationship of the new fast variable x^+ to the slow variable is

$$x^+ = \int_0^x \sqrt{\frac{q(\varepsilon z)}{p(\varepsilon z)}}\,dz = \frac{1}{\varepsilon}\int_0^{\tilde{x}} \sqrt{\frac{q(\xi)}{p(\xi)}}\,d\xi. \tag{3.3.78}$$

With this choice, we have

$$\psi''(x) = \frac{\varepsilon}{2}\sqrt{\frac{p}{q}}\left[\frac{1}{p}\frac{dq}{d\tilde{x}} - \frac{q}{p^2}\frac{dp}{d\tilde{x}}\right], \tag{3.3.79}$$

and Equation (3.3.74) is

$$\frac{d^2y}{dx^{+2}} + \varepsilon f(\tilde{x})\frac{dy}{dx^+} + \left\{1 - \varepsilon^2\frac{r(\tilde{x})}{p^2(\tilde{x})q(\tilde{x})}\right\}y = 0, \tag{3.3.80}$$

where

$$f(\tilde{x}) = \frac{1}{2}\frac{p^{1/2}}{q^{3/2}}\frac{dq}{d\tilde{x}} + \frac{1}{2}\frac{dp/d\tilde{x}}{p^{1/2}q^{1/2}}. \tag{3.3.81}$$

Now, for the two-variable expansion, we assume an expansion of the type

$$y(x^+;\varepsilon) = F_0(x^+, \tilde{x}) + \varepsilon F_1(x^+, \tilde{x}) + O(\varepsilon^2), \tag{3.3.82}$$

$$\frac{dy}{dx^+} = \frac{\partial F_0}{\partial x^+} + \frac{d\tilde{x}}{dx^+}\frac{\partial F_0}{\partial \tilde{x}} + \varepsilon\frac{\partial F_1}{\partial x^+} + O(\varepsilon^2), \tag{3.3.83}$$

$$\frac{d^2y}{dx^{+2}} = \frac{\partial^2 F_0}{\partial x^{+2}} + 2\frac{d\tilde{x}}{dx^+}\frac{\partial^2 F_0}{\partial x^+\,\partial\tilde{x}} + \varepsilon\frac{\partial^2 F_1}{\partial x^{+2}} + O(\varepsilon^2). \tag{3.3.84}$$

Note, from Equation (3.3.78), that

$$\frac{d\tilde{x}}{dx^+} = \varepsilon\sqrt{\frac{p(\tilde{x})}{q(\tilde{x})}}, \tag{3.3.85}$$

and $d^2\tilde{x}/dx^{+2} = O(\varepsilon^2)$. Thus, from Equation (3.3.80), the first two approximate equations are

$$O(1): \frac{\partial^2 F_0}{\partial x^{+2}} + F_0 = 0, \tag{3.3.86}$$

$$O(\varepsilon): \frac{\partial^2 F_1}{\partial x^{+2}} + F_1 = -2\sqrt{\frac{p}{q}}\frac{\partial^2 F_0}{\partial x^+\,\partial\tilde{x}} - f(\tilde{x})\frac{\partial F_0}{\partial x^+}. \tag{3.3.87}$$

Thus, we have

$$F_0(x^+, \tilde{x}) = A_0(\tilde{x})\cos x^+ + B_0(\tilde{x})\sin x^+. \tag{3.3.88}$$

Using the same argument as before, namely that fast growth (x^+ scale) is not permitted, we obtain differential equations for A_0, B_0 from the right-hand side of Equation (3.3.87). We have

$$\frac{\partial^2 F_1}{\partial x^{+2}} + F_1 = -2\sqrt{\frac{p}{q}}\left\{-\frac{dA_0}{d\tilde{x}}\sin x^+ + \frac{dB_0}{d\tilde{x}}\cos x^+\right\}$$
$$- f(\tilde{x})\{-A_0\sin x^+ + B_0\cos x^+\}. \tag{3.3.89}$$

Here A_0, B_0 satisfy the same differential equation (using the definition of f in Equation (3.3.81)):

$$2\frac{dA_0}{d\tilde{x}} + \frac{1}{2}\left(\frac{1}{q}\frac{dq}{d\tilde{x}} + \frac{1}{p}\frac{dp}{d\tilde{x}}\right)A_0 = 0. \tag{3.3.90}$$

The general solution of Equation (3.3.90) is, thus,

$$A_0(\tilde{x}) = a_0[p(\tilde{x})q(\tilde{x})]^{-1/4} \tag{3.3.91}$$

and the first approximation is the same as that usually found,

$$y(x^+; \varepsilon) = \frac{a_0 \cos x^+ + b_0 \sin x^+}{[p(\tilde{x})q(\tilde{x})]^{1/4}} + \varepsilon F_1(x^+, \tilde{x}) + \cdots, \tag{3.3.92}$$

where

$$F_1 = A_1(\tilde{x})\cos x^+ + B_1(\tilde{x})\sin x^+$$

and

$$x^+ = \frac{1}{\varepsilon}\int_0^{\tilde{x}} \sqrt{\frac{q(\xi)}{p(\xi)}}\, d\xi.$$

The same formulation can also be applied to an interval in which $p > 0$, $q < 0$, but $\lambda > 0$, and the sin and cos are replaced by exponential functions. In such a region, we have

$$y(x^+; \varepsilon) = \frac{c_0 e^{-x^+} + d_0 e^{x^+}}{[-p(\tilde{x})q(\tilde{x})]^{1/4}} + \varepsilon F_1(x^+, \tilde{x}) + \cdots, \tag{3.3.93}$$

where

$$x^+ = \frac{1}{\varepsilon}\int_0^{\tilde{x}} \sqrt{\frac{-q(\xi)}{p(\xi)}}\, d\xi.$$

Next we consider the behavior of the original Equation (3.3.70) near a simple turning point (say $\tilde{x} = 0$), where

$$q(\tilde{x}) = \alpha\tilde{x} + \cdots, \qquad p(\tilde{x}) = \beta + \beta_1\tilde{x} + \cdots, \qquad r(\tilde{x}) = \rho + \cdots. \tag{3.3.94}$$

The idea is to construct a limit-process expansion valid near $\tilde{x} = 0$. Introduce

$$x^* = \frac{\tilde{x}}{\delta(\varepsilon)}, \tag{3.3.95}$$

and consider an expansion procedure in which x^* is fixed and $\delta, \varepsilon \to 0$. The form of the expansion is

$$y(x^+; \varepsilon) = \sigma(\varepsilon)g(x^*) + \sigma_1(\varepsilon)g_1(x^*) + \cdots, \tag{3.3.96}$$

so that Equation (3.3.70) becomes

$$\varepsilon^2 \frac{p(\delta x^*)}{\delta^2}\frac{d^2 g}{dx^{*2}} + \varepsilon^2 \frac{p'(\delta x^*)}{\delta}\frac{dg}{dx^*} + [q(\delta x^*) - \varepsilon^2 r(\delta x^*)]g + \cdots = 0. \tag{3.3.97}$$

The dominant terms are

$$\frac{\varepsilon^2}{\delta^2}\beta\frac{d^2g}{dx^{*2}} + \delta(\varepsilon)\alpha x^* g = 0, \tag{3.3.98}$$

so that both terms are of the same order if $\varepsilon^2/\delta^2 = \delta$ or

$$\delta(\varepsilon) = \varepsilon^{2/3}. \tag{3.3.99}$$

Thus, the basic turning-point equation is obtained for $g(x^*)$,

$$\frac{d^2g}{dx^{*2}} + k^2x^*g = 0, \qquad k^2 = \frac{\alpha}{\beta}. \tag{3.3.100}$$

The problem is thus reduced to knowing the properties of the solution of Equation (3.3.100), and, since these solutions are expressed in terms of Airy functions or ordinary Bessel functions, further progress toward matching can be made. The general solution of Equation (3.3.100) can be written

$$g(x^*) = \sqrt{x^*}\{CJ_{-1/3}(\tfrac{2}{3}kx^{*3/2}) + DJ_{1/3}(\tfrac{2}{3}kx^{*3/2})\}, \tag{3.3.101}$$

with the corresponding analytic continuation to $x^* < 0$. The function $g(x^*)$ is well behaved at $x^* = 0$, the D term in Equation (3.3.101) varies like x^* and the C term is constant. (No branches!)

The two-variable expansion (Equation 3.3.92) is valid in some region $\tilde{x} > \tilde{x}_0$ excluding the turning point. For the matching of Equations (3.3.92) and (3.3.96), consider an intermediate limit x_η fixed, where

$$x_\eta = \frac{\tilde{x}}{\eta(\varepsilon)} \quad \text{such that } \eta \to 0, \frac{\eta}{\varepsilon^{2/3}} \to \infty. \tag{3.3.102}$$

Thus, we have

$$\tilde{x} = \eta x_\eta \to 0, \qquad x^* = \frac{\eta x_\eta}{\varepsilon^{2/3}} \to \infty, \tag{3.3.103}$$

and

$$x^+ = \frac{1}{\varepsilon}\int_0^{\eta x_\eta}\sqrt{\frac{q(\xi)}{p(\xi)}}\,d\xi \to \frac{1}{\varepsilon}\sqrt{\frac{\alpha}{\beta}}\int_0^{\eta x_\eta}\sqrt{\xi}\,d\xi + \cdots, \tag{3.3.104}$$

$$x^+ \to \frac{2}{3}\sqrt{\frac{\alpha}{\beta}}\frac{(\eta x_\eta)^{3/2}}{\varepsilon} + \cdots.$$

The two-variable expansion should contain all limit-process expansions valid in restricted x neighborhoods and so be able to be matched to Equation (3.3.101). Now the behavior of Equation (3.3.101) for large x^* is necessary for the matching, but the usual asymptotic expressions for $J_\nu(z)$ can be used:

$$J_\nu(z) = \sqrt{\frac{2}{\pi z}}\cos\left(z - \frac{\nu\pi}{2} - \frac{\pi}{4}\right) + O\left(\frac{1}{z}\right). \tag{3.3.105}$$

Thus, in terms of intermediate variables, we have the following.

Transition

$$y(x^+;\varepsilon) = \sigma(\varepsilon)\left\{\sqrt{\frac{3}{\pi k}}\frac{\varepsilon^{1/6}}{(\eta x_\eta)^{1/4}}\right\}\left\{C\cos\left(\frac{2}{3}k\frac{(\eta x_\eta)^{3/2}}{\varepsilon}-\frac{\pi}{12}\right)\right.$$
$$\left.+ D\cos\left(\frac{2}{3}k\frac{(\eta x_\eta)^{3/2}}{\varepsilon}-\frac{5\pi}{32}\right)\right\} + \cdots.$$

Two-Variable

$$y(x^+;\varepsilon) = \frac{a_0\cos(\frac{2}{3}\sqrt{\alpha/\beta}(\eta x_\eta)^{3/2}/\varepsilon) + b_0\sin(\frac{2}{3}\sqrt{\alpha/\beta}(\eta x_\eta)^{3/2}/\varepsilon)}{(\beta\alpha)^{1/4}(\eta x_\eta)^{1/4}}$$
$$+ \cdots + \varepsilon F_1 + \cdots.$$

Since $k = \sqrt{\alpha/\beta}$, it is seen that the dominant terms in these two expansions are matched if

$$\sigma(\varepsilon) = \varepsilon^{-1/6} \tag{3.3.106}$$

and, further, if

$$\sqrt{\frac{3}{\pi}}\beta^{1/4}\left\{C\cos\frac{\pi}{12} + D\cos\frac{5\pi}{32}\right\} = \frac{a_0}{\beta^{1/4}}, \tag{3.3.107}$$

$$\sqrt{\frac{3}{\pi}}\beta^{1/4}\left\{C\sin\frac{\pi}{12} + D\sin\frac{5\pi}{12}\right\} = \frac{b_0}{\beta^{1/4}}. \tag{3.3.108}$$

Equations (3.3.107) and (3.3.108) provide the basic relations for the constants in the solution. A uniformly valid first approximation including the transition point could be written down by adding Equations (3.3.101) and (3.3.92) and subtracting the common part. Further, the same procedure can be applied as $x^* \to -\infty$, so that ultimately the relationships between c_0, d_0 of Equation (3.3.93) and a_0, b_0, which provide the analytic continuation, are found.

When an eigenvalue problem is being considered, the asymptotic formulas are used, and the set of ε is determined by consideration of the homogeneous boundary conditions.

Problems

1. Consider the classical problem of beats for a linear oscillator in which the driver frequency ω is close to the natural frequency ω_N:

$$\frac{d^2y}{dt^2} + \omega_N^2 y = F_0\cos\omega t.$$

Here F_0 is a constant and the small parameter of the problem is

$$\varepsilon = \frac{\omega_N - \omega}{\omega_N}.$$

Nondimensionalize the problem and first solve it exactly for arbitrary initial values of y and dy/dt. Then compare this with the result of the formal application of the two-variable method.

2. Consider

$$\frac{d^2y}{dt^2} + (\tfrac{1}{4} + \varepsilon\delta_1 + \varepsilon^2\delta_2 + \varepsilon\cos t)y = 0$$

with the initial conditions $y(0) = a$, $\dot{y}(0) = b$. Assume that the solution is of the form

$$y = F_0(t, \tilde{t}) + \varepsilon F_1(t, \tilde{t}) + \varepsilon^2 F_2(t, \tilde{t}) + \cdots, \tilde{t} = \varepsilon t.$$

(a) The solution for F_0 was worked out in Section 3.3.1. Give the solution for F_1.
(b) Determine the stability boundaries (δ_2) to order ε^2.
(c) Determine, by a limiting procedure, the solution right on the stability boundaries. In particular, calculate the normalized solutions y_1 and y_2 satisfying the initial conditions $y_1(0) = 1$, $\dot{y}_1(0) = 0$, $y_2(0) = 0$, $\dot{y}_2(0) = 1$.
(d) With y_1 and y_2 as above, verify that the discriminant D, defined by $D = y_1(2\pi) + \dot{y}_2(2\pi)$, is equal to $-2 + O(\varepsilon^2)$. Also construct solutions y_1^*, y_2^* such that $y_1^*(t + 2\pi) = -y_1^*(t)$, $y_2^*(t + 2\pi) = -y_2^*(t) + y_1^*(t)$. Verify the general theorem that an arbitrary solution y has the property

$$y(t + 2\pi) = -y(t) + \theta p(t),$$

where θ is a constant and $p(t)$ is periodic.

3. Consider the system

$$\frac{d^2y}{dt^2} + \mu^2 y = 0$$

$$\frac{d\mu}{dt} = \varepsilon f(y, \mu)$$

analogous to the example discussed in Section 3.3.2 except that now μ is not given explicitly but is a slowly varying dependent variable coupled with y.

Make an exact change of variable from t to t^+ by setting

$$\frac{dt^+}{dt} = \mu$$

then solve the resulting system using a two variable expansion with t^+ and εt^+ as the fast and slow variables. Carry out the solution explicitly for the case $f = y^2$. In Section 3.5 an alternate approach will be presented for solving this type of problem.

4. Carry out the solution of the example of Section 3.3.2 to $O(\varepsilon)$ and deduce the form of the adiabatic invariant to $O(\varepsilon)$.

5. What is the effect of a weak nonlinear damping on the solution of the example of Section 3.3.2? For example, study the problem

$$\frac{d^2y}{dt^2} + \mu^2(\tilde{t})y + \varepsilon\left(\frac{dy}{dt}\right)^3 = 0.$$

Is there an adiabatic invariant for this problem?

6. We wish to examine whether the subharmonic response to the forced Duffing equation survives in the presence of damping. Study

$$\frac{d^2y}{dt^2} + y + \varepsilon\left(y^3 + \beta\frac{dy}{dt}\right) = \varepsilon\cos\Omega(\varepsilon)t$$

$$y(0) = a$$

$$\dot{y}(0) = b$$

$$\Omega = 3 + \varepsilon\omega_1 + \cdots,$$

where $\beta, a, b, \omega_1, \omega_2, \ldots$ are arbitrary constants independent of ε.

Choosing $t^+ = (3 + \varepsilon^2\omega_2 + \cdots)t$ and $\tilde{t} = \varepsilon t$ as fast and slow variables construct the solution to $O(1)$ and show that for certain values of a and b the steady-state solution is indeed a subharmonic oscillation.

7. A model equation for a resonance phenomenon which arises in celestial mechanics is given by

$$\frac{d^2y}{dt^2} + y + 2\varepsilon y(1 - 5\cos^2 R) = \varepsilon^2 R\cos t,$$

where $R^2 = y^2 + (dy/dt)^2$. The initial conditions are

$$y(0) = a\cos b$$

$$\frac{dy(0)}{dt} = a\sin b,$$

where a and b are constants independent of ε.

(a) Construct a two variable expansion for the solution in the form

$$y = F_0(t, \tilde{t}) + \varepsilon F_1(t, \tilde{t}) + \cdots$$

and show that the procedure works routinely [carry out the details of the solution to $O(\varepsilon)$] as long as $s \equiv 5\cos^2 a - 1 \neq 0$.

(b) Since s appears as a divisor in the solution for F_1, the procedure fails for initial values of a such that $s \approx 0$. Show that for s small it is incorrect to neglect the forcing term $\varepsilon^2 R\cos t$ in determining the solution for F_0. Retain this term formally in the equations governing the slowly varying amplitude and phase of F_0 and deduce the structure of the solution when $s \approx 0$.

(c) Guided by the results in (b) develop the solution when s is small in terms of appropriate time variables, to order ε. In particular show that the slow variable is now $\varepsilon^{3/2}t$, that the expansion should proceed in powers of $\varepsilon^{1/2}$ and that we must set $s = \varepsilon^{1/2}\bar{s}$ for some fixed $\bar{s}$ as $\varepsilon \to 0$.

(d) Match the solutions in (a) and (c) in some common overlap *domain in s*, and derive a result which is uniformly valid for all s to $O(\varepsilon)$.

References

3.3.1 N. W. McLachlan, *Theory and Application of Mathieu Functions*, Clarendon Press, Oxford, 1947.

3.3.2 C. S. Gardner, Adiabatic Invariants of Periodic Classical Systems, *Physics Reviews* **115**, 1959, pp. 791–794.

3.3.3 L. F. J. Broer and L. Wijngaarden, On the Motion of a Charged Particle in an Almost Homogeneous Magnetic Field, *Applied Scientific Research B* **8**, No. 3, 1960, pp. 159–176.

3.3.4 A. Erdelyi, *Asymptotic Expansions*, Dover Publications, New York, 1956.

3.4 Applications to Satellite Problems

The development of theories to describe adequately the motion of celestial bodies has been intimately tied with advances in mathematics from Newton's time to the turn of the century when Poincaré introduced our present notions of asymptotic expansions in connection with his works in celestial mechanics. More recently, the advent of artificial earth satellites, manned space flight, and interplanetary orbits has revitalized and broadened this area of study which until 1958 had almost exclusively been in the hands of a small number of celestial mechanicians.

In Section 2.7.1 we considered a simple illustrative model where different asymptotic expansions permit the description of a trajectory which passes close to two centers of gravity. The use of singular perturbations in earth-to-moon trajectories is an example wherein techniques developed in one field are successful in solving a long-standing problem in another field. In this section we study another class of problems, where the multiple variable procedure is very well adapted.

We are concerned with a class of motions which remain in a bounded region surrounding a gravitational center. These are called satellite motions and are distinguished by the fact that the dominant force is a spherically symmetric Newtonian gravitation perturbed by small effects. To account adequately for the cumulative effect of these small terms we need to use a multiple variable expansion.

The intimate connection between a weakly nonlinear oscillator and an almost Keplerian orbit was first noticed by Laplace (Reference 3.4.1), and it is a generalization of this idea which we will use to cast satellite problems into the same mathematical mold as the previous examples discussed in this chapter.

3.4.1 Satellite Equations in Local Orbital Plane

To fix ideas consider a planar motion which can be conveniently defined using the polar coordinates (R, θ) as follows

$$m\left[\frac{d^2R}{dT^2} - R\left(\frac{d\theta}{dT}\right)^2\right] = -\frac{GMm}{R^2} + F_0 f_1 \tag{3.4.1a}$$

$$m\left(R\frac{d^2\theta}{dT^2} + 2\frac{dR}{dT}\frac{d\theta}{dT}\right) = F_0 f_2. \tag{3.4.1b}$$

Here m is the mass of the satellite, and we recognize the left-hand sides of (3.4.1) as the mass times the acceleration expressed in radial (R) and tangential (θ) components. The term GmM/R^2 is the dominant gravitational attraction due to the planet of mass M located at $R = 0$ and G is the universal gravitational constant. The terms multiplied by the characteristic force F_0, which is intended to be small compared to the central gravitational force, are the radial and tangential perturbations. In general f_1 and f_2 may depend on both coordinates and their derivatives.

We introduce dimensionless variables by choosing some characteristic length L (say the initial value of R) and normalize the time by the characteristic time $(L^3/GM)^{1/2}$ and obtain the following equations in terms of $r = R/L$, θ, and $t = T/(L^3/GM)^{1/2}$.

$$\ddot{r} - r\dot{\theta}^2 = -\frac{1}{r^2} + \varepsilon f_1(r, \theta, \dot{r}, \dot{\theta}) \tag{3.4.2a}$$

$$r\ddot{\theta} + 2\dot{r}\dot{\theta} = \varepsilon f_2(r, \theta, \dot{r}, \dot{\theta}), \tag{3.4.2b}$$

where ε is the ratio of the small perturbation force F_0 to the gravitational force

$$\varepsilon = \frac{F_0}{(GmM/L^2)} \ll 1. \tag{3.4.3}$$

The general solution of (3.4.2) for $\varepsilon = 0$ can be found in any standard text in dynamics (e.g., Reference 3.4.2). This solution can be obtained by quadrature once we note that the nonlinear system (3.4.2) has the two integrals

$$r^2\dot{\theta} = l = \text{const.} = \text{angular momentum} \tag{3.4.4a}$$

$$\frac{1}{2}(\dot{r}^2 + r^2\dot{\theta}^2) - \frac{1}{r} = E = \text{const.} = \text{energy}. \tag{3.4.4b}$$

Here, we are only interested in the case of bounded orbits ($E < 0$) and this class of solutions is most conveniently expressed by the equation of the orbit

$$r = \frac{a(1 - e^2)}{1 + e\cos(\theta - \omega)} \tag{3.4.5a}$$

and Kepler's equation for the time history

$$t - \tau = a^{3/2}(\Phi - e\sin\Phi). \tag{3.4.5b}$$

The four constants of integration are the semi-major axis a, the eccentricity e, the argument of pericenter ω and the time of passage through pericenter τ. The relation between a, e and l, E is

$$l = \sqrt{a(1 - e^2)} \tag{3.4.6a}$$

$$E = -\frac{1}{2a} \tag{3.4.6b}$$

and the orbit is sketched in Figure 3.4.1.

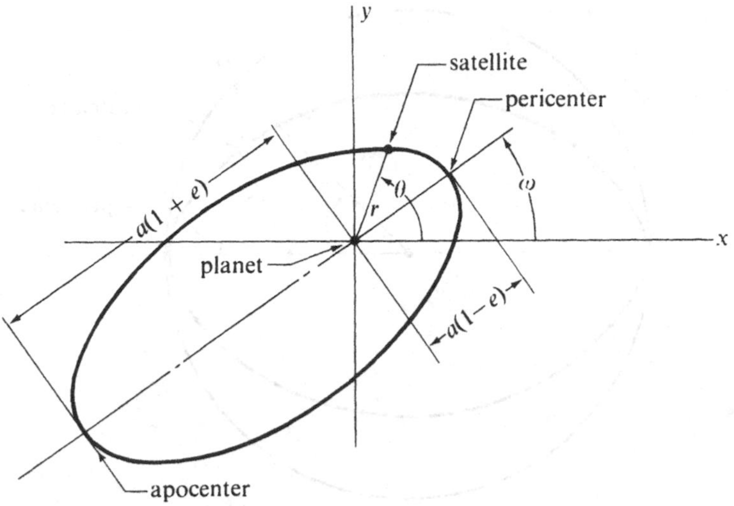

Figure 3.4.1 Kepler Ellipse

The time history of the particle in its orbit (3.4.5b) is compactly written using the so called "eccentric anomaly" Φ which is an angle related to θ according to

$$\sin \Phi = \frac{\sqrt{1 - e^2}\sin(\theta - \omega)}{1 + e\cos(\theta - \omega)}. \tag{3.4.7}$$

It has a simple geometrical interpretation derived by enclosing the ellipse inside a circle of radius a and drawing a line normal to the major axis of the ellipse from the particle to the circle. If we denote the intersection of this normal with the circle by Q, then Φ is the angle between the pericenter and Q as shown in Figure 3.4.2.

The principal result in the foregoing is the structure of (3.4.5a). We note that $u = 1/r$ is a harmonic function of θ. This observation led Laplace to propose the transformation of variables from r and θ in terms of t, to u and t in terms of θ. If we perform this transformation on (3.4.2) we find ($' = d/d\theta$)

$$u'' + u - u^4 t'^2 = -\varepsilon u^2 t'^2 \left(f_1 + \frac{u'}{u} f_2 \right) \tag{3.4.8a}$$

$$(u^2 t')' = -\varepsilon u^3 t'^3 f_2. \tag{3.4.8b}$$

Thus, if $\varepsilon = 0$, $u^2 t' = 1/l^2 = \text{const.}$, and we recover the solution (3.4.5). With $\varepsilon \neq 0$, u obeys the equation of a weakly nonlinear oscillator and this type of problem is well adapted for solution by the multiple variable method.

Actually, we can also handle three dimensional orbits in the same way by referring the motion to a local orbital plane determined by the instantaneous displacement and velocity vectors. This choice of variables was proposed

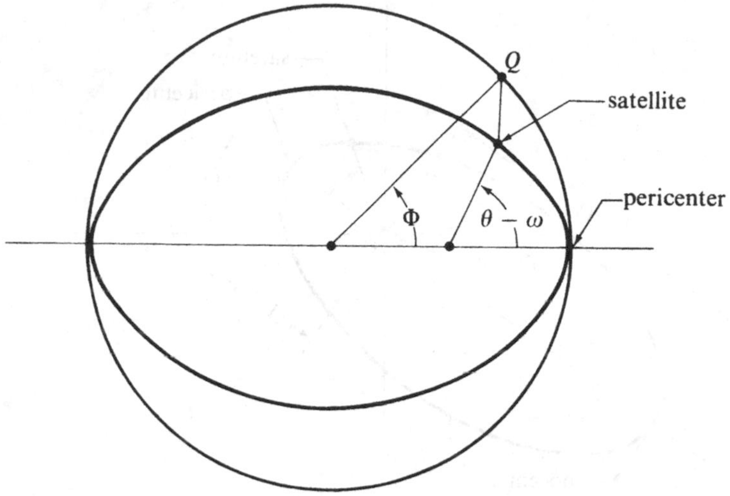

Figure 3.4.2 Eccentric Anomaly

by Struble in Reference 3.4.3 and the derivation of equations of motion for arbitrary three dimensional perturbations is given in Reference 3.4.4 in a form suitable for applying the multiple variable expansion technique. In this section we will only consider planar orbits for simplicity. More complicated examples of non-planar satellite problems using the multiple variable method can be found in Reference 3.4.5, 3.4.6 and 3.4.7.

3.4.2 Decay of Orbit Due to Drag

Consider the very idealised model of a spherical (hence non-lifting) satellite in orbit around a Newtonian gravitational center and perturbed by a thin, constant density atmosphere. Actually, a more realistic model of atmospheric density variation with altitude can also be used at the cost of some algebraic complexity. However, the main qualitative features of the solution are present in this simple model.

Since the drag force acts in the direction opposite to the velocity, we have the following equations of motion

$$m\frac{d^2R}{dT^2} - mR\left(\frac{d\theta}{dt}\right)^2 = -\frac{GmM}{R^2} - D\sin\gamma \tag{3.4.9a}$$

$$mR\frac{d^2\theta}{dT^2} + 2m\frac{dR}{dT}\frac{d\theta}{dT} = -D\cos\gamma, \tag{3.4.9b}$$

where D is the magnitude of the drag force and γ is the flight path angle as shown in Figure 3.4.3.

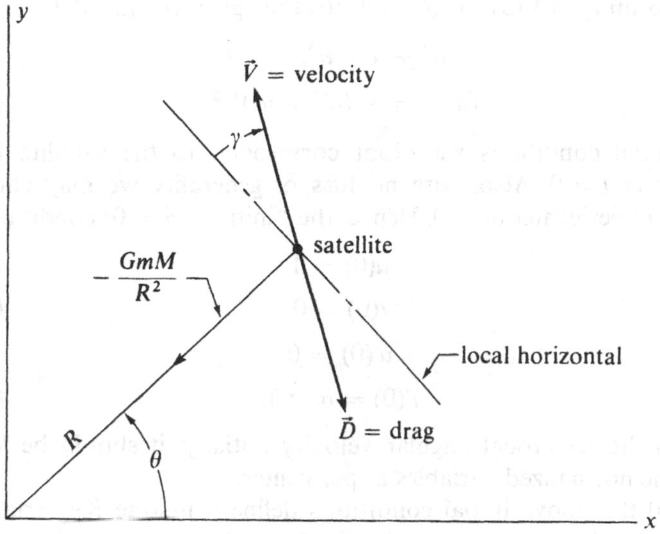

Figure 3.4.3 Drag on Satellite

Now D is given by

$$D = \tfrac{1}{2}\rho V^2 S C_D, \tag{3.4.10}$$

where ρ is the constant atmospheric density, S is the cross sectional area of the satellite, and C_D is the drag coefficient which we also assume to be constant. The magnitude of the velocity in polar coordinates is

$$V = \left[\left(\frac{dR}{dT}\right)^2 + R^2\left(\frac{d\theta}{dt}\right)^2\right]^{1/2} \tag{3.4.11}$$

and the flight path angle γ is simply

$$\gamma = \tan^{-1}\frac{dR/dT}{R\,d\theta/dT}. \tag{3.4.12}$$

If we use the dimensionless variables of the previous section with R normalized by R_0, its initial value, we obtain

$$\ddot{r} - r\dot{\theta}^2 = -\frac{1}{r^2} - \varepsilon\dot{r}(\dot{r}^2 + r^2\dot{\theta}^2)^{1/2} \tag{3.4.13a}$$

$$r\ddot{\theta} + 2\dot{r}\dot{\theta} = -\varepsilon r\dot{\theta}(\dot{r}^2 + r^2\dot{\theta}^2)^{1/2}, \tag{3.4.13b}$$

where the small parameter

$$\varepsilon = \frac{C_D \rho S R_0}{2m} \tag{3.4.14}$$

is the ratio of the drag force to the centrifugal force acting on the satellite.

Transforming (3.4.13) to $u(\theta)$ and $t(\theta)$ then gives [cf. (3.4.8)]

$$u'' + u - u^4 t'^2 = 0 \tag{3.4.15a}$$

$$(u^2 t')' = \varepsilon t'(u'^2 + u^2)^{1/2}. \tag{3.4.15b}$$

The initial conditions we adopt correspond to the satellite being at pericenter at $t = 0$. Also, with no loss of generality we may choose the argument of pericenter $\omega = 0$. Hence, the "initial" ($\theta = 0$) conditions are

$$u(0) = 1 \tag{3.4.16a}$$

$$t(0) = 0 \tag{3.4.16b}$$

$$u'(0) = 0 \tag{3.4.16c}$$

$$t'(0) = \sigma < 1. \tag{3.4.16d}$$

Since σ is the reciprocal angular velocity initially, it should be less than unity in the normalized variables at pericenter.

If $\varepsilon = 0$ the above initial conditions define a unique Keplerian ellipse with constant elements a, e, ω and τ. With $\varepsilon \neq 0$ the motion is more complicated. However to $O(1)$ it will still be in the form of a Keplerian orbit [cf. Equations 3.4.23] but with slowly varying elements. Therefore, it is convenient to express the initial conditions (3.4.16) in terms of equivalent conditions on the initial values of a, e, ω and τ.

First, we note that since the motion starts from the pericenter, Kepler's equation, Equation (3.4.5b), implies that

$$\tau(0) = 0. \tag{3.4.17a}$$

Moreover, we have chosen

$$\omega(0) = 0. \tag{3.4.17b}$$

Since the pericenter distance is $a(1 - e)$, Equation (3.4.16a) gives the following condition

$$a(0)[1 - e(0)] = 1. \tag{3.4.17c}$$

Finally, differentiating Kepler's Equation (3.4.5b) with respect to θ and using the definition (3.4.7) for the eccentric anomaly gives

$$\sigma = \frac{1}{\sqrt{1 + e(0)}}. \tag{3.4.17d}$$

Solving (3.4.17) gives

$$a(0) = \frac{\sigma^2}{2\sigma^2 - 1} \equiv a_0 \tag{3.4.18a}$$

$$e(0) = \frac{1 - \sigma^2}{\sigma^2} \equiv e_0 < 1 \tag{3.4.18b}$$

$$\tau(0) = 0 \tag{3.4.18c}$$

$$\omega(0) = 0. \tag{3.4.18d}$$

We note that the differential Equations (3.4.15) only involve t' and that the time t does not occur explicitly. Hence, we develop u and t' in two variable form as follows

$$u(\theta, \varepsilon) = u_0(\theta, \tilde{\theta}) + \varepsilon u_1(\theta, \tilde{\theta}) + \cdots \tag{3.4.19a}$$

$$t'(\theta, \varepsilon) = v_0(\theta, \tilde{\theta}) + \varepsilon v_1(\theta, \tilde{\theta}) + \cdots, \tag{3.4.19b}$$

where, as usual, $\tilde{\theta} = \varepsilon\theta$. Once the expansion (3.4.19b) for t' is known, t can be calculated by quadrature.

Substituting (3.4.19) and the corresponding formulas for the derivatives into (3.4.15) gives

$$\frac{\partial^2 u_0}{\partial\theta^2} + u_0 = u_0^4 v_0^2 \tag{3.4.20a}$$

$$u_0^2 \frac{\partial v_0}{\partial\theta} + 2u_0 \frac{\partial u_0}{\partial\theta} v_0 = 0 \tag{3.4.20b}$$

$$\frac{\partial^2 u_1}{\partial\theta^2} + u_1 = -2\frac{\partial^2 u_0}{\partial\theta\,\partial\tilde{\theta}} + 2u_0^4 v_0 v_1 + 4u_0^3 u_1 v_0^2 \tag{3.4.21a}$$

$$\begin{aligned} u_0^2\left(\frac{\partial v_1}{\partial\theta} + \frac{\partial v_0}{\partial\tilde{\theta}}\right) &+ 2u_0 u_1 \frac{\partial v_0}{\partial\theta} + 2u_1 \frac{\partial u_0}{\partial\theta} v_0 \\ &+ 2u_0\left[\frac{\partial u_0}{\partial\theta} v_1 + v_0\left(\frac{\partial u_1}{\partial\theta} + \frac{\partial u_0}{\partial\tilde{\theta}}\right)\right] \\ &= v_0\left[\left(\frac{\partial u_0}{\partial\theta}\right)^2 + u_0^2\right]^{1/2}. \end{aligned} \tag{3.4.21b}$$

The solution of Equations (3.4.20) is most conveniently expressed in terms of the three[7] slowly varying Keplerian elements a, e, and ω as follows. [Note: $l^2 = a^2(1 - e^2)$.]

$$u_0(\theta, \tilde{\theta}) = p^2[1 + e\cos(\theta - \omega)] \tag{3.4.22a}$$

$$v_0(\theta, \tilde{\theta}) = p^{-3}[1 + e\cos(\theta - \omega)]^{-2}, \tag{3.4.22b}$$

where p is the reciprocal angular momentum $p = 1/l = 1/\sqrt{a(1 - e^2)}$ and $a(\tilde{\theta})$, $e(\tilde{\theta})$, $\omega(\tilde{\theta})$ have initial values a_0, e_0, 0 defined in (3.4.18).

For simplicity we shall henceforth neglect terms of order e_0^2. This approximation will be justified by the fact that e is a monotone decreasing function of $\tilde{\theta}$.

Equation (3.4.21b) can be integrated once with respect to θ if we make use of the solution (3.4.22). We find

$$u_0^2 v_1 + 2p\frac{u_1}{u_0} + \frac{dp}{d\tilde{\theta}}\theta - \frac{1}{p}[\theta + e\sin(\theta - \omega)] = g_1(\tilde{\theta}) + O(e^2), \tag{3.4.23}$$

[7] The fourth element τ will only arise after the quadrature of v_0.

where g_1 is an unknown function of $\tilde{\theta}$. Using (3.4.23) in (3.4.21a) leads to

$$\begin{aligned}\frac{\partial^2 u_1}{\partial\theta^2} + u_1 = &\left(-2e + 2p^2\frac{de}{d\tilde{\theta}} + 4pe\frac{dp}{d\tilde{\theta}}\right)\sin(\theta - \omega)\\ &- 2ep^2\frac{d\omega}{d\tilde{\theta}}\cos(\theta - \omega) + 2\left(1 - p\frac{dp}{d\tilde{\theta}}\right)\theta\\ &+ 2pg_1 + O(e^2). \end{aligned} \tag{3.4.24}$$

Unless u_1 is bounded, the assumed expansion for u would be inconsistent. Therefore, we must set

$$1 - p\frac{dp}{d\tilde{\theta}} = 0 \tag{3.4.25a}$$

$$-e + p^2\frac{de}{d\tilde{\theta}} + 2pe\frac{dp}{d\tilde{\theta}} = 0 \tag{3.4.25b}$$

$$p^2 e\frac{d\omega}{d\tilde{\theta}} = 0 \tag{3.4.25c}$$

and these can easily be solved as follows

$$a = \frac{1}{1 + 2\tilde{\theta}} + O(e_0^2) \tag{3.4.26a}$$

$$e = \frac{e_0}{\sqrt{1 + 2\tilde{\theta}}} + O(e_0^2) \tag{3.4.26b}$$

$$\omega = 0.$$

This shows that to $O(1)$ the effect of drag leaves the pericenter fixed but produces an algebraic decay in the orbit and an algebraic decrease of the eccentricity. Thus, an initially elliptic orbit tends to spiral in and become more and more circular. Moreover, the behavior of e justifies expanding the solution in powers of e.

To determine the time history, i.e. to calculate $\tau(\tilde{\theta})$, is not a straight forward quadrature of v_0. We must keep in mind that a long periodic term (i.e. one depending on $\sin\tilde{\theta}$ or $\cos\tilde{\theta}$) in v_1 will, upon quadrature, drop by one order and contribute to $O(1)$ in t. But, in order to determine all the long periodic terms in v_1, we must study the solutions for v_2 and u_2 and require these to be bounded. Thus, we conclude that to determine $t(\theta)$ to $O(\varepsilon^n)$ we must know v_{n+1} completely, and this requires examining the equations for u_{n+2} and v_{n+2}. This question is explored in Problem 3.4.3.

The above feature is typical of the solution for the time history in all satellite problems where an exact integral of motion (e.g. an energy integral) is not available. If, however, we consider a problem where there is an integral, e.g. the examples in Section 3.4.3 and 3.4.4, one can use this integral to derive the missing long periodic terms to any order. This point is discussed in detail in Reference 3.4.8.

3.4.3 The Motion of a Close Satellite of the Moon in the Restricted Three-Body Problem

Consider the motion of three gravitational mass points (earth, moon, satellite) in the limit when one of the masses (satellite) is very much smaller than that of the other two. Clearly, in this limit the two large bodies will describe a Keplerian orbit about their common center of mass, while the satellite will move in this gravitational field, without influencing the motion of the two large bodies. If the two large bodies (called primaries) describe a circular orbit about their common mass center the satellite of negligible mass is governed by the following set of equations for motion in the orbital plane of the primaries

$$\frac{d^2\bar{x}}{dt^2} = -(1-\mu)\frac{(\bar{x}-\xi_1)}{\bar{r}_1^3} - \mu\frac{(\bar{x}-\xi_2)}{\bar{r}_2^3} \tag{3.4.27a}$$

$$\frac{d^2\bar{y}}{dt^2} = -(1-\mu)\frac{(\bar{y}-\eta_1)}{\bar{r}_1^3} - \mu\frac{(\bar{y}-\eta_2)}{\bar{r}_2^3}. \tag{3.4.27b}$$

The geometry is sketched in Figure 3.4.4.

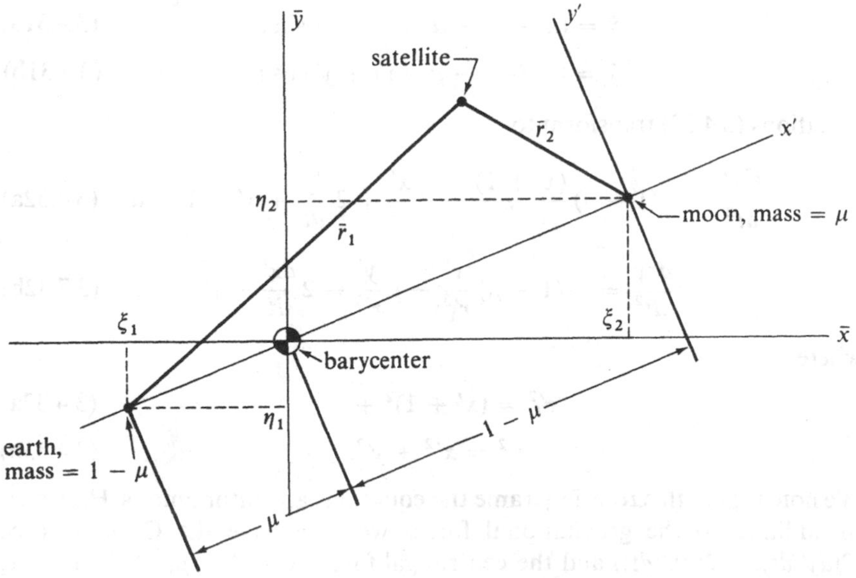

Figure 3.4.4 Motion in Barycentric Inertial Frame

The variables are normalized by choosing the earth-moon distance as the unit of length, and the reciprocal angular velocity of the earth moon orbit for the unit of time. The origin is located at the barycenter and the dimensionless masses μ and $1 - \mu$ are given by

$$\mu = \frac{m_2}{m_1 + m_2} \tag{3.4.28a}$$

$$1 - \mu = \frac{m_1}{m_1 + m_2}. \tag{3.4.28b}$$

Moreover, the coordinates of the earth and moon orbits are defined by

$$\xi_1 = -\mu \cos t \tag{3.4.29a}$$

$$\eta_1 = -\mu \sin t \tag{3.4.29b}$$

$$\xi_2 = (1 - \mu)\cos t \tag{3.4.29c}$$

$$\eta_2 = (1 - \mu)\sin t. \tag{3.4.29d}$$

The distances $\bar{r}_1$ and $\bar{r}_2$ between the satellite and the two primaries are

$$\bar{r}_1^2 = (\bar{x} - \xi_1)^2 + (\bar{y} - \eta_1)^2 \tag{3.4.30a}$$

$$\bar{r}_2^2 = (\bar{x} - \xi_2)^2 + (\bar{y} - \eta_2)^2. \tag{3.4.30b}$$

For the purpose of studying the motion of a lunar satellite it is more convenient to refer to the x', y' coordinates centered at the moon and rotating with the system, cf. Figure 3.4.4.

Since

$$\bar{x} = (x' - 1 - \mu)\cos t - y' \sin t \tag{3.4.31a}$$

$$\bar{y} = (x' + 1 - \mu)\sin t + y' \cos t \tag{3.4.31b}$$

Equations (3.4.27) transform to

$$\frac{d^2x'}{dt^2} = -(1 - \mu)\frac{(x' + 1)}{r_1'^3} - \mu\frac{x'}{r'^3} + 2\frac{dy'}{dt} + x' + 1 - \mu \tag{3.4.32a}$$

$$\frac{d^2y'}{dt^2} = -(1 - \mu)\frac{y'}{r_1'^3} - \mu\frac{y'}{r'^3} - 2\frac{dx'}{dt} - y', \tag{3.7.32b}$$

where

$$r_1'^2 = (x' + 1)^2 + y'^2 \tag{3.4.33a}$$

$$r'^2 = x'^2 + y'^2. \tag{3.4.33b}$$

We note that in this rotating frame the equations are autonomous. However, in addition to the gravitational forces we now have the Coriolis force $(2(dy'/dt), -2(dx'/dt))$ and the centrifugal force $(x' + 1 - \mu, y')$. Of course, we could have written down Equations (3.4.32) directly by taking proper account of the fictitious forces introduced by referring the motion to the rotating frame.

We wish to study the solution of (3.4.32) in the limit as $\mu \to 0$. (Actually, for the earth-moon system $\mu = 0.012$.) It is clear that for motion close to the moon it is inappropriate to use the x', y', t variables. In terms of these variables the moon's attraction will nominally occur only to order μ when in fact, for arbitrarily small μ there exists some neighborhood of the origin where the lunar gravitation dominates.

We therefore introduce inner variables

$$x^* = \frac{x'}{\delta(\mu)}; \qquad y^* = \frac{y'}{\delta(\mu)}; \qquad t^* = \frac{t}{\gamma(\mu)}, \tag{3.4.34}$$

where δ and γ are to be determined by an order of magnitude analysis. We anticipate for close satellites the period may also need to be rescaled as it may be significantly smaller than the earth-moon period which was used in normalising t.

Substituting the above into (3.4.32a) gives

$$\begin{aligned}\frac{\delta}{\gamma^2}\frac{d^2x^*}{dt^{*2}} &= -(1-\mu)\frac{(1+\delta x^*)}{[1+2\delta x^* + \delta^2(x^{*2}+y^{*2})]^{3/2}} \\ &\quad - \frac{\mu}{\delta^2}\frac{x^*}{[x^{*2}+y^{*2}]^{3/2}} + 2\frac{\delta}{\gamma}\frac{dy^*}{dt^*} + \delta x^* + 1 - \mu. \end{aligned} \tag{3.4.35}$$

Developing the right-hand side of (3.4.35) and multiplying by γ^2/δ gives

$$\begin{aligned}\frac{d^2x^*}{dt^{*2}} &= -\frac{\mu\gamma^2}{\delta^3}\frac{x^*}{[x^{*2}+y^{*2}]^{3/2}} + 2\gamma\frac{dy^*}{dt^*} \\ &\quad + 3\gamma^2x^* + O(\mu\gamma^2) + O(\delta\gamma^2). \end{aligned} \tag{3.4.36}$$

The fundamental criterion for a lunar satellite is that the lunar gravitation be dominant. This requires that we set $\mu\gamma^2/\delta^3 = 1$. With this premise, the richest equations result by setting $\gamma = 1$, $\delta = \mu^{1/3}$ and we have, in the limit as $\mu \to 0$,

$$\frac{d^2x^*}{dt^{*2}} = -\frac{x^*}{[x^{*2}+y^{*2}]^{3/2}} + 2\frac{dy^*}{dt^*} + 3x^* \tag{3.4.37a}$$

$$\frac{d^2y^*}{dt^{*2}} = -\frac{y^*}{[x^{*2}+y^{*2}]^{3/2}} - 2\frac{dx^*}{dt^*}. \tag{3.4.37b}$$

These equations were first derived by Hill using physical arguments [cf. Reference 3.4.9 for a comprehensive treatment]. They are not significantly simpler than the exact set (3.4.32) even though now the net effect of the earth's gravity plus centrifugal force is simply the term $3x^*$ in (3.4.37a).

In deriving (3.4.37) it was assumed that the lunar gravitation was comparable to the Coriolis force and to the net effect of the earth's gravity plus centrifugal forces. This would be true for orbits as far away as $O(\mu^{1/3})$ and having periods comparable to the earth-moon period. Such orbits are *not* perturbed Kepler ellipses and will not be considered here.[8] Rather, we are interested in closer satellites for which the lunar gravity dominates over the Coriolis force.

In this case any $\gamma(\mu)$ such that $\gamma \to 0$ as $\mu \to 0$ is appropriate. The smaller γ is the closer the orbit is to the moon. Setting $\gamma(\mu) = \mu^\alpha = \varepsilon$, $\alpha > 0$, we have the inner variables

$$x^{**} = \frac{x'}{\mu^{(1+2\alpha)/3}} \tag{3.4.38a}$$

$$y^{**} = \frac{y'}{\mu^{(1+2\alpha)/3}} \tag{3.4.38b}$$

$$t^{**} = \frac{t'}{\mu^\alpha} \tag{3.4.38c}$$

and the following equations of motion to order ε^2.

$$\frac{d^2x^{**}}{dt^{**2}} = -\frac{x^{**}}{r^{**3}} + 2\varepsilon\frac{dy^{**}}{dt^{**}} + 3\varepsilon^2 x^{**} \tag{3.4.39a}$$

$$\frac{d^2y^{**}}{dt^{**2}} = -\frac{y^{**}}{r^{**3}} - 2\varepsilon\frac{dx^{**}}{dt^{**}} \tag{3.4.39b}$$

$$r^{**2} = x^{**2} + y^{**2}. \tag{3.4.39c}$$

Transforming the above to the standard form, we set

$$x^{**} = r^{**}\cos\theta \tag{3.4.40a}$$

$$y^{**} = r^{**}\sin\theta \tag{3.4.40b}$$

and

$$u = \frac{1}{r^{**}} = u(\theta) \tag{3.4.41a}$$

$$p = u^{*2}\frac{dt^{**}}{d\theta} = p(\theta). \tag{3.4.41b}$$

Thus, p is the reciprocal angular momentum, and knowing p and u one can calculate the time history by quadrature.

Equations (3.4.39) become, $' = d/d\theta$,

$$u'' + u = p^2 - 2\varepsilon\left(\frac{p}{u} + \frac{p}{u^3}u'^2\right) - \frac{3}{2}\varepsilon^2\frac{p^2}{u^3}$$

$$+ \frac{3}{2}\varepsilon^2\frac{p^2}{u^4}u'\sin 2\theta - u\cos 2\theta) \tag{3.4.42a}$$

$$p' = -2\varepsilon u'\frac{p^2}{u^3} + \frac{3}{2}\varepsilon^2\frac{p^3}{u^4}\sin 2\theta. \tag{3.4.42b}$$

[8] Again we refer the reader to Reference 3.4.9 for a discussion of the solutions of Hill's equations.

We now assume a two variable expansion for u and p in the form

$$u = u_0(\theta, \tilde{\theta}) + \varepsilon u_1(\theta, \tilde{\theta}) + \varepsilon^2 u_2(\theta, \tilde{\theta}) + \cdots \tag{3.4.43a}$$

$$p = p_0(\theta, \tilde{\theta}) + \varepsilon p_1(\theta, \tilde{\theta}) + \varepsilon^2 p_2(\theta, \tilde{\theta}) + \cdots \tag{3.4.43b}$$

where the slow variable $\tilde{\theta}$ is now taken in the form

$$\tilde{\theta} = \varepsilon\theta(1 + \varepsilon\alpha_1 + \varepsilon^2\alpha_2 + \cdots). \tag{3.4.44}$$

The reason for expanding the slow variable here instead of the usual Lindstedt-type expansion of the fast variable is because θ occurs explicitly in the problem and it is inconvenient to transform it. The net result is, of course, the same.

We will only carry out the solution to $O(1)$ and the higher order calculations are left as an exercise (Problem 3.4.5). Substituting the expansions (3.4.43) into the differential Equations (3.4.42) gives

$$\frac{\partial^2 u_0}{\partial \theta^2} + u_0 = p_0^2 \tag{3.4.45a}$$

$$\frac{\partial p_0}{\partial \theta} = 0. \tag{3.4.45b}$$

To order ε, we find

$$\frac{\partial^2 u_1}{\partial \theta^2} + u_1 = -2\frac{\partial u_0}{\partial \theta\, \partial \tilde{\theta}} + 2p_0\left[\frac{1}{u_0} + \frac{1}{u_0^3}\left(\frac{\partial u_0}{\partial \theta}\right)^2\right] + 2p_0 p_1 \tag{3.4.46a}$$

$$\frac{\partial p_1}{\partial \theta} + \frac{\partial p_0}{\partial \tilde{\theta}} = -2\frac{\partial u_0}{\partial \theta}\frac{p_0^2}{u_0^3}. \tag{3.4.46b}$$

For reference to Problem (3.4.5), we also list the equations governing the terms of order ε^2.

$$\begin{aligned}\frac{\partial^2 u_2}{\partial \theta^2} + u_2 = {} & -2\frac{\partial^2 u_1}{\partial \theta\, \partial \tilde{\theta}} - \frac{\partial^2 u_0}{\partial \tilde{\theta}^2} - 2\alpha_1 \frac{\partial^2 u_0}{\partial \theta\, \partial \tilde{\theta}} \\ & + p_1^2 + 2p_0 p_2 - 2\left[\frac{p_1}{u_0} - \frac{p_0 u_1}{u_0^2} + \frac{p_1}{u_0^3}\left(\frac{\partial u_0}{\partial \theta}\right)^2\right. \\ & \left. - 3\frac{p_0}{u_0^4} u_1 \left(\frac{\partial u_0}{\partial \theta}\right)^2 - 2\frac{p_0}{u_0^3}\frac{\partial u_0}{\partial \theta}\left(\frac{\partial u_0}{\partial \tilde{\theta}} + \frac{\partial u_1}{\partial \theta}\right)\right] \\ & - \frac{3}{2}\frac{p_0^2}{u_0^3} + \frac{3}{2}\frac{p_0^2}{u_0^4}\left(\frac{\partial u_0}{\partial \theta}\sin 2\theta - u_0 \cos 2\theta\right)\end{aligned} \tag{3.4.47a}$$

$$\begin{aligned}\frac{\partial p_2}{\partial \theta} + \frac{\partial p_1}{\partial \tilde{\theta}} + \alpha_1 \frac{\partial p_0}{\partial \tilde{\theta}} = {} & -2\frac{p_0^2}{u_0^3}\left(\frac{\partial u_0}{\partial \tilde{\theta}} + \frac{\partial u_1}{\partial \theta}\right) \\ & - \frac{4p_0 p_1}{u_0^3}\frac{\partial u_0}{\partial \theta} + 6\frac{p_0^2 u_1}{u_0^4}\frac{\partial u_0}{\partial \theta} + \frac{3}{2}\frac{p_0^3}{u_0^4}\sin 2\theta.\end{aligned} \tag{3.4.47b}$$

The solution of the $O(1)$ terms is (cf. 3.4.22)

$$p_0 = p_0(\tilde{\theta}) \tag{3.4.48a}$$

$$u_0 = p_0^2(\tilde{\theta})\{1 + e(\tilde{\theta})\cos[\theta - \omega(\tilde{\theta})]\} \tag{3.4.48b}$$

and the slowly varying reciprocal angular momentum p_0, the eccentricity e and the argument of pericenter ω are to be determined by consistency of the solution to $O(\varepsilon)$.

Substituting the above into (3.4.46) gives

$$\begin{aligned}\frac{\partial^2 u_1}{\partial \theta^2} + u_1 = {} & -2p_0^2 e \frac{d\omega}{d\tilde{\theta}} \cos(\theta - \omega) - \frac{2}{p_0[1 + e\cos(\theta - \omega)]} \\ & - \frac{2e^2}{p_0} \frac{\sin^2(\theta - \omega)}{[1 + e\cos(\theta - \omega)]^3} + 2p_0 p_1 \\ & + 4p_0 \frac{dp_0}{d\tilde{\theta}} e \sin(\theta - \omega) + 2p_0^2 \frac{de}{d\tilde{\theta}} \sin(\theta - \omega)\end{aligned} \tag{3.4.49a}$$

$$\frac{\partial p_1}{\partial \theta} = -\frac{dp_0}{d\tilde{\theta}} + \frac{2e}{p_0^2} \frac{\sin(\theta - \omega)}{[1 + e\cos(\theta - \omega)]^3}. \tag{3.4.49b}$$

Since $dp_0/d\tilde{\theta}$ in (3.4.49b) depends only on $\tilde{\theta}$, integrating this term will give rise to an inconsistent secular term proportional to θ in p_1. We must therefore set $dp_0/d\tilde{\theta} = 0$, i.e., $p_0 =$ constant. Then, integrating (3.4.49b) gives

$$p_1 = \frac{1}{p_0^2[1 + e\cos(\theta - \omega)]^2} + \tilde{p}_1(\theta), \tag{3.4.50}$$

where $\tilde{p}_1$ is an unknown function of $\tilde{\theta}$.

We now substitute the above expression for p_1 into (3.4.49a) and can explicitly integrate the result. In addition to the secular terms $\psi \sin \psi$ and $\psi \cos \psi$ (with $\psi = \theta - \omega$), we encounter a term of the form $\sin \psi \cdot \sin^{-1}((e + \cos \psi)/(1 + e\cos \psi))$. The arc sine consists of a secular part equal to $-\psi$ plus a periodic part. In order to separate the periodic and secular terms in the solution we introduce the function

$$Z(\psi, e) = -\sin^{-1} \frac{e + \cos \psi}{1 + e\cos \psi} - \psi \tag{3.4.51}$$

which is strictly periodic. In fact, it is easy to show that Z has the Fourier series

$$Z = -\frac{\pi}{2} + 2\sum_{n=1}^{\infty} \frac{1}{n} \left(\frac{1 - \sqrt{1 - e^2}}{-e} \right)^n \sin n\psi. \tag{3.4.52}$$

Removing the secular terms in u_1 then requires that

$$\frac{d\omega}{d\tilde{\theta}} = -\frac{1}{p_0^3(1-e^2)^{3/2}}; \qquad \omega = -\frac{\tilde{\theta}}{p_0^3(1-e^2)^{3/2}} \tag{3.4.53a}$$

$$\frac{de}{d\tilde{\theta}} = 0; \qquad e = \text{const.} \tag{3.4.53b}$$

and the solution for u_1 can be calculated in the form

$$u_1 = 2p_0\tilde{p}_1 + \frac{1}{p_0(1-e^2)} + \left(A_1 - \frac{eZ}{p_0(1-e^2)^{3/2}}\right)\sin(\theta-\omega)$$
$$+ B_1\cos(\theta-\omega) - \frac{1}{p_0[1+e\cos(\theta-\omega)]}, \tag{3.4.54}$$

where the unknown functions A_1 and B_1 are to be found from the solution to $O(\varepsilon^2)$.

The principal result to $O(1)$ is that the Keplerian ellipse rotates within the x', y' coordinate system at a counterclockwise rate which is exactly equal to the clockwise rate of rotation of the coordinate system. (cf. Problem 3.4.5) Thus, the orbit to first order, preserves its orientation in space. This well-known result could have, of course, been derived more directly by noting that an orientation preserving ellipse is a solution of the differential equations to $O(\varepsilon)$.

This problem was studied in Reference 3.4.6 to $O(\varepsilon)$ for arbitrary orbital inclination and eccentricity. In this general case, the solution requires use of *two slow variables* $\varepsilon\theta$ as well as $\varepsilon^2\theta$ and the calculations are quite involved. The motion is also complicated as the orbital elements perform oscillatory or secular variations on both time scales.

3.4.4 The Planar Motion of a Trojan Asteroid

In this section we consider a class of planar orbits around the larger primary in the restricted circular three-body problem. We assume that the smaller primary (mass $= \mu$) is Jupiter and the larger primary (mass $1-\mu$) is the sun and neglect all other planets. Furthermore, we will consider asymptotic solutions in the limit $\mu \to 0$ [In the sun-Jupiter system $\mu = 9.53 \times 10^{-4}$] and assume that the asteroid moves in the sun-Jupiter orbital plane.

The Trojan asteroids are found clustered in neighborhoods of the triangular libration points [cf. Problem 3.4.7], and describe bounded orbits of small inclination relative to the sun-Jupiter plane. The model we adopt here is approximate in the sense that Jupiter's orbital eccentricity ($e = 0.05$) is assumed to be zero, the effect of Saturn is neglected, and the asteroids are assumed to move in the sun-Jupiter plane. Even though none of the actual Trojan asteroids depart from the libration point by more than 10°

of arc we will not restrict the solution in this respect. As a result, we will encounter the interesting possibility of asteroids coming very close to Jupiter.

For the purposes of this problem it is more convenient to choose sun-centered nonrotating coordinates which can easily be derived from equations (3.4.27) by a shift of the origin to the mass point μ. If $\hat{x}$ and $\hat{y}$ are nonrotating with respect to the $\bar{x}$, $\bar{y}$ inertial frame, and if we introduce polar coordinates r, and θ in the $\hat{x}$, $\hat{y}$ frame we calculate the following equations of motion from (3.4.27)

$$\ddot{r} - r\dot{\theta}^2 = -\frac{(1-\mu)}{r^2} + \mu\frac{[\cos(\theta - t) - r]}{r_J^3} - \mu\cos(\theta - t) \quad (3.4.55a)$$

$$r\ddot{\theta} + 2\dot{r}\dot{\theta} = \mu\sin(\theta - t)\left[1 - \frac{1}{r_J^3}\right], \quad (3.4.55b)$$

where r is the distance between sun and asteroid, r_J the distance between Jupiter and the asteroid

$$r_J^2 = 1 + r^2 - 2r\cos(\theta - t) \quad (3.4.56)$$

and the geometry is sketched in Figure 3.4.5.

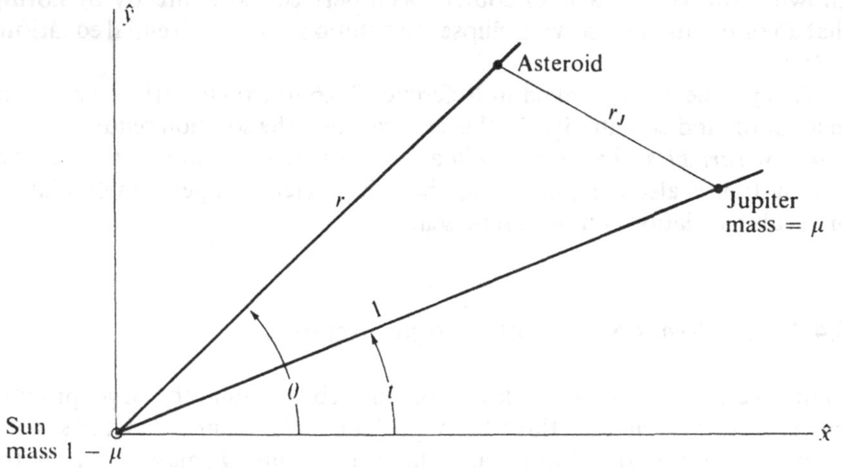

Figure 3.4.5 Sun-Centered Nonrotating Frame

As before we transform variables to $r = u(\theta)$ and $t(\theta)$ and obtain

$$u'' + u - u^4t'^2 = \mu\{ut'^2[u\cos(\theta - t) - u'\sin(\theta - t)]\}$$

$$+ \mu u^4 t'^2\left\{\frac{1 - u\cos(\theta - t) + u'\sin(\theta - t)}{[1 + u^2 - 2u\cos(\theta - t)]^{3/2}}\right\} \quad (3.4.57a)$$

$$(u^2t')' = \mu(ut')^3\sin(\theta - t)\left\{\frac{1}{[1 + u^2 - 2u\cos(\theta - t)]^{3/2}} - 1\right\}. \quad (3.4.57b)$$

The Jacobi integral for the system [cf. Problem 3.4.4] can also be written in terms of the $u(\theta)$, $t(\theta)$ variables as follows

$$\frac{1}{2u^2t'^2}\left(\frac{u'^2}{u^2}+1\right)-\frac{1}{u^2t'}+\frac{\mu}{u}\cos(\theta-t)-(1-\mu)u$$

$$-\mu\frac{u}{[1+u^2-2u\cos(\theta-t)]^{1/2}}=\frac{\mu^2}{2}-\frac{C}{2}. \tag{3.4.58}$$

We will use this *exact* integral of equations (3.4.57) as a partial check of our results.

In the usual application of the two variable method one assumes that the leading term in the asymptotic expansion for u and t are the Keplerian solutions [obtained by setting $\mu = 0$ in Equations (3.4.57)] with slowly varying elements depending on $\mu\theta$. It is easy to show that this approach fails, and leads to small divisors, for orbits which are nearly circular and close to the resonant values of the semi-major axis a given by

$$a_{\text{res}} = \left(\frac{n-m}{n}\right)^{2/3}, \tag{3.4.59}$$

where n and m are relatively prime integers [cf. Problem 3.4.8]. Since we are interested in the special case $n = 1$, $m = 0$ we cannot expand the solution in terms of θ and $\mu\theta$.

We assume the orbital elements for the first approximation in the form

$$a = 1 + \mu^\sigma \bar{a}(\tilde{\theta}) \tag{3.4.60a}$$

$$e = \mu^\sigma \bar{e}(\tilde{\theta}) \tag{3.4.60b}$$

$$\omega = \mu^\sigma \bar{\omega}(\tilde{\theta}) \tag{3.4.60c}$$

$$\tau = \mu^\sigma \bar{\tau}(\tilde{\theta}) \tag{3.4.60d}$$

with

$$\tilde{\theta} = (\theta - \theta_0)\mu^\lambda.$$

Here σ and λ are unknown constants and θ_0 is the initial ($t = 0$) value of θ. It can be shown [cf. Problem (3.4.9)] that the only consistent expansion corresponds to $\sigma = \lambda = \frac{1}{2}$.

Clearly, the elements defined by (3.4.60) correspond to initial conditions on u, t of the form

$$u(\theta_0) = 1 + \varepsilon\alpha \tag{3.4.61a}$$

$$t(\theta_0) = 0 \tag{3.4.61b}$$

$$u'(\theta_0) = \varepsilon\beta \tag{3.4.61c}$$

$$t'(\theta_0) = 1 + \varepsilon\gamma, \tag{3.4.61d}$$

where $\varepsilon = \mu^{1/2}$ and α, β and γ are arbitrary constants independent of ε.

We therefore expand u and t in two variable form as follows

$$u = 1 + \varepsilon u_1(\varphi, \tilde{\theta}) + \varepsilon^2 u_2(\varphi, \tilde{\theta}) + \cdots \tag{3.4.62a}$$

$$t = \varphi + \varepsilon t_1(\varphi, \tilde{\theta}) + \varepsilon^2 t_2(\varphi, \tilde{\theta}) + \cdots, \tag{3.4.62b}$$

where the fast variable is φ defined by

$$\frac{d\varphi}{d\theta} = 1 + \varepsilon f_1(\tilde{\theta}) + \varepsilon^2 f_2(\tilde{\theta}) + \varepsilon^3 f_3(\tilde{\theta}) + \cdots, \tag{3.4.63}$$

the slow variable is

$$\tilde{\theta} = \varepsilon(\theta - \theta_0),$$

and the f_n are unknown functions of $\tilde{\theta}$ to be determined by the requirement that the t_n be consistent.

The reason for choosing φ (which is a general version of a Lindstedt-type fast variable) instead of θ is because one of the effects of Jupiter on the asteroid is a net perturbation of the mean angular velocity. This quantity instead of being unity will, due to Jupiter's perturbation, contain long periodic terms of order ε, ε^2, etc. (cf. Equation (3.4.64). Unless such terms are accounted for in the definition of the fast variable one encounters inconsistent terms in t_1, t_2, etc. which cannot be removed [cf. Problem 3.4.9].

Integrating (3.4.63) defines φ in terms of θ once the f_n are known

$$\varphi = \theta - \theta_0 - F_1(\tilde{\theta}) - \varepsilon F_2(\tilde{\theta}) + \cdots, \tag{3.4.64a}$$

where

$$F_n(\tilde{\theta}) = -\int_{\theta_0}^{\tilde{\theta}} f_n(s)ds. \tag{3.4.64b}$$

In this section we only consider the solution to $O(\varepsilon)$ which contains most of the important features of the motion. The solution to $O(\varepsilon^2)$ is given in Reference 3.4.11. The case for the planar elliptic problem is discussed in Reference 3.4.12 and the extension of this to non-planar orbits can be found in Reference 3.4.13.

Substituting the expansions (3.4.62) into the differential equations (3.4.57) gives

$$\frac{\partial^2 u_1}{\partial \varphi^2} - 3u_1 - 2\frac{\partial t_1}{\partial \varphi} - 2f_1 = 0 \tag{3.4.65a}$$

$$\frac{\partial}{\partial \varphi}\left(2u_1 + \frac{\partial t_1}{\partial \varphi}\right) = 0 \tag{3.4.65b}$$

and to $O(\varepsilon^2)$ we find

$$\frac{\partial^2 u_2}{\partial\varphi^2} - 3u_2 + 2f_1\frac{\partial^2 u_1}{\partial\varphi^2} + 2\frac{\partial^2 u_1}{\partial\varphi\,\partial\tilde{\theta}} - 2\frac{\partial t_2}{\partial\varphi}$$

$$-2f_1\frac{\partial t_1}{\partial\varphi} - 2f_2 - 2\frac{\partial t_1}{\partial\tilde{\theta}} - \left(\frac{\partial t_1}{\partial\varphi} + f_1\right)^2$$

$$-8u_1\left(\frac{\partial t_1}{\partial\varphi} + f_1\right) - 6u_1^2 = \cos\xi_1 - 1 + \frac{1}{2^{3/2}(1-\cos\xi_1)^{1/2}} \tag{3.4.66a}$$

$$\frac{\partial}{\partial\varphi}\left(2u_2 + u_1^2 + 2u_1\frac{\partial t_1}{\partial\varphi} + 4u_1 f_1 + \frac{\partial t_2}{\partial\varphi} + 2f_1\frac{\partial t_1}{\partial\varphi} + \frac{\partial t_1}{\partial\tilde{\theta}}\right)$$

$$= -\frac{\partial}{\partial\tilde{\theta}}\left(2u_1 + \frac{\partial t_1}{\partial\varphi} + f_1\right) + \frac{\sin\xi_1}{[2(1-\cos\xi_1)]^{3/2}} - \sin\xi_1. \tag{3.4.66b}$$

In equations (3.4.66a) and (3.4.66b) we have introduced the notation

$$\xi_1(\tilde{\theta}) = \theta_0 - F_1(\tilde{\theta}) \tag{3.4.67}$$

and the meaning of ξ_1 will be discussed later.

The initial conditions in the solution to $O(\varepsilon)$ which result at $\theta = \theta_0$ from Equations (3.4.61) are

$$u_1 = \alpha \tag{3.4.68a}$$

$$\frac{\partial u_1}{\partial\varphi} = \beta \tag{3.4.68b}$$

$$t_1 = 0 \tag{3.4.68c}$$

$$\frac{\partial t_1}{\partial\varphi} = \gamma - f_1(0). \tag{3.4.68d}$$

Moreover, substituting the assumed expansions into the Jacobi integral (3.4.58) provides the following condition on the solution to $O(\varepsilon)$.

$$\frac{1}{2}\left[\left(\frac{\partial u_1}{\partial\varphi}\right)^2 + \left(\frac{\partial t_1}{\partial\varphi}\right)^2\right] + f_1\frac{\partial t_1}{\partial\varphi} + \frac{1}{2}f_1^2 - \frac{3}{2}u_1^2$$

$$+ (1+\cos\xi_1) - \frac{1}{\sqrt{2(1-\cos\xi_1)}} = \frac{\beta^2}{2} - \frac{3}{2}\alpha^2 - \frac{\gamma^2}{2}$$

$$+ 1 + \cos\theta_0 - \frac{1}{\sqrt{2(1-\cos\theta_0)}} = \text{const.} \tag{3.4.69}$$

After the solution for u_1 and t_1 is calculated we propose to substitute the results into (3.4.69) as a partial check on the calculations.

We now consider the solution to $O(\varepsilon)$. It follows from Equation (3.4.65b) that

$$2u_1 + \frac{\partial t_1}{\partial \varphi} = p_1(\tilde{\theta}). \quad (3.4.70)$$

Upon substitution of (3.4.70) into (3.4.65a) the following solution for u_1 can be calculated

$$u_1 = \rho_1 \cos(\varphi + \psi_1) + 2(p_1 + f_1), \quad (3.4.71)$$

where p_1, ρ_1 and ψ_1 are functions of $\tilde{\theta}$ to be determined together with $f_1(\tilde{\theta})$ by requiring that the expansion (3.4.62) be consistent.

Using (3.4.71) in (3.4.70) gives

$$\frac{\partial t_1}{\partial \varphi} = -(3p_1 + 4f_1) - 2\rho_1 \cos(\varphi + \psi_1). \quad (3.4.72)$$

Integration of (3.4.72) with respect to φ would result in a secular term in t_1 unless the first term is set equal to zero, and this gives the first relation governing the four unknown functions

$$p_1 = -\tfrac{4}{3} f_1. \quad (3.4.73)$$

Equation (3.4.72) is now integrated to give

$$t_1 = -2\rho_1 \sin(\varphi + \psi_1) + q_1, \quad (3.4.74)$$

where q_1 is a constant. In view of (3.4.73), u_1 becomes

$$u_1 = \rho_1 \cos(\varphi + \psi_1) - \tfrac{2}{3} f_1. \quad (3.4.75)$$

It is shown next that there is no loss of generality in taking q_1 constant as long as f_2 is allowed to depend on $\tilde{\theta}$. Consider the expression for φ in terms of θ and $\tilde{\theta}$ given in (3.4.64a).

Since the leading term in the expansion for t is φ, the final expression for t in terms of θ involves q_1 only through the difference $(q_1 - F_2)$. In fact, to any order ε^n in the expansion for t, only combinations of the form $(q_n - F_{n+1})$ appear. It therefore follows that allowing the q_n to vary will merely change the definitions of the F_{n+1} with no net effect on the final result.

Applying the initial conditions to the solutions for u_1 and t_1 leads to the following initial values (at $\theta = \theta_0$) for the four unknown functions and q_1.

$$f_1(0) = -3(2\alpha + \gamma) = -\tfrac{3}{4} p_1(0) \quad (3.4.76a)$$

$$\rho_1(0) \sin \psi_1(0) = -\beta \quad (3.4.76b)$$

$$\rho_1(0) \cos \psi_1(0) = -(3\alpha + 2\gamma) \quad (3.4.76c)$$

$$q_1 = -2\beta. \quad (3.4.76d)$$

Consider next the expression for $\partial t_2/\partial\varphi$ and u_2 governed by Equations (3.4.66). Substituting for u_1 and t_1 into the second bracketed term and use of (3.4.73) gives

$$\frac{\partial}{\partial\varphi}\left[2u_2 + u_1^2 + 2u_1\frac{\partial t_1}{\partial\varphi} + 4f_1u_1 + \frac{\partial t_2}{\partial\varphi} + 2f_1\frac{\partial t_1}{\partial\varphi}\right.$$
$$\left. + \frac{\partial t_1}{\partial\tilde{\theta}}\right] = \frac{1}{3}\xi_1'' + \frac{\sin\xi_1}{[2(1-\cos\xi_1)]^{3/2}} - \sin\xi_1, \qquad (3.4.77)$$

where the prime denotes differentiation with respect to $\tilde{\theta}$. In deriving (3.4.77) Equations (3.4.64b) and (3.4.67) have been used to set

$$f_1' = -\xi_1''. \qquad (3.4.78)$$

In order that the asymptotic expansion for u and t be consistent, u_2 and t_2 cannot contain terms proportional to φ. Therefore, the right-hand side of (3.4.77) must be set equal to zero; and this leads to the following differential equation governing ξ_1.

$$\xi_1'' + 3\left\{1 - \frac{1}{[2(1-\cos\xi_1)]^{3/2}}\right\}\sin\xi_1 = 0. \qquad (3.4.79)$$

In view of Equations (3.4.64a) and (3.4.62b), ξ_1 is the leading term in the mean motion of the asteroid in a frame rotating with Jupiter. This follows from the fact that the mean value of $\theta - t$ to $O(1)$ is merely ξ_1.

Equation (3.4.79) possesses the following integral which defines the evolution of the mean motion of the asteroid in the phase-plane of ξ_1 and ξ_1'.

$$\frac{1}{2}\xi_1'^2 + 3\left\{\frac{1}{\sqrt{2(1-\cos\xi_1)}} - \cos\xi_1\right\} = E \text{ const.}, \qquad (3.4.80a)$$

where the constant E may be expressed in terms of the initial parameters in the form

$$E = \frac{9}{2}(\gamma + 2\alpha)^2 + 3\left\{\frac{1}{\sqrt{2(1-\cos\theta_0)}} - \cos\theta_0\right\}. \qquad (3.4.80b)$$

Equation (3.4.80a) is in agreement with the first un-numbered equation in Section 9.14 of Reference 3.4.10.

Figure 3.4.6 is a scale drawing of curves of constant E in the phase plane ξ_1, ξ_1'. In view of the symmetry of (3.4.80a) only the quadrant $\xi_1' > 0$, $0 < \xi_1 < \pi$ need be exhibited. As expected, the equilateral libration points (L_4, L_5) appear as centers and the saddle corresponds to the straight-line equilibrium point (L_3) at opposition with Jupiter. [cf. Prob. 3.4.7].

For the special case of motion near these three equilibrium points, Equation (3.4.79) provides the leading term in the linearized long-periodic result. In

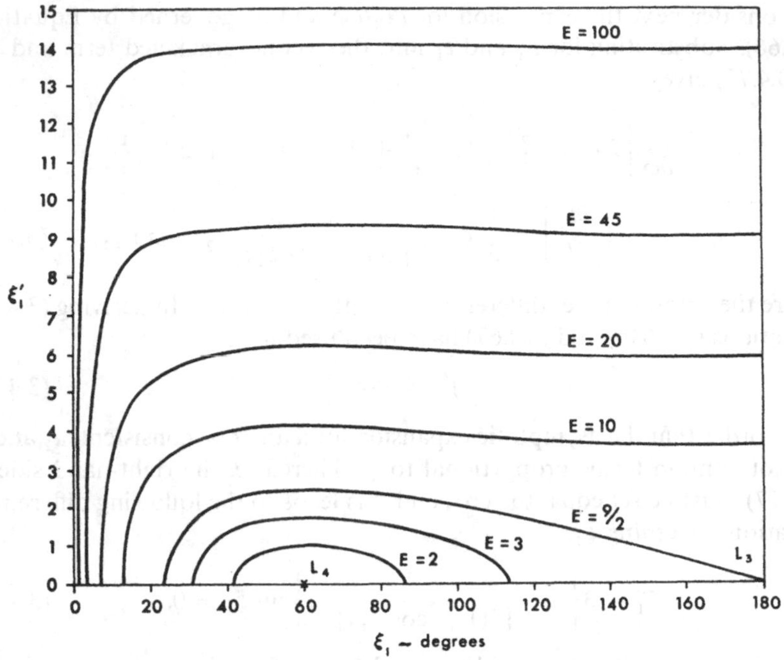

Figure 3.4.6 Phase Plane Solution for Mean Motion

particular, consider the motion near L_4, i.e., $\xi_1 \approx \pi/3$. Equation (3.4.79) may be linearized to

$$\xi_1'' + \frac{27}{4}\xi_1 = 0 \tag{3.4.81a}$$

according to which

$$\xi_1 = \xi_1^* \sin\left(\frac{\sqrt{27}}{2}\tilde{\theta} + \text{const.}\right) \tag{3.4.81b}$$

valid for sufficiently small ξ_1^*. Since, to first order $\theta = t$, Equation (3.4.81b) shows that near L_4, the long periodic librations have the frequency $\sqrt{27\mu}/2$ in agreement with the leading term in the expansion for the frequency s_1 in the linear solution [cf. Problem 3.4.7]

$$s_1 = \tfrac{1}{2}\sqrt{1 - [1 - 27\mu(1-\mu)]^{1/2}} = \frac{\sqrt{27\mu}}{2} + O(\mu^{3/2}). \tag{3.4.82}$$

An integral representation for ξ_1 in terms of $\tilde{\theta}$ follows immediately from (3.4.80); however, this is of little computational value. As suggested in Reference (3.4.10) one can construct the Fourier cosine series representation for ξ_1 in terms of $\tilde{\theta}$ easily as long as ξ_1 is near L_4 or L_5. This follows from the fact that for motion near L_4 or L_5 the coefficients of the Fourier series for ξ_1 proceed in decreasing powers of the amplitude in the phase plane (ξ_1, ξ_1') measured form L_4 and L_5. Clearly, for initial values such that E begins to approach $\frac{9}{2}$ (i.e., near or outside the separatrix) the above procedure fails and one must use the exact representation for ξ_1 to evaluate numerically the coefficients and period of the appropriate Fourier series.

Of course, even the exact result predicted by (3.4.80) is not actually valid for large values of E, since as ξ_1 approaches zero or 2π (i.e., as the asteroid approaches Jupiter) the expansions for u and t break down, as will be shown later. As an indication of the closeness of ξ_1 to zero with increasing E, consider Figure 3.4.7 drawn to scale using (3.4.80a) and showing $E - \xi_1'^2/2$ as a function of ξ_1 on the interval $0 < \xi_1 < \pi$. For example, the separatrix ($E = \frac{9}{2}$) in the phase-plane intercepts the ξ_1 axis at the two points with $\cos \xi_1 \approx 0.9$ which implies that the asteroid comes to within 24° of arc of Jupiter. Clearly, the assumption in this case that Jupiter's gravitational attraction is very small compared to the sun's is not strictly valid. We will return to this question later on.

To complete the solution to first order, it is necessary to evaluate ρ_1 and ψ_1 by requiring that u_2 be bounded. With the right-hand side equal to zero, Equation (3.4.77) integrates to

$$2u_2 + u_1^2 + 2u_1 \frac{\partial t_1}{\partial \varphi} + 4u_1 f_1 + \frac{\partial t_2}{\partial \varphi} + 2f_1 \frac{\partial t_1}{\partial \varphi} + \frac{\partial t_1}{\partial \tilde{\theta}} = p_2(\tilde{\theta}), \quad (3.4.83)$$

where p_2 is an unknown function of $\tilde{\theta}$.

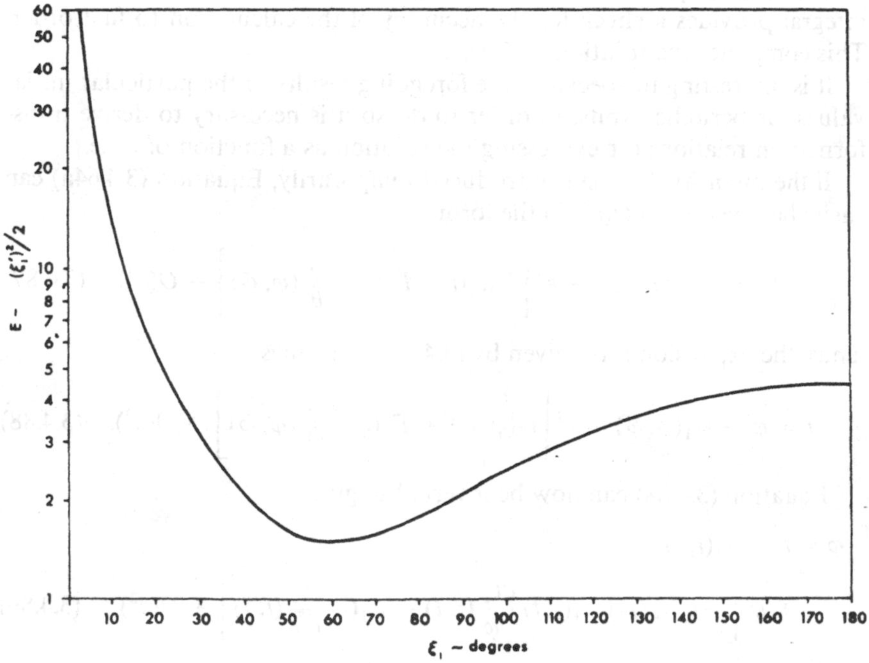

Figure 3.4.7 Energy Integral for Mean Motion

Substituting the value of $\partial t_2/\partial\varphi$ given by (3.4.83) into (3.4.66a), and use of the forms for u_1 and t_1 according to (3.4.71) and (3.4.74, gives

$$\frac{\partial^2 u_2}{\partial\varphi^2} + u_2 = 2\rho_1(f_1 + \psi_1')\cos(\varphi + \psi_1) + 2\rho_1' \sin(\varphi + \psi_1)$$
$$+ \frac{25}{9} f_1^2 + 2f_2 + 2p_2 - (1 - \cos\xi_1) + \frac{1}{\sqrt{8(1 - \cos\xi_1)}}. \quad (3.4.84)$$

In order to avoid mixed-secular terms in u_2, which would render the expansion for u inconsistent, one must set

$$\rho_1' = 0 \quad (3.4.85a)$$
$$\psi_1' + f_1 = 0. \quad (3.4.85b)$$

Equations (3.4.85) can be solved immediately to give

$$\rho_1 = \text{const.} = \sqrt{\beta^2 + (3\alpha + 2\gamma)^2} \quad (3.4.86a)$$
$$\psi_1(\tilde{\theta}) = \xi_1(\tilde{\theta}) - \theta_0 - \psi_0, \quad (3.4.86b)$$

where ψ_0 is defined in (3.4.76b, c).

Substitution of the above solution for u_1 and t_1 into the left-hand side of (3.4.69) shows that all the short periodic terms cancel *identically* and that the long periodic terms also cancel identically if ξ_1 obeys (3.4.80). The remaining constant terms on the left-hand side of (3.4.69) then add up precisely to the value required by the right-hand side of (3.4.69). Thus, the Jacobi integral provides a check for the accuracy of the calculation to first order. This completes the solution to $O(\varepsilon)$.

It is interesting to specialize the foregoing results to the particular initial values for periodic orbits. In order to do so it is necessary to derive transformation relations for expressing the solution as a function of time.

If the quantity $\tilde{\varphi} = \varepsilon\varphi$ is introduced temporarily, Equation (3.4.64a) can easily be inverted to $O(\varepsilon^2)$ in the form

$$\tilde{\theta} = \tilde{\varphi} + \varepsilon F_1(\tilde{\varphi}) + \varepsilon^2\left[F_2(\tilde{\varphi}) + F_1(\tilde{\varphi})\frac{\partial t_1}{\partial\tilde{\theta}}(\varphi, \tilde{\varphi})\right] + O(\varepsilon^3). \quad (3.4.87)$$

Thus, the expansion for t, given by (3.4.62b) becomes

$$t = \varphi + \varepsilon t_1(\varphi, \tilde{\varphi}) + \varepsilon^2\left[t_2(\varphi, \tilde{\varphi}) + F_1(\tilde{\varphi})\frac{\partial t_1}{\partial\tilde{\theta}}(\varphi, \tilde{\varphi})\right] + O(\varepsilon^3). \quad (3.4.88)$$

Equation (3.4.88) can now be inverted to give

$$\varphi = t - \varepsilon t_1(t, \tilde{t})$$
$$+ \varepsilon^2\left[-t_2(t, \tilde{t}) + t_1(t, \tilde{t})\frac{\partial t_1}{\partial\varphi}(t, \tilde{t}) - F_1(\tilde{t})\frac{\partial t_1}{\partial\tilde{\theta}}(t, \tilde{t})\right] + O(\varepsilon^3), \quad (3.4.89)$$

where $\tilde{t} = \varepsilon t$.

In the above t_1, t_2, and F_1 are the functions of φ and $\tilde{\theta}$ already defined; and the notation is self-explanatory.

If Equation (3.4.89) is multiplied by ε and the resulting expression for $\tilde{\varphi}$ in terms of t and $\tilde{t}$ is used in (3.4.87) one obtains the following relation between $\tilde{\theta}$ and the time:

$$\tilde{\theta} = \tilde{t} + \varepsilon F_1(\tilde{t}) + \varepsilon^2[F_2(\tilde{t}) - t_1(t, \tilde{t}) + F_1(\tilde{t})F_1'(\tilde{t})] + O(\varepsilon^3). \quad (3.4.90)$$

Equations (3.4.89) and (3.4.90) can now be used to write the solution as a function of time. The result for $\theta(t)$ is immediate and follows from substituting Equations (3.4.89) and (3.4.90) into (3.4.64a)

$$\theta = \theta_0 + t + F_1(\tilde{t}) + \varepsilon[F_2(\tilde{t}) + F_1(\tilde{t})F_1'(\tilde{t}) - t_1(t, \tilde{t})] + \varepsilon^2\left[F_3(\tilde{t}) + t_1(t, \tilde{t})\frac{\partial t_1}{\partial \varphi}(t, \tilde{t}) - F_1(\tilde{t})\frac{\partial t_1}{\partial \tilde{\theta}}(t, \tilde{t}) - t_2(t, \tilde{t})\right] + O(\varepsilon^3). \quad (3.4.91)$$

We see from the above that knowledge of F_3 is needed to define θ completely to $O(\varepsilon^2)$ as a function of time. This not surprising result is an inherent property of the solution and was discussed in the previous examples of this section. It is only in deriving explicit formulae for the coordinates as functions of time that one encounters the above situation. As pointed out earlier one can use a known exact integral of the motion (such as the Jacobi integral in this case) to calculate F_3 without considering the solution to the next order.

To calculate u as a function of time, one substitutes the expressions for φ and $\tilde{\theta}$ given by Equations (3.4.89) and (3.4.90) into the solution to find

$$u = 1 + \varepsilon u_1(t, \tilde{t}) + \varepsilon^2\left[u_2(t, \tilde{t}) + F_1(\tilde{t})\frac{\partial u_1}{\partial \tilde{\theta}}(t, \tilde{t}) - t_1(t, \tilde{t})\frac{\partial u_1}{\partial \varphi}(t, \tilde{t})\right] + O(\varepsilon^3). \quad (3.4.92)$$

Equations (3.4.91) and (3.4.92) define the solution completely as a function of time.

A necessary and sufficient condition for an orbit to be periodic in a coordinate system rotating with Jupiter is that *both* u and $\theta - t$ be periodic functions of time with the *same* period. Consider first the solution to order ε calculated earlier and expressed here as a function of time:

$$u = 1 + \varepsilon\{\rho_1 \cos[t + \xi_1'(\tilde{t})] + \tfrac{2}{3}\xi_1'(\tilde{t})\} \quad (3.4.93a)$$

$$\theta - t = \xi_1(\tilde{t}) + \varepsilon\{\xi_2(\tilde{t}) - \xi_1'(\tilde{t})[\xi_1(\tilde{t}) - \theta_0] + 2\rho_1 \sin[t + \xi_1'(\tilde{t})]\}. \quad (3.4.93b)$$

Clearly, if $\rho_1 = 0$, the functions above are periodic with the same period, since both ξ_1 and ξ_2 have the same period.[9] The condition $\rho_1 = 0$ implies that $\beta = 0$ and $\gamma = -3\alpha/2$. Thus, for varying values of α one has a one parameter family of long periodic orbits which are the continuations of the family

[9] See Reference 3.4.11 for the solution of $\xi_2 = F_2 - q_1$.

of long periodic orbits close to the triangular libration points [cf. Problem 3.4.7].

Since $r = 1 - \varepsilon u_1 + O(\varepsilon^2)$, the coordinates r and $\theta - t$ for the periodic solution are defined by

$$r = 1 - \tfrac{2}{3}\varepsilon\xi_1'(\tilde{t}) + O(\varepsilon^2) \tag{3.4.94a}$$

$$\theta - t = \xi_1(\tilde{t}) + \varepsilon\{\xi_2(\tilde{t}) - [\xi_1(\tilde{t}) - \theta_0]\xi_1'(\tilde{t})\} + O(\varepsilon^2). \tag{3.4.94b}$$

In view of the behavior of ξ_1 depicted in Figure 3.4.6 we obtain an evolution of periodic orbits from the two lenticular shapes near L_4 and L_5 to coalescence at L_3 and ultimately to horseshoe shaped orbits cnlosing L_3, L_4 and L_5. These orbits were obtained numerically in Reference 3.4.14 and are also reproduced in Figure 9.29 of Reference (3.4.9).

Consider now the behavior of the solution for values of the initial parameters such that ξ_1 approaches zero (or 2π). It is important to note that the condition of eventual close approach to Jupiter occurs, according to Equation (3.4.80a), even if θ_0 is not small as long as E is sufficiently large.

For example, if $\theta_0 = \pi$ and $\alpha = \gamma = 1$, one calculates $E = 45$ from Equation (3.4.80b); and according to Figure 3.4.7, $\xi_1' = 0$ occurs when $\xi_1 \approx 3.5°$ in this case. The choice $\alpha = \gamma = 1$ is certainly consistent with the order of magnitude assumed for the initial conditions in Equations (3.4.61). However, the results predicted for this case are dynamically impossible. To show this one merely need consider the motion relative to Jupiter resulting from the initial conditions near the turning point $\xi_1 = 3.5°$, $\xi_1' = 0$. Since the asteroid is within a distance of order $\mu^{1/2}$ of Jupiter, the gravitational attraction of Jupiter is now dominant and the orbit of the asteroid should describe at least one loop around Jupiter, contrary to the results given by Equations (3.4.87). A more direct evidence of the breakdown of the solution for large values of E is the occurrence of small divisors proportional to powers of $(1 - \cos \xi_1)$ in the solution for u_2 and t_2 in Equations (3.4.77) and (3.4.84). Thus, the asymptotic expansions in the form (3.4.62) are not uniformly valid for large E. This loss of validity for the case of close approach is to be expected since our expansions are constructed with the assumption that the attraction of Jupiter is of order μ and hence small regardless of the distance between Jupiter and the asteroid. A similar but somewhat simpler problem was first solved using matched asymptotic expansions in Reference 2.7.3 for the case of close approach to the moon in earth-to-moon trajectories. Unfortunately, matching of the present solution with one valid close to Jupiter is very difficult and has not been tackled. Solving this problem would answer the question of why the only asteroids observed in Jupiter's orbit are the Trojans. Did the others all collide with Jupiter?

3.4.5 Spin Stabilized Satellite

In a recent paper and thesis Gebman (References 3.4.15 and 3.4.16) has discussed, by combining perturbation theory and numerical solutions, the behavior of a spin-stabilized earth satellite. If the stabilizing rotor fails or is

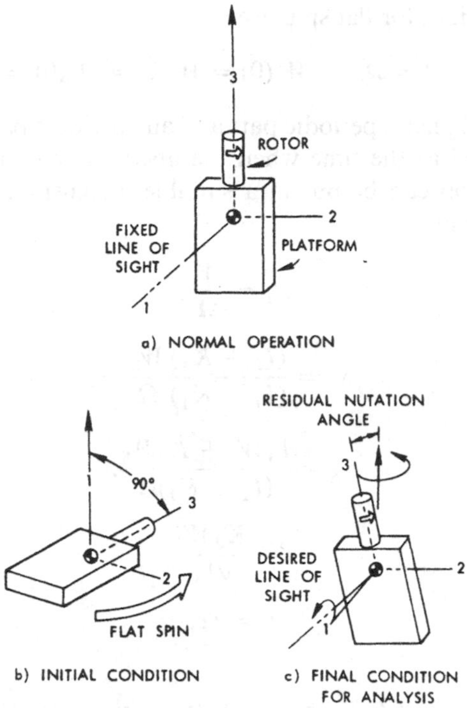

Figure 3.4.8 Illustration of the Relevant Operating States (Reference 3.4.15)

turned off the satellite reverts to an undesirable flat spin, but by restarting the rotor the spacecraft can be made to return to its upright position [cf. Figure 3.4.8]. A sketch of the perturbation analysis is presented here.

The principal moments of inertia for the platform are $I_1 > I_2 + I_3$ and for the rotor, $K_1 = K_2 > K_3$. Now, to write the equation of motion let

$W_{1,2,3}$ represent the components of the platform inertial angular velocity vector projected onto the spacecraft principal axes,

W_S be the bearing axis component of the rotor inertial angular velocity vector,

T mean physical time,

L be the torque, in excess of bearing friction, exerted by the motor on the rotor, and we have the following system:

$$(I_1 + K_1)\frac{dW_1}{dT} = (I_2 + K_1 - I_3)W_2 W_3 - K_3 W_S W_2 \tag{3.4.95a}$$

$$(I_2 + K_1)\frac{dW_2}{dT} = -(I_1 + K_1 - I_3)W_1 W_3 + K_3 W_S W_1 \tag{3.4.95b}$$

$$I_3 \frac{dW_3}{dT} = (I_1 - I_2)W_1 W_2 - L \tag{3.4.95c}$$

$$K_3 \frac{dW_S}{dT} = L. \tag{3.4.95d}$$

The initial conditions for flat spin are

$$W_1(0) = \Omega, \qquad W_2(0) = W_3(0) = W_S(0) = 0. \tag{3.4.96}$$

It turns out that W_3 has a periodic part and an aperiodic part. Recovery from flat spin is defined as the time when the aperiodic part reaches zero. The equation of motion can be put in a suitable dimensionless form with the following definitions

$$x_1 = \frac{W_1}{\Omega}, \tag{3.4.97a}$$

$$x_2 = \frac{(I_2 + K_1)}{(I_1 + K_1)} \frac{W_2}{\Omega}, \tag{3.4.97b}$$

$$x_3 = \frac{I_3 W_3 + K_3 W_S}{(I_1 + K_1)\Omega}, \tag{3.4.97c}$$

$$s = \frac{K_3 W_S}{\alpha I_3}, \tag{3.4.97d}$$

$$t = \alpha T, \tag{3.4.97e}$$

and

$$\alpha = \Omega \left\{ \left(\frac{I_1 + K_1}{I_3} - 1 \right) \left(\frac{I_1 + K_1}{I_2 + K_1} - 1 \right) \right\}^{1/2}. \tag{3.4.98}$$

With these definitions the equations of motion (3.4.95) become, when (3.4.95d) is integrated,

$$\frac{dx_1}{dt} = \frac{1 - \lambda^2}{\lambda} x_3 x_2 - \varepsilon t x_2 \tag{3.4.99a}$$

$$\frac{dx_2}{dt} = -\frac{1}{\lambda} x_3 x_1 + \varepsilon t x_1 \tag{3.4.99b}$$

$$\frac{dx_3}{dt} = \lambda x_1 x_2 \tag{3.4.99c}$$

$$s = \varepsilon t. \tag{3.4.99d}$$

The dimensionless parameters describing the system are

$$\lambda = \left\{ \frac{(I_1 + K_1)/(I_2 + K_1) - 1}{(I_1 + K_1)/I_3 - 1} \right\}^{1/2} \tag{3.4.100a}$$

and

$$\varepsilon = \frac{L/\Omega^2}{(I_1 + K_1 - I_3)\{(I_1 + K_1)/(I_2 + K_1) - 1\}}, \tag{3.4.100b}$$

where ε is the basic small parameter of the system. That is, the torque can be considered to be relatively small. The initial conditions (3.4.96) for this system become

$$x_1(0) = 1, \qquad x_2(0) = x_3(0) = 0. \tag{3.4.101}$$

Of course, the total angular momentum of the spacecraft is conserved so that

$$x_1^2 + x_2^2 + x_3^2 = 1 \tag{3.4.102}$$

is an integral of the motion, directly obvious in (3.4.99).

In order to construct an expansion valid for small ε in the initial phase of the motion a two variable expansion is necessary because of the slow drift of x_1 and the slow variations of frequency and amplitude for x_2 and x_3. After a little trial and error the following form can be arrived at:

$$x_1(t;\varepsilon) = F_{10}(\tilde{t}) + \varepsilon F_{11}(t^*, \tilde{t}) + \cdots \tag{3.4.103a}$$

$$x_2(t;\varepsilon) = \varepsilon F_{21}(t^*, \tilde{t}) + \cdots \tag{3.4.103b}$$

$$x_3(t;\varepsilon) = F_{30}(\tilde{t}) + \varepsilon F_{31}(t^*, \tilde{t}) + \cdots, \tag{3.4.103c}$$

where the slow time is

$$\tilde{t} = \varepsilon \lambda t \tag{3.4.103d}$$

and the fast time is found from

$$\frac{dt^*}{dt} = \psi(\tilde{t}). \tag{3.4.103e}$$

As a consequence, the only $O(1)$ terms come from the right-hand side of the x_2 equation, Equation (3.4.99b) and read

$$-F_{30}F_{10} + \tilde{t}F_{10} = 0. \tag{3.4.104}$$

The non-trivial solution has

$$F_{30} = \tilde{t} \tag{3.4.105}$$

and it follows from the momentum integral, Equation (3.4.102), that if $F_{10}(0) = 1$

$$F_{10} = (1 - \tilde{t}^2)^{1/2}. \tag{3.4.106}$$

Next the $O(\varepsilon)$ terms in the x_2, x_3 equations, Equations (3.4.99b, c) can be collected, using the results for F_{30} and F_{10}:

$$\frac{\partial F_{21}}{\partial t^*}\psi = -\frac{(1 - \tilde{t}^2)^{1/2}}{\lambda}F_{31} \tag{3.4.107a}$$

$$\frac{\partial F_{31}}{\partial t^*}\psi = \lambda(1 - \tilde{t}^2)^{1/2}F_{21} - \lambda. \tag{3.4.107b}$$

In order to have slowly modulated oscillations, the drift part of F_{21} must be removed from the right-hand side of (3.4.107b). Let

$$F_{21} = \frac{1}{(1-\tilde{t}^2)^{1/2}} + \bar{F}_{21}(t^*, \tilde{t}). \tag{3.4.108}$$

Then Equation (3.4.107b) becomes

$$\frac{\partial F_{31}}{\partial t^*}\psi = \lambda(1-\tilde{t}^2)^{1/2}\bar{F}_{21}. \tag{3.4.109}$$

We obtain an oscillator equation from Equations (3.4.107a) and (3.4.109)

$$\frac{\partial^2 F_{31}}{\partial t^{*2}} + \frac{(1-\tilde{t}^2)}{\psi^2}F_{31} = 0. \tag{3.4.110}$$

It is clear that in order to have constant frequency on the fast time scale we must set

$$\psi(\tilde{t}) = \frac{dt^*}{d\tilde{t}} = (1-\tilde{t}^2)^{1/2}. \tag{3.4.111}$$

Integration then defines t^* in the form

$$t^* = \frac{t}{2}\left\{(1-\tilde{t}^2)^{1/2} + \frac{1}{\tilde{t}}\sin^{-1}\tilde{t}\right\}. \tag{3.4.112}$$

Now, taking into account the initial conditions (3.4.101) we have

$$F_{11}(0,0) = 0, \qquad \bar{F}_{21}(0,0) = -1, \qquad F_{31}(0,0) = 0. \tag{3.4.113}$$

The solutions of Equations (3.4.107a) and (3.4.109) are of the form

$$\bar{F}_{21} = A_{21}(\tilde{t})\cos t^* + B_{21}(\tilde{t})\sin t^* \tag{3.4.114a}$$

$$F_{31} = \lambda A_{21}(\tilde{t})\sin t^* - \lambda B_{21}(\tilde{t})\cos t^*. \tag{3.4.114b}$$

It also follows from Equation (3.4.99a) that

$$\frac{\partial F_{11}}{\partial t^*}\psi = -\tilde{t}\lambda\bar{F}_{21}.$$

Thus,

$$F_{11} = -\frac{\tilde{t}}{(1-\tilde{t}^2)^{1/2}}\lambda A_{21}(\tilde{t})\sin t^* + \frac{\tilde{t}\lambda}{(1-\tilde{t}^2)^{1/2}}B_{21}(\tilde{t})\cos \tilde{t}^*. \tag{3.4.114c}$$

This also follows easily from the total momentum integral.

Now to find equations for $A_{21}(\tilde{t})$ and $B_{21}(\tilde{t})$ it is necessary to consider $O(\varepsilon^2)$ terms in the (x_1, x_3) equations, Equations (3.4.99b, c). The condition for non-occurrence of mixed secular terms involves only those forcing terms with $\sin t^*$ and $\cos t^*$. For example, the non-linear terms $F_{11}F_{21}$, $F_{11}F_{31}$

do not need to be considered. Thus, the significant terms are

$$\frac{\partial F_{22}}{\partial t^*}(1-\tilde{t}^2)^{1/2} = -\frac{(1-\tilde{t}^2)^{1/2}}{\lambda}F_{32} - \lambda\frac{\partial \bar{F}_{21}}{\partial \tilde{t}} + \cdots \qquad (3.4.115a)$$

$$\frac{\partial F_{32}}{\partial t^*}(1-\tilde{t}^2)^{1/2} = \lambda(1-\tilde{t}^2)^{1/2}F_{22} + \lambda\frac{F_{11}}{(1-\tilde{t}^2)^{1/2}} - \lambda\frac{\partial F_{31}}{\partial \tilde{t}}. \qquad (3.4.115b)$$

If we differentiate the first of these and use the second we find

$$\frac{\partial^2 F_{22}}{\partial t^{*2}} + F_{22} = \frac{1}{(1-\tilde{t}^2)^{1/2}}\frac{\partial F_{31}}{\partial \tilde{t}} - \frac{F_{11}}{1-\tilde{t}^2} - \frac{\lambda}{(1-\tilde{t}^2)^{1/2}}\frac{\partial^2 \bar{F}_{21}}{\partial t^*\,\partial \tilde{t}}. \qquad (3.4.116)$$

Referring to Equations (3.4.114) and setting the coefficients of $\cos t^*$, $\sin t^*$ equal to zero we find

$$2\frac{dA_{21}}{d\tilde{t}} = -\frac{\tilde{t}}{1-\tilde{t}^2}A_{21}, \qquad 2\frac{dB_{21}}{d\tilde{t}} = -\frac{\tilde{t}}{(1-\tilde{t}^2)}B_{21}. \qquad (3.4.117)$$

Thus,

$$A_{21} = a_{21}(1-\tilde{t}^2)^{1/4}, \qquad B_{21} = b_{21}(1-\tilde{t}^2)^{1/4}. \qquad (3.4.118)$$

The initial conditions of Equations (3.4.113) show that $a_{21} = -1$, $b_{21} = 0$. Collecting all the results in Equations (3.4.103) we find

$$x_1(t;\varepsilon) = (1-\tilde{t}^2)^{1/2} + \varepsilon\frac{\lambda\tilde{t}}{(1-\tilde{t}^2)^{1/4}}\sin t^* + \cdots \qquad (3.4.119a)$$

$$x_2(t;\varepsilon) = \frac{\varepsilon}{(1-\tilde{t}^2)^{1/2}} - \varepsilon(1-\tilde{t}^2)^{1/4}\cos t^* + \cdots \qquad (3.4.119b)$$

$$x_3(t;\varepsilon) = \tilde{t} - \varepsilon\lambda(1-\tilde{t}^2)^{1/4}\sin t^* + \cdots. \qquad (3.4.119c)$$

During the initial phase of motion just calculated the motion is dominated by coupling between the rotations around the x_2, x_3 axes. The x_1 angular-momentum decreases during the first phase. But it is clear from the form of the expansion for x_1 that the singularity at $\tilde{t} = 1$ makes the expansion invalid near $\tilde{t} = 1$. The rapid buildup of the x_2 angular momentum couples this with the x_1 momentum in the next phase of the motion, while $x_3 \to 1$. This turns out to be a relatively rapid phase on the $\tilde{t}$ scale and can be described by a limit process expansion. The general form of the expansion is

$$x_1(t;\varepsilon) = \mu_{11}(\varepsilon)G_{11}(\hat{t}) + \cdots \qquad (3.4.120a)$$

$$x_2(t;\varepsilon) = \mu_{21}(\varepsilon)G_{21}(\hat{t}) + \cdots \qquad (3.4.120b)$$

$$x_3(t;\varepsilon) = 1 + \mu_{31}(\varepsilon)G_{31}(\hat{t}) + \cdots, \qquad (3.4.120c)$$

where

$$\hat{t} = \frac{\tilde{t}-1}{\delta(\varepsilon)}. \qquad (3.4.121)$$

The limit process keeps $\hat{t}$ fixed as $\varepsilon \to 0$. The μ's and $\delta(\varepsilon)$ are to be found to give a distinguished limit, but we expect $\mu_{11}, \mu_{21} \gg \varepsilon$ because of the singularities in Equations (3.4.119a, b). The crucial equation for deciding about the orders of magnitude is the G_{21} equation, because the largest forcing terms cancel. First note the total angular momentum integral (3.4.102)

$$2G_{31} + G_{11}^2 = 0, \qquad \mu_{31} = \mu_{11}^2 \tag{3.4.122}$$

if we assume provisionally that $\mu_{11} \gg \mu_{21}$. Then the x_2, x_1 equations, Equations (3.4.99a, b), become

$$\varepsilon\lambda \frac{\mu_{11}}{\delta}\frac{dG_{11}}{d\hat{t}} = -\lambda\mu_{21}G_{21} \tag{3.4.123a}$$

$$\varepsilon\lambda \frac{\mu_{21}}{\delta}\frac{dG_{21}}{d\hat{t}} = \frac{\mu_{11}^3}{2\lambda}G_{11}^3 + \frac{\delta\mu_{11}}{\lambda}\hat{t}G_{11} \tag{3.4.123b}$$

where Equation (3.4.122) has been used. The distinguished limit has

$$\frac{\varepsilon\mu_{11}}{\delta} = \mu_{21}, \qquad \frac{\varepsilon\mu_{21}}{\delta} = \mu_{11}^3 = \delta\mu_{11}$$

so that

$$\mu_{11} = \varepsilon^{1/3}, \qquad \mu_{21} = \delta = \varepsilon^{2/3}. \tag{3.4.124}$$

The limiting equations are

$$\lambda\frac{dG_{11}}{d\hat{t}} = -G_{21} \tag{3.4.125a}$$

$$\lambda\frac{dG_{21}}{d\hat{t}} = \frac{1}{2\lambda}G_{11}^3 + \frac{1}{\lambda}\hat{t}G_{11}. \tag{3.4.125b}$$

Elimination yields the basic transition layer equation

$$\frac{d^2G_{11}}{d\hat{t}^2} - \frac{1}{\lambda^2}\{\tfrac{1}{2}G_{11}^3 + \hat{t}G_{11}\} = 0. \tag{3.4.126}$$

With the following change of scale,

$$G_{11} = \lambda^{1/3}y(\tau), \qquad \hat{t} = \lambda^{2/3}\tau,$$

this equation takes a canonical form

$$\frac{d^2y}{d\tau^2} + \frac{1}{2}y^3 + \tau y = 0. \tag{3.4.127}$$

The expansions in Equations (3.4.120) are of the form

$$x_1 = \varepsilon^{1/3}G_{11}(\hat{t}) + \cdots = \varepsilon^{1/3}\lambda^{1/3}y(\tau) + \cdots \tag{3.4.128a}$$

$$x_2 = \varepsilon^{2/3}G_{21}(\hat{t}) + \cdots \tag{3.4.128b}$$

$$x_3 = 1 - \frac{\varepsilon^{2/3}}{2}G_{11}^2(\hat{t}) + \cdots \tag{3.4.128c}$$

$$\tau = \frac{\hat{t}}{\lambda^{2/3}} = \frac{\tilde{t} - 1}{(\varepsilon\lambda)^{2/3}}. \tag{3.4.128d}$$

Equation (3.4.127) has the form of a nonlinear Airy equation. It brings the motion from one oscillatory regime to another. Not all solutions of this equation are of interest here but only that solution which matches to the earlier expansion as $\tau \to -\infty$ and carries the motion through to its next phase. The integration of Equation (3.4.127) is carried out numerically, but an advantage of the perturbation method is that only a single integration is needed to cover a set of parameter values. It is clear that as $\tau \to -\infty$, y must become large so that the $O(\varepsilon^{1/3})$ term in Equation (3.4.128a) can match with the $O(1)$ term in Equation (3.4.119a). As $\tau \to -\infty$ a solution of Equation (3.4.127) which is large balances $\frac{1}{2}y^3$ with $-\tau y$ and has the expansion

$$y(\tau) = \sqrt{-2\tau} + \frac{1}{(-2\tau)^{5/2}} + \cdots. \tag{3.4.129}$$

Formal matching can be carried out with the help of an intermediate limit (τ_ξ fixed)

$$\tau_\xi = \frac{\tilde{t} - 1}{\xi(\varepsilon)\lambda^{2/3}}, \tag{3.4.130}$$

where $\xi(\varepsilon)$ is such that

$$\varepsilon^{2/3} \ll \xi(\varepsilon) \ll 1.$$

Thus, under this limit

$$\tilde{t} = 1 + \lambda^{2/3}\xi(\varepsilon)\tau_\xi \to 1 \tag{3.4.131a}$$

$$\tau = \frac{\xi\tau_\xi}{\varepsilon^{2/3}} \to -\infty, \qquad \tau_\xi < 0. \tag{3.4.131b}$$

Comparing the initially valid expansion, Equation (3.4.119a), and the transition expansion, Equation (3.4.128a), we see:

Initially Valid

$$x_1 = \{1 - (1 + \lambda^{2/3}\xi\tau_\xi)^2\}^{1/2} + O\left(\frac{\varepsilon}{\sqrt{\xi}}\right)$$

$$x_1 \to \lambda^{1/3}\sqrt{-2\xi\tau_\xi} + O\left(\frac{\varepsilon}{\sqrt{\xi}}, \xi^{3/2}\right) \tag{3.4.132a}$$

Transition

$$x_1 = \varepsilon^{1/3} G_{11}\left(\frac{\lambda^{2/3}\xi\tau_\xi}{\varepsilon^{2/3}}\right) = \varepsilon^{1/3}\lambda^{1/3} y\left(\frac{\xi\tau_\xi}{\varepsilon^{2/3}}\right)$$

$$x_1 \to \varepsilon^{1/3}\lambda^{1/3}\left\{\sqrt{\frac{\xi 2(-\tau_\xi)}{\varepsilon^{2/3}}} + O\left(\frac{\varepsilon^{5/3}}{\xi^{5/2}}\right)\right\} \qquad (3.4.132b)$$

upon using Equation (3.4.130).

The $O(\sqrt{\xi})$ terms match identically. The special solution of the transition equation, Equation (3.4.127), defined by the behavior as $\tau \to -\infty$ can now be found by numerical integration. The behavior of this solution as $\tau \to +\infty$ can also be analyzed. The order of magnitude of the momentum about the x_1-axis continues to decrease as the satellite passes through the transition region. Thus as $\tau \to +\infty$ the transition equation is approximated by the Airy equation

$$\frac{d^2 y}{d\tau^2} + \tau y = 0. \qquad (3.4.133)$$

The solution eventually is given by the asymptotic form of the Airy function $Ai(-\tau)$

$$y \to \frac{A_0}{\tau^{1/4}} \cos(\tfrac{2}{3}\tau^{3/2} + \phi_0), \qquad (3.4.134)$$

where A_0, ϕ_0 are determined by the numerical integration. Numerical integration of Equation (3.4.127) starting from a large negative τ according to Equation (3.4.130) and proceeding to a large positive τ agrees with Equation (3.4.134) and gives (for details see References 3.4.15–16)

$$A_0 = 0.939, \qquad \phi_0 = 116°. \qquad (3.4.135)$$

After passing through the transition the satellite enters the final phase of motion. The asymptotic form of the transition expansion indicates that (see below) x_1, x_2 become of the same order and are coupled in the final phase. The form of the motion is again an oscillation with slowly varying phase. The $\tilde{t}$ coordinate again is the slow coordinate in the final phase. Matching between the transition expansion and the final expansion depends on the intermediate limit (τ_ζ fixed)

$$\tau_\zeta = \frac{\tilde{t} - 1}{\lambda^{2/3}\zeta(\varepsilon)}, \qquad (3.4.136)$$

where

$$\varepsilon^{2/3} \ll \zeta(\varepsilon) \ll 1.$$

Thus,

$$\tau = \frac{\zeta}{\varepsilon^{2/3}} \tau_\zeta \to \infty, \qquad \tau_\zeta > 0 \qquad (3.4.137a)$$

$$\tilde{t} = 1 + \lambda^{2/3}\zeta\tau_\zeta \to 1. \qquad (3.4.137b)$$

It follows from Equations (3.4.128a), (3.4.134) and (3.4.125a) that under this limit

$$x_1 = \varepsilon^{1/3}\lambda^{1/3}y\left(\frac{\zeta\tau_\zeta}{\varepsilon^{2/3}}\right) \to O\left(\frac{\varepsilon^{1/3}}{\left(\frac{\zeta\tau_\zeta}{\varepsilon^{2/3}}\right)^{1/4}}\right) = O(\varepsilon^{1/2}) \qquad (3.4.138a)$$

and

$$x_2 = \varepsilon^{2/3}G_{21}(\tilde{t}) \to O\left(\varepsilon^{2/3}\left(\frac{\zeta\tau_\zeta}{\varepsilon^{2/3}}\right)^{1/4}\right) = O(\varepsilon^{1/2}). \qquad (3.4.138b)$$

The order of x_3 follows from the total momentum integral.

We now summarize briefly the results for the final phase of motion. The general form of the expansion in the final region is

$$x_1 = \varepsilon^{1/2}H_{11}(t^+, \tilde{t}) + \varepsilon^{3/2}H_{12}(t^+, \tilde{t}) + \cdots \qquad (3.4.139a)$$

$$x_2 = \varepsilon^{1/2}H_{21}(t^+, \tilde{t}) + \varepsilon^{3/2}H_{22}(t^+\tilde{t}) + \cdots \qquad (3.4.139b)$$

$$x_3 = 1 + \varepsilon H_{31}(t^+, \tilde{t}) + \varepsilon^2 H_{32}(t^+, \tilde{t}) + \cdots. \qquad (3.4.139c)$$

The fast time is defined by

$$\frac{dt^+}{dt} = \chi(\tilde{t}), \qquad \tilde{t} = \varepsilon\lambda t \qquad (3.4.140)$$

and χ is found from the requirement of constant frequency on the fast scale. The equations for H_{11}, H_{21} are the basic coupled system and provide the general form of the oscillations

$$H_{11} = A_{11}(\tilde{t})\cos t^+ + B_{11}(\tilde{t})\sin t^+ \qquad (3.4.141)$$

etc.

Consideration of the equation for H_{12} provides the condition for non-occurrence of mixed secular terms. The equations are easily integrated with the help of the total momentum integral. The results are

$$H_{11} = -k_{11}\left(\frac{\Gamma_1}{\Gamma_2}\right)^{1/4} \cos(t^+ + \phi(\tilde{t})) \qquad (3.4.142a)$$

$$H_{21} = -k_{11}\left(\frac{\Gamma_2}{\Gamma_1}\right)^{1/4} \sin(t^+ + \phi(\tilde{t})) \qquad (3.4.142b)$$

$$H_{31} = -\frac{k_{11}^2}{4}\frac{\Gamma_1 + \Gamma_2}{\sqrt{\Gamma_1\Gamma_2}} - \frac{k_{11}^2\lambda^2}{\sqrt{\Gamma_1\Gamma_2}}\cos(2t^+ + 2\phi(\tilde{t})), \qquad (3.4.142c)$$

where

$$\Gamma_1 = \tilde{t} - 1 + \lambda^2, \qquad \Gamma_2 = \tilde{t} - 1 \qquad (3.4.143a)$$

and

$$\chi = \frac{1}{\lambda}(\Gamma_1\Gamma_2)^{1/2} \qquad (3.4.143b)$$

so that

$$t^+ = \frac{1}{\varepsilon\lambda^2}\left\{\frac{1}{2}\left(\Gamma_2 \frac{\lambda^2}{2}\right)(\Gamma_1\Gamma_2)^{1/2} - \frac{\lambda^4}{8}\log[2(\Gamma_1\Gamma_2)^{1/2} + 2\Gamma_2 + \lambda^2] + c_\omega\right\}. \tag{3.4.144}$$

The slowly varying phase $\phi(\tilde{t})$ is

$$\phi(\tilde{t}) = \frac{A_0^2}{8\lambda^2}\{\tfrac{3}{2}\lambda^2(\log\Gamma_2 - (1-\lambda^2)\log\Gamma_1) + (4 - 2\lambda^2)\tilde{t}\} + c_\phi. \tag{3.4.145}$$

In the above c_ω and c_ϕ are constants.

Now with the help of the intermediate limit, Equation (3.4.136), the matching between the final phase expansion, Equations (3.4.139), and the transition expansion, Equations (3.4.128), can be carried out. This results in the determination of the constants

$$c_\omega = \frac{\lambda^4}{\delta}\log\lambda^2, \qquad k_{11} = 0.939$$
$$c_\phi = \left(\frac{116°}{180°} - 1\right)\pi - \frac{k_{11}^2}{4}\{2 - \lambda^2 - \tfrac{3}{4}(1-\lambda^2)\log\lambda^2\}. \tag{3.4.146}$$

These results (in particular Equations (3.4.142)) are used in References (3.4.15) and (3.4.16) to estimate the final recovery time and nutation angle as well as to discover a recovery rule that

$$G = \text{const.} = \Omega T\eta^2. \tag{3.4.147}$$

The value of G depends on the inertia properties of the vehicle, Ω is the initial flat spin rate, T the time to final state, and η the residual nutation angle. The results of a complete numerical integration of the original system Equations (3.4.99) agree very well with the asymptotic analysis presented here as long as $\varepsilon < 0.1$ say and $\lambda \gg \varepsilon$.

Problems

1. Reconsider the problem of Section 3.4.2 for a lifting satellite with constant lift coefficient C_L. The lift force is given by

$$L = \tfrac{1}{2}\rho V^2 S C_L$$

and acts in the direction normal to the velocity vector. Show in particular, that to $O(1)$ the effect of lift is a slow motion of the perigee according to

$$\omega(\tilde{\theta}) = -\frac{C_L}{2C_D}\log(1 + 2\tilde{\theta}).$$

2. Consider the problem of a spherical satellite (no lift) in the presence of an exponentially varying atmosphere. Assume that the atmospheric density is given by

$$\rho = \rho_0 e^{(R-R_0)/H},$$

where H is the scale height of the atmosphere. Values for ρ_0 and H can be calculated from the following "typical" atmosphere $\rho = 3.65 \times 10^{-15}$ kg/m^3 at 1000 km altitude and $\rho = 5.604 \times 10^{-7}$ kg/m^3 at 100 km altitude.

Assume that initially the satellite is at apogee at 500 km with $e = 0.01$ and you wish to predict the decay of the orbit down to 100 km. Use appropriate dimensionless variables to model the density variation in the above altitude range and calculate the solution of the Keplerian elements a, e, and ω.

3. For the problem discussed in Section 3.4.2 ($C_L = 0$, $\rho =$ const.) calculate the time history to $O(1)$. In particular derive the expression for $\tau(\tilde{\theta})$ by considering the terms of $O(\varepsilon^2)$ in the solution.

4. Consider the restricted three-body problem defined by Equations (3.4.32). Transform the equations to barycentric rotating coordinates x, y defined by

$$x = x' + 1 - \mu$$

$$y = y'$$

to obtain

$$\frac{d^2x}{dt^2} = -\frac{(1-\mu)}{r_1^3}(x+\mu) - \mu\frac{(x-1+\mu)}{r_2^3} + 2\frac{dy}{dt} + x$$

$$\frac{d^2y}{dt^2} = -(1-\mu)\frac{y}{r_1^3} - \mu\frac{y}{r_2^3} - 2\frac{dx}{dt} + y,$$

where

$$r_1^2 = (x+\mu)^2 + y^2$$

$$r_2^2 = (x-1+\mu)^2 + y^2.$$

Multiply the first equation by dx/dt, the second equation by dy/dt and add the result to derive the Jacobi integral

$$\frac{1}{2}\left[\left(\frac{dx}{dt}\right)^2 + \left(\frac{dy}{dt}\right)^2\right] - \frac{1}{2}(x^2+y^2) - \frac{(1-\mu)}{r_1} - \frac{\mu}{r_2} = -\frac{C}{2},$$

where C is the Jacobi constant of integration.

By studying the curves of zero velocity, determine the limiting value of C for which the motion remains bounded within a region centered around either the earth or moon if it starts there.

Calculate the same result in moon-centered rotating coordinates x', y' as well as earth-centered rotating coordinates x'', y''. Use the former expression under limit process of Equation (3.4.38) to derive the limiting version of the Jacobi integral for motion close to the moon. and verify your result by direct derivation of this limiting form from Equations (3.4.39). Next use the form of the Jacobi integral in earth centered x'', y'' rotating coordinates to derive Equation (3.4.58).

5. (a) Introduce moon-centered nonrotating coordinates in Equations (3.4.39) and show that in terms of these the first perturbation term to Keplerian motion is $O(\varepsilon^2)$, thus directly verifying the result derived in Section 3.4.3 that for a close satellite of the moon the motion to $O(1)$ is a nonrotating Kepler ellipse.
   (b) Continue the solution of Section 3.4.3 to $O(\varepsilon)$ by considering boundedness of u_2 and p_2. Carry out the calculations for small e, correct to $O(e)$. In particular evaluate A_1, B_1 and $\tilde{p}_1$ as well as the constant α_1.

6. Calculate the time history of the motion correct to $O(1)$ for the problem in Section 3.4.3 in two ways.

   First derive the result by quadrature from the expression for p making use of the solution you derived in Problem 3.4.5 as well as a direct derivation of the *long periodic terms of order* ε^2. Next, make use of the Jacobi integral together with the solution you calculated in Problem 3.4.5 to derive the missing long periodic terms of order ε^2.

7. Consider the restricted three body problem in earth-centered rotating coordinates, i.e.,

$$\frac{d^2x''}{dt^2} = -(1-\mu)\frac{x''}{r''^3} - \mu\frac{(x''-1)}{r_2''^3} + 2\frac{dy''}{dt} + x'' - \mu$$

$$\frac{d^2y''}{dt^2} = -(1-\mu)\frac{y''}{r''^3} - \mu\frac{y''}{r_2''^3} - 2\frac{dx''}{dt} + y'',$$

where

$$x'' = x + \mu; \qquad y'' = y$$

$$r''^2 = x''^2 + y''^2; \qquad r_2''^2 = (x''-1)^2 + y''^2.$$

(a) Show that the points L_4 and L_5 with coordinates $x'' = \frac{1}{2}$, $y'' = \pm\sqrt{3}/2$ are equilibrium points of the preceding equations. These were discovered by Lagrange (hence the L notation) and form two equilateral triangles with the earth-moon axis as a common base.

(b) Show also that these equations have the straight line equilibrium points L_1, L_2, L_3 whose coordinates are defined by $y'' = 0$ and with x'' a solution of the equation

$$x'' - \mu + \varepsilon_1\frac{(1-\mu)}{x''^2} + \varepsilon_2\frac{\mu}{(x''-1)^2} = 0,$$

where

	L_1	L_2	L_3
ε_1	$+1$	-1	-1
ε_2	$+1$	$+1$	-1

(c) For $\mu \ll 1$ show that the x'' coordinates of L_1, L_2, L_3 are given by

$$L_1: x_1'' = -1 + \frac{7}{12}\mu - \frac{7}{144}\mu^2 + O(\mu^3)$$

$$L_2: x_2'' = 1 - \left(\frac{\mu}{3}\right)^{1/3} + \frac{3^{1/3}}{9}\mu^{2/3} + O(\mu)$$

$$L_3: x_3'' = 1 + \left(\frac{\mu}{3}\right)^{1/3} + \frac{3^{1/3}}{9}\mu^{2/3} + O(\mu).$$

Thus L_1 is located approximately one unit away from the earth in opposition to the moon, while L_2 and L_3 straddle the moon.

(d) Linearize the equations of motion in the neighborhood of $L_4: x'' = \frac{1}{2}, y'' = \sqrt{3}/2$ by introducing

$$\xi = x'' - \tfrac{1}{2}$$

$$\eta = y'' - \frac{\sqrt{3}}{2}$$

and show that for ξ, η small the motion is governed by

$$\frac{d^2\xi}{dt^2} - 2\frac{d\eta}{dt} = \frac{3}{4}\xi + \frac{3\sqrt{3}}{4}(1 - 2\mu)\eta$$

$$\frac{d^2y}{dt^2} + 2\frac{d\xi}{dt} = \frac{9}{4}\eta + \frac{3\sqrt{3}}{4}(1 - 2\mu)\xi.$$

Look for a solution in the form

$$\xi = Ae^{\lambda t}; \qquad \eta = Be^{\lambda t}$$

and show that the characteristic equation for λ obeys

$$\lambda_{1,2} = \tfrac{1}{2}\sqrt{-1 \pm \sqrt{1 - 27\mu(1 - \mu)}}.$$

Thus, the motion near L_4 is stable if

$$1 - 27\mu(1 - \mu) < 0; \qquad \text{i.e., } \mu < \tfrac{1}{2} - \sqrt{\tfrac{1}{4} - \tfrac{1}{27}} \approx 0.0385$$

which is the case for the earth-moon system where $\mu = 0.012$.

(e) For $\mu < 0.0385$ the four roots are

$$\lambda_{1,2} = \pm is_1, \qquad \lambda_{3,4} = \pm is_2,$$

with

$$s_1 = (\tfrac{27}{4}\mu)^{1/2} + O(\mu^{3/2}),$$

and

$$s_2 = 1 - \tfrac{27}{8}\mu + O(\mu^2).$$

Corresponding to each root one can find periodic orbits about L_4. Those with frequency s_1 are long periodic while those with frequency s_2 are short periodic. Calculate and sketch these two families of periodic orbits.

(f) Show that the motion near L_1, L_2 and L_3 are unstable for any $\mu > 0$.

8. Consider the perturbation expansion of Equations (3.4.57) in the form

$$u = u_0(\theta, \tilde{\theta}) + \varepsilon u_1(\theta, \tilde{\theta}) + \cdots$$

$$t = t_0(\theta, \tilde{\theta}) + \varepsilon t_1(\theta, \tilde{\theta}) + \cdots$$

with $\tilde{\theta} = \varepsilon(\theta - \theta_0)$.

Show that if in the first approximation u_0, t_0 we consider a nearly circular ($e \approx 0$) orbit with semi major axis $a = ((n - m)/n)^{2/3}$ where n and m are relatively prime integers, we will encounter small divisors in the solution to $O(\varepsilon)$.

9. (a) Show that the choice of $\sigma = \lambda = \frac{1}{2}$ in Equations (3.4.60) is the only consistent one for the given initial conditions. ($a \approx 1, e \approx 0$)

(b) Instead of the expansion (3.4.62) let u and t depend on θ, and $\tilde{\theta}$. Show by direct calculation why the result contains inconsistent terms which cannot be removed.

10. Show that the basic system Equations (3.4.99) yield the first approximation equations

$$\frac{\partial^2 H_{11}}{\partial t^{+2}} + \frac{\Gamma_1 \Gamma_2}{\lambda^2 \chi^2} H_{11} = 0$$

$$H_{21} = -\frac{\lambda\psi}{\Gamma_1} \frac{\partial H_{11}}{\partial t^+}.$$

Show also that the slowly varying parts of the oscillation can be found from the equations of the next order

$$2\frac{\partial^2 H_{12}}{\partial t^{+2}} + H_{12} = \frac{\lambda^2}{\Gamma_2} \frac{\partial H_{21}}{\partial \tilde{t}} + \frac{1}{\Gamma_2} H_{31} H_{11} - \frac{\lambda^2}{(\Gamma_1 \Gamma_2)^{1/2}} \frac{\partial^2 H_{11}}{\partial t^+ \, \partial \tilde{t}}$$

$$+ \frac{1 - \lambda^2}{(\Gamma_1 \Gamma_2)^{1/2}} \frac{\partial}{\partial t^+} (H_{31} H_{21}).$$

References

3.4.1 Marquis de Laplace, *Mécanique Céleste*, Translated by Nathaniel Bowditch, Vol. 1, Book II, Chap. 15, p. 517. Hillard, Gray, Little, and Wilkins, Boston, Mass, 1829.

3.4.2 J. M. A. Danby, *Fundamentals of Celestial Mechanics*, Chapter 6. The Macmillan Co., New York, 1962.

3.4.3 R. A. Struble, A Geometrical Derivation of the Satellite Equation, *J. Math. Anal. Appl.* **1** (1960), p. 300.

3.4.4 J. Kevorkian, *Lectures in Applied Mathematics*, Vol. 7, *Space Mathematics, Part III*, American Mathematical Society, Providence, Rhode Island, 1966, pp. 206–275.

3.4.5 M. C. Eckstein, Y. Y. Shi, and J. Kevorkian, Satellite motion for all inclinations around an oblate planet, *Proceedings of Symposium No. 25, International Astronomical Union*. Academic Press, New York, 1966, pp. 291–332.

3.4.6 M. C. Eckstein, Y. Y. Shi, and J. Kevorkian, Satellite motion for arbitrary eccentricity and inclination around the smaller primary in the restricted three-body problem, *The Astronomical Journal* **71**, 1966, pp. 248–263.

3.4.7 M. C. Eckstein, and Y. Y. Shi, Asymptotic solutions for orbital resonances due to the general geopotential, *The Astronomical Journal* **74**, 1969, pp. 551–562.

3.4.8 M. C. Eckstein, Y. Y. Shi, and J. Kevorkian, Use of the energy integral to evaluate higher-order terms in the time history of satellite motion, *The Astronomical Journal* **71**, 1966, pp. 301–305.

3.4.9 Victor Szebehely, *Theory of Orbits, The Restricted Problem of Three Bodies*, Academic Press, New York, 1967.

3.4.10 E. W. Brown, and C. A. Shook, *Planetary Theory*, Cambridge University Press, New York and London, 1933. Reprinted by Dover Publications, New York.

3.4.11 J. Kevorkian, The planar motion of a trojan asteroid, *Periodic Orbits, Stability, and Resonances*, G. E. O. Giacaglia, (editor), D. Reidel Publishing Company, Dordrecht, Holland 1970, pp. 283–303.

3.4.12 Balint Erdi, An asymptotic solution for the trojan case of the plane elliptic restricted problem of three bodies, *Celestial Mechanics*, Vol. 15, 1977, pp. 367–383.

3.4.13 Balint Erdi, The three-dimensional motion of trojan asteroids, *Celestial Mechanics* **18**, 1978, pp. 141–161.

3.4.14 E. Rabe, Determination and survey of periodic trojan orbits in the restricted problem of three bodies, *The Astronomical Journal* **66**, 1961, p. 500.

3.4.15 J. R. Gebman and D. L. Mingori, Perturbation solution for the flat spin recovery of a dual-spin spacecraft *AIAA J.* **14**, No. 7. July 1976, pp. 859–867.

3.4.16 J. R. Gebman, Perturbation analysis of the flat spin recovery of a dual-spin spacecraft, Ph.D. Dissertation. School of Engineering and Applied Science, University of California, Los Angeles, 1974.

3.5 Higher Order Problems

In the preceding sections of this chapter we have concentrated on second order equations. This was essentially true even in Section 3.4 where the availability of integrals of motion for Keplerian orbits reduced the calculation of planar motions to a second order equation for the orbit and a quadrature for the time history.

Higher order problems arise in many interesting applications and present several new features which will be illustrated by a series of examples in this section.

In Section 3.5.1 we consider a linear weakly coupled system of generalized oscillators and show how the ideas developed earlier now apply to each of the normal modes with no essential modification. However, when considering weakly nonlinear problems, as in Section 3.5.2, the use of one fast time and one slow time is insufficient to represent the solution and we will need to develop our results in terms of N times $t_0 = t, t_1 = \varepsilon t, t_2 = \varepsilon^2 t, \ldots, t_N = \varepsilon^N t$. The choice of the integer N will depend on the accuracy we wish to maintain in our solution. For example, in order to calculate the dependence of the solution to $O(1)$ on t_0, t_1 and t_2 we must take $N = 2$. In doing so, we will also calculate the dependence of the solution of order ε on t_0 and t_1 (but not t_2) and the dependence of the solution of order ε^2 on t_0 only, etc. The need for $N > 2$ times for systems of coupled oscillators was recognized by Kabakow in Reference 3.5.1 and the material in Section 3.5.2 is based on his thesis as well as the work in Reference 3.5.2.

The literature contains numerous investigations where the solution must be developed in terms of $N > 2$ times. We cite References 3.4.6, 3.5.3, and 3.5.4 as typical examples of early work using $N > 2$ times in other applications.

Another new feature associated with higher-order problems is the phenomenon of resonance which we first encountered in Section 3.4.4 and in Problem 3.4.8. Typically, resonance is manifested by the occurrence of zero divisors in some nominal representation of the solution. These divisors are usually in the form $m\omega_1 - n\omega_2$ where ω_1 and ω_2 are two natural frequencies and m and n are relatively prime integers.

The literature in celestial mechanics contains numerous studies of resonance where certain combinations of constants arising either in the differential equation, or initial conditions occur as zero divisors. We cite References 3.4.5, 6, 7, 11, 12, 13 as a representative, but by no means complete, collection of satellite problems exhibiting resonance that have been solved using the multiple variable expansion procedure. In all the above problems, as well as the one we consider in Section 3.5.2, the type of solution, i.e., whether it is non-resonant or resonant, is unaltered with time and is determined a priori by the constant parameters of the problem. In contrast, if the problem involves slowly varying parameters, e.g., if the natural frequencies ω_1 and ω_2 are given slowly varying functions of time, these parameters may evolve such that the solution passes through a state of resonance.

The problem of passage through resonance for systems with slowly varying parameters is significantly more complicated, and we devote Sections 3.5.3–3.5.5 to study of this case.

In Section 3.5.3 we establish the appropriate expansion procedure to represent the solution away from resonance for problems of this type. Actually, all the essential ideas of passage through resonance can be illustrated with the example of a single forced oscillator with slowly varying frequency as first discussed in Reference 3.5.5. Therefore, we digress from higher order problems in Section 3.5.4 to consider this example in some detail.

Applications to higher order systems can be modelled by the example of two oscillators with given slowly varying frequencies and weak nonlinear coupling. In Section 3.5.5 we follow the developments presented in the survey article (Reference 3.5.6) devoted to this problem.

Passage through resonance can also occur in more complicated dynamical systems where one (or more) of the slowly varying frequencies is an unknown obeying some differential equation. We will not consider this type of problem in this book. An example modelling a problem in flight mechanics is discussed in References 3.5.7 and 3.5.8. These references also give a discussion of the interesting phenomenon of sustained resonance where, under certain conditions, the unknown frequency will lock onto the given frequency indefinitely.

3.5.1 Linear Coupled System, Small Damping

Various physical systems lead to a system of differential equations formally similar to the linear oscillator equations studied in Section 3.2.1 when expressed in matrix notation

$$M\frac{d^2\mathbf{x}}{dt^2} + \varepsilon B\frac{d\mathbf{x}}{dt} + K\mathbf{x} = 0. \tag{3.5.1}$$

Here M, B, K are considered to be positive definite symmetric matrices, $\mathbf{x}$ an n-dimensional vector $\mathbf{x} = (x_1, x_2, \ldots, x_n)$. The two-variable formalism of Section 3.2.1 can be used to great advantage to discuss the general properties of the system represented by Equations (3.5.1).

One physical system which leads to Equations 3.5.1 is the system of masses, linear springs and linear dampers sketched in Figure 3.5.1. Every mass is coupled to every other mass with linear springs and dampers. The physical masses are M_j, springs K_{ij}, dampers B_{ij}, the deflection from equilibrium is $X_j(T)$, T = physical time, and $i, j = 1, \ldots, n$. Dimensionless variables can be

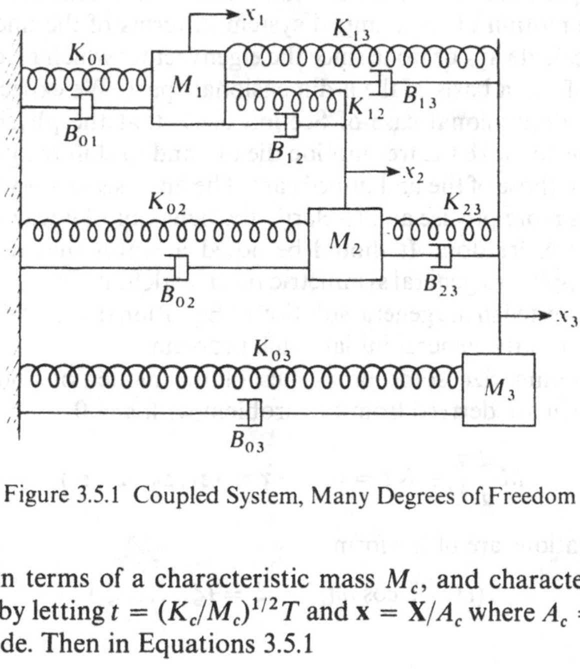

Figure 3.5.1 Coupled System, Many Degrees of Freedom

introduced in terms of a characteristic mass M_c, and characteristic spring constant K_c by letting $t = (K_c/M_c)^{1/2}T$ and $\mathbf{x} = \mathbf{X}/A_c$ where A_c = characteristic amplitude. Then in Equations 3.5.1

$$M = \frac{1}{M_c}\begin{pmatrix} M_1 & 0 & \cdots & 0 \\ 0 & M_2 & 0 & \cdots \\ \vdots & 0 & \ddots & \\ 0 & \vdots & & M_n \end{pmatrix} = (m_{ij}) \tag{3.5.2a}$$

$$B = \frac{1}{B_c}\begin{pmatrix} B_{01} + B_{12} + B_{13} + \cdots & -B_{12} & -B_{13} & \cdots \\ -B_{12} & B_{02} + B_{12} + B_{32} + \cdots & & \\ -B_{1n} & -B_{2n} & \cdots & \end{pmatrix} \equiv (b_{ij}) \tag{3.5.2b}$$

$$K = \frac{1}{K_c}\begin{pmatrix} K_{01} + K_{12} + K_{13} + \cdots & -K_{12} & -K_{13} & \cdots \\ -K_{12} & K_{02} + K_{12} + K_{32} + \cdots & & \\ -K_{1n} & -K_{2n} & \cdots & \end{pmatrix} \equiv (k_{ij}) \tag{3.5.2c}$$

and the small parameter ε is

$$\varepsilon = \frac{B_c}{(K_c M_c)^{1/2}}. \tag{3.5.3}$$

For example, average values can be used for the characteristic values. In order for an analysis based on small ε to give good results all the system constants should not deviate too much from their average values.

The motion of this model system for small damping should be close in some sense to the undamped case. The modes of free vibration of the undamped system are very convenient for representing the solution to damped initial value problems or forced motions. Hence it is reasonable to try to represent the motion of the damped system in terms of the undamped free modes. This can always be done since the eigenvectors which give the normal mode shapes form a basis of the n-dimensional space. We expect in analogy with the one-dimensional case of Section 3.2.1, that the physical effect of the damping is to make the free motion die out and to shift the frequencies of vibration from those of the undamped case. The analysis serves to make these intuitive ideas more precise and to clarify the behavior of linear systems with many degrees of freedom. It should be noted however that the analysis is general and applies to general symmetric positive definite matrices (M, B, K).

We are interested in the general solution of Equations (3.5.1) or equivalently in the solution to the general initial value problem.

First we summarize some properties of the modes of undamped free vibration which are derived from the problem with $\varepsilon = 0$:

$$M\frac{d^2\mathbf{z}}{dt^2} + K\mathbf{z} = 0, \qquad \mathbf{z} = (z_1, z_2, \ldots, z_n). \tag{3.5.4}$$

The free vibrations are of the form

$$\mathbf{z}(t) = \boldsymbol{\xi} \cos \omega t, \qquad \boldsymbol{\xi} = (\xi_1, \ldots, \xi_n) \tag{3.5.5}$$

so that

$$(K - M\omega^2)\boldsymbol{\xi} = 0 \tag{3.5.6}$$

is the system of n-linear equations for the ξ_j. The characteristic equation for the natural frequencies (dimensionless) of free vibration $\omega^{(j)}$ is

$$\det(K - M\omega^2) = 0. \tag{3.5.7}$$

We assume that the n-real roots are distinct and order them: $\omega^{(1)}, \omega^{(2)}, \ldots$. Thus, there is one eigenvector $\boldsymbol{\xi}^{(j)}$ corresponding to each frequency such that

$$(K - M\omega^{(j)^2})\boldsymbol{\xi}^{(j)} = 0, \qquad j = 1, \ldots, n. \tag{3.5.8}$$

These eigenvectors represent the free mode shapes and can be chosen to be an orthonormal set with respect to the weight function M

$$\boldsymbol{\xi}^{(i)} M \boldsymbol{\xi}^{(j)} = \delta_{ij}. \tag{3.5.9}$$

Now to study the damped problem we represent the motion in terms of the undamped modes

$$\mathbf{x}(t) = \sum_{i=1}^{n} \alpha_i(t)\xi^{(i)}. \tag{3.5.10}$$

$\alpha_i(t)$ are the mode coefficients to be found. Then (3.5.1) becomes

$$\sum_{i=1}^{n} M\xi^{(i)} \frac{d^2\alpha_i}{dt^2} + K\xi^{(i)}\alpha_i = -\varepsilon \sum_{i=1}^{n} B^{(i)}\xi^{(i)} \frac{d\alpha_i}{dt}. \tag{3.5.11}$$

Using (3.5.8) and multiplying by $\xi^{(j)}$ and using (3.5.10) we find

$$\frac{d^2\alpha_j}{dt^2} + \omega^{(j)^2}\alpha_j = -\varepsilon \sum_{i=1}^{n} \beta_{ij} \frac{d\alpha_i}{dt}. \tag{3.5.12}$$

β_{ij} is the quadratic form

$$\beta_{ij} = \xi^{(i)} B \xi^{(j)} \tag{3.5.13}$$

and is thus symmetric. The β_{ij} represent the coupling between the modes due to damping.

The solution of Equation (3.5.1) is now expressed in the form of a two variable expansion

$$\alpha_j(t) = F_{j0}(t^+, \tilde{t}) + \varepsilon F_{j1}(t^+, \tilde{t}) + \varepsilon^2 F_{j2}(t^+, \tilde{t}) + \cdots, \tag{3.5.14}$$

where

$$t^+ = t(1 + \varepsilon^2\sigma + \cdots), \qquad \text{“fast” time}$$

$$\tilde{t} = \varepsilon t, \qquad \text{“slow” time.}$$

Repeating the calculation of Section 3.2.1 we find

$$L(F_{j0}) \equiv \frac{\partial^2 F_{j0}}{\partial t^{+2}} + \omega^{(j)^2} F_{j0} = 0 \tag{3.5.15}$$

$$L(F_{j1}) = -2\frac{\partial^2 F_{j0}}{\partial t^+ \partial \tilde{t}} - \sum_{p=1}^{n} \beta_{jp} \frac{\partial F_{p0}}{\partial t^+} \tag{3.5.16}$$

$$L(F_{j2}) = -2\sigma \frac{\partial^2 F_{j0}}{\partial t^{+2}} - \frac{\partial^2 F_{j0}}{\partial \tilde{t}^2} - 2\frac{\partial^2 F_{j1}}{\partial t^+ \partial \tilde{t}}$$

$$- \sum_{p=1}^{n} \beta_{pj}\left(\frac{\partial F_{p0}}{\partial \tilde{t}} + \frac{\partial F_{p1}}{\partial t^+}\right). \tag{3.5.17}$$

The general solution of Equation (3.5.15) is

$$F_{j0} = A_{j0}(\tilde{t})\cos \omega^{(j)}t^+ + B_{j0}(\tilde{t})\sin \omega^{(j)}t^+. \tag{3.5.18}$$

Thus, from Equation (3.5.16)

$$L(F_{j1}) = 2\omega^{(j)} \frac{dA_{j0}}{d\tilde{t}} \sin \omega^{(j)}t^+ - 2\omega^{(j)} \frac{dB_{j0}}{d\tilde{t}} \cos \omega^{(j)}t^+$$

$$- \sum_{p=1}^{n} \beta_{jp}(-A_{p0}\omega^{(p)} \sin \omega^{(p)}t^+ + B_{p0}\omega^{(p)} \cos \omega^{(p)}t^+). \tag{3.5.19}$$

In order to eliminate mixed secular terms in F_{j1}, all driving terms with frequency $\omega^{(j)}$ on the right-hand side of (3.6.19) must vanish. Thus,

$$S(A_{j0}) \equiv 2\frac{dA_{j0}}{d\tilde{t}} + \beta_{jj}A_{j0} = 0 \tag{3.5.20a}$$

$$S(B_{j0}) = 0. \tag{3.5.20b}$$

The result thus far shows that to first order the modes remain uncoupled, each oscillating with its natural frequency, each mode having a damping coefficient $\frac{1}{2}\beta_{jj}$, i.e.,

$$A_{j0} = a_{j0}e^{-(1/2)\beta_{jj}\tilde{t}_j},\ B_{j0} = b_{j0}e^{-(1/2)\beta_{jj}\tilde{t}}. \tag{3.5.21}$$

This coefficient comes from the diagonal elements of the general damping matrix (Equation 3.5.13) and is always positive.

In order to calculate the first approximation to the frequency shift it is necessary to study the equation for F_{j2}. The solution for F_{j1} is now

$$F_{j1} = A_{j1}(\tilde{t})\cos\omega^{(j)}t^+ + B_{j1}(\tilde{t})\sin\omega^{(j)}t^+$$
$$+ \sum_{\substack{p=1\\ p\neq j}}^{n} \beta_{pj}\frac{\omega^{(p)}}{\omega^{(j)^2} - \omega^{(p)^2}}\left[A_{p0}\sin\omega^{(p)}t^+ - B_{p0}\cos\omega^{(p)}t^+\right]. \tag{3.5.22}$$

This shows that the $O(\varepsilon)$ terms of the jth mode contain all the frequencies of the whole system and that the mode is no longer "pure." It now follows from (3.5.17) that

$$L(F_{j2}) = 2\sigma\omega^{(j)^2}(A_{j0}\cos\omega^{(j)}t^+ + B_{j0}\sin\omega^{(j)}t^+)$$
$$- \frac{d^2A_{j0}}{d\tilde{t}^2}\cos\omega^{(j)}t^+ - \frac{d^2B_{j0}}{d\tilde{t}^2}\sin\omega^{(j)}t^+$$
$$- 2\frac{dA_{j1}}{dt^+}\omega^{(j)}\sin\omega^{(j)}t^+ - 2\frac{dB_{j1}}{dt^+}\omega^{(j)}\cos\omega^{(j)}t^+$$
$$-2\sum_{\substack{p=1\\ p\neq j}}^{n}\beta_{pj}\frac{\omega^{(p)^2}}{\omega^{(j)^2} - \omega^{(p)^2}}\left[\frac{dA_{p0}}{d\tilde{t}}\cos\omega^{(p)}t^+ + \frac{dB_{p0}}{d\tilde{t}}\sin\omega^{(p)}t^+\right]$$
$$- \sum_{q=1}^{n}\beta_{qj}\Bigg[\Bigg(\frac{dA_{q0}}{d\tilde{t}}\cos\omega^{(q)}t^+ + \frac{dB_{q0}}{d\tilde{t}}\sin\omega^{(q)}t^+$$
$$- \omega^{(q)}A_{q1}\sin\omega^{(q)}t^+ + \omega^{(q)}B_{q1}\cos\omega^{(q)}t^+\Bigg)$$
$$+ \sum_{\substack{p=1\\ p\neq q}}^{n}\beta_{pq}\frac{\omega^{(p)^2}}{\omega^{(q)^2} - \omega^{(p)^2}}(A_{p0}\cos\omega^{(p)}t^+ + B_{p0}\sin\omega^{(p)}t^+)\Bigg]. \tag{3.5.23}$$

Again, to avoid mixed secular terms the coefficients of $\cos\omega^{(j)}t^+$, $\sin\omega^{(j)}t^+$ in the right-hand side of Equation (3.5.23) must be set equal to zero. This gives, for $\cos \omega^{(j)}t^+$,

$$\omega^{(j)}S(B_{j1}) = \alpha_{jo}e^{-(1/2)\beta_{jj}\tilde{t}}\left[2\sigma\omega^{(q)^2} + \frac{1}{4}\beta_{jj}^2 - \sum_{\substack{p=1\\p\neq j}}^{n}\beta_{pj}^2\frac{\omega^{(j)^2}}{\omega^{(p)^2} - \omega^{(j)^2}}\right] \tag{3.5.24}$$

when Equation (3.5.21) is used.

An equivalent equation comes from the coefficient of $\sin \omega^{(j)}t^+$. As in Section 3.2.1 the right-hand side of Equation (3.5.24) must be set equal to zero, so that inconsistent terms, or non-uniform terms of the form

$$\tilde{t}\exp(-\tfrac{1}{2}\beta_{jj}\tilde{t})$$

do not appear. Thus the formula for the "shift" σ is

$$\sigma = -\frac{1}{8}\frac{\beta_{jj}^2}{\omega^{(j)^2}}\sum_{\substack{p=1\\p\neq j}}^{n}\frac{\beta_{pj}^2}{\omega^{(p)^2} - \omega^{(j)^2}} \quad \text{for the } j\text{th mode.} \tag{3.5.25}$$

From this a better approximation to the period of F_{j0} results. The frequency shift now depends on all damping coefficients as well as the natural frequencies of the undamped system. Thus, this section has shown how useful the two variable methods is in elucidating the behavior of a coupled system with small damping.

The old master Lord Rayleigh gave a succint discussion of this problem in the course of his general discussion of vibrating systems, Section 102 of Reference 3.5.13. The principal result of this section, the pure damping of each mode and the formula for damping coefficient (Equation 3.5.21), was given by Rayleigh. He also presented the equivalent of (Equation 3.5.22) and remarked that the $O(\varepsilon)$ modes excited in Equation 3.5.22 are all in phase with each other but "that phase differs by a quarter period from the phase of F_{j0}" (in our notation). This is manifest in Equation 3.5.22. Lastly, he gave in implicit form the recipe for the frequency shift Equation (3.5.25). Earlier in Reference 3.5.13 Rayleigh formulates conditions under which the three matrices M, B, K can be simultaneously diagonalized. For this case, which does not occur often in practical cases, each damped mode acts separately. Generalizations of Rayleigh's result appear in References 3.5.14, 3.5.15.

3.5.2 Two Weakly Nonlinear Oscillators, Constant Frequencies

Consider the system

$$\frac{d^2x}{dt^2} + a^2x = \varepsilon y^2 \tag{3.5.26a}$$

$$\frac{d^2y}{dt^2} + b^2y = 2\varepsilon xy, \tag{3.5.26b}$$

where a and b are arbitrary constants, and $0 < \varepsilon \ll 1$.

This problem was originally studied in References 3.5.9 and 3.5.10 using variants of the von Zeipel method (cf. Section 3.7.2); it models the motion of a star in a galaxy.

(a) *Necessary Conditions for Bounded Solutions*

It is instructive to examine the energy integral

$$\tfrac{1}{2}(\dot{x}^2 + \dot{y}^2) + \tfrac{1}{2}(a^2x^2 + b^2y^2) - \varepsilon xy^2 = E = \text{const.} \tag{3.5.27}$$

that is available for the system (3.5.26) when a and b are constants in order to establish the conditions under which solutions are always bounded. We take $\varepsilon > 0$, and construct the zero velocity curves that one obtains by setting $\dot{x} = \dot{y} = 0$ in (3.5.27) and considering the one-parameter family

$$V(x, y) \equiv \tfrac{1}{2}(a^2x^2 + b^2y^2) - \varepsilon xy^2 = E \tag{3.5.28}$$

corresponding to different values of E.

For any set of initial conditions: $x(0)$, $y(0)$, $\dot{x}(0)$, $\dot{y}(0)$, one calculates E from (3.5.27). The resulting curves $V(x, y) = E$ in the $x - y$ plane determines a "barrier" that cannot be crossed for motion evolving from these initial conditions, because $V > E$ on the other side of the zero velocity curve and, according to (3.5.27), this corresponds to an imaginary velocity. The exceptional case $V_x = V_y = 0$ on a zero velocity curve corresponds to an equilibrium solution of the system (3.5.26). It then follows that motion resulting from a given set of initial conditions is bounded if the corresponding zero velocity curve is closed and the initial point lies inside this closed curve.

To study (3.5.28), it is more convenient to rescale x and y and to consider the one-parameter family

$$\xi^2 + \eta^2 - \gamma\xi\eta^2 = C, \tag{3.5.29}$$

where

$$\xi = ax \tag{3.5.30a}$$

$$\eta = by \tag{3.5.30b}$$

$$\gamma = \frac{2\varepsilon}{ab^2} > 0 \tag{3.5.30c}$$

$$C = 2E. \tag{3.5.30d}$$

We have the equilibrium point ($V_\xi = V_\eta = 0$) at (0, 0) which is a center and the points $(\gamma^{-1}, \pm 2^{1/2}\gamma^{-1})$ which are saddles. Thus, motion with small initial values of x, $\dot{x}$, y, and $\dot{y}$ is stable, while motion with $x(0) \approx b^2/2\varepsilon$, $\dot{x}(0) \approx 0$, $y(0) \approx \pm ab/2^{1/2}\varepsilon$, $\dot{y}(0) \approx 0$ is unstable.

The zero velocity curves defined by (3.5.29) can be easily calculated and are shown in Figure 3.5.2. For $0 \le C < \gamma^{-2}$ one branch of the family consists of the nested set of closed curves surrounding the origin. As $C \to \gamma^{-2}$, these tend to the limiting curve bounded by the parabola $\eta^2 = \gamma^{-2}(1 + \gamma\xi)$ on the left and the vertical line $\xi = \gamma^{-1}$ on the right. The other branch for $0 < C < \gamma^{-2}$ consists of the set of curves (and their images for $\eta < 0$) evolving from D–E when $C = 0$ to the limiting curve A–B–D as $C \to \gamma^{-2}$.

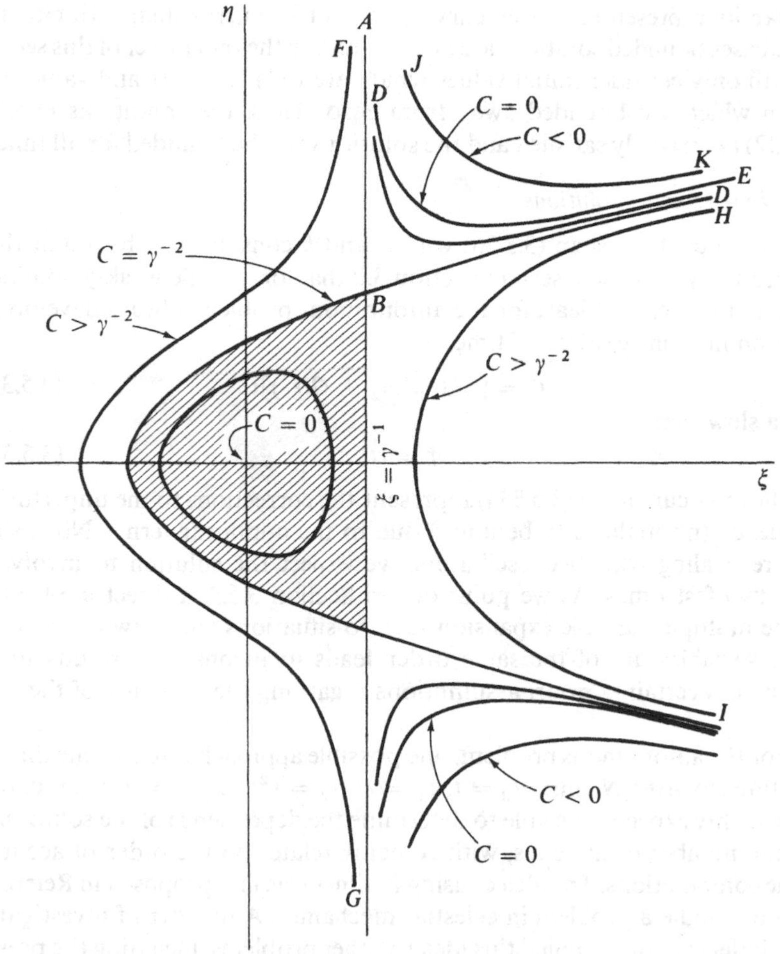

Figure 3.5.2 Zero Velocity Curves, Equation (3.5.29)

Thus, if $0 < C < \gamma^{-2}$, motion is bounded if the initial values of ξ and η lie inside the shaded region. Translating this to the original variables gives bounded motion as long as

$$0 \leq E < \frac{a^2 b^4}{8\varepsilon^2} \tag{3.5.31}$$

and

$$|y(0)| < \frac{a\sqrt{b^2 + 2\varepsilon x(0)}}{2\varepsilon} \tag{3.5.32a}$$

$$|x(0)| < \frac{b^2}{2\varepsilon}. \tag{3.5.32b}$$

For $C > \gamma^{-2}$, we have the two branches represented by the curves F–G and H–I in Figure (3.5.2). Finally, if $C < 0$, the zero velocity curves generate the family represented by the curve J–K and its mirror image. In either of these cases bounded solutions are not assured. In the remainder of this section we will only consider initial values which are $O(1)$ as $\varepsilon \to 0$, and values of a and b which are bounded away from zero. Thus, the conditions (3.5.31), (3.5.32) are trivially satisfied and the solutions will be bounded for all times.

(b) *Non-resonant Solutions*

We consider the system (3.5.26) with a and b constant (at this point these are arbitrary). We have seen in Section 3.2 that for a single weakly nonlinear oscillator one can, at least for the autonomous problem, always develop the solution in terms of a "fast" time

$$t^+ = [1 + \varepsilon^2\omega_2 + O(\varepsilon^3)]t \tag{3.5.33a}$$

and a slow time

$$\tilde{t} = \varepsilon t. \tag{3.5.33b}$$

The ω_i occurring in (3.5.33a) represent the corrections to the unperturbed frequency (normalized to be unity) due to the nonlinear terms. Now, since we are dealing with two oscillators, we expect the solution to involve at least two fast times. As we point out in Section 3.5.3, a direct application of the multiple variable expansion idea to situations where two or more of these variables are of the same order leads to inconsistent results unless one makes certain a priori assumptions regarding the structure of the solution.

For the autonomous problem, one possible approach is to assume that the solution involves N times $t_0 = t$, $t_1 = \varepsilon t$, $t_2 = \varepsilon^2 t, \ldots, t_N = \varepsilon^N t$. Of course, we can only expect to be able to determine the dependence of the solution on a finite number of these t_i's, with N being related to the order of accuracy of the computations. The idea of using N times was first proposed in Reference 3.4.6 to handle a problem in celestial mechanics. A number of investigators have independently applied this idea to other problems, including the present one discussed in Reference 3.5.1. Two other early examples are References 3.5.3 and 3.5.4.

We assume that the solution can be represented in the form

$$x(t; \varepsilon) = x_0(t_0, t_1, \ldots) + \varepsilon x_1(t_0, t_1, \ldots) + \varepsilon^2 x_2(t_0, t_1, \ldots) + \cdots \tag{3.5.34a}$$

$$y(t; \varepsilon) = y_0(t_0, t_1, \ldots) + \varepsilon y_1(t_0, t_1, \ldots) + \varepsilon^2 y_2(t_0, t_1, \ldots) + \cdots. \tag{3.5.34b}$$

We then compute

$$\frac{dx}{dt} = \frac{\partial x_0}{\partial t_0} + \varepsilon\left(\frac{\partial x_1}{\partial t_0} + \frac{\partial x_0}{\partial t_1}\right) + \varepsilon^2\left(\frac{\partial x_2}{\partial t_0} + \frac{\partial x_1}{\partial t_1} + \frac{\partial x_0}{\partial t_2}\right) + \cdots \tag{3.5.35a}$$

$$\begin{aligned}\frac{d^2x}{dt^2} &= \frac{\partial^2 x_0}{\partial t_0^2} + \varepsilon\left(\frac{\partial^2 x_1}{\partial t_0^2} + 2\frac{\partial^2 x_0}{\partial t_0\,\partial t_1}\right)\\ &\quad + \varepsilon^2\left(\frac{\partial^2 x_2}{\partial t_0^2} + 2\frac{\partial^2 x_1}{\partial t_0\,\partial t_1} + 2\frac{\partial^2 x_0}{\partial t_0\,\partial t_2} + \frac{\partial^2 x_0}{\partial t_1^2}\right) + \cdots,\end{aligned} \tag{3.5.35b}$$

and similar expressions for dy/dt and d^2y/dt^2. Note that to order ε the procedure gives the same formal results as if t_2 was not involved. Hence, to determine the dependence of x_0 on t_2 we need to consider the terms of order ε^2, etc.

Substituting the above expansions into (3.5.26) leads to the following sequence of pairs of equations for the x_i and y_i.

$$L_1(x_0) \equiv \frac{\partial^2 x_0}{\partial t_0^2} + a^2 x_0 = 0 \tag{3.5.36a}$$

$$L_2(y_0) \equiv \frac{\partial^2 y_0}{\partial t_0^2} + b^2 y_0 = 0 \tag{3.5.36b}$$

$$L_1(x_1) = y_0^2 - 2\frac{\partial^2 x_0}{\partial t_0\,\partial t_1} \tag{3.5.37a}$$

$$L_2(y_1) = 2x_0 y_0 - 2\frac{\partial^2 y_0}{\partial t_0\,\partial t_1} \tag{3.5.37b}$$

$$L_1(x_2) = 2y_0 y_1 - 2\frac{\partial^2 x_1}{\partial t_0\,\partial t_1} - 2\frac{\partial^2 x_0}{\partial t_0\,\partial t_2} - \frac{\partial^2 x_0}{\partial t_1^2} \tag{3.5.38a}$$

$$L_2(y_2) = 2(x_0 y_1 + x_1 y_0) - 2\frac{\partial^2 y_1}{\partial t_0\,\partial t_1} - 2\frac{\partial^2 y_0}{\partial t_0\,\partial t_2} - \frac{\partial^2 y_0}{\partial t_1^2}. \tag{3.5.38b}$$

Solving (3.5.36) gives

$$x_0(t_0, t_1, t_2, \ldots) = \alpha_0(t_1, t_2, \ldots)\cos\mu_0; \qquad \mu_0 = at_0 + \phi_0(t_1, t_2, \ldots) \tag{3.5.39a}$$

$$y_0(t_0, t_1, t_2, \ldots) = \beta_0(t_1, t_2, \ldots)\cos\nu_0; \qquad \nu_0 = bt_0 + \psi_0(t_1, t_2, \ldots) \tag{3.5.39b}$$

where α_0 and β_0 are unknown amplitudes and ϕ_0 and ψ_0 are unknown phases and all four functions may depend on $t_1, t_2, \ldots$ but not on t_0.

Using (3.2.39) to evaluate the right-hand sides of (3.5.37) gives

$$L_1(x_1) = \frac{\beta_0^2}{2}(1 + \cos 2\nu_0) + 2a\left(\frac{\partial \alpha_0}{\partial t_1}\sin\mu_0 + \alpha_0\frac{\partial \phi_0}{\partial t_1}\cos\mu_0\right) \tag{3.5.40a}$$

$$L_2(y_1) = \alpha_0\beta_0(\cos(\mu_0 + \nu_0) + \cos(\mu_0 - \nu_0)) + 2b\left(\frac{\partial \beta_0}{\partial t_1}\sin\nu_0 + \beta_0\frac{\partial \psi_0}{\partial t_1}\cos\nu_0\right). \tag{3.5.40b}$$

The trigonometric terms with μ_0 argument in (3.5.40a) and ν_0 argument in (3.5.40b) are homogeneous solutions and will therefore lead to unbounded contributions on the t_0 scale. Removing these inconsistent terms results in $\partial\alpha_0/\partial t_1 = \partial\phi_0/\partial t_1 = \partial\beta_0/\partial t_1 = \partial\psi_0/\partial t_1 = 0$. Thus,[10]

$$\alpha_0 = \alpha_0^{(2)}, \qquad \beta_0 = \beta_0^{(2)}, \qquad \phi_0 = \phi_0^{(2)}, \qquad \psi_0 = \psi_0^{(2)}. \tag{3.5.41}$$

What remains of (3.5.40) can now be solved to give

$$x_1 = \alpha_1^{(1)} \cos \mu_1 + \frac{\beta_0^2}{2a^2} + \frac{\beta_0^2}{2(a^2 - 4b^2)} \cos 2\nu_0 \tag{3.5.42a}$$

$$y_1 = \beta_1^{(1)} \cos \nu_1 - \frac{\alpha_0 \beta_0}{a(a + 2b)} \cos(\mu_0 + \nu_0) - \frac{\alpha_0 \beta_0}{a(a - 2b)} \cos(\mu_0 - \nu_0), \tag{3.5.42b}$$

where

$$\mu_1 = at_0 + \phi_1^{(1)} \tag{3.5.43a}$$

$$\nu_1 = bt_0 + \psi_1^{(1)}. \tag{3.5.43b}$$

We note the occurrence of the divisors a, a^2, $(a + 2b)$ and $(a - 2b)$ in our results. By hypothesis, for bounded solutions, a and b are bounded away from zero and positive (cf. the discussion following (3.5.32)). Hence, $a > 0$, $a + 2b > 0$ and these divisors are not troublesome. However, if $a = 2b$, our results become singular even though the solution must be bounded. Clearly, this is a reflection of the inadequacy of the assumed expansion for values of $a \approx 2b$ which is called the "first resonance" condition for (3.5.26). We will return to this case presently. For the time being we assume $a \neq 2b$ and proceed with the calculations for the next order.

Using the known solutions for x_0, y_0, x_1, and y_1 in (3.5.38) gives

$$L_1(x_2) = \beta_0 \beta_1[\cos(\nu_1 + \nu_0) + \cos(\nu_1 - \nu_0)] - \frac{\alpha_0 \beta_0^2}{a(a + 2b)} \cos(2\nu_0 + \mu_0) - \frac{\alpha_0 \beta_0^2}{a(a - 2b)} \cos(2\nu_0 - \mu_0) + S \tag{3.5.44a}$$

$$\begin{aligned} L_2(y_2) = {} & \alpha_0 \beta_1[\cos(\mu_0 + \nu_1) + \cos(\mu_0 - \nu_1)] \\ & - \alpha_0^2 \beta_0[\cos(2\mu_0 + \nu_0) + \cos(2\mu_0 - \nu_0)] \\ & + \alpha_1 \beta_0[\cos(\nu_0 + \mu_1) + \cos(\nu_0 - \mu_1)] + \frac{\beta_0^3}{2(a^2 - 4b^2)} \cos 3\nu_0 + T, \end{aligned} \tag{3.5.44b}$$

[10] Henceforth, we will adopt the superscript notation $f^{(n)}$ to denote functions $f^{(n)}(t_n, t_{n+1}, \ldots)$ which do not depend on $t_0, t_1, \ldots, t_{n-1}$.

where S and T denote the following terms which are solutions of the homogeneous equations $L_1(x_2) = 0$ and $L_2(y_2) = 0$ respectively.

$$S = 2\left(a\frac{\partial}{\partial t_1}\alpha_1\cos(\phi_1 - \phi_0) + a\frac{\partial \alpha_0}{\partial t_2}\right)\sin\mu_0$$
$$+ 2\left(a\frac{\partial}{\partial t_1}\alpha_1\sin(\phi_1 - \phi_0) + \alpha_0 a\frac{\partial \phi_0}{\partial t_2} - \frac{\alpha_0\beta_0^2}{a^2 - 4b^2}\right)\cos\mu_0 \quad (3.5.45a)$$

$$T = 2\left(b\frac{\partial}{\partial t_1}\beta_1\cos(\psi_1 - \psi_0) + b\frac{\partial \beta_0}{\partial t_2}\right)\sin\nu_0$$
$$+ 2\left\{b\frac{\partial}{\partial t_1}\beta_1\sin(\psi_1 - \psi_0) + \beta_0\frac{\partial \psi_0}{\partial t_2}b\right.$$
$$\left. + \frac{\beta_0[\beta_0^2(3a^2 - 8b^2) - 4a^2\alpha_0^2]}{4a^2(a^2 - 4b^2)}\right\}\cos\nu_0. \quad (3.5.45b)$$

Clearly, we must set $S = T = 0$ in order to insure that x_2 and y_2 be bounded on the t_0 scale. Now, examining the coefficients of the $\sin\mu_0$ and $\sin\nu_0$ terms in (3.5.45) we see that α_1 and β_1 will be *unbounded on the* t_1 *scale* unless we set $\partial\alpha_0/\partial t_2 = 0$, $\partial\beta_0/\partial t_2 = 0$. Therefore, the amplitudes of the $O(1)$ solution do not depend on t_2 either. We next examine the coefficients of the $\cos\mu_0$ and $\cos\nu_0$ term and note that boundedness of α_1 and β_1 on the t_1 scale also requires (with α_0 and β_0 independent of t_2) that we take

$$\frac{\partial \phi_0}{\partial t_2} = \frac{\beta_0^2}{a(a^2 - 4b^2)} \quad (3.5.46a)$$

$$\frac{\partial \psi_0}{\partial t_2} = -\frac{\beta_0^2(3a^2 - 8b^2) - 4a^2\alpha_0^2}{4a^2(a^2 - 4b^2)b} \quad (3.5.46b)$$

which we can immediately integrate. The solution to $O(1)$ is now determined up to terms involving t_2 and is summarized below

$$\alpha_0 = \alpha_0^{(3)} \quad (3.5.47a)$$

$$\beta_0 = \beta_0^{(3)} \quad (3.5.47b)$$

$$\phi_0 = \frac{\beta_0^2}{a(a^2 - 4b^2)}t_2 + k_1^{(3)} \quad (3.5.47c)$$

$$\psi_0 = \frac{\beta_0^2(3a^2 - 8b^2) - 4a^2\alpha_0^2}{4a^2(a^2 - 4b^2)b}t_2 + k_2^{(3)}, \quad (3.5.47d)$$

where the functions $\alpha_0^{(3)}$, $\beta_0^{(3)}$, $k_1^{(3)}$ and $k_2^{(3)}$ arise after integration with respect to t_2 and therefore depend only on $t_3, t_4, \ldots$. Actually, if we stop our calculations at this stage, we can regard these functions as constants and evaluate them using the initial values of x, y, dx/dt and dy/dt. We also have a strong suspicion that α_0, β_0 will turn out to be pure constants, while ϕ_0, ψ_0 will depend linearly on the t_2, t_3, etc. This is borne out by the calculations, at least to the next order, which we do not give. (cf. Problem 3.5.2).

The need for $N > 2$ time variables is now apparent since the phases for the x and y solutions have *different* corrections and could not have been uniformly represented by a single t^+ variable.

Having eliminated the inconsistent terms with respect to t_2 from (3.5.45), these reduce to the statement that α_1, β_1, ϕ_1 and ψ_1 are independent of t_1, i.e.,

$$\alpha_1 = \alpha_1^{(2)}; \quad \beta_1 = \beta_1^{(2)}; \quad \phi_1 = \phi_1^{(2)}; \quad \psi_1 = \psi_1^{(2)}. \tag{3.5.48}$$

Now, we can also integrate what remains of (3.5.44) to calculate x_2 and y_2 in the form

$$\begin{aligned} x_2 = {} & \alpha_2^{(1)} \cos \mu_2 + \frac{\beta_0 \beta_1}{a^2 - 4b^2} \cos(\nu_1 + \nu_0) + \frac{\beta_0 \beta_1}{a^2} \cos(\nu_1 - \nu_0) \\ & + \frac{\alpha_0 \beta_0^2}{4ab(a+b)(a+2b)} \cos(2\nu_0 + \mu_0) \\ & + \frac{\alpha_0 \beta_0^2}{4ab(a-b)(a-2b)} \cos(2\nu_0 - \mu_0) \end{aligned} \tag{3.5.49a}$$

$$\begin{aligned} y_2 = {} & \beta_2^{(1)} \cos \nu_2 - \frac{\alpha_0 \beta_0}{a(a+2b)} [\cos(\mu_0 + \nu_1) + \cos(\mu_0 - \nu_1)] \\ & + \frac{\alpha_0^2 \beta_0}{4a^2(a+b)^2} \cos(2\mu_0 + \nu_0) + \frac{\alpha_0^2 \beta_0}{4a^2(a-b)^2} \cos(2\mu_0 - \nu_0) \\ & - \frac{\alpha_1 \beta_0}{a(a+2b)} \cos(\nu_0 + \mu_1) - \frac{\alpha_1 \beta_0}{a(a-2b)} \cos(\nu_0 - \mu_1) \\ & - \frac{\beta_0^3}{16b^2(a^2 - 4b^2)} \cos 3\nu_0, \end{aligned} \tag{3.5.49b}$$

where

$$\mu_2 = at_0 + \phi_2^{(2)} \tag{3.5.50a}$$

$$\nu_2 = bt_0 + \psi_2^{(2)}. \tag{3.5.50b}$$

The reason we calculated the solution for x_2 and y_2 is to exhibit the second resonance for this problem associated with the small divisor $a - b$ when $a = b$. We conclude that to each higher order in ε the solution will contain a new divisor which can vanish for a certain ratio of a/b.

These higher order resonances become gradually weaker in the sense that for any given small but non-vanishing value of the small divisor, the corresponding singularity is strongest for the first resonance and decreases by one order in ε for each succeeding resonance. Actually, these higher resonances are not interesting because they do not lead to an exchange of energy between modes as is the case for the first resonance. In the next subsection we concentrate on the solution for the case $a \approx 2b$.

Finally, we note that if we truncate the procedure at the stage where the form of the $O(\varepsilon^N)$ solution is determined, we have already defined the dependence of the $O(1)$ solution on $t_0, t_1, \ldots, t_N$, the dependence of the $O(\varepsilon)$ solution on $t_0, t_1, \ldots, t_{N-1}$, etc.

(c) *Solution near the First Resonance*

The procedure for handling the case when $a \approx 2b$ in the previous subsection is quite straightforward and merely involves setting $b = (a/2) + s$ (where s is a small constant) explicitly in the equation. It is easy to verify that the richest equations to $O(\varepsilon)$ result by choosing $as + s^2 = \varepsilon\kappa$, where κ is an arbitrary $O(1)$ constant.

We can then rescale t and ε so that with no loss of generality we need only study the system

$$\frac{d^2x}{dt^2} + x = \varepsilon y^2 \tag{3.5.51a}$$

$$\frac{d^2y}{dt^2} + \frac{1}{4}y = 2\varepsilon xy - \varepsilon\kappa y. \tag{3.5.51b}$$

We will only study the dependence of the solution on t_0 and t_1 as the calculations to higher order are difficult. Therefore, we ignore the dependence of x and y on t_2, t_3, etc., and consider a development in the form

$$x = x_0(t_0, t_1) + \varepsilon x_1(t_0, t_1) + \cdots \tag{3.5.52a}$$

$$y = y_0(t_0, t_1) + \varepsilon y_1(t_0, t_1) + \cdots. \tag{3.5.52b}$$

The differential equations governing x_0, y_0, x_1 and y_1 are then easily derived in the form

$$L_1(x_0) \equiv \frac{\partial^2 x_0}{\partial t_0^2} + x_0 = 0 \tag{3.5.53a}$$

$$L_2(y_0) \equiv \frac{\partial^2 y_0}{\partial t_0^2} + \frac{1}{4}y_0 = 0 \tag{3.5.53b}$$

$$L_1(x_1) = -2\frac{\partial^2 x_0}{\partial t_0\,\partial t_1} + y_0^2 \tag{3.5.54a}$$

$$L_2(y_1) = -2\frac{\partial^2 y_0}{\partial t_0\,\partial t_1} - 2x_0 y_0 - \kappa y_0. \tag{3.5.54b}$$

The solution of (3.5.53) is

$$x_0 = \alpha_0(t_1)\cos\mu_0, \qquad \mu_0 = t_0 + \phi_0(t_1), \tag{3.5.55a}$$

$$y_0 = \beta_0(t_1)\cos\nu_0, \qquad \nu_0 = \frac{t_0}{2} + \psi_0(t_1). \tag{3.5.55b}$$

We substitute these into the right-hand sides of (3.5.54) and isolate homogeneous solutions to obtain (primes denote d/dt_1):

$$L_1(x_1) = \left(2\alpha_0' - \frac{\beta_0^2}{2}\sin(2\psi_0 - \phi_0)\right)\sin\mu_0 + \left(2\alpha_0\phi_0'\frac{\beta_0^2}{2}\cos(2\psi_0 - \phi_0)\right)\cos\mu_0 + \frac{\beta_0^2}{2} \tag{3.5.56a}$$

$$L_2(y_1) = [\beta_0' + \alpha_0\beta_0\sin(2\psi_0 - \phi_0)]\sin\nu_0 + [\beta_0\psi_0' + \alpha_0\beta_0\cos(2\psi_0 - \phi_0) - \kappa\beta_0]\cos\nu_0 + \alpha_0\beta_0\cos(\mu_0 + \nu_0). \tag{3.5.56b}$$

The bracketed terms in (3.5.56) multiply solutions of the homogeneous equations $L_1(x_1) = 0$ and $L_2(y_1) = 0$, and must be set equal to zero. Note that the terms which produced small divisors in the preceding subsection are now homogeneous solutions and will be eliminated.

The resulting equations governing the four unknowns α_0, β_0, ϕ_0 and ψ_0 are the coupled nonlinear system

$$2\alpha_0' - \frac{\beta_0^2}{2}\sin(2\psi_0 - \phi_0) = 0 \tag{3.5.57a}$$

$$2\alpha_0\phi_0' + \frac{\beta_0^2}{2}\cos(2\psi_0 - \phi_0) = 0 \tag{3.5.57b}$$

$$\beta_0' + \alpha_0\beta_0\sin(2\psi_0 - \phi_0) = 0 \tag{3.5.57c}$$

$$\beta_0\psi_0' + \alpha_0\beta_0\cos(2\psi_0 - \phi_0) - \kappa\beta_0 = 0. \tag{3.5.57d}$$

Despite the forbidding nature of these equations they can be solved because there exist two integrals for the system. The absence of t_1 then reduces the solution to quadrature.

Before tackling this solution we integrate what is left of (3.5.56) and calculate x_1 and y_1 in the following form free of small divisors

$$x_1 = \alpha_1(t_1)\cos\mu_1 + \frac{\beta_0^2}{2}, \qquad \mu_1 = t_0 + \phi_1(t_1); \tag{3.5.58a}$$

$$y_1 = \beta_1(t_1)\cos\nu_1 - \tfrac{16}{9}\alpha_0\beta_0\cos(\mu_0 + \nu_0), \qquad \nu_1 = \frac{t_0}{2} + \psi_1(t_1). \tag{3.5.58b}$$

A first integral for the system (3.5.57) can be derived by multiplying (3.5.57a) by α_0, (3.5.57c) by β_0 and adding the result. We find

$$\alpha_0^2 + \frac{\beta_0^2}{4} = \text{const.} = 2E_0. \tag{3.5.59}$$

It is easy to verify that E_0 is the leading term of the total energy of the system (3.5.51)

$$E = \frac{1}{2}\left[\left(\frac{dx}{dt}\right)^2 + \left(\frac{dy}{dt}\right)^2\right] + \frac{1}{2}\left(x^2 + \frac{y^2}{4}\right) - \varepsilon\left(xy^2 + \frac{\kappa y^2}{2}\right) = \text{const.} \quad (3.5.60)$$

which is an exact integral.

The second integral of the system (3.5.57) is more subtle. One way of obtaining it is to differentiate (3.5.57b) solved for ϕ_0. Letting $\xi = 2\psi_0 - \phi_0$ temporarily, we obtain

$$\phi_0'' = \frac{\beta_0^2 \xi'}{4\alpha_0} \sin \xi - \frac{\beta_0 \beta_0'}{2\alpha_0} \cos \xi + \frac{\beta_0^2 \alpha_0'}{4\alpha_0^2} \cos \xi. \quad (3.5.61)$$

Now, using (3.5.57a, b and d) to eliminate ϕ_0', β_0' and ψ_0' we obtain

$$\phi_0'' = -\frac{2\alpha_0'}{\alpha_0}(\phi_0' - \kappa) \quad (3.5.62)$$

which integrates to give

$$\phi_0' = \frac{\lambda_0}{\alpha_0^2} + \kappa, \quad (3.5.63)$$

where λ_0 is a constant of integration. It is interesting to note that λ_0 corresponds to the leading term of the so called "adelphic" integral discussed by Whittaker in Reference 3.5.11. Procedures for calculating such formal[11] integrals directly are discussed in References 3.5.9 and 3.5.10 and numerous other papers dealing with Hamiltonian systems in celestial mechanics (see also Section 3.7.2).

Now, to reduce the solution to quadrature, we square and add (3.5.57a, b), then use the energy integral to find

$$4\alpha_0'^2 + 4\alpha_0^2 \phi_0'^2 = \frac{\beta_0^4}{2} = 4(2E_0 - \alpha_0^2)^2. \quad (3.5.64)$$

Using (3.5.63) to express ϕ_0' in terms of α_0 then gives

$$\alpha_0'^2 + \frac{\lambda_0^2}{\alpha_0^2} + p^2\alpha_0^2 - \alpha_0^4 = q, \quad (3.5.65)$$

where

$$p^2 = 4E_0 + \kappa^2 \quad (3.5.66a)$$

$$q = 2E_0^2 - 2\kappa\lambda_0. \quad (3.5.66b)$$

[11] We use this terminology because one cannot prove the convergence in general for the series representation of such integrals. At any rate, their asymptotic validity is more easily established.

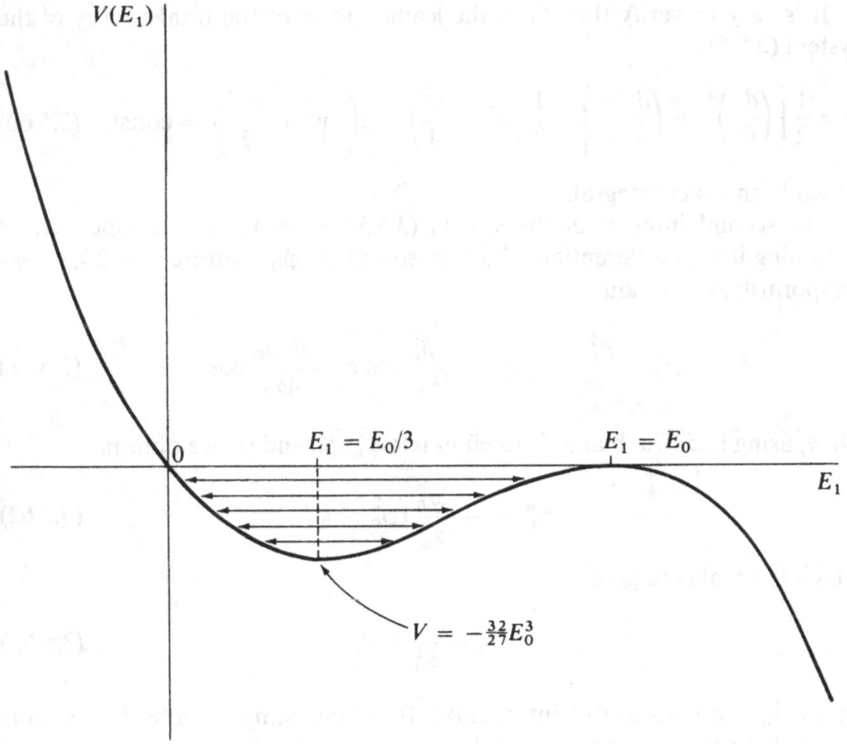

Figure 3.5.3 V as a Function of E_1, Equation (3.5.69)

Equation (3.5.65) is an integral involving the amplitude of the x oscillator and can be solved by quadrature. Actually, in view of the apparent[12] singularity in (3.5.65) when $\alpha_0 \to 0$, $\lambda_0 \neq 0$, it is more convenient to express this result in terms of the energy E_1 of the x oscillator to $O(1)$. With

$$E_1 = \frac{\alpha_0^2}{2} \tag{3.5.67}$$

(3.5.65) becomes

$$E_1'^2 - 8E_1(E_0 - E_1)^2 + 4\kappa E_1(\lambda_0 + \kappa E_1) = -\lambda_0^2. \tag{3.5.68}$$

The solution of (3.5.68) can be carried out using elliptic functions. However, it is more instructive to study the qualitative behavior of E_1 using energy arguments. To fix ideas, let $\kappa = 0$ and denote

$$V(E_1) = -8E_1(E_0 - E_1)^2. \tag{3.5.69}$$

For some fixed E_0, $V(E_1)$ is shown in Figure 3.5.3.

[12] The singularity is not worrisome because α_0 is never equal to zero unless $\lambda_0 = 0$ also, and this limiting case will be considered in our study of (3.5.68)

In view of the definition (3.5.67), $E_1 \geq 0$ and also from (3.5.59) we must have $E_1 \leq E_0$. Therefore, solutions of (3.5.69) only exist ($E'^2 \geq 0$) and are consistent with the other integrals of motion for values of λ_0 such that

$$0 \leq \lambda_0^2 \leq \tfrac{32}{27}E_0^3. \tag{3.5.70}$$

For this range of values of λ_0, E_1 oscillates between $E_{1_{\min}}$ and $E_{1_{\max}}$ where $E_{1_{\min}}$ is the smallest root of $V(E_1) = 0$ and $E_{1_{\max}}$ is the second root. We note from Figure 3.5.3 (or the calculation of the roots of the cubic $V(E_1) = 0$) that the third root is larger than E_0 and must be excluded.

The curves in the $(E_1 - E_1')$ plane are the ovals sketched in Figure 3.5.4 for a fixed E_0. Thus, we have the interesting phenomenon of periodic energy exchange on the t_1 scale between the two oscillators at resonance.

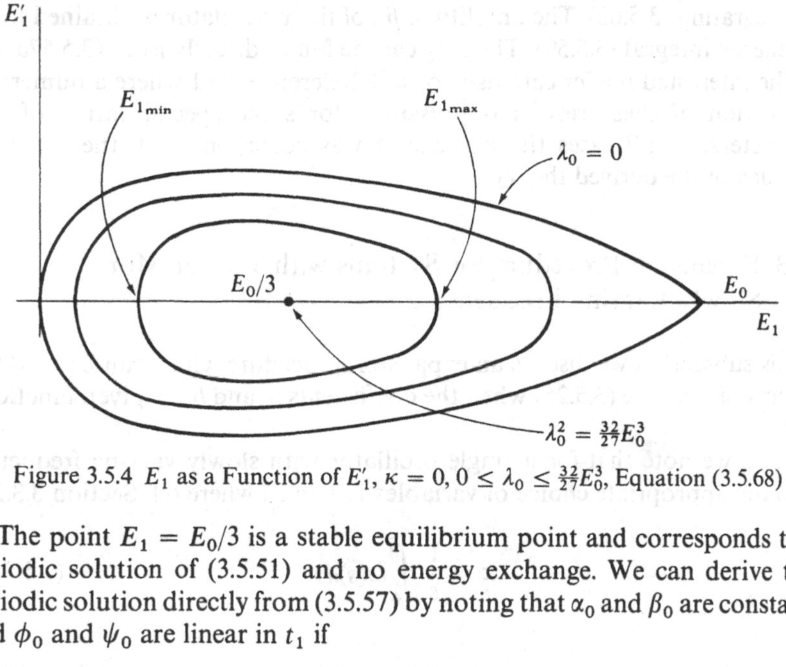

Figure 3.5.4 E_1 as a Function of E_1', $\kappa = 0$, $0 \leq \lambda_0 \leq \frac{32}{27}E_0^3$, Equation (3.5.68)

The point $E_1 = E_0/3$ is a stable equilibrium point and corresponds to a periodic solution of (3.5.51) and no energy exchange. We can derive this periodic solution directly from (3.5.57) by noting that α_0 and β_0 are constants and ϕ_0 and ψ_0 are linear in t_1 if

$$2\psi_0 - \phi_0 = 0 \tag{3.5.71a}$$

$$\phi_0' = \frac{-\beta_0^2}{4\alpha_0} = \text{const.} \tag{3.5.71b}$$

$$\psi_0' = \kappa - \alpha = \text{const.} = \frac{\phi_0'}{2}. \tag{3.5.71c}$$

Thus, given some α_0, we calculate the following values for β_0, ϕ_0 and ψ_0

$$\beta_0 = \sqrt{8\alpha_0(\alpha_0 - \kappa)} \tag{3.5.72a}$$

$$\phi_0 = 2(\kappa - \alpha_0)t_1 + \text{const.} \tag{3.5.72b}$$

$$\psi_0 = (\kappa - \alpha_0)t_1 + \text{const.} \tag{3.5.72c}$$

and the expressions for x and y which are periodic, at least to $O(1)$, are

$$x = \alpha_0 \cos\{[1 + 2\varepsilon(\kappa - \alpha_0) + O(\varepsilon^2)]t + \text{const.}\} + O(\varepsilon) \tag{3.5.73a}$$

$$y = \sqrt{8\alpha_0(\alpha_0 - \kappa)}\cos\left\{[1 + 2\varepsilon(\kappa - \alpha_0) + O(\varepsilon^2)]\frac{t}{2} + \text{const.}\right\} + O(\varepsilon). \tag{3.5.73b}$$

These results agree with the expressions given in Reference 3.5.10 to $O(1)$ in amplitude and to $O(\varepsilon)$ in frequency. Note that this periodic solution corresponds to very special initial conditions, and could have also been derived by using the method of strained coordinates (cf. Problem 3.5.4).

The situation for $\kappa \neq 0$ in (3.5.68) is not much different from the above and is not presented here. Finally, once E_1 (or α_0) is known, we calculate ϕ_0 by integrating (3.5.63). The amplitude β_0 of the y oscillator is obtained from the energy integral (3.5.59). Then ψ_0 can be found directly from (3.5.57a, c).

The interested reader can also consult Reference 3.5.1 where a numerical verification of these results is presented for some special values of the parameters. In all cases the agreement was consistent with the order of accuracy of the derived theory.

3.5.3 Expansion Procedure for Systems with Two or More Slowly Varying Frequencies

In this subsection we discuss an expansion procedure which can be used for systems of the type (3.5.26) when the coefficients a, and b are given functions of εt.

First, we note that for a single oscillator with slowly varying frequency $\omega(\varepsilon t)$ the appropriate choice of variables is τ and $\tilde{t}$ where (cf. Section 3.3.2)

$$\tau = \frac{1}{\varepsilon}\int_0^{\tilde{t}} \omega(s)ds \tag{3.5.74a}$$

and

$$\tilde{t} = \varepsilon t. \tag{3.5.74b}$$

It is therefore very tempting, in the case of two coupled oscillators, to assume that x and y depend on $\tilde{t}$ and the two fast times μ and ν defined by

$$\mu = \frac{1}{\varepsilon}\int_0^{\tilde{t}} a(s)ds; \qquad \nu = \frac{1}{\varepsilon}\int_0^{\tilde{t}} b(s)ds \tag{3.5.75}$$

and to seek expansions of the form

$$x(t, \varepsilon) = x_0(\mu, \nu, \tilde{t}) + \varepsilon x_1(\mu, \nu, \tilde{t}) + \cdots \tag{3.5.76a}$$

$$y(t, \varepsilon) = y_0(\mu, \nu, \tilde{t}) + \varepsilon y_1(\mu, \nu, \tilde{t}) + \cdots. \tag{3.5.76b}$$

We will now exhibit some of the difficulties inherent in this assumption. We calculate the following expressions for the first and second derivatives

$$\frac{d}{dt} = a\frac{\partial}{\partial \mu} + b\frac{\partial}{\partial \nu} + \varepsilon\frac{\partial}{\partial \tilde{t}} \tag{3.5.77a}$$

$$\frac{d^2}{dt^2} = L + \varepsilon M + \varepsilon^2\frac{\partial^2}{\partial \tilde{t}^2}, \tag{3.5.77b}$$

where L and M are the operators

$$L = a^2\frac{\partial^2}{\partial \mu^2} + 2ab\frac{\partial^2}{\partial \mu\,\partial \nu} + b^2\frac{\partial^2}{\partial \nu^2} \tag{3.5.78a}$$

$$M = 2a\frac{\partial^2}{\partial \mu\,\partial \tilde{t}} + 2b\frac{\partial^2}{\partial \nu\,\partial \tilde{t}} + a'\frac{\partial}{\partial \mu} + b'\frac{\partial}{\partial \nu} \tag{3.5.78b}$$

and $' = d/d\tilde{t}$.

Substituting the assumed expansions (3.5.76) for x and y together with the above expression for the second derivative into (3.5.26) gives the sequence of equations:

$$L(x_0) + a^2x_0 = 0 \tag{3.5.79a}$$

$$L(y_0) + b^2y_0 = 0 \tag{3.5.79b}$$

$$L(x_1) + a^2x_1 = -M(x_0) + y_0^2 \tag{3.5.80a}$$

$$L(y_1) + b^2y_1 = -M(y_0) + 2x_0y_0. \tag{3.5.80b}$$

The equations governing x_n and y_n will have the form

$$L(x_n) + a^2x_n = -M(x_{n-1}) + P_n \tag{3.5.81a}$$

$$L(y_n) + b^2y_n = -M(y_{n-1}) + Q_n, \tag{3.5.81b}$$

where P_n and Q_n depend on x_{n-1}, y_{n-1} as well as *known* functions of μ, ν and $\tilde{t}$ (because at this stage $x_0, \ldots, x_{n-2}, y_0, \ldots, y_{n-2}$ have been determined.)

Actually, we need not study the partial differential operator L, as the transformation[13]

$$\xi = b\mu - a\nu; \qquad \eta = a\mu + b\nu \tag{3.5.82}$$

will reduce L to a second derivative with respect to η, and (3.5.79–81) will reduce to

$$\bar{L}(x_0) + \bar{a}^2x_0 = 0 \tag{3.5.83a}$$

$$\bar{L}(y_0) + \bar{b}^2y_0 = 0 \tag{3.5.83b}$$

$$\bar{L}(x_1) + \bar{a}^2x_1 = -\bar{M}(x_0) + (a^2+b^2)^{-2}y_0^2 \tag{3.5.84a}$$

$$\bar{L}(y_1) + \bar{b}^2y_1 = -\bar{M}(y_0) + 2(a^2+b^2)^{-2}x_0y_0 \tag{3.5.84b}$$

$$\bar{L}(x_n) + \bar{a}^2x_n = -\bar{M}(x_{n-1}) + \bar{P}_n \tag{3.5.85a}$$

$$\bar{L}(y_n) + \bar{b}^2y_n = -\bar{M}(y_{n-1}) + \bar{Q}_n \tag{3.5.85b}$$

[13] We note that for a and b constant, ξ is identically equal to zero.

where $\bar{a}^2$, $\bar{b}^2$, $\bar{P}_n$ and $\bar{Q}_n$ are simply equal to the corresponding unbarred quantities divided by $(a^2 + b^2)^2$ and $\bar{L}$ and $\bar{M}$ are the operators

$$\bar{L} = \frac{\partial^2}{\partial \eta^2} \tag{3.5.86a}$$

$$\begin{aligned} \bar{M} = \frac{1}{(a^2 + b^2)^2} \bigg\{ & 2(a^2 + b^2) \frac{\partial^2}{\partial \eta \, \partial \tilde{t}} + 2[(aa' + bb')\xi \\ & + (ab' - ba')\eta] \frac{\partial^2}{\partial \xi \, \partial \eta} + 2[(aa' + bb')\eta - (ab' - ba')\xi] \frac{\partial^2}{\partial \eta^2} \\ & + (ab' - ba') \frac{\partial}{\partial \xi} + 3(aa' + bb') \frac{\partial}{\partial \eta} \bigg\}. \end{aligned} \tag{3.5.86b}$$

Consider now the homogeneous solution of the $(n - 1)$st equation,

$$x_{n-1} = A_{n-1}(\xi, \tilde{t}) \sin \bar{a}\eta + B_{n-1}(\xi, \tilde{t}) \cos \bar{a}\eta \tag{3.5.87a}$$

$$y_{n-1} = C_{n-1}(\xi, \tilde{t}) \sin \bar{b}\eta + D_{n-1}(\xi, \tilde{t}) \cos \bar{b}\eta. \tag{3.5.87b}$$

We wish to determine the $A_{n-1}, \ldots, D_{n-1}$ by requiring the solution to $O(\varepsilon^n)$ to be bounded. The right-hand sides of (3.5.85) involve $\bar{M}(x_{n-1})$ and $\bar{M}(y_{n-1})$, which according to Equations (3.5.87) and (3.5.86b) contain the two terms which are proportional to $\eta \sin \bar{a}\eta$, $\eta \cos \bar{a}\eta$ in (3.5.85a) and the two terms proportional to $\eta \sin \bar{b}\eta$, $\eta \cos \bar{b}\eta$ in (3.5.85b). In each of (3.5.85a) and (3.5.85b) we also have present two homogeneous solutions $\sin \bar{a}\eta$, $\cos \bar{a}\eta$ and $\sin \bar{b}\eta$, $\cos \bar{b}\eta$. It is clear that we cannot eliminate the eight unbounded contributions resulting from the above terms by any choice of only four functions $A_{n-1}, \ldots, D_{n-1}$. This situation is the analog to what would happen in a single oscillator problem with slowly varying frequency if one chooses t as the fast variable. Thus, assuming that the solution depends on μ, ν, and $\tilde{t}$ and regarding these to be independent leads to inconsistent results.

In Reference 3.5.12 an expansion of the type (3.5.76) is used with x_0 depending only on μ and $\tilde{t}$ and y_0 depending only on ν and $\tilde{t}$ which is certainly the case since the coupling terms are small. However, the inconsistency we pointed out above will occur to $O(\varepsilon^2)$ when deriving conditions on the coefficients A_1, B_1, C_1 and D_1 of the $O(\varepsilon)$ solution. This difficulty is avoided in Reference 3.5.12 by the implicit assumption [cf. (2.5) of Reference 3.5.12] that the homogeneous solution of (3.5.85a) depends only on μ, $\tilde{t}$ and that the homogeneous solution of (3.5.85b) depends only on ν, $\tilde{t}$ for each $n = 0, 1, \ldots$; the coupling terms only arise as particular solutions. This is a plausible assumption and was essentially the one proposed explicitly in the discussion given in section 5 of Reference 3.5.5. We devote the remainder of this subsection to a detailed study of the implications of this assumption. In Section 3.5.4, we will demonstrate that it is correct for the special example of Reference 3.5.5 if the resulting multi-variable expansion is derived with due care. In

Section 3.5.5 we will apply these ideas to the system (3.5.26) for the case where a and b are given functions of $\tilde{t}$.

In order to avoid having to deal with an operator L involving the two fast times, it is assumed [cf. (5.6a) of Reference 3.5.5] that the solution can be expanded in the form

$$x(t, \varepsilon) = x_0(\mu, \tilde{t}) + \varepsilon x_1(\mu, \tilde{t}) + \varepsilon^2 x_2(\mu, \tilde{t}) + \cdots \tag{3.5.88a}$$

$$y(t, \varepsilon) = y_0(v, \tilde{t}) + \varepsilon y_1(v, \tilde{t}) + \varepsilon^2 y_2(v, \tilde{t}) + \cdots \tag{3.5.88b}$$

and that coupling terms depending on v arising in the solution for x will be regarded as functions of μ and vice versa.

Let us now examine the implications of this assumption on the representation of a given function F of μ, v and $\tilde{t}$. By definition, if

$$x(t, \varepsilon) = F(\mu, v, \tilde{t}) \tag{3.5.89}$$

with

$$\frac{d\mu}{dt} = a(\tilde{t}) \tag{3.5.90a}$$

$$\frac{dv}{dt} = b(\tilde{t}) \tag{3.5.90b}$$

$$\tilde{t} = \varepsilon t \tag{3.5.90c}$$

we have

$$\frac{dx}{dt} = a\frac{\partial F}{\partial \mu} + b\frac{\partial F}{\partial v} + \varepsilon\frac{\partial F}{\partial \tilde{t}} \tag{3.5.91a}$$

and

$$\frac{d^2x}{dt^2} = a^2\frac{\partial^2 F}{\partial \mu^2} + 2ab\frac{\partial F}{\partial \mu\,\partial v} + b^2\frac{\partial^2 F}{\partial v^2} + \varepsilon\left(2a\frac{\partial^2 F}{\partial \mu\,\partial \tilde{t}} + 2b\frac{\partial^2 F}{\partial v\,\partial \tilde{t}} + a'\frac{\partial F}{\partial \mu} + b'\frac{\partial F}{\partial v}\right) + \varepsilon^2\frac{\partial^2 F}{\partial \tilde{t}^2}. \tag{3.5.91b}$$

Now, if we wish to, for example, regard v as a function of μ, we can do so as follows. We solve the expression

$$\varepsilon\mu = \int_0^{\tilde{t}} a(s)ds \equiv A(\tilde{t}) \tag{3.5.92}$$

obtained by integrating (3.5.90a) for $\tilde{t}$ as a function of $\varepsilon\mu$ (this can always be done if $a \neq 0$) to obtain

$$\tilde{t} = A^{-1}(\varepsilon\mu). \tag{3.5.93}$$

Now,

$$\varepsilon v = \int_0^{\tilde{t}} b(s)ds \equiv B(\tilde{t}). \tag{3.5.94}$$

Therefore,

$$v = \frac{1}{\varepsilon} B(A^{-1}(\varepsilon\mu)) = \frac{M(\varepsilon\mu)}{\varepsilon} \tag{3.5.95}$$

relates v to μ. We use this in (3.5.89) to define x as a function of μ, and $\tilde{t}$ only as follows

$$x = F\left\{\mu, \frac{1}{\varepsilon} B[A^{-1}(\varepsilon\mu)], \tilde{t}\right\} \equiv G(\mu, \tilde{t}). \tag{3.5.96}$$

It then follows that the derivatives of x have the following expressions in terms of G

$$\frac{dx}{dt} = a \frac{\partial G}{\partial \mu} + \varepsilon \frac{\partial G}{\partial \tilde{t}} \tag{3.5.97a}$$

$$\frac{d^2x}{dt^2} = a^2 \frac{\partial G}{\partial \mu^2} + 2a\varepsilon \frac{\partial^2 G}{\partial \mu\, \partial \tilde{t}} + a'\varepsilon \frac{\partial G}{\partial \mu} + \varepsilon^2 \frac{\partial^2 G}{\partial \tilde{t}^2}. \tag{3.5.97b}$$

To verify that (3.5.97) agree with (3.5.91), we use the identity (3.5.96) to relate F and G and regard v as a function of μ. We calculate

$$\frac{\partial G}{\partial \mu} = \frac{\partial F}{\partial \mu} + m \frac{\partial F}{\partial v}, \qquad \frac{\partial G}{\partial \tilde{t}} = \frac{\partial F}{\partial \tilde{t}}, \tag{3.5.98}$$

where

$$m(\varepsilon\mu) = \frac{1}{\varepsilon} \frac{dM}{d\mu}.$$

Moreover,

$$\frac{\partial^2 G}{\partial \mu^2} = \frac{\partial^2 F}{\partial \mu^2} + 2m \frac{\partial^2 F}{\partial \mu\, \partial v} + \varepsilon \frac{dm}{d(\varepsilon\mu)} \frac{\partial F}{\partial v} + m^2 \frac{\partial^2 F}{\partial v^2} \tag{3.5.99a}$$

$$\frac{\partial^2 G}{\partial \mu\, \partial \tilde{t}} = \frac{\partial^2 F}{\partial \mu\, \partial \tilde{t}} + m \frac{\partial^2 F}{\partial v\, \partial \tilde{t}} \tag{3.5.99b}$$

$$\frac{\partial^2 G}{\partial \tilde{t}^2} = \frac{\partial^2 F}{\partial \tilde{t}^2}. \tag{3.5.99c}$$

We note that

$$m = \frac{b}{a} \tag{3.5.100a}$$

and therefore

$$\frac{dm}{dt} = \varepsilon\left(\frac{b}{a}\right)'.$$

But

$$\frac{dm}{dt} = \frac{dm}{d(\varepsilon\mu)}\frac{d(\varepsilon\mu)}{dt} = \varepsilon a \frac{dm}{d(\varepsilon\mu)}.$$

Hence

$$\frac{dm}{d(\varepsilon\mu)} = \frac{1}{a}\left(\frac{b}{a}\right)', \tag{3.5.100b}$$

where (3.5.93) is to be used in computing the right-hand side of (3.5.100b) in terms of $\varepsilon\mu$.

Using (3.5.100) and the expressions relating the partial derivatives of G and F in (3.5.97) reproduces (3.5.91). Thus, it is correct to regard x as a function of μ and $\tilde{t}$ as long as derivatives are computed according to (3.5.97). Note in particular that M, m, and $dm/d(\varepsilon\mu)$ *must all be regarded as functions of* $\varepsilon\mu$. If one yields to the temptation of regarding m as the function of $\tilde{t}$ defined in (3.5.100b), the term $\varepsilon[dm/d(\varepsilon\mu)](\partial F/\partial v)$ appearing in (3.5.99a) would be missing and the result for d^2x/dt^2 would be incorrect.[14] The author of Reference 3.5.5 overlooked this, and as a result, the last term on the right-hand side of (2.6) of Reference 3.5.5 is incorrect. We will return to the specific example of Reference 3.5.5 in Section 3.5.4 to illustrate the proposed expansion procedure. It will become evident that the need for regarding M, m and $dm/d(\varepsilon\mu)$ as functions of $(\varepsilon\mu)$ is notationally very cumbersome, particularly since our results must also involve the original slow variable $\tilde{t} = \varepsilon t$. A short-cut proposed in Reference 3.5.7 avoids this requirement at the cost of distinguishing between the mixed partials $\partial^2 G/\partial\mu\,\partial\tilde{t}$ and $\partial^2 G/\partial\tilde{t}\,\partial\mu$. We consider this next.

Suppose that we regard m, whenever it occurs, as the function of $\tilde{t}$ given in (3.5.100a). Clearly, now $\partial^2 G/\partial\mu\,\partial\tilde{t} \neq \partial^2 G/\partial\tilde{t}\,\partial\mu$. In fact, (3.5.98) for the first partial derivatives of G are unchanged, so is (3.5.99c) for $\partial^2 G/\partial\tilde{t}^2$, however, we now have the following expressions to replace (3.5.99a), (3.5.99b) and (3.5.97b).

$$\frac{\partial^2 G}{\partial\mu^2} = \frac{\partial^2 F}{\partial\mu^2} + 2m\frac{\partial^2 F}{\partial\mu\,\partial v} + m^2\frac{\partial^2 F}{\partial v^2} \tag{3.5.101a}$$

$$\frac{\partial^2 G}{\partial\mu\,\partial\tilde{t}} \equiv \frac{\partial}{\partial\mu}\left(\frac{\partial G}{\partial\tilde{t}}\right) = \frac{\partial^2 F}{\partial\mu\,\partial\tilde{t}} + m\frac{\partial^2 F}{\partial\tilde{t}\,\partial v} \tag{3.5.101b}$$

$$\frac{\partial^2 G}{\partial\tilde{t}\,\partial\mu} \equiv \frac{\partial}{\partial\tilde{t}}\left(\frac{\partial G}{\partial\mu}\right) = \frac{\partial^2 F}{\partial\mu\,\partial\tilde{t}} + m'\frac{\partial F}{\partial v} + m\frac{\partial^2 F}{\partial v\,\partial\tilde{t}} \tag{3.5.101c}$$

$$\frac{d^2x}{dt^2} = a^2\frac{\partial^2 G}{\partial\mu^2} + a\varepsilon\left[\frac{\partial}{\partial\mu}\left(\frac{\partial G}{\partial\tilde{t}}\right) + \frac{\partial}{\partial\tilde{t}}\left(\frac{\partial G}{\partial\mu}\right)\right] + a'\varepsilon\frac{\partial G}{\partial\mu} + \varepsilon^2\frac{\partial^2 G}{\partial\tilde{t}^2}. \tag{3.5.101d}$$

[14] The authors are indebted to Prof. P. A. Lagerstrom for pointing this out.

The term we missed in (3.5.101a) by regarding m to be a function of $\tilde{t}$ now shows up in (3.5.101c) and the resulting expressions for dx/dt and d^2x/dt^2 are also correct. Use of this unconventional expansion procedure simplifies the calculations considerably. We illustrate ideas using the model problem of Reference 3.5.5 in the next subsection.

3.5.4 Passage Through Resonance, Single Forced Oscillator

Consider the problem of Section 3.3.2 with a forcing term in the form:

$$\frac{d^2y}{dt^2} + \mu^2(\tilde{t})y = \alpha \cos(t + \beta). \tag{3.5.102}$$

Here α and β are constants and μ is a given function of $\tilde{t}$ such that

$$\mu(\tilde{t}) > 0, \qquad 0 \leq \tilde{t} < \infty.$$

We also assume that near some given positive value $\tilde{t}_0$, μ has the following behavior:

$$\mu(\tilde{t}) = 1 + a_1(\tilde{t} - \tilde{t}_0) + a_2(\tilde{t} - \tilde{t}_0)^2 + O[(\tilde{t} - \tilde{t}_0)^3] \quad \text{as } \tilde{t} \to \tilde{t}_0, \tag{3.5.103}$$

where $a_1 > 0$ and a_2 are given constants. Thus, near $\tilde{t} = \tilde{t}_0$, μ is approximately equal to unity and coincides with the forcing frequency. Therefore, we expect some sort of local resonant behavior for the solution. The question we would like to answer is how does the solution pass through the resonant stage and what happens for times $\tilde{t} \gg \tilde{t}_0$?

This problem was first studied in Reference 3.5.5 as a mathematical model for more complicated nonlinear resonance problems examples of which are discussed in References 3.5.6–8. Thus, even Equation (3.5.102) is linear and can perhaps be solved more efficiently by a WKB type approach we will tackle it, as we did the example of Section 3.3.3, by combining multiple scale and limit process expansions, because these do more readily generalize to nonlinear problems.

(a) *Solution before Resonance* (*Outer Expansion*)

First, we consider the solution for times $\tilde{t} < \tilde{t}_0$, and denote the corresponding expansion the preresonance expansion. As in Section 3.3.2, we represent the solution in terms of the two times

$$t^+ = \frac{1}{\varepsilon}\int_0^{\tilde{t}} \mu(s)ds = \frac{1}{\varepsilon}M(\tilde{t}) \tag{3.5.104a}$$

and

$$\tilde{t} = \varepsilon t \tag{3.5.104b}$$

in the form

$$y(t;\varepsilon) = y_0(t^+, \tilde{t}) + \varepsilon y_1(t^+, \tilde{t}) + \cdots. \tag{3.5.105}$$

It is important to note that even though Equation (3.5.102) is of second order, we must account for the two distinct fast times t and t^+ in the solution. We will calculate the solution using the expansion procedure discussed in the latter part of Section 3.5.3, where we regard t as a function of t^+, but let the expression for

$$\frac{dt}{dt^+} = \frac{1}{\mu(\tilde{t})} \tag{3.5.106}$$

be a function of $\tilde{t}$ instead of the function of εt^+ obtained by inverting Equation (3.5.104a). Thus, we must distinguish between the mixed second partial derivatives $\partial^2 y_0/\partial t^+ \partial \tilde{t}$ and $\partial^2 y_0/\partial \tilde{t}\,\partial t^+$.

We compute [cf. Eqs. (3.5.97a) and (3.5.101d)].

$$\frac{dy}{dt} = \mu \frac{\partial y_0}{\partial t^+} + \varepsilon\left(\mu \frac{\partial y_1}{\partial t^+} + \frac{\partial y_0}{\partial \tilde{t}}\right) + O(\varepsilon^2) \tag{3.5.107}$$

$$\frac{d^2 y}{dt^2} = \mu^2 \frac{\partial^2 y_0}{\partial t^{+2}} + \varepsilon\left[\mu^2 \frac{\partial^2 y_1}{\partial t^{+2}} + \mu\left(\frac{\partial^2 y_0}{\partial t^+ \partial \tilde{t}} + \frac{\partial^2 y_0}{\partial \tilde{t}\,\partial t^+}\right) + \mu' \frac{\partial y_0}{\partial t^+}\right] + O(\varepsilon^2), \tag{3.5.108}$$

where a prime denotes $d/d\tilde{t}$.

The equations governing y_0 and y_1 are obtained by substituting the assumed expansions into (3.5.102). We find

$$L(y_0) \equiv \frac{\partial^2 y_0}{\partial t^{+2}} + y_0 = \frac{\alpha}{\mu^2}\cos(t + \beta) \tag{3.5.109}$$

$$L(y_1) = -\frac{1}{\mu}\left(\frac{\partial^2 y_0}{\partial t^+ \partial \tilde{t}} + \frac{\partial^2 y_0}{\partial \tilde{t}\,\partial t^+}\right) - \frac{\mu'}{\mu^2}\frac{\partial y_0}{\partial t^+}, \tag{3.5.110}$$

where t in Equation (3.5.109) is given by (cf. Equation 3.5.104a)

$$t = \frac{1}{\varepsilon} M^{-1}(\varepsilon t^+) \tag{3.5.111}$$

and M^{-1} always exists since $\mu > 0$.

We assume a solution of (3.5.109) in the form

$$y_0(t^+, \tilde{t}) = \rho_0(\tilde{t})\cos[t^+ + \phi_0(\tilde{t})] + A_0(\tilde{t})\cos(t + \beta) + B_0(\tilde{t})\sin(t + \beta) \tag{3.5.112}$$

and proceed to calculate the particular solution $A_0 \sin(t + \beta) + B_0 \cos(t + \beta)$ by substitution into Equation (3.5.109).

If we may regard dt/dt^+ to be the function of $\tilde{t}$ given by $\mu^{-1}(\tilde{t})$, it is easy to show that

$$A_0(\tilde{t}) = \frac{\alpha}{\mu^2 - 1} \tag{3.5.113a}$$

$$B_0(\tilde{t}) = 0. \tag{3.5.113b}$$

If on the other hand we insist that dt/dt^+ be the function of εt^+ given by differentiating Equation (3.5.111), we compute

$$A_0(\tilde{t}) = \frac{\alpha}{(\mu^2 - 1) + \varepsilon^2\mu'^2/\mu^2(\mu^2 - 1)} \tag{3.5.114a}$$

$$B_0(\tilde{t}) = \frac{-\varepsilon A(\tilde{t})\mu'}{\mu(\mu^2 - 1)}. \tag{3.5.114b}$$

The result in (3.5.114) is somewhat unpleasant in that y_0 now contains higher order terms in ε. Moreover, the calculations are unnecessarily cumbersome and can be simplified by using the short-cut which leads to (3.5.113). It is easy to show that the higher order terms occurring in (3.5.114) will be properly taken into account in the higher order expressions (cf. Problem 3.5.6).

The cost of the short-cut is modest—we must distinguish between the mixed partial derivatives $\partial^2 y_0/\partial t^+ \partial\tilde{t}$ and $\partial^2 y_0/\partial\tilde{t}\,\partial t^+$. In fact, we compute

$$\frac{\partial^2 y_0}{\partial t^+ \partial \tilde{t}} = -\rho_0' \sin(t^+ + \phi_0) - \rho_0\phi_0' \cos(t^+ + \phi_0) + \frac{2\alpha\mu'}{(\mu^2 - 1)^2}\sin(t + \beta) \tag{3.5.115}$$

$$\frac{\partial^2 y_0}{\partial \tilde{t}\,\partial t^+} = -\rho_0' \sin(t^+ + \phi_0) - \rho_0\phi_0' \cos(t^+ + \phi_0) + \frac{\alpha\mu'(3\mu^2 - 1)}{\mu^2(\mu^2 - 1)^2}\sin(t + \beta) \tag{3.5.116}$$

and the expression for $L(y_1)$ in Equation (3.5.110) becomes

$$L(y_1) = \frac{2\rho_0}{\mu}\phi_0' \cos(t^+ + \phi_0) + \left(\frac{2}{\mu}\rho_0' + \frac{\mu'}{\mu^2}\rho_0\right)\sin(t^+ + \phi_0) - \frac{4\alpha\mu'}{\mu(\mu^2 - 1)^2}\sin(t + \beta). \tag{3.5.117}$$

The last term in (3.5.117) differs from the one appearing in Equation (2.6) of Reference 3.5.5 because the noncommutativity of the mixed second partial derivative was not observed there.

The homogeneous solutions appearing in the right-hand side of Equation (3.5.117) will lead to mixed secular terms on the t^+ scale in y_1 and we eliminate them by setting

$$\frac{d\phi_0}{d\tilde{t}} = 0 \tag{3.5.118a}$$

$$\frac{d\rho_0}{d\tilde{t}} + \frac{1}{2\mu}\frac{d\mu}{d\tilde{t}}\rho_0 = 0 \tag{3.5.118b}$$

and this gives us the same result that we calculated in Section 3.3.2 for the homogeneous equation, i.e.,

$$\rho_0(\tilde{t}) = \tilde{\rho}_0\mu^{-1/2}, \qquad \tilde{\rho}_0 = \rho_0(0)\mu(0)^{1/2} = \text{const.}; \tag{3.5.119a}$$

$$\phi_0(\tilde{t}) = \phi_0, \qquad \phi_0 = \phi_0(0) = \text{const.} \tag{3.5.119b}$$

What remains of (3.5.117) can now be solved (again letting $dt/dt^+ =$ function of $\tilde{t}$ in calculating a particular solution) and gives

$$y_1 = \rho_1(\tilde{t})\cos[t^+ + \phi_1(\tilde{t})] - \frac{4\alpha\mu\mu'}{(\mu^2 - 1)^3}\sin(t + \beta). \tag{3.5.120}$$

We note the worsening singularities in the solution for $\mu = 1$, and cannot use the results for values of $\tilde{t} \approx \tilde{t}_0$. The nature of the singularity is significantly different now from the case of constant μ. If μ were constant, say $\mu = 1 + \sigma$ one can calculate the exact solution of Equation (3.5.102) for some given initial conditions. Then letting $\sigma \to 0$ one observes that the singular term in the particular solution cancels with a corresponding singular term in the homogeneous solution and the limiting result is the well-known mixed secular solution. Now, however, resonance occurs at some $\tilde{t}_0 \neq 0$ and since there is no singularity in the homogeneous solution at $\tilde{t} = \tilde{t}_0$ our results predict that $y \to \infty$.

Clearly, the assumed form of the expansion in Equation (3.5.105) is not valid for $\tilde{t} \approx \tilde{t}_0$. Accordingly, we introduce a new slow variable

$$\hat{t} = \frac{\tilde{t} - \tilde{t}_0}{\delta(\varepsilon)}, \tag{3.5.121}$$

where δ is an unknown function of ε such that $\delta \to 0$ as $\varepsilon \to 0$. The duration of the resonance is thus assumed to be of order δ on the $\tilde{t}$ scale which means it is of order δ/ε on the t scale. Now, if it turns out that $\delta/\varepsilon \to \infty$ as $\varepsilon \to 0$ the duration of resonance will be large on the t scale and therefore *the solution in this domain must also be developed as a two variable expansion.*

In anticipation of this possibility, and since we expect y to become large as $\tilde{t} \to \tilde{t}_0$ we assume a two variable resonance expansion in the form

$$y(t, \varepsilon) = \frac{1}{\nu_0(\varepsilon)} G_0(t, \hat{t}) + \nu_1(\varepsilon)G_1(t, \hat{t}) + \cdots, \tag{3.5.122}$$

where $\nu_0 \to 0$, $\nu_1\nu_0 \to 0$, etc. as $\varepsilon \to 0$. We choose t as the fast variable because $\mu \approx 1$ during resonance. Of course, at this point, all the above are assumptions, plausible though they may be; the final justification depends on being able to derive a consistent resonance expansion which matches with the pre-resonance solution.

Substituting Equation (3.5.122) into (3.5.102) and using Equation (3.5.103) to develop μ gives

$$\nu_0^{-1}\left(\frac{\partial^2 G_0}{\partial t^2} + G_0\right) + \nu_1\left(\frac{\partial^2 G_1}{\partial t^2} + G_1\right) - \alpha\cos(t+\beta)$$
$$+ \frac{2\varepsilon}{\nu_0\delta}\frac{\partial^2 G_0}{\partial t\,\partial\tilde{t}} + \frac{2\delta}{\nu_0}a_1\tilde{t}G_0 + \cdots = 0. \qquad (3.5.123)$$

Clearly, the first term proportional to ν_0^{-1} always dominates, and the richest equation for G_1 results if all the remaining terms are $O(1)$ (since we have the $O(1)$ forcing function appearing). This means that

$$\nu_1 = 1, \qquad \frac{\varepsilon}{\nu_0\delta} = 1, \qquad \frac{\delta}{\nu_0} = 1 \qquad (3.5.124a)$$

or

$$\nu_0 = \delta = \varepsilon^{1/2}. \qquad (3.5.124b)$$

Thus, the duration of the resonance region is $O(\varepsilon^{1/2})$ on the $\tilde{t}$ scale and $O(\varepsilon^{-1/2})$ on the t scale. Furthermore, during resonance the amplitude can be as large as $O(\varepsilon^{-1/2})$.[15]

With the above choices of δ, ν_0, ν_1, we have the resonance expansion

$$y(t,\varepsilon) = \varepsilon^{-1/2}G_0(t,\tilde{t}) + G_1(t,\tilde{t}) + \varepsilon^{1/2}G_2(t,\tilde{t}) + \cdots \qquad (3.5.125)$$

and the following sequence of equations to be solved for the G_i.

$$L(G_0) \equiv \frac{\partial^2 G_0}{\partial t^2} + G_0 = 0 \qquad (3.5.126a)$$

$$L(G_1) = \alpha\cos(t+\beta) - 2a_1\tilde{t}G_0 - 2\frac{\partial^2 G_0}{\partial t\,\partial\tilde{t}} \qquad (3.5.126b)$$

$$L(G_2) = -2a_1\tilde{t}G_1 - (2a_2 + a_1^2)\tilde{t}^2 G_0 - 2\frac{\partial^2 G_1}{\partial t\,\partial\tilde{t}} - \frac{\partial^2 G_0}{\partial\tilde{t}^2}. \qquad (3.5.126c)$$

We use the solution of Equation (3.3.126a) in the form

$$G_0(t,\tilde{t}) = A(\tilde{t})\sin t + B(\tilde{t})\cos t \qquad (3.5.127)$$

[15] With n equal to the first nonvanishing exponent of $(\tilde{t} - \tilde{t}_0)$ in the expansion (3.5.103) we can show that the corresponding values of δ, ν_0 and ν_1 are $\delta = \varepsilon^{1/(1+n)}$, $\nu_0 = \varepsilon^{n/(1+n)}$, $\nu_1 = 1$ (cf. Problem 3.5.9).

in Equation (3.5.126b) and find that in order to remove inconsistent terms proportional to t sin t and t cos t from the solution for G_1 we must set

$$\frac{dA}{d\bar{t}} + a_1\bar{t}B = \frac{\alpha}{2}\cos\beta \tag{3.5.128a}$$

$$\frac{dB}{d\bar{t}} - a_1\bar{t}A = \frac{\alpha}{2}\sin\beta. \tag{3.5.128b}$$

As a result, the right-hand side of Equation (3.5.126b) vanishes and G_1 is simply

$$G_1(t, \bar{t}) = C(\bar{t})\sin t + D(\bar{t})\cos t. \tag{3.5.129}$$

A convenient way to solve Equations (3.5.128) is to introduce the complex amplitude-phase function

$$z(\bar{t}) = A(\bar{t}) + iB(\bar{t}) \tag{3.5.130}$$

and note that Equations (3.5.128) reduce to the simple equation

$$\frac{dz}{d\bar{t}} - ia_1\bar{t}z = \frac{\alpha}{2}e^{i\beta} \tag{3.5.131}$$

for z which can be immediately integrated in the form

$$z = z_{-\infty}e^{ia_1\bar{t}^2/2} + \frac{\alpha}{2}e^{ia_1\bar{t}^2/2}\int_{-\infty}^{\bar{t}} e^{i(\beta - a_1\sigma^2/2)}\,d\sigma. \tag{3.5.132}$$

Here $z_{-\infty}$ is a complex constant determined by the value of z as $\bar{t} \to -\infty$, which will be given by the matching. The particular solution given by the integral vanishes as $\bar{t} \to -\infty$.

The contribution of the homogeneous solution $z_{-\infty}e^{ia_1\bar{t}^2/2}$ in G_1 will be an oscillatory term proportional to $e^{i(t + a_1\bar{t}^2/2)}$. Clearly, there is no such term in the pre-resonance solution so that without carrying out the matching in detail we can set $z_{-\infty} = 0$.

The real and imaginary parts of the integral representation for z can be expressed in terms of Fresnel integrals

$$c(\xi) = \int_{\xi}^{\infty} \cos r^2\,dr \tag{3.5.133a}$$

$$s(\xi) = \int_{\xi}^{\infty} \sin r^2\,dr, \tag{3.5.133b}$$

where

$$\xi^2 = \zeta = \frac{a_1\bar{t}^2}{2} \tag{3.5.133c}$$

and for $\bar{t} < 0$, we have (with $z_{-\infty} = 0$)

$$A(\bar{t}) = \frac{\alpha}{(2a)^{1/2}} [\cos(\zeta + \beta)c(-\xi) + \sin(\zeta + \beta)s(-\xi)] \quad (3.5.134a)$$

$$B(\bar{t}) = \frac{\alpha}{(2a_1)^{1/2}} [\sin(\zeta + \beta)c(-\xi) - \cos(\zeta + \beta)s(-\xi)]. \quad (3.5.134b)$$

For $\bar{t} > 0$, we write Equation (3.5.132) as

$$z = \frac{\alpha}{2} e^{ia_1\bar{t}^2/2} \int_{-\infty}^{\infty} e^{i(\beta - a_1\sigma^2/2)}\, d\sigma - \frac{\alpha}{2} e^{ia_1\bar{t}^2/2} \int_{\bar{t}}^{\infty} e^{i(\beta - a_1\sigma^2/2)}\, d\sigma. \quad (3.5.135)$$

Using the fact that

$$\int_0^{\infty} \cos r^2\, dr = \int_0^{\infty} \sin r^2\, dr = \frac{1}{2}\sqrt{\frac{\pi}{2}} \quad (3.5.136)$$

we find that for $\bar{t} > 0$, A and B are given by

$$A = \frac{\alpha}{2}\sqrt{\frac{\pi}{a_1}} [(\cos\beta - \sin\beta)\sin\zeta + (\sin\beta + \cos\beta)\cos\zeta]$$
$$- \frac{\alpha}{(2a_1)^{1/2}} [\cos(\zeta + \beta)c(\xi) + \sin(\zeta + \beta)s(\xi)] \quad (3.5.137a)$$

$$B = \frac{\alpha}{2}\sqrt{\frac{\pi}{a_1}} [(-\cos\beta + \sin\beta)\cos\zeta + (\sin\beta + \cos\beta)\sin\zeta]$$
$$- \frac{\alpha}{(2a_1)^{1/2}} [\sin(\zeta + \beta)c(\xi) - \cos(\zeta + \beta)s(\xi)]. \quad (3.5.137b)$$

Note that for $\bar{t} > 0$, A and B have two contributions. The first is a homogeneous solution proportional to $\sin\zeta$ and $\cos\zeta$ arising from the first term in Equation (3.5.135). In addition, there is the particular solution which has the opposite sign as for the case $\bar{t} < 0$. In Reference (3.5.5) the homogeneous solution was overlooked for the case $\bar{t} > 0$ and as we shall see presently, this has a crucial effect in determining the post resonance behavior.

The above determines G_0 completely. In order to derive a uniformly valid solution to $O(1)$ it is also necessary to calculate G_1, which means we must examine the right-hand side of Equation (3.5.126c). Removal of mixed secular terms in G_2 leads to the following equation for $w = C + iD$

$$\frac{dw}{d\bar{t}} - ia_1\bar{t}w = \left(ia_2\bar{t}^2 - \frac{a_1}{2}\right)z - \frac{a_1\bar{t}}{4}\alpha e^{i\beta}$$
$$\equiv h(\bar{t}). \quad (3.5.138)$$

This can also be easily integrated as follows

$$w = w_{-\infty} e^{i\zeta} - \frac{i\alpha e^{i\beta}}{4a_1^2} \int_0^{\bar{t}} h(\sigma) e^{-ia_1\sigma^2/2}\, d\sigma, \tag{3.5.139}$$

where as before $\zeta = a_1 \bar{t}^2/2$ and we have the complex constant of integration w_0 which will later be connected to constants w_∞ and $w_{-\infty}$. To be consistent with the notation of Reference (3.5.5) we let

$$w_{-\infty} = \delta - i\gamma, \tag{3.5.140}$$

where γ and δ are real constants.

In anticipation of the matching we now derive the asymptotic expansions of G_0 and G_1 as $|\bar{t}| \to \infty$. One could always use the integral representations for z and w and repeated integrations by parts to do this. But this is quite laborious. A much easier approach discussed in Section 2.3.2 is to proceed directly from the differential equations for z and w. The particular solution for z [i.e., the integral in Equation (3.5.132)] must clearly have an expansion of the form

$$z_p = \frac{c_1}{\bar{t}} + \frac{c_2}{\bar{t}^3} + O(\bar{t}^{-5}) \quad \text{as } \bar{t} \to -\infty \tag{3.5.141}$$

deduced by noting that the term $-a_1 \bar{t} z$ always dominates $dz/d\bar{t}$ in Equation (3.5.131). Thus, the leading term obtained by setting $-ia_1 \bar{t} z = (\alpha/2)e^{i\beta}$ is $O(\bar{t}^{-1})$. Differentiation will introduce higher negative powers of $\bar{t}$ with all the even powers being set equal to zero. After a simple calculation, we find

$$c_1 = \frac{i\alpha}{2a_1} e^{i\beta} \tag{3.5.142a}$$

$$c_2 = -\frac{c_1}{ia_1} = -\frac{\alpha}{2a_1^2} e^{i\beta}. \tag{3.5.142b}$$

Now for $\bar{t} > 0$, Equation (3.5.135) can be written

$$z = \alpha \sqrt{\frac{\pi}{2a_1}} \exp\left[i\left(\frac{a_1 \bar{t}^2}{2} + \beta - \frac{\pi}{4} \right) \right] - z_p(-\bar{t}), \tag{3.5.143}$$

where z_p is the particular integral defined in (3.5.132). Since the expansion (3.5.141) gives an odd function of $\bar{t}$, we calculate

$$z = c_0 e^{i\zeta} + \frac{c_1}{\bar{t}} + \frac{c_2}{\bar{t}^3} + O(\bar{t}^{-5}) \quad \text{as } \bar{t} \to \infty, \tag{3.5.144a}$$

where

$$c_0 = \alpha \sqrt{\frac{\pi}{2a_1}} e^{i(\beta - \pi/4)}. \tag{3.5.144b}$$

Using Equations (3.5.141) and (3.5.144) we calculate the following expansions for G_0 as $|\bar{t}| \to \infty$

$$G_0 = \frac{\alpha}{2a_1\bar{t}} \cos(t + \beta) - \frac{\alpha}{2a_1^2\bar{t}^3} \sin(t + \beta) + O(\bar{t}^{-5}) \quad \text{as } \bar{t} \to -\infty \tag{3.5.145a}$$

$$G_0 = \frac{\alpha}{2}\sqrt{\frac{\pi}{a_1}}\left[\sin\left(\frac{a_1}{2}\bar{t}^2 + t + \beta\right) - \cos\left(\frac{a_1}{2}\bar{t}^2 + t + \beta\right)\right]$$
$$+ \frac{\alpha}{2a_1\bar{t}} \cos(t + \beta) - \frac{\alpha}{2a_1^2\bar{t}^3} \sin(t + \beta) + O(\bar{t}^{-5}) \quad \text{as } \bar{t} \to \infty. \tag{3.5.145b}$$

A calculation similar to the one used in obtaining Equation (3.5.144) can be used for w. The result is

$$w_p = -\frac{i\alpha e^{i\beta}}{4a_1^2}(2a_2 + a_1^2) + O(\bar{t}^{-2}) \quad \text{as } \bar{t} \to -\infty. \tag{3.5.146}$$

It then follows that, as $\bar{t} \to -\infty$, G_1 is given by

$$G_1 = \delta \sin(\zeta + t) - \gamma \cos(\zeta + t) - \frac{\alpha(2a_2 + a_1^2)}{4a_1^2} \cos(t + \beta). \tag{3.5.147}$$

However, when $\bar{t} \to \infty$ the right-hand side of Equation (3.5.138) will contain terms proportional to $\bar{t}^n e^{i\zeta}$ with $n = 0, 2$, arising from the fact that z now contains a homogeneous solution [cf. Equation (3.5.143)]. Since these terms are unbounded as $\bar{t} \to \infty$ we must exercise more caution in developing the asymptotic expansion in this limit. Using the expansion given by Equation (3.5.144) for z when $\bar{t} \to \infty$, we calculate the following equation for w:

$$\frac{dw}{d\bar{t}} - ia_1\bar{t}w = ia_2c_0\bar{t}^2e^{i\zeta} + ia_2c_1\bar{t} - \frac{a_1}{2}c_0e^{i\zeta} - \frac{a_1}{4}\alpha e^{i\beta}\bar{t} + O\left(\frac{1}{\bar{t}}\right). \tag{3.5.148}$$

Clearly, the particular solution corresponding to these terms must be of the form

$$w_p(\bar{t}) = d_0\bar{t}^3e^{i\zeta} + d_1\bar{t}^2e^{i\zeta} + d_2\bar{t}e^{i\zeta} + d_3 + \frac{d_4}{\bar{t}} + O(\bar{t}^{-2}) \quad \text{as } \bar{t} \to \infty. \tag{3.5.149}$$

Substituting Equation (3.5.149) into Equation (3.5.148) and equating coefficients of like terms gives

$$d_0 = \frac{ia_2\alpha}{3}\sqrt{\frac{\pi}{2a_1}}\, e^{i(\beta-\pi/4)} \tag{3.5.150a}$$

$$d_1 = 0 \tag{3.5.150b}$$

$$d_2 = -\frac{a_1\alpha}{2}\sqrt{\frac{\pi}{2a_1}}\, e^{i(\beta-\pi/4)} \tag{3.5.150c}$$

$$d_3 = -\frac{i\alpha}{4a_1^2}(2a_2 + a_1^2)e^{i\beta} \tag{3.5.150d}$$

$$d_4 = 0. \tag{3.5.150e}$$

All that remains is the determination of the homogeneous solution corresponding to the initial condition $w_{-\infty} = \delta - i\gamma$.

We first write Equation (3.5.138) in the following form where all the singular terms are isolated on the right-hand side

$$\frac{dw}{d\bar{t}} - ia_1\bar{t}w = k(\bar{t}) + l(\bar{t}), \tag{3.5.151a}$$

where

$$k(\bar{t}) = ia_2\bar{t}^2 z = -\frac{a_1}{2}z - ia_2 c_0 \bar{t}^2 e^{i\zeta} - ia_2 c_1 \bar{t} + \frac{a_1}{2}c_0 e^{i\zeta} \tag{3.5.151b}$$

$$l(\bar{t}) = ia_2 c_0 \bar{t}^2 e^{i\zeta} = +ia_2 c_1 \bar{t} - \frac{a_1}{2}c_0 e^{i\zeta} - \frac{a_1}{4}\alpha\bar{t}e^{i\beta}. \tag{3.5.151c}$$

Clearly $k + l$ equals the right-hand side of Equation (3.5.138). Moreover using k for h in Equation (3.5.139) when $\bar{t}$ is positive allows one to write the solution for w in the form (valid for $\bar{t} > 0$)

$$w = \left[w_0 - d_3 + \int_0^\infty k(\sigma)e^{-ia_1\sigma^2/2}\,d\sigma\right]e^{i\zeta}$$
$$- e^{i\zeta}\int_{\bar{t}}^\infty k(\sigma)e^{-ia_1\sigma^2/2}\,d\sigma + (d_0\bar{t}^3 + d_2\bar{t})e^{i\zeta} + d_3. \tag{3.5.152}$$

Now, the asymptotic behavior of w as $\bar{t} \to \infty$ follows immediately

$$w = d_0\bar{t}^3 e^{i\zeta} + d_2\bar{t}e^{i\zeta} + d_3 + w_\infty e^{i\zeta} + O(\bar{t}^{-2}), \tag{3.5.153}$$

where the constant w_∞ is given by

$$w_\infty = w_0 - d_3 + \int_0^\infty k(\sigma)e^{-ia_1\sigma^2/2}\,d\sigma. \tag{3.5.154}$$

A similar calculation gives $w_{-\infty} = w_\infty$.

Using the above expression for w as $\tilde{t} \to \infty$ in Equation (3.5.129) defines G_1 in this limit.

This completes the resonance expansion and its asymptotic behavior as $|\tilde{t}| \to \infty$ explicitly to $O(1)$. Next, we consider the matching of the pre-resonance and resonance expansions to relate the unknown integration constants in the resonance solution with the initial values of y and dy/dt.

As discussed in Chapter 2, we introduce the intermediate variable

$$t_\nu = \frac{\tilde{t} - \tilde{t}_0}{\varepsilon^\nu}, \tag{3.5.155}$$

where ν is some constant[16] in the interval $0 \leq \nu_1 < \nu \leq \nu_2 < \frac{1}{2}$. The constants ν_1 and ν_2 which determine the overlap domain will be given by the matching. Matching to $O(1)$ will consist of requiring that the preresonance and resonance solutions, when written in terms of t_ν, agree to $O(1)$ in the limit as $\varepsilon \to 0$ with t_ν fixed. In this problem, there are trigonometric terms in both expansions. These terms may not possess a limit as $\varepsilon \to 0$ with t_ν fixed. However they can be directly identified and related so we avoid expanding the arguments of trigonometric terms for simplicity.

Consider first the behavior of the pre-resonance expansion in the overlap domain. It is clear from Equation (3.5.155) that functions of $\tilde{t}$ must be developed around $\tilde{t}_0$. Therefore, the fast variable t^+ can be expanded using the identity

$$t^+ = \frac{1}{\varepsilon}\int_0^{\tilde{t}_0} \mu(s)ds + \frac{1}{\varepsilon}\int_0^{\tilde{t}-\tilde{t}_0} \mu(\tilde{t}_0 + \sigma)d\sigma \tag{3.5.156}$$

obtained from Equation (3.5.104a)

If we denote the constant

$$\tau_0 = \int_0^{\tilde{t}_0} \mu(s)ds \tag{3.5.157}$$

we calculate the following result by expanding the second integral in Equation (3.5.156)

$$t^+ = \frac{\tau_0}{\varepsilon} + \varepsilon^{\nu-1}t_\nu + \varepsilon^{2\nu-1}\frac{a_1}{2}t_\nu^2 + O(\varepsilon^{3\nu-1}). \tag{3.5.158}$$

Using the above and the fact that

$$\tilde{t} = \tilde{t}_0 + \varepsilon^\nu t_\nu$$

[16] Actually, Equation (3.5.155) is a short-cut in the sense that we should match with respect to a variable $t_\eta = (\tilde{t} - \tilde{t}_0)/\eta(\varepsilon)$ with $\eta(\varepsilon)$ in some domain $\varepsilon^{1/2} \lll \eta_1(\varepsilon) \ll \eta(\varepsilon) \ll \eta_2(\varepsilon) \lll 1$. However, the simplification (3.5.155) does not affect the results.

we calculate the following form of the pre-resonance expansion in the overlap domain

$$y = \tilde{\rho}_0 \cos\left(\frac{\tau_0}{\varepsilon} + \varepsilon^{\nu-1} t_\nu + \varepsilon^{2\nu-1}\frac{a_1}{2} t_\nu^2 + \tilde{\phi}_0\right)$$
$$+ \varepsilon^{-\nu}\frac{\alpha}{2a_1 t_\nu}\cos\left(\frac{\bar{t}_0}{\varepsilon} + \varepsilon^{\nu-1} t_\nu + \beta\right)$$
$$- \left(\frac{2a_2 + a_1^2}{4a_1^2}\right)\alpha \cos\left(\frac{\bar{t}_0}{\varepsilon} + \varepsilon^{\nu-1} t_\nu + \beta\right)$$
$$- \varepsilon^{1-3\nu}\frac{\alpha}{2a_1^2 t_\nu^3}\sin\left(\frac{\bar{t}_0}{\varepsilon} + \varepsilon^{\nu-1} t_\nu + \beta\right)$$
$$+ O(\varepsilon^{\nu}) + O(\varepsilon^{1-2\nu}) + O(\varepsilon^{3\nu-1}). \tag{3.5.159}$$

In terminating the expansion of t^+ with the term of order $\varepsilon^{2\nu-1}$ we have assumed that $\varepsilon^{3\nu-1} \to 0$, i.e., that $\nu_1 = \frac{1}{3}$. This will be justified by the fact that all the terms in Equation (3.5.159) will match with corresponding terms in the resonance expansion. In this connection it is noted that the singular term of order $\varepsilon^{1-3\nu}$ in Equation (3.5.159) is contributed by the particular solution of y_1, and that the homogeneous solution in y_1 is not needed for the matching to $O(1)$.

In order to express the resonance expansion in terms of t_ν we note that

$$\bar{t} = \varepsilon^{-1/2+\nu} t_\nu$$

and

$$\zeta = \varepsilon^{-1+2\nu}\frac{a_1}{2} t_\nu^2. \tag{3.5.160}$$

Use of the above in Equations (3.5.145a) and (3.5.147) gives, after some algebra,

$$y = \frac{\varepsilon^{-\nu}}{2a_1 t_\nu}\alpha\cos\left(\frac{\bar{t}_0}{\varepsilon} + \varepsilon^{\nu-1} t_\nu + \beta\right) - \frac{\varepsilon^{1-3\nu}}{2a_1^2 t_\nu^3}\alpha\sin\left(\frac{\bar{t}_0}{\varepsilon} + \varepsilon^{\nu-1} t_\nu + \beta\right)$$
$$- \frac{\alpha(2a_2 + a_1^2)}{4a_1^2}\cos\left(\frac{\bar{t}_0}{\varepsilon} + \varepsilon^{\nu-1} t_\nu + \beta\right) - \gamma\cos\left(\frac{\bar{t}_0}{\varepsilon} + \varepsilon^{\nu-1} t_\nu + \varepsilon^{2\nu-1}\frac{a_1}{2} t_\nu^2\right)$$
$$+ \delta\sin\left(\frac{\bar{t}_0}{\varepsilon} + \varepsilon^{\nu-1} t_\nu + \varepsilon^{2\nu-1}\frac{a_1}{2} t_\nu^2\right) + O(\varepsilon^{2-5\nu}). \tag{3.5.161}$$

Comparing Equation (3.5.159) with Equation (3.5.161) we see that all the singular terms match identically. In order to match the terms of order unity arising from the homogeneous solution in G_1, we must set

$$\gamma = -\tilde{\rho}_0 \cos\left(\frac{\tau_0 - \bar{t}_0}{\varepsilon} + \tilde{\phi}_0\right) \tag{3.5.162a}$$

$$\delta = -\tilde{\rho}_0 \sin\left(\frac{\tau_0 - \bar{t}_0}{\varepsilon} + \tilde{\phi}_0\right) \tag{3.5.162b}$$

and since the largest neglected term in Equation (3.5.161) is $O(\varepsilon^{2-5\nu})$, we must choose $\nu_2 = \frac{2}{5}$. Moreover, the absence of terms of order $\varepsilon^{-1/2}$ in Equation (3.5.159) confirms our earlier statement that $z_{-\infty} = 0$.

All the constants of the resonance solution are thus determined. To predict what happens after passage through resonance we must construct a post resonance expansion and match it with the resonance solution. The details of this calculation are left as an exercise (cf. Problem 3.5.7). We simply note here that since G_0 does not vanish as $t \to \infty$ (as was incorrectly assumed in Reference 3.5.5) it must match with a post resonance expansion of order $\varepsilon^{-1/2}$! Thus, after resonance the amplitude remains large and is slowly modulated by the variations in μ.

3.5.5 Passage of Two Weakly Nonlinear Coupled Oscillators Through Resonance

In this subsection we return to the system

$$\frac{d^2x}{dt^2} + a^2(\tilde{t})x = \varepsilon y^2 \tag{3.5.163a}$$

$$\frac{d^2y}{dt^2} + b^2(\tilde{t})y = 2\varepsilon xy \tag{3.5.163b}$$

that we considered in Section 3.5.2 for the case a and b constant.

Since Equations (3.5.163) are Hamiltonian, the procedure to be discussed in Section 3.7.2 can be used to first transform the problem to a basic *second-order* one exhibiting passage through resonance. This reduced problem can then be solved routinely by the techniques discussed in Section 3.5.4. A general treatment can be found in Reference 3.5.16 and the essential ideas are outlined in Problem 3.7.9. However, in Reference 3.5.6 (which we follow in this section) the Hamiltonian structure of Equations (3.5.163) and the possibility of order reduction are not exploited to illustrate the techniques needed to solve *general* weakly nonlinear systems of order higher than two.

(a) *Solution before Resonance* (*Outer Expansion*)

As in the autonomous problem, the solution to order ε will contain the divisor $a - 2b$, and we will concentrate on the question of how the solution passes through this condition, which we take to occur at $\tilde{t} = \tilde{t}_0$.

Another similarity with the autonomous problem is the need to account for corrections to the unperturbed frequencies to higher order. Since we are taking into account the dependence of the solution on $\tilde{t}$ we only need correction terms of order ε^2 and higher and we assume that the two fast variables are given by[17]

$$\mu = \frac{1}{\varepsilon}\int_0^{\tilde{t}} [a(s) + \varepsilon^2 k_2(s) + \cdots]ds \quad (3.5.164a)$$

$$\nu = \frac{1}{\varepsilon}\int_0^{\tilde{t}} [b(s) + \varepsilon^2 l_2(s) + \cdots]ds, \quad (3.5.164b)$$

where k_2 and l_2 are unknown functions of $\tilde{t}$ to be determined by consistency of the solution to $O(\varepsilon^2)$.

In view of the discussion in Section 3.5.3, we assume that x and y can be expanded as follows

$$x = x_0(\mu, \tilde{t}) + \varepsilon x_1(\mu, \tilde{t}) + \varepsilon^2 x_2(\mu, \tilde{t}) + \cdots \quad (3.5.165a)$$

$$y = y_0(\nu, \tilde{t}) + \varepsilon y_1(\nu, \tilde{t}) + \varepsilon^2 y_2(\nu, \tilde{t}) + \cdots, \quad (3.5.165b)$$

where μ and ν are the fast variables defined in (3.5.164).

The equations that result for the x_i and y_i using the expansion procedure discussed in Section 3.5.3 are listed below

$$a^2 L_\mu(x_0) \equiv a^2\left(\frac{\partial^2 x_0}{\partial \mu^2} + x_0\right) = 0 \quad (3.5.166a)$$

$$b^2 L_\nu(y_0) \equiv b^2\left(\frac{\partial^2 y_0}{\partial \nu^2} + y_0\right) = 0 \quad (3.5.166b)$$

$$a^2 L_\mu(x_1) = -aM_\mu(x_0) - a'\frac{\partial x_0}{\partial \mu} + y_0^2 \quad (3.5.167a)$$

$$b^2 L_\nu(y_1) = -bM_\nu(y_0) - b'\frac{\partial y_0}{\partial \nu} + 2x_0 y_0 \quad (3.5.167b)$$

$$a^2 L_\mu(x_2) = -aM_\mu(x_1) - a'\frac{\partial x_1}{\partial \mu} + 2ak_2 x_0 - \frac{\partial^2 x_0}{\partial \tilde{t}^2} + 2y_0 y_1 \quad (3.5.168a)$$

$$b^2 L_\nu(y_2) = -bM_\nu(y_1) - b'\frac{\partial y_1}{\partial \nu} + 2bl_2 y_0 - \frac{\partial^2 y_0}{\partial \tilde{t}^2} + 2(x_0 y_1 + y_1 x_0), \quad (3.5.168b)$$

where the operators M_μ and M_ν are simply (note that we are distinguishing $\partial^2/\partial\mu\,\partial\tilde{t}$ and $\partial^2/\partial\tilde{t}\,\partial\mu$)

$$M_\mu = \frac{\partial^2}{\partial \mu\,\partial \tilde{t}} + \frac{\partial^2}{\partial \tilde{t}\,\partial \mu} \quad (3.5.169a)$$

$$M_\nu = \frac{\partial^2}{\partial \nu\,\partial \tilde{t}} + \frac{\partial^2}{\partial \tilde{t}\,\partial \nu}. \quad (3.5.169b)$$

[17] Actually one can prove that the final result for x and y is unchanged if one introduces arbitrary functions εk_1 and εl_1 in the definitions (3.5.164) (cf. the discussion following (3.5.172)).

We use the solutions

$$x_0 = \alpha_0(\tilde{t})\cos[\mu + \phi_0(\tilde{t})] \tag{3.5.170a}$$

$$y_0 = \beta_0(\tilde{t})\cos[\nu + \psi_0(\tilde{t})] \tag{3.5.170b}$$

of (3.5.166) to calculate the right-hand sides of (3.5.167) and remove inconsistent terms. We find

$$\alpha_0(\tilde{t}) = \frac{\alpha_0(0)}{\sqrt{a(\tilde{t})/a(0)}} \tag{3.5.171a}$$

$$\beta_0(\tilde{t}) = \frac{\beta_0(0)}{\sqrt{b(\tilde{t})/b(0)}} \tag{3.5.171b}$$

$$\phi_0(\tilde{t}) = \phi_0(0) = \text{const.} \tag{3.5.172a}$$

$$\psi_0(\tilde{t}) = \psi_0(0) = \text{const.} \tag{3.5.172b}$$

Had we assumed that the definitions of μ and ν in (3.5.164) contained terms εk_1 and εl_1 respectively, the results for ϕ_0 and ψ_0 would have read

$$\phi_0(\tilde{t}) = -\int_0^{\tilde{t}} k_1(s)ds \tag{3.5.173a}$$

$$\psi_0(\tilde{t}) = -\int_0^{\tilde{t}} l_1(s)ds. \tag{3.5.173b}$$

It is easily seen that the expressions for $\mu + \phi_0$ and $\nu + \psi_0$ are unaltered by the choice of the functions k_1 and l_1 since these cancel out in the final result. Thus, the simple choice $k_1 = l_1 = 0$ in (3.5.164) is justified.

The expressions we calculate for x_1 and y_1 are identical to those for the autonomous problem [cf. Equations (3.5.42)]

$$x_1 = \alpha_1(\tilde{t})\cos[\mu + \phi_1(\tilde{t})] + \frac{\beta_0^2}{2a^2} + \frac{\beta_0^2}{2(a^2 - 4b^2)}\cos 2(\nu + \psi_0) \tag{3.5.174a}$$

$$y_1 = \beta_1(\tilde{t})\cos[\nu + \psi_1(\tilde{t})] - \frac{\alpha_0\beta_0}{a(a+2b)}\cos(\mu + \nu + \phi_0 + \psi_0)$$
$$- \frac{\alpha_0\beta_0}{a(a-2b)}\cos(\mu - \nu + \phi_0 - \psi_0). \tag{3.5.174b}$$

This is as far as we need to go in order to exhibit all the essential features of passage through resonance. However, to demonstrate the procedure for determining k_2 and l_2 we also consider the solution to $O(\varepsilon^2)$.

After much tedious but straightforward algebra, one can derive the following expressions governing the unknowns α_1, β_1, ϕ_1 and ψ_1 such that the $O(\varepsilon^2)$ terms in (3.5.165) are bounded functions of μ and ν.

$$2a\frac{dA_1}{d\tilde{t}} + a'A_1 = 0 \tag{3.5.175a}$$

$$2a\frac{dB_1}{d\tilde{t}} + a'B_1 = -\left(2ak_2\alpha_0 - \alpha_0'' - 2\frac{\alpha_0\beta_0^2}{a^2 - 4b^2}\right) \tag{3.5.175b}$$

$$2b\frac{dC_1}{d\tilde{t}} + b'C_1 = 0 \tag{3.5.175c}$$

$$2b\frac{dD_1}{d\tilde{t}} + b'D_1 = -\left\{2bl_2\beta_0 - \beta_0'' + \frac{2\beta_0[\beta_0^2(3a^2 - 8b^2) - 4a^2\alpha_0^2]}{4a^2(a^2 - 4b^2)}\right\}, \tag{3.5.175d}$$

where

$$A_1 = \alpha_1 \cos(\phi_1 - \phi_0) \tag{3.5.176a}$$

$$B_1 = \alpha_1 \sin(\phi_1 - \phi_0) \tag{3.5.176b}$$

$$C_1 = \beta_1 \cos(\psi_1 - \psi_0) \tag{3.5.176c}$$

$$D_1 = \beta_1 \sin(\psi_1 - \psi_0). \tag{3.5.176d}$$

Now, the right-hand sides of (3.5.175b, d), which are known functions of $\tilde{t}$, will lead to unbounded terms on the $\tilde{t}$ scale, and we eliminate these by choosing

$$k_2 = \frac{\beta_0^2}{a(a^2 - 4b^2)} + \frac{\alpha_0''}{2a\alpha_0} \tag{3.5.177}$$

$$l_2 = -\frac{\beta_0''}{2b\beta_0} - \frac{\beta_0^2(3a^2 - 8b^2) - 4a^2\alpha_0^2}{4a^2b(a^2 - 4b^2)} \tag{3.5.178}$$

and these define k_2 and l_2 in terms of known quantities. Comparing these expressions with those given in Equations (3.5.46) shows agreement in the limiting case of constant frequencies a and b, i.e., $\alpha_0'' = \beta_0'' = 0$.

What remains of (3.5.175) gives the same structure for α_1, β_1, ϕ_1, and ψ_1 as we calculated for the corresponding $O(1)$ quantities [cf. (3.5.171, 172)], and this completes the solution to $O(\varepsilon)$. We note that in addition to the singularities appearing in the amplitudes of the x_1, y_1 solution when $a \to 2b$, we also have singularities in the slowly varying phases k_2 and l_2. Thus, our results break down for values of $\tilde{t}$ close to the critical value $\tilde{t}_0$ where $a(\tilde{t}_0) = 2b(\tilde{t}_0)$. The nature of this singularity in the outer solution is significantly different for the present problem in comparison with the autonomous case discussed in Section 3.5.2. There, if we take the limit as $a \to 2b$, the outer solution (when initial conditions are properly taken into account) does not become infinite but merely exhibits mixed secular terms of the form $t \sin t$ or $t \cos t$. Now, however, the singularity as $\tilde{t} \to \tilde{t}_0$ does not cancel and the solution grows without bound. Clearly, this result is incorrect and we consider the appropriate representation of the solution near $\tilde{t}_0$ next.

(b) *Solution during Resonance* (*Inner Expansion*)

For simplicity, we assume that a and b have the following behavior near $\tilde{t} = \tilde{t}_0$

$$a = 2b_0 + a_1(\tilde{t} - \tilde{t}_0) + a_2(\tilde{t} - \tilde{t}_0)^2 + \cdots \tag{3.5.179a}$$

$$b = b_0 + b_1(\tilde{t} - \tilde{t}_0) + b_2(\tilde{t} - \tilde{t}_2)^2 + \cdots, \tag{3.5.179b}$$

where the constants above are subject to the conditions that $b_0 > 0$, $a_1 \neq 0$, $b_1 \neq 0$, $a_1 \neq 2b_1$[18] and a_2, b_2, etc. are arbitrary. Other more complicated behaviors near resonance can also be handled.

For the choice $a_1 \neq 0, b_1 \neq 0$, it is easy to show by considering an arbitrary function of ε in the denominator of (4.2) that the appropriate slow variable is (cf. the discussion following Equation (3.5.121) in Section 3.5.4)

$$\bar{t} = \frac{\tilde{t} - \tilde{t}_0}{\varepsilon^{1/2}} \tag{3.5.180}$$

and the solution is expanded in multiple variable form in powers of $\varepsilon^{1/2}$.

$$x = \bar{x}_0(t, \bar{t}) + \varepsilon^{1/2}\bar{x}_1(t, \bar{t}) + \varepsilon\bar{x}_2(t, \bar{t}) + \cdots \tag{3.5.181a}$$

$$y = \bar{y}_0(t, \bar{t}) + \varepsilon^{1/2}\bar{y}_1(t, \bar{t}) + \varepsilon\bar{y}_2(t, \bar{t}) + \cdots. \tag{3.5.181b}$$

The reason for seeking a multi-variable expansion of the solution here is because resonance lasts for a period of $O(\varepsilon^{-1/2})$ on the fast scale t. Since in this period the leading terms in the frequencies of the x and y oscillators are constants, the solution (3.5.181) is developed as a conventional two variable expansion.

We calculate the following equations for the $\bar{x}_i$ and $\bar{y}_i$

$$\bar{L}_1(\bar{x}_0) \equiv \frac{\partial^2 \bar{x}_0}{\partial t^2} + 4b_0^2 \bar{x}_0 = 0 \tag{3.5.182a}$$

$$\bar{L}_2(\bar{y}_0) \equiv \frac{\partial^2 \bar{y}_0}{\partial t^2} + b_0^2 \bar{y}_0 = 0 \tag{3.5.182b}$$

$$\bar{L}_1(\bar{x}_1) = -2\frac{\partial^2 \bar{x}_0}{\partial t\, \partial \bar{t}} - 4b_0 a_1 \bar{t}\bar{x}_0 \tag{3.5.183a}$$

$$L_2(\bar{y}_1) = -2\frac{\partial^2 \bar{y}_0}{\partial t\, \partial \bar{t}} - 2b_0 b_1 \bar{t}\bar{y}_0 \tag{3.5.183b}$$

$$\bar{L}_1(\bar{x}_2) = -2\frac{\partial^2 \bar{x}_1}{\partial t\, \partial \bar{t}} - (a_1^2 + 4b_0 a_2)\bar{t}^2 x_0 - \frac{\partial^2 \bar{x}_0}{\partial \bar{t}^2} - 4b_0 a_1 \bar{t}\bar{x}_1 + \bar{y}_0^2 \tag{3.5.184a}$$

$$\bar{L}_2(\bar{y}_2) = -2\frac{\partial^2 \bar{y}_1}{\partial t\, \partial \bar{t}} - (b_1^2 + 2b_0 b_2)\bar{t}^2 \bar{y}_0 - \frac{\partial^2 \bar{y}_0}{\partial \bar{t}^2} - 2b_0 b_1 \bar{t}\bar{y}_1 + 2\bar{x}_0 \bar{y}_0. \tag{3.5.184b}$$

[18] If $a_1 - 2b_1 = 0$, the singularity in $(a - 2b)^{-1}$ is $O[(\tilde{t} - \tilde{t}_0)^{-2}]$ and the inner variable $\bar{t}$ as well as the expansions (3.5.181) must be modified (cf. Problem 3.5.16).

We write the solution of (3.5.182) in the form

$$\bar{x}_0 = \bar{\alpha}_0(\bar{t})\cos[2b_0 t + \bar{\Phi}_0(\bar{t})] \tag{3.5.185a}$$

$$\bar{y}_0 = \bar{\beta}_0(\bar{t})\cos(b_0 t + \bar{\Psi}_0(\bar{t})] \tag{3.5.185b}$$

and find by examining (3.5.183) that we must set

$$\bar{\alpha}_0(\bar{t}) = \bar{\alpha}_0(0) = \text{const.} \tag{3.5.186a}$$

$$\bar{\beta}_0(\bar{t}) = \bar{\beta}_0(0) = \text{const.} \tag{3.5.186b}$$

$$\bar{\Phi}_0(\bar{t}) = \bar{\Phi}_0(0) + a_1\bar{t}^2/2 \tag{3.5.186c}$$

$$\bar{\Psi}_0(\bar{t}) = \bar{\Psi}_0(0) + b_1\bar{t}^2/2. \tag{3.5.186d}$$

The four constants $\bar{\alpha}_0(0)$, $\bar{\beta}_0(0)$, $\bar{\Phi}_0(0)$ and $\bar{\Psi}_0(0)$ are to be determined by matching with the pre-resonance expansion.

The above choices eliminate all the terms on the right-hand sides of (3.5.183) and we use the Cartesian form of the slowly varying coefficients to express $\bar{x}_1$ and $\bar{y}_1$ as

$$\bar{x}_1 = \bar{A}_1(\bar{t})\cos 2b_0 t + \bar{B}_1(\bar{t})\sin 2b_0 t \tag{3.5.187a}$$

$$\bar{y}_1 = \bar{C}_1(\bar{t})\cos b_0 t + \bar{D}_1(\bar{t})\sin b_0 t. \tag{3.5.187b}$$

The conditions on $\bar{A}_1$, $\bar{B}_1$, $\bar{C}_1$ and $\bar{D}_1$ that result from requiring that $\bar{x}_2$ and $\bar{y}_2$ be bounded on the t scale are found to be

$$\frac{d\bar{A}_1}{d\bar{t}} - a_1\bar{t}\bar{B}_1 = -\frac{a_1\bar{\alpha}_0}{4b_0}\cos\bar{\Phi}_0 + \frac{\bar{\beta}_0^2}{8b_0}\sin 2\bar{\Psi}_0 - a_2\bar{\alpha}_0\bar{t}^2\sin\bar{\Phi}_0 \tag{3.5.188a}$$

$$\frac{d\bar{B}_1}{d\bar{t}} + a_1\bar{t}\bar{A}_1 = \frac{a_1\bar{\alpha}_0}{4b_0}\sin\bar{\Phi}_0 + \frac{\bar{\beta}_0^2}{8b_0}\cos 2\bar{\Psi}_0 - a_2\bar{\alpha}_0\bar{t}^2\cos\bar{\Phi}_0 \tag{3.5.188b}$$

and

$$\frac{d\bar{C}_1}{d\bar{t}} - b_1\bar{t}\bar{D}_1 = -\frac{b_1\bar{\beta}_0}{2b_0}\cos\bar{\Psi}_0 + \frac{\bar{\alpha}_0\bar{\beta}_0}{2b_0}\sin(\bar{\Phi}_0 - \bar{\Psi}_0) - b_2\bar{t}^2\bar{\beta}_0\sin\bar{\Psi}_0 \tag{3.5.189a}$$

$$\frac{d\bar{D}_1}{d\bar{t}} + b_1\bar{t}\bar{C}_1 = \frac{b_1\bar{\beta}_0}{2b_0}\sin\bar{\Psi}_0 + \frac{\bar{\alpha}_0\bar{\beta}_0}{2b_0}\cos(\bar{\Phi}_0 - \bar{\Psi}_0) - b_2\bar{t}^2\bar{\beta}_0\cos\bar{\Psi}_0. \tag{3.5.189b}$$

To solve (3.5.188, 189) we introduce the complex variables

$$z = \bar{A}_1 + i\bar{B}_1 \tag{3.5.190a}$$

$$w = \bar{C}_1 + i\bar{D}_1 \tag{3.5.190b}$$

and find that z and w obey the simple equations

$$\frac{dz}{d\bar{t}} + ia_1\bar{t}z = (p + \bar{t}^2 q)e^{-ia_1\bar{t}^2/2} + re^{-ib_1\bar{t}^2} \tag{3.5.191a}$$

$$\frac{dw}{d\bar{t}} + ib_1\bar{t}w = (c + \bar{t}^2 d)e^{-ib_1\bar{t}^2/2} + he^{-i(a_1-b_1)\bar{t}^2/2}, \tag{3.5.191b}$$

where p, q, r, c, d, h are the complex constants

$$p = -\frac{\bar{\alpha}_0 a_1}{4b_0} e^{-i\bar{\phi}_0(0)} \equiv p_1 + ip_2 \tag{3.5.192a}$$

$$q = -ia_2\bar{\alpha}_0 e^{-i\bar{\phi}_0(0)} \equiv q_1 + iq_2 \tag{3.5.192b}$$

$$r = i\frac{\bar{\beta}_0^2}{8b_0} e^{-2i\bar{\psi}_0(0)} \equiv r_1 + ir_2 \tag{3.5.192c}$$

$$c = -\frac{\bar{\beta}_0 b_1}{2b_0} e^{-i\bar{\psi}_0(0)} \equiv c_1 + ic_2 \tag{3.5.193a}$$

$$d = i\bar{\beta}_0 b_2 e^{-i\bar{\psi}_0(0)} \equiv d_1 + id_2 \tag{3.5.193b}$$

$$h = i\frac{\bar{\alpha}_0\bar{\beta}_0}{2b_0} e^{-(\bar{\phi}_0(0)-\bar{\psi}_0(0))} \equiv h_1 + ih_2. \tag{3.5.193c}$$

The solutions of (3.5.191) are then easy to calculate in the form

$$z = z(0)e^{-ia_1\bar{t}^2/2} + (p\bar{t} + q\bar{t}^3/3)e^{-ia_1\bar{t}^2/2} + re^{-ia_1\bar{t}^2}\int_0^{\bar{t}} e^{i(a_1-2b_1)s^2/2}\,ds \tag{3.5.194}$$

$$w = w(0)e^{-ib_1\bar{t}^2/2} + (c\bar{t} + d\bar{t}^3/3)e^{-ib_1\bar{t}^2/2} + he^{-ib_1\bar{t}^2/2}\int_0^{\bar{t}} e^{-i(a_1-2b_1)s^2/2}\,ds \tag{3.5.195}$$

and these define $\bar{x}_1$ and $\bar{y}_1$ explicitly. The four additional constants $z(0) = \bar{A}_1(0) + i\bar{B}_1(0)$, $w(0) = \bar{C}_1(0) + i\bar{D}_1(0)$ will also be determined by the matching.

This completes the inner solution to $O(\varepsilon^{1/2})$, and is as far as we need to proceed in order to derive the solution beyond resonance to $O(\varepsilon^{1/2})$ also. In preparation for the matching of the inner and outer expansions we must calculate the behavior of z and w as $|\bar{t}| \to \infty$. We denote the integrals appearing in (3.5.194, 195) by z_p and w_p, and for $\bar{t} > 0$, we write these in the form

$$z_p = \int_0^\infty e^{i(a_1-2b_1)s^2/2}\,ds - \int_{\bar{t}}^\infty e^{i(a_1-2b_1)s^2/2}\,ds \equiv I - J(\bar{t}) \tag{3.5.196a}$$

$$w_p = z_p^*, \tag{3.5.196b}$$

where * denotes the complex conjugate.

For $\bar{t} < 0$, we write z_p in the form

$$z_p = \int_{-\infty}^{\bar{t}} e^{i(a_1 - 2b_1)s^2/2}\, ds - \int_{-\infty}^{0} e^{i(a_1 - 2b_1)s^2/2}\, ds \equiv J(-\bar{t}) - I. \quad (3.5.197)$$

Now it is easy to show that the constant I reduces to

$$I = \frac{1}{2}\sqrt{\frac{\pi}{|a_1 - 2b_1|}}[1 + i\,\mathrm{sgn}(a_1 - 2b_1)] \equiv I_1 + iI_2 \quad (3.5.198)$$

and that J has the following asymptotic behavior as $|\bar{t}| \to \infty$.

$$J = i\frac{e^{i(a_1 - 2b_1)\bar{t}^2/2}}{(a_1 - 2b_1)|\bar{t}|} + O(\bar{t}^{-3}). \quad (3.5.199)$$

With the above, we calculate the following asymptotic expressions for x and y as $|\bar{t}| \to \infty$.

$$\begin{aligned} x = {} & \bar{\alpha}_0(0)\cos\bar{\phi}_0(0)\cos\rho - \bar{\alpha}_0(0)\sin\bar{\phi}_0(0)\sin\rho \\ & + \varepsilon^{1/2}\Big\{[\bar{A}_1(0) + (\mathrm{sgn}\,\bar{t})(r_1 I_1 - r_2 I_2)]\cos\rho \\ & + [\bar{B}_1(0) + (\mathrm{sgn}\,\bar{t})(r_2 I_1 + r_1 I_2)]\sin\rho \\ & + p_1\bar{t}\cos\rho + p_2\bar{t}\sin\rho + q_1\frac{\bar{t}^3}{3}\cos\rho + q_2\frac{\bar{t}^3}{3}\sin\rho \\ & - \frac{1}{(a_1 - 2b_1)\bar{t}}(r_1\sin 2\theta - r_2\cos 2\theta)\Big\} + O(\varepsilon) + O(\varepsilon^{1/2}\bar{t}^{-3}) \quad \text{as } |\bar{t}| \to \infty \end{aligned} \quad (3.5.200)$$

$$\begin{aligned} y = {} & \bar{\beta}_0(0)\cos\bar{\psi}_0(0)\cos\theta - \bar{\beta}_0(0)\sin\bar{\psi}_0(0)\sin\theta \\ & + \varepsilon^{1/2}\Big\{[\bar{C}_1(0) + (\mathrm{sgn}\,\bar{t})(I_1 h_1 + I_2 h_2)]\cos\theta \\ & + [\bar{D}_1(0) + (\mathrm{sgn}\,\bar{t})(I_1 h_2 - I_2 h_1)]\sin\theta \\ & + c_1\bar{t}\cos\theta + c_2\bar{t}\sin\theta + d_1\frac{\bar{t}^3}{3}\cos\rho + d_2\frac{\bar{t}^3}{3}\sin\rho \\ & + \frac{1}{(a_1 - 2b_1)\bar{t}}[h_1\sin(\rho - \theta) - h_2\cos(\rho - \theta)]\Big\} \\ & + O(\varepsilon) + O(\varepsilon^{1/2}\bar{t}^{-3}) \quad \text{as } |\bar{t}| \to \infty, \end{aligned} \quad (3.5.201)$$

where we have introduced the notation

$$\rho = 2b_0 t + a_1\bar{t}^2/2 \quad (3.5.202a)$$

$$\theta = b_0 t + b_1\bar{t}^2/2. \quad (3.5.202b)$$

We remark that the inner solution is by construction only valid for times of order $\varepsilon^{-1/2}$ around the critical value $t_c = \tilde{t}_0/\varepsilon$. Thus, the mixed secular terms proportional to $\tilde{t}$ and $\tilde{t}^3$ exhibited in (3.5.194, 195) are not troublesome because $\tilde{t} = O(1)$ during resonance. Furthermore, we will demonstrate in part (c) that in the overlap domains for the matching $t - t_c$ is never as large as $O(\varepsilon^{-1/2})$. Therefore, the above inner representation is adequate for our purposes.

(c) *Matching of Pre-resonance and Resonance Solutions*

We adopt the procedure discussed in Section 3.5.4 for matching two multiple variable expansions. In particular, we express all the variables in terms of the intermediate variable t_η

$$t_\eta = \frac{\tilde{t} - \tilde{t}_0}{\varepsilon^\eta}, \qquad 0 \le \eta_1 < \eta < \eta_2 \le \tfrac{1}{2}, \tag{3.5.203}$$

where the constants η_1, η_2 to be determined by the matching, define the overlap domain. Moreover, we must match the pre-resonance and resonance solutions to $O(\varepsilon^{1/2})$ in order to exhibit the cancellation of the singular divisors and to derive the effect of passage through resonance on the solution for $\tilde{t} > \tilde{t}_0$.

With the basic definition of $\tilde{t}$ we derive the following expressions for μ and ν

$$\mu = \frac{\tau_0}{\varepsilon} + \varepsilon^{\eta-1} 2b_0 t_\eta + \frac{a_1}{2}\varepsilon^{2\eta-1} t_\eta^2 + \frac{a_2}{3}\varepsilon^{3\eta-1} t_\eta^3 + O(\varepsilon) + O(\varepsilon^{4\eta-1}) \tag{3.5.204a}$$

$$\nu = \frac{\kappa_0}{\varepsilon} + \varepsilon^{\eta-1} b_0 t_\eta + \frac{b_1}{2}\varepsilon^{2\eta-1} t_\eta^2 + \frac{b_2}{3}\varepsilon^{3\eta-1} t_\eta^3 + O(\varepsilon) + O(\varepsilon^{4\eta-1}), \tag{3.5.204b}$$

where the constants τ_0 and κ_0 are given by

$$\tau_0 = \int_0^{\tilde{t}_0} a(s)\, ds \tag{3.5.205a}$$

$$\kappa_0 = \int_0^{\tilde{t}_0} b(s)\, ds. \tag{3.5.205b}$$

The $O(\varepsilon^{4\eta-1})$ remainder terms in (3.5.204) arise from truncating the power series expansions of a and b near $\tilde{t}_0$ while neglecting the k_2 and l_2 terms in the definitions of μ and ν results in the $O(\varepsilon)$ remainders.

Since the inner expansion is expressed in terms of the variables ρ and θ, [cf. (3.5.200, 201, 202)] it is convenient to express these in terms of t_η. We find

$$\rho = \frac{2b_0\tilde{t}_0}{\varepsilon} + \varepsilon^{\eta-1}2b_0 t_\eta + \frac{a_1}{2}\varepsilon^{2\eta-1}t_\eta^2 \tag{3.5.206a}$$

$$\theta = \frac{b_0\tilde{t}_0}{\varepsilon} + \varepsilon^{\eta-1}2b_0 t_\eta + \frac{b_1}{2}\varepsilon^{2\eta-1}t_\eta^2. \tag{3.5.206b}$$

Comparing (3.5.206) with (3.5.204) we see that

$$\mu = \frac{\tau_0 - 2b_0\tilde{t}_0}{\varepsilon} + \rho + \frac{a_2}{3}\varepsilon^{3\eta-1}t_\eta^3 + O(\varepsilon) + O(\varepsilon^{4\eta-1}) \tag{3.5.207a}$$

$$\nu = \frac{\kappa_0 - b_0\tilde{t}_0}{\varepsilon} + \theta + \frac{b_2}{3}\varepsilon^{3\eta-1}t_\eta^3 + O(\varepsilon) + O(\varepsilon^{4\eta-1}). \tag{3.5.207b}$$

Consider now the expression x^0 for the outer expansion of x in the matching domain. Using (3.5.206a) for μ and expanding functions of $\tilde{t}$ by setting

$$\tilde{t} = \tilde{t}_0 + \varepsilon^\eta t_\eta \tag{3.5.208}$$

in (3.5.170a) and (3.5.174a), we calculate

$$\begin{aligned} x^0(t_\eta, \varepsilon) = {} & \frac{\alpha_0(0)\sqrt{a(0)}}{\sqrt{2b_0}}\left[\cos\left(\rho + \frac{\tau_0 - 2b_0\tilde{t}_0}{\varepsilon} + \phi_0(0)\right)\right. \\ & - \frac{a_1}{4b_0}\varepsilon^\eta t_\eta \cos\left(\rho + \frac{\tau_0 - 2b_0\tilde{t}_0}{\varepsilon} + \phi_0(0)\right) \\ & \left. - \frac{a_2}{3}\varepsilon^{3\eta-1}t_\eta^3 \sin\left(\rho + \frac{\tau_0 - 2b_0\tilde{t}_0}{\varepsilon} + \phi_0(0)\right)\right] \\ & + \frac{\varepsilon^{1-\eta}\beta_0^2(0)b(0)}{8b_0(a_1 - 2b_1)t_\eta}\cos 2\left(\frac{(\kappa_0 - b_0\tilde{t}_0)}{\varepsilon} + \psi_0(0) + \theta\right) \\ & - O(\varepsilon^{4\eta-1}) + O(\varepsilon). \end{aligned} \tag{3.5.209}$$

The singular term in x_1 contributes the last term in (3.5.209) and the rest of the εx_1 solution is $O(\varepsilon)$.

The solution for x in the inner region, denoted by x^i, is also expanded in terms of t_η. Since $|\bar{t}| = |\varepsilon^{\eta-1/2}t_\eta| \to \infty$ as $\varepsilon \to 0$, we use (3.5.200) and calculate (with $\bar{t} < 0$)

$$\begin{aligned} x^i(t_\eta, \varepsilon) = {} & \bar{\alpha}_0(0)\cos[\rho + \bar{\phi}_0(0)] + \varepsilon^{1/2}[(\bar{A}_1(0) + r_2 I_2 - r_1 I_1)\cos\rho \\ & + (\bar{B}_1(0) - r_2 I_1 - r_1 I_2)\sin\rho] + \varepsilon^\eta p_1 t_\eta \cos\rho + \varepsilon^\eta p_2 t_\eta \sin\rho \\ & + \varepsilon^{3\eta-1}\frac{q_1}{3}t_\eta^3\cos\rho + \varepsilon^{3\eta-1}\frac{q_2}{3}t_\eta^3\sin\rho \\ & - \frac{\varepsilon^{1-\eta}}{(a_1 - 2b_1)t_\eta}(r_1\sin 2\theta - r_2\cos 2\theta) + O(\varepsilon^{2-3\eta}). \end{aligned} \tag{3.5.210}$$

The remainder term of order $\varepsilon^{2-3\eta}$ in (3.5.210) represents terms of order $\varepsilon^{1/2}\bar{t}^{-3}$ which were neglected in deriving (3.5.200).

In order to match the results to $O(\varepsilon^{1/2})$ we must show that

$$\lim_{\substack{\varepsilon\to 0\\ t_\eta \text{ fixed}}} \frac{x^0(t_\eta, \varepsilon) - x^i(t_\eta, \varepsilon)}{\varepsilon^{1/2}} = 0. \tag{3.5.211}$$

Comparing (3.5.209) and (3.5.210) we see that the terms of order unity in x^0 and x^i match if we set

$$\bar{\alpha}_0(0) = \frac{\alpha_0(0)\sqrt{a(0)}}{\sqrt{2b_0}} \tag{3.5.212a}$$

$$\bar{\phi}_0(0) = \frac{\tau_0 - 2b_0\bar{t}_0}{\varepsilon} + \phi_0(0). \tag{3.5.212b}$$

The above choice then implies that both the $O(\varepsilon^\eta)$ and $O(\varepsilon^{3\eta-1})$ terms match identically, and this provides a partial check on the results.

Since the outer expansion prior to resonance does not contain terms of order $\varepsilon^{1/2}$, we must set these in x^i equal to zero, and we obtain

$$\bar{A}_1(0) = r_1 I_1 - r_2 I_2 \tag{3.5.213a}$$

$$\bar{B}_1(0) = r_2 I_1 + r_1 I_2. \tag{3.5.213b}$$

The singular $O(\varepsilon^{1-\eta})$ terms match if we set

$$\bar{\beta}_0(0) = \beta_0(0)\sqrt{\frac{b(0)}{b_0}} \tag{3.5.214a}$$

$$\bar{\psi}_0(0) = \frac{\kappa_0 - b_0\bar{t}_0}{\varepsilon} + \psi_0(0). \tag{3.5.214b}$$

Finally, we note that the remainder terms in (3.5.209) divided by $\varepsilon^{1/2}$ vanish if $\eta > 3/8$, while the remainder term in (3.5.210) divided by $\varepsilon^{1/2}$ vanishes if $\eta < 1/2$. In fact, this result would have also followed with the more general choice $t_\lambda = (\bar{t} - \bar{t}_0)/\lambda(\varepsilon)$ of intermediate variable. We conclude that the overlap domain is the class of functions $\lambda(\varepsilon)$ such that $\varepsilon^{1/2} \ll \lambda(\varepsilon) \ll \varepsilon^{3/8}$.

The details of the matching for y are similar to the above and will not be repeated. Matching of the $O(1)$ solutions reproduces (3.5.214) and also verifies the cancellation of all singular terms. The removal of terms of order $\varepsilon^{1/2}$ in y^i gives the additional conditions (cf. (3.5.201))

$$\bar{C}_1(0) = I_1 h_1 + I_2 h_2 \tag{3.5.215a}$$

$$\bar{D}_1(0) = I_1 h_2 - I_2 h_1. \tag{3.5.215b}$$

Now, the inner solution is completely defined and can be used to determine a post-resonance outer expansion.

(d) *Outer Expansion after Resonance*

The outer solution appropriate for $\tilde{t} > \tilde{t}_0$ is dictated by the behavior of the inner expansion as $\bar{t} \to \infty$. Using (3.5.200) and (3.5.201) we conclude that the expressions for x^i and y^i for the $t_\eta > 0$ matching are as before except for the sign changes indicated for the r_1, r_2 and h_1, h_2 terms. This means that x^i and y^i will now contain the homogeneous solutions sin ρ, cos ρ and sin θ, cos θ, respectively, to $O(\varepsilon^{1/2})$ and matching will require the presence of corresponding terms in the outer expansion.

Therefore, we seek a post-resonance expansion in the form

$$x = x_0^*(\mu, \tilde{t}) + \varepsilon^{1/2} x_{1/2}^*(\mu, \tilde{t}) + \varepsilon x_1^*(\mu, \tilde{t}) + \cdots \tag{3.5.216a}$$

$$y = y_0^*(v, \tilde{t}) + \varepsilon^{1/2} y_{1/2}^*(v, \tilde{t}) + \varepsilon y_1^*(v, \tilde{t}) + \cdots. \tag{3.5.216b}$$

It is easily seen by paralleling the expansion procedure discussed in part (a) that to $O(\varepsilon^{1/2})$ the form of the solutions is unchanged and we have

$$x_{i/2}^* = \alpha_{i/2}^*(\tilde{t})\cos[\mu + \phi_{i/2}^*(\tilde{t})], \qquad i = 0, 1 \tag{3.5.217a}$$

$$y_{i/2}^* = \beta_{i/2}^*(\tilde{t})\cos[v + \psi_{i/2}^*(\tilde{t})], \qquad i = 0, 1 \tag{3.5.217b}$$

with the $\alpha_{i/2}^*$, $\phi_{i/2}^*$, $\beta_{i/2}^*$, $\psi_{i/2}^*$ defined as before, i.e.,

$$\alpha_{i/2}^*(\tilde{t}) = \alpha_{i/2}^*(\tilde{t}_1)\left(\frac{a(\tilde{t}_1)}{a(\tilde{t})}\right)^{1/2}, \qquad i = 0, 1 \tag{3.5.218a}$$

$$\beta_{i/2}^*(\tilde{t}) = \beta_{i/2}^*(\tilde{t}_1)\left(\frac{b(\tilde{t}_1)}{b(\tilde{t})}\right)^{1/2}, \qquad i = 0, 1 \tag{3.5.218b}$$

$$\phi_{i/2}^*(\tilde{t}) = \phi_{i/2}^*(\tilde{t}_1) = \text{const.}, \qquad i = 0, 1 \tag{3.5.218c}$$

$$\psi_{i/2}^*(\tilde{t}) = \psi_{i/2}^*(\tilde{t}_1) = \text{const.}, \qquad i = 0, 1 \tag{3.5.218d}$$

and we are referring the initial conditions to some time $\tilde{t}_1 > \tilde{t}_0$.

Corresponding to (3.5.209) we now have an expression of the form

$$\begin{aligned} x^{*0} = {} & \alpha_0^*(\tilde{t}_1)\left(\frac{a(\tilde{t}_1)}{2b_0}\right)^{1/2}\left[\cos\left(\rho + \frac{\tau_0 - 2b_0\tilde{t}_0}{\varepsilon} + \phi_0^*(\tilde{t}_1)\right)\right] \\ & + \varepsilon^{1/2}\alpha_{1/2}^*(\tilde{t}_1)\left(\frac{a(\tilde{t}_1)}{2b_0}\right)^{1/2}\left[\cos\left(\rho + \frac{\tau_0 - 2b_0\tilde{t}_0}{\varepsilon} + \phi_{1/2}^*(\tilde{t}_1)\right)\right] + T, \end{aligned} \tag{3.5.219}$$

where T indicates terms which are identical to those in (3.5.209) and which will obviously also match with corresponding terms in the inner expansion.

If we now compare (3.5.219) with (3.5.210) in which r_1 and r_2 have the opposite sign, we find that matching requires

$$\bar{\alpha}_0(0) = \alpha_0^*(\tilde{t}_1)\left(\frac{a(\tilde{t}_1)}{2b_0}\right)^{1/2} \tag{3.5.220a}$$

$$\bar{\phi}_0(0) = \frac{\tau_0 - 2b_0\tilde{t}_0}{\varepsilon} + \phi_0^*(\tilde{t}_1) \tag{3.5.220b}$$

$$\bar{A}_1(0) + (r_1 I_1 - r_2 I_2) = \alpha_{1/2}^*(\tilde{t}_1)\left(\frac{a(\tilde{t}_1)}{2b_0}\right)^{1/2} \cos\left(\frac{\tau_0 - 2b_0\tilde{t}_0}{\varepsilon} + \phi_{1/2}^*(\tilde{t}_1)\right) \tag{3.5.220c}$$

$$\bar{B}_1(0) + (r_2 I_1 + r_1 I_2) = -\alpha_{1/2}^*(\tilde{t}_1)\left(\frac{a(\tilde{t}_1)}{2b_0}\right)^{1/2} \sin\left(\frac{\tau_0 - 2b_0\tilde{t}_0}{\varepsilon} + \phi_{1/2}^*(\tilde{t}_1)\right). \tag{3.5.220d}$$

Using the known values of $\bar{\alpha}_0(0)$, $\bar{\phi}_0(0)$, etc., we calculate

$$\alpha_0^*(\tilde{t}) = \alpha_0(\tilde{t}) = \alpha_0(0)\left(\frac{a(0)}{a(\tilde{t})}\right)^{1/2} \tag{3.5.221a}$$

$$\phi_0^*(\tilde{t}_1) = \phi_0(0) \tag{3.5.221b}$$

$$\alpha_{1/2}^*(\tilde{t}_1) = \frac{\beta_0^2(0)b(0)}{[8b_0^3 a(\tilde{t}_1)]^{1/2}}(I_1^2 + I_2^2)^{1/2} \tag{3.5.221c}$$

$$\phi_{1/2}^*(\tilde{t}_1) = \frac{2\kappa_0 - \tau_0}{\varepsilon} + 2\psi_0(0) + \pi + \tan^{-1}\frac{I_1}{I_2}. \tag{3.5.221d}$$

Thus, as expected, the solution to $O(1)$ is identical to the preresonance value, and the $O(\varepsilon^{1/2})$ solution is explicitly defined by the slowly varying parameters $\alpha_{1/2}^*$ and $\beta_{1/2}^*$.

Analogous calculations for y show that

$$\beta_0^*(\tilde{t}) = \beta_0(\tilde{t}) = \beta_0(0)\left(\frac{b(0)}{b(t)}\right)^{1/2} \tag{3.5.222a}$$

$$\psi_0^*(\tilde{t}_1) = \psi_0(0) \tag{3.5.222b}$$

$$\beta_{1/2}^*(\tilde{t}_1) = \frac{\alpha_0(0)\beta_0(0)[a(0)b(0)]^{1/2}[I_1^2 + I_2^2]^{1/2}}{[2b(\tilde{t}_1)b_0^3]^{1/2}} \tag{3.5.222c}$$

$$\psi_{1/2}^*(\tilde{t}_1) = \frac{\tau_0 - 2\kappa_0}{\varepsilon} + \phi_0(0) - \psi_0(0) - \tan^{-1}\frac{I_1}{I_2}. \tag{3.5.222d}$$

To proceed further with the outer expansion for $\tilde{t} > \tilde{t}_0$ we need to know the inner solution to $O(\varepsilon)$, which we have not done. However, the results to $O(\varepsilon^{1/2})$ already contain the main effects of passage through resonance, and this is discussed in Problems 3.5.14, 15 and Reference 3.5.6. The reader is also referred to Problem 3.7.9 and Reference 3.5.16 where a general version of this problem is studied by canonical perturbation theory.

Problems

1. Apply the results of Section 3.5.1 to the simple system of two equal masses (M) with equal springs ($K = K_{01} = K_{12} = K_{02}$) but general damping. Verify that $\beta_{11} < \beta_{22}$ so that the fundamental mode decays more slowly than the second mode. Calculate the approximate period of each mode.

2. Carry out the calculations of the example in Section 3.5.2b and determine the dependence of α_0, β_0, ϕ_0 and ψ_0 on t_3.

3. Using the results calculated in Section 3.5.2b verify that the energy integral (3.5.27) is satisfied to order ε.

4. Derive the periodic solution (3.5.73) to order ε using the method of strained coordinates of Section 3.1.

5. Discuss the solution for E_1 given by (3.5.68) for the case $\kappa \neq 0$.

6. Calculate the outer solution of Equation (3.5.102) correct to order ε using the alternate (cumbersome) approach of regarding t to be the function of εt^+ given by (3.5.111). Show that your results agree with those in Section 3.5.4 when the expressions for A_0, B_0, etc. are expanded out. Also, substitute the expression for y given by $y = y_0 + \varepsilon y_1$ (where y_0 is defined by Equations. (3.5.112) and (3.5.113), and y_1 is defined by Equation (3.5.120)) into Equation (3.5.102) and show that it is a solution to order ε.

7. For the example of Section 3.5.4, assume an outer expansion valid for $\tilde{t} > \tilde{t}_0$ in the form

$$y = \frac{y^*_{-1/2}(t^+, \tilde{t})}{\varepsilon^{1/2}} + y^*_0(t^+, \tilde{t}) + \varepsilon^{1/2} y^*_{1/2}(t^+, \tilde{t}) + \cdots$$

and calculate the dependence of $y^*_{-1/2}$ and y^*_0 on t^+ and $\tilde{t}$ explicitly in terms of four integration constants. Match this post resonance solution with the resonance solution derived in Section 3.5.4 to determine the integration constants. Use these results to calculate a uniformly valid composite solution for

$$0 \leq \tilde{t} \leq \tilde{t}_1, \qquad \text{where } \tilde{t}_1 > \tilde{t}_0.$$

8. Modify the inner expansion to handle the case $a_1 = 0, a_2 \neq 0$ in Equation (3.5.103) of Section 3.5.4.

9. Parallel the analysis of Section 3.5.4 for the problem of the forced Duffing equation

$$\frac{d^2y}{dt^2} + \mu^2(\varepsilon t)y + \varepsilon y^3 = \alpha \cos(t + \beta)$$

with μ as defined by Equation (3.5.103).

10. Reconsider the special case of Problem 3.3.3 with a forcing function, i.e.,

$$\frac{d^2y}{dt^2} + \mu^2 y = \cos 2t \qquad \frac{d\mu}{dt} = \varepsilon y^2$$

with initial conditions

$$y(0) = 1 \qquad \dot{y}(0) = 0 \qquad \mu(0) = 1.$$

(a) Make the exact change of variables $t \to t^+$ suggested in Problem 3.3.3 and derive a *fourth* order system for the dependent variables y, μ, and t as functions of t^+.
(b) Solve this system by using a two variable expansion with t^+ and $\tilde{t} = \varepsilon t^+$ as the fast and slow variables, and show that resonance occurs at some finite time t_c^+.
(c) Derive an appropriate resonance expansions valid near t_c^+, and solve it to $O(1)$.
(d) Match the expansions (b) and (c) in their common overlap domain for values of $t^+ < t_c^+$.
(e) Calculate the post resonance expansion and match it also with the expansion in (c). Hence, derive a uniformly valid representation for the solution to $O(1)$ for values of t^+ beyond t_c^+.

11. Derive the results in Equations (3.5.175).

12. Derive the results in Equations (3.5.188) and (3.6.189).

13. Carry out the matching for x and y explicitly for the problem of Section 3.5.5 and derive the results in Equations (3.5.212), (3.5.213), (3.5.214) and (3.5.215).

14. Carry out the details of the matching between the resonance and post-resonance solutions and derive the results in Equations (3.5.221) and (3.5.222).

15. Define a normalized action for each oscillator in the example of Section 3.5.5 by

$$R_x = \frac{\dot{x}^2 + a^2(\tilde{t})x^2}{a}; \qquad R_y = \frac{\dot{y}^2 + b^2(\tilde{t})y^2}{b}.$$

(a) Calculate R_x and R_y to $O(\varepsilon)$ using the pre-resonance solution. To what order are these quantities conserved?
(b) Calculate R_x and R_y to $O(\varepsilon^{1/2})$ using the post-resonance solution. Show that for $\tilde{t} > \tilde{t}_0$ these reduce to

$$R_x = \alpha_0^2(0)a(0) - \varepsilon^{1/2}R_{1/2} + O(\varepsilon)$$
$$R_y = \beta_0^2(0)b(0) + 2\varepsilon^{1/2}R_{1/2} + O(\varepsilon),$$

where

$$R_{1/2} = \alpha_0(0)\beta_0^2(0)\left[\frac{a(0)b^2(0)\pi}{8|a_1 - 2b_1|b_0^3}\right]$$
$$\cos\left[\frac{\tau_0 - 2\kappa_0}{\varepsilon} + \phi_0(0) - 2\psi_0(0) - \tan^{-1}\frac{1}{\operatorname{sgn}(a_1 - 2b_1)}\right].$$

Thus, R_x and R_y are constants to $O(\varepsilon^{1/2})$, but these constants change by $O(\varepsilon^{1/2})$ after passage through resonance. However, $2R_x + R_y$ remains constant to $O(\varepsilon^{1/2})$ through resonance (cf. Problem 3.7.9).

16. Modify the resonance solution of Section 3.5.5 to the case

$$a_1 - 2b_1 = 0, \qquad a_2 - 2b_2 \neq 0.$$

17. Reconsider the example of Problem 3.3.3

$$\frac{d^2y}{dt^2} + \mu^2 y = 0 \qquad \frac{d\mu}{dt} = \varepsilon y^2$$

and solve it using the expansion procedure of Section 3.5.5. In particular assume that y and μ have the expansions

$$y = y_0(\tau, \tilde{t}) + \varepsilon y_1(\tau, \tilde{t}) + \cdots$$

$$\mu = \mu_0(\tilde{t}) + \varepsilon \mu_1(\tau, \tilde{t}) + \cdots,$$

where

$$\frac{d\tau}{dt} = \mu_0(\tilde{t}) + \varepsilon k_1(\tilde{t}) + \cdots$$

and k_1 is unknown. Note that since the slowly varying frequency μ is an unknown we need a correction term of order ε in the fast time τ. Calculate y_0, μ_0 and k_1 explicitly according to the procedure discussed in Section 3.5.5. Show that these results agree with those found in Problem 3.3.3.

References

3.5.1 H. Kabakow, A perturbation procedure for nonlinear oscillations, Ph.D. Thesis, California Institute of Technology, Pasadena, California, 1968.

3.5.2 I. Gros, Resonance in a coupled oscillatory system, M.S. Thesis, University of Washington, Seattle, Washington, 1973.

3.5.3 G. Sandri, A new method of expansion in mathematical physics, Nuovo Cimento, B36, 1965, pp. 67–93.

3.5.4 W. Lick, Two-variable expansions and singular perturbation problems, *S.I.A.M.J. on Applied Mathematics*, **17**, 1969, pp. 815–825.

3.5.5 J. Kevorkian, Passage through resonance for a one-dimensional oscillator with slowly varying frequency, *S.I.A.M. Journal on Applied Mathematics* **20**, 1971, pp. 364–373. See also Errata in Vol. 26, 1974, p. 686.

3.5.6 J. Kevorkian, Resonance in a weakly nonlinear system with slowly varying parameters, *Studies in Applied Mathematics* **62**, 1980, pp. 23–67.

3.5.7 J. Kevorkian, On a model for reentry roll resonance, *S.I.A.M. Journal on Applied Mathematics* **26**, 1974, pp. 638–669.

3.5.8 L. Lewin and J. Kevorkian, On the problem of sustained resonance, *S.I.A.M. Journal on Applied Mathematics* **35**, 1978, pp. 738–754.

3.5.9 G. Contopoulos, A third integral of motion in a galaxy, *Z. Astrophys.* **49**, 1960, p. 273.

3.5.10 G.-I. Hori, Nonlinear coupling of two harmonic oscillations, *Publications of the Astronomical Society of Japan* **19**, 1967, pp. 229–241.

3.5.11 E. T. Whittaker, *Analytical Dynamics*, Cambridge University Press, London and New York, 1904.

3.5.12 M. J. Ablowitz, B. A. Funk, and A. C. Newell, Semi resonant interactions and frequency dividers, *Studies in Applied Mathematics* **L11**, 1973, pp. 51–74.

3.5.13 Lord Rayleigh, Theory of Sound, Second Edition, Dover reprint, New York, 1945.
3.5.14 T. K. Caughey, Classical normal modes in damped linear systems, *J. Appl. Mech.* **82**, Series E, 1960, pp. 269–271.
3.5.15 T. K. Caughey and M. E. J. O'Kelly, Classical normal modes in damped linear dynamic systems, *J. Appl. Mech.* **32**, Series E, 1965, pp. 583–588.
3.5.16 J. Kevorkian, Adiabatic invariance and passage through resonance for nearly periodic Hamiltonian systems, *Studies in Applied Mathematics*, **66**, 1982, pp. 95–119.

3.6 Strongly Nonlinear Oscillations, The Method of Kuzmak-Luke

In this section we will generalize the ideas of multiple variable expansions to strictly nonlinear second order equations. That is to say to equations which, as $\varepsilon \to 0$, remain nonlinear.

The basic technique was discussed in Reference 3.6.1, by Kuzmak who studied equations of the form

$$\frac{d^2y}{dt^2} + \varepsilon h(y, \tilde{t})\frac{dy}{dt} + g(y, \tilde{t}) = 0, \tag{3.6.1}$$

where $0 < \varepsilon \ll 1$, $\tilde{t} = \varepsilon t$ and h and g are arbitrary analytic functions of their arguments.

The only restriction is that for $\varepsilon = 0$, the reduced problem for the nonlinear oscillator

$$\frac{d^2y}{dt^2} + g(y, 0) = 0 \tag{3.6.2}$$

have periodic solutions.

Kuzmak's calculations are concise and straightforward, but only provide the solution to $O(1)$ partially; the equation governing the slowly varying phase is not derived. In Reference 3.6.2 Luke, in his study of nearly uniform nonlinear dispersive waves, extends Kuzmak's method to higher orders. As we shall see in Chapter 4, the basic problem for nearly uniform waves is essentially the same as given in (3.6.1), and Luke derives the missing formula for the slowly varying phase, and shows that it is constant to $O(1)$. Thus, Kuzmak's theory is indeed complete to $O(1)$. We will follow Luke's procedure as it applies to Equation (3.6.1) in this section and will discuss applications to nonlinear waves in Chapter 4.

The basic idea dictating the choice of a fast variable is identical to that for the linear problem where, [cf. Equation (3.3.47)] $g = \mu^2(\tilde{t})y$, i.e., the fast time t^+ must be chosen such that the period is independent of $\tilde{t}$ when measured on the t^+ scale. For the linear case, the choice was, a priori, obvious and depended only on μ in the form

$$\frac{dt^+}{dt} = \mu. \tag{3.6.3}$$

Now, however, the period is amplitude (or energy) dependent and the expression equivalent to Equation (3.6.3) is more complicated. Also, for the linear problem, we obtained conditions governing the slowly varying amplitude and phase to any order by requiring the solution to the next order to be periodic in the fast variable. This precise condition also applies to the nonlinear problem, although the calculations are now more difficult.

3.6.1 General Theory

We assume that the solution of Equation (3.6.1) can be developed in the form

$$y(t, \varepsilon) = F(t^+, \tilde{t}; \varepsilon) = y_0(t^+, \tilde{t}) + \varepsilon y_1(t^+, \tilde{t}) + \cdots, \tag{3.6.4}$$

where

$$\frac{dt^+}{dt} = \omega(\tilde{t}) \tag{3.6.5}$$

which is an unknown, to be determined by the requirement that y_0 be periodic in t^+ with a period which does not depend on $\tilde{t}$.

As usual, we calculate the following expressions for the derivatives

$$\frac{dy}{dt} = \omega(\tilde{t})\frac{\partial F}{\partial t^+} + \varepsilon\frac{\partial F}{\partial \tilde{t}}$$

$$= \omega(\tilde{t})\frac{\partial y_0}{\partial t^+} + \varepsilon\left(\omega\frac{\partial y_1}{\partial t^+} + \frac{\partial y_0}{\partial \tilde{t}}\right) + O(\varepsilon^2) \tag{3.6.6a}$$

$$\frac{d^2y}{dt^2} = \omega^2\frac{\partial^2 F}{\partial t^{+2}} + 2\varepsilon\omega\frac{\partial^2 F}{\partial t^+\partial \tilde{t}} + \varepsilon\frac{d\omega}{d\tilde{t}}\frac{\partial F}{\partial t^+} + \varepsilon^2\frac{\partial^2 F}{\partial \tilde{t}^2}$$

$$= \omega^2\frac{\partial^2 y_0}{\partial t^{+2}} + \varepsilon\left(\omega^2\frac{\partial^2 y_1}{\partial t^{+2}} + 2\omega\frac{\partial^2 y_0}{\partial t^+\partial \tilde{t}} + \frac{d\omega}{d\tilde{t}}\frac{\partial y_0}{\partial t^+}\right) + O(\varepsilon^2). \tag{3.6.6b}$$

We also develop h and g as follows

$$h(y, \tilde{t}) = h(y_0, \tilde{t}) + O(\varepsilon) \tag{3.6.7a}$$

$$g(y, \tilde{t}) = g(y_0, \tilde{t}) + \varepsilon g_y(y_0, \tilde{t})y_1 + O(\varepsilon^2) \tag{3.6.7b}$$

to obtain the following equations for y_0 and y_1

$$\omega^2\frac{\partial^2 y_0}{\partial t^{+2}} + g(y_0, \tilde{t}) = 0 \tag{3.6.8a}$$

$$\omega^2\frac{\partial^2 y_1}{\partial t^{+2}} + g_y(y_0, \tilde{t})y_1 = -2\omega\frac{\partial^2 y_0}{\partial t^+\partial \tilde{t}}$$

$$-\left[\frac{d\omega}{d\tilde{t}} + \omega h(y_0, \tilde{t})\right]\frac{\partial y_0}{\partial t^+} \equiv p_1. \tag{3.6.8b}$$

The solution of Equation (3.6.8a) can be carried out by quadrature. We multiply it by $\partial y_0/\partial t^+$ and observe that the result can be integrated with respect to t^+ to yield the "energy" integral

$$\frac{\omega^2}{2}\left(\frac{\partial y_0}{\partial t^+}\right)^2 + V(y_0, \tilde{t}) = E_0(\tilde{t}), \tag{3.6.9a}$$

where $E_0(\tilde{t})$ is the slowly varying energy and V is the potential defined by

$$V(y_0, \tilde{t}) = \int_0^{y_0} g(\eta, \tilde{t})d\eta. \tag{3.6.9b}$$

Now, by the periodicity hypothesis, the curves in the $(\partial y_0/\partial t^+, y_0)$ plane for fixed E_0, ω, and $\tilde{t}$ are ovals which are symmetric with respect to the y_0 axis as sketched in Figure 3.6.1. Note that since $\tilde{t}$ is held fixed, the closed curve in Figure 3.6.1 is not an actual integral curve of Equation (3.6.9a). However,

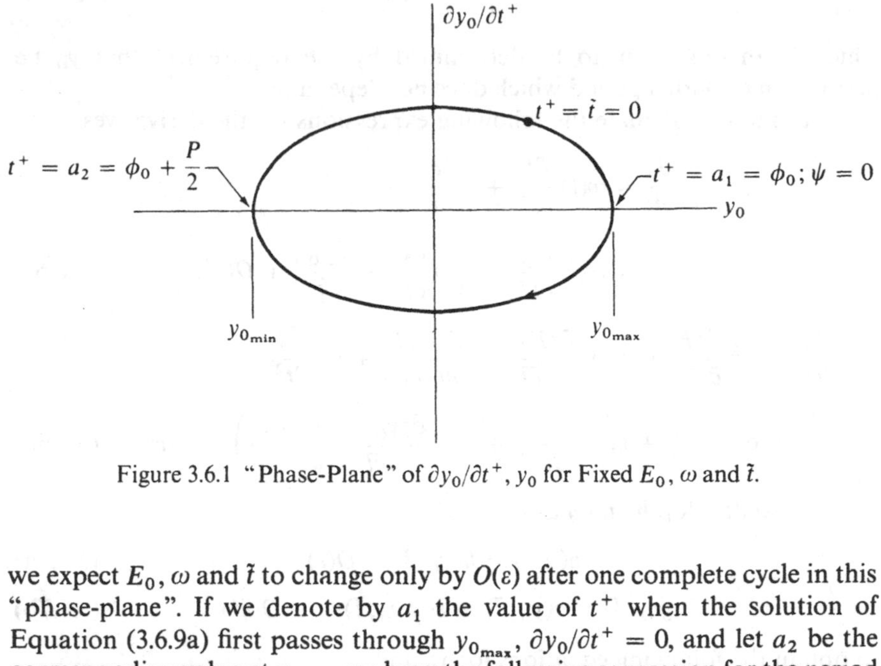

Figure 3.6.1 "Phase-Plane" of $\partial y_0/\partial t^+$, y_0 for Fixed E_0, ω and $\tilde{t}$.

we expect E_0, ω and $\tilde{t}$ to change only by $O(\varepsilon)$ after one complete cycle in this "phase-plane". If we denote by a_1 the value of t^+ when the solution of Equation (3.6.9a) first passes through $y_{0_{max}}$, $\partial y_0/\partial t^+ = 0$, and let a_2 be the corresponding value at $y_{0_{min}}$, we have the following expression for the period P by quadrature from Equation (3.6.9a).

$$P = 2\omega \int_{y_{0_{min}}}^{y_{0_{max}}} \frac{dy_0}{\sqrt{2[E_0(\tilde{t}) - V(y_0, \tilde{t})]}} = 2(a_2 - a_1). \tag{3.6.10a}$$

In general, for a given V, Equation (3.6.10a) defines P as a function of E_0, ω, and $\tilde{t}$:

$$P = P(E_0, \omega, \tilde{t}). \tag{3.6.10b}$$

However, we insist that P be a constant, which we may choose for convenience to be 2π. Then, Equation (3.6.10b) can be solved for ω in the form

$$\omega = \omega(E_0, \tilde{t}). \tag{3.6.11}$$

In the linear case, $g = \mu^2(\tilde{t})y$, the choice $P = 2\pi$ gives $\omega = \mu$. Note that the normalization of the period is arbitrary since any constant times ω can be used in Equation (3.6.10b) without affecting the constancy of P.

We next integrate Eq. (3.6.9a) solved for ∂t^+ and invert the result to find y_0 in the form

$$\begin{aligned} y_0 &= f[t^+ - \phi_0(\tilde{t}), E_0, \omega, \tilde{t}] \\ &\equiv f[\psi, E_0, \omega, \tilde{t}]; \ \psi = t^+ - \phi_0(\tilde{t}), \end{aligned} \tag{3.6.12a}$$

where the additive slowly varying function ϕ_0 is the phase. If we expressed ω in terms of E_0 and $\tilde{t}$ using Equation (3.6.11), the above would reduce to

$$y_0 = f[\psi, E_0, \omega(E_0, \tilde{t}), \tilde{t}] \equiv f^*(\psi, E_0, \tilde{t}). \tag{3.6.12b}$$

We now need to look at the solution to $O(\varepsilon)$ in order to determine the two unknowns E_0 and ϕ_0.

Since Equation (3.6.8b) for y_1 is linear, we can calculate the general solution once one solution is known. Kuzmak pointed out that $\partial f/\partial\psi$ is a solution of the homogeneous Equation (3.6.8b). To see this we take the partial derivative of Equation (3.6.8a) [for $y_0 = f(\psi, E_0, \omega, \tilde{t})$] with respect to ψ, holding $E_0, \omega, \tilde{t}$ fixed. We find

$$\omega^2 \frac{\partial^2}{\partial t^{+2}} f_\psi + g_y(y_0, \tilde{t}) f_\psi = 0 \tag{3.6.13a}$$

and this is just the homogeneous Equation (3.6.8b).

Kuzmak then uses this result to construct a particular solution of Equation (3.6.8b) from which the periodicity of y_1 gives an equation defining E_0. This is a straightforward calculation which we do not repeat because it provides no information on ϕ_0. To calculate ϕ_0 and to indicate how the procedure can be carried out to any order in ε, we derive the general solution of Equation (3.6.8b) and follow the steps outlined by Luke in Reference 3.6.2.

In particular, Luke points out that $\partial f/\partial E_0$, calculated by holding ψ, ω and $\tilde{t}$ fixed in Equation (3.6.12a) is a second linearly independent solution of the homogeneous Equation (3.6.8b). This again follows by taking the partial derivative of Equation (3.6.8a) with respect to E_0 holding ψ, ω, and $\tilde{t}$ fixed, i.e.,

$$\omega^2 \frac{\partial^2}{\partial t^{+2}} f_{E_0} + g_y(y_0, \tilde{t}) f_{E_0} = 0. \tag{3.6.13b}$$

Note that in deriving Equation (3.6.13b) it is essential *not* to use Equation (3.6.11) relating ω to E_0 and $\tilde{t}$, because it is easily verified that $\partial f^*/\partial E_0$ is *not* a solution of the homogeneous Equation (3.6.8b).

To prove that $\partial f/\partial\psi$ and $\partial f/\partial E_0$ are linearly independent, we calculate the Wronskian

$$W(f_\psi, f_{E_0}) \equiv f_\psi f_{E_0\psi} - f_{E_0} f_{\psi\psi} \tag{3.6.14a}$$

then take the partial derivative of W with respect to ψ

$$\frac{\partial W}{\partial \psi} = f_\psi f_{E_0\psi\psi} - f_{E_0} f_{\psi\psi\psi} \tag{3.6.14b}$$

and use Equations (3.6.13) to find $\partial W/\partial\psi = 0$.

It then follows that we can calculate the general solution of the inhomogeneous Equation (3.6.8b) by variation of parameters in the form

$$y_1 = \alpha f_\psi + \beta f_{E_0}. \tag{3.6.15}$$

As usual, α and β depend on ψ but we require

$$f_\psi \alpha_\psi + f_{E_0}\beta_\psi = 0 \tag{3.6.16}$$

in order to simplify the expression for $\partial y_1/\partial\psi$ to be

$$y_{1\psi} = \alpha f_{\psi\psi} + \beta f_{E_0\psi}. \tag{3.6.17}$$

Using Equation (3.6.17), Equation (3.6.8b) becomes

$$\omega^2(f_{\psi\psi}\alpha_\psi + f_{E_0\psi}\beta_\psi) = p_1 \tag{3.6.18}$$

and we solve Equations (3.6.16) and (3.6.18) for α_ψ and β_ψ, then integrate the results with respect to ψ and substitute back into Equation (3.6.15).

This defines the general solution of y_1:

$$y_1 = \phi_1(\tilde{t})f_\psi + E_1(\tilde{t})f_{E_0} - \frac{f_\psi}{\omega^2 W}\int_0^\psi p_1 f_{E_0}\,d\psi' + \frac{f_{E_0}}{\omega^2 W}\int_0^\psi p_1 f_\psi\,d\psi'. \tag{3.6.19}$$

The first two terms on the right hand side with the unknown coefficients $\phi_1(\tilde{t})$ and $E_1(\tilde{t})$ represent the homogeneous solution while the integrals define a particular solution.

In Reference (3.6.1), only the last term in (3.6.19) was derived, and as we shall see presently, periodicity of this term determines E_0.

In the linear case, [cf. Section 3.6.2] with $\omega = \mu$, $g = \mu^2 y$, we have $y_0 = \sqrt{2E_0}\cos\psi$ and boundedness (or periodicity) of y_1 on the t^+ scale reduces to the requirement that p_1 be orthogonal to f_{E_0} and f_ψ, i.e.,

$$\int_0^{2\pi} p_1 \sin\psi\,d\psi = \int_0^{2\pi} p_1\cos\psi\,d\psi = 0. \tag{3.6.20}$$

One of these conditions, namely that

$$\int_0^{2\pi} p_1 f_\psi\,d\psi' = 0 \tag{3.6.21}$$

still holds for the nonlinear case and can be deduced most directly as follows. Since f and hence f_ψ are 2π-periodic functions of ψ, it follows from the definition of p_1 in Equation (3.6.8b) that p_1 is also a 2π-periodic function of ψ. The product $p_1 f_\psi$ is then 2π-periodic and therefore $\int_0^\psi p_1 f_\psi \, d\psi'$ will have a linear term in ψ, *unless its average value over one period vanishes*, i.e., Equation (3.6.21) must hold.

Unfortunately, the same argument does not apply to the second integral, $\int_0^\psi p_1 f_{E_0} \, d\psi'$, because f_{E_0} is not periodic in ψ for the nonlinear case. To see this we note that the periodicity in f implies that

$$f[\psi + P(E_0, \omega, \tilde{t}), E_0, \omega, \tilde{t}] = f(\psi, E_0, \omega, \tilde{t}). \tag{3.6.22}$$

Since f_{E_0} must be calculated before the constancy condition on P is imposed, we must use Equation (3.6.10a) for P. If we now take the partial derivative of Equation (3.6.22) with respect to E_0, i.e.,

$$f_\psi(\psi + P, E_0, \omega, \tilde{t})P_{E_0} + f_{E_0}(\psi + P, E_0, \omega, \tilde{t}) = f_{E_0}(\psi, E_0, \omega, \tilde{t}) \tag{3.6.23}$$

and use the partial derivative of Equation (3.6.22) with respect to ψ, i.e.,

$$f_\psi(\psi + P, E_0, \omega, \tilde{t}) = f_\psi(\psi, E_0, \omega, \tilde{t}), \tag{3.6.24}$$

we find

$$f_{E_0}(\psi + P, E_0, \omega, \tilde{t}) = f_{E_0}(\psi, E_0, \omega, \tilde{t}) - f_\psi(\psi, E_0, \omega, \tilde{t})P_{E_0}. \tag{3.6.25}$$

This exhibits the fact that f_{E_0} is not periodic. However, its nonperiodic component is easily isolated by noting that the function k defined by

$$k(\psi, E_0, \omega, \tilde{t}) = f_{E_0}(\psi, E_0, \omega, \tilde{t}) + \frac{\psi P_{E_0}}{P} f_\psi(\psi, E_0, \omega, \tilde{t}) \tag{3.6.26}$$

is periodic. In fact, direct substitution of $\psi + P$ for ψ in (3.6.26) shows that $k(\psi + P, E_0, \omega, \tilde{t}) = k(\psi, E_0, \omega, \tilde{t})$.

We can now use Equation (3.6.26) to express f_{E_0} as the combination of a periodic term k and a mixed secular term in the form[19]

$$f_{E_0} = k - \frac{P_{E_0}}{P} \psi f_\psi. \tag{3.6.27}$$

Substituting this expression f_{E_0} into Equation (3.6.19) gives

$$y_1 = f_\psi\left[\phi_1 - \frac{1}{\omega^2 W}\int_0^\psi p_1\left(k - \frac{\psi' P_{E_0}}{P} f_\psi\right) d\psi'\right] + \left(k - \frac{\psi P_{E_0}}{P} f_\psi\right)\left(\int_0^\psi \frac{f_\psi p_1}{\omega^2 W}\, d\psi' + E_1\right). \tag{3.6.28}$$

[19] Note that in the linear problem P does not depend on E_0, therefore Equation (3.6.27) reduces to $f_{E_0} = k$ which is periodic.

We remove ψ from the integrand in the second term on the right-hand side by integrating this term by parts. The result then reduces to

$$y_1 = f_\psi\left\{\phi_1 - \int_0^\psi\left[\frac{p_1 k}{\omega^2 W} + \frac{P_{E_0}E_1}{P} + \frac{P_{E_0}}{P\omega^2 W}\int_0^{\psi'} p_1 f_\psi\, d\psi''\right]d\psi'\right\} + k\left[E_1 + \frac{1}{\omega^2 W}\int_0^\psi p_1 f_\psi\, d\psi'\right]. \quad (3.6.29)$$

The second periodicity condition for y_1 is then

$$\int_0^{2\pi}\left[\frac{p_1 k}{\omega^2 W} + \frac{P_{E_0}E_1}{P} + \frac{P_{E_0}}{P\omega^2 W}\int_0^{\psi'} p_1 f_\psi\, d\psi''\right]d\psi' = 0. \quad (3.6.30)$$

The two conditions (3.6.21) and (3.6.30) ensure the periodicity of y_1 and it is clear that this procedure can be repeated at each succeeding order in ε. In our formulas, we simply need to replace f by f_N, p_1 by p_{N+1}, etc. to derive the formal periodicity conditions for the solution to $O(\varepsilon^N)$.

Let us now consider how the availability of conditions analogous to Equations (3.6.21) and (3.6.30) to some order ε^N can be used to calculate the E_n, ϕ_n for $n < N$. In the linear case, since $P_{E_0} = 0$ and $k = f_{E_0}$, Equation (3.6.30) is simply the orthogonality condition

$$\int_0^{2\pi} p_1 f_{E_0}\, d\psi = \int_0^{2\pi} p_1 \cos\psi\, d\psi = 0 \quad (3.6.31a)$$

while Equation (3.6.21) reduces to the second orthogonality condition

$$\int_0^{2\pi} p_1 f_\psi\, d\psi = \int_0^{2\pi} p_1 \sin\psi\, d\psi = 0. \quad (3.6.31b)$$

We will see presently that Equation (3.6.31a) determines ϕ_0 and Equation (3.6.31b) determines E_0, and in general for the linear problem, periodicity of the $O(\varepsilon^{N+1})$ solution determines the two unknown slowly varying functions of the $O(\varepsilon^N)$ solution.

In contrast, for the nonlinear problem, Equation (3.6.30) *involves* E_1, and this must be first determined by considering Equation (3.6.21) with p_2 replacing p_1, and f_1 replacing f, i.e., by periodicity considerations on y_2. Luke points out that to any order ε^{N+1} the equation corresponding to (3.6.21) does not involve ϕ_{N+1}. Therefore, it can be solved first for E_N, and this result can then be used in the equation corresponding to (3.6.30) for $O(\varepsilon^N)$ to determine ϕ_{N-1}. He also points out that in the second approximation, the condition

$$\int_0^{2\pi} p_2 y_{1\psi}\, d\psi = 0$$

gives $E_1 = 0$, and the condition (3.6.30) gives $\phi_0 = $ const. Thus, Kuzmak's theory, which did not include Equation (3.6.30), is, in fact, complete to $O(1)$.

Equations (3.6.21) and (3.6.30) can be made more explicit and we only consider the first one as (3.6.30) is not interesting for $N = 1$.

Since f is an even function of ψ, Equation (3.6.21) reduces to

$$\int_0^\pi \left\{ 2\omega f_{\psi\tilde{t}} f_\psi + \left[\frac{d\omega}{d\tilde{t}} + \omega h(f, \tilde{t}) \right] f_\psi^2 \right\} d\psi = 0 \tag{3.6.32}$$

and this can be written as

$$\frac{d}{d\tilde{t}} \left[\omega(\tilde{t}) \int_0^\pi f_\psi^2 \, d\psi \right] + \omega(\tilde{t}) \int_0^\pi h(f, \tilde{t}) f_\psi^2 \, d\psi = 0. \tag{3.6.33}$$

For the special case where the damping coefficient is independent of the amplitude ($h = h(\tilde{t})$) Equation (3.6.33) is a linear differential equation

$$\frac{d\lambda}{d\tilde{t}} + h(\tilde{t})\lambda = 0 \tag{3.6.34}$$

for

$$\lambda(\tilde{t}) = \omega(\tilde{t}) \int_0^\pi f_\psi^2 \, d\psi = \frac{2}{\omega} \int_0^{2\pi} [E_0 - V(f, \tilde{t})] d\psi, \tag{3.6.35a}$$

where the last part of Equation (3.6.35) follows from Equation (3.6.9a).

Solving Equation (3.6.34) gives

$$\lambda(\tilde{t}) = \lambda(0) \exp\left[- \int_0^{\tilde{t}} h(s) ds \right] \tag{3.6.35b}$$

and for the special case $h = 0$, λ is a constant. This is just the adiabatic invariant to $O(1)$, [cf. Sections 3.3.2 and 3.7.3].

3.6.2 Linear Problem

To illustrate the ideas we first consider the simple example of a weakly damped linear oscillator with slowly varying damping and spring constants:

$$\frac{d^2 y}{dt^2} + \varepsilon h(\tilde{t}) \frac{dy}{dt} + \mu^2(\tilde{t}) y = 0. \tag{3.6.36}$$

Equation (3.6.8a) becomes

$$\omega^2 \frac{\partial^2 y_0}{\partial t^{+2}} + \mu^2 y_0 = 0 \tag{3.6.37}$$

with solution

$$y_0 = \frac{\sqrt{2E_0}}{\mu(\tilde{t})} \cos \frac{\mu}{\omega} \psi; \qquad \psi = t^+ - \phi_0(\tilde{t}). \tag{3.6.38}$$

Clearly, we must set $\mu/\omega = \text{const.}$ to have a constant period on the t^+ scale, the choice $\omega = \mu$ normalizes this period to be 2π. This result also follows from the more general relation (3.6.10a) which is valid for nonlinear problems.

With $V(y_0, \tilde{t}) = \mu^2(\tilde{t})y_0^2/2$, we first solve Equation (3.6.9a) for the zeros of $(\partial y_0/\partial t^+)$

$$y_{0_{\text{max, min}}} = \pm\sqrt{2E_0}/\mu(\tilde{t}). \tag{3.6.39a}$$

Now, Equation (3.6.10a) with $P = 2\pi$ becomes

$$\pi = \omega \int_{\sqrt{2E_0}/\mu}^{\sqrt{2E_0}/\mu} \frac{dy_0}{\sqrt{2E_0 - \mu^2 y_0^2}}. \tag{3.6.39b}$$

We change variables of integration to $\eta = \mu y_0/\sqrt{2E_0}$ and obtain

$$\pi = \frac{\omega}{\mu}\int_{-1}^{1} \frac{d\eta}{\sqrt{1-\eta^2}} = \frac{\omega}{\mu}\pi \tag{3.6.40}$$

and this gives $\omega = \mu(\tilde{t})$.

In our case $\partial y_0/\partial t^+ = -(\sqrt{2E_0}/\mu)\sin\psi$, therefore

$$p_1 = \mu\sqrt{2E_0}\left(\frac{1}{E_0}\frac{dE_0}{d\tilde{t}} + \frac{1}{\mu}\frac{d\mu}{d\tilde{t}} + h\right)\sin\psi - 2\mu\sqrt{2E_0}\,\frac{d\phi_0}{d\tilde{t}}\cos\psi. \tag{3.6.41}$$

So far, we have duplicated all the steps of our usual multiple-variable procedure, and at this point we would set the coefficients of $\sin\psi$ and $\cos\psi$ equal to zero in Equation (3.6.41) in order to avoid mixed secular terms in y_1. The resulting equations for E_0 and ϕ_0 according to the usual procedure are then

$$\frac{1}{E_0}\frac{dE_0}{d\tilde{t}} + \frac{1}{\mu}\frac{d\mu}{d\tilde{t}} + h = 0 \tag{3.6.42a}$$

$$\frac{d\phi_0}{d\tilde{t}} = 0. \tag{3.6.42b}$$

These are precisely the results we obtain by invoking Equations (3.6.31). The solutions for E_0 and ϕ_0 are

$$\frac{E_0}{\mu} = \frac{E_0(0)}{\mu(0)}\exp\left(-\int_0^{\tilde{t}} h(s)\,ds\right) \tag{3.6.43a}$$

$$\phi_0 = \phi_0(0) \tag{3.6.43b}$$

and Equation (3.6.43a) also follows from Equation (3.6.35).

3.6.3 Undamped Nonlinear Oscillator with Slowly Varying Parameters

We now consider the nonlinear problem discussed in Reference 3.6.1

$$\frac{d^2y}{dt^{+2}} + a(\tilde{t})y + b(\tilde{t})y^3 = 0, \tag{3.6.44}$$

where a and b are given functions.

The equation corresponding to (3.6.8a) is

$$\omega^2(\tilde{t})\frac{\partial^2 y_0}{\partial t^{+2}} + a(\tilde{t})y_0 + b(\tilde{t})y_0^3 = 0 \tag{3.6.45}$$

and the "energy" integral is

$$\frac{\omega^2(\tilde{t})}{2}\left(\frac{\partial y_0}{\partial t^+}\right)^2 + V(y_0, a, b) = E_0(\tilde{t}), \tag{3.6.46}$$

where the potential V is

$$V(y_0, a, b) = a(\tilde{t})y_0^2/2 + b(\tilde{t})y_0^4/4. \tag{3.6.47}$$

Examining V for the different possible combinations of the signs of a and b will determine the cases for which Equation (3.6.45) can have periodic solutions.

(i) If $a > 0$, $b > 0$, y_0 is periodic for any positive E_0, (and Equation (3.6.46) shows that E_0 cannot be negative) because as shown in Figure 3.6.2a V is concave.

(ii) If $a < 0$, $b < 0$, V has the opposite sign as in case (i) and is convex. Therefore, none of the solutions are periodic.

(iii) If $a < 0$, $b > 0$, V is "W-shaped" as seen in Figure 3.6.2b and we have two families of periodic solutions centered about $y_0 = \pm\sqrt{-a/b}$ for negative values of E_0. When E_0 becomes positive the two periodic solutions coalesce to one centered about $y_0 = 0$.

(iv) Finally if $a > 0$, $b < 0$, is "M-shaped" and is given by the reflection of Figure 3.6.2b about the y_0 axis. Therefore, we have periodic solutions around $y_0 = 0$ as long as $0 < E_0 < a^2/(-4b)$.

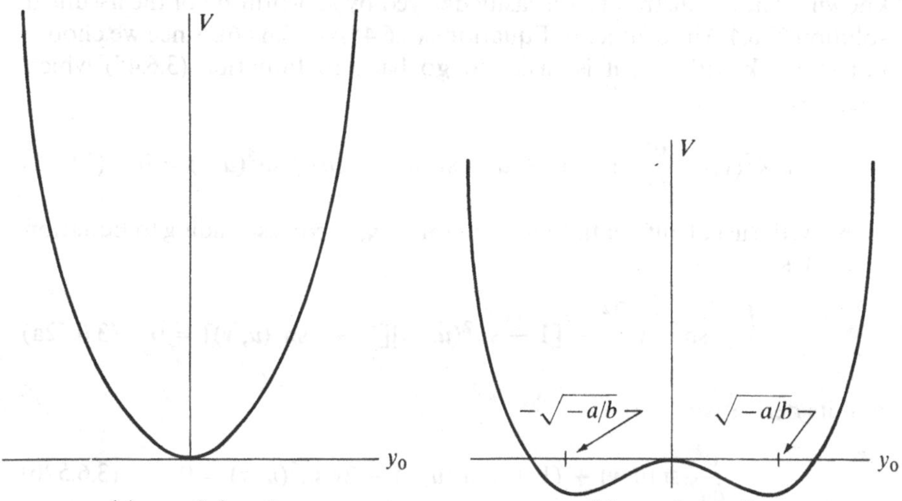

(a) $a > 0, b > 0$ (b) $a < 0, b > 0$

Figure 3.6.2 $V(y_0, a, b)$ for Different Signs of a, b

We will restrict our attention to those cases which admit periodic solutions around $y_0 = 0$.

The solution of Equation (3.6.45), or (3.6.46), is conveniently expressed in terms of the elliptic sine function,[20] sn:

$$y_0 = A_0(\tilde{t})\text{sn}[K(v(\tilde{t}))t^+, v(\tilde{t})], \tag{3.6.48}$$

where K is the complete elliptic integral of the first kind with modulus $\sqrt{v}$. The function $A_0(\tilde{t})$ is a slowly varying amplitude, related to $E_0(\tilde{t})$, and in accordance with the discussion in Section 3.6.1, we have set the constant value of the phase equal to zero. Thus, we have

$$K(v) = \int_0^1 \frac{d\xi}{\sqrt{(1-\xi^2)(1-v\xi^2)}}. \tag{3.6.49}$$

Let us denote $t^+K(v) = u$, then $\psi = t^+ = u/K$

$$\begin{aligned}\frac{\partial y_0}{\partial \psi} &= A_0(\tilde{t})K(v(\tilde{t}))\frac{\partial}{\partial u}\text{sn}(u, v)\\ &= A_0(\tilde{t})K(v(\tilde{t}))\text{cn}(u, v)\text{dn}(u, v).\end{aligned} \tag{3.6.50}$$

Using Equations (3.6.50) and (3.6.48) in (3.6.46) relates A_0 to E_0 and the other parameters. Henceforth, we will use A_0 instead of E_0 in our calculations.

The condition that the period be independent of $\tilde{t}$ has already been imposed by the choice of ψ since [cf. Reference 3.6.3] the period of sn(u, v) is $4K$. Thus, we have chosen to normalize the period to be $P = 4$ with respect to t^+.

At this point we have three unknowns A_0, v, and ω but we should only have one (two if we had retained ϕ_0).

This means that there must exist two constraints relating our three unknowns. These constraints are easily derived by substitution of the assumed solution (3.6.48) into either of Equations (3.6.45) or (3.6.46). Since we choose not to work with E_0, it is easier to go back to Equation (3.6.45) which becomes

$$\omega^2 K^2(v) A_0 \frac{\partial^2}{\partial u^2}\text{sn}(u, v) + aA_0\,\text{sn}(u, v) + bA_0^3\,\text{sn}^3(u, v) = 0. \tag{3.6.51}$$

Now, the usual differential equation for sn(u, v) corresponding to Equation (3.6.46) is

$$\left[\frac{\partial}{\partial u}\text{sn}(u, v)\right]^2 - [1 - \text{sn}^2(u, v)][1 - v\,\text{sn}^2(u, v)] = 0 \tag{3.6.52a}$$

and its derivative

$$\frac{\partial^2}{\partial u^2}\text{sn}(u, v) + (1 + v)\text{sn}(u, v) - 2v\,\text{sn}^3(u, v) = 0 \tag{3.6.52b}$$

[20] The reader will find Reference 3.6.3 useful for definitions and formulas for elliptic functions.

must correspond to Equation (3.6.51) if Equation (3.6.48) is to be a solution. This means that substituting the expression in Equation (3.6.52b) for $\partial^2 \operatorname{sn}(u, v)/\partial u^2$ into Equation (3.6.52a) must give an identity. In other words, the expression

$$\omega^2 K^2(v)[-(1 + v)\operatorname{sn}(u, v) + 2v \operatorname{sn}^3(u, v)] + a \operatorname{sn}(u, v) + bA_0^2 \operatorname{sn}^3(u, v) = 0 \tag{3.6.53}$$

must hold identically. Therefore the coefficients of sn and sn^3 must vanish individually, i.e.,

$$-(1 + v)\omega^2(\tilde{t})K^2(v(\tilde{t})) + a(\tilde{t}) = 0 \tag{3.6.54a}$$

$$2\omega^2(\tilde{t})v(\tilde{t})K^2(v(\tilde{t})) + b(\tilde{t})A_0^2(\tilde{t}) = 0. \tag{3.6.54b}$$

Equations (3.6.54) provide two relations for the three quantities A_0, v, and ω; the third relation is provided by the periodicity condition on y_1, $\lambda = \text{const.} = 2B$, which, using Equations (3.6.35) and (3.6.50) can be written

$$\begin{aligned} 2B &= \omega \int_1^3 A_0^2(\tilde{t})K^2(v(\tilde{t}))\operatorname{cn}^2(K\psi, v)\operatorname{dn}^2(K\psi, v)d\psi \\ &= \omega K A_0^2 \int_K^{3K} \operatorname{cn}^2(u, v)dn^2(u, v)du \\ &= 2\omega K A_0^2 \int_0^K \operatorname{cn}^2(u, v)\operatorname{dn}^2(u, v)du. \end{aligned} \tag{3.6.55}$$

Thus, the third relation can be written

$$K\omega A_0^2 L(v) = B = \text{const.}, \tag{3.6.56}$$

where

$$L(v) = \int_0^K \operatorname{cn}^2(u, v)\operatorname{dn}^2(u, v)du$$

and according to Equation (361.03) of Reference 3.6.3 this gives

$$L(v) = \frac{1}{3v}[(1 + v)E(v) - (1 - v)K(v)]. \tag{3.6.57a}$$

In the above $E(v)$ is the complete elliptic integral of the second kind:

$$E(v) = \int_0^1 \sqrt{\frac{1 - v\xi^2}{1 - \xi^2}}\, d\xi. \tag{3.6.57b}$$

The three equations governing A_0, v, and ω can be solved numerically at any time $\tilde{t}$ and depend only on $a(\tilde{t})$ and $b(\tilde{t})$.

The results can be expressed in the form

$$\frac{4L^2(v)v^2}{(1+v)^3} = \frac{B^2b^2}{a^3} \equiv c(\tilde{t}) \tag{3.6.58a}$$

$$\omega(\tilde{t}) = \frac{1}{K(v)}\sqrt{\frac{a}{1+v}} \tag{3.6.58b}$$

$$A_0(\tilde{t}) = \sqrt{\frac{-2av}{b(1+v)}}, \tag{3.6.58c}$$

where $c(\tilde{t})$ is known once we use the definitions for B, ω, A_0, and L.

A graph of the solution for v as a function of c, taken from Reference 3.6.1, is shown below. Various combinations of solutions can occur depending on the signs of a and b. Note that the sign of c is the same as the sign of a.

(i) If $a > 0$, $b > 0$ [cf. Fig. 3.6.2a], the solution for v lies in the fourth quadrant and exists for all positive c.

(ii) If $a < 0$, $b < 0$, [cf. Figure 3.6.2a reflected about y_0], there are no real solutions for v, i.e., the solution is not periodic as anticipated earlier.

(iii) If $a < 0$, $b > 0$ [cf. Figure 3.6.2b], the curve for v lies in the third quadrant and exists for $-\infty < c < -4/9$.

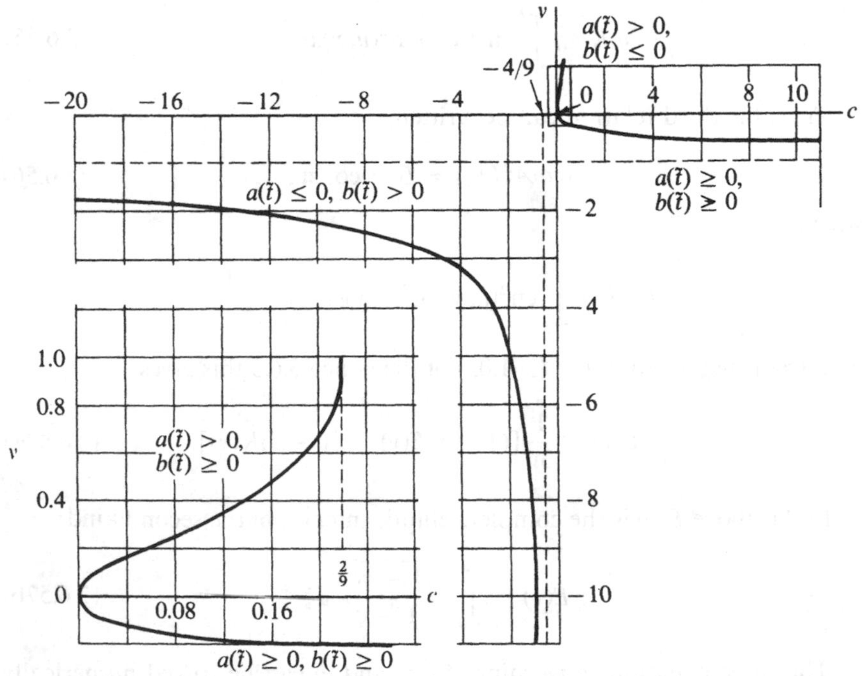

Figure 3.6.3 Numerical Results for v vs c

(iv) If $a > 0$, $b < 0$ [cf. Figure 3.6.2b reflected about y_0], the curve for v lies in the first quadrant since Equation (3.6.54b) shows that $v > 0$. The solution exists only in the range $0 < c < \frac{2}{9}$, and if $c > \frac{2}{9}$ there are no periodic solutions. An enlargement of cases (i) and (iv) appears on the lower left-hand corner of Figure 3.6.3.

As pointed out earlier for case (iii), we also have the two families of periodic solutions centered about $y_0 = \pm\sqrt{-a/b}$. The calculations for this case are essentially similar to those given above, and the results are summarized in Reference 3.6.1.

For the cases where a small damping term appears in the equation, the solution can still be expressed in terms of elliptic functions since y_0 satisfies the same equations. But now the slowly varying modulus $v(\tilde{t})$ is found as the solution of an ordinary differential equation, as is usual in the two variable method.

It should be noted that the results for this example include as special cases both the results of the linear problem, Equation (3.6.48), and the weakly nonlinear spring, Equation (3.1.1). In particular the results for Equation (3.3.47) are approached as $v, b \to 0$.

Finally, we point out that for the undamped problem $h = 0$, Equation (3.6.1) is derivable from a Hamiltonian, and the techniques of canonical perturbations discussed in Sections 3.7.2 and 3.7.3 are also applicable.

Problems

1. Specialize the results of this section to the case [cf. Equation (3.2.60)]

$$\frac{d^2y}{dt^2} + y - y^3 - \varepsilon\frac{dy}{dt} = 0, \qquad 0 < \varepsilon \ll 1$$

$$y(0) = \tfrac{1}{2}, \qquad \dot{y}(0) = 0.$$

Thus, the motion is initially periodic, but the effect of the negative damping is to gradually raise the motion out of the "potential-well". In particular, derive an expression for the time elapsed when the motion ceases to be periodic.

2. Consider the problem of a pendulum with slowly varying length undergoing large amplitude oscillations. In suitable dimensionless variables the governing equation is

$$\frac{d^2y}{dt^2} + a^2(\tilde{t})\sin y = 0.$$

In a sense, Equation (3.6.44) is an approximation to the above where only two terms of the Taylor expansion for sin y are retained.

Apply the technique in this section to derive an approximate solution to $O(1)$ in the oscillatory regime.

3. In many problems it is not possible to express the first approximation y_0 in terms of special functions, yet y_0 is periodic with period 2π; therefore it can be developed in a

Fourier series. Consider $\phi_0 = 0$ and truncate the series after N terms, i.e.,

$$y_0(t^+, \tilde{t}) = \frac{a_0}{2} + \sum_{n=1}^{N} a_n(\tilde{t})\cos nt^+.$$

Show by substitution into Equation (3.6.8a), and neglecting harmonics higher than N that

$$-\omega^2(\tilde{t})a_n(\tilde{t})n^2 + g_n[a_0(\tilde{t}), a_1(\tilde{t}), \ldots, a_N(\tilde{t}), \tilde{t}] = 0$$

for each $n = 0, 1, 2, \ldots, N$, where

$$g_n(a_0, a_1, \ldots, a_N, \tilde{t}) = \frac{2}{\pi}\int_0^{\pi} g\left(\sum_{n=0}^{N} a_n \cos nt^+, \tilde{t}\right)\cos nt^+ \, dt^+.$$

The above defines a system of $N + 1$ equations for the $N + 2$ unknowns, ω, $a_0, \ldots, a_N$. Derive the additional equation needed to close the system by substituting the assumed truncated Fourier series into the periodicity condition (3.6.33) to obtain

$$\frac{d}{d\tilde{t}}\left[\frac{\omega\pi}{2}\sum_{n=0}^{N} n^2 a_n^2\right] + \omega\int_0^{\pi} h\left[\sum_{n=0}^{N} a_n \cos nt^+, \tilde{t}\right]\left[\sum_{n=0}^{N} na_n \sin nt^+\right]^2 dt^+ = 0.$$

Thus, the functions $\omega, a_0, \ldots, a_N$ can be computed in principle.

Specialize your results to the problem of Section 3.6.3 with $a > 0$ and arbitrary b. Take $N = 1$, and compute ω, a_0 and a_1 and compare your results with the "exact" solution given in Section 3.6.3.

References

3.6.1 G. E. Kuzmak, Asymptotic solutions of nonlinear second order differential equations with variable coefficients, *P.M.M.* **23**, 3 (1959), pp. 515–526. Also appears in English translation.

3.6.2 J. C. Luke, A perturbation method for nonlinear dispersive wave problems, *Proc. Roy, Soc. Ser. A* **292** (1966), pp. 403–412.

3.6.3 P. F. Byrd and M. D. Friedman, *Handbook of Elliptic Integrals for Engineers and Scientists*, 2nd Ed., Springer-Verlag, New York, 1971.

3.7 Summary of Other Solution Techniques

To give a detailed survey of all the methods that have been proposed in the literature for solving the type of problems with which we have been occupied in this chapter would require a separate volume. Up till now we have concentrated on the multiple variable approach and its ramifications, primarily because experience has shown that it is simple and efficient without sacrificing generality of application.

From a practical standpoint the simplicity of the multiple variable method is evidenced by the fact that the approximate solution is derived directly as an explicit function of the independent variable. In contrast, the methods that we survey in this section provide the solution in implicit form, and further manipulations are required to derive explicit results.

We admit that this judgment might be considered unduly harsh and therefore we devote this final section of Chapter 3 to a brief survey of competing procedures.

In each case, we present the basic ideas without attempting to give the most general version. Moreover, it would indeed be a difficult task to track down the original contribution of each idea and this is generally not attempted. Rather, we refer the reader to standard texts and whenever possible to survey articles for detailed comparisons among the various procedures.

3.7.1 The Method of Averaging

This method is a generalization of the idea proposed by Krylov and Bogolinbov in Reference 3.7.1. The principal text on the subject including numerous examples and rigorous proofs of asymptotic validity is Reference 3.7.2. More recent results, primarily for non autonomous problems are discussed in Reference 3.7.3. The two survey articles by Morrison, References 3.2.5 and 3.7.6, are particularly recommended as they provide detailed expositions as well as critical comparisons of the averaging method with both the multiple variable procedure and the von Zeipel procedure of Section 3.7.2.

In discussing the method of averaging it is convenient to start with the differential equations in "standard" form. A large class of problems can be represented by the system:

$$\frac{d\mathbf{x}}{dt} = \varepsilon\mathbf{F}(\mathbf{x}, \boldsymbol{\theta}; \varepsilon) \tag{3.7.1a}$$

$$\frac{d\boldsymbol{\theta}}{dt} = \boldsymbol{\omega}(\mathbf{x}) + \varepsilon\mathbf{u}(\mathbf{x}, \boldsymbol{\theta}, \varepsilon), \tag{3.7.1b}$$

where $\mathbf{x}$ is the n-vector $(x_1, x_2, \ldots, x_n)$, $\boldsymbol{\theta}$ is the m-vector $(\theta_1, \theta_2, \ldots, \theta_m)$ and $\mathbf{F}$, $\boldsymbol{\omega}$ and $\mathbf{u}$ are vector valued functions of dimension n, m and m respectively. It is further assumed that $\mathbf{F}$ and $\mathbf{u}$ are 2π-periodic functions of θ_i, and that no resonances occur in the solution. Thus, $\mathbf{x}$ is a slowly varying vector while $\boldsymbol{\theta}$ is a rapidly varying one.

As an example of a system which can be brought to the form of Equations (3.7.1) consider the model problem of Reference 3.5.7, i.e.,

$$\frac{d^2x}{dt^2} + [\omega^2(\varepsilon t) + p^2]x = 0 \tag{3.7.2a}$$

$$\frac{dp}{dt} = \varepsilon\omega^2 x \sin\psi \tag{3.7.2b}$$

$$\frac{d\psi}{dt} = \sqrt{2}p \tag{3.7.2c}$$

with

$$\omega^2 = \omega_0^2 e^{\varepsilon t}, \qquad \omega_0 = \text{const.} \tag{3.7.2d}$$

Thus, ω is a given slowly varying function of time, while the slowly varying function p is governed by Equation (3.7.2b).

We introduce the *exact* transformation of variables

$$x = \rho(t, \varepsilon)\cos[\tau - \phi(t, \varepsilon)] \tag{3.7.3a}$$

$$\dot{x} = -\rho(t, \varepsilon)k \sin[\tau - \phi(t, \varepsilon)], \tag{3.7.3b}$$

where ρ and ϕ are the two new dependent variables; τ is defined by

$$\dot{\tau} = k \equiv (\omega^2 + p^2)^{1/2}. \tag{3.7.4}$$

We note, as in the method of variation of parameters for calculating a particular solution, that the expression for $\dot{x}$ given by (3.7.3b) implies that we have set

$$\dot{\rho} \cos(\tau - \phi) + \rho\dot{\phi} \sin(\tau - \phi) = 0. \tag{3.7.5}$$

We now use Equations (3.7.3) to calculate Equation (3.7.2a). We find

$$\ddot{x} + k^2 x = -\dot{\rho}k \sin(\tau - \phi) - \rho\dot{k} \sin(\tau - \phi) + \rho k\dot{\phi} \cos(\tau - \phi) = 0. \tag{3.7.6}$$

Solving this together with Equation (3.7.5) for $\dot{\rho}$ and $\rho\dot{\phi}$ gives

$$\dot{\rho} = -\frac{\rho\dot{k}}{k} \sin^2(\tau - \phi) \tag{3.7.7a}$$

$$\dot{\phi} = \frac{\dot{k}}{k} \sin(\tau - \phi)\cos(\tau - \phi). \tag{3.7.7b}$$

Using the definition (3.7.4) for k and Equations (3.7.2b), (3.7.2d) we can express $\dot{k}/k$ in terms of the new variables as

$$\frac{\dot{k}}{k} = \varepsilon\omega^2\left[\frac{1 + 2p\rho \cos(\tau - \phi)\sin(\tau - \phi)}{2(\omega^2 + p^2)}\right]. \tag{3.7.8}$$

If we now regard ω as a new dependent variable defined by (cf. Equation (3.7.2d)]

$$\dot{\omega} = \varepsilon\omega/2 \tag{3.7.9}$$

we can avoid dealing with a non-autonomous problem at the cost of adding one variable.

It then follows that Equations (3.7.7) (with $\dot{k}/k$ defined by Equation (3.7.8)) Equation (3.7.2b) and Equation (3.7.9) define the slowly varying vector $\mathbf{x} = (\rho, \phi, p, \omega)$, while Equations (3.7.4) and (3.7.2c) define the rapidly varying vector $\boldsymbol{\theta} = (\tau, \psi)$. Note that in this example $\mathbf{u} = 0$.

The method of averaging applied to systems with one fast variable is discussed in Reference 3.7.2, the extension to more than one fast variable is given in Reference 3.7.3 and Reference 3.7.6. Here we will not consider the general case of Equations (3.7.1), but rather we will adopt the special case

$$\frac{d\mathbf{x}}{dt} = \varepsilon \mathbf{F}(\mathbf{x}, t; \varepsilon) \tag{3.7.10}$$

in order to illustrate the ideas. In the above as in Equation (3.7.1) it is assumed that $\mathbf{F}$ is a 2π-periodic function of t. Here again, we find the discussion given by Morrison in Reference 3.2.5 particularly instructive and will follow it closely.

To fix ideas, we note that the general weakly nonlinear second order equation

$$\frac{d^2y}{dt^2} + y + \varepsilon f\left(y, \frac{dy}{dt}\right) = 0 \tag{3.7.11}$$

that we studied in Section 3.2.5 can be transformed to the form (3.7.10). In fact, if we introduce the transformations analogous to Eqs. (3.7.3)

$$y = \rho \cos(t + \phi) \tag{3.7.12a}$$

$$\frac{dy}{dt} = -\rho \sin(t + \phi) \tag{3.7.12b}$$

we find that Equation (3.7.11) reduces to

$$\dot{\rho} = \varepsilon \sin(t + \phi) f[\rho \cos(t + \phi), -\rho \sin(t + \phi)] \tag{3.7.13a}$$

$$\dot{\phi} = \frac{\varepsilon}{\rho} \cos(t + \phi) f[\rho \cos(t + \phi), -\rho \sin(t + \phi)]. \tag{3.7.13b}$$

The basic idea of the method of averaging is the observation that the long term behavior of functions such as ρ and ϕ which have oscillatory derivatives of order ε, must be predominantly contributed by the average value of Equations (3.7.13) over one period in t. This is easy to see if we were to develop the right-hand sides of Equations (3.7.13) in their Fourier series with respect to t. Suppose this resulted in expressions of the form

$$\dot{\rho} = \varepsilon\left[\frac{a_0}{2} + \sum_{n=1}^{\infty} a_n \cos nt + b_n \sin nt\right] \tag{3.7.14a}$$

$$\dot{\phi} = \varepsilon\left[\frac{c_0}{2} + \sum_{n=1}^{\infty} c_n \cos nt + d_n \sin nt\right] \tag{3.7.14b}$$

with the a_i, b_i, c_i, d_i depending on ρ and ϕ and defined in the usual way. Since the oscillatory terms have little net effect in the long term, the dominant contribution to ρ and ϕ is described by the "averaged" equations

$$\dot{\rho} \simeq \varepsilon \frac{a_0}{2} \tag{3.7.15a}$$

$$\dot{\phi} \simeq \varepsilon \frac{c_0}{2} \tag{3.7.15b}$$

hence the name "method of averaging".

It is not too difficult to see that the multiple variable expansion of Equations (3.7.14) leads to precisely this result to $O(1)$, with the understanding that Equations (3.7.15) define the first terms in the expansions for ρ and ϕ and that these first terms are functions of $\tilde{t} = \varepsilon t$ only.

It is in the calculations of the higher order terms that we see a significant departure in the structure of the solution according to the two methods. Whereas in the multiple variable method we assume an expansion in terms explicitly depending on $\tilde{t}$ and t, the role of the independent variable is not specified in the method of averaging. In fact, returning to the general form of Equation (3.7.10) one assumes an expansion (Let $\mathbf{x} = x_1, x_2, \ldots, x_n$, $\boldsymbol{\xi} = \xi_1, \xi_2, \ldots, \xi_n$, $\mathbf{F} = F_1, F_2, \ldots, F_n$, etc.) in the form

$$x_i = \xi_i + \varepsilon \eta_i^{(1)}(\boldsymbol{\xi}, t) + \varepsilon^2 \eta_i^{(2)}(\boldsymbol{\xi}, t) + \cdots, \tag{3.7.16a}$$

where the η_i are 2π-periodic functions of t and the ξ_i are governed by the averaged equations

$$\frac{d\xi_i}{dt} = \varepsilon M_i^{(1)}(\boldsymbol{\xi}) + \varepsilon^2 M_i^{(2)}(\boldsymbol{\xi}) + \cdots \tag{3.7.16b}$$

and we anticipate that

$$M_i^{(1)}(\boldsymbol{\xi}) = \frac{1}{2\pi} \int_0^{2\pi} F_i(\boldsymbol{\xi}, t) dt. \tag{3.7.17}$$

One can derive the equations governing the $M_i^{(j)}$ and $\eta_i^{(j)}$ systematically by substituting the assumed form (3.7.16) into Equation (3.7.10). First, we calculate dx_i/dt by differentiating Equation (3.7.16a). With the convention of summing over repeated indices, we find

$$\frac{dx_i}{dt} = \frac{d\xi_i}{dt} + \varepsilon\left(\frac{\partial \eta_i^{(1)}}{\partial \xi_j}\frac{d\xi_j}{dt} + \frac{\partial \eta_i^{(1)}}{\partial t}\right) + \varepsilon^2\left(\frac{\partial \eta_i^{(1)}}{\partial \xi_j}\frac{d\xi_j}{dt} + \frac{\partial \eta_i^{(2)}}{\partial t}\right) + O(\varepsilon^3) \tag{3.7.18a}$$

and using the expansion given by Equation (3.7.16b), this reduces to

$$\frac{dx_i}{dt} = \varepsilon\left(M_i^{(1)} + \frac{\partial \eta_i^{(1)}}{\partial t}\right) + \varepsilon^2\left(M_i^{(2)} + \frac{\partial \eta_i^{(1)}}{\partial \xi_j} M_j^{(1)} + \frac{\partial \eta_i^{(2)}}{\partial t}\right) + O(\varepsilon^3). \tag{3.7.18b}$$

The right-hand side of Equation (3.7.10) can also be expanded. First, we assume the ε dependence of the F_i can be developed in the form

$$F_i(\mathbf{x}, t; \varepsilon) = F_i^{(1)}(\mathbf{x}, t) + \varepsilon F_i^{(2)}(\mathbf{x}, t) + O(\varepsilon^2). \tag{3.7.19a}$$

Then, when we use Equation (3.7.16a) to expand the first argument we find

$$F_i(\mathbf{x}, t; \varepsilon) = F_i^{(1)}(\boldsymbol{\xi}, t) + \varepsilon \left[\frac{\partial F_i^{(1)}}{\partial x_j}(\boldsymbol{\xi}, t)\eta_j^{(1)} + F_i^{(2)}(\boldsymbol{\xi}, t) \right] + O(\varepsilon^2). \tag{3.7.19b}$$

Therefore, equating terms of like powers in ε gives the following sequence of equations

$$M_i^{(1)}(\boldsymbol{\xi}) + \frac{\partial \eta_i^{(1)}}{\partial t}(\boldsymbol{\xi}, t) = F_i^{(1)}(\boldsymbol{\xi}, t) \tag{3.7.20a}$$

$$M_i^{(2)}(\boldsymbol{\xi}) + \frac{\partial \eta_i^{(1)}}{\partial \xi_j}(\boldsymbol{\xi}, t)M_j^{(1)}(\boldsymbol{\xi}) + \frac{\partial \eta_i^{(2)}}{\partial t}(\boldsymbol{\xi}, t)$$
$$= F_i^{(2)}(\boldsymbol{\xi}, t) + \frac{\partial F_i^{(1)}}{\partial x_j}(\boldsymbol{\xi}, t)\eta_j^{(1)}(\boldsymbol{\xi}, t). \tag{3.7.20b}$$

Consider first Equation (3.7.20a). Since $\eta_i^{(1)}$ is periodic in t, $\partial \eta_i^{(1)}/\partial t$ is also periodic in t with zero average value.

Now, $F_i^{(1)}$ is also periodic in t therefore the average value of $F_i^{(1)}$ must equal $M_i^{(1)}$. This is just Equation (3.7.17) and it defines $M_i^{(1)}$ as a function of $\boldsymbol{\xi}$. Let us denote this function (for consistency with later notation) by $R_i^{(1)}(\boldsymbol{\xi})$, i.e.,

$$M_i^{(1)}(\boldsymbol{\xi}) = \frac{1}{2\pi}\int_0^{2\pi} F_i^{(1)}(\boldsymbol{\xi}, t) \equiv R_i^{(1)}(\boldsymbol{\xi}). \tag{3.7.21}$$

What remains of Equation (3.7.20a) once Equation (3.7.21) is imposed has zero average value and determines $\eta_i^{(1)}$ by quadrature in the form

$$\eta_i^{(1)}(\boldsymbol{\xi}, t) = x_i^{(1)}(\boldsymbol{\xi}, t) + q_i^{(1)}(\boldsymbol{\xi}), \tag{3.7.22}$$

where $x_i^{(1)}$ is the oscillatory part of $\eta_i^{(1)}$ defined by

$$\frac{\partial x_i^{(1)}}{\partial t}(\boldsymbol{\xi}, t) = F_i^{(1)}(\boldsymbol{\xi}, t) - R_i^{(1)}(\boldsymbol{\xi}) \tag{3.7.23a}$$

and has zero average value, i.e.,

$$\int_0^{2\pi} x_i^{(1)}(\boldsymbol{\xi}, t)dt = 0. \tag{3.7.23b}$$

The arbitrary function of $\boldsymbol{\xi}$ which arises in the quadrature is denoted by $q_i^{(1)}(\boldsymbol{\xi})$.

A similar argument determines $M_i^{(2)}$ and $\eta_i^{(2)}$. If we average Equation (3.7.20b) in which we use the known expressions $\eta_i^{(1)}$ and $M_i^{(1)}$, we have

$$M_i^{(2)}(\xi) = \frac{1}{2\pi}\int_0^{2\pi}\left[F_i^{(2)}(\xi, t) + \frac{\partial F_i^{(1)}}{\partial x_j}(\xi, t)(x_j^{(1)} + q_j^{(1)}) - \left(\frac{\partial x_i^{(1)}}{\partial \xi_j} + \frac{\partial q_i}{\partial \xi_j}\right)R_j^{(1)}(\xi)\right]dt. \tag{3.7.24}$$

We note that

$$\int_0^{2\pi}\frac{\partial x_i^{(1)}}{\partial \xi_j}R_j^{(1)}\,dt = R_j^{(1)}(\xi)\frac{\partial}{\partial \xi_j}\int_0^{2\pi}x_i^{(1)}(\xi, t)dt = 0. \tag{3.7.25}$$

Let us define $R_i^{(2)}(\xi)$ by

$$R_i^{(2)}(\xi) \equiv \frac{1}{2\pi}\int_0^{2\pi}\left[F_i^{(2)}(\xi, t) + \frac{\partial F_i^{(1)}}{\partial x_j}(\xi, t)x_j^{(1)}(\xi, t)\right]dt \tag{3.7.26}$$

so that it is uniquely determined by $\mathbf{F}^{(1)}$ and $\mathbf{F}^{(2)}$. Now, Equation (3.7.24) defines $M_i^{(2)}(\xi)$ in the form

$$M_i^{(2)}(\xi) = R_i^{(2)}(\xi) + q_j^{(1)}(\xi)\frac{\partial R_i^{(1)}}{\partial \xi_j}(\xi) - R_j^{(1)}(\xi)\frac{\partial q_i^{(1)}}{\partial \xi_j}(\xi). \tag{3.7.27}$$

Knowing $M_i^{(2)}$, we calculate $\eta_i^{(2)}(\xi)$ from what remains of Equation (3.7.20b) by quadrature in terms of an arbitrary function $q_i^{(2)}(\xi)$ etc.

The functions $q_i^{(j)}(\xi)$ which arise at each order are usually chosen such that the $\eta_i^{(j)}$ have zero average values. In particular, this would mean that $q_i^{(1)}(\xi) = 0$. However, they are quite arbitrary in the sense that the final result for the approximation of $\mathbf{x}$ should not depend on the choice of the $q_i^{(1)}(\xi)$. Morrison in Reference 3.2.5 points out that by a judicious choice of the $q_i^{(j)}$ one can directly exhibit the equivalence of the multiple variable results of Section 3.2 with the results given by the method of averaging, and we consider this question next.

Let us calculate the solution of Equations (3.7.13) by the method of averaging and compare the results with those of Section 3.2. We assume expansions for ρ and ϕ in the form [cf. Equations (3.7.16a) and (3.7.22)]

$$\rho = \xi_1 + \varepsilon[x_1^{(1)}(\xi_1, t + \xi_2) + q_1(\xi_1)] + O(\varepsilon^2) \tag{3.7.28a}$$

$$\phi = \xi_2 + \varepsilon[x_2^{(1)}(\xi_1, t + \xi_2) + q_2(\xi_1)] + O(\varepsilon^2) \tag{3.7.28b}$$

and the dependence of the solution on $t + \xi_2$ follows from the structure of Equations (3.7.13). Note that we are assuming that $q_i^{(1)} \equiv q_i(\xi_1)$ depend only on ξ_1, and the choice of these arbitrary functions will be made later on.

Equations (3.7.28) determine ρ and ϕ to order ε^2 for the averaged variables ξ_1 and ξ_2, and to order ε for the oscillatory terms when we use the results derived above. Thus, using the result for $M_i^{(1)}$ in Equation (3.7.21) and the fact [cf. Equations (3.7.13)] that

$$F_1^{(1)} = f[\xi_1 \cos(t + \xi_2), -\xi_1 \sin(t + \xi_2)]\sin(t + \xi_2) \quad (3.7.29a)$$

$$F_2^{(1)} = \frac{1}{\xi_1} f[\xi_1 \cos(t + \xi_2), -\xi_1 \sin(t + \xi_2)]\cos(t + \xi_2) \quad (3.7.29b)$$

we see, once we make the change of variable $t + \xi_2 = \psi$ that

$$M_1^{(1)} = P_1(\xi_1) \quad (3.7.30a)$$

$$M_2^{(1)} = Q_1(\xi_1)/\xi_1, \quad (3.7.30b)$$

where P_1 and Q_1 were defined in Equations (3.2.92) of Section 3.2.5.

To second order, the definition (3.7.26) for the $R_i^{(2)}$ gives

$$R_1^{(2)} = R_2(\xi_1) \equiv \frac{1}{2\pi} \int_0^{2\pi} \left\{ x_1^{(1)}(\xi_1, \psi) \frac{\partial}{\partial \xi_1} [f_0(\xi_1, \psi)\sin \psi] + x_2^{(1)}(\xi_1, \psi) \frac{\partial}{\partial \psi} [f_0(\xi_1, \psi)\sin \psi] \right\} d\psi \quad (3.7.31a)$$

$$R_2^{(2)} = S_2(\xi_1) \equiv \frac{1}{2\pi} \int_0^{2\pi} \left\{ x_1^{(1)}(\xi_1, \psi) \frac{\partial}{\partial \xi_1} \left[f_0(\xi_1, \psi) \frac{\cos \psi}{\xi_1} \right] + x_2^{(1)}(\xi_1, \psi) \frac{\partial}{\partial \psi} \left[f_0(\xi_1, \psi) \frac{\cos \psi}{\xi_1} \right] \right\} d\psi, \quad (3.7.31b)$$

where, as defined in Equation (3.2.99), we have set

$$f_0(\xi_1, \psi) = f(\xi_1 \cos \psi, -\xi_1 \sin \psi). \quad (3.7.32)$$

Evaluating Equation (3.7.27) for the $M_i^{(2)}$ now gives

$$M_1^{(2)} = M_1^{(2)}(\xi_1) = R_2 + q_1 \frac{dP_1}{d\xi_1} - P_1 \frac{dq_1}{d\xi_1} \quad (3.7.33a)$$

$$M_2^{(2)} = M_2^{(2)}(\xi_1) = S_2 + q_1 \frac{d}{d\xi_1} \left(\frac{Q_1}{\xi_1} \right) - P_1 \frac{dq_2}{d\xi_1} \quad (3.7.33b)$$

and this is as far as we need to proceed with the solution of Equations (3.7.13) in order to compare with the results of Section 3.2.5.

We now identify the periodic functions $x_1^{(1)}$ and $x_2^{(1)}$ appearing in Equations (3.7.28) with λ_1 and $-\mu_1$ defined by Equations (3.2.103, 4) of Section 3.2.5, i.e.,

$$x_1^{(1)}(\xi_1, \psi) = \lambda_1(\xi_1, \psi) \quad (3.7.34a)$$

$$x_2^{(1)}(\xi_1, \psi) = -\mu_1(\xi_1, \psi). \quad (3.7.34b)$$

Using Equations (3.2.102–104), it is easy to show that

$$\int_0^{2\pi} \mu_1(\alpha_0, \psi) f_0(\alpha_0, \psi) \cos \psi \, d\psi$$
$$= \int_0^{2\pi} \mu_1(\alpha_0, \psi)\left[Q_1(\alpha_0) - \alpha_0 \frac{\partial \mu_1}{\partial \psi}\right] d\psi = 0 \tag{3.7.35a}$$

$$\int_0^{2\pi} [\lambda_1(\alpha_0, \psi)\cos \psi - \alpha_0 \mu_1(\alpha_0, \psi)\sin \psi] f_0(\alpha_0, \psi) d\psi$$
$$= \int_0^{2\pi}\left\{\lambda_1(\alpha_0, \psi)\left[Q_1 - \alpha_0 \frac{\partial \mu_1}{\partial \psi}\right] - \alpha_0 \mu_1(\alpha_0, \psi)\left[\frac{\partial \lambda_1}{\partial \psi} + P_1(\alpha_0)\right]\right\} d\psi = 0. \tag{3.7.35b}$$

It then follows from Equations (3.7.33) and the definitions of P_2 and Q_2 given by Equations (3.2.113) that

$$R_2(\xi_1) = P_2(\xi_1) \tag{3.7.36a}$$

$$S_2(\xi_1) = \frac{Q_2(\xi_1)}{\xi_1}. \tag{3.7.36b}$$

To correspond with the results of Section 3.2.5 we identify ξ_1 with α_0 and ξ_2 with $-\phi_0 + \varepsilon^2\omega_2 t$. Then, using Equations (3.2.93) with $\omega_1 = 0$ we want, if possible, to have

$$\frac{d\xi_1}{dt} = \varepsilon P_1(\xi_1) \tag{3.7.37a}$$

$$\xi_1 \frac{d}{dt}(\xi_2 - \varepsilon^2\omega_2 t) = \varepsilon Q_1(\xi_1). \tag{3.7.37b}$$

In view of the averaged equations, Equations (3.7.16b), for ξ_1 and ξ_2 to order ε^2, and the result in Equations (3.7.33) for $M_1^{(2)}$ and $M_2^{(2)}$, we can satisfy Equations (3.7.37) by choosing q_1 and q_2 to be solutions of

$$P_1 \frac{dq_1}{d\xi_1} - \frac{dP_1}{d\xi_1} q_1 = P_2(\xi_1) \tag{3.7.38a}$$

$$P_1 \frac{dq_2}{d\xi_1} = \frac{Q_2(\xi_1)}{\xi_1} + q_1(\xi_1)\frac{d}{d\xi_1}\left[\frac{Q_1(\xi_1)}{\xi_1}\right] - \omega_2. \tag{3.7.38b}$$

The first equation is linear and can be solved for $q_1(\xi_1)$, then the second equation defines $q_2(\xi_1)$ by quadrature.

We now identify q_1 with A_1 and q_2 with $-B_1/\alpha_0$ of Section 3.2 and the correspondence between the two expansions is complete to order ε explicitly, because Equations (3.7.38) reduce to Equations (3.2.126).

Having established the correspondence between the results by the two methods we now state without proof a theorem proven in Reference 3.7.2

regarding the asymptotic validity of the first approximation ξ for the system (3.7.10) under fairly general assumptions.

Let $\mathbf{F}(\mathbf{x}, t)$, independent of ε, satisfy the following conditions in some domain D of $\mathbf{x}$ and $t \geq 0$.

(a)
$$|\mathbf{F}(\mathbf{x}\ t)| = O(1) \tag{3.7.39}$$
for all $\mathbf{x}$ in D.

(b) Given any two vectors $\mathbf{x}_1$, $\mathbf{x}_2$ in D, there exists a positive constant δ such that for all $\mathbf{x}$ in D
$$|\mathbf{F}(\mathbf{x}_1, t) - \mathbf{F}(\mathbf{x}_2, t)| \leq \delta|\mathbf{x}_1 - \mathbf{x}_2|. \tag{3.7.40}$$

(c) For all $\mathbf{x}$ in D, the limit
$$\lim_{T\to\infty} \frac{1}{T}\int_0^T \mathbf{F}(\mathbf{x}, t)dt = \mathbf{M}(\mathbf{x}) \tag{3.7.41}$$
exists.

It then follows that the solution $\xi(t)$ of the system
$$\frac{d\xi}{dt} = \varepsilon\mathbf{M}(\xi) \tag{3.7.42}$$
is the first term of the asymptotic expansion for $\mathbf{x}$ as $\varepsilon \to 0$, and is uniformly valid for all times in an interval $0 \leq t \leq L/\varepsilon$ for any given constant L. Similar statements can also be proved for the higher approximations.

In order to appreciate the relative simplicity of the multiple variable method, the reader is urged to use the method of averaging to calculate explicit solutions to $O(\varepsilon)$ for the various examples discussed in Section 3.2.4. [cf. Problems 3.7.1–3.7.3].

The main advantage of the method of averaging is its generality and the fact that one need make no a priori assumptions regarding the structure of the solution once the governing equations are transformed to the standard form. For example, many nonautonomous problems can still be brought to the form (3.7.10) yet the choice of fast and slow variables is not evident a priori as it is for Equation (3.7.11). [cf. Problems 3.7.4].

3.7.2 Canonical Perturbation Theory; The von Zeipel Method

This method applies to certain periodic Hamiltonian systems and was used extensively by workers in celestial mechanics until the advent of more general and efficient methods. Since a large body of work in the field of celestial mechanics relies on this method we will review it briefly in order to provide the interested reader some background for studying the literature.

The original source, Reference 3.7.4, is somewhat obscure. In fact some authors refer to the Bohlin-von Zeipel method and give equal credit to,

Bohlin, another astronomer for the basic idea. At any rate, a modern review and a survey of the older references can be found in Reference 3.7.5. A discussion of the method can also be found in Reference 3.7.6 where Morrison shows that this is a special case of the averaging method of Section 3.7.1.

(a) *Hamiltonian Systems, Canonical Transformations, Action-Angle Variables*

A discussion of von Zeipel's method requires knowledge of Hamiltonian mechanics and we must digress briefly here to review some of the basic concepts. This review is directed strictly toward the reader who has some familiarity with the subject. For the uninitiated, we urge studying first a standard text on dynamics, e.g., Reference 3.7.7, where this material is covered in several chapters.

Consider a system of $2n$ first-order differential equations in the form

$$\frac{dq_i}{dt} = \frac{\partial H}{\partial p_i}; \qquad i = 1, 2, \ldots, n \tag{3.7.43a}$$

$$\frac{dp_i}{dt} = -\frac{\partial H}{\partial q_i}; \qquad i = 1, 2, \ldots, n. \tag{3.7.43b}$$

We will refer to the q_i as the "coordinates" and to the p_i as the "momenta" conjugate to the q_i; the function H is called the Hamiltonian and depends on the q_i, p_i and t. Thus,[21]

$$\begin{aligned} H &= H(q_1, q_2, \ldots, q_n, p_1, p_2, \ldots, p_n, t) \\ &\equiv H(\mathbf{q}, \mathbf{p}, t). \end{aligned} \tag{3.7.44}$$

A system of differential equations which is expressible in the special form of Equations (3.7.43) is called a Hamiltonian system of differential equations.

Many dynamical and other systems can be represented in Hamiltonian form, and often one arrives at the system (3.7.43) via Lagrange's equations

$$\frac{d}{dt}\left(\frac{\partial L}{\partial \dot{q}_i}\right) - \frac{\partial L}{\partial q_i} = 0; \qquad i = 1, \ldots, n \tag{3.7.45}$$

defined for a Lagrangian function

$$L(q_1, q_2, \ldots, q_n, \dot{q}_1, \dot{q}_2, \ldots, \dot{q}_n, t) \equiv L(\mathbf{q}, \dot{\mathbf{q}}, t) \tag{3.7.46}$$

by eliminating the $\dot{q}_i$ in favor of the p_i through a Legendre transformation:

$$p_i = \frac{\partial L}{\partial \dot{q}_i}; \qquad i = 1, \ldots, n \tag{3.7.47}$$

[21] We will use the abbreviated notation $F(\mathbf{q}, \mathbf{p}, t)$ to denote a function $F(q_1, q_2, \ldots, q_n, p_1, p_2, \ldots, p_n, t)$.

and using the definition of H as follows

$$H = \sum_{i}^{n} p_i \dot{q}_i - L(\mathbf{q}, \dot{\mathbf{q}}, t). \tag{3.7.48}$$

It is important to note that not all dynamical systems can be expressed in Hamiltonian form. For example, for a dissipative system neither a Lagrangian nor a Hamiltonian function exist. In this sense, the technique we discuss here is somewhat special.

The system (3.7.43) is a direct consequence of the variational principle, (Hamilton's modified principle)

$$\delta \int_{t_1}^{t_2} [\mathbf{p} \cdot \dot{\mathbf{q}} - H(\mathbf{q}, \mathbf{p}, t)]dt = 0 \tag{3.7.49}$$

subject to the requirements that t_1, t_2, as well as the values of $\mathbf{q}$, $\dot{\mathbf{q}}$ and $\mathbf{p}$ at t_1, t_2 be fixed. We will use this variational principle to generate canonical transformations. We define a canonical transformation as a change of the variables $\mathbf{q}$, $\mathbf{p}$ to $\mathbf{Q}$, $\mathbf{P}$ in the form

$$Q_i = Q_i(\mathbf{q}, \mathbf{p}, t); \qquad i = 1, \ldots, n \tag{3.7.50a}$$

$$P_i = P_i(\mathbf{q}, \mathbf{p}, t); \qquad i = 1, \ldots, n \tag{3.7.50b}$$

and such that the Hamiltonian structure of Equations (3.7.43) is preserved. More precisely, let the inverse transformations to Equations (3.7.50) be defined in the form

$$q_i = q_i(\mathbf{Q}, \mathbf{P}, t); \qquad i = 1, \ldots, n \tag{3.7.50c}$$

$$p_i = p_i(\mathbf{Q}, \mathbf{P}, t); \qquad i = 1, \ldots, n. \tag{3.7.50d}$$

If there exists a function $\mathscr{H}$ of the new variables, i.e.,

$$\mathscr{H} = \mathscr{H}(\mathbf{Q}, \mathbf{P}, t) \tag{3.7.51}$$

such that Equations (3.7.43) transform to

$$\frac{dQ_i}{dt} = \frac{\partial \mathscr{H}}{\partial P_i}; \qquad i = 1, \ldots, n \tag{3.7.52a}$$

$$\frac{dP_i}{dt} = -\frac{\partial \mathscr{H}}{\partial Q_i}; \qquad i = 1, \ldots, n \tag{3.7.52b}$$

then the transformations (3.7.50) are said to be canonical.

The reader can easily verify that an arbitrary transformation of the form (3.7.50a, b) is not canonical.

Since Equations (3.7.43) are a consequence of the variational principle (3.7.49), and since a canonical transformation preserves the Hamiltonian form of the differential equations, it must be true that Equations (3.7.52) also result from the variational principle

$$\delta \int_{t_1}^{t_2} [\mathbf{P} \cdot \dot{\mathbf{Q}} - \mathscr{H}(\mathbf{Q}, \mathbf{P}, t)]dt = 0. \tag{3.7.53}$$

Subtracting Equation (3.7.49) from Equation (3.7.53) then implies that the difference of their integrands is the differential of some arbitrary function S. We will refer to S as the generating function and it may depend on the $4n + 1$ variable $\mathbf{q}, \mathbf{p}, \mathbf{Q}, \mathbf{P}$ and t. However, since a canonical transformation defines $2n$ relations, we consider generating functions depending only on $2n + 1$ variables consisting of t and any of the four possible combinations of pairs $(\vec{\mathbf{p}}, \vec{\mathbf{Q}})(\vec{\mathbf{q}}, \vec{\mathbf{P}})$, $(\mathbf{Q}, \mathbf{p})$ and $(\mathbf{p}, \mathbf{P})$ of "old" and "new" variables.

For example, if we select S to be a function $S_1(\mathbf{q}, \mathbf{Q}, t)$, we have as a consequence of subtracting Equations (3.7.49) and (3.7.53):

$$\sum_{i=1}^{n} p_i \dot{q}_i - H(\mathbf{q}, \mathbf{p}, t) - \sum_{i=1}^{n} P_i \dot{Q}_i + \mathscr{H}(\mathbf{Q}, \mathbf{P}, t) = \frac{d}{dt} S_1(\mathbf{q}, \mathbf{Q}, t)$$

$$= \sum_{i=1}^{n} \frac{\partial S_1}{\partial q_i} \dot{q}_i + \sum_{i=1}^{n} \frac{\partial S_1}{\partial Q_i} \dot{Q}_i + \frac{\partial S_1}{\partial t}. \tag{3.7.54}$$

It then follows that we must set

$$p_i = \frac{\partial S_1}{\partial q_i}; \qquad i = 1, \ldots, n \tag{3.7.55a}$$

$$P_i = -\frac{\partial S_1}{\partial Q_i}; \qquad i = 1, \ldots, n \tag{3.7.55b}$$

$$\mathscr{H} = H + \frac{\partial S_1}{\partial t}. \tag{3.7.55c}$$

To show how Equations (3.7.55) generate a canonical transformation once an S_1 is chosen, consider first the n equations defined by (3.7.55a). We solve these n equations for the n new coordinates Q_i, defining these in the form of Equations (3.7.50a). When this result is used in the right hand sides of Equations (3.7.55b) we obtain the new momenta P_i in the form of Equations (3.7.50b). Finally, Equation (3.7.55c) can be used to calculate $\mathscr{H}(\mathbf{Q}, \mathbf{P}, t)$ by substitution, i.e.,

$$\mathscr{H}(\mathbf{Q}, \mathbf{P}, t) = H[\mathbf{q}(\mathbf{Q}, \mathbf{P}, t), \mathbf{p}(\mathbf{Q}, \mathbf{P}, t), t]$$

$$+ \frac{\partial S_1}{\partial t} [\mathbf{q}(\mathbf{Q}, \mathbf{P}, t), \mathbf{Q}, t]. \tag{3.7.56}$$

Note that $\partial S_1/\partial t$ is evaluated with its first two arguments fixed and then $\mathbf{q}$ is expressed in terms of $\mathbf{q}, \mathbf{P}, t$. Thus, if $\partial S_1/\partial t = 0$, the time will also not occur in the canonical transformations (3.7.50). Moreover, if H is time independent then so is $\mathscr{H}$ for all transformations with $\partial S_1/\partial t = 0$.

It is easy to show that corresponding to each of the remaining three choices of generating functions, there result analogous transformation relations. For example with the choice of a generating function $S_2(\mathbf{q}, \mathbf{P}, t)$ we derive

$$p_i = \frac{\partial S_2}{\partial q_i}; \qquad i = 1, \ldots, n \tag{3.7.57a}$$

$$Q_i = \frac{\partial S_2}{\partial P_i}; \qquad i = 1, \ldots, n \tag{3.7.57b}$$

$$\mathscr{H} = H + \frac{\partial S_2}{\partial t}, \tag{3.7.57c}$$

etc. We will use the generating function S_2 in developing the von Zeipel method.

The fundamental role of canonical transformations is exhibited by the following observation. If we can find a canonical transformation generated by a function, say, $S_2(\mathbf{q}, \mathbf{P}, t)$ and such that $\mathscr{H}$ vanishes identically, then the Hamilton differential Equations (3.7.52) have zero right hand sides, and thus, the new coordinates and momenta are *constants*. In this case, the transformation relations (3.7.50c, d) define the solution of the original system in the form of $2n$ functions of time involving $2n$ arbitrary constants $Q_1, \ldots, Q_n, P_1, \ldots, P_n$. This is the general solution of the system of Equations (3.7.43).

If we now ask how one would go about finding such an S_2, we see immediately from Equations (3.7.57) that S_2 must be a solution of the first order partial differential equation

$$\frac{\partial S_2}{\partial t} + H\left(q_1, \ldots, q_n, \frac{\partial S_2}{\partial q_1}, \ldots, \frac{\partial S_2}{\partial q_n}, t\right) = 0. \tag{3.7.58}$$

This is known as the Hamilton-Jacobi equation and solution techniques for this and similar nonlinear first order equations are discussed in standard texts on partial differential equations, e.g., Reference 3.7.8 or dynamics, e.g., Reference 3.7.7.

A review of the main results of Hamilton-Jacobi theory would cause too much of a digression here, particularly since we will only be concerned with time independent Hamiltonians for which a parallel theory is appropriate.

We now show that one can always formally eliminate the time from H by regarding t as the $n + 1$st variable and introducing an appropriate P_{n+1}. In fact, consider the following generating function S_2 depending on the n coordinates q_i, the n new momenta P_i, the time t and the $n + 1$st new momentum P_{n+1}:

$$S_2 = \sum_{i=1}^{n} q_i P_i + P_{n+1} t. \tag{3.7.59}$$

The series $\sum_{i=1}^{n} q_i P_i$ generates the identity transformation and applying Equations (3.7.57) gives

$$p_i = P_i; \qquad i = 1, \ldots, n \tag{3.7.60a}$$

$$Q_i = q_i; \qquad i = 1, \ldots, n \tag{3.7.60b}$$

$$Q_{n+1} = t \tag{3.7.60c}$$

$$\mathscr{H} = H(Q_1, \ldots, Q_n, P_1, \ldots, P_n, t) + P_{n+1}. \tag{3.7.60d}$$

Consequently given a Hamiltonian H depending on the $2n$ coordinates and momenta q_i, p_i and t, we can formally associate with it a time independent Hamiltonian $\mathscr{H}$ of the $m = n + 1$ coordinates $q_1, \ldots, q_n$, t, and $m = n + 1$ momenta $p_1, \ldots, p_n, p_{n+1}$.

Thus, there is no loss of generality in studying time independent Hamiltonians. For such a Hamiltonian assume that we have found a canonical transformation which renders $\mathscr{H}$ independent of the Q_i. It then follows from Equations (3.7.52b) that the new momenta P_i are constants, and since the $\partial\mathscr{H}/\partial P_i$ depend on the P_i alone these also are constants, say

$$\frac{\partial \mathscr{H}}{\partial P_i} = v_i = \text{const.}, \qquad i = 1, \ldots, n. \tag{3.7.61}$$

Equation (3.7.52a) then implies that the solution for the Q_i is simply

$$Q_i = v_i t + Q_i(0), \tag{3.7.62}$$

where the $Q_i(0)$ are also constants.

So in this case the general solution of Equations (3.7.43) for the q_i and p_i in terms of t and the $2n$ constants follows immediately from the transformation relations (3.7.50c, d).

Also, we note that, in general, for any $H(\mathbf{q}, \mathbf{p}, t)$

$$\frac{dH}{dt} = \sum_{i=1}^{n} \left(\frac{\partial H}{\partial q_i} \dot{q}_i + \frac{\partial H}{\partial p_i} \dot{p}_i \right) + \frac{\partial H}{\partial t} \tag{3.7.63a}$$

and using Equations (3.7.43) to eliminate $\dot{q}_i$ and $\dot{p}_i$, we find

$$\frac{dH}{dt} = \frac{\partial H}{\partial t}. \tag{3.7.63b}$$

Now if $\partial H/\partial t = 0$ it follows that H is an integral of the system (3.7.43), say

$$H(\mathbf{q}, \mathbf{p}) = \alpha_1 = \text{const.} \tag{3.7.64}$$

If we use the formulas (3.7.57) to characterize the generating function $W(\mathbf{q}, \mathbf{P})$ of a transformation from $H(\vec{\mathbf{q}}, \vec{\mathbf{p}})$ to $\mathscr{H}(\mathbf{Q}, \mathbf{P})$ such that $\mathscr{H}$ is independent of the Q_i, we find that W obeys

$$H\left(q_1, q_2, \ldots, q_n, \frac{\partial W}{\partial q_1}, \frac{\partial W}{\partial q_2}, \ldots, \frac{\partial W}{\partial q_n}\right) = \alpha_1. \tag{3.7.65}$$

This is the time-independent form of the Hamilton-Jacobi equation; it also follows from Equation (3.7.58), for time-independent Hamiltonians, from the substitution $S_2 = W - \alpha_1 t$ [cf. the discussion following Equation (3.7.83)].

The general solution of Equation (3.7.65) is the so called complete integral in the form[22]

$$W = W(\mathbf{q}, \alpha_1, \alpha_2, \ldots, \alpha_{n-1}) + \alpha_n \tag{3.7.66}$$

involving the n constants, $\alpha_1, \alpha_2, \ldots, \alpha_n$. The new momenta P_i can be chosen as any n arbitrary functions of the n constants α_i.

Finding the solution (3.7.66) is particularly simple if the Hamiltonian function H is separable for the given choice of variables in the following sense. Assume a solution of Equations (3.7.65) for W in the form

$$W = \sum_{i=1}^{n} W_i(q_i, \alpha_1, \ldots, \alpha_n), \tag{3.7.67}$$

where each W_i involves only the one coordinate q_i. If substitution of Equation (3.7.67) into Equation (3.7.65) decomposes the latter into a sequence of n ordinary differential equations of the form

$$H_i\left(q_i, \frac{\partial W_i}{\partial q_i}, \alpha_1, \ldots, \alpha_n\right) = \alpha_i, \qquad i = 1, \ldots, n, \tag{3.7.68}$$

then these can be individually solved by quadrature. Note that each H_i only depends on one q_i and the corresponding $\partial W_i/\partial q_i$. An example of a separable system is the motion of a particle in a central force field, and is discussed in Reference 3.7.7.

Whether a given Hamiltonian is separable in a given set of coordinates is a straightforward question to answer. However, the basic question of whether there exist appropriate coordinates with respect to which a given Hamiltonian system is separable is more difficult and in general not possible to answer definitively. Since separability of the Hamiltonian implies the explicit solvability of Equations (3.7.43), the answer to the question of separability is quite important. For systems which are close to a sólvable one, we will attempt to derive successively more accurate canonical transformations to variables such that the transformed Hamiltonian is independent of the coordinates (hence solvable) to any desired degree of accuracy. This, in essence, is the von Zeipel procedure.

In many practical applications, particularly in celestial mechanics, the unperturbed system is periodic and is described most concisely by action and angle variables and we will establish what these are next.

Assume that we have a time independent Hamiltonian which is separable in some system of coordinates and momenta $\{q_i\}$, $\{p_i\}$. We call this system

[22] Note that since Equation (3.6.65) does not involve W explicitly we only need to find a solution involving the $n-1$ independent constants $\alpha_1, \ldots, \alpha_{n-1}$ then the nth constant is additive.

periodic, if the motion in *each* of the 2-dimensional planes $(q_i, p_i), i = 1, \ldots, n$ describes either a simple closed curve (libration) or corresponds to the p_i being a periodic function of q_i (rotation). Thus, for the case of libration both q_i and p_i must be periodic functions of t with the same period.

For the case of rotation, only p_i is periodic in t with some given period τ_i, while q_i has a secular component added to a τ_i-periodic function, i.e., $\dot{q}_i$ is τ_i-periodic in t with a non-zero average value. Note also that the periods involved in each (q_i, p_i) plane need not be the same.

Since the system is assumed to be separable, one can determine a priori what the projected motions in each q_i, p_i plane are without solving the problem.

In fact, using Equation (3.7.67) and (3.7.57a) with $S_2 = W$, and $\partial/\partial t = 0$ we have the following relations

$$p_i = \frac{\partial W_i}{\partial q_i}(q_i, \alpha_1, \ldots, \alpha_n); \qquad i = 1, \ldots, n \tag{3.7.69}$$

linking each pair (p_i, q_i) with the n constants $\alpha_1, \ldots, \alpha_n$. One could choose the new momenta P_i as n independents functions of the α_i. For this choice, Equations (3.7.57), with $S_2 = W$ and $\partial/\partial t = 0$, define a canonical transformation to a new Hamiltonian which does not involve the Q_i.

A particular choice of the P_i in terms of the α_i with some useful properties consists of letting $P_i = J_i$, the "action", defined by

$$J_i = \oint p_i \, dq_i; \qquad i = 1, \ldots, n, \tag{3.7.70}$$

where the integral is taken over one complete cycle in p_i, q_i plane for fixed values of $\alpha_1, \ldots, \alpha_n$. Thus, for the case of libration J_i is the area inside the closed (p_i, q_i) curve, while for the case of rotation it is simply the area for one period under the p_i vs. q_i curve. Using Equation (3.7.69) J_i can be computed by

$$J_i = \oint \frac{\partial W_i}{\partial q_i}(q_i, \alpha_1, \ldots, \alpha_n) dq_i; \qquad i = 1, \ldots, n \tag{3.7.71}$$

and this defines each J_i as a function of $\alpha_1, \ldots, \alpha_n$. The J_i as functions of $\alpha_1, \ldots, \alpha_n$ are independent and these functions can be inverted to express each α_i in terms of $J_1, \ldots, J_n$. Substituting this result into Equation (3.7.67) for W defines this in the form

$$W = W(\mathbf{q}, \mathbf{J}), \tag{3.7.72}$$

which is one of the standard forms [cf. Equations (3.7.57)] for a generating function. The new coordinates corresponding to the choice (3.7.70) are the "angle" variables w_i defined according to Equation (3.7.57b) by

$$w_i = \frac{\partial W}{\partial J_i}; \qquad i = 1, \ldots, n, \tag{3.7.73}$$

where Equation (3.7.72) is to be used in computing the partial derivatives.

The transformed Hamiltonian $\mathscr{H}$ is now a function of only the J_i,

$$\mathscr{H} = \mathscr{H}(\mathbf{J}). \quad (3.7.74)$$

Hence, Hamilton's equations (3.7.52a) for the w_i reduce to

$$\frac{dw_i}{dt} = \frac{\partial \mathscr{H}}{\partial J_i} \equiv v_i(\mathbf{J}); \qquad i = 1, \dots, n \quad (3.7.75a)$$

and can be solved to give

$$w_i(t) = v_i t + w_i(0); \qquad i = 1, \dots, n. \quad (3.7.75b)$$

If we denote by $\Delta_j w_i$ the change in a given w_i after the coordinate q_j has gone through a complete cycle (with the remaining coordinates held fixed), it is easy to show that $\Delta_j w_i = \delta_{ij}$ (Kronecker-delta). Thus, w_i changes by a unit amount only if q_i varies through a complete cycle; w_i returns to its original value if any of the other coordinates are varied. It is also easy to show that $\Delta_j W = J_j$ and this property of the action variable will be used in the next subsection.

It follows from $\Delta_i w_i = 1$ that the angle variables have unit period and that the v_i in Equation (3.7.75b) are the reciprocal periods.

We now illustrate the foregoing ideas with some simple examples. Consider first the one-dimensional linear oscillator defined in dimensionless form by

$$\frac{d^2q}{dt^2} + \omega^2 q = 0, \quad (3.7.76)$$

where $\omega = \text{const}$.

The kinetic energy is $T = \dot{q}^2/2$ and the potential energy is $V(q) = \omega^2 q^2/2$. Therefore, the Lagrangian $L \equiv T - V = \dot{q}^2/2 - \omega^2 q^2/2$. It is easily seen that Lagrange's equation (3.7.45) gives Equation (3.7.76). In order to cast Equation (3.7.76) into Hamiltonian form, we introduce the momentum p, by Equation (3.7.47)

$$p = \frac{\partial L}{\partial \dot{q}} = \dot{q}. \quad (3.7.77)$$

Therefore, the Hamiltonian, defined by Equation (3.7.48), is

$$H = p\dot{q} - L = p^2 - p^2/2 + \omega^2 q^2/2 = p^2/2 + \omega^2 q^2/2 = H(q, p). \quad (3.7.78)$$

Hamilton's equation (3.7.43) are

$$\dot{q} = \partial H/\partial p = p \quad (3.7.79a)$$

$$\dot{p} = -\partial H/\partial q = -\omega^2 q \quad (3.7.79b)$$

and these are clearly equivalent to Equation (3.7.76).

In this example, since H is independent of t, $dH/dt = 0$, [cf. Equation (3.7.63b)] and the constant value of H is simply the total energy. The reader can also verify that all the above results up to Equations (3.7.79) remain true if ω depends on t. In this case, however, $dH/dt = \partial H/\partial t = \omega\dot{\omega}q^2 \neq 0$. We will consider the case of slowly varying ω in Section 3.7.3.

As a second example consider the motion in three dimensions of a particle in the force field of Newtonian center of gravitation. Using suitable dimensionless variables we can express the kinetic energy in terms of the spherical polar coordinates as follows

$$T = \tfrac{1}{2}(\dot{r}^2 + r^2\dot{\phi}^2 + r^2\dot{\theta}^2 \sin^2 \phi) \tag{3.7.80}$$

and thc potential energy is simply

$$V(r) = -\frac{1}{r}. \tag{3.7.81}$$

Letting $q_1 = r$, $q_2 = \phi$, $q_3 = \theta$, we calculate the momenta p_1, p_2, p_3 from the Lagrangian $L = T - V$ in the form

$$p_1 = \frac{\partial L}{\partial \dot{q}_1} = \dot{q}_1 \tag{3.7.82a}$$

$$p_2 = \frac{\partial L}{\partial \dot{q}_2} = q_1^2\dot{q}_2 \tag{3.7.82b}$$

$$p_3 = \frac{\partial L}{\partial \dot{q}_3} = q_1^2\dot{q}_3 \sin^2 q_2.$$

It then follows that the Hamiltonian is

$$H = \tfrac{1}{2}[p_1^2 + p_2^2/q_1^2 + p_3^2/q_1^2 \sin^2 q_2] - 1/q_1 \tag{3.7.83}$$

and Hamilton's equations, which we will not write down, reduce to the familiar equations of motion.

To illustrate some of the ideas of canonical transformations, let us return to Equation (3.7.78) for the oscillator and seek a canonical transformation to a new Hamiltonian $\mathscr{H}$ which is identically zero. The transformation S which effects this obeys [cf. Equation (3.7.58)]

$$\frac{\partial S}{\partial t} + \frac{1}{2}\left(\frac{\partial S}{\partial q}\right)^2 + \frac{\omega^2}{2} q^2 = 0. \tag{3.7.84}$$

Let us assume that $S(q, t)$ can be solved in the separated form

$$S(q, t) = S_1(q) + S_2(t). \tag{3.7.85}$$

Substituting Equation (3.7.85) into Equation (3.7.84) gives

$$\frac{1}{2}\left(\frac{\partial S_1}{\partial q}\right)^2 + \frac{\omega^2}{2} q^2 = -\frac{\partial S_2}{\partial t}. \tag{3.7.86}$$

Therefore, both sides of Equation (3.7.86) equal a constant, say α_1, and we have, setting $S_1 = W$,

$$S = W(q, \alpha_1) - \alpha_1 t + \alpha_2, \tag{3.7.87}$$

where the additive constant α_2 may be set equal to zero with no loss of generality. Now W obeys

$$\frac{1}{2}\left(\frac{\partial W}{\partial q}\right)^2 + \frac{\omega^2}{2} q^2 = \alpha_1 \tag{3.7.88}$$

which is the result we would have obtained had we started from the time independent Hamilton-Jacobi equation (3.7.68) for this example.

The solution of Equation (3.7.88) for W is

$$\begin{aligned} W &= \int^q [2\alpha_1 - \omega^2 q'^2]^{1/2}\, dq' \\ &= \frac{\alpha_1}{\omega}\left[\sin^{-1}\frac{\omega q}{\sqrt{2\alpha_1}} + \frac{\omega q}{\sqrt{2\alpha_1}}\left(1 - \frac{\omega^2 q^2}{2\alpha_1}\right)^{1/2}\right] \end{aligned} \tag{3.7.89}$$

$$S = \int^q [2\alpha_1 - \omega^2 q'^2]^{1/2} dq' - \alpha_1 t. \tag{3.7.90}$$

If we regard α_1 as a new momentum, $\alpha_1 = P$. The transformation relations (3.7.57) give

$$Q = \frac{\partial S}{\partial P} = \int^q [2P - \omega^2 q'^2]^{-1/2}\, dq' - t, \tag{3.7.91}$$

where Q is a constant.

The integral in Equation (3.7.91) can be evaluated, and we find

$$Q + t = \frac{1}{\omega}\cos^{-1}\frac{q\omega}{\sqrt{2P}}. \tag{3.7.92a}$$

Solving Equation (3.7.92a) for q defines this quantity in terms of t and the two constants P, Q

$$q = \frac{\sqrt{2P}}{\omega}\cos\omega(t + Q). \tag{3.7.92b}$$

We can use Equation (3.7.57a) and Equation (3.7.92) to evaluate the constants P and Q in terms of the initial values $q_0 = q(0)$ and $p_0 = p(0)$ in the form

$$Q = \frac{1}{\omega}\cos^{-1}[q_0\omega/(p_0^2 + \omega^2 q_0^2)^{1/2}] \tag{3.7.93a}$$

$$P = \tfrac{1}{2}(p_0^2 + \omega^2 q_0^2). \tag{3.7.93b}$$

Thus, the new coordinate Q is the phase and the new momentum P is the energy, the two natural constants of integration for the oscillator equation (3.7.76).

Instead of α_1, we may choose any function of α_1 as a new momentum. The action J is one such choice. A direct derivation of J follows from the observation that Equation (3.7.78) with $\omega =$ const. and $H = \alpha_1$ represents a one-parameter family of ellipses in the q, p plane. We write Equation (3.7.78) in the form

$$\frac{p^2}{(\sqrt{2\alpha_1})^2} + \frac{q^2}{(\sqrt{2\alpha_1}/\omega)^2} = 1. \tag{3.7.94}$$

Since the area of the ellipse $(p/a)^2 + (q/b)^2 = 1$ is πab, we find

$$J = \pi(\sqrt{2\alpha_1})(\sqrt{2\alpha_1}/\omega) = 2\pi\alpha_1/\omega = \pi(p^2 + \omega^2 q^2)/\omega. \tag{3.7.95}$$

To calculate w, we express W in terms of J as follows

$$W = \frac{J}{2\pi}\left[\sin^{-1}\left(\frac{\pi\omega}{J}\right)^{1/2} q + \left(\frac{\pi\omega}{J}\right)^{1/2} q\left(1 - \frac{\pi\omega q^2}{J}\right)^{1/2}\right]. \tag{3.7.96}$$

Hence

$$w = \frac{\partial W}{\partial J} = \frac{1}{2\pi}\sin^{-1}\left(\frac{\pi\omega}{J}\right)^{1/2} q = \frac{1}{2\pi}\tan^{-1}\frac{\omega q}{p} \tag{3.7.97}$$

and the transformed Hamiltonian becomes

$$\mathscr{H} = \omega J/2\pi. \tag{3.7.98}$$

Assume now that instead of the harmonic oscillator we have the perturbed equation

$$\frac{d^2 q}{dt^2} + q + \varepsilon f(q, t; \varepsilon) = 0, \tag{3.7.99}$$

where f is 2π-periodic in t. Note that this is a special case of Equation (3.7.11) in that f does not depend on $\dot{q}$.

Equation (3.7.99) is derivable from the time dependent Hamiltonian

$$H_1(q, p, t; \varepsilon) = \tfrac{1}{2}(p^2 + q^2) + \varepsilon g(q, t; \varepsilon) = E(t), \tag{3.7.100}$$

where

$$p = \dot{q} \tag{3.7.101a}$$

$$g = \int_0^q f(y, t; \varepsilon)dy \tag{3.7.101b}$$

and g is also 2π-periodic in t.

We now transform the problem to a two-dimensional one by regarding the time as a second coordinate and introducing its conjugate momentum, [cf. Equation (3.7.59)]. The transformed Hamiltonian H_2 will then be formally time independent but two-dimensional:

$$H_2(q_1, q_2, p_1, p_2; \varepsilon) = \tfrac{1}{2}(p_1^2 + q_1^2) + \varepsilon g(q_1, q_2; \varepsilon) + p_2 = C = \text{const.}, \tag{3.7.102}$$

where

$$q_1 = q; \qquad p_1 = p; \qquad q_2 = t; \qquad p_2 = C - E(t) \tag{3.7.103}$$

and C is an arbitrary constant.

We next introduce the canonical transformation to action and angle variables in the (q_1, p_1) and (q_2, p_2) planes by [cf. Equation (3.7.95) and (3.7.97)]:

$$\tilde{J}_1 = \pi(q_1^2 + p_1^2) \tag{3.7.104a}$$

$$\tilde{w}_1 = \frac{1}{2\pi}\tan^{-1}(q_1/p_1) \tag{3.7.104b}$$

$$\tilde{J}_2 = 2\pi p_2 \tag{3.7.104c}$$

$$\tilde{w}_2 = q_2/2\pi. \tag{3.7.104d}$$

The last two equations follow from the definitions (3.7.70) and (3.7.73) by noting that the motion in the q_2, p_2 plane consists of $p_2 = \text{const.}$ which is a rotation.

The Hamiltonian associated with the new variables is

$$\begin{aligned}\tilde{H}(\tilde{w}_1, \tilde{w}_2, \tilde{J}_1, \tilde{J}_2; \varepsilon) &= (\tilde{J}_1 + \tilde{J}_2)/2\pi \\ &\quad + \varepsilon g[(\tilde{J}_1/\pi)^{1/2} \sin 2\pi\tilde{w}_1, 2\pi\tilde{w}_2; \varepsilon].\end{aligned} \tag{3.7.105}$$

With $\varepsilon = 0$, $\tilde{w}_1$ and $\tilde{w}_2$ have the same frequency $1/2\pi$ and we remove this degeneracy with the canonical transformation

$$w_1 = \tilde{w}_1 - \tilde{w}_2; \qquad J_1 = \tilde{J}_1 \tag{3.7.106a}$$

$$w_2 = \tilde{w}_2; \qquad J_2 = \tilde{J}_1 + \tilde{J}_2. \tag{3.7.106b}$$

The Hamiltonian now becomes

$$H(w_1, w_2, J_1, J_2; \varepsilon) = J_2/2\pi + \varepsilon g[(J_1/\pi)^{1/2} \sin 2\pi(w_1 + w_2), 2\pi w_2; \varepsilon] \tag{3.7.107}$$

and this is a special case of a two-dimensional Hamiltonian that we will consider later on in this section. Note that the Hamiltonian to $O(1)$ involves only the action J_2 for which the corresponding frequency is non-zero.

The interested reader can find a discussion in Reference 3.7.7 of how the Hamiltonian (3.7.83) for motion about a gravitational center can be transformed to action-angle variables. As above, the Hamiltonian in terms of the final set of variables involves only one action variable corresponding to a nonzero frequency.

These variables are known as the "Delaunay elements" and are often used in satellite problems in cases where the perturbation terms are derivable from a Hamiltonian. The resulting Hamiltonian is then of the form

$$H(w_1, w_2, w_3, J_1, J_2, J_3; \varepsilon) = -\frac{2\pi^2}{J_2^2} + \varepsilon F(w_1, w_2, w_3, J_1, J_2, J_3, \varepsilon),$$

where $-2\pi^2/J_2^2$ is the Hamiltonian corresponding to the $-1/r^2$ force field in Delaunay elements, and F represents perturbations to this force field. For further discussion of Delaunay elements and examples of perturbed satellite problems using canonical transformations, we refer the reader to Reference 3.7.5.

(b) *The von Zeipel Method*[23]

We will only consider two dimensional Hamiltonians of the form

$$H(w_1, w_2, J_1, J_2; \varepsilon) = J_2/2\pi + \varepsilon g(w_1, w_2, J_1; \varepsilon) \tag{3.7.108}$$

where w_1 and w_2 are angle variables conjugate to the action variables J_1 and J_2. The function g is periodic with respect to w_1 and w_2 with unit period, and ε is a small positive parameter. Thus, the perturbed oscillator problem of Equations (3.7.99), (3.7.107) is a special case of Equation (3.7.108). An example of a more general two dimensional Hamiltonian is considered in Problem (3.7.5). The corresponding problem with slowly varying frequencies is studied in Problem 3.7.9.

Since g is periodic in both w_1 and w_2, it may be uniquely decomposed into two functions

$$g = \sigma(w_1, w_2, J_1; \varepsilon) + \lambda(w_1, J_1; \varepsilon), \tag{3.7.109}$$

where σ is periodic in w_1 and w_2 with unit period and has a zero average value over one period of w_2. We will use the notation

$$\langle\sigma\rangle \equiv \int_0^1 \sigma(w_1, w_2, J_1; \varepsilon)dw_2 \tag{3.7.110}$$

for averages over one period in w_2. Thus, $\langle\sigma\rangle = 0$.

In view of the fact that Hamilton's differential equations for Equation (3.7.108) imply that [cf. Equations (3.7.43)] $w_2 = t/2\pi + \text{const.}$, and $w_1 = O(\varepsilon)$, we denote w_2 and w_1 as the short and long period variables respectively. Thus, σ represents short period perturbation while λ represents long period ones.

The von Zeipel procedure consists of the transformation of H to any given order in ε to a form free of the angle variables. In this case, only one variable

[23] The following discussion is taken from Reference 3.7.9.

need be removed to reduce the solution to quadrature. In general, however, one can, in principle, eliminate all the angle variables simultaneously by an appropriate canonical transformation. As a result, the transformed action variables are constants to the order of accuracy retained.

We denote the new variables by an asterisk and introduce a generating function $S(w_1, w_2, J_1^*. J_2^*; \varepsilon)$ which depends on the original angle variables w_1, w_2 and new action variables J_1^*, J_2^*, in order to transform H to a form free of w_2^*. In view of the fact that the unperturbed Hamiltonian is already in the proper form, i.e., free of w_1 and w_2, S must reduce to the generating function for the identity transformation when $\varepsilon \to 0$, and we assume that it may be developed asymptotically in powers of ε as follows:

$$S = w_1 J_1^* + w_2 J_2^* + \varepsilon S_1 + \varepsilon^2 S_2 + O(\varepsilon^3), \tag{3.7.111}$$

where the S_i do not depend on ε.

The new variables are related to the original according to [cf. Equations (3.7.57) with $\partial/\partial t = 0$]

$$J_i = \frac{\partial S}{\partial w_i} = J_i^* + \varepsilon \frac{\partial S_1}{\partial w_i} + \varepsilon^2 \frac{\partial S_2}{\partial w_i} + O(\varepsilon^3), \tag{3.7.112a}$$

$$w_i^* = \frac{\partial S}{\partial J_i^*} = w_i + \varepsilon \frac{\partial S_1}{\partial J_i^*} + \varepsilon^2 \frac{\partial S_2}{\partial J_i^*} + O(\varepsilon^3), \tag{3.7.112b}$$

and the new Hamiltonian H^* is obtained from the original Hamiltonian by substituting for the original variables in H the transformed values involving starred quantities. Before considering the conditions which define the S_i explicitly, it is useful to develop the detailed relationship between the original and new variables correct to $O(\varepsilon^2)$, and subject to the transformation formulas in Equations (3.7.112).

When the following expansions:

$$J_i = J_i^* + \varepsilon F_i^{(1)} + \varepsilon^2 F_i^{(2)} + O(\varepsilon^3), \tag{3.7.113a}$$

$$w_i = w_i^* + \varepsilon G_i^{(1)} + \varepsilon^2 G_i^{(2)} + O(\varepsilon^3), \tag{3.7.113b}$$

with the $F_i^{(j)}$ and $G_i^{(j)}$ depending upon w_1^*, w_2^*, J_1^*, and J_2^*, are substituted into Equations (3.7.112) and terms of equal orders are identified, one finds

$$F_i^{(1)} = \partial S_1/\partial w_i, \tag{3.7.114a}$$

$$G_i^{(1)} = -\partial S_1/\partial J_i^*, \tag{3.7.114b}$$

$$F_i^{(2)} = -\frac{\partial^2 S_1}{\partial w_i\, \partial w_1} \frac{\partial S_1}{\partial J_1^*} - \frac{\partial^2 S_1}{\partial w_i\, \partial w_2} \frac{\partial S_1}{\partial J_2^*} + \frac{\partial S_2}{\partial w_i}, \tag{3.7.114c}$$

$$G_i^{(2)} = \frac{\partial^2 S_1}{\partial J_i^*\, \partial w_1} \frac{\partial S_1}{\partial J_1^*} + \frac{\partial S_1}{\partial J_i^*\, \partial w_2} \frac{\partial S_1}{\partial J_2^*} - \frac{\partial S_2}{\partial J_i^*}, \tag{3.7.114d}$$

where all the partial derivatives appearing on the right-hand sides are evaluated at $w_i = w_i^*$.

Use of Equations (3.7.112) in Equation (3.7.108) and the ordering of the results according to powers of ε yields the following expression for the new Hamiltonian H^*.

$$H^* = H_0^* + \varepsilon H_1^* + \varepsilon^2 H_2^* + O(\varepsilon^3), \tag{3.7.115a}$$

$$H_0^* = J_2^*/2\pi, \tag{3.7.115b}$$

$$H_1^* = (1/2\pi)(\partial S_1/\partial w_2) + \sigma + \lambda, \tag{3.7.115c}$$

$$\begin{aligned} H_2^* = {} & \frac{1}{2\pi}\left(\frac{\partial S_2}{\partial w_2} - \frac{\partial^2 S_1}{\partial w_1\,\partial w_2}\frac{\partial S_1}{\partial J_1^*} - \frac{\partial^2 S_1}{\partial w_2^2}\frac{\partial S_1}{\partial J_2^*}\right) \\ & - \frac{\partial\sigma}{\partial w_1}\frac{\partial S_1}{\partial J_1^*} - \frac{\partial\sigma}{\partial w_2}\frac{\partial S_1}{\partial J_2^*} + \frac{\partial\sigma}{\partial J_1}\frac{\partial S_1}{\partial w_1} \\ & - \frac{\partial\lambda}{\partial w_1}\frac{\partial S_1}{\partial J_1^*} + \frac{\partial\lambda}{\partial J_1}\frac{\partial S_1}{\partial w_1} + \frac{\partial\sigma}{\partial\varepsilon} + \frac{\partial\lambda}{\partial\varepsilon}, \end{aligned} \tag{3.7.115d}$$

where all the terms on the right-hand sides in the above are evaluated at $w_i = w_i^*$, $J_i = J_i^*$, and $\varepsilon = 0$.

Clearly, the requirement that H_1^* be independent of w_2^* is accomplished by setting

$$(1/2\pi)(\partial S_1(w_1, w_2, J_1^*, J_2^*)/\partial w_2) + \sigma(w_1, w_2, J_1^*; 0) = 0, \tag{3.7.116a}$$

in which case H_1^* is simply

$$H_1^* = \lambda(w_1^*, J_1^*; 0). \tag{3.7.116b}$$

Equation (3.7.116a) defines S_1 to within an arbitrary long periodic term in the form

$$S_1 = \hat{S}_1(w_1, w_2, J_1^*) + S_1^*(w_1, J_1^*, J_2^*), \tag{3.7.117a}$$

where

$$\hat{S}_1 = -2\pi\int_a^{w_2}\sigma(w_1, \eta, J_1^*; 0)\,d\eta, \tag{3.7.117b}$$

a is an arbitrary function of w_1 and J_1^*, and S_1^* is arbitrary. We note that since $\langle\sigma\rangle = 0$, it is possible to choose the lower limit a in Equation (3.7.117b) such that $\langle S_1\rangle = 0$ too. Use of Equations (3.7.116) in Equation (3.7.115d) simplifies the expression for H_2^* to

$$\begin{aligned} H_2^* = {} & \frac{1}{2\pi}\frac{\partial S_2}{\partial w_2} + \frac{\partial\sigma}{\partial J_1}\frac{\partial S_1}{\partial w_1} - \frac{\partial\lambda}{\partial w_1}\frac{\partial S_1}{\partial J_1^*} \\ & + \frac{\partial\lambda}{\partial J_1}\frac{\partial S_1}{\partial w_1} + \frac{\partial\sigma}{\partial\varepsilon} + \frac{\partial\lambda}{\partial\varepsilon}. \end{aligned} \tag{3.7.118}$$

This result could perhaps have been derived more directly by expressing the new Hamiltonian, which involves J_1^*, J_2^*, and w_1^*, in terms of J_1^*, J_2^*, w_1, and w_2 and identifying this result with the original Hamiltonian [cf. Reference 3.7.6)].

Removal of the short-periodic terms from H_2^* requires that

$$\frac{1}{2\pi}\frac{\partial S_2}{\partial w_2} + \left(\frac{\partial \sigma}{\partial J_1}\frac{\partial \hat{S}_1}{\partial w_1} - \left\langle \frac{\partial \sigma}{\partial J_1}\frac{\partial \hat{S}_1}{\partial w_1}\right\rangle\right) + \frac{\partial \sigma}{\partial J_1}\frac{\partial S_1^*}{\partial w_1}$$
$$- \frac{\partial \lambda}{\partial w_1}\frac{\partial \hat{S}_1}{\partial J_1^*} + \frac{\partial \lambda}{\partial J_1}\frac{\partial \hat{S}_1}{\partial w_1} + \frac{\partial \sigma}{\partial \varepsilon} = 0, \tag{3.7.119}$$

which in turn defines S_2 by quadrature to within an arbitrary long-period function S_2^*. The role of these arbitrary long-period functions is discussed in Reference (3.7.6) in detail. For the purpose of the present comparison, the choice of these functions is immaterial as may be seen subsequently. (However, we only consider the case $\partial S_i^*/\partial J_2^* = 0$, which implies that $w_2 = w_2^*$.) We note in passing that the judicious choice of S_1^* could simplify the calculation of S_2, etc.

The second-order term in the transformed Hamiltonian now becomes

$$H_2^* = \left\langle \frac{\partial \sigma}{\partial J_1}\frac{\partial \hat{S}_1}{\partial w_1}\right\rangle - \frac{\partial \lambda}{\partial w_1}\frac{\partial S_1^*}{\partial J_1^*} + \frac{\partial \lambda}{\partial J_1}\frac{\partial S_1^*}{\partial w_1} + \frac{\partial \lambda}{\partial \varepsilon}. \tag{3.7.120}$$

This is as far as the von Zeipel procedure is carried out.

The new Hamiltonian for this simple case can now be solved by quadrature, as shown presently. In general, the entire procedure may be repeated at this, or a subsequent, stage in order to eliminate another variable or group of variables with the next order of magnitude in period. Herein lies an advantage and the elegance of the von Zeipel method. One has an almost automatic means for the isolation and analysis of the various time scales that are characteristic for a given Hamiltonian system.

With H^* now completely defined to $O(\varepsilon^2)$ we find from Hamilton's differential equations for J_1^* and w_1^* that these quantities only depend on $\tilde{t} = \varepsilon t$ to $O(\varepsilon)$ since

$$\frac{dw_1^*}{d\tilde{t}} = \frac{\partial H_1^*}{\partial J_1^*} + \varepsilon\frac{\partial H_2^*}{\partial J_1^*} + O(\varepsilon^2), \tag{3.7.121a}$$

$$\frac{dJ_1^*}{d\tilde{t}} = -\frac{\partial H_1^*}{\partial w_1^*} - \varepsilon\frac{\partial H_2^*}{\partial w_1^*} + O(\varepsilon^2). \tag{3.7.121b}$$

Furthermore, since $H_1^* + \varepsilon H_2^*$ is also independent of J_2^*, the dependence on w_2^* and J_2^* is completely isolated. The Hamiltonian $H_1^* + \varepsilon H_2^*$ is an integral to $O(\varepsilon)$ of the system (3.7.121), thus reducing the solution for w_1^* and J_1^* in terms of $\tilde{t}$ from (3.7.121) to quadrature.

The solutions for w_2^* and J_2^* are simply

$$w_2^* = (t/2\pi) + \text{const.} + O(\varepsilon^3) \tag{3.7.122a}$$

$$J_2^* = \text{const.} + O(\varepsilon^3) \tag{3.7.122b}$$

and the above together with the solutions one could obtain for w_1^* and J_1^* in terms of $\tilde{t}$, represent a complete solution to the problem when used in Equations (3.7.113).

(c) *Two Variable Solution*

For the purposes of comparison of the methods and results let us develop the solution of Hamilton's differential equations

$$\frac{dw_i}{dt} = \partial H/\partial J_i; \qquad i = 1, 2 \tag{3.7.123a}$$

$$\frac{dJ_i}{dt} = -\partial H/\partial w_i; \qquad i = 1, 2 \tag{3.7.123b}$$

associated with the Hamiltonian H of Equation (3.7.108) by the multiple variable method.

For the problem at hand it is sufficient to use t itself as the "fast variable" and to choose $\tilde{t} = \varepsilon t$ as the slow variable. Moreover, since Equation (3.7.123a) with $i = 2$ implies that $w_2 = t/2\pi$ + constant, we need only expand w_1, J_1, and J_2 as functions of t and $\tilde{t}$ in the following form:

$$w_1 = \sum_{n=0}^{N} w_1^{(n)}(t, \tilde{t})\varepsilon^n + O(\varepsilon^{N+1}) \tag{3.7.124a}$$

$$J_i = \sum_{n=0}^{N} J_i^{(n)}(t, \tilde{t})\varepsilon^n + O(\varepsilon^{N+1}), \qquad i = 1, 2. \tag{3.7.124b}$$

With the assumed explicit dependence of the expressions for w_1 and J_i upon both t and $\tilde{t}$ in Equation (3.7.124a), derivatives with respect to t become

$$\begin{aligned}\frac{dw_1}{dt} &= \frac{\partial w_1}{\partial t} + \varepsilon\frac{\partial w_1}{\partial \tilde{t}} \\ &= \frac{\partial w_1^{(0)}}{\partial t} + \sum_{n=1}^{N}\left[\frac{\partial w_1^{(n)}}{\partial t} + \frac{\partial w_1^{(n-1)}}{\partial \tilde{t}}\right]\varepsilon^n + O(\varepsilon^{N+1})\end{aligned} \tag{3.7.125a}$$

$$\begin{aligned}\frac{dJ_i}{dt} &= \frac{\partial J_i}{\partial t} + \varepsilon\frac{\partial J_i}{\partial \tilde{t}} \\ &= \frac{\partial J_i^{(0)}}{\partial t} + \sum_{n=1}^{N}\left[\frac{\partial J_i^{(n)}}{\partial t} + \frac{\partial J_i^{(n-1)}}{\partial \tilde{t}}\right]\varepsilon^n + O(\varepsilon^{N+1}).\end{aligned} \tag{3.7.125b}$$

When the expressions given in Equations (3.7.124) and (3.7.125) are substituted into Equations (3.7.123) and the latter developed in powers of ε, the following sets of equations are obtained for the terms up to order ε^2.

$$\partial w_1^{(0)}/\partial t = 0, \tag{3.7.126a}$$

$$\partial J_1^{(0)}/\partial t = 0, \tag{3.7.126b}$$

$$\partial J_2^{(0)}/\partial t = 0, \tag{3.7.126c}$$

$$\frac{\partial w_1^{(1)}}{\partial t} = -\frac{\partial w_1^{(0)}}{\partial \tilde{t}} + \frac{\partial \lambda}{\partial J_1} + \frac{\partial \sigma}{\partial J_1}, \tag{3.7.127a}$$

$$\frac{\partial J_1^{(1)}}{\partial t} = -\frac{\partial J_1^{(0)}}{\partial \tilde{t}} - \frac{\partial \lambda}{\partial w_1} - \frac{\partial \sigma}{\partial w_1}, \tag{3.7.127b}$$

$$\frac{\partial J_2^{(1)}}{\partial t} = -\frac{\partial J_2^{(0)}}{\partial \tilde{t}} - \frac{\partial \sigma}{\partial w_2}; \tag{3.7.127c}$$

$$\frac{\partial w_1^{(2)}}{\partial t} = -\frac{\partial w_1^{(1)}}{\partial \tilde{t}} + \frac{\partial^2(\lambda + \sigma)}{\partial J_1\, \partial w_1} w_1^{(1)} + \frac{\partial^2(\lambda + \sigma)}{\partial J_1^2} J_1^{(1)} + \frac{\partial^2(\lambda + \sigma)}{\partial J_1\, \partial \varepsilon}, \tag{3.7.128a}$$

$$\frac{\partial J_1^{(2)}}{\partial t} = -\frac{\partial J_1^{(1)}}{\partial \tilde{t}} - \frac{\partial^2(\lambda + \sigma)}{\partial w_1^2} w_1^{(1)} - \frac{\partial^2(\lambda + \sigma)}{\partial J_1\, \partial w_1} J_1^{(1)} - \frac{\partial^2(\lambda + \sigma)}{\partial w_1\, \partial \varepsilon}, \tag{3.7.128b}$$

$$\frac{\partial J_2^{(2)}}{\partial t} = -\frac{\partial J_2^{(1)}}{\partial \tilde{t}} - \frac{\partial^2 \sigma}{\partial w_1\, \partial w_2} w_1^{(1)} - \frac{\partial^2 \sigma}{\partial J_1\, \partial w_2} J_1^{(1)} - \frac{\partial^2 \sigma}{\partial w_2\, \partial \varepsilon}. \tag{3.7.128c}$$

The terms on the right-hand sides of Equations (3.7.127) and (3.7.128) are evaluated at the following arguments:

$$w_1 = w_1^{(0)}, \tag{3.7.129a}$$

$$w_2 = t/2\pi, \tag{3.7.129b}$$

$$J_1 = J_1^{(0)}, \tag{3.7.129c}$$

$$J_2 = J_2^{(0)}, \tag{3.7.129d}$$

$$\varepsilon = 0, \tag{3.7.129e}$$

where in Equation (3.7.129b) and subsequently the constant of integration for w_2 has been set equal to zero with no loss of generality.

Equations (3.7.126) imply that $w_1^{(0)}$, $J_1^{(0)}$, and $J_2^{(0)}$ depend on $\tilde{t}$ alone and are so far arbitrary. We note that since λ does not depend on w_2, $\partial\lambda/\partial J_1$, in Equation (3.7.127a) and $\partial\lambda/\partial w_1$ in Equation (3.7.127b) do not involve t explicitly. Thus, unless we set

$$\frac{dw_1^{(0)}}{d\tilde{t}} = \frac{\partial\lambda}{\partial J_1} \tag{3.7.130a}$$

$$\frac{dJ_1^{(0)}}{d\tilde{t}} = -\frac{\partial\lambda}{\partial w_1} \tag{3.7.130b}$$

$$\frac{dJ_2^{(0)}}{d\tilde{t}} = 0 \tag{3.7.130c}$$

integration of Equations (3.7.127) with respect to t will lead to inconsistent terms proportional to t.

Equation (3.7.130c) implies that $J_2^{(0)}$ is a constant, and since $\lambda(w_1^{(0)}, J_1^{(0)}; 0) = \text{const.}$ is an integral of Equations (3.7.130a, b), these can be solved for $w_1^{(0)}$ and $J_1^{(0)}$ by quadrature. We denote these solutions by

$$w_1^{(0)} = \tilde{w}_1^{(0)}(\tilde{t}) \tag{3.7.131a}$$

$$J_1^{(0)} = \tilde{J}_1^{(0)}(\tilde{t}). \tag{3.7.131b}$$

Equations (3.7.127) may now be integrated with respect to t and give

$$\begin{aligned} w_1^{(1)} &= 2\pi \int_0^{w_2} \frac{\partial\sigma}{\partial J_1}(w_1^{(0)}, \eta, J_1^{(0)}; 0)d\eta + \tilde{w}_1^{(1)}(\tilde{t}) \\ &= -\frac{\partial\hat{S}_1}{\partial J_1^*}(w_1^{(0)}, w_2, J_1^{(0)}; 0) + \tilde{w}_1^{(1)}(\tilde{t}) \end{aligned} \tag{3.7.132a}$$

$$\begin{aligned} J_1^{(1)} &= -2\pi \int_0^{w_2} \frac{\partial\sigma}{\partial w_1}(w_1^{(0)}, \eta, J_1^{(0)}; 0)d\eta + \tilde{J}_1^{(1)}(\tilde{t}) \\ &= \frac{\partial\hat{S}_1}{\partial w_1}(w_1^{(0)}, w_2, J_1^{(0)}; 0) + \tilde{J}_1^{(1)}(\tilde{t}) \end{aligned} \tag{3.7.132b}$$

$$J_2^{(1)} = -2\pi\sigma(w_1^{(0)}, w_2, J_1^{(0)}; 0) + \tilde{J}_2^{(1)}(\tilde{t}), \tag{3.7.132c}$$

where $\tilde{w}_1^{(1)}$, $\tilde{J}_1^{(1)}$, and $\tilde{J}_2^{(1)}$ are functions of $\tilde{t}$ alone to be determined by requirements upon the terms of $O(\varepsilon^2)$ in Equations (3.7.128). Furthermore, the use of $\hat{S}_1$, in Equations (3.7.132a, b) is justified [cf. Equation (3.7.117b)] and will be elaborated later.

Before we consider Equations (3.7.128) it is useful to note the following identities from Equation (3.7.132) and (3.7.130):

$$\frac{\partial w_1^{(1)}}{\partial\tilde{t}} = \frac{d\tilde{w}_1^{(1)}}{d\tilde{t}} - \frac{\partial^2\hat{S}_1}{\partial J_1^* \partial w_1}\frac{\partial\lambda}{\partial J_1} + \frac{\partial^2\hat{S}_1}{\partial J_1^{*2}}\frac{\partial\lambda}{\partial w_1} \tag{3.7.133a}$$

$$\frac{\partial J_1^{(1)}}{\partial\tilde{t}} = \frac{d\tilde{J}_1^{(1)}}{d\tilde{t}} + \frac{\partial^2\hat{S}_1}{\partial w_1^2}\frac{\partial\lambda}{\partial J_1} - \frac{\partial^2\hat{S}_1}{\partial J_1^* \partial w_1}\frac{\partial\lambda}{\partial w_1} \tag{3.7.133b}$$

$$\frac{\partial J_2^{(1)}}{\partial\tilde{t}} = \frac{d\tilde{J}_2^{(1)}}{d\tilde{t}} - 2\pi\frac{\partial\sigma}{\partial w_1}\frac{\partial\lambda}{\partial J_1} + 2\pi\frac{\partial\sigma}{\partial J_1}\frac{\partial\lambda}{\partial w_1}, \tag{3.7.133c}$$

where the arguments of all the partial derivatives occurring above are evaluated according to Equations (3.7.129).

All the terms appearing on the right-hand sides of Equations (3.7.128) have now been defined. The consistency of the expansions to $O(\varepsilon^2)$ and their uniformity for times of order ε^{-1} requires the elimination of all the terms which are independent of w_2. It then follows from Equations (3.7.128) with the use of Equations (3.7.132) and (3.7.133) that we must set

$$\frac{d\tilde{w}_1^{(1)}}{d\tilde{t}} = \frac{\partial^2\lambda}{\partial J_1^2}\tilde{J}_1^{(1)} + \frac{\partial^2\lambda}{\partial J_1\,\partial w_1}\tilde{w}_1^{(1)} + \frac{\partial^2\lambda}{\partial J_1\,\partial\varepsilon} + \left\langle\frac{\partial^2\sigma}{\partial J_1^2}\frac{\partial\hat{S}_1}{\partial w_1} - \frac{\partial^2\sigma}{\partial J_1\,\partial w_1}\frac{\partial\hat{S}_1}{\partial J_1^*}\right\rangle, \tag{3.7.134a}$$

$$\frac{d\tilde{J}_1^{(1)}}{d\tilde{t}} = -\frac{\partial^2\lambda}{\partial w_1\,\partial J_1}\tilde{J}_1^{(1)} - \frac{\partial^2\lambda}{\partial w_1^2}\tilde{w}_1^{(1)} - \frac{\partial^2\lambda}{\partial w_1\,\partial\varepsilon} - \left\langle\frac{\partial^2\sigma}{\partial w_1\,\partial J_1}\frac{\partial\hat{S}_1}{\partial w_1} - \frac{\partial^2\sigma}{\partial w_1^2}\frac{\partial\hat{S}_1}{\partial J_1^*}\right\rangle, \tag{3.7.134b}$$

$$\frac{d\tilde{J}_2^{(1)}}{d\tilde{t}} = \left\langle\frac{\partial^2\sigma}{\partial w_1\,\partial w_2}\frac{\partial\hat{S}_1}{\partial J_1^*} - \frac{\partial^2\sigma}{\partial J_1\,\partial w_2}\frac{\partial\hat{S}_1}{\partial w_1}\right\rangle, \tag{3.7.134c}$$

where, again, the partial derivatives on the right-hand sides are all evaluated according to Equations (3.7.129).

The equations for $w_1^{(2)}$, $J_1^{(2)}$, and $J_2^{(2)}$, which are not given here, are the remainders of Equations (3.7.128) after cancellation of the long-period terms according to Equations (3.7.134). This is as far as the procedure is carried out. The pattern of the development is quite straightforward, but leads to rather tedious calculations if explicit results are desired to any higher order.

(d) *Comparison of the Results by the Two Methods*

We will find it useful to recast the solution in part (b) into an expansion analogous to Equations (3.7.124).

Actually, it is easier and more informative to derive expressions for the derivatives of the long-period terms given by the von Zeipel method and then compare these derivatives with their counterparts given in Equations (3.7.130) and (3.7.134) rather than attempting to convert the results of part (c) into the form of part (b). Furthermore, since the starred variables whose solutions are governed by Equations (3.7.121) involve ε implicitly, the transformations Equations (3.7.113), linking the starred and original variables must be reordered accordingly.

If we introduce the asymptotic expansions for the solutions of Equations (3.7.121) in the form

$$w_1^* = w_1^{*(0)}(\tilde{t}) + \varepsilon w_1^{*(1)}(\tilde{t}) + O(\varepsilon^2), \tag{3.7.135a}$$

$$J_1^* = J_1^{*(0)}(\tilde{t}) + J_1^{*(1)}(\tilde{t}) + O(\varepsilon^2), \tag{3.7.135b}$$

the following equations for the derivatives of the various terms on the right-hand side of Equations (3.7.135) follow directly from Equations (3.7.121):

$$dw_1^{*(0)}/d\tilde{t} = \partial\lambda/\partial J_1, \tag{3.7.136a}$$

$$dJ_1^{*(0)}/d\tilde{t} = -\partial\lambda/\partial w_1; \tag{3.7.136b}$$

$$\begin{aligned}\frac{dw_1^{*(1)}}{d\tilde{t}} &= \frac{\partial^2\lambda}{\partial J_1\,\partial w_1}\left(w_1^{*(1)} - \frac{\partial S_1^*}{\partial J_1^*}\right) + \frac{\partial^2\lambda}{\partial J_1^2}\left(J_1^{*(1)} + \frac{\partial S_1^*}{\partial w_1}\right)\\ &\quad + \left\langle\frac{\partial^2\sigma}{\partial J_1^2}\frac{\partial\hat{S}_1}{\partial w_1}\right\rangle + \left\langle\frac{\partial\sigma}{\partial J_1}\frac{\partial^2\hat{S}_1}{\partial w_1\,\partial J_1^*}\right\rangle - \frac{\partial\lambda}{\partial w_1}\frac{\partial^2 S_1^*}{\partial J_1^{*2}}\\ &\quad + \frac{\partial\lambda}{\partial J_1}\frac{\partial^2 S_1^*}{\partial w_1\,\partial J_1^*} + \frac{\partial^2\lambda}{\partial\varepsilon\,\partial J_1},\end{aligned} \tag{3.7.137a}$$

$$\begin{aligned}\frac{dJ_1^{*(1)}}{d\tilde{t}} &= -\frac{\partial^2\lambda}{\partial w_1^2}\left(w_1^{*(1)} - \frac{\partial S_1^*}{\partial J_1^*}\right) - \frac{\partial^2\lambda}{\partial w_1\,\partial J_1}\left(J_1^{*(1)} + \frac{\partial S_1^*}{\partial w_1}\right)\\ &\quad - \left\langle\frac{\partial^2\sigma}{\partial J_1\,\partial w_1}\frac{\partial\hat{S}_1}{\partial w_1}\right\rangle - \left\langle\frac{\partial\sigma}{\partial J_1}\frac{\partial^2\hat{S}_1}{\partial w_1^2}\right\rangle + \frac{\partial\lambda}{\partial w_1}\frac{\partial^2 S_1^*}{\partial J_1^*\,\partial w_1}\\ &\quad - \frac{\partial\lambda}{\partial J_1}\frac{\partial^2 S_1^*}{\partial w_1^2} - \frac{\partial^2\lambda}{\partial\varepsilon\,\partial w_1},\end{aligned} \tag{3.7.137b}$$

where all the partial derivatives on the right-hand sides in the above are evaluated at the following arguments:

$$w_1^* = w_1^{*(0)} \tag{3.7.138a}$$

$$J_1^* = J_1^{*(0)}, \tag{3.7.138b}$$

$$\varepsilon = 0. \tag{3.7.138c}$$

The exact correspondence of the long-periodic terms to order unity is immediately apparent by comparing Equations (3.7.136a, b) with Equations (3.7.130a, b) where we identify the following pairs of terms:

$$w_1^{(0)} = w_1^{*(0)}, \tag{3.7.139a}$$

$$J_1^{(0)} = J_1^{*(0)}. \tag{3.7.139b}$$

The constancy of $J_2^{(0)}$ according to Equation (3.7.130c) is also in agreement with Equation (3.7.122b) and we identify

$$J_2^{(0)} = \lim_{\varepsilon\to 0} J_2^*. \tag{3.7.139c}$$

We next consider the short-periodic contributions to $O(\varepsilon)$. According to the von Zeipel method [cf. Equations (3.7.113), (3.7.114) and the definition of S_1 in Equation (3.7.117)], these consist of $-\partial\hat{S}_1/\partial J_1^*$, $\partial\hat{S}_1/\partial w_1$, and $\partial\hat{S}_1/\partial\tilde{w}_2$ for w_1, J_1, and J_2, respectively. This, again, is in exact agreement with Equations (3.7.132) for the short-periodic parts of $w_1^{(1)}$, $J_1^{(1)}$, and $J_2^{(1)}$.

The long-periodic terms of order ε are to be considered next. These terms consisting of $\tilde{w}_1^{(1)}$, $\tilde{J}_1^{(1)}$, and $\tilde{J}_2^{(1)}$ arise naturally and quite distinctly in Equations (3.7.132) and are defined by Equations (3.7.134). Isolation of the corresponding terms according to the von Zeipel method, however, requires further calculations which are presented next. The derivatives denoted by $d\tilde{w}_1^{(1)}/d\tilde{t}$, $d\tilde{J}_1^{(1)}/d\tilde{t}$, and $d\tilde{J}_2^{(1)}/d\tilde{t}$ in the two variable method and defined in Equations (3.7.134) are the terms of order ε in the expansions of the *averages of* $dw_1/d\varepsilon t$, $dJ_1/d\varepsilon t$, and $dJ_2/d\varepsilon t$ as can be seen from Equations (3.7.124) (3.7.131) and (3.7.132), i.e.,

$$\left\langle \frac{dw_1}{d\varepsilon t} \right\rangle = \frac{d\tilde{w}_1^{(0)}}{d\tilde{t}} + \varepsilon \frac{d\tilde{w}_1^{(1)}}{d\tilde{t}} + O(\varepsilon^2) \tag{3.7.140a}$$

$$\left\langle \frac{dJ_1}{d\varepsilon t} \right\rangle = \frac{d\tilde{J}_1^{(0)}}{d\tilde{t}} + \varepsilon \frac{d\tilde{J}_1^{(1)}}{d\tilde{t}} + O(\varepsilon^2) \tag{3.7.140b}$$

$$\left\langle \frac{dJ_2}{d\varepsilon t} \right\rangle = \frac{d\tilde{J}_2^{(0)}}{d\tilde{t}} + \varepsilon \frac{d\tilde{J}_2^{(1)}}{d\tilde{t}} + O(\varepsilon^2). \tag{3.7.140c}$$

Thus, in calculating the corresponding terms given by the von Zeipel solution we must first calculate derivatives with respect to εt then average these over one period of w_2. It will be crucial in the comparison of the two methods to higher orders to recognize that in the solution of the w_i and J_i as functions of time by the von Zeipel method *not all the long-periodic terms are contained in the solution of the new Hamiltonian H^**. This is due to the presence, in the transformation relations (3.7.113) of products of short-periodic terms which have, in general non-zero averages. Furthermore, since we propose to compare averages of derivatives with respect to εt, we must also anticipate the fact that certain terms of order ε^2 in Equations (3.7.113) have a contribution to $O(\varepsilon)$ in Equations (3.7.140).

Therefore, in evaluating the derivatives to be averaged in forming the left-hand sides of Equations (3.7.140), we use the transformation relations (3.7.113) and (3.7.114) to $O(\varepsilon^2)$, and neglect only those terms which, after differentiation with respect to εt, have a zero average over one period of w_2. This gives

$$\left\langle \frac{dw_1}{d\varepsilon t} \right\rangle = \frac{dw_1^{*(0)}}{d\varepsilon t} + \varepsilon \left[\frac{dw_1^{*(1)}}{d\varepsilon t} - \frac{d}{d\varepsilon t}\left(\frac{\partial S_1^*}{\partial J_1^*}\right) + \frac{1}{2\pi}\left\langle \frac{\partial}{\partial w_2}\left(\frac{\partial^2 \hat{S}_1}{\partial w_1\, \partial J_1^*}\frac{\partial \hat{S}_1}{\partial J_1^*}\right)\right\rangle \right] + O(\varepsilon^2) \tag{3.7.141a}$$

$$\left\langle \frac{dJ_1}{d\varepsilon t} \right\rangle = \frac{dJ_1^{*(0)}}{d\varepsilon t} + \varepsilon \left[\frac{dJ_1^{*(1)}}{d\varepsilon t} + \frac{d}{d\varepsilon t}\left(\frac{\partial S_1^*}{\partial w_1}\right) - \frac{1}{2\pi}\left\langle \frac{\partial}{\partial w_2}\left(\frac{\partial^2 \hat{S}_1}{\partial w_1^2}\frac{\partial \hat{S}_1}{\partial J_1^*}\right)\right\rangle \right] + O(\varepsilon^2) \tag{3.7.141b}$$

$$\left\langle \frac{dJ_2}{d\varepsilon t} \right\rangle = -\frac{\varepsilon}{2\pi}\left\langle \frac{\partial}{\partial w_2}\left(\frac{\partial^2 \hat{S}_1}{\partial w_1\, \partial w_2}\frac{\partial \hat{S}_1}{\partial J_1^*}\right) + 2\pi \frac{\partial}{\partial w_2}\left(\frac{\partial \sigma}{\partial J_1}\frac{\partial \hat{S}_1}{\partial w_1}\right)\right\rangle + O(\varepsilon^2), \tag{3.7.141c}$$

where the last term on the right-hand side of Equation (3.7.141c) is obtained by using Equation (3.7.119) for $\partial S_2/\partial w_2$ in Equation (3.7.114c).

Next, we use Equations (3.7.136) to calculate the derivatives of $\partial S_1^*/\partial J_1^*$ and $\partial S_1^*/\partial w_1$ and find

$$\frac{d}{d\varepsilon t}\left(\frac{\partial S_1^*}{\partial J_1^*}\right) = \frac{\partial^2 S_1^*}{\partial w_1\, \partial J_1^*}\frac{\partial \lambda}{\partial J_1} - \frac{\partial^2 S_1^*}{\partial J_1^{*2}}\frac{\partial \lambda}{\partial w_1} \tag{3.7.142a}$$

$$\frac{d}{d\varepsilon t}\left(\frac{\partial S_1^*}{\partial w_1}\right) = \frac{\partial^2 S_1^*}{\partial w_1^2}\frac{\partial \lambda}{\partial J_1} - \frac{\partial^2 S_1^*}{\partial w_1\, \partial J_1^*}\frac{\partial \lambda}{\partial w_1} \tag{3.7.142b}$$

and the other terms to be evaluated in Equations (3.7.141) become

$$\frac{1}{2\pi}\left\langle\frac{\partial}{\partial w_2}\left(\frac{\partial^2 \hat{S}_1}{\partial w_1\, \partial J_1^*}\frac{\partial \hat{S}_1}{\partial J_1^*}\right)\right\rangle = - \left\langle\frac{\partial^2 \sigma}{\partial w_1\, \partial J_1}\frac{\partial \hat{S}_1}{\partial J_1^*} + \frac{\partial^2 \hat{S}_1}{\partial w_1\, \partial J_1^*}\frac{\partial \sigma}{\partial J_1}\right\rangle \tag{3.7.143a}$$

$$\frac{1}{2\pi}\left\langle\frac{\partial}{\partial w_2}\left(\frac{\partial^2 \hat{S}_1}{\partial w_1^2}\frac{\partial S_1}{\partial J_1^*}\right)\right\rangle = - \left\langle\frac{\partial^2 \sigma}{\partial w_1^2}\frac{\partial \hat{S}_1}{\partial J_1^*} + \frac{\partial^2 \hat{S}_1}{\partial w_1^2}\frac{\partial \sigma}{\partial J_1}\right\rangle \tag{3.7.143b}$$

$$-\frac{1}{2\pi}\left\langle\frac{\partial}{\partial w_2}\left(\frac{\partial^2 \hat{S}_1}{\partial w_1\, \partial w_2}\frac{\partial \hat{S}_1}{\partial J_1^*} + 2\pi\frac{\partial \sigma}{\partial J_1}\frac{\partial \hat{S}_1}{\partial w_1}\right)\right\rangle = \left\langle\frac{\partial^2 \sigma}{\partial w_1\, \partial w_2}\frac{\partial \hat{S}_1}{\partial J_1^*} - \frac{\partial^2 \sigma}{\partial w_2\, \partial J_1}\frac{\partial \hat{S}_1}{\partial w_1}\right\rangle. \tag{3.7.143c}$$

When Equations (3.7.137), (3.7.142) and (3.7.143) are substituted into Equations (3.7.141), the terms of order ε agree identically with their counterparts given by Equations (3.7.132) and we identify the following pairs of corresponding quantities by the two methods:

$$\frac{d}{d\tilde{t}}(\tilde{w}_1^{(0)} + \varepsilon\tilde{w}_1^{(1)}) = \left\langle\frac{dw_1}{d\varepsilon t}\right\rangle + O(\varepsilon^2) \tag{3.7.144a}$$

$$\frac{d}{d\tilde{t}}(\tilde{J}_1^{(0)} + \varepsilon\tilde{J}_1^{(1)}) = \left\langle\frac{dJ_1}{d\varepsilon t}\right\rangle + O(\varepsilon^2) \tag{3.7.144b}$$

$$\frac{d}{d\tilde{t}}(J_2^{(0)} + \varepsilon\tilde{J}_2^{(1)}) = \left\langle\frac{dJ_2}{d\varepsilon t}\right\rangle + O(\varepsilon^2) \tag{3.7.144c}$$

$$\tilde{w}_1^{(1)} = w_1^{*(1)} - \frac{\partial S_1^*}{\partial J_1^*} \tag{3.7.144d}$$

$$\tilde{J}_1^{(1)} = J_1^{*(1)} + \frac{\partial S_1^*}{\partial w_1} \tag{3.7.144e}$$

$$\tilde{J}_2^{(1)} = \lim_{\varepsilon\to 0}(J_2^* - J_2^{(0)})/\varepsilon. \tag{3.7.144f}$$

We note by comparing Equations (3.7.144d, e) and (3.7.113a, b) in which we set $S_1 = \hat{S}_1 + S_1^*$, that the choice of S_1^* is irrelevant to what are actually the long-periodic parts of w_1 and J_1.

This completes the detailed demonstration that the two methods studied are identical at least to $O(\varepsilon)$ for the Hamiltonian system adopted, provided the functions of $\tilde{t}$ arising in the two variable method are properly interpreted in terms of the results of the von Zeipel solution. In fact, we note in Equations (3.7.141) that in recovering the results of the two variable solution to $O(\varepsilon)$, it was necessary to include in the von Zeipel solution averages of certain derivatives of products of short-periodic terms of order ε^2. This means that truncating the von Zeipel procedure at a given order *does not necessarily determine the solution to that order.*

It is evident from the above discussion that if one seeks to determine the solution in the explicit form of q as a function of t, the two variable approach is far more efficient when starting either from Equations (3.7.123) or from the basic second-order equation which had to be transformed to the Hamiltonian in Equation (3.7.108). The main advantage presented by the von Zeipel procedure seems to be the ability to derive formal integrals of the type $J_2^* = \text{const.} + O(\varepsilon^3)$ obtained by expressing J_2^* in terms of the original coordinates.

The von Zeipel procedure, which does not directly provide explicit results, is related to a procedure based on Lie transforms which is explicit. Discussion of this approach is beyond the scope of this text and we refer the reader to References 3.7.10 and 3.7.11 and the literature cited there. A concise description and survey of some previous work appears in Reference 3.7.10 and the author of Reference 3.7.11 compares the Lie transform approach to the von Zeipel method.

3.7.3 Adiabatic Invariance[24]

We introduced the idea of adiabatic invariance in Section 3.3.2 in connection with the example of a linear oscillator with slowly várying frequency. As mentioned there, an adiabatic invariant is characterized by the property that its derivative is small and oscillatory. Therefore, assuming that such a quantity is constant does not lead to a cumulative error as $t \to \infty$.

(a) *One-Dimensional Nearly Periodic Hamiltonian*

To fix ideas, we consider a one-dimensional Hamiltonian system with a slowly varying parameter ω in the form

$$H(q, p, \omega) = \alpha_1; \qquad p = \dot{q}. \tag{3.7.145a}$$

[24] Some of the material in this subsection is based on a series of lectures given by Prof. P. A. Lagerstrom at Caltech during 1960–1961.

In the above, ω is a given slowly varying function of εt:

$$\omega = \omega(\varepsilon t); \qquad 0 < \varepsilon \ll 1 \tag{3.7.145b}$$

and consequently α_1 is also time dependent [cf. Equation (3.7.63b)]. Assume further, that with $\omega = \text{const.}$, the system is periodic according to the definition in part (a) of Section 3.7.2. Thus, Equation (3.7.145a) is a generalization of the oscillator problem defined by Equation (3.7.78).

For such a system, a function $A_N(q, p, \omega; \varepsilon)$ is called an adiabatic invariant to $O(\varepsilon^N)$ if

$$\frac{dA_N}{dt} = \varepsilon^{N+1}\phi_{N+1}(q, p, \omega; \varepsilon), \tag{3.7.146a}$$

where ϕ_{N+1} has zero average value over one cycle in the q, p plane (with $\omega = \text{const.}$). Since $\alpha_1 = \text{const.}$ if $\omega = \text{const.}$, Equation (3.7.145a) defines a one-parameter (α_1) family of curves in the q, p plane. These curves are either closed and nested (libration) or periodic in q (rotation) but are *not* integral curves for the actual time dependent Hamiltonian. Nevertheless, we can explicitly calculate these curves from Equation (3.7.145a) and we denote one complete cycle along one of them for a fixed α_1 by $C(\alpha_1)$. The statement that ϕ_{N+1} has zero average value simply means that the line integral of ϕ_{N+1} along $C(\alpha_1)$ vanishes

$$\oint_{C(\alpha_1)} \phi_{N+1}(q, p, \omega; \varepsilon)dq = 0. \tag{3.7.146b}$$

The basic idea of the technique we present in this section was discussed in Reference 3.7.12 where it was shown that by iteration of the canonical transformations to action and angle variables one generates adiabatic invariants to any desired degree of accuracy.

We begin our study by examining the properties of the canonical transformation $W(q, \alpha_1, \omega)$ obtained by solving the *time-independent* Hamilton-Jacobi equation

$$H\left(q, \frac{\partial W}{\partial q}, \omega\right) = \alpha_1, \tag{3.7.147}$$

where we regard ω and α_1 *as constants.* We must keep in mind that since ω (and hence α_1) are actually time-dependent, Equation (3.7.147) is not the appropriate Hamilton-Jacobi equation for solving the problem *exactly* [cf. the discussion following Equation (3.7.58)]. However, we are completely at liberty to use Equation (3.7.147) for the purpose of generating a canonical transformation.

The solution of Equation (3.7.147), which can always be derived by quadrature, defines W in terms of q, ω and α_1 as follows:

$$W = W(q, \alpha_1, \omega). \tag{3.7.148}$$

Let us regard P as a new momentum and choose it, for the time being, as some unspecified function of α_1 and ω:

$$P = P(\alpha_1, \omega). \tag{3.7.149}$$

Using Equation (3.7.149) to eliminate α_1 in favor of P in Equation (3.7.148) defines W as a function of q, P and ω:

$$W = W[q, \alpha_1(P, \omega), \omega] \equiv F(q, P, \omega). \tag{3.7.150}$$

Having derived F in the above form, by treating ω (and α_1) to be constant, we now change our mind and let ω be the slowly varying function given in Equation (3.7.145b), and study the canonical transformation generated by F.

Using Equations (3.7.57), defining a canonical transformation with a generating function, such as F, of the old coordinate q, the new momentum P and the time, we have the following relations linking the old and new variables:

$$p = \frac{\partial F}{\partial q} \tag{3.7.151a}$$

$$Q = \frac{\partial F}{\partial P}. \tag{3.7.151b}$$

The new Hamiltonian $\mathscr{H}$ involving Q, P and ω is defined by Equation (3.7.57c) and this reduces, in our special case, to

$$\mathscr{H} = H_0(P, \omega) + \varepsilon H_1(Q, P, \varepsilon t), \tag{3.7.152a}$$

where H_0 is the solution of α_1 in terms of P and ω from Equation (3.7.149), and εH_1 is $\partial F/\partial t$, i.e.,

$$H_1 = \frac{\partial F}{\partial \omega}\omega', \qquad ' = \frac{d}{d\varepsilon t}. \tag{3.7.152b}$$

Hamilton's differential equations associated with $\mathscr{H}$ are then

$$\frac{dQ}{dt} = \frac{\partial \mathscr{H}}{\partial P} \tag{3.7.153a}$$

$$\frac{dP}{dt} = -\frac{\partial \mathscr{H}}{\partial Q} = -\varepsilon\frac{\partial H_1}{\partial Q}. \tag{3.7.153b}$$

The time derivative of $\mathscr{H}$ [cf. Equation (3.7.63b)] is

$$\frac{d\mathscr{H}}{dt} = \frac{\partial \mathscr{H}}{\partial t} = \varepsilon\omega'\frac{\partial H_0}{\partial \omega} + \varepsilon^2\frac{\partial H_1}{\partial \varepsilon t}. \tag{3.7.153c}$$

Note that although $\mathscr{H}$ is an invariant to $O(1)$ (i.e., $d\mathscr{H}/dt = O(\varepsilon)$) it is not necessarily adiabatic since the average value of $\partial H_0/\partial\omega$ over one cycle in the q-p plane need not be zero.

On the other hand, if we are able to render the right-hand side of Equation (3.7.153b) oscillatory, P would be an adiabatic invariant. We proceed to do this by the appropriate choice of P.

According to Equation (3.7.153b), a necessary and sufficient condition for P to the adiabatic to $O(1)$ is that

$$\oint_{C(\alpha_1)} \frac{\partial H_1}{\partial Q}\, dq = 0. \tag{3.7.154}$$

Recalling thc definition of H_1, we conclude that we must set $\Delta(\partial F/\partial\omega) = 0$ [cf. the definition of Δ following Equation (3.7.75)]. Now $\partial F/\partial\omega$ is computed with both q and P fixed. It then follows that we must require

$$\Delta \frac{\partial F}{\partial\omega} = \frac{\partial}{\partial\omega}(\Delta F) = 0. \tag{3.7.155}$$

In general ΔF will depend on both ω and P and $\partial(\Delta F)/\partial\omega \neq 0$ unless ΔF is a function of P alone. The simplest choice is $\Delta F = P$ in which case $P = J$, the action [cf. the discussion following Equation (3.7.75)].

With the choice $P = J$, the generating function is now $F(q, J, \omega)$ where

$$J = \oint_{C(\alpha_1)} p\, dq = J(q, p, \varepsilon t) \tag{3.7.156a}$$

and Equation (3.7.151b) defines the angle variable w in the form:

$$w = \frac{\partial F}{\partial J} = w(q, p, t). \tag{3.7.156b}$$

The transformed Hamiltonian will become

$$\mathscr{H}^{(1)}(w, J, \varepsilon t; \varepsilon) = H_0^{(1)}(J, \omega) + \varepsilon H_1^{(1)}(w, J, \varepsilon t), \tag{3.7.157}$$

where the terms on the right-hand side of Equation (3.7.157) are those in Equation (3.7.152a) with $P = J$ and $Q = w$.

We have shown that J is adiabatic to $O(1)$. In order to calculate an adiabatic invariant to $O(\varepsilon)$ we repeat the previous process starting with Equation (3.7.157), i.e., we seek a canonical transformation from w, J to $w^{(1)}$, $J^{(1)}$ such that $dJ^{(1)}/dt = O(\varepsilon^2)$ and has a zero average computed over one cycle in the $w^{(1)}$, $J^{(1)}$ plane with ω held fixed. To effect this transformation we solve the Hamilton-Jacobi equation associated with $\mathscr{H}^{(1)}$, i.e.,

$$H_0^{(1)}\left(\frac{\partial W^{(1)}}{\partial J}, \omega\right) + \varepsilon H_1^{(1)}\left(w, \frac{\partial W^{(1)}}{\partial J}, \varepsilon t\right) = \alpha_1^{(1)}, \tag{3.7.158}$$

where again we regard ω, εt, $\alpha_1^{(1)}$ as constants.

As before, the proper choice for $J^{(1)}$ turns out to be

$$J^{(1)} = \oint_{C(\alpha_1^{(1)})} J\, dw, \tag{3.7.159}$$

where $C(\alpha_1^{(1)})$ is the curve in the w, J plane that one obtains from Equation (3.7.158) by regarding ω, εt, and $\alpha_1^{(1)}$ as constants. This procedure can be repeated indefinitely to derive an adiabatic invariant to any desired order ε^N.

It also was pointed out in Reference 3.7.12 that if at any time all the derivatives of H with respect to εt are zero then $J^{(N)} = J$ for any integer N, and we have the following result.

Assume that at $\varepsilon t = 0$ and $\varepsilon t = 1$ the derivatives of all orders of H with respect to εt are equal to zero. Let the solution be defined for the initial values q_0, p_0 (both independent of ε) at $\varepsilon t = 0$. If q_1 and p_1 are the values of q and p at $\varepsilon t = 1$ we have

$$J(q_1, p_1, 1) = J(q_0, p_0, 0) + O(\varepsilon^{N-1}) \tag{3.7.160}$$

for any integer N.

(b) *Oscillator with Slowly Varying Frequency*

To illustrate the ideas, we consider as a first example the problem of Section 3.3.2

$$\frac{d^2q}{dt^2} + \omega^2(\varepsilon t)q = 0. \tag{3.7.161}$$

With $\dot{q} = p$, the Hamiltonian associated with Equation (3.7.161) is

$$H(q, p, \omega) = \tfrac{1}{2}p^2 + \tfrac{1}{2}\omega^2 q^2. \tag{3.7.162}$$

Regarding ω as a constant, we solve the Hamilton-Jacobi equation

$$\frac{1}{2}\left(\frac{\partial W}{\partial q}\right)^2 + \frac{1}{2}\omega^2 q^2 = \alpha_1 \tag{3.7.163}$$

and calculate W as [cf. Equations (3.7.89) and (3.7.148)]

$$\begin{aligned} W &= \int^q \sqrt{2\alpha_1 - \omega^2 q'^2}\, dq' \\ &= \frac{\alpha_1}{\omega}\left[\sin^{-1}\frac{\omega q}{\sqrt{2\alpha_1}} + \frac{\omega q}{\sqrt{2\alpha_1}}\left(1 - \frac{\omega^2 q^2}{2\alpha_1}\right)^{1/2}\right] \\ &= W(q, \alpha_1, \omega). \end{aligned} \tag{3.7.164}$$

Next, we eliminate α_1 in favor of J, where

$$J = \oint_{C(\alpha_1)} p\, dq. \tag{3.7.165a}$$

Clearly, $C(\alpha_1)$ is the ellipse defined by Equation (3.7.94) and J is the area of this ellipse

$$J = 2\pi\alpha_1/\omega. \tag{3.7.165b}$$

The expression for the generating function $F(q, J, \omega)$ is then obtained by substituting $\alpha_1 = \omega J/2\pi$ in Equation (3.7.164). This gives Equation (3.7.96) which we repeat below

$$F(q, J, \omega) = \frac{J}{2\pi}\left[\sin^{-1}\left(\frac{\pi\omega}{J}\right)^{1/2} q + \left(\frac{\pi\omega}{J}\right)^{1/2} q\left(1 - \frac{\pi\omega q^2}{J}\right)^{1/2}\right]. \tag{3.7.166}$$

Note that so far our results are identical to those in Section 3.7.2 for $\omega = $ const.

At this point we regard Equation (3.7.166) as a *time-dependent* generating function for a canonical transformation from q, p to w, J.

Clearly, the relations between the old and new variables are as before, i.e.,

$$J = \pi(p^2 + \omega^2 q^2)/\omega \tag{3.7.167a}$$

$$w = \frac{1}{2\pi}\tan^{-1} \omega q/p. \tag{3.7.167b}$$

However, the transformed Hamiltonian obeys Equation (3.7.157) which, for our example, reduces to the "standard" form:

$$\mathscr{H}^{(1)}(w, J, \varepsilon t; \varepsilon) = \frac{\omega J}{2\pi} + \frac{\varepsilon J \omega'}{4\pi\omega}\sin 4\pi w = \alpha_1^{(1)}. \tag{3.7.168}$$

In analogy with the discussion in Section 3.7.2, we will call this the standard form for the Hamiltonian [cf. Equations (3.7.108)]. It is characterized by the fact that for $\varepsilon = 0$, the angle variables are absent, therefore the actions are invariants. Moreover, the Hamiltonian is periodic with respect to the angle variables and these occur to higher orders only.

For this example, the reader may verify from Equation (3.7.166) that

$$\frac{\partial F}{\partial \omega} = \frac{q}{2\omega}\left(\frac{\omega J}{\pi} - \omega^2 q^2\right)^{1/2}. \tag{3.7.169a}$$

Thus, eliminating q in favor of w according to Equation (3.7.167b) gives

$$\frac{\partial F}{\partial \omega} = \frac{J}{4\pi\omega}\sin 4\pi w \tag{3.7.169b}$$

and this is the main ingredient in $H_1^{(1)}$ occurring in Equation (3.7.168).

The canonical transformation leading to Equations (3.7.157) and (3.7.158) is exact and the problem is now also exactly defined by the associated Hamilton differential equations

$$\frac{dw}{dt} = \frac{\partial \mathscr{H}^{(1)}}{\partial J} = \frac{\omega}{2\pi} + \frac{\varepsilon\omega'}{4\pi\omega}\sin 4\pi w \tag{3.7.170a}$$

$$\frac{dJ}{dt} = -\frac{\partial \mathscr{H}^{(1)}}{\partial w} = -\frac{\varepsilon J \omega'}{\omega}\cos 4\pi w. \tag{3.7.170b}$$

Equation (3.7.170b) exhibits explicitly the property that J is an adiabatic invariant to $O(1)$ since the right-hand side has a zero average value over one cycle.

Let us now calculate the adiabatic invariant to $O(\varepsilon)$. The Hamilton-Jacobi equation associated with Equation (3.7.168) gives

$$\frac{\partial W^{(1)}}{\partial w} = \frac{2\pi\alpha_1^{(1)}}{\omega}\left(1 - \frac{\varepsilon\omega'}{2\omega^2}\sin 4\pi w\right) \tag{3.7.171a}$$

and this can be integrated to define $W^{(1)}$:

$$W^{(1)} = \frac{2\pi\alpha_1^{(1)}}{\omega}\left(w + \frac{\varepsilon\omega'}{8\pi\omega^2}\cos 4\pi w\right). \tag{3.7.171b}$$

This is a function of $\alpha_1^{(1)}$ and w, and will define a canonical transformation once we select the new angle variable $J_1^{(1)}$ as an appropriate function of $\alpha_1^{(1)}$.

In analogy with the discussion leading to the choice of J, we must choose $J^{(1)}$ according to Equation (3.7.159) and this reduces to

$$J^{(1)} = \int_{w=0}^{w=1} J\,dw, \tag{3.7.172}$$

where J is expressed in terms of w for $\alpha_1^{(1)} = \text{const.}$ through Equation (3.7.168) or Equation (3.7.171a):

$$J = \frac{\partial W^{(1)}}{\partial w} = \frac{2\pi\alpha_1^{(1)}}{\omega}\left(1 - \frac{\varepsilon\omega'}{2\omega^2}\sin 4\pi w\right). \tag{3.7.173}$$

Therefore, integrating Equation (3.7.172) with respect to w gives

$$\begin{aligned} J^{(1)} &= \frac{2\pi\alpha_1^{(1)}}{\omega} = J\Bigg/\left(1 - \frac{\varepsilon\omega'}{2\omega^2}\sin 4\pi w\right) \\ &= J\left[1 + \frac{\varepsilon\omega'}{2\omega^2}\sin 4\pi w + O(\varepsilon^2)\right]. \end{aligned} \tag{3.7.174}$$

Thus, $J^{(1)}$ can be computed in terms of q, p, and w using the definition (3.7.167) for w and J.

In order to exhibit the adiabatic invariance of $J^{(1)}$ to $O(\varepsilon)$ we must calculate the transformed Hamiltonian in terms of $w^{(1)}$ and $J^{(1)}$.

First, we use the result $J^{(1)} = 2\pi\alpha_1^{(1)}/\omega$ to express $W^{(1)}$ in terms of $J^{(1)}$ and w as follows

$$W^{(1)} = F^{(1)}(w, J^{(1)}; \varepsilon t) = J^{(1)}\left(w + \frac{\varepsilon\omega'}{8\pi\omega^2}\cos 4\pi w\right) \tag{3.7.175}$$

and use $F^{(1)}$ as a generating function for a canonical transformation from w, J, to $w^{(1)}, J^{(1)}$.

In particular,

$$w^{(1)} = \frac{\partial F^{(1)}}{\partial J^{(1)}} = w + \frac{\varepsilon\omega'}{8\pi\omega^2}\cos 4\pi w \tag{3.7.176}$$

and

$$\mathscr{H}^{(2)} = \mathscr{H}^{(1)} + \frac{\partial F^{(1)}}{\partial t} = \alpha_1^{(1)} + \varepsilon \frac{\partial F^{(1)}}{\partial \varepsilon t}$$

$$= \frac{\omega J^{(1)}}{2\pi} + \frac{\varepsilon^2}{8\pi}\left(\frac{\omega'}{\omega^2}\right)' J^{(1)} \cos 4\pi w. \tag{3.7.177}$$

Since, $w = w^{(1)} + O(\varepsilon)$ according to Equation (3.7.176), we can write the new Hamiltonian $\mathscr{H}^{(2)}$ in the form

$$\mathscr{H}^{(2)}(w^{(1)}, J^{(1)}, \varepsilon t; \varepsilon) = \frac{\omega J^{(1)}}{2\pi} + \frac{\varepsilon^2}{8\pi}\left(\frac{\omega'}{\omega^2}\right)' J^{(1)} \cos 4\pi w^{(1)} + O(\varepsilon^3). \tag{3.7.178}$$

The adiabatic invariance of $J^{(1)}$ to $O(\varepsilon)$ is now exhibited by Hamilton's equation

$$\frac{dJ^{(1)}}{dt} = -\frac{\partial \mathscr{H}^{(2)}}{\partial w^{(1)}} = \frac{\varepsilon^2}{2}\left(\frac{\omega'}{\omega^2}\right)' J^{(1)} \sin 4\pi w^{(1)} + O(\varepsilon^3) \tag{3.7.179}$$

and this procedure can be repeated indefinitely.

It is important to note that once the Hamiltonian is transformed to the standard form (3.7.168), where it is periodic in w, *the von Zeipel procedure of Section* 3.7.2 *and the approach we have discussed in this section are equivalent.*

To demonstrate this equivalence, we return to Equation (3.7.168), and seek a canonical transformation from w, J to $w^{(1)}, J^{(1)}$ such that $w^{(1)}$ is absent from the transformed Hamiltonian to $O(\varepsilon)$. Since $\mathscr{H}^{(1)}$ depends on t, we seek a time dependent generating function S in the form [cf. Equations (3.7.57)].

$$S(w, J^{(1)}, \tilde{t}; \varepsilon) = wJ^{(1)} + \varepsilon S_1(w, J^{(1)}, \tilde{t}) + \cdots. \tag{3.7.180}$$

Proceeding as in Section 3.7.2, we calculate the following expression for the transformed Hamiltonian (Note that $\partial S/\partial t = O(\varepsilon^2)$):

$$\mathscr{H}^{(1)}(w^{(1)}, J^{(1)}, \varepsilon t) = \frac{\omega}{2\pi}\left[J^{(1)} + \varepsilon \frac{\partial S_1}{\partial w}(w^{(1)}, J^{(1)}) + \cdots\right] + \frac{\varepsilon J^{(1)}\omega'}{4\pi\omega} \sin 4\pi w^{(1)} + \cdots. \tag{3.7.181}$$

Thus, we can eliminate the term of $O(\varepsilon)$ depending on $w^{(1)}$ by setting

$$\frac{\partial S_1}{\partial w} = -\frac{J^{(1)}\omega'}{2\omega^2} \sin 4\pi w \tag{3.7.182a}$$

and this defines S_1 in the form

$$S_1 = \frac{J^{(1)}\omega'}{8\pi\omega^2}\cos 4\pi w. \tag{3.7.182b}$$

Comparing Equations (3.7.180), (3.7.182b) with Equation (3.7.175) we see that the generating functions S and W are identical. Thus, the von Zeipel procedure is equivalent to the one proposed in Reference 3.7.12 for this example.

(c) *Charged Particle in Slowly Varying Magnetic Field*

The equivalence of the two procedures is actually more general and we illustrate the ideas with the example of the motion of a changed particle in a magnetic field which varies slowly in the x direction and which is cylindrically symmetric about the x-axis.

For simplicity we assume that the mass m of the particle, the charge e and the speed of light c are unity. (This can be accomplished by choosing appropriate dimensionless variables). We assume that the electric field $\mathbf{E}$ is zero and the magnetic vector potential $\mathbf{A}$ is cylindrically symmetric and is of the special form[25]

$$\mathbf{A} = \frac{r}{2} f(\varepsilon x)\mathbf{e}_\theta, \tag{3.7.183}$$

where

$$r = \sqrt{y^2 + z^2}, \qquad 0 < \varepsilon \ll 1.$$

Here f is an arbitrary non-vanishing function which describes the slow axial variation of $\mathbf{A}$, and $\mathbf{e}_\theta$ is a unit vector in the tangential (to surfaces r = const.) direction.

Using Cartesian x, y, z coordinates, the Lagrangian has the form

$$L(x, y, z, \dot{x}, \dot{y}, \dot{z}) = \frac{1}{2}(\dot{x}^2 + \dot{y}^2 + \dot{z}^2) + \frac{f}{2}(y\dot{z} - z\dot{y}), \tag{3.7.184a}$$

where a dot denotes d/dt.

If we introduce cylindrical polar coordinates (x, r, θ) we have

$$L(x, r, \theta, \dot{x}, \dot{r}, \dot{\theta}) = \frac{1}{2}(\dot{x}^2 + \dot{r}^2 + r^2\dot{\theta}^2) + \frac{f}{2}r^2\dot{\theta} \tag{3.7.184b}$$

and the absence of θ from L implies that

$$\frac{\partial L}{\partial \dot{\theta}} = p_\theta = r^2[\dot{\theta} + f(\varepsilon x)/2] = l = \text{const.} \tag{3.7.185}$$

[25] The reader may refer to Reference 3.7.7 for notation and a derivation of the Lagrangian formulation of the motion of a charged particle in an electromagnetic field.

This is an exact invariant and its significance will become clear once we have studied the $\varepsilon = 0$ problem.

A straightforward formulation of our problem results from Lagrange's equations in Cartesian coordinates. These are [cf. Equation (3.7.45)]

$$\ddot{x} = \varepsilon f'(y\dot{z} - z\dot{y})/2; \qquad ' = \frac{d}{d\varepsilon x} \tag{3.7.186a}$$

$$\ddot{y} - f\dot{z} = \varepsilon f'\dot{x}z/2 \tag{3.7.186b}$$

$$\ddot{z} + f\dot{y} = -\varepsilon f'\dot{x}y/2. \tag{3.7.186c}$$

When $\varepsilon = 0$, we can immediately integrate these and find

$$x = x(0) + \dot{x}(0)t \tag{3.7.187a}$$

$$y = y_g + \rho\cos(ft - \phi) \tag{3.7.187b}$$

$$z = z_g - \rho\sin(ft - \phi), \tag{3.7.187c}$$

where $x(0)$ and $\dot{x}(0)$ are the initial values of x and $\dot{x}$, while y_g, z_g, ρ and ϕ are the four constants of the motion in the y, z plane. We see that the motion has two distinct components: a uniform translation in the x direction superposed on a uniform clockwise rotation with constant angular velocity f around the "guiding-center" located at y_g, z_g. See Figure 3.7.1 for a projection of the motion onto the y, z plane.

Thus, the composite motion is a spiralling trajectory around the axis $y = y_g$, $z = z_g$. The constants y_g, z_g, ρ and ϕ can be related to the initial values of y, $\dot{y}$, z and $\dot{z}$ by

$$\rho = \sqrt{\dot{y}(0)^2 + \dot{z}(0)^2}/f \tag{3.7.188a}$$

$$\sin\phi = \dot{y}(0)/\rho f; \qquad \cos\phi = -\dot{z}(0)/\rho f \tag{3.7.188b}$$

$$y_g = y(0) + \dot{z}(0)/f \tag{3.7.188c}$$

$$z_g = z(0) - \dot{y}(0)/f. \tag{3.7.188d}$$

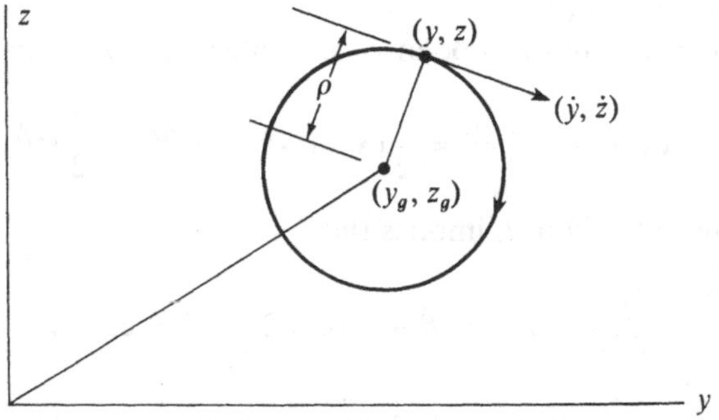

Figure 3.7.1 Circular Motion around the Guiding-Center

Note that $r_g \equiv \sqrt{y_g^2 + z_g^2} \geq \rho$, and the equal sign only occurs if the particle passes through the origin.

The invariant p_θ reduces to

$$p_\theta = \frac{f}{2}(y_g^2 + z_g^2 - \rho^2) = \text{const.} \qquad (3.7.189)$$

which is true even when f depends on x. Since r_g is constant if f is constant, Equation (3.7.189) reduces, in this case, to the statement that the angular momentum about the guiding center is constant. For the case of slowly varying f, we expect both ρ and r_g to also vary with x and Equation (3.7.189) shows that if f is such that the particle's clockwise angular momentum about the guiding center increases then $r_g^2 f$ must decrease and vice versa.

In order to analyze the problem for the case of slow axial variations of f, by canonical perturbations, we must first express the Hamiltonian for the problem in standard form. This task is not trivial and will occupy us next.

We start with the definition of the system (3.7.184a) in Lagrangian form using x, y, z as the coordinates. Next we calculate [cf. Equations (3.7.47) and (3.7.48)] the following expressions for the momenta and the Hamiltonian

$$p_x = \dot{x} \qquad (3.7.190a)$$

$$p_y = \dot{y} - fz/2; \qquad \dot{y} = p_y + fz/2 \qquad (3.7.190b)$$

$$p_z = \dot{z} + fy/2; \qquad \dot{z} = p_z - fy/2 \qquad (3.7.190c)$$

$$H = \tfrac{1}{2}v^2 = \tfrac{1}{2}[p_x^2 + (p_y + fz/2)^2 + (p_z - fy/2)^2] = \text{const.} \qquad (3.7.190d)$$

The constancy of H (even for variable f) follows from the fact that H is time independent. Thus, the speed of the particle is an exact invariant. Incidentally, this result also follows less directly from Equations (3.7.186). First we note that the last two equations (or Equation (3.7.185)) imply that $y\dot{z} - z\dot{y} = l - (f/2)(y^2 + z^2)$. We then write Equation (3.7.186a) in the form

$$\ddot{x} = \frac{\varepsilon f'}{2}\left[l - \frac{f}{2}(y^2 + z^2)\right]. \qquad (3.7.191)$$

Now, multiplying Eq. (3.7.191) by $\dot{x}$, Equation (3.7.186b) by $\dot{y}$ and Equation (3.7.186c) by $\dot{z}$ and adding gives $v^2/2 = \text{const.}$

It is convenient to make a preliminary canonical transformation to variables $q_1, q_2, q_3, p_1, p_2, p_3$ which exhibit some of the features of the unperturbed motion. Since the axial motion is rectilinear we need not transform x and we choose $q_1 = x$. If we choose p_2 and q_2 proportional to $\dot{y}$ and $\dot{z}$ we will be able to depict the nearly uniform and circular motion of the particle around the guiding center by introducing action and angle variables in the p_2, q_2 plane .Finally, since the coordinates of the guiding center are constant for $\varepsilon = 0$, we choose q_3 and p_3 proportional to these.

With the observation that

$$\dot{z} = p_z - fy/2, \qquad \dot{y} = p_y + fz/2 \tag{3.7.192}$$

$$y_g = \frac{1}{f}\left(p_z + \frac{fy}{2}\right); \qquad z_g = -\frac{1}{f}\left(p_y - \frac{fz}{2}\right) \tag{3.7.193}$$

in the unperturbed problem, we seek a canonical transformation from x, y, z, p_x, p_y, p_z to $q_1, q_2, q_3, p_1, p_2, p_3$ in the form

$$q_1 = x; \qquad p_1 = p_x + F \tag{3.7.194a}$$

$$q_2 = A(p_z - fy/2); \qquad p_2 = B(p_y + fz/2) \tag{3.7.194b}$$

$$q_3 = D(p_y - fz/2); \qquad p_3 = C(p_z + fy/2), \tag{3.7.194c}$$

where F is unknown and we expect it to be $O(\varepsilon)$, and A, B, C, D are unknown functions of εx.

In order that Equations (3.7.194) be canonical, we must have [cf. the discussion in Section 3.7.2]

$$\mathbf{p} \cdot d\mathbf{q} \equiv p_1\, dq_1 + p_2\, dq_2 + p_3\, dq_3 = p_x\, dx + p_y\, dy + p_z\, dz + dS. \tag{3.7.195}$$

We will use this criterion, rather than an explicit derivation by a generating function, to determine the unknowns A, B, C, D, F.

The right-hand side of Equation (3.7.195), when we use Equations (3.7.194), becomes

$$\begin{aligned}\mathbf{p} \cdot d\mathbf{q} = p_1\, dx &+ B(p_y + fz/2)d[A(p_z + fy/2) - Afy] \\ &+ D(p_z + fy/2)d[C(p_y + fz/2) - Cfz].\end{aligned} \tag{3.7.196a}$$

Clearly, we must set $A = D$, $B = C$ and the above simplifies to

$$\begin{aligned}\mathbf{p} \cdot d\mathbf{q} = [p_1 &- \varepsilon B(Af)'yp_y - \varepsilon A(Bf)'zp_z]\, dx \\ &- ABf\, p_y\, dy - ABf\, p_z\, dz + dS,\end{aligned} \tag{3.7.196b}$$

where dS denotes an exact differential, and therefore does not interest us. Comparing this result with the right-hand side of Equation (3.7.195) shows that we must set $ABf = -1$ in order to have a canonical transformation.

It is convenient to choose $A = -1/\sqrt{f}$, $B = 1/\sqrt{f}$, in which case the following is *an exact canonical transformation*

$$q_1 = x; \qquad p_1 = p_x - \varepsilon f'(yp_y + zp_z)/2f \tag{3.7.197a}$$

$$q_2 = -(p_z - fy/2)/\sqrt{f}; \qquad p_2 = (p_y + fz/2)/\sqrt{f} \tag{3.7.197b}$$

$$q_3 = -(p_y - fz/2)/\sqrt{f}; \qquad p_3 = (p_z + fy/2)/\sqrt{f}. \tag{3.7.197c}$$

The transformed Hamiltonian corresponding to Equation (3.7.190d) becomes

$$\mathscr{H} = \frac{1}{2}\left[p_1 + \frac{\varepsilon f'}{2f}(p_2 p_3 - q_2 q_3)\right]^2 + \frac{f}{2}(q_2^2 + p_2^2) = \text{const.} \tag{3.7.198}$$

Notice that x, p_3 and q_3 only occur in the form εx, εp_3 and εq_3.

We can now introduce action and angle variables in both the q_2, p_2 and q_3, p_3 planes. In order to avoid the occurrence of unnecessary factors π in the definitions of the action and angle variables, we adopt the following normalized variables:

$$p_2 = -\sqrt{2J_2^*}\sin w_2^*, \qquad q_2 = \sqrt{2J_2^*}\cos w_2^*; \tag{3.7.199a}$$

$$p_3 = \sqrt{2J_3^*}\cos w_3^*, \qquad q_3 = \sqrt{2J_3^*}\sin w_3^*. \tag{3.7.199b}$$

The reader can verify by direct calculation that $(p_2\,dq_2 + p_3\,dq_3 - J_2^*\,dw_2^* - J_3^*\,dw_3^*)$ is an exact differential, hence the transformations (3.7.199) are canonical. Moreover Equations (3.7.199a) are orientation preserving in the sense that the direction of increasing w_2^* is along the motion (clockwise) in the q_2, p_2 plane.

In terms of the new variables, we calculate the following expression for the Hamiltonian

$$\mathscr{H}^*(q_1, w_1^*, w_2^*, p_1, J_1^*, J_2^*)$$
$$= \frac{1}{2}\left[p_1 - \frac{\varepsilon f'}{f}\sqrt{J_2^* J_3^*}\sin(w_2^* + w_3^*)\right]^2 + fJ_2^* = \text{const.} \tag{3.7.200}$$

Finally, introduce the canonical transformation

$$w_2 = w_2^* + w_3^*, \qquad J_2 = J_2^* \tag{3.7.201a}$$

$$w_3 = w_3^* \qquad J_3 = J_3^* - J_2^* \tag{3.7.201b}$$

to write the Hamiltonian in the standard form

$$\mathscr{H}^{(1)} = \frac{1}{2}\left[p_1 - \frac{\varepsilon f'}{f}\sqrt{J_2(J_2 + J_3)}\sin w_2\right]^2 + fJ_2 \tag{3.7.201c}$$
$$= \text{const.}$$

The relationship between these variables and the Cartesian coordinates is easy to calculate and is summarized in Figure 3.7.2.

We immediately conclude that in addition to the exact invariants $\mathscr{H}^{(1)}$ and J_3 (which are just H and p_θ as defined earlier) we have J_2 as adiabatic invariant to $O(1)$. The momentum p_1 is an invariant to $O(1)$ also but is not necessarily adibatic unless f is periodic in εx.

Let us focus attention on the adiabatic invariant J_2. If we want to calculate an adiabatic invariant to $O(\varepsilon^2)$, we can follow the von Zeipel procedure discussed in the preceding section, or alternately the procedure outlined earlier in this section. At any rate, these are equivalent and result in removing the w_2 term from the Hamiltonian to $O(\varepsilon)$. The details of the calculation are straightforward and are outlined in Problem 3.7.7.

We conclude this subsection by emphasizing that once a problem has been reduced to the standard form, successively more accurate adiabatic

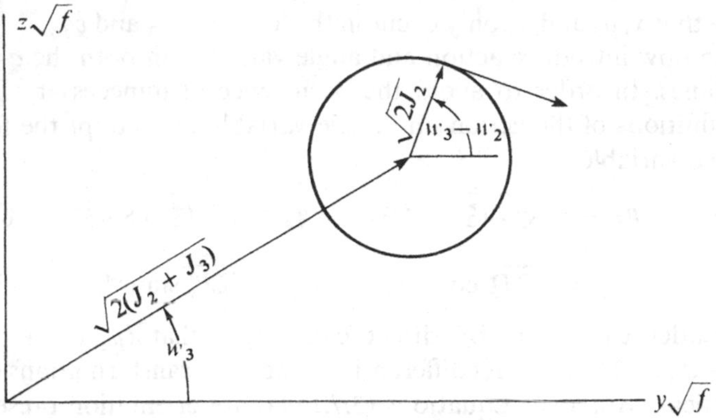

Figure 3.7.2 Action and Angle Variables

invariants can be computed routinely by the von Zeipel procedure. The crucial step for problems with several degrees of freedom is finding the appropriate action and angle variables for the first step.

PROBLEMS

1. Calculate the solution of

$$\frac{d^2y}{dt^2} + y + \varepsilon\left(\frac{dy}{dt}\right)^3 = 0$$

by the method of averaging to $O(\varepsilon)$ and compare the results with Equations (3.2.36) and (3.2.39).

2. Recall that a direct application of the multiple variable method to the problem

$$\frac{d^2y}{dt^2} + y + \varepsilon\left(\frac{dy}{dt}\right)^3 + 3\nu\varepsilon^2\frac{dy}{dt} = 0, \ \nu = \text{const.}$$

gave inconsistent results, [cf. Equations (3.2.61) and (3.2.67)]. It was necessary to rescale y and t before applying the multiple variable method. Use the method of averaging directly on Equation (3.2.61) to calculate the solution to $O(\varepsilon)$ and compare your results with Equations (3.2.74).

3. By paralleling the procedure in Section 3.2.5 calulate the multiple variable expansion of Equation (3.2.75) starting from the standard form given by Equations (3.7.13). Compare your results with those given in Section 3.2.5 and 3.7.1.

4. Study Mathieu's equation

$$\frac{d^2y}{dt^2} + (\delta + \varepsilon\cos t)y = 0$$

for the case

$$\delta = 1 + \varepsilon\delta_1 + \varepsilon^2\delta_2 + \cdots, \qquad \delta_i = \text{const.}$$

by the method of averaging. Deduce the conditions on δ_1 and δ_2 for stable solutions, and compare your results with those in Section 3.3.1.

5. We wish to study the problem of Section 3.5.2 by the von Zeipel method. We have the system

$$\frac{d^2x}{dt^2} + a^2x = \varepsilon y^2$$

$$\frac{d^2y}{dt^2} + b^2y = 2\varepsilon xy,$$

where a and b are constants.

(a) Derive the Hamiltonian in the form

$$H_1(q_1, q_2, p_1, p_2, \varepsilon) = \tfrac{1}{2}(p_1^2 + p_2^2) + \tfrac{1}{2}(a_1^2q_1^2 + a_2^2q_2^2) - \varepsilon q_1 q_2^2 = \alpha_1 = \text{const.},$$

where $a_1 = a$ and $a_2 = b$.

(b) Consider the problem fo $\varepsilon = 0$. Clearly, the Hamiltonian is separable and the motion is periodic in each $q_i - p_i$ plane (uncoupled oscillators). Thus, we can define action and angle variables, [cf. Equations (3.7.70–73)] by the generating function

$$W(q_1, q_2, J'_1, J'_2) = \sum_{i=1}^{2} \frac{J_i'}{2\pi}\left[\sin^{-1}\left(\frac{\pi a_i}{J'_i}\right)^{1/2} q_i + q_i\left(\frac{\pi a_i}{J'_i}\right)^{1/2}\left(1 - \frac{\pi a_i q_i^2}{J'_i}\right)^{1/2}\right]$$

which corresponds to the following explicit canonical transformation to the action (J'_i), angle (w'_i) variables:

$$J'_i = \pi(p_i^2 + a_i^2q_i^2)/a_i; \qquad i = 1, 2$$

$$w'_i = \frac{1}{2\pi}\tan^{-1} a_i q_i/p_i; \qquad i = 1, 2.$$

We can use the above W to carry out a canonical transformation for the $\varepsilon \neq 0$ problem. Since H is independent of t Equations (3.7.57) imply that the w'_i and J'_i are as defined above, and the new Hamiltonian is obtained by expressing the q_i, p_i in terms of the w'_i, J'_i. Show that this gives

$$\begin{aligned} H'(w'_1, w'_2, J'_1, J'_2, \varepsilon) &= (a_1J'_1 + a_2J'_2)/2\pi \\ &\quad -\varepsilon\left(\frac{J'_2}{\pi a_2}\right)\sin^2 2\pi w'_2 \cdot \left(\frac{J'_1}{\pi a_1}\right)^{1/2} \sin 2\pi w'_1 \\ &= (a_1J'_1 + a_2J'_2)/2\pi - \frac{\varepsilon}{2}\left(\frac{J'_1 J'^2_2}{\pi^3 a_1 a_2^2}\right)^{1/2} \\ &\quad \times \left[\sin 2\pi w'_1 - \frac{1}{2}\sin 2\pi(w'_1 + 2w'_2) - \frac{1}{2}\sin 2\pi(w'_1 - 2w'_2)\right]. \end{aligned}$$

Now introduce the canonical transformation defined by the generating function

$$S = J_1\left(\frac{w'_1}{a_1} - \frac{w'_2}{a_2}\right) + J_2\frac{w'_2}{a_2}$$

to the new angle variables w_i and action variables J_i in the form

$$J_1 = a_1 J_1'$$
$$J_2 = a_2 J_2' + a_1 J_1'$$
$$w_1 = w_1'/a_1$$
$$w_2 = w_2'/a_2$$

and the new Hamiltonian H given by

$$H(w_1, w_2, J_1, J_2; \varepsilon) = J_2/2\pi - \varepsilon\left(\frac{J_1}{\pi^3}\right)^{1/2} \frac{(J_1 - J_2)}{a_1 a_2^2}$$

$$\times \left[\sin 2\pi a_1 w_1 - \frac{1}{2}\sin 2\pi(a_1 w_1 + 2a_2 w_2) - \frac{1}{2}\sin 2\pi(a_1 w_1 - 2a_2 w_2)\right].$$

(c) Note that since $\partial H/\partial J_1 = O(\varepsilon)$, w_1 is a long periodic variable (slow variable) while $\partial H/\partial J_2 = 1/2\pi + O(\varepsilon)$ implies that w_2 is a short periodic variable (fast variable).

The Hamiltonian H is slightly more general than the one considered in Section 3.7.2 in that the perturbation term also involves J_2.

Carry out the von Zeipel procedure and remove w_2 from H to $O(\varepsilon)$. Where does the first resonance condition, $a_1 = 2a_2$ first occur? Compare your results with those in Section 3.5.2 for the case $a_1 \neq 2a_2$. Discuss the meaning of the formal integral which results for the momentum conjugate to the new w_2.

(d) An equivalent more efficient approach for solving this problem is to transform H_1 to a new Hamiltonian which is separable.

This is the approach in Reference 3.5.10.

Assume a generating function $S(q_1, q_2, P_1, P_2, \varepsilon)$ from the (q_i, p_i) to (Q_i, P_i) in the form

$$S = q_1 P_1 + q_2 P_2 + \varepsilon S_1(q_1, q_2, P_1, P_2) + \varepsilon^2 S_2(q_1, q_2, P_1, P_2).$$

Deduce the conditions on S_1 and S_2 such that the new Hamiltonian is separable (i.e., corresponds to two uncoupled oscillators for the transformed variables).

6. Study the motion of a particle of unit mass in the plane in the field of force with potential $-k(\varepsilon t)/r$, where k is a given slowly varying function which might represent the mass loss of the central star.

(a) Express the Hamiltonian

$$H = \frac{1}{2}\left[p_r^2 + \frac{p_\theta^2}{r^2}\right] - \frac{k(\varepsilon t)}{r}$$

in terms of action and angle variables.

(b) Show that $J_r + J_\theta$ is an adiabatic invariant to $O(1)$ and use this result to study the behavior of the orbit as k varies.

7. (a) Continue the solution of Equation (3.7.201) by seeking a canonical transformation from $q_1, w_2, w_3, p_1, J_2, J_3$ to the variables $q_1, w_2^{(1)}, w_3^{(1)}, p_1, J_2^{(1)}, J_3^{(1)}$ such that the

new Hamiltonian $\mathscr{H}^{(2)}$ is free of $w_2^{(1)}$ to $O(\varepsilon)$. Proceed as in Section 3.7.2 by introducing the generating function

$$S = w_2 J_2^{(1)} + w_3 J_3^{(1)} + \varepsilon S_1(w_2, w_3, J_2^{(1)}, J_3^{(1)}) + O(\varepsilon^2)$$

and determine S_1.

(b) Relate the $J_2^{(1)}$, which is an adiabatic invariant to $O(\varepsilon)$ to x, y, z, $\dot{x}$, $\dot{y}$, $\dot{z}$ and discuss the physical significance of $J_2^{(1)} = \text{const.}$

8. Guided by the nature of the solution of Equations (3.7.186) by canonical perturbations, deduce an appropriate multiple variable expansion procedure which gives x, y, z directly in the form of multiple variable expansions. Compare the efficiency of your derivation by this approach with that using canonical perturbations.

9. Consider the Hamiltonian

$$H(q_1, q_2, p_1, p_2, \tilde{t}; \varepsilon) = \frac{p_1^2 + p_2^2}{2} + \frac{\omega_1^2 q_1^2 + \omega_2^2 q_2^2}{2} - \varepsilon q_1 q_2^2,$$

corresponding to Equations (3.5.163) with the following notation

$$\tilde{t} = \varepsilon t, x = q_1, y = q_2, \dot{x} = p_1, \dot{y} = q_2, a(\tilde{t}) = \omega_1; \; b(\tilde{t}) = \omega_2.$$

(a) Solve the time-independent Hamilton–Jacobi equation (cf. Equation 3.7.65) associated with the unperturbed ($\varepsilon = 0$, $\omega_i = \text{const.}$) problem to obtain W in the form

$$W = \sum_{i=1}^{2} P_i \left[\sin^{-1}\left(\frac{\omega_i}{2P_i}\right)^{1/2} q_i + \left(\frac{\omega_i}{2P_i}\right)^{1/2} q_i \left(1 - \frac{\omega_i q_i^2}{2P_i}\right)^{1/2} \right].$$

Here we have set the new momenta P_i equal to E_i/ω_i, where E_i is the energy of the ith oscillator and is constant if $\varepsilon = 0$.

(b) Now let ω_i depend on $\tilde{t}$ in the above expression for W. Regarding W as a *time-dependent* generating function of a canonical transformation $\{q_i, p_i\} \to \{Q_i, P_i\}$, show that the new coordinates Q_i, momenta P_i, and Hamiltonian $\mathscr{H}$ are given by

$$P_i = \frac{p_i^2 + \omega_i^2 q_i^2}{2\omega_i}, \qquad i = 1, 2$$

$$Q_i = \tan^{-1}\left(\frac{\omega_i q_i}{p_i}\right), \qquad i = 1, 2.$$

$$\begin{aligned}\mathscr{H}(Q_1, Q_2, P_1, P_2, \tilde{t}; \varepsilon) = {} & \omega_1 P_1 + \omega_2 P_2 + \varepsilon \frac{P_2}{2\omega_2}\left(\frac{2P_1}{\omega_1}\right)^{1/2} \sin(Q_1 - 2Q_2) \\ & + \varepsilon \frac{P_2}{2\omega_2}\left(\frac{2P_1}{\omega_1}\right)^{1/2} [\sin(Q_1 + 2Q_2) - 2\sin Q_1] \\ & + \frac{\varepsilon}{2}\left[\frac{P_1}{\omega_1}\frac{d\omega_1}{d\tilde{t}} \sin 2Q_1 + \frac{P_2}{\omega_2}\frac{d\omega_2}{d\tilde{t}} \sin 2Q_2\right].\end{aligned}$$

Note that the P_i and Q_i are normalized action and angle variables $P_i = J_i'/2\pi$, $Q_i = 2\pi w_i'$ (cf. the definitions of J_i' and w_i' in Problem 5.7.5b). Also, because the ω_i are now time-dependent we have the extra two last terms in $\mathscr{H}$ as compared with the expression for H' in Problem 5.7.5b.

(c) Show that a conventional von Zeipel transformation which removes both Q_1 and Q_2 from $\mathscr{H}$ to order ε leads to a singularity in the generating function to $O(\varepsilon)$ at resonance, i.e., for the value of $\tilde{t}$ when $\omega_1 = 2\omega_2$.

(d) Isolate the critical combination of variables $Q_1 - 2Q_2$ by the time-independent canonical transformation $\{Q_i, P_i\} \leftrightarrow \{q_i', p_i'\}$, $\mathscr{H} \leftrightarrow \mathscr{H}'$ generated by

$$V = (p_1' + p_2')Q_1 - 2p_2'Q_2$$

and use a von Zeipel transformation $\mathscr{H}' \leftrightarrow \mathscr{H}''$ to remove only $q_1' = Q_1$ to order ε from $\mathscr{H}'$. Deduce from the form of $\mathscr{H}''$ that

$$2P_1 + P_2 + \varepsilon\left\{\frac{P_2}{\omega_2}\left(\frac{2P_1}{\omega_1}\right)^{1/2}\left[\frac{2}{\omega_1 + 2\omega_2}\sin(Q_1 + 2Q_2) - \frac{2}{\omega_1}\sin Q_1\right]\right.$$
$$\left. + \frac{P_1}{\omega_1^2}\frac{d\omega_1}{d\tilde{t}}\sin 2Q_1 + \frac{P_2}{2\omega_2^2}\frac{d\omega_2}{d\tilde{t}}\sin 2Q_2\right\}$$

is an adiabatic invariant to order ε which remains valid through resonance.

(e) Solve the two first-order differential equations associated with the reduced Hamiltonian $\mathscr{H}''$ by constructing and matching appropriate two variable expansions before, during, and after resonance as in Section 3.5.3.

References

3.7.1 N. M. Krylov and N. N. Bogoliubov, *Introduction to Nonlinear Mechanics*, Acad. Sci., Ukrain. S.S.R, 1937. Translated by S. Lefschetz, Princeton University Press, Princeton, N.J., 1947.

3.7.2 N. N. Bogoliubov and Y. A. Mitropolski, *Asymptotic Methods in the Theory of Nonlinear Oscillations*, Hindustan Publishing Corp., Delhi, India, 1961.

3.7.3 Y. A. Mitropolski, *Problèmes de la Théorie Asymptotique des Oscillations Non Stationnaires*, Translated by G. Carvallo, Gauthier–Villars, Paris, France, 1966.

3.7.4 H. von Zeipel, Recherche sur le movement des petites planètes, *Ark. Astron. Mat. Fys.*, 1916, **11–13**.

3.7.5 D. Brouwer and G. M. Clemence, *Methods of Celestial Mechanics*, Academic Press, New York, 1961.

3.7.6 J. A. Morrison, Generalized method of averaging and the von Zeipel method, *Progress in Astronautics and Aeronautics* **17**, *Methods in Astrodynamics and Celestial Mechanics*, Ed. by R. L. Duncombe and V. G. Szebehely, Academic Press, N.Y., 1966, pp. 117–138.

3.7.7 H. Goldstein, *Classical Mechanics*, Addison Wesley, Reading, Massachusetts, 1950.

3.7.8 R. Courant and D. Hilbert, *Methods of Mathematical Physics*, Vol. 2, Interscience, New York, 1961.

3.7.9 J. Kevorkian, von Zeipel method and the two variable expansion procedure, *The Astron. J.* **71**, 1966, pp. 878–885.

3.7.10 A. A. Kamel, Perturbation method in the theory of nonlinear oscillations, *Celestial Mechanics* **3**, 1970, pp. 90–106.

3.7.11 W. A. Mersman, Explicit recursive algorithms for the construction of equivalent canonical transformations, *Celestial Mechanics* **3**, 1971, pp. 384–389.

3.7.12 C. S. Gardner, Adiabatic invariants of periodic classical systems, *The Physical Review* **115**, 1959, pp. 791–794.

Chapter 4

Applications to Partial Differential Equations

In this chapter, the methods developed previously are applied to partial differential equations. The plan is the same as for the cases of ordinary differential equations discussed earlier. First, the very simplest case is discussed, in which a singular perturbation problem arises. This is a second-order equation which becomes a first-order one in the limit $\varepsilon \to 0$. Following this, various more complicated physical examples of singular perturbations and boundary-layer theory are discussed. Next, the ideas of matching and inner and outer expansions are applied in some problems that are analogous to the singular boundary problems of Section 2.7. The final section deals with multiple variable expansions for partial differential equations, and several applications dealing with different aspects of the procedure are discussed.

4.1 Limit-Process Expansions for Second-Order Partial Differential Equations

In this section, a study is made of the simplest problems for partial differential equations that lead to boundary layers; that is, which are singular in the sense of Section 2.2 and 2.7. The aim is to base the discussion as much as possible on the mathematical situation. The simplest nontrivial case is that of a second-order equation which drops to a first-order equation as the small parameter $\varepsilon \to 0$. It is clear that some of the boundary data cannot be satisfied by the limit equation, so that boundary layers (in general) occur. The analogous case of a first-order partial differential equation reducing to a zero-order equation as $\varepsilon \to 0$, is, by the theory of characteristics for first-order equations, equivalent to a problem in ordinary differential equations, and is not discussed here.

Consider

$$\varepsilon\left\{\alpha_{11}\frac{\partial^2 u}{\partial x^2} + 2\alpha_{12}\frac{\partial^2 u}{\partial x\,\partial y} + \alpha_{22}\frac{\partial^2 u}{\partial y^2}\right\} = a\frac{\partial u}{\partial x} + b\frac{\partial u}{\partial y} \tag{4.1.1}$$

with constant coefficients. In the case, where the coefficients are functions of (x, y) the solutions can be expected to behave locally in the same way as the constant coefficient equation. However, nonlinearities especially in the lower-order operator usually introduce new effects.

A boundary or initial-value problem for Equation (4.1.1) which leads to a unique solution $u(x, y; \varepsilon)$ is considered. The kind of boundary-value problem that makes sense for Equation (4.1.1) depends on the type of the equation which is a property only of the highest-order differential operator appearing in that equation:

$$L_2 u \equiv \alpha_{11}\frac{\partial^2 u}{\partial x^2} + 2\alpha_{12}\frac{\partial^2 u}{\partial x\,\partial y} + \alpha_{22}\frac{\partial^2 u}{\partial y^2}. \tag{4.1.2}$$

The main type classifications of the operator L_2 and some significant properties are as follows.

I. *Elliptic Type*

No real characteristics. A point-disturbance solution of $L_2 u = 0$ influences the entire space. Thus, one boundary condition u, or a normal derivative, or combination is prescribed on a closed boundary. The simplest case and canonical form is the Laplace equation $u_{xx} + u_{yy} = 0$.

II. *Hyperbolic Type*

Real characteristic curves exist in the xy-plane and form a coordinate system. A point disturbance solution of $L_2 u = 0$ influences a restricted portion, bounded by the characteristic curves, of the (x, y) space. The direction of propagation (future) is not necessarily implicit in the equation and then must be assigned. Two initial conditions on a space-like arc define a solution in a domain bounded by characteristics. One boundary condition is assigned on a time-like arc. A characteristic value problem assigns a compatible boundary condition on a characteristic curve. The simplest case is the wave equation $u_{xx} - u_{yy} = 0$.

III. *Parabolic Type*

A dividing case between I and II, in which the real characteristics coalesce. Half the space is influenced by a point disturbance. One initial condition on a characteristic arc or one boundary condition is prescribed. The simplest case is $u_{xx} = u_y$.

If $\phi(x, y) = \text{const.}$ is the equation of a family of characteristic curves for L_2, the equation of these curves is

$$\alpha_{11}\left(\frac{\partial\phi}{\partial x}\right)^2 + 2\alpha_{12}\left(\frac{\partial\phi}{\partial x}\right)\left(\frac{\partial\phi}{\partial y}\right) + \alpha_{22}\left(\frac{\partial\phi}{\partial y}\right)^2 = 0 \tag{4.1.3}$$

or, if the slope of a characteristic $\zeta = dy/dx$ on $(\phi = \text{const.}) = -\phi_x/\phi_y$ is introduced, we have

$$\alpha_{11}\zeta^2 - 2\alpha_{12}\zeta + \alpha_{22} = 0. \tag{4.1.4}$$

The type of classification of Equation (4.1.1) thus depends only on the discriminant and we have three possibilities:

I. Elliptic: $\alpha_{12}^2 - \alpha_{11}\alpha_{22} < 0$, no real roots,
II. Hyperbolic: $\alpha_{12}^2 - \alpha_{11}\alpha_{22} > 0$, two real directions $\zeta_{\pm}$,
III. Parabolic: $\alpha_{12}^2 - \alpha_{11}\alpha_{22} = 0$, double root $\zeta = \alpha_{12}/\alpha_{11}$.

A separate discussion is given for each type of equation.

4.1.1 Linear Elliptic Equations, $\alpha_{12}^2 - \alpha_{11}\alpha_{22} < 0$

Since α_{11}, α_{22} must be of the same algebraic sign, let $\alpha_{11} > 0$, $\alpha_{22} > 0$, and note that

$$|\alpha_{12}| < \sqrt{\alpha_{11}\alpha_{22}}; \qquad \alpha_{11}\alpha_{22} > 0. \tag{4.1.5}$$

Consider, at first, an interior boundary-value problem with $u = u_B(P_B)$, a prescribed function of position on a closed smooth boundary curve (Fig. 4.1.1). Here u_B is independent of ε.

This set of boundary conditions defines a unique regular solution $u(x, y; \varepsilon)$ at all points interior to and on the boundary. As $\varepsilon \to 0$ with (x, y) fixed in the

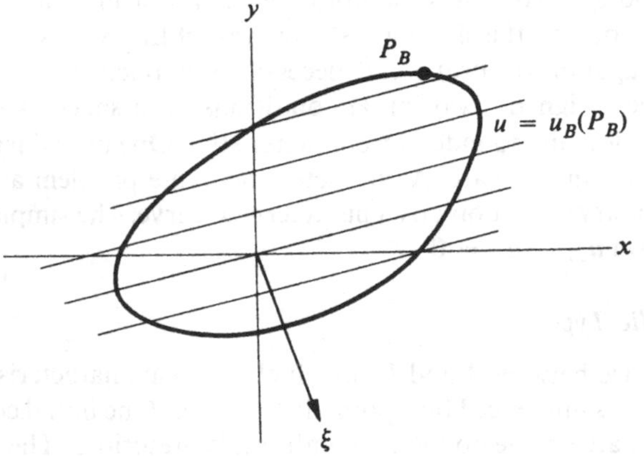

Figure 4.1.1 Boundary-Value Problem

interior of the domain, the exact solution approaches the outer solution, which can be thought of as the first term of an outer expansion (valid off the boundary):

$$\lim_{\substack{\varepsilon \to 0 \\ x,\, y \text{ fixed}}} u(x, y; \varepsilon) = u_0(x, y). \tag{4.1.6}$$

Here $u_0(x, y)$ satisfies the limit equation of first-order,

$$a \frac{\partial u_0}{\partial x} + b \frac{\partial u_0}{\partial y} = 0. \tag{4.1.7}$$

Solutions of Equation (4.1.7) are expressed in terms of the characteristic curves:

$$\xi = bx - ay = \text{const.}$$

These are defined parametrically by $dx/ds = a$, $dy/ds = b$; $du/ds = 0$ in accord with the familiar interpretation of Eq. (4.1.7) as a derivative in the direction with slope b/a.

We call the lines $\xi = \text{const.}$ subcharacteristics of the original equation (4.1.1). The main underlying structure of the solution is given by these subcharacteristics since $u_0 = u_0(\xi)$. The subcharacteristic curves are sketched in Figure 4.1.1. It is clear that the boundary condition on one side of the domain is sufficient to define $u_0(x, y) = u_0(\xi)$ in the whole domain. Also, $u_0(\xi)$, in general, does not satisfy the boundary condition on the other side of the domain, so that a boundary layer is needed.

In order to study this boundary layer and its matching to $u_0(\xi)$, it is convenient to introduce an orthogonal coordinate system (ξ, η) based on the subcharacteristic ξ. The coordinate η is chosen here arbitrarily, for convenience. Different choices of η do not affect the essential boundary-layer character but may influence the form of higher-order corrections.

$$\xi = bx - ay, \qquad \eta = ax + by. \tag{4.1.8}$$

The transformation formulas are

$$\begin{aligned}
&\frac{\partial u}{\partial x} = b \frac{\partial u}{\partial \xi} + a \frac{\partial u}{\partial \eta}, \qquad \frac{\partial u}{\partial y} = -a \frac{\partial u}{\partial \xi} + b \frac{\partial u}{\partial \eta}, \\
&\frac{\partial^2 u}{\partial x^2} = b^2 \frac{\partial^2 u}{\partial \xi^2} + 2ab \frac{\partial^2 u}{\partial \xi\, \partial \eta} + a^2 \frac{\partial^2 u}{\partial \eta^2}, \\
&\frac{\partial^2 u}{\partial x\, \partial y} = -ab \frac{\partial^2 u}{\partial \xi^2} + (b^2 - a^2) \frac{\partial^2 u}{\partial \xi\, \partial \eta} + ab \frac{\partial^2 u}{\partial \eta^2}, \\
&\frac{\partial^2 u}{\partial y^2} = a^2 \frac{\partial^2 u}{\partial \xi^2} - 2ab \frac{\partial^2 u}{\partial \xi\, \partial \eta} + b^2 \frac{\partial^2 u}{\partial \eta^2}.
\end{aligned} \tag{4.1.9}$$

The original equation (4.1.1) now reads

$$\varepsilon\left[A_{11}\frac{\partial^2 u}{\partial\xi^2} + 2A_{12}\frac{\partial^2 u}{\partial\xi\,\partial\eta} + A_{22}\frac{\partial^2 u}{\partial\eta^2}\right] = \frac{\partial u}{\partial\eta}, \tag{4.1.10}$$

where the coefficients A_{ij} are given by

$$\begin{aligned}(a^2 + b^2)A_{11} &= \alpha_{11}b^2 - 2\alpha_{12}ab + \alpha_{22}a^2,\\ (a^2 + b^2)A_{12} &= \alpha_{11}ab + \alpha_{12}(b^2 - a^2) - \alpha_{22}ab,\\ (a^2 + b^2)A_{22} &= \alpha_{11}a^2 + 2\alpha_{12}ab + \alpha_{22}b^2.\end{aligned} \tag{4.1.11}$$

Since the equation is still elliptic, we know that the discriminant

$$A_{12}^2 - A_{11}A_{22} < 0 \tag{4.1.12}$$

and, further, that

$$(a^2 + b^2)A_{22} = (\sqrt{\alpha_{11}}a \pm \sqrt{\alpha_{22}}b)^2 \mp 2ab(\sqrt{\alpha_{11}\alpha_{22}} \mp \alpha_{12}) > 0 \tag{4.1.13}$$

if we choose the upper sign for $ab < 0$, and the lower for $ab > 0$.

The original domain has an image denoted by $\eta = \eta_B(\xi)$ in the $\xi\eta$-plane (Fig. 4.1.2). Let the assigned boundary values be denoted $u = u_U(\xi)$ on the upper part of the domain between A and B, and $u = u_L(\xi)$ on the lower part.

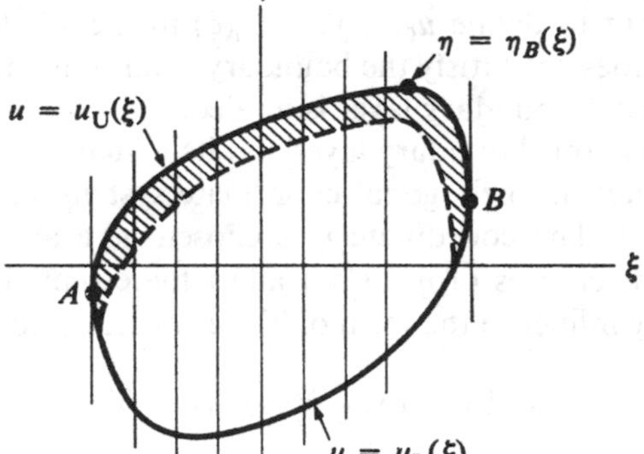

Figure 4.1.2 Subcharacteristic Coordinates

In the region of the boundary layer, we expect $\partial/\partial\eta$ to be large, so that a boundary-layer coordinate

$$\eta^* = \frac{\eta - \eta_B(\xi)}{\delta(\varepsilon)} \tag{4.1.14}$$

and an associated limit process can be introduced ($\varepsilon \to 0$, η^*, ξ fixed). The largest order terms on the left-hand side of equation (4.1.10) are, then, $O(\varepsilon/\delta^2)$, and the term on the right is $O(1/\delta)$. Thus, the boundary layer has thickness ε, and it follows that

$$\eta^* = \frac{\eta - \eta_B(\xi)}{\varepsilon}. \tag{4.1.15}$$

The boundary-layer solution is represented as the first term of an asymptotic expansion

$$u(x, y; \varepsilon) = u_{\mathrm{BL}}(\xi, \eta^*) + \cdots. \tag{4.1.16}$$

Note that, under the transformation from (ξ, η) to (ξ, η^*), we have

$$\frac{\partial u}{\partial \eta} \to \frac{1}{\varepsilon}\frac{\partial u}{\partial \eta^*}, \qquad \frac{\partial u}{\partial \xi} \to -\frac{1}{\varepsilon}\left(\frac{d\eta_B}{d\xi}\right)\frac{\partial u}{\partial \eta^*} + \cdots,$$

so that the boundary-layer equation derived from Equation (4.1.10) is

$$K(\xi)\frac{\partial^2 u_{\mathrm{BL}}}{\partial \eta^{*2}} = \frac{\partial u_{\mathrm{BL}}}{\partial \eta^*}, \tag{4.1.17}$$

where

$$K(\xi) = A_{11}\left(\frac{d\eta_B}{d\xi}\right)^2 - 2A_{12}\left(\frac{d\eta_B}{d\xi}\right) + A_{22}. \tag{4.1.18}$$

Notice that the boundary-layer equation is an ordinary differential equation in this case. The assumed orders of magnitude are certainly all right, unless $d\eta_B/d\xi \to \infty$. The important case where $d\eta_B/d\xi = \infty$ along an arc, that is, the case of subcharacteristic boundary, will be discussed later. The location of the boundary layer is now decided by the criterion that it must match with the outer solution $u_0(\xi)$.

This means (exponential) decay of the boundary-layer dependence on η^* in this case. The solution of Equation (4.1.17) is

$$u_{\mathrm{BL}}(\xi, \eta^*) = A(\xi) + B(\xi)e^{\eta^*/K(\xi)}. \tag{4.1.19}$$

Everything depends on the sign of $K(\xi)$. Since the equation is elliptic, there are no real roots $d\eta_B/d\xi$ of Equation (4.1.18), which is nothing but the characteristic form (Equation 4.1.4) expressed in A_{ij}. Thus, $K(\xi)$ does not vanish, and sign K = sign $A_{22} > 0$ from Equation (4.1.13).

Exponential decay occurs as $\eta^* \to -\infty$; the boundary layer appears on the upper boundary in Fig. 4.1.2. With matching, the solution is thus

$$u_{\mathrm{BL}}(\xi, \eta^*) = u_{\mathrm{L}}(\xi) + \{u_{\mathrm{U}}(\xi) - u_{\mathrm{L}}(\xi)\}e^{\eta^*/K(\xi)}, \qquad u_0(\xi) = u_{\mathrm{L}}(\xi). \tag{4.1.20}$$

The boundary-layer solution in this case is also the first term of a uniformly valid composite expansion. The location of the boundary layer in the original figure (Figure 4.1.1) depends on the orientation of the coordinates (ξ, η); that is, only on the signs of (a, b). A qualitative sketch of the possibilities is given in Figure 4.1.3.

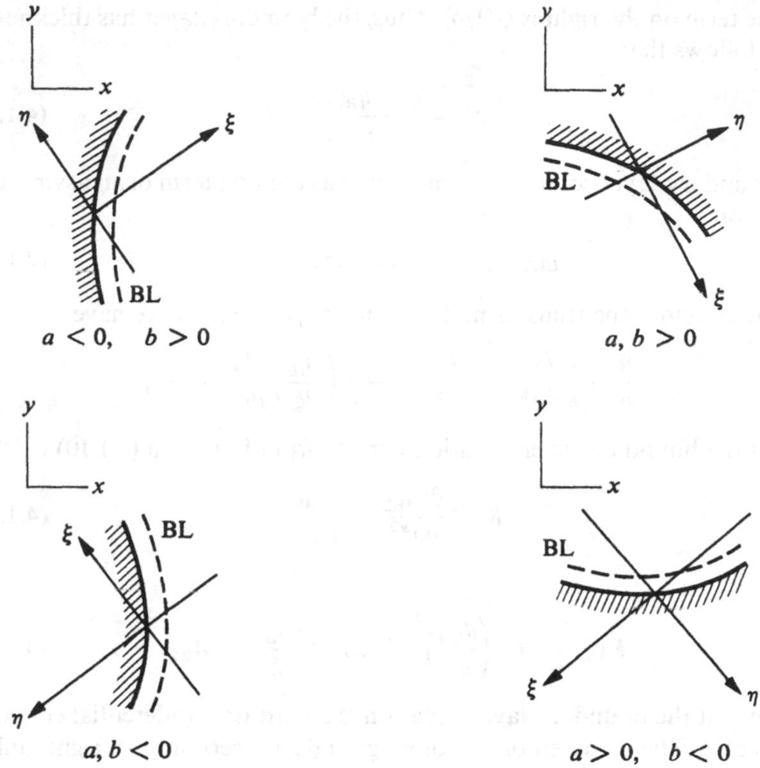

Figure 4.1.3 Various Locations of Boundary Layers

We still have to discuss the case where a segment of the boundary is subcharacteristic, say $\xi = \xi_S =$ const. as in Figure 4.1.4.

The outer solution carries the constant $u_L(\xi_S)$ as the boundary is approached, so that the boundary condition $u = u_S(\eta)$ is violated. In this case, it can be expected that $\partial/\partial\xi$ is large in the layer, so that the suitable boundary-

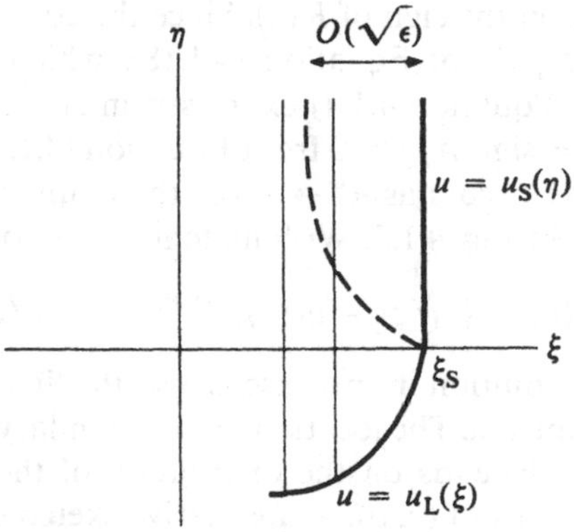

Figure 4.1.4 Boundary Layer on a Subcharacteristic

layer coordinate (balancing $\partial^2/\partial\xi^2$ and $\partial/\partial\eta$) is

$$\xi^* = \frac{\xi - \xi_S}{\sqrt{\varepsilon}}, \tag{4.1.21}$$

and the boundary-layer thickness is $O(\sqrt{\varepsilon})$. The boundary-layer solution is the first term of an asymptotic expansion

$$u(\xi, \eta; \varepsilon) = u^*_{BL}(\xi^*, \eta) + \cdots \tag{4.1.22}$$

associated with the limit ($\varepsilon \to 0$, ξ^*, η fixed). Thus, Equation (4.1.10) yields the boundary-layer equation

$$A_{11}\frac{\partial^2 u^*_{BL}}{\partial \xi^{*2}} = \frac{\partial u^*_{BL}}{\partial \eta}. \tag{4.1.23}$$

Here A_{11} = const., so that the boundary-layer equation is the heat equation, a partial differential equation in this case.

Also $A_{11} > 0$, just as A_{22} (cf. Equation 4.1.13), so that $+\eta$ is the timelike coordinate. The solution of Equation (4.1.23) is to be found satisfying the boundary condition $u^*_{BL} = u_S(\eta)$ on $\xi^* = 0$ and matching to $u_0(\xi)$. The matching as usual can be carried out by an intermediate limit and results in

$$u^*_{BL}(-\infty, \eta) = u_L(\xi_S). \tag{4.1.24}$$

Various representations can be used for the solution of the heat equation. For example, if a source representation is used on $\xi^* = 0$, we have

$$u^*_{BL}(\xi^*, \eta) = u_L(\xi_S) + \frac{1}{2\sqrt{\pi A_{11}}}\int_0^{\eta} \exp\frac{-\xi^{*2}}{4A_{11}(\eta - \bar{\eta})}. \tag{4.1.25}$$

This matches u_0 as $\varepsilon^* \to -\infty$, and the source strength $Q(\eta)$ is found from the solution of Abel's integral equation on the boundary $\xi^* = 0$,

$$u_S(\eta) - u_L(\xi_S) = \frac{1}{2\sqrt{\pi A_{11}}}\int_0^{\eta}\frac{Q(\bar{\eta})}{\sqrt{\eta - \bar{\eta}}}\,d\bar{\eta}. \tag{4.1.26}$$

The solution of the integral equation is

$$Q(\eta) = \frac{1}{\pi}\frac{d}{d\eta}\int_0^{\eta}\frac{2[u_S(\bar{\eta}) - u_L(\xi_S)]\sqrt{\pi A_{11}}}{\sqrt{\eta - \bar{\eta}}}\,d\bar{\eta}. \tag{4.1.27}$$

Alternatively, a doublet representation could be used directly. In any case, the boundary-layer spreads upwards in Figure 4.1.4 from its origin at ($\eta = 0$, $\xi = \xi_S$). For example, if u_S = const., the boundary-layer solution is

$$\begin{aligned} u^*_{BL}(\xi^*, \eta) &= u_S + [u_L(\xi_S) - u_S)]\operatorname{erf}\frac{\xi^*}{2\sqrt{\eta A_{11}}} \\ &= u_S + [u_L(\xi_S) - u_S]\operatorname{erf}\frac{\xi - \xi_S}{2\sqrt{\varepsilon\eta A_{11}}}. \end{aligned} \tag{4.1.28}$$

It has been tacitly assumed here that the "initial condition" ($\eta = 0$) for the boundary-layer equation is the same as the condition at the edge of the boundary layer [$u^*_{BL} \to u_L(\xi_S)$ either as ($\xi^* \to -\infty$, η fixed) or as ($\eta \to 0$, ξ^* fixed]. A rigorous treatment of the initial condition demands consideration of an approximation valid near $\xi = \xi_S$, $\eta = 0$, which we do not give.

It can be expected that more general elliptic cases have a similar structure. For example, for strictly linear equations where

$$a = a(x, y) \quad \text{and} \quad b = b(x, y), \tag{4.1.29}$$

the subcharacteristics are not straight but form a family of curves defined by the differential equation

$$\left(\frac{dy}{dx}\right) = \frac{b(x, y)}{a(x, y)}. \tag{4.1.30}$$

If no singular points of this equation occur in the domain of interest, then the behavior is qualitatively the same as before. If singular points do occur, then certain complications may arise. An example will be considered presently.

When the lower-order operator is quasi-linear, that is, when

$$a = a(x, y, u), \qquad b = b(x, y, u).$$

the solution of the limit equation $\varepsilon = 0$ (Equation 4.1.7) can become many-valued (steepening of fronts), and jumps in u (shock waves) may have to be inserted in order to obtain a one-valued solution of the lower order operator. As long as the second order operator is linear, jumps in u (or its first derivatives) can only occur for the case of weak solutions of the hyperbolic problem, and in this case, these jumps are only permitted along characteristics of the second order operator. Thus, in all of the three types of problems associated with the linear second-order operator (4.1.2), the shocks inserted to obtain a one-valued solution of the outer problem must be smoothed out by introducing suitable interior layer equations. Since the basic problem is the same in all three classes of equations, we will defer our study of this question to Section 4.1.3 where we consider a parabolic equation with a quasi-linear first order operator.

Finally, if the higher-order operator is quasi-linear, that is,

$$\alpha_{ij} = \alpha_{ij}(x, y, u), \tag{4.1.31}$$

then the type of the equation is not known in advance in a given (x, y) domain but depends on the solution u. It is clear that various complications can arise, although a knowledge of the outer expansion should enable the overall-type structure to be sketched. We do not consider any examples of this situation.

We now study an elliptic problem with variable coefficients in the operator for the outer limit.

$$\varepsilon(u_{xx} + u_{yy}) = xu_x + yu_y. \tag{4.1.32a}$$

The subcharacteristics are the family of straight lines $y/x = \text{const}$. Therefore, the origin is a source point for this family. Let us solve the interior problem of Equation (4.1.32a) in the unit circle $x^2 + y^2 = 1$ with prescribed boundary values of u.

It is more convenient to transform the problem to polar coordinates r, θ, and we have to satisfy

$$\varepsilon\left(u_{rr} + \frac{u_r}{r} + \frac{u_{\theta\theta}}{r^2}\right) = ru_r \tag{4.1.32b}$$

inside $0 \leq r \leq 1$, with boundary condition

$$u(1, \theta) = f(\theta), \tag{4.1.33}$$

prescribed.

The outer limit is $u_r = 0$, i.e., $u_0 = A(\theta)$. This means that u is constant along rays $y/x = \text{constant}$. It follows by consideration of the solution near $r = 0$, where there is no singularity for $\varepsilon > 0$, that A must be a constant, say $A = \alpha$.

In order to satisfy the boundary condition at $r = 1$, we introduce a boundary layer there. It is easily seen that

$$r^* = \frac{r-1}{\varepsilon} \tag{4.1.34}$$

is the appropriate boundary layer variable. Hence Eq. (4.1.32b) in the limit $\varepsilon \to 0$, r^* fixed, reduces to

$$\frac{\partial^2 u}{\partial r^{*2}} - \frac{\partial u}{\partial r^*} = O(\varepsilon) \tag{4.1.35}$$

and we have the boundary-layer limit

$$u_{\text{BL}} = a(\theta) + [f(\theta) - a(\theta)]e^{r^*}. \tag{4.1.36}$$

Matching with the outer limit u_0 shows that

$$a(\theta) = \alpha = \text{const.} \tag{4.1.37}$$

but there is no information on the value of this constant.

This is the analog, for partial differential equations, of the problem we encountered in Section 2.3.4, and is also discussed in Reference 2.3.1. A boundary layer is possible everywhere along the circle $r = 1$, and matching does not determine the unknown constant α.

Motivated by the resolution of the difficulty for the ordinary differential equation case, we seek a variational principle for which Equation (4.1.32) is the associated Euler equation. We will then use the variational principle to calculate the unknown constant α.

Thus, we seek a Lagrangian $L(x, y, u, u_x, u_y)$ such that Equation (4.1.32a) corresponds to Euler's equation [cf. Reference 4.1.1]

$$\frac{\partial}{\partial x}\left(\frac{\partial L}{\partial u_x}\right) + \frac{\partial}{\partial y}\left(\frac{\partial L}{\partial u_y}\right) - \frac{\partial L}{\partial u} = 0 \tag{4.1.38}$$

associated with the variational principle

$$\delta J \equiv \delta\left[\iint_{\sqrt{x^2+y^2}\le 1} L(x, y, u, u_x, u_y)dx\, dy\right] = 0 \tag{4.1.39}$$

with u prescribed on the boundary.

For the Laplacian operator, interpreted as defining the equilibrium deflection u of a membrane, it is well known that $L = (u_x^2 + u_y^2)/2$, and the variational principle (4.1.39) is the principle of least potential energy.

Guided by this result, we attempt to calculate a Lagrangian associated with Equation (4.1.32b). As, in Section 2.3.4, we first eliminate the first derivative term on the right hand side by introducing the transformation

$$v(r, \theta) = u(r, \theta)e^{-r^2/4\varepsilon}.$$

Equation (4.1.32b) becomes

$$\varepsilon\left(v_{rr} + \frac{v_r}{r} + \frac{v_{\theta\theta}}{r^2}\right) + \left(1 - \frac{r^2}{4\varepsilon}\right)v = 0. \tag{4.1.40}$$

Now, since the Lagrangian associated with the Laplacian-operator is known, we can easily extend the definition to include the added linear term in v in Equation (4.1.40).

We find

$$L(r, \theta, v, v_r, v_\theta) = \frac{\varepsilon}{2}\left(v_r^2 + \frac{v_\theta^2}{r^2}\right) - \frac{1}{2}\left(1 - \frac{r^2}{4\varepsilon}\right)v^2. \tag{4.1.41a}$$

Thus, Equation (4.1.40) is the Euler equation associated with

$$\delta\left\{\frac{1}{2}\int_0^1\int_0^{2\pi}\left[\varepsilon\left(v_r^2 + \frac{v_\theta^2}{r^2}\right) - \left(1 - \frac{r^2}{4\varepsilon}\right)v^2\right]r\, dr\, d\theta\right\} = 0. \tag{4.1.41b}$$

Hence, Equation (4.1.32b) is the Euler equation for the variational principle

$$\delta J = 0, \qquad \delta u = 0 \quad \text{on } r = 1, \tag{4.1.42a}$$

where

$$J = \frac{1}{2}\int_0^1\int_0^{2\pi}\left[\varepsilon\left(u_r^2 + \frac{u_\theta^2}{r^2}\right) + \left(\frac{r^2u^2}{2\varepsilon} - u^2 - ruu_r\right)\right]e^{-r^2/2\varepsilon}r\, dr\, d\theta. \tag{4.1.42b}$$

Again, we point out that the form of the Lagrangian is not unique [cf. Problem 2.3.3], and Equation (4.1.42b) is somewhat simpler than the expression given in Reference 2.3.1.

It is easily verified that the Euler equation

$$\frac{\partial}{\partial r}\left(r\frac{\partial L_1}{\partial u_r}\right)+\frac{\partial}{\partial \theta}\left(r\frac{\partial L_1}{\partial u_\theta}\right)-\frac{\partial}{\partial u}(rL_1)=0 \tag{4.1.43a}$$

with

$$L_1(r,\theta,u,\mu_r,u_\theta)=\left[\frac{\varepsilon}{2}\left(u_r^2+\frac{u_\theta^2}{r^2}\right)+\frac{1}{2}\left(\frac{r^2u^2}{2\varepsilon}-u^2-ruu_r\right)\right]e^{-r^2/2\varepsilon} \tag{4.1.43b}$$

gives Eq. (4.1.32b).

What we need to do now is to evaluate L_1 along the "solution"

$$u=\alpha+[f(\theta)-\alpha]e^{-(1-r)/\varepsilon} \tag{4.1.44}$$

which is uniformly valid to $O(1)$ in the interior of the unit circle. The calculations are left as an exercise [cf. Problem 4.1.4]. Knowing L_1 to leading order then defines J to leading order and α is the solution of

$$\frac{\partial J}{\partial \alpha}=0. \tag{4.1.45}$$

The result is

$$\alpha=\frac{1}{2\pi}\int_0^{2\pi}f(\theta)d\theta, \tag{4.1.46}$$

i.e., α is the average value of the boundary data.

The situation when the singular point is not at the center and for a more general boundary is explored in Problem 4.1.5. Another example for which the subcharacteristics have a saddle point in the interior of the domain is given in Problem 4.1.6.

4.1.2 Linear Hyperbolic Equations, $\alpha_{12}^2-\alpha_{11}\alpha_{22}>0$

Typically, linear hyperbolic equations describe perturbations about a uniform solution for a quasi-linear hyperbolic system. The essential features are illustrated by considering L_2 to be the simple wave operator, and t a coordinate corresponding to time. Assuming a suitable length scale, and normalizing the time by (length/signal speed), a dimensionless equation can be written

$$\varepsilon\left\{\frac{\partial^2 u}{\partial x^2}-\frac{\partial^2 u}{\partial t^2}\right\}=a\frac{\partial u}{\partial x}+b\frac{\partial u}{\partial t}. \tag{4.1.47}$$

Equation (4.1.47) has real characteristics (r, s)

$$r=t-x,\qquad s=t+x. \tag{4.1.48}$$

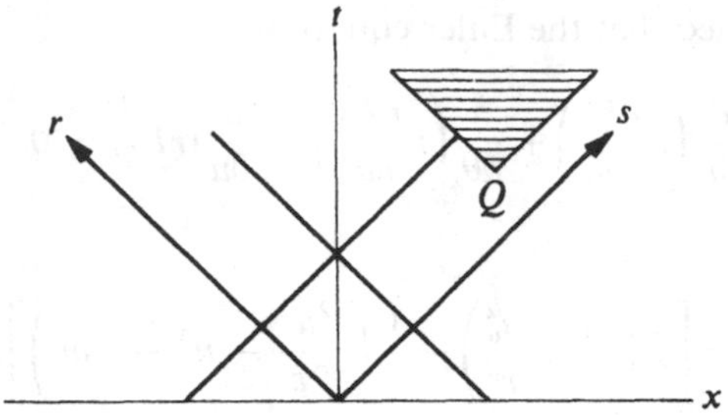

Figure 4.1.5 Characteristic Coordinate System

The characteristics serve to define the region of influence, propagating into the future, of a disturbance at a point Q (Figure 4.1.5). For the specification of a boundary-value problem for Equation (4.1.47), the number of boundary conditions to be specified on an arc depends on the nature of the arc with respect to the characteristic directions of propagation. For example, on $t = 0$ typical initial conditions (u, u_t) must be given in order to find the solution for $t > 0$. For $x = 0$, one boundary condition (for example, u) must be given to define the solution (signal propagating from the boundary) for $x > 0$. Along $t = 0$, two characteristics lead from the boundary into the region of interest, but for $x = 0$, only one does. Generalizing this idea, the directions of an arc, with respect to the characteristic directions and the future directions, can be classified as timelike, spacelike, or characteristic. (See Figure 4.1.6, where the characteristics have arrows pointing to the future.)

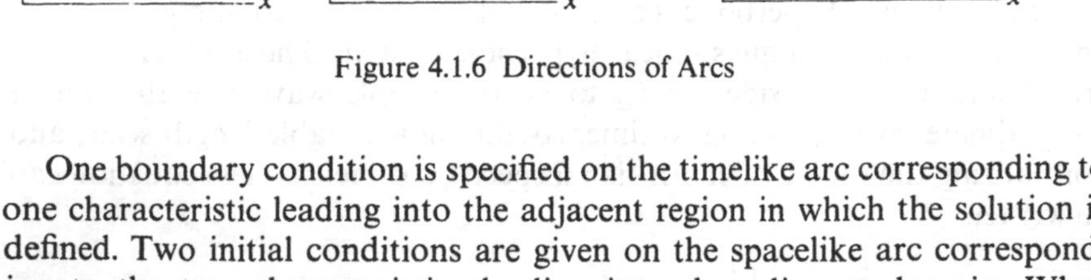

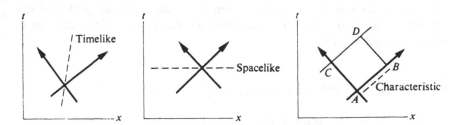

Figure 4.1.6 Directions of Arcs

One boundary condition is specified on the timelike arc corresponding to one characteristic leading into the adjacent region in which the solution is defined. Two initial conditions are given on the spacelike arc corresponding to the two characteristics leading into the adjacent domain. When the boundary curves are characteristic, only one condition can be prescribed, and the characteristic relations must hold. The characteristic initial-value problem prescribes one condition on AB and on AC to define the solution in $ABCD$.

Now consider the initial-value problem in $-\infty < x < \infty$ for Equation (4.1.47):

$$u(x, 0) = F(x), \tag{4.1.49}$$

$$u_t(x, 0) = G(x). \tag{4.1.50}$$

(see Figure 4.1.7).

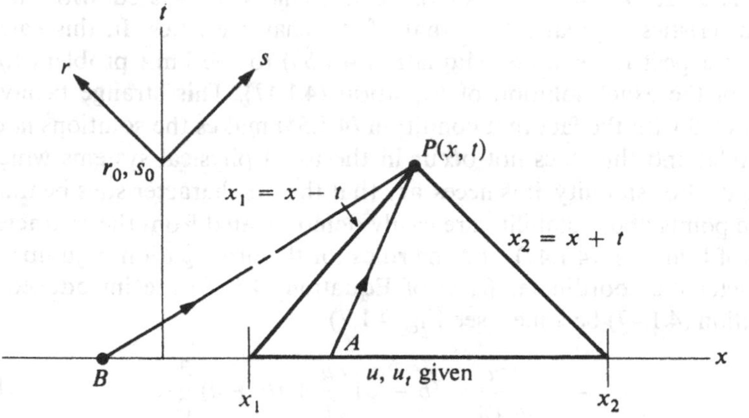

Figure 4.1.7 Initial Value Problem

According to the general theory, the solutions at a point $F(x, t)$ can depend only on that part of the initial data which can send a signal to P, the part cut out of the initial line by the backward running characteristics through P, $(x_1 < x < x_2)$. Now consider what happens as $\varepsilon \to 0$. In particular, consider the behavior of the limit equation

$$a\frac{\partial u}{\partial x} + b\frac{\partial u}{\partial t} = 0. \tag{4.1.51}$$

The solution in this case depends only on the data connected to P along a subcharacteristic of the original equation

$$bx - at = \text{const.} \tag{4.1.52}$$

The general solution of Equation (4.1.51) has the form

$$u(x, t) = f\left(x - \frac{a}{b}t\right). \tag{4.1.53}$$

Now if the subcharacteristic originates at point A between x_1 and x_2, that is if

$$\left|\frac{b}{a}\right| > 1, \tag{4.1.54}$$

it is reasonable to conceive of Equation (4.1.53) as a limiting form of the exact solution. In this case, as $\varepsilon \to 0$, it is only the data at point A which affects the solution in the limit. However, if

$$\left|\frac{a}{b}\right| > 1, \tag{4.1.55}$$

the subcharacteristic to P lies outside the usual domain of influence and originates at B. The speed of the disturbances associated with the subcharacteristics is greater than that of the characteristics. In this case, one cannot expect the solution (Equation 4.1.53) to the limit problem to be a limit of the exact solution of Equation (4.1.47). This strange behavior is connected with the fact that condition (4.1.55) makes the solutions $u(x, t; \varepsilon)$ unstable; and this does not occur in the usual physical systems which are studied.[1] For stability, it is necessary that the subcharacteristics be timelike. These points about stability are easily demonstrated from the characteristic form of Equation (4.1.47), and the rules for the propagation of jumps. If the characteristic coordinates (r, s) of Equation (4.1.48) are introduced, then Equation (4.1.47) becomes (see Fig. 4.1.7)

$$-4\varepsilon \frac{\partial^2 u}{\partial r\, \partial s} = (b - a)\frac{\partial u}{\partial r} + (b + a)\frac{\partial u}{\partial s}, \tag{4.1.56a}$$

and

$$\frac{\partial u}{\partial x} = -\frac{\partial u}{\partial r} + \frac{\partial u}{\partial s}, \qquad \frac{\partial u}{\partial t} = \frac{\partial u}{\partial r} + \frac{\partial u}{\partial s}. \tag{4.1.56b}$$

Consider now the propagation along $(r = r_0 = \text{const.})$ of a jump in the derivative $(\partial u/\partial r)$. Let

$$\kappa = \left[\frac{\partial u}{\partial r}\right]_{r=r_0} \equiv \frac{\partial u}{\partial r}(r_0^+, s) - \frac{\partial u}{\partial r}(r_0^-, s). \tag{4.1.57}$$

Assuming that u and $\partial u/\partial s$ are continuous across $r = r_0$, we can evaluate Equation (4.1.56a) at r_0^+ and r_0^- and form the difference to obtain

$$-4\varepsilon \frac{\partial \kappa}{\partial s} = (b - a)\kappa. \tag{4.1.58}$$

The solution has the form

$$\kappa = \kappa_0 \exp\left[-\frac{b - a}{4s}(s - s_0)\right]. \tag{4.1.59}$$

[1] In particular, the small perturbation assumption, which is implicit in the derivation of the linear problem (4.1.47), is violated if u is unstable.

A jump across a characteristic propagates to infinity along that characteristic. If

$$b - a > 0, \quad \text{we have exponential decay and stability;}$$

if

$$b - a < 0, \quad \text{we have exponential growth and instability.}$$

A parallel discussion for jumps in $\partial u/\partial s$ across a characteristic $s = s_0$ gives the following results:

$$b + a > 0 \quad \text{implies exponential decay and stability;}$$
$$b + a < 0 \quad \text{implies exponential growth and instability.}$$

Combining these relations, we see that for stability we must have

$$\frac{b}{|a|} > 1. \tag{4.1.60}$$

Thus, we restrict the further discussion to the stable case in which Equation (4.1.60) is satisfied and study first the initial-value problem specified by Equations (4.1.49) and (4.1.50) as $\varepsilon \to 0$.

Initial-Value Problem

Since the limit solution (4.1.53) can only satisfy one initial condition, the existence of an initial boundary layer, analogous to that discussed in Section 2.2, can be expected. An initially valid expansion can be expressed in the coordinates (t^*, x), where

$$t^* = \frac{t}{\delta(\varepsilon)}, \qquad \delta \to 0. \tag{4.1.61}$$

The associated limit process has (x, t^* fixed) and consists of a "vertical" approach to the initial line. The expansion has the form

$$u(x, t; \varepsilon) = U_0(x, t^*) + \beta_1(\varepsilon)U_1(x, t^*) + \cdots. \tag{4.1.62}$$

In order to satisfy the initial conditions independent of ε, we need

$$\beta_1(\varepsilon) = \delta,$$

and then we obtain

$$\frac{\partial u}{\partial t}(x, t; \varepsilon) = \frac{1}{\delta}\frac{\partial U_0}{\partial t^*} + \frac{\partial U_1}{\partial t^*} + \cdots. \tag{4.1.63}$$

Thus, to take care of the initial conditions (Equations 4.1.49, 4.1.50), we need

$$U_0(x, 0) = F(x), \qquad U_1(x, 0) = U_2(x, 0) = \cdots = 0, \tag{4.1.64}$$

$$\frac{\partial U_0}{\partial t^*}(x, 0) = 0, \qquad \frac{\partial U_1}{\partial t^*}(x, 0) = G(x), \qquad \frac{\partial U_2}{\partial t^*}(x, 0) = \cdots = 0. \tag{4.1.65}$$

Using the inner expansion in Equation (4.1.47), we have

$$\varepsilon\left\{\frac{\partial^2 U_0}{\partial x^2} + \delta\frac{\partial^2 U_1}{\partial x^2} + \cdots - \frac{1}{\delta^2}\frac{\partial^2 U_0}{\partial t^{*2}} - \frac{1}{\delta}\frac{\partial^2 U_1}{\partial t^{*2}} + \cdots\right\}$$

$$= a\frac{\partial U_0}{\partial x} + \cdots + \frac{b}{\delta}\frac{\partial U_0}{\partial t^*} + b\frac{\partial U_1}{\partial t^*} + \cdots.$$

A second-order equation results for U_0 only if

$$\delta = \varepsilon$$

and, with this choice, the following sequence of approximate equations is obtained:

$$\frac{\partial^2 U_0}{\partial t^{*2}} + b\frac{\partial U_0}{\partial t^*} = 0, \tag{4.1.66}$$

$$\frac{\partial^2 U_1}{\partial t^{*2}} + b\frac{\partial U_1}{\partial t^*} = -a\frac{\partial U_0}{\partial x}. \tag{4.1.67}$$

In accordance with the general ideas of the first part of this Section, the boundary-layer equations are ordinary differential equations, since the boundary layer does not occur on a subcharacteristic. This must be true for any hyperbolic initial-value problem, since a spacelike arc can never be subcharacteristic. Equations (4.1.64) and (4.1.65) provide the initial conditions for the initial-layer equations, and the solutions are easily found:

$$U_0(x, t^*) = F(x), \tag{4.1.68}$$

$$U_1(x, t^*) = \left[G(x) + \frac{a}{b}F'(x)\right]\left[\frac{1 - e^{-bt^*}}{b}\right] - \frac{a}{b}t^*F'(x). \tag{4.1.69}$$

Thus, finally, we have the initially valid expansion (Equation 4.1.62)

$$u(x, t; \varepsilon) = F(x) + \varepsilon\left\{\left[G(x) + \frac{a}{b}F'(x)\right]\left[\frac{1 - e^{-bt^*}}{b}\right] - \frac{a}{b}t^*F'(x)\right\} + \cdots. \tag{4.1.70}$$

These solutions contain persistent terms as well as typical boundary-layer decay terms with a time scale $t = \varepsilon$. The behavior of Equation (4.1.70) provides initial conditions for an outer expansion. Next, we construct the outer expansion, based on the limit process ($\varepsilon \to 0$, x, t fixed), the first term of which is the limit solution (4.1.53). The orders of the various terms are evident from the orders in Equation (4.1.70). Thus, we have

$$u(x, t; \varepsilon) = u_0(x, t) + \varepsilon u_1(x, t) + \varepsilon^2 u_2(x, t) + \cdots. \tag{4.1.71}$$

In the outer expansion, the higher-order derivatives are small and the lower-order operator dominates. The sequence of equations which approximates

Equation (4.1.47) is

$$a\frac{\partial u_0}{\partial x} + b\frac{\partial u_0}{\partial t} = 0, \tag{4.1.72}$$

$$a\frac{\partial u_1}{\partial x} + b\frac{\partial u_1}{\partial t} = \left(\frac{\partial^2 u_0}{\partial x^2} - \frac{\partial^2 u_0}{\partial t^2}\right). \tag{4.1.73}$$

The general solutions can all be expressed in terms of arbitrary functions:

$$u_0 = f(\xi), \qquad \xi = x - \frac{a}{b}t. \tag{4.1.74}$$

The equation for u_1 becomes

$$a\frac{\partial u_1}{\partial x} + b\frac{\partial u_1}{\partial t} = \left(1 - \frac{a^2}{b^2}\right)f''(\xi). \tag{4.1.75}$$

This is easily solved by introducing ($\tau = t, \xi$) as coordinates, and the result is

$$u_1(x, t) = \left(1 - \frac{a^2}{b^2}\right)\frac{t}{b}f''(\xi) + f_1(\xi). \tag{4.1.76}$$

Thus, the outer expansion is

$$u(x, t; \varepsilon) = f(\xi) + \varepsilon\left[\left(1 - \frac{a^2}{b^2}\right)\frac{t}{b}f''(\xi) + f_1(\xi)\right] + \cdots. \tag{4.1.77}$$

The arbitrary functions, f, f_1, etc., in the outer expansion (Equation (4.1.77) must be determined by matching Equation (4.1.70). A limit intermediate to the two already considered has (x, t_η fixed) as $\varepsilon \to 0$:

$$t_\eta = \frac{t}{\eta(\varepsilon)}, \qquad \varepsilon \ll \eta \ll 1. \tag{4.1.78}$$

Thus, we have

$$t = \eta t_\eta \to 0, \qquad t^* = \frac{\eta}{\varepsilon}t_\eta \to \infty.$$

Under this limit, the initially valid expansion (Equation (4.1.70) becomes

$$u(x, t; \varepsilon) = F(x) + \varepsilon\left\{\frac{G(x)}{b} + \frac{a}{b^2}F'(x) - \frac{a}{b}\frac{\eta}{\varepsilon}t_\eta F'(x)\right\} + O(\varepsilon^2) + \text{T.S.T.} \tag{4.1.79}$$

In the outer expansion, we have

$$f(\xi) = f\left(x - \frac{a}{b}t\right) = f\left(x - \frac{a}{b}\eta t_\eta\right) = f(x) - \frac{a}{b}\eta t_\eta f'(x) + O(\eta^2).$$

The outer expansion approaches

$$u(x, t; \varepsilon) = f(x) - \frac{a}{b}\eta t_\eta f'(x) + \varepsilon f_1(x) + O(\varepsilon\eta) + O(\varepsilon^2). \quad (4.1.80)$$

The terms of order one in Equations (4.1.79) and (4.1.80) can be matched, and then the terms of order η are matched identically; terms of order ε can also be matched. Thus, we have

$$f(x) = F(x), \quad (4.1.81)$$

$$f_1(x) = \frac{G(x)}{b} + \frac{a}{b^2}F'(x), \quad (4.1.82)$$

as long as $\varepsilon|\log \varepsilon| \ll \eta \ll 1$.

The final result expresses the outer expansion in terms of the given initial values,

$$u(x, t; \varepsilon) = F\left(x - \frac{a}{b}t\right) + \varepsilon\left\{\frac{1}{b}\left(1 - \frac{a^2}{b^2}\right)tF''\left(x - \frac{a}{b}t\right) + \frac{a}{b^2}F'\left(x - \frac{a}{b}t\right) + \frac{1}{b}G\left(x - \frac{a}{b}t\right)\right\} + O(\varepsilon^2). \quad (4.1.83)$$

We can see that after a little while, the solution is dominated by the given initial value of u which propagates along the subcharacteristic. We also note that the outer expansion is not uniformly valid in the far field ($t = O(\varepsilon^{-1})$). The solution in the far-field requires a different expansion [cf. Section 5.3].

Signaling Problems

We next consider a signaling or radiation problem in which boundary conditions are prescribed on a timelike arc and propagate into the quiescent medium in $x > 0$. For the first problem, the boundary condition is prescribed at $x = 0$, and we have to distinguish two cases according to the slope of the subcharacteristics, that is, whether they run into or out of the boundary $x = 0$. (See Figure 4.1.8). The subcharacteristics are given by

$$\xi = x - \frac{a}{b}t = \text{const.} \quad (4.1.84)$$

and for $a > 0$, outgoing, and $a < 0$, incoming, the boundary condition is

$$u(0, t) = F(t), \qquad t > 0. \quad (4.1.85)$$

There is a real discontinuity in the function and in its derivative along the characteristic curve $x = t$, but the intensity of the jump decays exponentially ($b > |a|$) according to the considerations of Equation (4.1.59). The solution is identically zero for $x > t$.

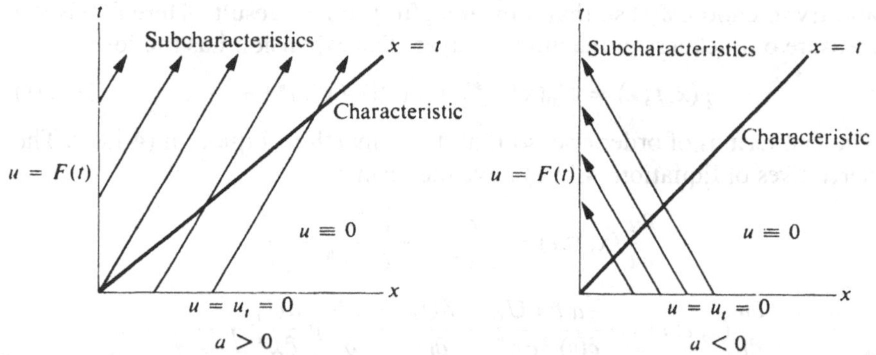

Figure 4.1.8 Signaling Problems

$a > 0$: *Outgoing Subcharacteristics*

The "outer" solution is an asymptotic expansion of the same form as before:

$$u(x, t; \varepsilon) = u_0(x, t) + \varepsilon u_1(x, t) + \cdots. \tag{4.1.86}$$

The sequence of equations satisfied by u_i is the same as before. Equations (4.1.72) and (4.1.73) and the solution u_0 can be written

$$u_0 = f(\zeta), \quad \text{where } \zeta = t - \frac{b}{a}x. \tag{4.1.87}$$

The boundary condition (Equation 4.1.85) can be satisfied by identifying $f = F$, so that the first-approximation outer solution is

$$u_0 = \begin{cases} 0, & t < \dfrac{b}{a}x, \\ F\left(t - \dfrac{b}{a}x\right), & t > \dfrac{b}{a}x. \end{cases} \tag{4.1.88}$$

This solution, however, has a discontinuity on the particular subcharacteristic through the origin. Such a discontinuity is not permitted in the solution to the exact Equation (4.1.47) with $\varepsilon > 0$, since any discontinuities can appear only on characteristics. Thus, to obtain a uniformly valid solution, a suitable interior layer must be introduced on the particular subcharacteristic $\zeta = 0$ which supports the discontinuity in the outer solution. In order to derive the interior-layer equations, we consider a limit process in which (x^*, t^*) are fixed, where

$$x^* = \frac{x - (a/b)t}{\delta(\varepsilon)}, \qquad \delta(\varepsilon) \to 0; \tag{4.1.89}$$

$$t^* = t, \tag{4.1.90}$$

and try to choose $\delta(\varepsilon)$ so that a meaningful problem results. Here $\delta(\varepsilon)$ is the measure of thickness of the interior layer. The expansion has the form

$$u(x, t; \varepsilon) = U_0(x^*, t^*) + \mu(\varepsilon)U_1(x^*, t^*) + \cdots. \tag{4.1.91}$$

The first term is of order one, so that it can match to Equation (4.1.88). The derivatives of Equation (4.1.91) have the form

$$\frac{\partial u}{\partial x}(x, t; \varepsilon) = \frac{1}{\delta}\frac{\partial U_0}{\partial x^*} + \frac{\mu}{\delta}\frac{\partial U_1}{\partial x^*} + \cdots,$$

$$\frac{\partial u}{\partial t}(x, t; \varepsilon) = \frac{-a/b}{\delta(\varepsilon)}\frac{\partial U_0}{\partial x^*} + \frac{\partial U_0}{\partial t^*} - \frac{a/b}{\delta}\mu\frac{\partial U_1}{\partial x^*} + \cdots.$$

The operator on the right-hand side of Equation (4.1.47) now has the form

$$a\frac{\partial u}{\partial x} + b\frac{\partial u}{\partial t} = b\left[\frac{\partial U_0}{\partial t^*} + \mu(\varepsilon)\frac{\partial U_1}{\partial t^*} + \cdots\right]. \tag{4.1.92}$$

The dominant terms of the wave operator are $\partial^2 U_0/\partial x^{*2}$, so that Equation (4.1.47) becomes

$$\frac{\varepsilon}{\delta^2}\left\{\frac{\partial^2 U_0}{\partial x^{*2}} + \cdots - \frac{a^2}{b^2}\frac{\partial^2 U_0}{\partial x^{*2}} + \cdots\right\} = b\frac{\partial U_0}{\partial t^*} + \cdots. \tag{4.1.93}$$

The distinguished limiting case, which results in a nontrivial equation, has

$$\delta = \sqrt{\varepsilon}. \tag{4.1.94}$$

The interior-layer thickness here is an order of magnitude larger than in the initial boundary layer. The interior-layer equation is, then, a partial differential equation with one coordinate along a subcharacteristic. From Equation (4.1.93), we obtain

$$\kappa\frac{\partial^2 U_0}{\partial x^{*2}} = \frac{\partial U_0}{\partial t^*}, \tag{4.1.95}$$

where $\kappa = (1 - a^2/b^2)/b > 0$. Here $\kappa > 0$ assures that $t^* = t$ is a positive timelike direction and that Equation (4.1.95) is an ordinary heat or diffusion equation which describes the resolution of the discontinuity of the outer expansion on $\zeta = 0$. The boundary conditions for Equation (4.1.95) have to come from matching with the outer expansion. Matching for this problem is carried out with a class of intermediate limits (x_η, t fixed), where

$$x_\eta = \frac{x - (a/b)t}{\eta(\varepsilon)}, \qquad \sqrt{\varepsilon} \ll \eta \ll 1. \tag{4.1.96}$$

In the limit, then, we have

$$x - \frac{a}{b}t = \eta x_\eta \to 0, \qquad x^* = \frac{x - (a/b)t}{\sqrt{\varepsilon}} = \frac{\eta}{\sqrt{\varepsilon}}x_\eta \to \pm\infty.$$

Under this limit, the outer and interior-layer expansions behave as follows:

$$\text{Outer:} \quad u(x, t; \varepsilon) = u_0 + \cdots \begin{cases} \to 0, & t < \dfrac{b}{a}x, \\ \to F(0+), & t > \dfrac{b}{a}x; \end{cases} \tag{4.1.97}$$

$$\text{Interior layer:} \quad u(x, t; \varepsilon) = U_0(x^*, t) + \cdots \to U_0(\pm\infty, t) + \cdots. \tag{4.1.98}$$

For matching, the terms in Equations (4.1.97) and (4.1.98) must be the same, and this provides the boundary conditions illustrated in Figure 4.1.9 for the interior-layer equation (Equation (4.1.95)). Initial conditions are chosen here consistent with the boundary and in such a way that the physical process

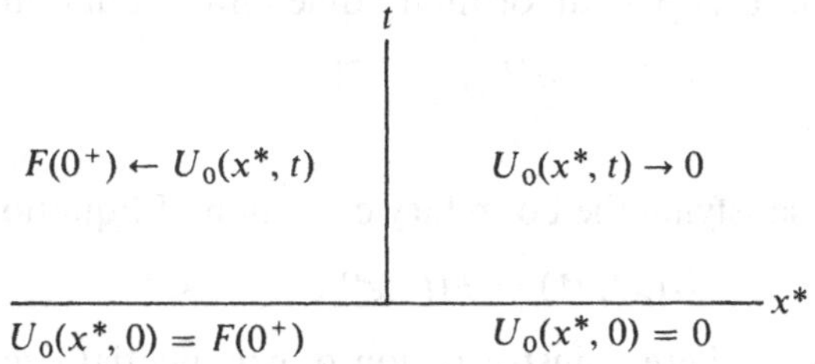

Figure 4.1.9 Initial Condition and Limiting Values of U_0

represents the resolution of the discontinuity at $\zeta = 0$. A rigorous treatment of the initial conditions demands a discussion of initially valid expansions, which is not given here. The main point here is the nature of the interior-layer equation. The solution corresponding to the conditions stated is

$$U_0(x^*, t) = \frac{F(0+)}{2} \operatorname{erfc} \frac{x^*}{2\sqrt{\kappa t}}. \tag{4.1.99}$$

The discontinuity of the outer solution is replaced here by the diffusive solution of the heat equation.

$a < 0$: *Incoming Subcharacteristics*

The qualitative difference between this case and the case just discussed is striking. If an outer expansion is contemplated of the form of Equation (4.1.86), the only reasonable solution of the form $f(t - (b/a)x)$ is zero, since now disturbances propagate along the subcharacteristics from the quiescent region to the boundary.

Thus, we have

$$u_0 = 0. \tag{4.1.100}$$

The discontinuity occurs at the boundary $x = 0$, and the boundary layer occurs at $x = 0$. The boundary-layer equations should be ordinary differential equations, since again in this case the boundary layer is not on a subcharacteristic. To derive these equations, consider (x^*, t^* fixed), where

$$x^* = \frac{x}{\delta(\varepsilon)}, \qquad t^* = t, \quad \text{as } \varepsilon \to 0. \tag{4.1.101}$$

The expansion is of the usual form,

$$u(x, t; \varepsilon) = U_0(x^*, t^*) + \nu_1(\varepsilon)U_1(x^*, t^*) + \cdots. \tag{4.1.102}$$

The basic equation (Equation 4.1.47) takes the form

$$\frac{\varepsilon}{\delta^2}\frac{\partial^2 U_0}{\partial x^{*2}} + \cdots = \frac{a}{\delta}\frac{\partial U_0}{\partial x^*} + b\frac{\partial U_0}{\partial t^*} + \cdots. \tag{4.1.103}$$

Again, $\delta = \varepsilon$ is a distinguished case for the boundary layer not on a subcharacteristic and U_0 obeys an ordinary differential equation

$$\frac{\partial^2 U_0}{\partial x^{*2}} = a\frac{\partial U_0}{\partial x^*}. \tag{4.1.104}$$

The solution satisfying the boundary condition of Equation (4.1.85) is

$$U_0(x^*, t^*) = F(t^*)e^{ax^*}, \qquad a < 0. \tag{4.1.105}$$

The boundary layer here is just a region of exponential decay adjacent to the boundary.

4.1.3 A Quasi-linear Problem for the Parabolic Case

In this section we illustrate some of the consequences of having a quasi-linear lower order operator and we choose an equation which can be solved exactly as our model.

Exact Solution

Burgers' equation

$$\varepsilon u_{xx} = u_t + uu_x \tag{4.1.106}$$

was originally proposed by Burgers to model turbulence. It was later shown in Reference 4.1.2 that Equation (4.1.106) is derivable from the Navier–Stokes equations in the limit of a weak shock layer. It was also observed in Reference 4.1.2, and independently in Reference 4.1.3, that Equation (4.1.106) can be transformed to the diffusion equation. We consider this transformation next.

As a preliminary step, let

$$u = \psi_x \tag{4.1.107}$$

then Equation 4.1.106 can be integrated to

$$\varepsilon\psi_{xx} = \psi_t + \tfrac{1}{2}\psi_x^2. \tag{4.1.108}$$

We now eliminate the nonlinear term by setting

$$\psi = -2\varepsilon \log v. \tag{4.1.109a}$$

Then

$$\psi_t = -\frac{2\varepsilon}{v} v_t \tag{4.1.109b}$$

$$\psi_x = -\frac{2\varepsilon}{v} v_x \tag{4.1.109c}$$

$$\psi_{xx} = -\frac{2\varepsilon v_{xx}}{v} + \frac{2\varepsilon}{v^2} v_x^2 \tag{4.1.109d}$$

and Equation (4.1.108) transforms to

$$\varepsilon v_{xx} = v_t. \tag{4.1.110}$$

The combined transformation $u \to v$ is

$$u = -2\varepsilon \frac{v_x}{v}. \tag{4.1.111}$$

Thus, if we wish to solve the initial value problem

$$u(x, 0) = F(x) \tag{4.1.112}$$

for Equation (4.1.106) on the infinite interval, $-\infty < x < \infty$, we need to solve the linear first-order equation (4.1.111) to calculate $v(x, 0)$.

$$v(x, 0) = \exp\left(-\frac{1}{2\varepsilon}\int_0^x F(s)ds\right) \equiv G(x) \tag{4.1.113}$$

where with no loss of generality we have set $v(0, 0) = 1$. This arbitrary constant will cancel out in the final result for u.

Now, using the well known result for the solution of the initial value problem for v:

$$v(x, t) = \frac{1}{\sqrt{4\pi\varepsilon t}} \int_{-\infty}^{\infty} G(s)e^{-(x-s)^2/4\varepsilon t}\, ds \tag{4.1.114}$$

we calculate u from Equation (4.1.111) in the form

$$u(x, t) = \frac{\displaystyle\int_{-\infty}^{\infty} G(s)\,\frac{(x-s)}{t}\,e^{-(x-s)^2/4\varepsilon t}\, ds}{\displaystyle\int_{-\infty}^{\infty} G(s)e^{-(x-s)^2/4\varepsilon t}\, ds}. \tag{4.1.115}$$

This gives u explicitly for a given $F(x)$.

The key to calculating this result once the transformation (4.1.111) was obtained lies in the fact that the appropriate initial condition for v can be easily calculated.

In contrast, if we wish to solve the signaling problem

$$u(x, 0) = 0; \qquad u(0, t) = H(t), \qquad x > 0 \tag{4.1.116}$$

we still have Equation (4.1.110) for v. However, the calculation of the boundary condition for v at $x = 0$ is complicated and involves the solution of an integral equation.

To see this, assume that the boundary condition on v is

$$v(0, t) = K(t). \tag{4.1.117}$$

Then the solution of Equation (4.1.110) subject to $v(x, 0) = 1$ and $v(0, t) = K(t)$ is [cf. Equations (4.1.26)–(4.1.27)].

$$v(x, t) = 1 + \frac{1}{\pi}\int_0^t \frac{e^{-x^2/4\varepsilon(t-\tau)}}{(t-\tau)^{1/2}}\left[\frac{d}{d\tau}\int_0^\tau \frac{K(s) - 1}{(\tau - s)^{1/2}}\,ds\right]d\tau. \tag{4.1.118}$$

Therefore, u is of the form

$$u(x, t) = \frac{\displaystyle\int_0^t \frac{xe^{-x^2/4\varepsilon(t-\tau)}}{(t-\tau)^{3/2}}\left[\frac{d}{d\tau}\int_0^\tau \frac{K(s) - 1}{(\tau - s)^{1/2}}\,ds\right]d\tau}{\displaystyle 1 + \frac{1}{\pi}\int_0^t \frac{e^{-x^2/4\varepsilon(t-\tau)}}{(t-\tau)^{1/2}}\left[\frac{d}{d\tau}\int_0^\tau \frac{K(s) - 1}{(\tau - s)^{1/2}}\,ds\right]d\tau}. \tag{4.1.119}$$

In the limit as $x \to 0^+$ Equation (4.1.119) gives a complicated integral equation for K. This result may be simplified for special choices of $H(t)$. Otherwise, even though the problem is solvable in principle, not much progress can be made unless $K(t)$ is known.

In this section we will study two simple examples using perturbations as $\varepsilon \to 0$. The first will be an initial value problem and will illustrate how shock or corner layers can be inserted to smooth out discontinuities in u or its first partial derivatives as computed from the $\varepsilon = 0$ problem. The second example will deal with the signalling problem, and will exhibit boundary layers as well as shock, corner and transition layers in exact analogy with the problem discussed in Section 2.5.

Weak Solutions

Since the outer limit in our example consists of

$$u_t + uu_x = 0, \tag{4.1.120}$$

we review briefly the appropriate results for constructing a "weak" solution. This is a solution of Equation (4.1.120) everywhere in some domain of the (x, t) plane except along a certain curve where u or its first partial derivatives may be discontinuous. The reader will find a general discussion in Reference 4.1.4.

If we associate the divergence form

$$u_t + \left(\frac{u^2}{2}\right)_x = 0 \tag{4.1.121}$$

with Equation (4.1.120), we can integrate Equation (4.1.121) between two fixed points x_1 and x_2 to derive the conservation law

$$\frac{d}{dt}\int_{x_1}^{x_2} u(x, t)dx = \tfrac{1}{2}[u^2(x_1, t) - u^2(x_2, t)]. \tag{4.1.122}$$

Now, if u is a continuously differentiable solution of Equation (4.1.121) everywhere in some domain D, then Equation (4.1.122) holds for any $x_1 < x_2$ in D. Conversely, if Equation (4.1.122) holds for any $x_1 < x_2$ in D and u is continuously differentiable, then Equation (4.1.121) is implied.[2]

A weak solution of Equation (4.1.121) is one where u is continuously differentiable everywhere in D except along some curve $x_s(t)$, and where the conservation law (4.1.122) holds *even if* $x_1 < x_s < x_2$.

It then follows in the limit $(x_2 - x_1) \to 0$ that the shock must travel with speed dx_s/dt given by

$$\frac{dx_s}{dt} = \frac{u(x_s^+, t) + u(x_s^-, t)}{2}. \tag{4.1.123}$$

Moreover, an allowable shock must satisfy the inequalities

$$u(x_s^+, t) \le \frac{dx_s}{dt} \le u(x_s^-, t). \tag{4.1.124}$$

This is sometimes called the "entropy condition" because its generalization to the case of compressible flow implies that the entropy must rise across a shock.

Note that the conservation law, Equation (4.1.122), and hence the rule for shock propagation, Equation (4.1.123), is directly related to the divergence form, Equation (4.1.121).

We note that a given first-order equation such as Equation (4.1.120) *does not have a unique divergence form.* For example, we could also have used the divergence relation

$$(\log u)_t + (u)_x = 0 \tag{4.1.125}$$

which is equivalent to Equation (4.1.121) if $v = \log u$ is continuously differentiable. However, the conservation law and shock condition associated

[2] A similar statement can be made for general divergence forms

$$\frac{\partial}{\partial t} p(t, x, u) + \frac{\partial}{\partial x} q(t, x, u) + r(t, x, u) = 0.$$

with (4.1.125) are quite different from those derived earlier. We obtain

$$\frac{d}{dt}\int_{x_1}^{x_2} \log u(x, t)dx = u(x_1, t) - u(x_2, t) \tag{4.1.126a}$$

$$\frac{dx_s}{dt} = \frac{u(x_s^+, t) - u(x_s^-, t)}{\log[u(x_s^+, t)] - \log[u(x_s^-, t)]}. \tag{4.1.126b}$$

In general, the choice of the correct conservation law and allowable shocks are derived either from physical considerations or knowledge of the limiting form, as $\varepsilon \to 0$, of the exact solution of the higher order problem. In our example we must use Equations (4.1.123) and (4.1.124) as they correspond to the limiting form as $\varepsilon \to 0$ of the exact solutions of Equation (4.1.106). The details of this calculation can be found in Reference 4.1.5.

Initial Value Problem

A simple way to generate a discontinuity in the solution of the lower order operator is by considering discontinuous initial data. Let us study the solution of Equation (4.1.106) with initial conditions

$$u(x, 0) = \begin{cases} 1, & x < 0, \\ -1, & 0 < x < 1, \\ 0, & 1 < x < \infty. \end{cases} \tag{4.1.127}$$

Since the characteristics of the lower order operator, Equation (4.1.120) are straight lines with slope equal to $u(x, 0)$, we have the pattern of subcharacteristics sketched in Figure 4.1.10.

Clearly, we need to introduce a shock at the point $t = 0$, $x = 0$, and since the values of u on either side of this shock are constant (± 1) for $0 \leq t \leq 1$, we conclude from Equation (4.1.123) that $x_s = 0$ for $0 \leq t \leq 1$.

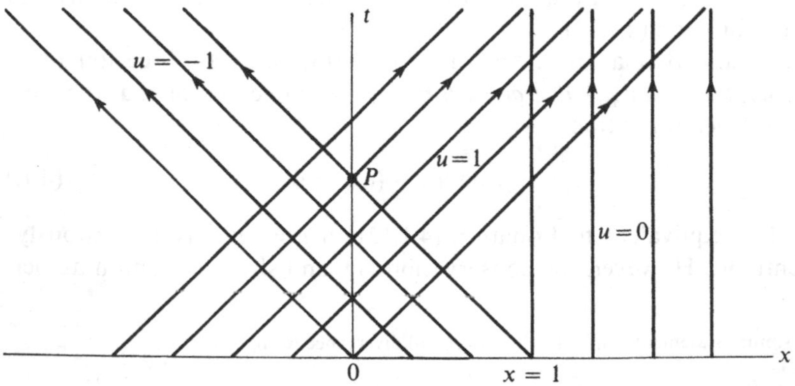

Figure 4.1.10 Subcharacteristics for the Initial Values in Equation (4.1.127)

Next, we consider the gap left in the t, x plane because of the divergence of the subcharacteristics emanating from $t = 0$, $x = 1$. It is easy to verify that one cannot satisfy Equation (4.1.124) with any finite number of discontinuities passing through $t = 0$, $x = 1$. In fact, we must introduce the "simple wave" solution

$$u = \frac{x-1}{t}; \qquad -1 < \frac{x-1}{t} < 0, \qquad t < 1. \tag{4.1.128}$$

This is a weak solution since u_x (and u_t) are discontinuous along $x = 1 - t$ and $x = 1$. One may regard the simple wave solution as an infinite number of infinitesimal discontinuities along the lines $(x - 1)/t = \text{const}$.

Beyond the point $P = (1, 0)$ the solution above the shock OP will be affected by the simple wave solution. Thus, to continue the shock beyond P we must solve

$$\frac{dx_s}{dt} = \frac{1}{2}\left[\frac{x_s - 1}{t} + 1\right] \tag{4.1.129a}$$

$$x_s = 0 \quad \text{at } t = 1. \tag{4.1.129b}$$

This solution is easy to calculate and we obtain

$$x_s = t - 2\sqrt{t} + 1; \qquad 1 \le t \le 4. \tag{4.1.130}$$

The reason that the above is only valid up to $t = 4$ is because at $Q = (4, 1)$, the shock intersects the last subcharacteristic of the simple wave solution, and begins to be influenced by the solution $u = 0$. See Figure 4.1.11.

Therefore, beyond Q, the shock is again straight and propagates with velocity $\frac{1}{2}$.

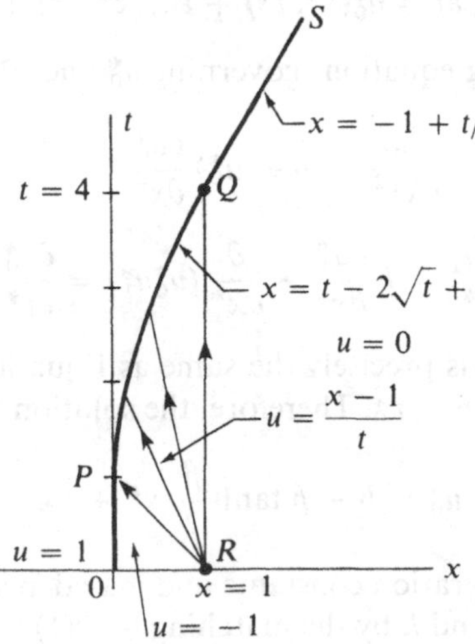

Figure 4.1.11 Outer Limit, Shock and Simple Wave

The solution summarized in Figure 4.1.11 is the leading term in an outer expansion (t, x fixed, $\varepsilon \to 0$) of Equation (4.1.106). In this example the outer expansion terminates with the outer limit when this is a constant. What remains to $O(1)$ is the introduction of appropriate shock layers along OP, PQ and QS to smooth out the discontinuities in u, and of corner-layers along RP and RQ to smooth out the discontinuities in u_x (or u_t) there.

Consider first neighborhoods of straight shocks such as OP, QS which divide the t, x plane into regions of uniform u. Such shocks are of the form

$$x_s = a + bt, \tag{4.1.131a}$$

where a and b are constants.

We introduce the inner variables

$$x^* = \frac{x - a - bt}{\delta(\varepsilon)}; \qquad t^* = t \tag{4.1.131b}$$

and calculate $\delta(\varepsilon)$ by substitution into Equation (4.1.106). We find

$$\frac{\varepsilon}{\delta^2}\frac{\partial^2 u}{\partial x^{*2}} = \frac{\partial u}{\partial t^*} + \frac{1}{\delta}(u - b)\frac{\partial u}{\partial x^*}. \tag{4.1.132a}$$

Therefore, we must choose $\delta = \varepsilon$ and the equation governing the solution near a straight shock is

$$\frac{\partial^2 u}{\partial x^{*2}} + (b - u)\frac{\partial u}{\partial x^*} = \varepsilon\frac{\partial u}{\partial t^*}. \tag{4.1.132b}$$

If we assume a shock layer expansion of the form

$$u(x, t, \varepsilon) = u_0^*(x^*, t^*) + \varepsilon u_1^*(x^*, t^*) + \cdots \tag{4.1.133}$$

we obtain the following equations governing u_0^* and u_1^*:

$$\frac{\partial^2 u_0^*}{\partial x^{*2}} + (b - u_0^*)\frac{\partial u_0^*}{\partial x^*} = 0 \tag{4.1.134a}$$

$$\frac{\partial^2 u_1^*}{\partial x^{*2}} + b\frac{\partial u_1^*}{\partial x^*} - \frac{\partial}{\partial x^*}(u_0^* u_1^*) = \frac{\partial u_0^*}{\partial t^*}. \tag{4.1.134b}$$

Equation (4.1.134a) is precisely the same as Equation (2.5.11) of Section 2.5 if we interpret $g = b - u_0^*$. Therefore, the solution for u_0^* is

$$u_0^* = b - \beta \tanh\frac{\beta}{2}(x^* + k), \tag{4.1.135}$$

where β and k are integration constants and may depend on t^*.

We can evaluate β and k by the matching to $O(1)$ with the outer limit on either side of the shock. Since in our case the shock is straight, it borders on two regions of constant u. In fact, we know that the outer solution is of the form

$$u = u_1 = \text{const.} \qquad x > a + bt \tag{4.1.136a}$$

$$u = u_2 = \text{const.} \qquad x < a + bt \tag{4.1.136b}$$

to all orders on either side of the shock, and

$$b = \frac{u_1 + u_2}{2} \tag{4.1.137}$$

according to Equation (4.1.123).

The inner limit u_0^*, when evaluated for $|x^*| \to \infty$ gives (with $k = 0$):

$$u = b - \beta + \text{T.S.T.} \quad \text{as } x^* \to \infty \tag{4.1.138a}$$

$$u = b + \beta + \text{T.S.T.} \quad \text{as } x^* \to -\infty. \tag{4.1.138b}$$

Matching to $O(1)$ to the right of the shock gives

$$\beta = b - u_1 \tag{4.1.139}$$

and this also satisfies the matching to the left since $u_2 = 2b - u_1$ according to Equation (4.1.137).

In this case (straight shock) the inner expansion also terminates with u_0^* as we can match u_0^* with the outer limit to all orders. In fact, u_0^* is the uniformly valid solution for the case of a straight shock to all orders. This result is easily confirmed by considering the exact solution for the initial value problem

$$u(x, 0) = \begin{cases} u_1; & x > 0 \\ u_2 > u_1; & x < 0. \end{cases} \tag{4.1.140}$$

Thus, in order to have a non-trivial situation we need to consider a curved shock bounding adjacent domains of varying u. The simplest such example is when u is constant on one side and is a simple wave solution on the other side as is the case for the shock between P and Q in Figure 4.1.11.

In this case, the inner variables are

$$x^* = \frac{x - x_0(t) - \varepsilon x_1(t)}{\varepsilon}; \qquad x_0(t) = t - 2\sqrt{t} + 1 \tag{4.1.141a}$$

$$t^* = t, \tag{4.1.141b}$$

where we envisage the possibility that the shock is displaced to $O(\varepsilon)$ and include the unknown $x_1(t)$ in the definition of x^*.

The equations corresponding to Equations (4.1.134) now become

$$\frac{\partial^2 u_0^*}{\partial x^{*2}} + (\dot{x}_0 - u_0^*)\frac{\partial u_0^*}{\partial x^*} = 0 \tag{4.1.142a}$$

$$\frac{\partial^2 u_1^*}{\partial x^{*2}} + \dot{x}_0 \frac{\partial u_1^*}{\partial x^*} - \frac{\partial}{\partial x^*}(u_0^* u_1^*) = -\dot{x}_1 \frac{\partial u_0^*}{\partial x^*} + \frac{\partial u_0^*}{\partial t^*}. \tag{4.1.142b}$$

The solution for u_0^* is

$$u_0^* = \dot{x}_0 - \beta \tanh \beta(x^* + k)/2 \tag{4.1.143}$$

and the matching to $O(1)$ proceeds as follows.

We introduce the intermediate variable

$$x_\eta = \frac{x - x_0 - \varepsilon x_1}{\eta(\varepsilon)}; \qquad \varepsilon \ll \eta \ll 1. \tag{4.1.144}$$

Thus,

$$x = x_0 + \eta(\varepsilon) x_\eta + \varepsilon x_1 \to x_0 \tag{4.1.145a}$$

and

$$x^* = \frac{\eta(\varepsilon)}{\varepsilon} x_\eta \to \pm\infty. \tag{4.1.145b}$$

The simple wave solution to the right of the shock, $u = (x - 1)/t$ has the form

$$u = \frac{x_0 - 1}{t} + \frac{\eta x_\eta}{t} + \frac{\varepsilon x_1}{t} + O(\varepsilon^2) \tag{4.1.146}$$

the term of order ε^2 is included to account for a possible displacement of the shock to $O(\varepsilon^2)$. To the left of the shock, we have $u = 1$ to all orders.

Now, the inner expansion to the right of the shock ($x^* \to \infty$) gives, with $k = 0$

$$u = \dot{x}_0 - \beta + O(\varepsilon) + \text{T.S.T.} \tag{4.1.147}$$

Thus, matching to $O(1)$ requires that we set

$$\beta = \dot{x}_0 - \frac{x_0 - 1}{t^*} = \frac{1}{\sqrt{t^*}} \tag{4.1.148}$$

and u_0^* is defined in the form

$$u_0^* = \left(1 - \frac{1}{\sqrt{t^*}}\right) - \frac{1}{\sqrt{t^*}} \tanh \frac{x^*}{2\sqrt{t^*}}. \tag{4.1.149}$$

The calculation of u_1^* and the matching to $O(\varepsilon)$ is left as an exercise, (Problem 4.1.8). The main points to note are that u_1^* obeys a linear equation and can be calculated explicitly. The matching is *not sufficient* to determine the two unknown functions of t^* which arise in the integration of Equation (4.1.142b); one of these, say $C(t^*)$, multiplies a homogeneous solution: $u^* = C(t^*)\text{sech}^2\, x^*/2\sqrt{t^*}$ which is transcendentally small in the matching on *either side of the shock*! In order to determine $C(t^*)$ one must match $u_0^* + \varepsilon u_1^*$ with a solution valid near the point P.

If we introduce the local variables near P

$$\bar{x} = \frac{x}{\gamma(\varepsilon)}; \qquad \bar{t} = \frac{t-1}{\delta(\varepsilon)} \tag{4.1.150}$$

Equation (4.1.106) transforms to

$$\frac{\varepsilon}{\gamma^2}\frac{\partial^2 u}{\partial \bar{x}^2} = \frac{1}{\delta}\frac{\partial u}{\partial \bar{t}} + \frac{1}{\gamma} u \frac{\partial u}{\partial \bar{x}}. \tag{4.1.151}$$

Clearly, we must set $\gamma = \delta = \varepsilon$, and the result is the full equation. This same situation is also true in neighborhoods of the points O, R and Q.

In the case of Equation (4.1.106) the exact solution is available, and one could derive from it appropriate limiting forms required for the matching. We do not consider these calculations. Fortunately, knowledge of the exact solution is only required to calculate the asymptotic results to $O(\varepsilon)$. The $O(1)$ results are complete.

To complete our study of the initial value problem we must also seek solutions of Equation 4.1.106 which are capable of smoothing out the discontinuity in u_x (or u_t) along the lines RP or RQ in Figure 4.1.11. The situation is similar in either case, so if we examine the solution near the line $x = 1 - t$, we note that we need to find a corner-layer solution of Equation (4.1.106) for $u \approx -1$ and such that it matches with $u_x = 0$ to the left and with $u_x = 1/t$ to the right of the line $x = 1 - t$.

If we scale the variables as follows

$$u_c = \frac{u+1}{\mu(\varepsilon)}; \qquad x_c = \frac{x-1+t}{\lambda(\varepsilon)}, \qquad t_c = t \tag{4.1.152}$$

Equation (4.1.106) transforms to

$$\frac{\varepsilon\mu}{\lambda^2}\frac{\partial^2 u_c}{\partial x_c^2} = \mu\frac{\partial u_c}{\partial t_c} + \frac{\mu^2}{\lambda} u_c \frac{\partial u_c}{\partial x_c}. \tag{4.1.153a}$$

We must choose $\mu = \lambda = \varepsilon^{1/2}$ and again the result is the full equation. Now, we need to demonstrate that

$$\frac{\partial^2 u_c}{\partial x_c^2} = \frac{\partial u_c}{\partial t_c} + u_c \frac{\partial u_c}{\partial x_c} \tag{4.1.153b}$$

has a solution with the following behavior

$$\frac{\partial u_c}{\partial x_c} \to 0 \quad \text{as } x_c \to -\infty \tag{4.1.154a}$$

$$\frac{\partial u_c}{\partial x_c} \to \frac{1}{t_c} \quad \text{as } x_c \to \infty. \tag{4.1.154b}$$

Equations (4.1.154) give the boundary conditions at infinity for the solution of Equation (4.1.153). The initial conditions would follow from matching with another exact solution of the type (4.1.151) near the point R.

We will not exhibit these exact solutions as the calculations become quite involved.

The conclusion one draws from the above discussion is that any calculation beyond the $O(1)$ solution will usually depend on information that can only be provided by local exact solutions.

Signaling Problem

Actually, a more general problem consists of solving Equation (4.1.106) over a finite interval $0 \leq x \leq a$ with initial condition

$$u(x, 0) = F(x) \tag{4.1.155a}$$

and boundary conditions

$$u(0, t) = G(t) \tag{4.1.155b}$$

$$u(a, t) = H(t). \tag{4.1.155c}$$

The essential new feature now is how various possible outer limits can be reconciled with the prescribed boundary conditions. All possibilities can be illustrated with the signaling problem (one boundary condition) with constant boundary and initial data.

We propose to solve Equation 4.1.106 in $0 \leq x < \infty$, $0 < t < \infty$, with the initial condition

$$u(x, 0) = A \tag{4.1.156a}$$

and the boundary condition

$$u(0, t) = B, \tag{4.1.156b}$$

where A and B are constants which do not depend on ε.

The simplest choice is $A = B$, in which case the exact solution is $u = A = B$. This gives us a starting point for examining various possible combina-

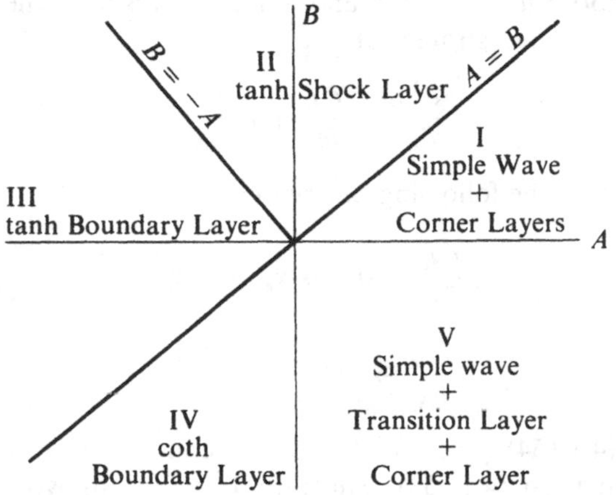

Figure 4.1.12 Possible Solutions in the $A - B$ Plane

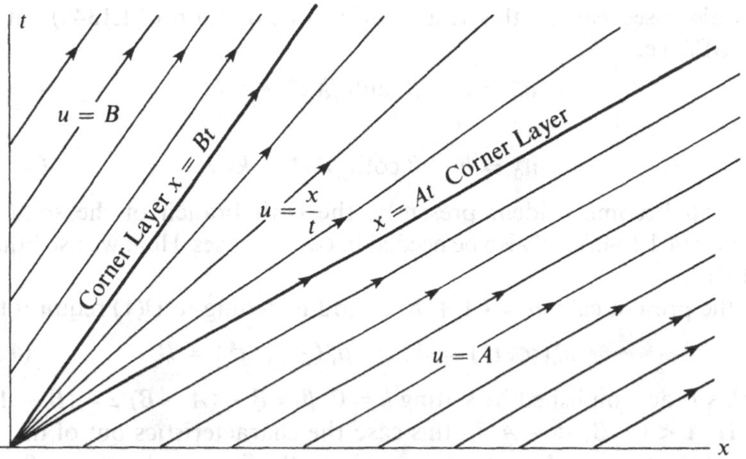

Figure 4.1.13 Solutions for $0 < B < A$

tions of A and B as we did in the example of Section 2.5. We consider first the case (see Figure 4.1.12):

(I) $0 < B < A$. In this case, the outer limit consists of a simple wave $u = x/t$ between the two uniform states $u = A$, $u = B$ as shown in Figure 4.1.13.

To smooth out the discontinuities in u_x along the rays $x = Bt$, $x = At$ we introduce corner layers similar to the ones given by Equation (4.1.153).

(II) $0 < |A| < B$. In this case, the subcharacteristics flowing out of the axes intersect and we must introduce a shock with slope $dx/dt = (A + B)/2 > 0$. The shock divides the first quadrant of the $x - t$ plane into two regions of uniform u as shown in Figure 4.1.14.

To smooth out the discontinuity in u along the shock, we introduce a shock layer of order ε thickness along $x = (A + B)t/2$.

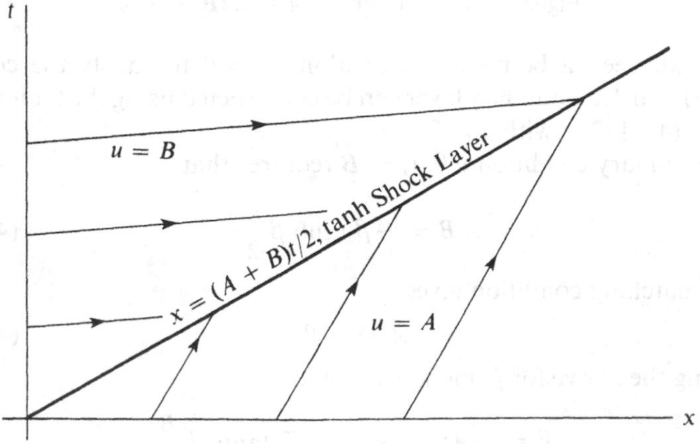

Figure 4.1.14 Solution for $0 < |A| < B$

As discussed earlier, this is a solution of Equation (4.1.134a) with $b = (A + B)/2$, i.e.,

$$u_0^* = b - \beta \tanh \beta(x^* + k)/2 \qquad (4.1.157a)$$

or

$$u_0^* = b - \beta \coth \beta(x^* + k)/2. \qquad (4.1.157b)$$

As will become evident presently, the coth branch of the solution of Equation (4.1.134a) will also be needed in certain cases. Here we use Equation (4.1.157a).

In the present case, $b = (A + B)/2$, and matching to $O(1)$ requires that

$$u_0^*(\infty, t^*) = A; \qquad u_0^*(-\infty, t^*) = B \qquad (4.1.158)$$

and this is accomplished by setting $k = 0$, $\beta = B - (A + B)/2 = (B - A)/2$.

(III) $A < 0$, $|B| < -A$. In this case the characteristics out of the x axis intersect the t axis and no shock can exist in the first quadrant ($t > 0, x > 0$) as shown in Figure 4.1.15.

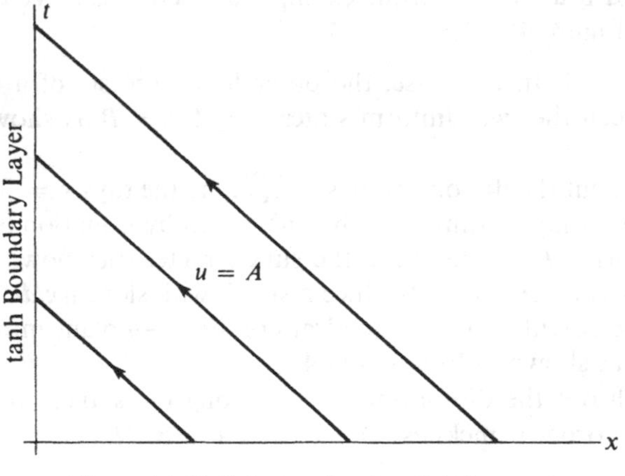

Figure 4.1.15 Solution for $A < 0$, $|B| < -A$

Thus, we need a boundary layer along $x = 0$ to satisfy the condition $u(0, t) = B$. Such a boundary layer can be constructed using the tanh solution, Equation (4.1.157a), with $b = 0$.

The boundary condition $u(0, t) = B$ requires that

$$B = -\beta \tanh \beta \frac{k}{2} \qquad (4.1.159a)$$

and the matching condition gives

$$A = -\beta. \qquad (4.1.159b)$$

Solving the above for β and k, we obtain

$$\beta = -A; \qquad k = -\frac{2}{A} \tanh^{-1} \frac{B}{A}. \qquad (4.1.160)$$

Note that $\tanh^{-1} B/A$ is real only if $|B/A| \leq 1$, so that $B = A < 0$ is the limit beyond which we cannot use a tanh solution. However, we do have the coth solution, Equation (4.1.157b) and this is needed in case IV.

(IV) $A < 0$, $B < 0$, $|B/A| > 1$. Now, the outer solution is as in case III, and we use a coth boundary layer, and we calculate

$$\beta = -A \tag{4.1.161a}$$

$$k = -\frac{2}{A}\coth^{-1} B/A. \tag{4.1.161b}$$

The above cases exhaust the possibilities of using layers of order ε thickness, but we have not been able to handle the case $A > 0$, $B < 0$.

(V) $A > 0$, $B < 0$. Now, the outer solution consists of a uniform region and a simple wave as sketched in Figure 4.1.16.

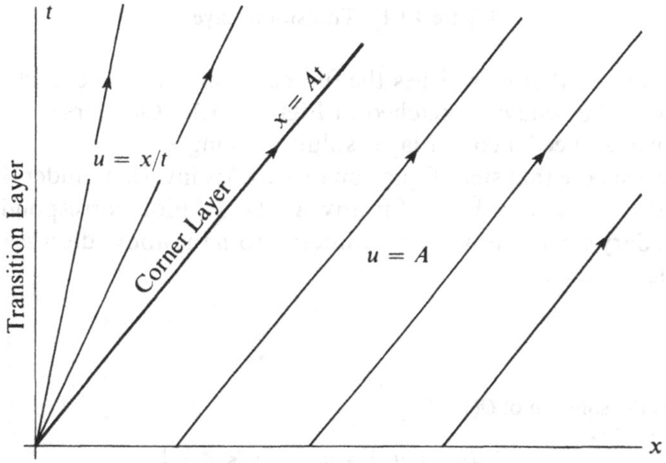

Figure 4.1.16 Solution for $A > 0$, $B < 0$

Thus, on the boundary $x = 0$, the outer limit, $u = 0$, disagrees with the boundary condition $u = B < 0$. It is clear that neither the tanh nor the coth solutions can be used as they do not increase from a negative value to zero as $x^* \to \infty$.

The situation here is analogous to case (V) of Section 2.5. We need to introduce a transition layer using the variables

$$x_T = \frac{x - \varepsilon\xi_T(t, \varepsilon)}{\sqrt{\varepsilon}} \tag{4.1.162a}$$

$$u_T = u/\sqrt{\varepsilon}, \tag{4.1.162b}$$

where $\xi_T < 0$ is the location of the transition layer.

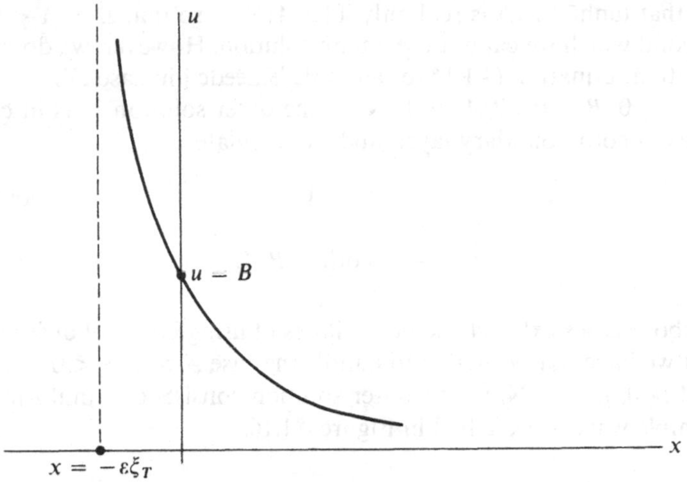

Figure 4.1.17 Transition Layer

Again, we find that u_T satisfies the full equation, and we expect an exact solution with the behavior sketched in Figure 4.1.17. Of course, in addition to the above, we need a corner layer solution along $x = At$.

Finally, we note that since Equation (4.1.106) is invariant under the transformation $u \to -u$, $x \to \bar{X} - x$ for any $\bar{X}$, the solution corresponding to a right boundary can be derived by symmetry from the above discussion of the case of a left boundary.

Problems

1. What is the solution of $O(1)$ of

$$\varepsilon(u_{xx} + u_{yy}) = u_y, \qquad 0 < \varepsilon \ll 1$$

in the interior of the unit square, $0 < x < 1$, $0 < y < 1$, with boundary conditions

$$u_y(x, 0) = 0; \qquad u(1, y) = -1$$
$$u(0, y) = 1; \qquad u(x, 1) = 0.$$

2. An incompressible fluid (density ρ, specific heat c, thermal conductivity k) flows through a circular grid of radius L located at $x = 0$. The velocity of the fluid is assumed always to be U in the $+x$-direction. The temperature of the grid is maintained at $T = T_B =$ const., and the temperature of the fluid at upstream infinity is T_∞. Thus, the differential equation for the temperature field is

$$\kappa\left\{\frac{\partial^2 T}{\partial x^2} + \frac{\partial^2 T}{\partial r^2} + \frac{1}{r}\frac{\partial T}{\partial r}\right\} = U\frac{\partial T}{\partial x},$$

where

$$\kappa = k/\rho c.$$

Write the problem in suitable dimensionless coordinates. [Use

$$x^* = x/L, \qquad r^* = r/L, \qquad T(x, r) = T_\infty + (T_B - T_\infty)\theta(x^*, r^*).]$$

Study the behavior of the solution for ε small, where

$$\varepsilon = \kappa/UL.$$

Find the "outer" solution and the necessary boundary layers. In particular show that the rate of heat transfer to the fluid is independent of k as $\varepsilon \to 0$.

Discuss the validity of these solutions for the regions where $x^* \to \infty$.

3. Consider heat transfer to a viscous incompressible fluid flowing steadily in a circular pipe of radius R. The equation for the temperature distribution is

$$\rho c u(r) \frac{\partial T}{\partial x} = k\left\{\frac{\partial^2 T}{\partial r^2} + \frac{1}{r}\frac{\partial T}{\partial r} + \frac{\partial^2 T}{\partial x^2}\right\},$$

where ρ = fluid density, c = fluid specific heat = const., k = thermal conductivity = const. For laminar flow, the velocity distribution in the pipe is parabolic:

$$\frac{u(r)}{U} = 1 - \left(\frac{r}{R}\right)^2.$$

Let the temperature be raised at the wall from the constant value T_0 for $x < 0$ to T_1 for $x > 0$. (See Fig. 4.1.18.)

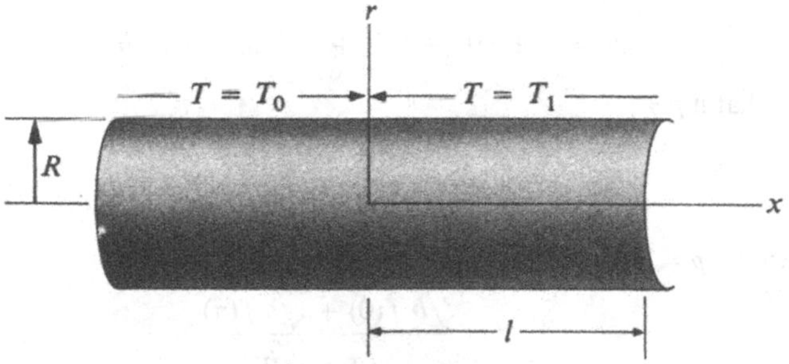

Figure 4.1.18 Heat Transfer in Pipe Flow.

For

$$\varepsilon = \frac{k/\rho c}{UR} \ll 1 \quad \left(\text{note that } \varepsilon = \frac{1}{\text{Re } Pr}\right),$$

construct a suitable boundary-layer theory. Thus, show that the heat transferred to the fluid $Q(\imath)$ in the length $\imath$ is approximately

$$\frac{Q}{([k(T_i - T_0)]/R)(2\pi R\imath)} = K^*\left(\frac{R}{\imath}\right)^{1/3},$$

and evaluate the constant K^*.

[Hint: Use similarity methods or Laplace transforms on the boundary-layer equation.]

Indicate the mathematical problem to be solved for the next higher-order approximation. Is there a singularity of heat transfer at $x = 0$ in the higher approximations?

4. Carry out the details of the calculations leading to Eq. (4.1.46).

5. Study Eq. (4.1.32) inside a general bounded domain D surrounding the origin. Let the boundary be given by

$$r = r_B(\theta)$$

and assume that the boundary condition is specified in the form

$$u(r_B, \theta) = f(\theta).$$

(a) Assume that the point $P: r = 1, \theta = 0$, is a unique closest point to the origin, and that near this point

$$r_B(\theta) = 1 + a\theta^2 + \cdots, \quad \text{as } \theta \to 0; \ a > 0.$$

Thus, at P, r_B has a first order contact with the unit circle. Show that in this case

$$\alpha = f(0).$$

(b) Let the boundary have *two* distinct points ($P_1: r = 1, \theta = 0$ and $P_2: r = 1, \theta = \pi$) nearest the origin. Also assume that r_B has the following behavior near these points

$$r_B(\theta) = 1 + a\theta^{2p} + \cdots \quad \text{as } \theta \to 0; a > 0$$

$$r_B(\theta) = 1 + b(\theta - \pi)^{2q} + \cdots \quad \text{as } \theta \to \pi; \ b > 0.$$

Show that if $p \neq q$

$$\alpha = \begin{cases} f(0); & \text{if } p > q \\ f(\pi); & \text{if } p < q \end{cases}$$

and that if $p = q$

$$\alpha = \frac{\sqrt{b}\, f(0) + \sqrt{a}\, f(\pi)}{\sqrt{a} + \sqrt{b}}.$$

6. Study the problem

$$\varepsilon(u_{xx} + u_{yy}) = yu_x + xu_y$$

in the interior of the unit circle, $x^2 + y^2 = 1$, with boundary condition: u specified on the circumference.

For this problem, the subcharacteristics have a saddle point at the origin. Determine the locations of allowable boundary layers and discuss the nature of the solution near the origin.

7. Calculate the solution to $O(1)$ of

$$\varepsilon u_{xx} = u_t + \cos t u_x, \qquad 0 \leq x, 0 \leq t$$

with initial condition

$$u(x, 0) = 0$$

and boundary condition

$$u(0, t) = 1.$$

Note the locations of shock and boundary layers and calculate the inner limit in these layers. Also note the points near which exact solutions are needed.

8. Derive the solution of Eq. (4.1.142b) for u_1^* then carry out the matching between the inner expansion, $u = u_0^* + \varepsilon u_1^*$, and the two outer expansions $u = (x - 1)/t + \varepsilon u_1(x, t)$ and $u = 1$ on either side. Does the matching determine $x_1(t)$ and the two unknown functions of t which arise in the solution?

9. Construct the Riemann function R for

$$\varepsilon(u_{xx} - u_{tt}) = au_x + bu_t,$$

so that the solution of the initial value problem with

$$u(x, 0) = 0, \qquad u_t(x, 0) = V(x)$$

is represented as

$$u(x, t; \varepsilon) = \int_{x-t}^{x+t} R(x - \xi, t; \varepsilon)V(\xi)d\xi.$$

Show how inner and outer expansions as $\varepsilon \to 0$ can be obtained from the exact solution.

10. Consider the wave equation

$$\varepsilon\left(\frac{\partial}{\partial t} + c_1\frac{\partial}{\partial x}\right)\left(\frac{\partial}{\partial t} + c_2\frac{\partial}{\partial x}\right)u + \left(\frac{\partial}{\partial t} + b\frac{\partial}{\partial x}\right)u = 0,$$

where $0 < \varepsilon \ll 1$, $c_1 > 0$, $c_2 < 0$, and c_1, c_2 and b are constants.

(a) What is the condition on a for stable solutions?
(b) For the stable case, study the signaling problem.

$$u(x, 0) = 0$$

$$u(0, t) = F(t).$$

(c) For the special case $F = 1$ when $0 < t \leq 1$, $F = 0$ for $t > 1$, study the behavior of the signal for large t.

References

4.1.1 R. Courant and D. Hilbert, *Methods of Mathematical Physics*, Vol. 1, Interscience Publishers, Inc., New York, 1953.

4.1.2 Cole, J. D., On a quasilinear parabolic equation occurring in aerodynamics, *Q. Appl. Math.* **9**, 1951, pp. 225–236.

4.1.3 Hopf, E., The partial differential equation $u_t + uu_x = \mu u_{xx}$, *Comm. Pure Appl. Math.* **3**, 1950, pp. 201–230.

4.1.4 Lax, P. D., Hyperbolic Systems of Conservation Laws and the Mathematical Theory of Shock Waves, *S.I.A.M. Regional Conference Series in Applied Mathematics*, Vol. 11, 1973.

4.1.5 Whitham, G. B., *Linear and Nonlinear Waves*, John Wiley and Sons, New York, 1974.

4.2 Boundary-Layer Theory in Viscous Incompressible Flow

The original physical problem from which the ideas of mathematical boundary-layer theory originated was the problem of viscous, incompressible flow past an object. The aim was to explain the origin of the resistance in a slightly viscous fluid. By the use of physical arguments, Prandtl deduced that for small values of the viscosity, a thin region near the solid boundary (where the fluid is brought to rest) is described by approximate boundary-layer equations, and the flow outside this region is essentially inviscid. These ideas find a natural mathematical expression in terms of the ideas of singular perturbation problems discussed in Sections 2.1 through 2.6. In this Section, we show how the external inviscid flow is associated with an outer-limit process and the boundary layer with an inner-limit process. The boundary condition of no slip is lost, and the order of the equations is lowered in the outer limit, so that the problem is indeed singular in the terminology used previously.

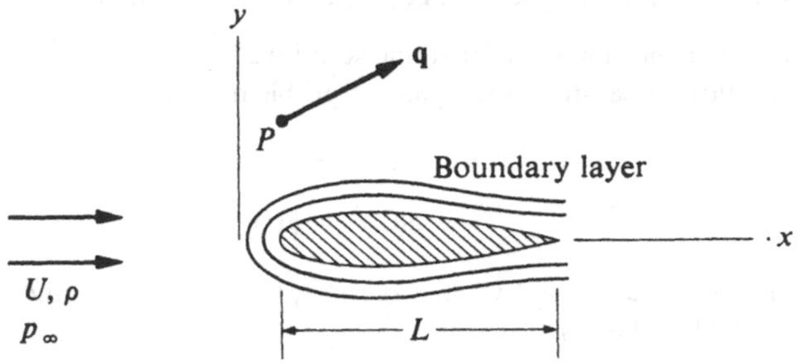

Figure 4.2.1 Flow Past a Body

In order to illustrate these ideas explicitly, the entire discussion should be carried out in dimensionless variables. Consider uniform flow with velocity U past an object with characteristic length L (Figure 4.2.1). Given the fluid density ρ and viscosity coefficient μ, there is one overall dimensionless number, the Reynolds number, Re:

$$\mathrm{Re} = UL\rho/\mu = UL/\nu. \tag{4.2.1}$$

Pressure does not enter a dimensionless parameter since the level of pressure has no effect on the flow.[3] The limit processes are all concerned with Re $\to \infty$ or

$$\varepsilon = 1/\text{Re} \to 0. \tag{4.2.2}$$

Now, make all velocities dimensionless with U, all lengths with L, and the pressure with ρU^2. The full problem is expressed in the continuity and momentum equations, written in an invariant vector form:

Navier-Stokes Equations

$$\operatorname{div} \mathbf{q} = 0, \tag{4.2.3a}$$

$$\underbrace{\mathbf{q}\cdot\nabla\mathbf{q} \equiv \nabla\left(\frac{q^2}{2}\right) - (\mathbf{q}\times\boldsymbol{\omega})}_{\text{transport or inertia}} = \underbrace{-\nabla p}_{\text{pressure}} \underbrace{- \,\varepsilon \operatorname{curl}\boldsymbol{\omega}}_{\text{viscous body force}}, \tag{4.2.3b}$$

where

$$\boldsymbol{\omega} = \text{vorticity} = \operatorname{curl}\mathbf{q}. \tag{4.2.3c}$$

Vorticity represents the angular velocity of a fluid element. Also it can be shown that the viscous force per area on a surface is

$$\boldsymbol{\tau}_v = -\varepsilon(\boldsymbol{\omega}\times\mathbf{n}), \quad \mathbf{n} = \text{outward normal (dimensionless)}, \tag{4.2.4}$$

$$\mathbf{q}(\infty) = \mathbf{i}, \quad \text{uniform flow}, \tag{4.2.5}$$

$$\mathbf{q} = \mathbf{q}_b = 0 \quad \text{on the body} \quad \text{(no slip condition)}. \tag{4.2.6}$$

The outer expansion is carried out, keeping the representative point P fixed and letting $\varepsilon \to 0$. The expansion has the form of an asymptotic expansion in terms of a suitable sequence $\alpha_i(\varepsilon)$:

$$\left.\begin{aligned} \mathbf{q}(P;\varepsilon) &= \mathbf{q}_0(P) + \alpha_1(\varepsilon)\mathbf{q}_1(P) + \cdots, & \varepsilon &\to 0;\\ p(P;\varepsilon) &= p_0(P) + \alpha_1(\varepsilon)p_1(P) + \cdots, & P &\text{ fixed.} \end{aligned}\right. \tag{4.2.7}$$

Here $(\mathbf{q}_0, p_0)$ represents an inviscid flow. In many cases of interest, this flow is irrotational. This fact can be demonstrated by considering the equation for vorticity propagation obtained from curl (4.2.3b),

$$\operatorname{curl}(\mathbf{q}\times\boldsymbol{\omega}) = \varepsilon \operatorname{curl}\operatorname{curl}\boldsymbol{\omega}. \tag{4.2.8}$$

For plane flow, the vorticity vector is normal to the xy-plane and can be written

$$\boldsymbol{\omega} = \mathbf{k}\omega(P), \quad \text{plane flow}. \tag{4.2.9}$$

Thus, Equation (4.2.8) can be written

$$\mathbf{q}\cdot\nabla\omega = \varepsilon\nabla^2\omega. \tag{4.2.10}$$

[3] That is, the Mach number is always zero.

The physical interpretation of Equation (4.2.10) is that the vorticity is transported along the streamlines but diffuses (like heat) due to the action of viscosity. The solid boundary is the only source of vorticity; as $\varepsilon \to 0$, the diffusion is small and is confined to a narrow boundary layer close to the body (except if the flow separates). Under the outer limit, Equation (4.2.10) becomes

$$\mathbf{q} \cdot \nabla\omega = 0, \tag{4.2.11}$$

so that vorticity is constant along a streamline.[4] For uniform flow, $\omega = 0$ at ∞, and hence $\omega = 0$ throughout. We obtain, in terms of the outer expansion,

$$\boldsymbol{\omega}_0 = \operatorname{curl} \mathbf{q}_0 = 0. \tag{4.2.12}$$

Thus, the basic outer flow $(\mathbf{q}_0, p_0)$ is a potential flow. For example, in cartesian components, $\mathbf{q}_0 = u_0 \mathbf{i} + v_0 \mathbf{j}$, and the components can be expressed by an analytic function,

$$u_0 - iv_0 = F'(z), \qquad z = x + iy. \tag{4.2.13}$$

The problem is thus purely kinematic. Integration of the limit form of Equations (4.2.3) along a streamline yields Bernoulli's law:

$$p_0 = \tfrac{1}{2}(1 - q_0^2), \qquad p_0(\infty) = 0, \tag{4.2.14}$$

and accounts for all the dynamics of potential flow. In addition, stream (ψ) and potential (ϕ) functions exist

$$\phi + i\psi = F(z), \qquad u_0 = \phi_x = \psi_y, \qquad v_0 = \phi_y = -\psi_x. \tag{4.2.15}$$

An important question, which can not be answered by the present approach, is what potential flow $F(z)$ to choose for the problem of high Re flow past a given body. Real flows tend to separate toward the rear of a closed body and to generate a viscous wake. This implies that the correct limiting potential flow separates from the body. Furthermore, if Re is sufficiently high, turbulence sets in, so that a description under the steady Navier–Stokes equations is not valid. Thus, for our purposes we consider the simplest potential flow, for example, that which closes around the body. The approximation is understood to be valid only in a region of limited extent near the nose of the body.

Now, with respect to the higher terms of the outer expansion, the following observation can be made. Every $(\mathbf{q}_i, p_i)$ is a potential flow. This follows by induction from the fact that the viscous body force is zero in a potential flow:

$$\operatorname{curl} \boldsymbol{\omega}_0 = 0 \quad \text{or} \quad \nabla^2_{(x,y)}\mathbf{q}_0 = 0. \tag{4.2.16}$$

The inner expansion is derived from a limit process in which a representative point P^* approaches the boundary as $\varepsilon \to 0$. The boundary layer in this problem is along a streamline of the inviscid flow, a subcharacteristic of the full problem. Characteristic surfaces in general are the loci of possible

[4] $\mathbf{q} \cdot \nabla$ is the operator of differentiation along a streamline.

discontinuities, and streamlines of an inviscid flow can support a discontinuity in vorticity. In the inviscid limit in which the external flow is potential flow, this discontinuity is only at the solid surface where the tangential velocity jumps (if the boundary condition of zero velocity on the surface is enforced). Now the vorticity equation (Equation 4.2.10) has the same structure as the general partial differential equation discussed in Section 4.1. The boundary layer resolving the vorticity occurs on a subcharacteristic and, hence, should be of $O(\sqrt{\varepsilon})$ in thickness. Thus, symbolically,

$$P^* = \frac{P - P_b}{\sqrt{\varepsilon}}, \qquad P_b = \text{point of the boundary,} \tag{4.2.17}$$

is held fixed as $\varepsilon \to 0$. This order of magnitude can also be checked explicitly as below.

Before considering flow past a body, however, it is worthwhile to outline the simpler problem of purely radial flow in a wedge-shaped sector. (Figure 4.2.2). For this simple geometry, the Navier–Stokes equations simplify

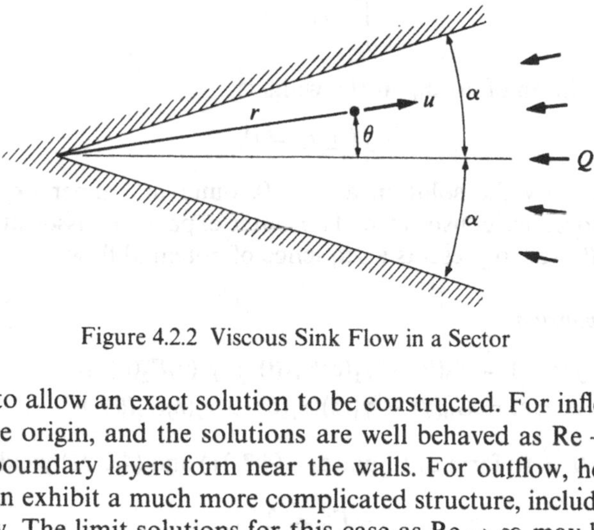

Figure 4.2.2 Viscous Sink Flow in a Sector

sufficiently to allow an exact solution to be constructed. For inflow, there is a sink at the origin, and the solutions are well behaved as $\mathrm{Re} \to \infty$ in the sense that boundary layers form near the walls. For outflow, however, the solutions can exhibit a much more complicated structure, including regions of back-flow. The limit solutions for this case as $\mathrm{Re} \to \infty$ may have vortex sheets in the interior of the channel. Only the case of inflow is considered here.

4.2.1 Radial Viscous Inflow

The mass flux, per unit width of channel, Q = mass/sec-length, is prescribed. The overall Reynolds number is, thus,

$$\mathrm{Re} = \frac{Q}{\mu} = \frac{1}{\varepsilon}. \tag{4.2.18}$$

By dimensional reasoning, the radial velocity and pressure must be of the form

$$\text{outward radial velocity} = \frac{Q}{\rho r} f(\theta), \tag{4.2.19}$$

$$\text{pressure} = \frac{Q^2}{\rho r^2} g(\theta). \tag{4.2.20}$$

Thus, the full Navier–Stokes equations become ordinary differential equations:

$$\text{radial momentum,} \qquad -f^2(\theta) = 2g(\theta) + \varepsilon \frac{d^2 f}{d\theta^2}; \tag{4.2.21}$$

$$\text{tangential momentum,} \qquad 0 = -\frac{dg}{d\theta} + 2\varepsilon \frac{df}{d\theta}. \tag{4.2.22}$$

Mass conservation leads to the normalization

$$-\int_{-\alpha}^{\alpha} f(\theta) d\theta = 1, \tag{4.2.23}$$

while the condition of no slip at the walls is

$$f(\pm\alpha) = 0. \tag{4.2.24}$$

In order to study the solution as $\varepsilon \to 0$, outer and inner expansions, as indicated above, are constructed. The outer expansion, associated with the limit $\varepsilon \to 0$, θ fixed, represents a sequence of potential flows.

Outer Expansion

$$\begin{aligned} f(\theta; \varepsilon) &= F_0(\theta) + \gamma_1(\varepsilon)F_1(\theta) + \gamma_2(\varepsilon)F_2(\theta) + \cdots, \\ g(\theta; \varepsilon) &= G_0(\theta) + \gamma_1(\varepsilon)G_1(\theta) + \gamma_2(\varepsilon)G_2(\theta) + \cdots. \end{aligned} \tag{4.2.25}$$

The limit ($\varepsilon = 0$) form of Equations (4.2.21) and (4.2.22) yields

$$-F_0^2 = 2G_0, \tag{4.2.26}$$

$$0 = -\frac{dG_0}{d\theta}. \tag{4.2.27}$$

The no-slip condition is given up so that $F_0 = G_0 = \text{const.}$, and the normalization (Equation 4.2.23) yields

$$F_0 = -\frac{1}{2\alpha}, \qquad G_0 = -\frac{1}{8\alpha^2}. \tag{4.2.28}$$

Now, in order to have a balance of viscous forces and inertia near the walls ($\theta = \pm\alpha$), it is necessary that the viscous layer have a thickness $O(\sqrt{\varepsilon})$.

An inner limit $\varepsilon \to 0$, $\theta^* = (\theta \pm \alpha)/\sqrt{\varepsilon}$ fixed, is considered. It follows that the inner expansion is of the form (valid near each wall).

Inner Expansion

$$f(\theta;\varepsilon) = f_0(\theta^*) + \beta(\varepsilon)f_1(\theta^*) + \cdots, \qquad g(\theta;\varepsilon) = g_0(\theta^*) + \beta(\varepsilon)g_1(\theta^*) + \cdots,$$
$$\theta^* = (\theta \pm \alpha)/\sqrt{\varepsilon}. \tag{4.2.29}$$

The equations of motion reduce to

$$O(1): \qquad -f_0^2 = 2g_0 + \frac{d^2 f_0}{d\theta^{*2}}. \tag{4.2.30}$$

$$O\left(\frac{1}{\sqrt{\varepsilon}}\right): \qquad 0 = -\frac{dg_0}{d\theta^*}. \tag{4.2.31}$$

The solutions to these boundary-layer equations should satisfy the no-slip condition, so that

$$f_0(0) = 0. \tag{4.2.32}$$

The other boundary conditions for Equations (4.2.30) and (4.2.31) are found by matching with the outer solution. Equation (4.2.31) states that, to this order, there is no pressure gradient across the thin viscous layer adjacent to the wall. The matching, thus, fixes the level of pressure in the boundary layer. An intermediate limit is obtained by holding θ_η fixed as $\varepsilon \to 0$, where

$$\theta_\eta = \frac{(\theta \pm \alpha)}{\eta(\varepsilon)}, \qquad \sqrt{\varepsilon} \ll \eta \ll 1. \tag{4.2.33}$$

In this limit, we have

$$\theta = \mp\alpha + \eta\theta_\eta \to \mp\alpha, \tag{4.2.34}$$

$$\theta^* = \frac{\eta}{\sqrt{\varepsilon}}\theta_\eta \to \pm\infty. \tag{4.2.35}$$

It is assumed that the inner and outer expansions are matched by an intermediate limit. Consider now only the lower wall $\theta = -\alpha$; the solution at the upper wall is found by symmetry. The matching of pressures to first order is

$$\lim_{\substack{\varepsilon\to 0\\ \theta_\eta \text{ fixed}}} \left[G_0(-\alpha + \eta\theta_\eta) + \cdots - g_0\left(\frac{\eta}{\sqrt{\varepsilon}}\theta_\eta\right) + \cdots\right] = 0. \tag{4.2.36}$$

In this case, we know that

$$G_0 = g_0 = \text{const.} = -\frac{1}{8\alpha^2}, \tag{4.2.37}$$

Thus, Equation (4.2.30) becomes

$$\frac{d^2 f_0}{d\theta^{*2}} + f_0^2(\theta^*) = \frac{1}{4\alpha^2}. \tag{4.2.38}$$

Matching of the velocities now gives

$$\lim_{\substack{\varepsilon \to 0 \\ \theta_\eta \text{ fixed}}} \left\{ F_0(-\alpha + \eta\theta_\eta) - f_0\left(\frac{\eta}{\sqrt{\varepsilon}}\theta_\eta\right)\right\} = 0. \tag{4.2.39}$$

The velocity of the potential flow at the wall is matched to the velocity of the boundary-layer flow at infinity in this case:

$$f_0(\infty) \to F_0(-\alpha) = -1/2\alpha. \tag{4.2.40}$$

Thus, the solution of Equation (4.2.38) for $0 \le \theta^* < \infty$ must be found satisfying the conditions of Equations (4.2.32) and (4.2.40). It is clear that the boundary condition (Equation 4.2.40) is consistent with Equation (4.2.38).

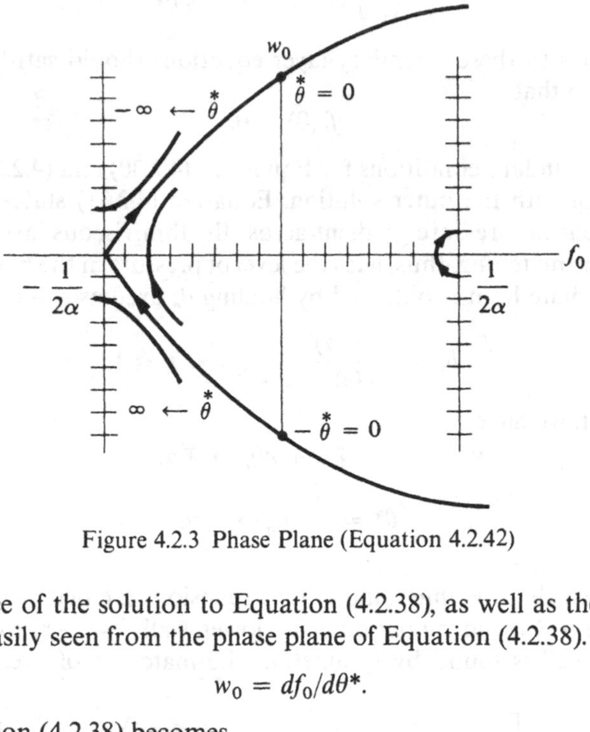

Figure 4.2.3 Phase Plane (Equation 4.2.42)

The existence of the solution to Equation (4.2.38), as well as the form near infinity, is easily seen from the phase plane of Equation (4.2.38). Let

$$w_0 = df_0/d\theta^*. \tag{4.2.41}$$

Then Equation (4.2.38) becomes

$$\frac{dw_0}{df_0} = \frac{(1/4\alpha^2) - f_0^2}{w_0}. \tag{4.2.42}$$

The paths of the integral curves are indicated in Figure 4.2.3. The arrows indicate the direction of increasing θ^* according to Equation (4.2.41). The singularity at $w_0 = 0, f_0 = -1/2\alpha$ is a saddle point whose paths are

$$w_0^2 - \frac{1}{\alpha}\left(f_0 + \frac{1}{2\alpha}\right)^2 = \text{const.} \tag{4.2.43}$$

Two exceptional paths,

$$w_0 = \pm \frac{1}{\sqrt{\alpha}}\left(f_0 + \frac{1}{2\alpha}\right), \tag{4.2.44}$$

enter the saddle points, the (−) sign corresponding to the boundary layer at the lower wall. The value of θ^* along the path is found by integration of Equation (4.2.42) along the path from ($f_0 = 0$, $\theta^* = 0$). Near the singular point, integration of Equation (4.2.41) shows that

$$f_0 = -\frac{1}{2\alpha} + k_0 e^{-\theta^*/\sqrt{\alpha}} + \cdots, \tag{4.2.45}$$

where k_0 is known from integration along the path. Equation (4.2.45) shows that the boundary layer approaches its limiting value with an error which is transcendentally small for intermediate limits.

The need for higher-order terms arises because of the mass-flow defect in the boundary layer. The first term of a uniformly valid ($-\alpha \le \theta \le \alpha$) composite expansion of the form

$$f(\theta; \varepsilon) = \mathscr{F}_0(\theta; \varepsilon) + \gamma_1(\varepsilon)\mathscr{F}_1(\theta; \varepsilon) + \cdots \tag{4.2.46}$$

can be found by adding f_0 for both walls to F_0 and subtracting the common part $-1/2\alpha$:

$$\mathscr{F}_0(\theta; \varepsilon) = f_0\left(\frac{\theta + \alpha}{\sqrt{\varepsilon}}\right) + f_0\left(\frac{\theta - \alpha}{\sqrt{\varepsilon}}\right) + \frac{1}{2\alpha}. \tag{4.2.47}$$

The mass-flow integral (Equation 4.2.23) is

$$-\int_{-\alpha}^{\alpha}\left[f_0\left(\frac{\theta + \alpha}{\sqrt{\varepsilon}}\right) + f_0\left(\frac{\theta - \alpha}{\sqrt{\varepsilon}}\right) + \frac{1}{2\alpha}\right]d\theta$$

$$\to 1 - 2\sqrt{\varepsilon}\int_0^{\infty}\left[\frac{1}{2\alpha} + f_0(\theta^*)\right]d\theta^* + \cdots.$$

The error is thus $O(\sqrt{\varepsilon})$, and this has to be made up by the next term in the outer expansion. Hence, we have $\gamma_1(\varepsilon) = \sqrt{\varepsilon}$ and

$$\int_{-\alpha}^{\alpha} F_1(\theta)d\theta = -2\int_0^{\infty}\left[\frac{1}{2\alpha} + f_0(\theta^*)\right]d\theta^*. \tag{4.2.48}$$

Once the (F_1, G_1) are found, the next boundary-layer terms (f_1, g_1) can be constructed and the procedure repeated.

We next apply these ideas to flow past a body.

4.2.2 Flow Past a Body

Consider steady plane flow past a body of the general form indicated in Figure 4.2.4. In order to carry out the boundary-layer and outer expansions, it is convenient to choose a special coordinate system. In a very interesting

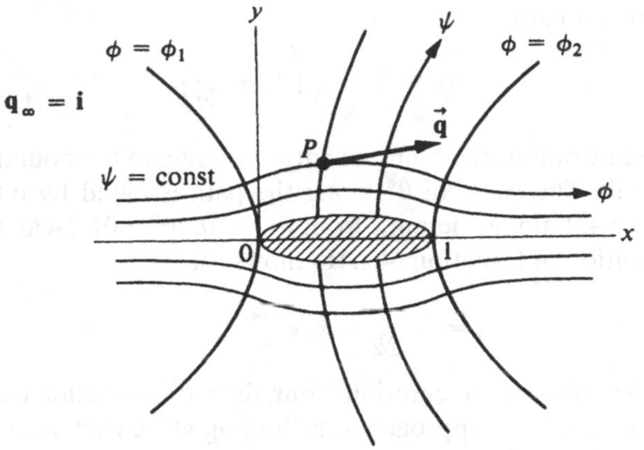

Figure 4.2.4 Streamline Coordinates

paper (Reference 4.2.1), Kaplun discussed the choice of "optimal" coordinates. He shows that it is possible to find certain coordinates in which the boundary-layer equations and solutions are uniformly valid first approximations in the entire flow field, including the so-called flow due to displacement thickness. However, the construction of such coordinates is, in general, just as difficult as it is to proceed directly in any convenient system. The first approximation to the skin-friction is independent of the coordinates. Here we express the Navier–Stokes equations in terms of a network of potential lines (ϕ = const.) and streamlines (ψ = const.), which represent the idealized inviscid flow around the object.

This choice of coordinates at least has the advantage of allowing the ideas of boundary-layer theory to be expressed independent of the body shape and of having a simple representation of the first-order outer flow. Thus, $\psi = 0$ is always the bounding streamline along which the boundary layer appears. In addition to the basic definitions (Equations 4.2.13, 4.2.14, and 4.2.15), we note the following expressions for the velocity components, (q_ϕ, q_ψ), vorticity, etc., in the *viscous* flow. The results follow from general vector formulas in orthogonal curvilinear coordinates. (See Figure 4.2.5.)

$$dz = \frac{dF}{F'}, \qquad (dx)^2 + (dy)^2 = \frac{(d\phi)^2 + (d\psi)^2}{|F'|^2} = \frac{(d\phi)^2 + (d\psi)^2}{q_0^2}, \quad (4.2.49)$$

$$\omega = q_0^2\left\{\frac{\partial}{\partial\phi}\left(\frac{q_\psi}{q_0}\right) - \frac{\partial}{\partial\psi}\left(\frac{q_\phi}{q_0}\right)\right\}, \qquad \boldsymbol{\omega} = \mathbf{k}\omega, \quad (4.2.50)$$

$$\mathbf{q} \times \boldsymbol{\omega} = (q_\phi \mathbf{i}_\phi + q_\psi \mathbf{i}_\psi) \times (\omega\mathbf{k}) = \omega q_\psi \mathbf{i}_\phi - \omega q_\phi \mathbf{i}_\psi, \quad (4.2.51)$$

$$\nabla = \left(q_0\frac{\partial}{\partial\phi}, q_0\frac{\partial}{\partial\psi}\right), \quad (4.2.52)$$

$$\operatorname{curl}\boldsymbol{\omega} = \mathbf{i}_\phi q_0\frac{\partial\omega}{\partial\psi} - \mathbf{i}_\psi q_0\frac{\partial\omega}{\partial\phi}. \quad (4.2.53)$$

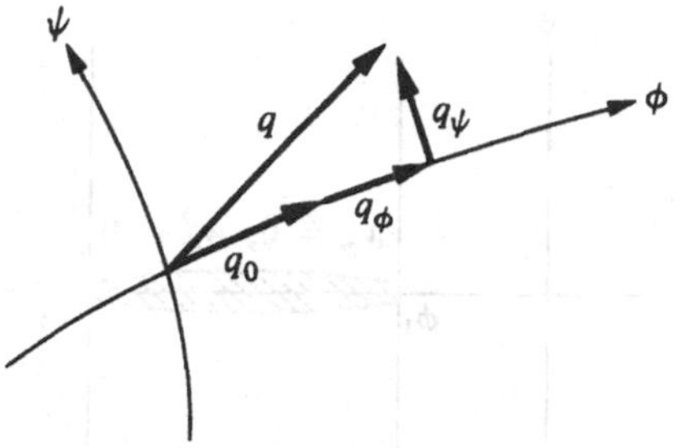

Figure 4.2.5 Detail of Velocity Components

Thus, the basic Navier–Stokes equations (4.2.3a, b) become

$$\text{continuity,}\quad \frac{\partial}{\partial\phi}\left(\frac{q_\phi}{q_0}\right) + \frac{\partial}{\partial\psi}\left(\frac{q_\psi}{q_0}\right) = 0; \tag{4.2.54}$$

$$\phi\text{-momentum,}\quad q_0\frac{\partial}{\partial\phi}\left(\frac{q_\phi^2 + q_\psi^2}{2}\right) - q_\psi\omega = -q_0\frac{\partial p}{\partial\phi} - \varepsilon q_0\frac{\partial\omega}{\partial\psi}; \tag{4.2.55}$$

$$\psi\text{-momentum,}\quad q_0\frac{\partial}{\partial\psi}\left(\frac{q_\phi^2 + q_\psi^2}{2}\right) + q_\phi\omega = -q_0\frac{\partial p}{\partial\psi} + \varepsilon q_0\frac{\partial\omega}{\partial\phi}. \tag{4.2.56}$$

The viscous stress is now $\tau_v = \varepsilon q_0^2(\partial/\partial\psi)(q_\phi/q_0)$. From the form of these equations, it seems clear that a small simplification can be achieved by measuring the velocities at a point P relative to the inviscid velocity q_0 at that point. Let

$$w_\phi = \frac{q_\phi}{q_0}, \qquad w_\psi = \frac{q_\psi}{q_0}. \tag{4.2.57}$$

It follows that, for the vorticity, we have

$$\omega = q_0^2\left\{\frac{\partial w_\psi}{\partial\phi} - \frac{\partial w_\phi}{\partial\psi}\right\}, \tag{4.2.58}$$

and Equations (4.2.54), (4.2.55), and (4.2.56) become

$$\frac{\partial w_\phi}{\partial\phi} + \frac{\partial w_\psi}{\partial\psi} = 0,$$

$$w_\phi\frac{\partial w_\phi}{\partial\phi} + w_\psi\frac{\partial w_\phi}{\partial\psi} + (w_\phi^2 + w_\psi^2)\frac{\partial}{\partial\phi}(\log q_0)$$
$$= -\frac{1}{q_0^2}\frac{\partial p}{\partial\phi} + \varepsilon\left\{\frac{\partial^2 w_\phi}{\partial\phi^2} + \frac{\partial^2 w_\phi}{\partial\psi^2} - 2\frac{\partial\log q_0}{\partial\psi}\left(\frac{\partial w_\psi}{\partial\phi} - \frac{\partial w_\phi}{\partial\psi}\right)\right\}, \tag{4.2.59}$$

$$w_\phi\frac{\partial w_\psi}{\partial\phi} + w_\psi\frac{\partial w_\psi}{\partial\psi} + (w_\phi^2 + w_\psi^2)\frac{\partial}{\partial\psi}(\log q_0)$$
$$= -\frac{1}{q_0^2}\frac{\partial p}{\partial\psi} + \varepsilon\left\{\frac{\partial^2 w_\psi}{\partial\phi^2} + \frac{\partial^2 w_\psi}{\partial\psi^2} + 2\frac{\partial\log q_0}{\partial\phi}\left(\frac{\partial w_\psi}{\partial\phi} - \frac{\partial w_\phi}{\partial\psi}\right)\right\}. \tag{4.2.60}$$

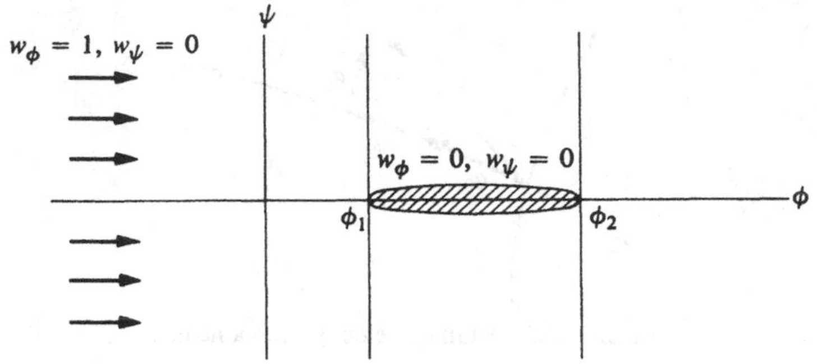

Figure 4.2.6 Boundary-Value Problem in the (ϕ, ψ)-Plane

The viscous surface stress, correspondingly, is $\tau_v = \varepsilon q_0^2(\partial w_\phi/\partial\psi)$. The domain of the problem is sketched in Figure 4.2.6. The body occupies the slit $\psi = 0$, $\phi_1 < \phi < \phi_2$, for symmetric flow. For unsymmetric flow $\phi_2(\psi = 0^+) \neq \phi_2(\psi = 0^-)$. The boundary conditions are as follows:

$$\text{uniform flow at infinity upstream,} \quad w_\psi \to 0, \quad w_\phi \to 1, \quad \phi \to -\infty; \tag{4.2.61}$$

$$\text{no-slip at the body surface,} \quad w_\phi = w_\psi = 0, \quad \psi = 0, \quad \phi_1 < \phi < \phi_2. \tag{4.2.62}$$

Now, in order to construct the expansions, we consider first the outer or Euler limit ($\varepsilon \to 0$, ϕ, ψ fixed). This represents inviscid and, in this case, irrotational, flow around the object. The limit flow is, thus,

$$w_\phi \to 1, \quad w_\psi \to 0, \quad p = \tfrac{1}{2}(1 - q_0^2) \quad \text{(Bernoulli equation).} \tag{4.2.63}$$

As an outer expansion, we have the limit flow as the first term, and corrections due to the inner solution appearing as higher terms. The general form of the outer expansion is, thus,

$$w_\phi(\phi, \psi; \varepsilon) = 1 + \beta(\varepsilon)w_\phi^{(1)}(\phi, \psi) + \cdots, \tag{4.2.64}$$

$$w_\psi(\phi, \psi; \varepsilon) = \beta(\varepsilon)w_\psi^{(1)}(\phi, \psi) + \cdots, \tag{4.2.65}$$

$$p(\phi, \psi; \varepsilon) = \tfrac{1}{2}(1 - q_0^2) + \beta(\varepsilon)p^{(1)}(\phi, \psi) + \cdots. \tag{4.2.66}$$

All corrections with superscript (1) vanish at upstream infinity. However, other boundary conditions for these correction terms cannot be found without discussing the inner viscous boundary layer. To construct the boundary layer and correction equations, we consider an inner-limit process and associated expansion where

$$\psi^* = \frac{\psi}{\delta(\varepsilon)}, \quad \phi \text{ fixed as } \varepsilon \to 0. \tag{4.2.67}$$

The expansion has the following form:

$$w_\phi(\phi, \psi; \varepsilon) = W_\phi(\phi, \psi^*) + \cdots, \tag{4.2.68}$$

$$w_\psi(\phi, \psi; \varepsilon) = \delta(\varepsilon) W_\psi(\phi, \psi^*) + \cdots, \tag{4.2.69}$$

$$p(\phi, \psi; \varepsilon) = P(\phi, \psi^*) + \cdots. \tag{4.2.70}$$

The form is deduced from the following considerations, in addition to those which indicated that the boundary layer occupies a thin region close to $\psi = 0$. The first term of the expression for the velocity component w_ϕ along the streamline is of order one, so that it can be matched to the outer expansion (Equation 4.2.64). The first term in the expansion for w_ψ must, then, be of the order $\delta(\varepsilon)$, so that a nontrivial continuity equation results:

$$\frac{\partial W_\phi}{\partial \phi} + \frac{\partial W_\psi}{\partial \psi} = 0. \tag{4.2.71}$$

The first term in the pressure is also $O(1)$ in order to match to Equation (4.2.66). Under the boundary-layer limit, the inviscid velocity field which occurs in the coefficient approaches the surface distribution of inviscid velocity according to

$$q_0(\phi, \psi) = q_0(\phi, \delta(\varepsilon)\psi^*) = q_0(\phi, 0) + \delta(\varepsilon)\psi^* \frac{\partial q_0}{\partial \psi}(\phi, 0) + \cdots \tag{4.2.72}$$

or

$$q_0(\phi, \psi) = q_B(\phi) + O(\delta(\varepsilon)),$$

where $q_B(\phi)$ = inviscid surface-velocity distribution. The inertia and pressure terms in Equation (4.2.59) are both of $O(1)$, while $\varepsilon(\partial^2 W_\phi/\partial \psi^2)$ is of the order ε/δ^2. The distinguished limiting case has

$$\delta = \sqrt{\varepsilon}. \tag{4.2.73}$$

It is only this case that allows a nontrivial system of boundary-layer equations capable of satisfying the boundary conditions and being matched to the outer flow. With this assumption, the first-approximation momentum equations are

$$W_\phi \frac{\partial W_\phi}{\partial \phi} + W_\psi \frac{\partial W_\phi}{\partial \psi^*} + \frac{W_\phi^2}{q_B(\phi)} \frac{dq_B}{d\phi} = -\frac{1}{q_B^2} \frac{\partial P}{\partial \phi} + \frac{\partial^2 W_\phi}{\partial \psi^{*2}} \tag{4.2.74}$$

$$0 = -\frac{1}{q_B^2} \frac{\partial P}{\partial \psi^*}. \tag{4.2.75}$$

Equation (4.2.75) tells us that the layer is so thin that the pressure does not vary across the layer and, rather, that

$$P = P(\phi). \tag{4.2.76}$$

Hence, the pressure is easily matched to the pressure in the outer solution. All the matching is carried out with the help of an intermediate limit in which the representative point approaches the wall but not as fast as it does in the distinguished limit. In the intermediate limit, we have

$$\psi_\eta = \frac{\psi}{\eta(\varepsilon)}, \qquad \phi \text{ fixed}, \tag{4.2.77}$$

where

$$\sqrt{\varepsilon} \ll \eta \ll 1.$$

Thus, in the intermediate limit, we see that

$$\psi = \eta\psi_\eta \to 0, \qquad \psi^* = \frac{\eta}{\sqrt{\varepsilon}}\psi_\eta \to \infty. \tag{4.2.78}$$

It is sufficient to consider only positive ψ to illustrate the ideas. Matching of pressure (cf. Equations 4.2.66, 4.2.70) takes the form

$$\lim_{\varepsilon\to 0}\{\tfrac{1}{2}(1 - q_0^2(\phi, \eta\psi_\eta)) + \beta(\varepsilon)p^{(1)}(\phi, \eta\psi_\eta) + \cdots - P(\phi) - \cdots\} = 0.$$

Hence, to first order, we have

$$P(\phi) = \tfrac{1}{2}(1 - q_B^2(\phi)) = P_B(\phi) \quad \text{(say)}. \tag{4.2.79}$$

The pressure distribution on the body is that of the inviscid flow, if we neglect the boundary layer. Thus, the system of boundary-layer equations (Equations 4.2.71, 4.2.74) is

$$\frac{\partial W_\phi}{\partial \phi} + \frac{\partial W_\psi}{\partial \psi^*} = 0,$$

$$W_\phi \frac{\partial W_\phi}{\partial \phi} + W_\psi \frac{\partial W_\phi}{\partial \psi^*} = \frac{1 - W_\phi^2}{q_B}\frac{dq_B}{d\phi} + \frac{\partial^2 W_\phi}{\partial \psi^{*2}}, \tag{4.2.80}$$

a system for (W_ϕ, W_ψ). The boundary conditions to be satisfied are no-slip,

$$W_\phi(\phi, 0) = W_\psi(\phi, 0) = 0, \qquad \phi_1 < \phi < \phi_2, \tag{4.2.81}$$

and matching. The system (Equation 4.2.80) is parabolic, so that only the interval $\phi_1 < \phi < \phi_2$ need be considered at first. The next quantity to be matched is the velocity component along a streamline, which also contains an order-one term. Inner and outer expansions (Equations 4.2.68 and 4.2.64) must match in terms of intermediate variables, so that

$$\lim_{\varepsilon\to 0}\{1 + \beta(\varepsilon)w_\phi^{(1)}(\phi, \eta\psi_\eta) + \cdots - W_\phi(\phi, (\eta/\sqrt{\varepsilon})\psi_\eta) - \cdots\} = 0. \tag{4.2.82}$$

Thus, the boundary condition is obtained:

$$\lim_{\psi^*\to\infty} W_\phi(\phi, \psi^*) = 1. \tag{4.2.83}$$

This is usually interpreted by saying that the velocity at the outer edge of the boundary layer is that of the inviscid flow adjacent to the body. Since the system (Equations 4.2.80) is parabolic, there is no upstream influence, so that the solution again must match the undisturbed flow:

$$W_\phi(\phi_1, \psi^*) = 1. \tag{4.2.84}$$

The conditions of Equations (4.2.84), (4.2.83), and (4.2.81) serve to define a unique solution in the strip $\phi_1 < \phi < \phi_2$. The solution downstream of the body, $\phi > \phi_2$, should really be discussed also. The boundary-layer equations and expansion are the same, but the boundary conditions corresponding to the wake are different. Now the upstream boundary-layer solution just calculated provides initial conditions on $\phi = \phi_2$ for $-\infty < \psi^* < \infty$; and the initial-value problem can be solved to find the flow downstream.

Assume now that the solution of Equation (4.2.80) has been found for all $\phi > \phi_1$, so that W_ϕ, W_ψ are known functions. The matching of the normal component of velocity along the potential lines W_ψ can be discussed next, and this provides a boundary condition which defines the correction in the outer flow due to the presence of the boundary layer. The intermediate limit of inner and outer expansions is

$$\lim_{\varepsilon \to 0} \left\{ \beta(\varepsilon) w_\psi^{(1)}(\phi, \eta\psi_\eta) + \cdots - \sqrt{\varepsilon}\, W_\psi\left(\phi, \frac{\eta\psi_\eta}{\sqrt{\varepsilon}}\right) - \cdots \right\} = 0. \tag{4.2.85}$$

Matching is achieved to first order, provided the limits exist, if first of all

$$\beta(\varepsilon) = \sqrt{\varepsilon} \tag{4.2.86}$$

and

$$w_\psi^{(1)}(\phi, 0) = W_\psi(\phi, \infty), \qquad \phi > \phi_1. \tag{4.2.87}$$

Equation (4.2.87) has the form of a boundary condition for $w_\psi^{(1)}$, which can be interpreted as an effective thin body added to the original body; it defines the flow due to displacement thickness. The limit in Equation (4.2.87) exists since the solutions for $[1 - W_\phi]$ can be shown to decay exponentially as $\psi \to \infty$ and

$$W_\psi(\phi, \psi) = \int_0^\psi \left[-\frac{\partial W_\phi}{\partial \phi}(\phi, \lambda) \right] d\lambda. \tag{4.2.88}$$

Thus, the outer flow $w_\phi^{(1)}$, $w_\psi^{(1)}$, $p^{(1)}$ can, in principle, be computed, and further matching of p, w_ϕ can be used to define the second-order boundary layer. Various local nonuniformities can develop, such as near sharp, leading, or trailing edges, corners, etc., which make it unwise to attempt to carry the procedure very far. Analogous procedures can be carried out for compressible flow where the energy balance of the flow must also be considered. The subject is discussed with a point of view similar to that given here by P. A. Lagerstrom in Reference 4.2.2. Further complications occur when different types of interaction with shock waves in the outer flow have to be considered, but the ideas behind these methods seem capable of handling all cases that arise.

The quantity of most physical interest from the boundary layer theory is the skin-friction on the surface, which now is represented by

$$\tau_v = \sqrt{\varepsilon} q_B^2(\phi)(\partial W_\phi/\partial \psi^*)(\phi, 0). \qquad (4.2.89)$$

When the boundary-layer solution is found, $(\partial W_\phi/\partial\psi^*)(\phi, 0)$ can be calculated, and the estimate of the skin friction is obtained. The classical result is given here—the skin-friction coefficient is proportional to $1/\sqrt{\text{Re}}$.

Unfortunately, no elementary solutions of the boundary-layer system (Equation 4.2.80) under boundary conditions (Equations 4.2.81, 4.2.84) exist. The only cases in which substantial simplifications can be achieved are cases of similarity when the problem can be reduced to ordinary differential equations. Otherwise, numerical integration of the system must be relied on, although some rough approximate methods can also be derived.

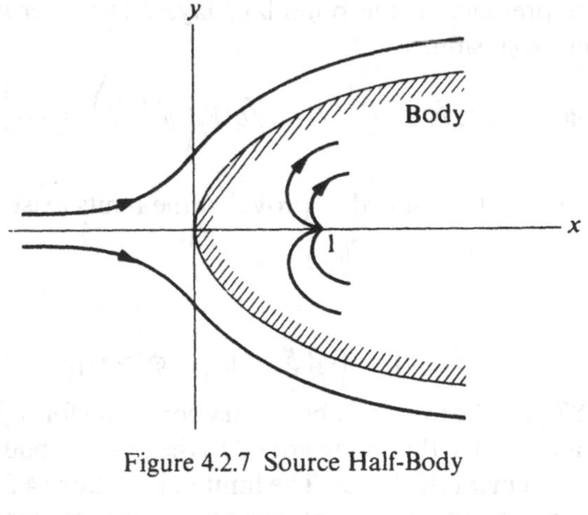

Figure 4.2.7 Source Half-Body

The cases of similarity can either be interpreted as local approximations or as solutions which are really asymptotic to solutions of the Navier–Stokes equations in a sense different from having $\varepsilon \to 0$. That is, the characteristic length L used to define ε really drops out of the problem, and the expansion is really in terms of the coordinates (x, y) or (ϕ, ψ). For example, consider the flow past a semi-infinite body generated by a source at $z = 1$ in a free stream (see Figure 4.2.7). The inviscid flow and coordinates are given by

$$\phi + i\psi = z + \log(z - 1) - i\pi = -\frac{z^2}{2} + \cdots, \qquad z \to 0. \qquad (4.2.90)$$

Thus, near the origin we have the stagnation-point flow

$$\phi = \frac{y^2 - x^2}{2}, \qquad \psi = -xy. \qquad (4.2.91)$$

The body is, at $x = 0$,

$$q_B = \frac{\partial \phi}{\partial y} = y = \sqrt{2\phi}. \tag{4.2.92}$$

The boundary-layer equations (Equation 4.2.80) are, for this case,

$$\frac{\partial W_\phi}{\partial \phi} + \frac{\partial W_\psi}{\partial \psi^*} = 0, \tag{4.2.93}$$

$$W_\phi \frac{\partial W_\phi}{\partial \phi} + W_\psi \frac{\partial W_\phi}{\partial \psi^*} = \frac{1 - W_\phi^2}{2\phi} + \frac{\partial^2 W_\phi}{\partial \psi^{*2}} \tag{4.2.94}$$

with boundary conditions

$$W_\phi(\phi, 0) = W_\psi(\phi, 0) = 0, \qquad W_\phi(\phi, \infty) = 1. \tag{4.2.95}$$

The system (Equations 4.2.93, 4.2.94, 4.2.95) has similarity, which enables the problem to be reduced to ordinary differential equations. The form is

$$W_\phi = F(\eta), \tag{4.2.96}$$

where

$$\eta = \frac{\psi^*}{2\sqrt{\phi}},$$

$$W_\psi = \frac{1}{\sqrt{\phi}} G(\eta), \tag{4.2.97}$$

$$\frac{dG}{d\eta} - \eta \frac{dF}{d\eta} = 0, \qquad G(0) = F(0) = 0, \tag{4.2.98}$$

$$\frac{d^2F}{d\eta^2} + 2(\eta F - G)\frac{dF}{d\eta} + 2(1 - F^2) = 0, \qquad F(\infty) = 1. \tag{4.2.99}$$

According to Equation (4.2.89), the skin-friction is obtained once this system is solved from

$$\tau_v = \sqrt{\phi \varepsilon} F'(0). \tag{4.2.100}$$

The existence of the solution to the problem posed in Equations (4.2.98) and (4.2.99) as well as to the more general class of similar problems in which $q_B = c\phi^m$ (the form 4.2.96, 4.2.97 is the same) is proved by H. Weyl in Reference 4.2.3. For the special case of the stagnation-point flow, the similar solution can be interpreted as the local solution near the origin. It turns out that in this case the solution to the boundary-layer equation (Equations 4.2.98, 4.2.99) is also a solution to the full Navier–Stokes equations. The parameter ε really drops out of the local solution, since the local solution cannot depend on the length L. The length L drops out of the similarity variable $\psi^*/\sqrt{\phi}$ when dimensional coordinates are re-introduced as follows:

$$\psi^* = UL\sqrt{\varepsilon}\Psi, \qquad \phi = UL\Phi, \tag{4.2.101}$$

where Φ, Ψ are dimensional,

$$\frac{\psi^*}{\sqrt{\phi}} = \frac{UL}{\sqrt{UL}}\sqrt{\frac{\mu}{\rho UL}}\frac{\Psi}{\sqrt{\Phi}} = \Psi\sqrt{\frac{\mu}{\rho\Phi}}. \tag{4.2.102}$$

Similar considerations apply to the velocity components to show that expansion is really in terms of Φ or (X) and is valid near the origin.

The same remarks apply to another classical case that is usually discussed, namely the flow past a semi-infinite flat plate, in which case we have

$$\phi = x, \qquad \psi = y, \tag{4.2.103}$$

$$q_B(\phi) = 1. \tag{4.2.104}$$

The similarity form is the same (Equations 4.2.96, 4.2.97), and the equations are a simplified version of Equations (4.2.98) and (4.2.99) with the $(1 - F^2)$ term missing. There is no characteristic length L in the problem, so that the parameter ε is artificial. If an arbitrary length is used for L (and this can be done), it must drop out of the answer. When similarity is combined with the artificial expansion in terms of ε, the expansion corresponding to boundary-layer theory becomes an expansion in terms of the space coordinates. For example, in dimensional coordinates (X, Y), the boundary-layer expansion (Equations 4.2.68, 4.2.69, 4.2.70) and outer expansions (Equations 4.2.64, 4.2.65, 4.2.66) take the form

$$\frac{q_x}{U} = U_0(\zeta) + \sqrt{\frac{\nu}{UX}}\,U_1(\zeta) + \cdots$$

$$\frac{q_y}{U} = \sqrt{\frac{\nu}{UX}}\,V_0(\zeta) + \cdots, \quad \text{where } \zeta = \frac{Y}{\sqrt{X}}\sqrt{\frac{U}{\nu}}, \tag{4.2.105}$$

$$\frac{p - p_\infty}{\rho U^2} = \sqrt{\frac{\nu}{UX}}\,P_1(\zeta) + \cdots, \text{ (boundary layer)},$$

$$\frac{q_x}{U} = 1 + \sqrt{\frac{\nu}{UX}}\,u_1(\zeta) + \cdots,$$

$$\frac{q_y}{U} = \sqrt{\frac{\nu}{UX}}\,v_1(\zeta) + \cdots \tag{4.2.106}$$

$$\frac{p - p_\infty}{\rho U^2} = \sqrt{\frac{\nu}{UX}}\,p_1(\zeta) + \cdots \text{ (outer expansion)}.$$

These expansions are seen to be valid for small ν/UX and are thus nonuniform near the nose, where a more complete treatment of the Navier–Stokes equations is needed. However, the skin-friction has a singularity only like $1/\sqrt{x}$ which is integrable at the nose. This indicates that probably a first approximation to the total drag can be found as $\varepsilon \to 0$.

A general result can be proved: If a problem with a parameter has similarity, then the approximate solution in terms of this parameter can not be uniformly valid, unless the approximate solution turns out to be the exact solution (as in the stagnation-point case). By similarity, here we mean the fact, for example, that if a solution depends on coordinates and a parameter $(x, y; \varepsilon)$, the solution must depend on two combinations of these due to invariance. In the case of the semi-infinite flat plate, the Navier–Stokes solution $u(x, y; \varepsilon) = fn(x/\varepsilon, y/\varepsilon)$ and the boundary-layer solution is not uniformly valid. The proof of this theorem, as well as much detailed discussion of expansions for both ε small and ε large in special problems for the Navier–Stokes equations, is given in Reference 4.2.4.

References

4.2.1 S. Kaplun, The role of coordinate systems in boundary layer theory, *Z.A.M.P.* V, **2** (1954) 111–135.

4.2.2 P. A. Lagerstrom, Laminar flow theory, *High Speed Aerodynamics and Jet Propulsion*, IV, Princeton Univ. Press, 1964, 20–282.

4.2.3 H. Weyl, On the simplest differential equations of boundary layer theory. *Annals of Math.* **43**, 2, 381–407.

4.2.4 P. A. Lagerstrom, and J. D. Cole, Examples illustrating expansion procedures for the Navier–Stokes equations, *Journal of Rational Mechanics and Analysis* **4**, 6 (Nov. 1955), 817–882.

4.3 Singular Boundary-Value Problems

Just as we found for ordinary differential equations discussed in Section 2.7, there are problems for partial differential equations in which various asymptotic expansions are constructed in different regions, but where the order of the system does not change in the limit of vanishing of the small parameter. In these problems the form of the expansions is dominated by the boundary conditions and usually, in one limit or other, a region degenerates to a line or a point, and may thus be singular. Narrow domains, slender bodies, and disturbances of small spatial extent are examples. In these cases it is often useful, although not always necessary, to construct different expansions in different regions. Expansions valid near the singularity can be matched with expansions valid far away. Several such examples are now considered.

4.3.1 Flow Due to a Stationary Deformable Slender-Body in a Uniform Stream

Consider the incompressible flow of a perfect fluid past a body of revolution capable of arbitrary radial deformations. This is a generalization of classical slender-body theory. The body is held fixed in a flow which has a velocity U_∞ along the body axis at infinity. We denote the body length by L and let the deformations be defined by

$$R = R_{\max} F\left(\frac{X}{L}, \frac{T}{\tau}\right) \tag{4.3.1}$$

where capital letters denote dimensional variables and τ is the characteristic time for the deformations (see Figure 4.3.1).

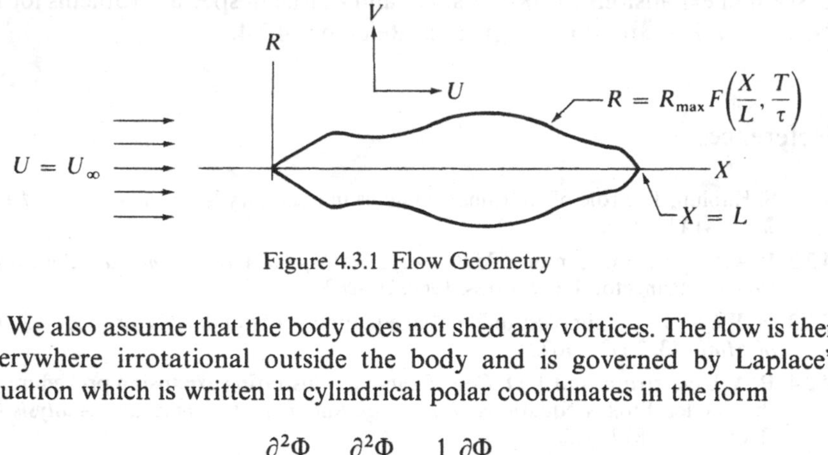

Figure 4.3.1 Flow Geometry

We also assume that the body does not shed any vortices. The flow is then everywhere irrotational outside the body and is governed by Laplace's equation which is written in cylindrical polar coordinates in the form

$$\frac{\partial^2 \Phi}{\partial X^2} + \frac{\partial^2 \Phi}{\partial R^2} + \frac{1}{R}\frac{\partial \Phi}{\partial R} = 0. \tag{4.3.2}$$

Here Φ is the velocity potential $\Phi = \Phi(X, R, T)$ which determines the axial and radial components U and V of the flow field according to

$$U = \frac{\partial \Phi}{\partial X} \tag{4.3.3a}$$

$$V = \frac{\partial \Phi}{\partial R}. \tag{4.3.3b}$$

We note that the time occurs only as a parameter in Equation (4.3.2). Once the solution for Φ is known, one can calculate the pressure field from Bernoulli's equation

$$P - P_\infty = -\rho \frac{\partial \Phi}{\partial T} + \frac{1}{2}\rho\left[U_\infty^2 - \left(\frac{\partial \Phi}{\partial X}\right)^2 - \left(\frac{\partial \Phi}{\partial R}\right)^2\right],$$

where P_∞ is the ambient pressure of the undisturbed flow at infinity, P is the pressure in the flow field and ρ is the constant density of the fluid.

The boundary conditions to be imposed are that the potential tends to the undisturbed free-stream value at infinity, and that the component of the flow velocity normal to the body must be equal to the normal component of the body velocity on the surface.

Now, the outward normal to the body $\mathbf{n}$ has components

$$\mathbf{n} = \left[-R_{\max}\frac{\partial F}{\partial X}, 1\right];$$

the velocity components of the body surface are defined by

$$\mathbf{V}_B = \left[0, R_{\max}\frac{\partial F}{\partial T}\right]$$

and the velocity components of the fluid are given by Equations (4.3.3). Thus, the boundary condition

$$\mathbf{n}\cdot\operatorname{grad}\Phi = \mathbf{n}\cdot\mathbf{V}_B \tag{4.3.4}$$

becomes

$$\frac{\partial\Phi}{\partial R}(X, R_{\max}F, T) = R_{\max}\left[\frac{\partial F}{\partial X}\frac{\partial\Phi}{\partial X} + \frac{\partial F}{\partial T}\right]. \tag{4.3.5}$$

To calculate the horizontal force on the body, we consider an annular element of thickness dX as shown in Figure (4.3.2) below.

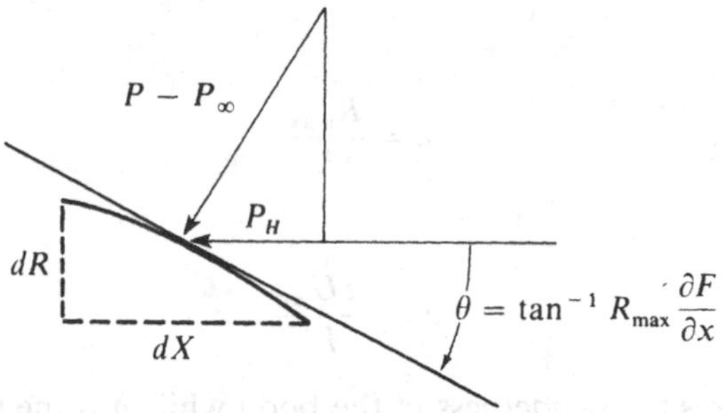

Figure 4.3.2 Pressure on the Surface

The net pressure acting on the surface is $P - P_\infty$ and is normal to the surface. Thus, the horizontal component of the pressure is given by

$$P_H = (P - P_\infty)\sin\theta, \tag{4.3.6}$$

where θ is the angle between the tangent and the horizontal; i.e.,

$$\theta = \tan^{-1} R_{\max}\frac{\partial F}{\partial X}. \tag{4.3.7}$$

We note that P_H is negative i.e., propulsive if, say $P > P_\infty$ and

$$\frac{\partial F}{\partial X} < 0$$

as drawn in Figure (4.3.2).

The horizontal force on the annulus is thus

$$dD = 2\pi(P - P_\infty)(\tan\theta)R_{\max}F\,dX \tag{4.3.8a}$$

and the total force D on the body is therefore

$$D(T) = 2\pi R_{\max}^2 \int_0^L (P - P_\infty)F\frac{\partial F}{\partial X}\,dX, \tag{4.3.8b}$$

where we have used the identity (4.3.7).

It is convenient to introduce dimensionless variables (denoted by lower case letters) and defined as follows

$$r = \frac{R}{L}, \qquad x = \frac{X}{L}, \qquad t = \frac{T}{\tau}, \qquad \phi = \frac{\Phi}{LU_\infty}. \tag{4.3.9}$$

Then, the governing equation (4.3.2) becomes

$$\phi_{xx} + \phi_{rr} + \frac{1}{r}\phi_r = 0. \tag{4.3.10}$$

The boundary condition on the surface reduces to

$$\phi_r[x, \varepsilon F(x,t), t] = \varepsilon\left\{\phi_x[x, \varepsilon F(x,t), t]F_x(x,t) + \frac{1}{\delta}F_t(x,t)\right\}, \tag{4.3.11a}$$

where

$$\varepsilon = \frac{R_{\max}}{L} \tag{4.3.11b}$$

and

$$\delta = \frac{\tau U_\infty}{L}. \tag{4.3.11c}$$

Thus, ε measures the slenderness of the body while δ is the ratio of the two characteristic times in the problem, and we assume that $\delta \ll 1$ or $\delta = O(1)$.

If we introduce the pressure coefficient C_p according to the usual convention

$$C_p = \frac{(P - P_\infty)}{\rho U_\infty^2} \tag{4.3.12}$$

Bernoulli's equation becomes

$$C_p = -\frac{1}{\delta}\phi_t + \frac{1}{2}[1 - \phi_x^2 - \phi_r^2]. \tag{4.3.13}$$

Finally, the force on the body can be expressed in terms of an axial force coefficient C_D nondimensionalizing Equation (4.3.8b) as follows

$$C_D = 2 \int_0^1 C_p(x, t) G_x(x, t) dx, \tag{4.3.14}$$

where

$$C_D = \frac{2D}{\rho U_\infty^2 \pi R_{\max}^2} \tag{4.3.15}$$

and

$$G = F^2.$$

Thus, the classical (rigid) slender-body theory corresponds to the special case $F_t = 0$.

For future reference, it is important to point out that if the potential ϕ is discontinuous with respect to the time (as would be the case if the velocity of surface deformations were discontinuous in time) Equation (4.3.13) remains valid as long as we interpret ϕ_t in the sense of distributions. More precisely, if at some time $t = t_0$, ϕ has a finite jump discontinuity of the form

$$[\phi]_{t_0} \equiv \phi(x, r, t_0^+) - \phi(x, r, t_0^-) \tag{4.3.16a}$$

then Equation (4.3.13) holds as long as we interpret

$$\phi_t = \{\phi_t\} + [\phi]_{t_0} \delta(t - t_0), \tag{4.3.16b}$$

where { } denotes the continuous part of ϕ_t and δ is the Dirac delta function.

This result follows directly by integration of the Euler equations

$$\operatorname{grad} p = -\frac{\partial}{\partial t} \operatorname{grad} \phi - \operatorname{grad}[\tfrac{1}{2}(\operatorname{grad} \phi)^2] \tag{4.3.17}$$

since it is easy to show that

$$\frac{\partial}{\partial t} \operatorname{grad} \phi = \operatorname{grad}\left[\left\{\frac{\partial \phi}{\partial t}\right\} + [\phi]_{t_0} \delta(t - t_0)\right]. \tag{4.3.18}$$

It is well known that the outer expansion for the potential ϕ in the form

$$\phi = \phi_0(x, r, t) + \mu_1(\varepsilon) \phi_1(x, r, t) + \cdots \tag{4.3.19}$$

becomes singular near the boundary $r = \varepsilon F(x, t)$, because in effect the boundary condition (4.3.11a) must be applied at $r = 0$ and the developments of ϕ_i as $r \to 0$ are singular. Thus, it is natural to seek an expansion of Equation (4.3.10) which is valid near $r = 0$ by introducing the inner variable

$$r^* = \frac{r}{\varepsilon} \tag{4.3.20}$$

and letting r^* be fixed in the limit as $\varepsilon \to 0$. We develop ϕ in the form

$$\phi = \phi_0^*(x, r^*, t) + \mu_1^*(\varepsilon) \phi_1^*(x, r^*, t) + \cdots. \tag{4.3.21}$$

Consider the leading term of this expansion. It is governed by

$$\frac{\partial^2 \phi_0^*}{\partial r^{*2}} + \frac{1}{r^*}\frac{\partial \phi_0^*}{\partial r^*} = 0 \tag{4.3.22}$$

subject to the boundary condition on the body [cf. Equation (4.3.11a)]

$$\frac{\partial \phi_0^*}{\partial r^*}(x, F(x, t), t) = \frac{\varepsilon^2}{\delta}\frac{\partial F}{\partial t}(x, t) + \varepsilon^2 \frac{\partial \phi_0^*}{\partial x}(x, F(x, t), t)\frac{\partial F}{\partial x}(x, t) + \cdots. \tag{4.3.23}$$

Thus, we must distinguish the two cases

$$\delta = 1, \quad \text{and} \quad \delta = \varepsilon^2.$$

Consider first the case $\delta = 1$. It is reasonable to assume that the outer expansion starts with the potential for the undisturbed flow $\phi_0 = x$ (this assumption will be justified by the matching). The solution of Equation (4.3.22) for ϕ_0^* subject to the boundary condition $\partial\phi_0^*/\partial r^* = 0$, of Equation (4.3.23) is

$$\phi_0^* = B_0(x, t)$$

and matching with the outer expansion to $O(1)$ then gives $B_0(x, t) = x$. To proceed to higher orders, we note that each term in the outer expansion obeys the full Laplace equation. In particular, ϕ_1 can be represented as the potential due to the superposition of sources and sinks of unknown strength S_1 along the axis on the unit interval. Thus,

$$\phi_1(x, r, t) = -\frac{1}{4\pi}\int_0^1 \frac{S_1(\xi, t)d\xi}{\sqrt{(x-\xi)^2 + r^2}}. \tag{4.3.24}$$

To calculate, ϕ_1^*, we substitute Equation (4.3.21) into Equation (4.3.10) and find that ϕ_1^* also obeys Equation (4.3.22).

Therefore,

$$\phi_1^*(x, r^*, t) = A_1(x, t)\log r^* + B_1(x, t). \tag{4.3.25}$$

The boundary condition on $r^* = F(x, t)$ to this order becomes

$$\mu_1^*(\varepsilon)\frac{\partial \phi_1^*}{\partial r^*} = \varepsilon^2[F_x + F_t]. \tag{4.3.26}$$

Thus, we must choose $\mu_1^*(\varepsilon) = \varepsilon^2$, and using Equation (4.3.25) in Equation (4.3.26) gives

$$A_1(x, t) = \tfrac{1}{2}(G_x + G_t). \tag{4.3.27}$$

To determine $B_1(x, t)$, we must match with the outer expansion to $O(\varepsilon^2)$. In particular, we must calculate the behavior of the singular integral (4.3.24) as $r \to 0$.

To accomplish this, we note the identities

$$\frac{1}{\sqrt{(x-\xi)^2+r^2}} = -\frac{\partial}{\partial \xi}\log[x-\xi+\sqrt{(x-\xi)^2+r^2}], \qquad \xi \le x \tag{4.3.28a}$$

$$\frac{1}{\sqrt{(x-\xi)^2+r^2}} = +\frac{\partial}{\partial \xi}\log[\xi-x+\sqrt{(x-\xi)^2+r^2}], \qquad \xi \ge x. \tag{4.3.28b}$$

Thus, Equation (4.3.24) can be split into two parts as follows

$$\begin{aligned}\phi_1(x,r,t) = {}& \frac{1}{4\pi}\int_0^x S_1(\xi,t)\frac{\partial}{\partial \xi}\log[x-\xi+\sqrt{(x-\xi)^2+r^2}]d\xi \\ & -\frac{1}{4\pi}\int_x^1 S_1(\xi,t)\frac{\partial}{\partial \xi}\log[\xi-x+\sqrt{(x-\xi)^2+r^2}]d\xi. \end{aligned}\tag{4.3.29}$$

Integrating by parts then gives the following exact result

$$\begin{aligned}\phi_1(x,r,t) = {}& \frac{1}{2\pi}S_1(x,t)\log r - \frac{1}{4\pi}S_1(0,t)\log(x+\sqrt{x^2+r^2}) \\ & -\frac{1}{4\pi}S_1(1,t)\log[1-x+\sqrt{(1-x)^2+r^2}] \\ & -\frac{1}{4\pi}\int_0^x \frac{\partial S_1}{\partial \xi}\log[x-\xi+\sqrt{(x-\xi)^2+r^2}]d\xi \\ & +\frac{1}{4\pi}\int_x^1 \frac{\partial S_1}{\partial \xi}\log[\xi-x+\sqrt{(x-\xi)^2+r^2}]d\xi. \end{aligned}\tag{4.3.30}$$

The limiting form of the above as $r \to 0$ is

$$\begin{aligned}\phi_1(x,r,t) = {}& \frac{1}{2\pi}S_1(x,t)\log r - \frac{1}{4\pi}S_1(0,t)\log 2x \\ & -\frac{1}{4\pi}S_1(1,t)\log 2(1-x) \\ & -\frac{1}{4\pi}\int_0^1 \frac{\partial S_1}{\partial \xi}\operatorname{sgn}(x-\xi)\log 2|x-\xi|d\xi + O(r^2). \end{aligned}\tag{4.3.31}$$

Let us assume that near $x = 0$ and $x = 1$ the source strength tends to zero faster that $[\log x]^{-1}$ or $[\log(1-x)]^{-1}$. This condition will, once S_1 is determined in terms of $F(x, t)$, impose a restriction on the allowable nose and tail shapes of the body. To carry out the matching, we introduce the intermediate variable r_η defined by

$$r_\eta = \frac{r}{\eta(\varepsilon)} \tag{4.3.32}$$

for functions $\eta(\varepsilon)$ such that $\varepsilon \ll \eta \ll 1$. The outer expansion then becomes:

$$\phi(x, r, t, \varepsilon) = x + \mu_1(\varepsilon)\left[\frac{S_1(x, t)}{2\pi} \log \eta r_\eta + T_1(x, t) + O(\eta^2)\right] + o(\mu_1), \tag{4.3.33a}$$

where

$$T_1(x, t) = -\frac{1}{4\pi}\int_0^1 \frac{\partial S_1}{\partial \xi} \operatorname{sgn}(x - \xi)\log 2|x - \xi|\, d\xi. \tag{4.3.33b}$$

The inner expansion, written in terms of r_η gives

$$\phi(x, r, t, \varepsilon) = x + \varepsilon^2\left[A_1(x, t)\log \frac{\eta r_\eta}{\varepsilon} + B_1(x, t)\right] + o(\varepsilon^2). \tag{4.3.34}$$

We see that in order to match we must set

$$\mu_1(\varepsilon) = \varepsilon^2 \tag{4.3.35a}$$

$$A_1(x, t) = \frac{S_1(x, t)}{2\pi} \tag{4.3.35b}$$

$$B_1(x, t) = T_1(x, t). \tag{4.3.35c}$$

However, the term $-A_1(x, t)\varepsilon^2 \log \varepsilon$ in the inner expansion cannot be matched.

As is in the example of Section 2.7.2 matching this term requires introducing a homogeneous solution of order $\varepsilon^2 \log \varepsilon$ in the inner expansion. Thus, Equation (4.3.21) must now be modified to read

$$\phi = x + A_1(x, t)\varepsilon^2 \log \varepsilon + [A_1(x, t)\log r^* + B_1(x, t)] + o(\varepsilon^2) \tag{4.3.36}$$

and the matching is demonstrated to $O(\varepsilon^2)$ with the definitions in Equation (4.3.35).

We note that the source strength S_1 is related to the body shape according to Equation (4.3.35b) and Equation (4.3.27) by

$$S_1(x, t) = \pi(G_x + G_t). \tag{4.3.37}$$

Since $G_t = 0$ at $x = 0$ and $x = 1$, we have the classical restrictions on the nose and tail shapes defined by

$$\lim_{x\to 0} G_x \log x = 0, \qquad \lim_{x\to 1} G_x \log(1 - x) = 0. \tag{4.3.38}$$

For example, if the nose and tail are given by a power law $r \sim x^n$ or $(1 - x)^n$ we must restrict our attention to values of $n > \frac{1}{2}$.

Consider now the remaining case when $\delta = \varepsilon^2$. The inner solution to $O(1)$ is

$$\phi_0^* = A_0(x, t)\log r^* + B_0(x, t) \tag{4.3.39}$$

subject to the boundary condition

$$\frac{\partial \phi_0^*}{\partial r^*} = \frac{A_0(x, t)}{r^*} = F_t. \tag{4.3.40}$$

Therefore,

$$A_0(x, t) = \frac{G_t}{2}. \tag{4.3.41}$$

It is clear that the leading term of the outer expansion is now no longer just the undisturbed potential, as the radial dependence from the inner solution must be accounted for. In fact, the outer solution must include, to first order, the contribution of the source distribution and this is physically reasonable since for δ sufficiently small, the rapid deformations of the boundary must have a first-order effect in the outer region.

We then seek an outer expansion with leading term ϕ_0 given by

$$\phi_0 = x - \frac{1}{4\pi} \int_0^1 \frac{S_0(\xi, t)d\xi}{\sqrt{(x - \xi)^2 + r^2}} \tag{4.3.42}$$

and the matching to $O(1)$ can be carried out as before giving

$$B_0(x, t) = x + T_1(x, t) \tag{4.3.43a}$$

with T_1 as defined in Equation (4.3.33b) and

$$S_0(x, t) = 2\pi A_0(x, t) = \pi G_t(x, t). \tag{4.3.43b}$$

Thus, the results for $\delta = \varepsilon^2$ are formally identical to the previous case except for the orders of magnitude of the various terms appearing in the formulas.

We can now calculate the pressure coefficient C_p on the body by substituting Equation (4.3.36) evaluated on $r^* = F(x, t)$ into Equation (4.3.13). The result is easily derived in the form

$$C_p = -\frac{1}{2}\varepsilon^2 \log \varepsilon H^2(G) - \varepsilon^2 \left\{ \frac{1}{4} H^2(G) \log G + H(T_1) + \frac{1}{8} \frac{[H(G)]^2}{G} \right\} + O(\varepsilon^4 \log \varepsilon), \tag{4.3.44a}$$

where H is the operator

$$H \equiv \frac{\partial}{\partial t} + \frac{\partial}{\partial x}. \tag{4.3.44b}$$

If $\delta = \varepsilon^2$, we find

$$C_p = -\frac{1}{2}\varepsilon^2 \log \varepsilon G_{tt} - \frac{\varepsilon^2}{2}\left(\frac{1}{2} G_{tt} \log G + 2T_{1t} + \frac{1}{4G} G_t^2\right) + O(\varepsilon^4 \log \varepsilon). \tag{4.3.44c}$$

In the example of this section, the use of an inner expansion is not strictly necessary in the sense that the inner expansion is really completely contained in the outer expansion [cf. Equations (4.3.33a) and (4.3.34)]. However, it is useful in making explicit the behavior near the boundary (particularly in the calculation of the pressure coefficient) and in emphasizing the nature of the different expansions as $\varepsilon \to 0$ for a point fixed on the boundary as compared to a point fixed in space. For more complicated differential equations, the idea of local behavior near a singular line or point can be essential.

(a) *Force on the Body.* Consider now the axial force coefficient defined by Equation (4.3.14). In all the cases which arise in Equation (4.3.44) we only encounter the following three integrals

$$I_1(t) = \int_0^1 L^2(G)G_x\,dx \tag{4.3.45a}$$

$$I_2(t) = \int_0^1 \left\{L^2(G)\log G + \frac{1}{2G}[L(G)]^2\right\}G_x\,dx \tag{4.3.45b}$$

$$I_3(t) = \int_0^1 L(T_1)G_x\,dx, \tag{4.3.45c}$$

where L is either equal to $H \equiv (\partial/\partial t) + (\partial/\partial x)$ or $L \equiv (\partial/\partial t)$, and T_1 is defined by Equations (4.3.33b), (4.3.35b) and (4.3.27) in the form

$$T_1(x, t) = -\frac{1}{4}\int_0^1 \frac{\partial}{\partial \xi} L(G)\operatorname{sgn}(x - \xi)\log 2|x - \xi|\,d\xi. \tag{4.3.46}$$

In evaluating these integrals, it is useful to note the identity for any two functions $A(x, t)$, $B(x, t)$

$$L(AB) = AL(B) + BL(A) \tag{4.3.47}$$

valid whether $L = H$ or $L = \partial/\partial t$. Consider the following identity for I_1

$$I_1 = \int_0^1 L^2(G)G_x\,dx = \int_0^1 L[L(G)G_x]dx - \frac{1}{2}\int_0^1 \frac{\partial}{\partial x}[L(G)]^2\,dx. \tag{4.3.48a}$$

Since $L(G) = 2FL(F)$ and $F(0, t) = F(1, t) = 0$, the second integral on the right hand side of (4.3.48) vanishes and the first integral reduces to

$$I_1 = \int_0^1 \frac{\partial}{\partial t}[L(G)G_x]dx \tag{4.3.48b}$$

because $L(G)(\partial G/\partial x) = 0$ at $x = 0$ and $x = 1$. Similar calculations give the following results for I_2 and I_3

$$I_2 = \int_0^1 \frac{\partial}{\partial t}[L(G)G_x \log G]dx \tag{4.3.49}$$

$$I_3 = \int_0^1 \frac{\partial}{\partial t}[T_1 G_x]dx. \tag{4.3.50}$$

Thus, for the case where $G_t = 0$, i.e., a rigid body, we recover the classical result $C_D = 0$ (D'Alembert paradox).

Now consider a periodic deformation $G(x, t)$ with period λ, i.e.,

$$G(x, t + \lambda) = G(x, t)$$

and let G and its partial derivatives be continuous. It then follows that C_D is of the form

$$C_D = \frac{d}{dt} h(t, \varepsilon), \tag{4.3.51}$$

where $h(t, \varepsilon)$ is the total contribution of all integrals in C_p.

Thus, C_D is also periodic with period λ and with zero average, i.e., C_D can be represented in a Fourier series without a constant term in the form:

$$C_D = \sum_{n=1}^{\infty} a_n(\varepsilon)\sin\left[\frac{2n\pi t}{\lambda} + \beta_n(\varepsilon)\right], \tag{4.3.52}$$

where the a_n and β_n can be determined by quadrature once the deformation G is known. It is particularly important to note that the average value of C_D over one period is zero. In other words *the net impulse over one period is zero.*

This situation is unaltered if we allow G_t to have any number of finite discontinuities over a period with G and G_x continuous. To prove this, we note that since the Bernoulli equation is still valid in this case (as long as time derivatives are interpreted in the sense of distributions) we have to consider in the calculation for C_D integrals of the form

$$I = \int_0^1 \frac{\partial}{\partial t}\left[K(x, t)\frac{\partial G}{\partial x}(x, t)\right]dx, \tag{4.3.53}$$

where both K and G_t have discontinuities. It is sufficient to consider the case of one discontinuity at $t = t_0$ in the interval $0 \le t \le \lambda$.

The impulse during one period is the integral of I which has the form

$$J = \int_0^\lambda \int_0^1 \frac{\partial}{\partial t}\left[K(x, t)\frac{\partial G}{\partial x}(x, t)\right]dx\,dt. \tag{4.3.54}$$

If we now split the integral as indicated in Equation (4.3.16b) we have two contributions denoted by J_1 and J_2:

$$J = J_1 + J_2, \tag{4.3.55}$$

where

$$J_1 = \int_0^\lambda \int_0^1 \left\{\frac{\partial}{\partial t}\left[K(x, t)\frac{\partial G}{\partial x}\right]\right\}dx\,dt \tag{4.3.56a}$$

$$J_2 = \int_0^\lambda \int_0^1 [K(x, t)]_{t_0}\delta(t - t_0)\frac{\partial G}{\partial x}\,dx\,dt \tag{4.3.56b}$$

and { } denotes the continuous part of the quantity.

We calculate J_1, in two parts as follows

$$J_1 = \int_0^{t_0}\int_0^1 \left\{\frac{\partial}{\partial t}\left[K(x,t)\frac{\partial G}{\partial x}\right]\right\}dx\,dt + \int_{t_0}^{\lambda}\int_0^1 \left\{\frac{\partial}{\partial t}\left[K(x,t)\frac{\partial G}{\partial x}\right]\right\}dx\,dt. \tag{4.3.57}$$

Since the integrands are continuous in both open intervals, we can immediately integrate with respect to t and obtain

$$J_1 = \int_0^1 K(x, t_0^-)\frac{\partial G}{\partial x}(x, t_0)dx - \int_0^1 K(x, t_0^+)\frac{\partial G}{\partial x}(x, t_0)dx \tag{4.3.58a}$$

or

$$J_1 = -\int_0^1 [K(x,t)]_{t_0}\frac{\partial G}{\partial x}(x, t_0)dx. \tag{4.3.58b}$$

Integrating J_2 gives

$$J_2 = \int_0^1 [K(x,t)]_{t_0}\frac{\partial G}{\partial x}(x, t_0)dx. \tag{4.3.59}$$

Therefore

$$J = 0.$$

Actually, the above conclusion regarding the impulse is valid even if the motion is non-periodic. It is sufficient that the body return to some given configuration after an interval of time λ for the impulse during this interval to vanish.

To show an example of a deformation with non-vanishing impulse let us consider only the leading term in the expansion for C_p. We then have

$$C_D = -\varepsilon^2 \log \varepsilon \frac{d}{dt}\int_0^1 H(G)G_x\,dx + O(\varepsilon^2). \tag{4.3.60}$$

Assume now that G has the form

$$G(x,t) = f(x)g(t) \tag{4.3.61}$$

with $f(0) = f(1) = 0$. Thus, the body undergoes geometrically similar deformations. Equation (4.3.60) can then be easily integrated to give

$$C_D = -\varepsilon^2(\log \varepsilon)\alpha^2 \frac{d(g^2)}{dt} + O(\varepsilon^2), \tag{4.3.62a}$$

where α^2 is the positive constant

$$\alpha^2 = \int_0^1 \left[\frac{df}{dx}\right]^2 dx. \tag{4.3.62b}$$

We note that in order to have a propulsive force, ($C_D < 0$), we must let $dg^2/dt < 0$. Hence the body must "collapse." For example, if we wish to maintain a contant propulsive force $C_D = C < 0$ we calculate (Note: $C/\varepsilon^2 \log \varepsilon > 0$)

$$g(t) = \sqrt{1 - \frac{Ct}{\alpha^2\varepsilon^2 \log \varepsilon}}.$$

Thus, the body will have collapsed to a "needle" after an interval of time equal to $\alpha^2\varepsilon^2 \log \varepsilon/C$. It is also interesting to note (cf. Problem 4.3.1) that for a body starting from rest and moving under the influence of forces generated by periodic surface deformations, the best average velocity that can be achieved is a constant.

(b) *Nonuniformity Near the Nose.* Let us now consider the local non-uniformity of the slender-body expansion near a blunt nose. We will consider the case of a rigid body for simplicity and outline a method based on a local solution for eliminating the difficulty. As we saw earlier [cf. the discussion following Equation (4.3.38)] some difficulty occurs for a nose (or tail) which is so blunt that $F(x) \sim \sqrt{x}$, but if $F(x) \sim x^n, n < \frac{1}{2}$, the pressure force, at least in the first approximation, is integrable. Thus, we consider here a slender-body whose shape function $F(x)$ has the following behavior near $x = 0$

$$F(x) = (2ax)^{1/2}(1 + bx + \cdots) \tag{4.3.63a}$$

$$F'(x) = (a/2x)^{1/2} + O(x^{1/2}) \tag{4.3.63b}$$

$$F''(x) = -(2a)^{1/2}/4x^{3/2} + O(x^{-1/2}). \tag{4.3.63c}$$

The radius of curvature r_c at the nose is

$$r_c \to \frac{\varepsilon^3(F'^3)}{\varepsilon F''} \to \varepsilon^2 a \quad \text{as } x \to 0. \tag{4.3.64}$$

The slender-body inner expansion for the potential is given by Equation (4.3.36), but near the nose we find that the source strength is

$$A_1(x) = FF' = a + O(x). \tag{4.3.65}$$

Furthermore, in order to determine the function $B_1(x)$ we have to return to Equation (4.3.31) and retain the boundary term $-(1/4\pi)S_1(0)\log 2x$ which was omitted for a sufficiently pointed nose. Now, the matching gives

$$B_1(x) = -\frac{1}{2}A_1(0)\log x + \cdots$$
$$-\frac{1}{2}\int_0^1 A_1'(\xi)\text{sgn}(x - \xi)\log 2|x - \xi|\,d\xi \tag{4.3.66}$$

or, as $x \to 0$,

$$B_1(x) \to -(a/2)\log 2 - (a/2)\log x + \cdots. \tag{4.3.67}$$

Thus, the potential has the behavior

$$\phi = x + a\varepsilon^2 \log \varepsilon + \varepsilon^2[a \log r^* - a \log 2 - (a/2)\log x + \cdots] \text{ as } x \to 0. \tag{4.3.68}$$

On the body surface, $r^* = (2ax)^{1/2}$ and ϕ is finite. However, the velocity is given by

$$u = \phi_x = 1 - a\varepsilon^2/2x + \cdots \tag{4.3.69a}$$

$$v = \frac{1}{\varepsilon}\,\phi_{r^*} = a\varepsilon/r^* \tag{4.3.69b}$$

so that the local surface pressure coefficient, C_{p_b} given by the Bernoulli equation (4.3.13) with $\phi_t = 0$ is

$$\begin{aligned} 2C_{p_b} &= 1 - (1 - a\varepsilon^2/2x + \cdots)^2 - \left(\frac{a\varepsilon}{\sqrt{2ax}} + \cdots\right)^2 \\ &= \varepsilon^2 a/2x + O(\varepsilon^4). \end{aligned} \tag{4.3.70}$$

The term of $O(\varepsilon^2)$ shows large physically unrealistic compression and, in fact, the total force on the nose which is proportional to $\int C_{pb}FF'\,dx$ is infinite. In order to give a better representation of the flow near the nose, we can try to find a local expansion based on a limit process which preserves the structure of the flow near the nose. Since we are interested in the neighborhood of a point, both x and r must tend to zero in the limit, and it is clear that all terms in the basic equations (Equation 4.3.10) should be retained. Thus, the general form has

$$\bar{x} = \frac{x}{\alpha(\varepsilon)}, \bar{r} = \frac{r}{\alpha(\varepsilon)}$$

in the limit. But now considering that $r_b \sim \varepsilon x^{1/2}$ as $x \to 0$, we see that $\alpha = \varepsilon^2$ in order to keep the typical body structure near the surface. Thus, let

$$\bar{x} = \frac{x}{\varepsilon^2}, \qquad \bar{r} = \frac{r}{\varepsilon^2}. \tag{4.3.71}$$

The asymptotic expansion of the potential near the nose is assumed in the form

$$\phi = x + \bar{\mu}_1(\varepsilon)\bar{\Phi}_1(\bar{x}, \bar{r}) + \cdots, \tag{4.3.72}$$

where $\bar{\Phi}_1$ obeys Equation (4.3.10).

As far as the first approximation goes, the body is represented by

$$\bar{r}_b = (2a\bar{x})^{1/2}. \tag{4.3.73}$$

Then the problem is one of flow past a paraboloid. The surface boundary condition, Equation (4.3.11a) with $\partial/\partial t = 0$, becomes

$$\frac{\bar{\mu}_1}{\varepsilon^2}\frac{\partial\bar{\Phi}_1}{\partial\bar{r}}(\bar{x}, \sqrt{2a\bar{x}}) + \cdots = \sqrt{\frac{a}{2\bar{x}}}\left[1 + \frac{\bar{\mu}_1}{\varepsilon^2}\frac{\partial\bar{\Phi}_1}{\partial\bar{x}}(\bar{x}, \sqrt{2a\bar{x}}) + \cdots\right] \tag{4.3.74}$$

from which we see that the proper choice for $\bar{\mu}_1$ is

$$\bar{\mu}_1 = \varepsilon^2. \tag{4.3.75}$$

Note that the free-stream term x in Equation (4.3.72) is just the same order ($x = \varepsilon^2\bar{x}$) as the $\bar{\mu}_1\bar{\Phi}_1$ term; we had a similar situation in Equation (4.3.42). The potential for the paraboloid with the boundary condition of Equation (4.3.74) can be written

$$\bar{\Phi}_1 = \frac{1}{2}a\log\left\{\left[\left(\bar{x}-\frac{a}{2}\right)^2+\bar{r}^2\right]^{1/2}-\left(\bar{x}-\frac{a}{2}\right)\right\}. \tag{4.3.76}$$

There is no arbitrary constant here due to the form of Equation (4.3.74). The x term in Equation (4.3.72) is already matched, and the ϕ_1 term can be matched to the previously calculated inner expansion to remove the singularity at the nose and enable a uniformly valid approximation to be constructed. For the matching, an intermediate limit, with r^* and x_η fixed, can be used where

$$x_\eta = \frac{x}{\eta(\varepsilon)}, \qquad \varepsilon^2 \ll \eta \ll \varepsilon. \tag{4.3.77}$$

Under this limit, we have

$$\bar{x} = \frac{\eta}{\varepsilon^2}x_\eta \to \infty, \qquad x = \eta x_\eta \to 0, \qquad \bar{r} = \frac{r^*}{\varepsilon} \to \infty. \tag{4.3.78}$$

Thus, we obtain

$$\left[\left(\bar{x}-\frac{a}{2}\right)^2+\bar{r}^2\right]^{1/2} = \frac{\eta}{\varepsilon^2}x_\eta\left(1+\frac{1}{2}\frac{\varepsilon^2}{\eta^2}\frac{r^{*2}}{x_\eta 2}+\cdots\right) \tag{4.3.79}$$

if $\varepsilon^2 \ll \eta$, and therefore

$$\bar{\Phi}_1 = \frac{a}{2}\log\frac{r^{*2}}{2\eta x_\eta}+\cdots. \tag{4.3.80}$$

By adding suitable constants which do not affect the velocity, it is seen that the potential in Equation (4.3.68) matches with the $\log r^*$ and $\log x$ terms in Equation (4.3.88). Thus, near the nose the pressure should be computed from the velocity components as found from Equation (4.3.76),

$$\frac{\partial\phi}{\partial r} = \frac{a}{2}\frac{1}{\sqrt{(\bar{x}-a/2)^2+\bar{r}^2}-(\bar{x}-a/2)}\cdot\frac{\bar{r}}{\sqrt{(\bar{x}-a/2)^2+\bar{r}^2}}$$

$$\to \frac{(2a\bar{x})^{1/2}}{2\bar{x}+a} \quad \text{on the surface} \tag{4.3.81}$$

$$\frac{\partial\phi}{\partial x} = 1+\frac{a}{2}\frac{1}{\sqrt{(\bar{x}-a/2)^2+\bar{r}^2}-(\bar{x}-a/2)}\left\{-1+\frac{\bar{x}-a/2}{\sqrt{(\bar{x}-a/2)^2+\bar{r}^2}}\right\}$$

$$\to \frac{\bar{x}}{\bar{x}+a/2} \quad \text{on the surface.} \tag{4.3.82}$$

This is a typical example of how a local solution, in this case flow past a paraboloid, can be used to improve the representation of the solution near a singularity. A composite expansion can be written by adding the local and outer expansions and subtracting the common part.

4.3.2 Low Reynolds-Number Viscous Flow Past a Circular Cylinder

For this problem, the Navier–Stokes equations (Equations 4.2.3) are again considered to describe the flow. There is uniform flow at infinity, and the body is at the origin. Since the size of the body was used as the characteristic length in writing the system (Equations 4.2.3), the body diameter is one (see Figure 4.3.3). The boundary condition of no-slip,

$$q = 0 \quad \text{on} \quad r = \sqrt{x^2 + y^2} = \tfrac{1}{2}, \tag{4.3.83}$$

and conditions at infinity serve to define the problem. We are interested in a low Reynolds number, so that in Equation (4.2.3b) we have

$$\varepsilon = \frac{1}{\mathrm{Re}} = \frac{\nu}{UL} \to \infty. \tag{4.3.84}$$

The variables based on L are inner variables (Stokes variables in the notation of References 4.3.2, 4.3.3, 4.3.4), since the boundary remains fixed in the limit. As it turns out, the inner problem which is Stokes flow cannot satisfy the complete boundary condition at infinity, so that some suitable outer expansion, valid near infinity, must also be constructed. Both inner and outer expansions, which can be identified with the usual Stokes and Oseen flow approximations, respectively, are described here, and the matching is carried out. The model example corresponding to the kind of singular boundary-value problem that occurs here has already been discussed in Section 2.7.2.

The inner expansion is based on $\mathrm{Re} \to 0$, $\varepsilon \to \infty$ in Equation (4.2.3), but if pressure is measured in units of ρU^2, both inertia and pressure terms drop

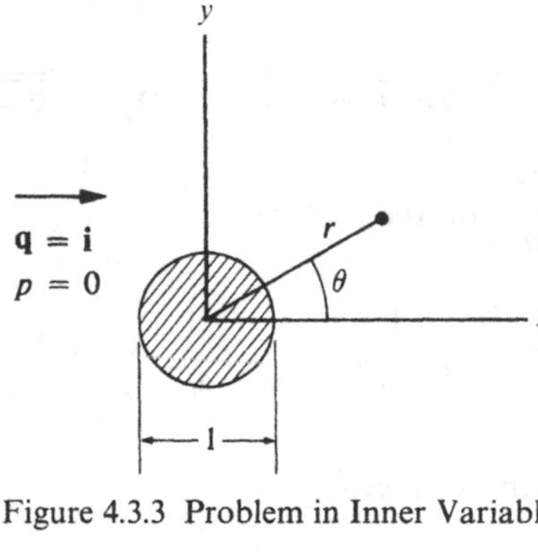

Figure 4.3.3 Problem in Inner Variables

out of the limiting momentum equations. There are not enough variables if continuity is to be considered, so that the physical pressure (difference from infinity) should be measured in terms of $U\mu/L$. Or let

$$p^*(x, y) = \frac{1}{\varepsilon} p(x, y) = (\text{Re})p(x, y). \tag{4.3.85}$$

This is in accord with Stokes' idea of a balance between viscous stresses and pressure forces, at least near the body, for slow flow. Thus, in inner variables, the Navier–Stokes system can be written

$$\operatorname{div} \mathbf{q} = 0, \tag{4.3.86a}$$

$$\text{Re}(\mathbf{q} \cdot \nabla \mathbf{q}) = -\operatorname{grad} p^* - \operatorname{curl} \boldsymbol{\omega}, \qquad \boldsymbol{\omega} = \operatorname{curl} \mathbf{q}. \tag{4.3.86b}$$

The inner expansion, associated with the limit process (Re → 0, x, y fixed) has the form

$$\mathbf{q}(x, y; \text{Re}) = \alpha_0(\text{Re})\mathbf{q}_0(x, y) + \alpha_1(\text{Re})\mathbf{q}_1(x, y) + \cdots, \tag{4.3.87a}$$

$$p^*(x, y; \text{Re}) = \alpha_0(\text{Re})p_0^*(x, y) + \alpha_1(\text{Re})p_1^*(x, y) + \cdots. \tag{4.3.87b}$$

Taking the limit of the Navier–Stokes equations (Equation 4.3.86) shows that the first term of the inner expansion satisfies the usual Stokes equations:

$$\operatorname{div} \mathbf{q}_0 = 0, \tag{4.3.88a}$$

$$0 = -\operatorname{grad} p_0^* - \operatorname{curl} \boldsymbol{\omega}_0, \qquad \boldsymbol{\omega}_0 = \operatorname{curl} \mathbf{q}_0. \tag{4.3.88b}$$

A fairly general discussion of the solutions to Equation (4.3.88) can be given while taking account of the boundary conditions on the surface, so that the behavior at infinity can be ascertained. It is convenient to introduce the stream function $\psi(x, y)$ satisfying continuity identically by

$$\mathbf{q} = \operatorname{curl} \boldsymbol{\psi}, \qquad \boldsymbol{\psi} = \psi(x, y)\mathbf{k}, \qquad q_x = \frac{\partial \psi}{\partial y}, \qquad q_y = -\frac{\partial \psi}{\partial x}. \tag{4.3.89}$$

The equation for the vorticity $\boldsymbol{\omega}_0 = \omega_0(x, y)\mathbf{k}$ is

$$\omega_0(x, y) = -\nabla^2 \psi_0(x, y). \tag{4.3.90}$$

Further, taking the curl of the momentum equation (Equation 4.3.88b) shows that

$$\operatorname{curl}\operatorname{curl} \boldsymbol{\omega}_0 = 0 \quad \text{or} \quad \nabla^2 \omega_0 = 0. \tag{4.3.91}$$

Thus, the vorticity field is a harmonic function and can be represented, in general, outside the circular cylinder $r = \frac{1}{2}$ by a series with unknown coefficients. The velocity field must be symmetric with respect to the x-axis, and the vorticity field antisymmetric. Thus, the general form is the familiar solution of Laplace's equation in cylindrical coordinates.

$$\omega_0(r, \theta) = \sum_{n=0}^{\infty} (a_n r^n + b_n r^{-n}) \sin n\theta. \tag{4.3.92}$$

The general solution for $\psi_0(r, \theta)$ satisfying Equation (4.3.90) is

$$\frac{\partial^2 \psi_0}{\partial r^2} + \frac{1}{r}\frac{\partial \psi_0}{\partial r} + \frac{1}{r^2}\frac{\partial^2 \psi_0}{\partial \theta^2} = -\sum_{n=0}^{\infty} (a_n r^n + b_n r^{-n}) \sin n\theta, \tag{4.3.93}$$

and the boundary condition of no-slip

$$\psi_0\left(\frac{1}{2}, \theta\right) = \frac{\partial \psi_0}{\partial r}\left(\frac{1}{2}, \theta\right) = 0, \tag{4.3.94}$$

can now be found. Thus, let

$$\psi_0(r, \theta) = \sum_{n=1}^{\infty} \Psi^{(n)}(r) \sin n\theta, \tag{4.3.95}$$

so that we have

$$L^{(n)}\Psi^{(n)} \equiv \frac{d^2\Psi^{(n)}}{dr^2} + \frac{1}{r}\frac{d\Psi^{(n)}}{dr} - \frac{n^2}{r^2}\Psi^{(n)} = -a_n r^n - b_n r^{-n}. \tag{4.3.96}$$

It can be verified that

$$L^{(n)} r^m = (m^2 - n^2) r^{m-2}, \tag{4.3.97}$$

so that by choosing $m = n + 2$, we have

$$L^{(n)} r^{n+2} = (4n + 4) r^n, \tag{4.3.98}$$

which is good for all $n \neq -1$. Further, for $n = -1$, we obtain

$$L^{(n)} r \log r = 2/r. \tag{4.3.99}$$

Thus, introducing new constants, the general solution (Equation 4.3.96) is

$$\Psi^{(1)} = A_1 r^3 + B_1 r \log r + C_1 r + \frac{D_1}{r}, \tag{4.3.100a}$$

$$\Psi^{(n)} = A_n r^{2+n} + B_n r^{2-n} + C_n r^n + \frac{D_n}{r^n}, \qquad n = 2, 3, \ldots \tag{4.3.100b}$$

and

$$\frac{d\Psi^{(1)}}{dr} = 3A_1 r^2 + B_1(\log r + 1) + C_1 - \frac{D_1}{r^2}, \tag{4.3.101a}$$

$$\frac{d\Psi^{(n)}}{dr} = (2 + n)A_n r^{1+n} + (2 - n)B_n r^{1-n} + nC_n r^{n-1} - n\frac{D_n}{r^{n+1}},$$
$$n = 2, 3, \ldots. \tag{4.3.101b}$$

By applying the boundary condition at the body surface $r = \frac{1}{2}$ (Equation 4.3.94), we obtain two relations between the four constants A_n, B_n, C_n, D_n.

Further determination of the solution must come from the boundary conditions at infinity, which would read

$$q_x = \sin\theta \frac{\partial\psi}{\partial r}(r,\theta) + \frac{\cos\theta}{r}\frac{\partial\psi}{\partial\theta}(r,\theta) = 1 \quad \text{as} \quad r \to \infty, \tag{4.3.102}$$

$$q_y = -\cos\theta \frac{\partial\psi}{\partial r}(r,\theta) + \frac{\sin\theta}{r}\frac{\partial\psi}{\partial\theta}(r,\theta) = 0 \quad \text{as} \quad r \to \infty \tag{4.3.103}$$

or

$$\frac{\partial\psi}{\partial r}(r,\theta) \to \sin\theta \quad \text{as} \quad r\to\infty; \qquad \frac{1}{r}\frac{\partial\psi}{\partial\theta}(r,\theta) \to \cos\theta \quad \text{as} \quad r\to\infty. \tag{4.3.104}$$

If the condition of Equation (4.3.104) is imposed to fix $C_1 = 1, A_1 = B_1 = A_n = C_n = 0$, then the two boundary conditions at the wall cannot be satisfied. Thus, the condition has to be given up and replaced by a condition of matching at infinity. The inner expansion is not uniform at infinity. The situation is possibly a little clearer for the corresponding problem for a sphere, where, although the first term of the inner expansion can satisfy the conditions at infinity, the second can not and becomes larger than the first at some distance from the origin. In general, only one more constant B_1 is needed, so that we can choose

$$B_1 \neq 0, \qquad A_1 = 0, \qquad A_n = 0, \qquad C_n = 0, \qquad n = 2, 3, \ldots \tag{4.3.105}$$

and obtain the weakest possible divergence of the solution at infinity. This has to be verified by matching. Thus, from the boundary condition at the surface (Equation 4.3.94) applied to Equations (4.3.100) and (4.3.101) we obtain

$$\begin{aligned}
\Psi^{(1)}\left(\frac{1}{2}\right) &= 0 = \frac{B_1}{2}\log\frac{1}{2} + \frac{C_1}{2} + 2D_1.\\
\Psi^{(n)}\left(\frac{1}{2}\right) &= 0 = B_n 2^{n-2} + 2^n D_n, \qquad n = 2, 3, \ldots;\\
\frac{d\Psi^{(1)}}{dr}\left(\frac{1}{2}\right) &= 0 = B_1\left(\log\frac{1}{2} + 1\right) + C_1 - 4D_1.\\
\frac{d\Psi^{(n)}}{dr}\left(\frac{1}{2}\right) &= 0 = (2-n)\beta_n 2^{n-1} - nD_n 2^{n+1}, \qquad n = 2, 3, \ldots.
\end{aligned} \tag{4.3.106}$$

Thus, from the conditions at the surface of the circular cylinder, we have

$$B_n = D_n = 0, \qquad n = 2, 3, \ldots, \tag{4.3.107}$$

and we are left with two relations between the three constants B_1, C_1, and D_1:

$$(\tfrac{1}{2}\log\tfrac{1}{2})B_1 + \tfrac{1}{2}C_1 + 2D_1 = 0, \qquad (\log\tfrac{1}{2} + 1)B_1 + C_1 - 4D_1 = 0. \tag{4.3.108}$$

Thus, the first term of the inner expansion becomes

$$\psi_0(x, r) = [B_1 r \log r + C_1 r + (D_1/r)]\sin\theta, \tag{4.3.109}$$

and

$$q_x = \alpha_0(\mathrm{Re})\{B_1 \log r + C_1 + B_1 \sin^2\theta + (D_1/r^2)\cos 2\theta\} + \alpha_1(\mathrm{Re})q_{1_x} + \cdots. \tag{4.3.110}$$

Now, in order to construct the outer expansions, a suitable outer variable has to be chosen. It was a basic idea of Kaplun to use the characteristic length ν/U for defining the expansion in the far field. Thus, in these units the body radius is very small and approaches zero in the limit. It can then be anticipated that the first term of the outer expansion is the undisturbed stream, since the body of infinitesimal size has no arresting power. Compare this procedure with that in the model example in Section 2.7.2. The formalities involve a limit procedure with $\tilde{x}$, $\tilde{y}$ fixed, Re $\to 0$, where

$$\tilde{x} = (\mathrm{Re})x, \qquad \tilde{y} = (\mathrm{Re})y \tag{4.3.111}$$

since x, y are based on the diameter. In these units, the body surface itself is

$$\tilde{r} = \sqrt{\tilde{x}^2 + \tilde{y}^2} = \tfrac{1}{2}\mathrm{Re}, \quad \mathrm{Re} \to 0. \tag{4.3.112}$$

If the Navier–Stokes equations (Equation 4.3.86) are written in these units, the parameter Re disappears. The pressure is again based on ρU^2:

$$\widetilde{\mathrm{div}}\, \mathbf{q} = 0, \tag{4.3.113a}$$

$$\mathbf{q} \cdot \tilde{\nabla}\mathbf{q} + \tilde{\nabla}p = -\widetilde{\mathrm{curl}}\, \boldsymbol{\omega}, \qquad \boldsymbol{\omega} = \widetilde{\mathrm{curl}}\, \mathbf{q}, \tag{4.3.113b}$$

where ($\tilde{\ }$) means space derivatives with respect to ($\tilde{x}$, $\tilde{y}$). The form of the outer expansion is thus assumed to be a perturbation about the uniform free stream. Thus, we have

$$q_x(x, y; \mathrm{Re}) = 1 + \beta(\mathrm{Re})u(\tilde{x}, \tilde{y}) + \beta_1(\mathrm{Re})u_1(\tilde{x}, \tilde{y}) + \cdots, \tag{4.3.114a}$$

$$q_y(x, y; \mathrm{Re}) = \beta(\mathrm{Re})v(\tilde{x}, \tilde{y}) + \cdots, \tag{4.3.114b}$$

$$p(x, y; \mathrm{Re}) = \beta(\mathrm{Re})\tilde{p}(\tilde{x}, \tilde{y}) + \cdots. \tag{4.3.114c}$$

From this expansion, it is clear that the first approximation equation is linearized about the free stream. The transport operator is

$$\mathbf{q} \cdot \tilde{\nabla} = \frac{\partial}{\partial \tilde{x}} + \beta(\mathrm{Re})\left(u\frac{\partial}{\partial \tilde{x}} + v\frac{\partial}{\partial \tilde{y}}\right) + \cdots. \tag{4.3.115}$$

Thus, we have

$$\frac{\partial u}{\partial \tilde{x}} + \frac{\partial v}{\partial \tilde{y}} = 0, \tag{4.3.116a}$$

$$\frac{\partial u}{\partial \tilde{x}} + \frac{\partial \tilde{p}}{\partial \tilde{x}} = \frac{\partial^2 u}{\partial \tilde{x}^2} + \frac{\partial^2 u}{\partial \tilde{y}^2}, \tag{4.3.116b}$$

$$\frac{\partial v}{\partial \tilde{x}} + \frac{\partial \tilde{p}}{\partial \tilde{y}} = \frac{\partial^2 v}{\partial \tilde{x}^2} + \frac{\partial^2 v}{\partial \tilde{y}^2}. \tag{4.3.116c}$$

These are the equations proposed by Oseen as a model for high Reynolds-number flow, but they appear here as part of an actual approximation scheme for low Re. The idea of the matching of the two expansions can now be discussed. If the methods used previously are followed, a class of intermediate limits is considered in which x_η, y_η are held fixed and in which

$$x_\eta = \eta(\text{Re})x, \qquad y_\eta = \eta(\text{Re})y, \qquad \eta \to 0 \tag{4.3.117}$$

and

$$\text{Re} \ll \eta(\text{Re}) \ll 1.$$

Therefore, in this limit, we know that

$$r = \frac{r_\eta}{\eta} \to \infty, \qquad \tilde{r} = \frac{\text{Re}}{\eta} r_\eta, \qquad r_\eta \to 0. \tag{4.3.118}$$

Assuming that the two expansions are valid in an overlap domain, we can compare the intermediate forms of inner and outer expansions (Equations 4.3.110, 4.3.114a) for q_x:

$$\lim_{\substack{\varepsilon \to 0,\\ r_\eta \text{ fixed}}} \left\{ \alpha_0(\text{Re}) \left[B_1 \log \frac{r_\eta}{\eta(\text{Re})} + C_1 + B_1 \sin^2\theta + D_1 \frac{\eta^2(\text{Re})}{r_\eta^2} \cos 2\theta \right] \right.$$

$$\left. + \alpha_1(\text{Re}) q_1 + \cdots - 1 - \beta(\text{Re}) u\left(\frac{\text{Re}}{\eta} x_\eta, \frac{\text{Re}}{\eta} y_\eta \right) - \cdots \right\} = 0. \tag{4.3.119}$$

It is clear from Equation (4.3.119) that a solution of the Oseen equation (Equation 4.3.116) must be found, in which

$$u(\tilde{x}, \tilde{y}) \to a \log \tilde{r} + \cdots \quad \text{as} \quad \tilde{r} \to 0;$$

$$u(\tilde{x}, \tilde{y}) \to a \log \frac{\text{Re}\, r_\eta}{\eta} + \cdots \quad \text{under the intermediate limit} \tag{4.3.120}$$

if the expansions are to be matched and the dominant remaining terms are

$$-\alpha_0(Re) B_1 \log \eta(\text{Re}) + \cdots - 1 + \beta(\text{Re}) a \log \eta(\text{Re}) + \beta(\text{Re}) a \log\left(\frac{1}{\text{Re}} \right) + \cdots.$$

Matching is accomplished if

$$\alpha_0(\text{Re}) = \beta(\text{Re}), \qquad B_1 = a, \tag{4.3.121}$$

and

$$\beta(\text{Re}) = \frac{1}{\log(1/\text{Re})}, \qquad a = 1. \tag{4.3.122}$$

Thus, Equation (4.3.120) provides the necessary boundary condition for the solution of the first outer approximation. The stream function, pressure, and other velocity component can also be considered and matched. In this way, the complete first approximation to the flow near the body is found (cf. Equation 4.3.108):

$$B_1 = 1, \qquad C_1 = -\tfrac{1}{2}\log\tfrac{1}{2} - \tfrac{1}{2}, \qquad D_1 = \tfrac{1}{8}. \tag{4.3.123}$$

The continuation of this procedure enables the various higher approximations to be carried out.

Note that the nonlinear terms of the Navier–Stokes equations, when expressed in inner variables (Equation 4.3.86b), are $O(\text{Re})$, and hence transcendentally small compared to $\beta(\text{Re}) = 1/[\log(1/\text{Re})]$ as $\text{Re} \to 0$. In particular, when successive terms of the inner expansions are constructed, each satisfies the same Stokes equations (Equation 4.3.88). The nonlinear effects appear only explicitly in the outer equation and outer expansion. Thus, the nonlinearity indicates the existence of terms,

$$\beta_1(\text{Re}) = \beta^2(\text{Re}) = \log^2\left(\frac{1}{\text{Re}}\right), \tag{4.3.124}$$

so that $(u_1, v_1, \tilde{p}_1)$ satisfy nonhomogeneous Oseen equations. Of course, terms of intermediate order satisfying the homogeneous Oseen equations may appear between (u, u_1) to complete the matching. For the incompressible case, it turns out that the outer expansion includes the inner expansion, and that a uniformly valid solution is found from the outer expansion with a boundary condition satisfied on $\tilde{r} = \text{Re}/2$. Such a result can not be expected in the more general compressible case.

A much more sophisticated version of this problem and the general problem of low Re flow appears in References 4.3.2, 4.3.3, and 4.3.4.

Independently of the above work similar results were also obtained in Reference 4.3.5.

4.3.3 One-Dimensional Heat Conduction

In many examples, the geometrical shape of the domain of the problem introduces a small parameter. For such thin domains, it is often possible to introduce various asymptotic expansions based on the limit $\varepsilon \to 0$. The terms in these asymptotic expansions can correspond to simplified models for the physical process. In this Section, one-dimensional heat conduction is derived from a three-dimensional equation. In Section 4.3.4, elastic shell theory is derived from the three-dimensional linear elasticity equations, and there are many other examples. In general, the boundary conditions for the simplified equations have to be derived from matching with a more complicated boundary layer involving more independent variables.

Consider steady heat-conduction in a long rod of circular cross section whose shape is given by

$$S(X, R) = 0 = R - BF(X/L), \qquad 0 \le X \le L \tag{4.3.125}$$

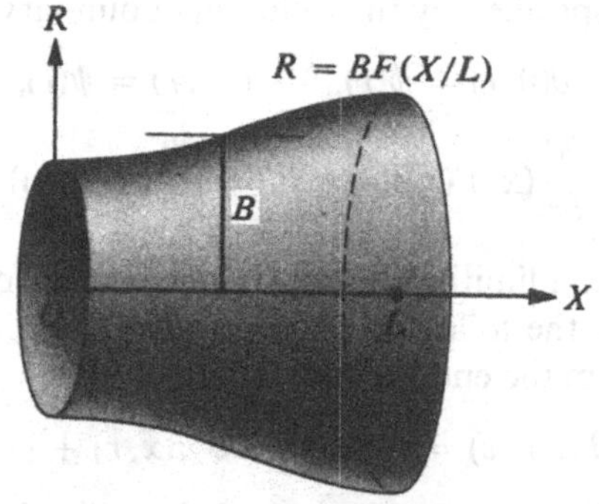

Figure 4.3.4 Quasi-One-Dimensional Heat Conduction

(see Figure 4.3.4). Assume that the side of the rod is insulated, so that $\partial T/\partial n = 0$, and assume that the temperature $T(X, R)$ is prescribed on the ends and is written in the form

$$T(0, R) = T^*\phi\left(\frac{R}{B}\right), \qquad T(L, R) = T^*\psi\left(\frac{R}{B}\right), \tag{4.3.126}$$

so that heat flows down the rod. We are interested in the case where $B/L \ll 1$. Here T^* is a characteristic temperature, and the equation for steady heat flow with constant thermal properties is Laplace's equation with axial symmetry,

$$\frac{\partial^2 T}{\partial R^2} + \frac{1}{R}\frac{\partial T}{\partial R} + \frac{\partial^2 T}{\partial X^2} = 0. \tag{4.3.127}$$

The boundary condition on the insulated surface $S = 0$ can be expressed as

$$\nabla T \cdot \nabla S = 0 \qquad \text{on } S = 0$$

or

$$\frac{\partial T}{\partial R} = \frac{B}{L} F'\left(\frac{X}{L}\right)\frac{\partial T}{\partial X} \quad \text{on } R = BF\left(\frac{X}{L}\right). \tag{4.3.128}$$

The entire problem can be expressed in the following suitable dimensionless coordinates:

$$r = \frac{R}{B}, \qquad x = \frac{X}{L}, \qquad \theta(x, r; \varepsilon) = \frac{T(X, R)}{T^*},$$

where $\varepsilon = B/L$. In terms of these variables, Equation (4.3.127) becomes

$$\frac{\partial^2 \theta}{\partial r^2} + \frac{1}{r}\frac{\partial \theta}{\partial r} + \varepsilon^2 \frac{\partial^2 \theta}{\partial x^2} = 0, \tag{4.3.129}$$

and the problem is specified by the following boundary conditions:

$$\text{ends} \qquad \theta(0, r) = \phi(r), \qquad \theta(1, r) = \psi(r), \tag{4.3.130}$$

$$\text{side} \qquad \frac{\partial \theta}{\partial r}(x, F(x)) = \varepsilon^2 F'(x) \frac{\partial \theta}{\partial x}(x, F(x)) \tag{4.3.131}$$

We now assume that a limiting solution, independent of ε, appears as $\varepsilon \to 0$ and represents θ by the following asymptotic expansion, which we expect to be valid away from the ends of the rod:

$$\theta(x, r; \varepsilon) = \theta_0(x, r) + \varepsilon^2 \theta(x, r) + \cdots. \tag{4.3.132}$$

The corresponding limit process has $\varepsilon \to 0$, (x, r) fixed.

In general, an arbitrary order could be chosen for the θ_0 term, but matching would show it to be of $O(1)$. The second term is of order ε^2, so that a nonhomogeneous equation for θ_1 results. Terms of intermediate order could be inserted if needed. The sequence of equations approximating Equation (4.3.129) is

$$\frac{\partial^2 \theta_0}{\partial r^2} + \frac{1}{r} \frac{\partial \theta_0}{\partial r} = 0, \tag{4.3.133}$$

$$\frac{\partial^2 \theta_1}{\partial r^2} + \frac{1}{r} \frac{\partial \theta_1}{\partial r} = -\frac{\partial^2 \theta_0}{\partial x^2}. \tag{4.3.134}$$

All subsequent equations are of the form of Equation (4.3.134). The boundary condition (Equation 4.3.131) has the expansion

$$\frac{\partial \theta_0}{\partial r}(x, F(x)) + \varepsilon^2 \frac{\partial \theta_1}{\partial r}(x, F(x)) + \cdots = \varepsilon^2 F'(x) \left\{ \frac{\partial \theta_0}{\partial x}(x, F(x)) + \cdots \right\},$$

so that, on the insulated boundary, we have

$$\frac{\partial \theta_0}{\partial r}(x, F(x)) = 0, \tag{4.3.135}$$

$$\frac{\partial \theta_1}{\partial r}(x, F(x)) = F'(x) \frac{\partial \theta_0}{\partial x}(x, F(x)). \tag{4.3.136}$$

Next the solution must be investigated. We have

$$\theta_0(x, r) = A_0(x) + B_0(x) \log r. \tag{4.3.137}$$

If we require finite temperature at the axis, then $B_0 = 0$, and the basic approximation is a one-dimensional temperature distribution:

$$\theta_0(x, r) = A_0(x). \tag{4.3.138}$$

This distribution automatically satisfies the boundary condition (Equation 4.3.135). In this case, further information about $A_0(x)$ cannot be found without considering the equation for θ_1 and its boundary condition. Equation (4.3.134) is now

$$\frac{\partial^2\theta_1}{\partial r^2} + \frac{1}{r}\frac{\partial\theta_1}{\partial r} = -\frac{d^2 A_0}{\partial x^2}, \tag{4.3.139}$$

which, if we disregard the log r term, has the solution

$$\theta_1(x, r) = A_1(x) - \frac{r^2}{4}\frac{d^2 A_0}{dx^2}. \tag{4.3.140}$$

Information about A_1 is found from the equation for θ_2, etc. Now the boundary condition on the insulated surface (Equation 4.3.136) becomes

$$-\frac{F(x)}{2}\frac{d^2 A_0}{dx^2} = F'(x)\frac{dA_0}{dx} \tag{4.3.141}$$

or

$$\frac{d}{dx}\left(F^2(x)\frac{dA_0}{dx}\right) = 0. \tag{4.3.142}$$

Remembering that $F(x)$ is proportional to the radius of a cross section, we see that Equation (4.3.142) is the equation for one-dimensional heat conduction. It arises here as a formal consequence of the insulation boundary condition. For the uniform accuracy of this approximation over the center section of the rod, $F(x)$ has to be sufficiently smooth.

Now, boundary conditions for Equation (4.3.129) as $x \to 0, 1$ have to be found by studying the solution in the neighborhood of the ends. Near $x = 0$, the only distinguished limit which preserves enough structure in (4.3.129) to allow for boundary conditions and matching is one in which $x^* = x/\varepsilon$ is fixed. Thus, consider the asymptotic expansion valid near $x = 0$,

$$\theta(x, r; \varepsilon) = \vartheta(x^*, r) + \cdots, \qquad x^* = x/\varepsilon. \tag{4.3.143}$$

Then the full equation results for ϑ:

$$\frac{\partial^2\vartheta}{\partial r^2} + \frac{1}{r}\frac{\partial\vartheta}{\partial r} + \frac{\partial^2\vartheta}{\partial x^{*2}} = 0, \tag{4.3.144}$$

but the boundary condition on the insulated surface is somewhat simplified. Equation (4.3.131) becomes

$$\frac{\partial\vartheta}{\partial r}(x^*, F(\varepsilon x^*)) + \cdots = \varepsilon^2 F'(\varepsilon x^*)\frac{1}{\varepsilon}\frac{\partial\vartheta}{\partial x^*}(x^*, F(\varepsilon x^*)) + \cdots. \tag{4.3.145}$$

Thus, as $\varepsilon \to 0$, we have

$$\frac{\partial\vartheta}{\partial r}(x^*, r_0) = 0, \tag{4.3.146}$$

where $r_0 = F(0)$. Again the assumption that F is smooth has been used. It can be seen from Equation (4.3.145) that the next term in the boundary-layer expansion (Equation 4.3.143) is $O(\varepsilon)$, but this is not considered here. At the end $x = 0$, we have

$$\vartheta(0, r) = \phi(r). \qquad 0 \le r \le r_0. \tag{4.3.147}$$

Thus, the problem to be solved is that of heat flow in an insulated cylinder. The extent in the x^* direction is infinite, since matching according to any intermediate limit makes $(x^* \to \infty,\ x \to 0)$. The matching condition here takes the simple form

$$\vartheta(\infty, r) = A_0(0). \tag{4.3.148}$$

It remains to be shown that $\vartheta(x^*, r) \to \text{const.}$ as $x^* \to \infty$ and to evaluate the constant. The solution to the problem for ϑ can be expressed, by separation of variables in terms of functions like

$$e^{-\lambda x^*} J_0(\lambda r), \qquad \lambda \ge 0,$$

and the transcendental equation for the eigenvalues λ_n follows from Equation (4.3.146):

$$\lambda_n J_0'(\lambda_n r) = 0. \tag{4.3.149}$$

There are an infinite set of roots starting with $\lambda_0 = 0, \lambda_1, \lambda_2, \lambda_3, \ldots$, and an infinite complete set of eigenfunctions. Thus, we have

$$\vartheta(x^*, r) = a_0 + \sum_{n=1}^{\infty} a_n e^{-\lambda_n x^*} J_0(\lambda_n r). \tag{4.3.150}$$

From the equation for $J_0(\lambda r)$,

$$\frac{d}{dr}\left\{r \frac{dJ_0}{dr}\right\} + \lambda^2 J_0(\lambda r) = 0 \tag{4.3.151}$$

it follows by integration from 0 to r_0 that

$$\int_0^{r_0} J_0(\lambda_n r) r\, dr = 0. \tag{4.3.152}$$

Thus, the constant a_0 is determined from

$$\int_0^{r_0} \vartheta(x^*, r) r\, dr = \frac{r_0^2}{2} a_0 = \int_0^{r_0} \vartheta(0, r) r\, dr$$

or

$$a_0 = \frac{2}{r_0^2} \int_0^{r_0} \phi(r) r\, dr. \tag{4.3.153}$$

Thus, the matching condition (Equation 4.3.148) states that the (weighted) average temperature at the end should be used as the boundary condition for the one-dimensional heat flow:

$$A(0) = (2/r_0^2) \int_0^{r_0} \phi(r) r\, dr, \qquad r_0 = F(0). \tag{4.3.154}$$

Similar considerations apply near $x = 1$, so that we have

$$A(1) = (2/r_1^2) \int_0^{r_1} \psi(r) r \, dr, \qquad r_1 = F(1), \tag{4.3.155}$$

and the net heat flow can then be calculated to $O(\varepsilon)$.

The other coefficients in the expansion (Equation 4.3.150) can be calculated from the usual orthogonality properties of the eigenfunctions:

$$\int_0^{r_0} J_0(\lambda_n r) J_0(\lambda_m r) r \, dr = \begin{cases} 0, & n \neq m, \\ \gamma_m^2, & n = m. \end{cases} \tag{4.3.156}$$

The correctness of the one-dimensional approximation depended to a large extent on the type of boundary conditions. Problems 4.3.2 and 4.3.3 illustrate this point.

4.3.4 Elastic-Shell Theory. Spherical Shell

By an elastic shell, we mean a thin region of elastic material which responds to a load in a special way due to its geometrical properties. The theory of elastic shells can be derived in a systematic way, by the use of perturbation expansions, from the three-dimensional equations of elasticity. This is not the method usually followed in various books. Rather, shell equations are derived from overall assumptions about the total forces and moments acting on an infinitesimal element. These forces and moments are often thought of as average across the cross section of a shell or else corresponding strain-energy methods are used. (See References 4.3.6, 4.3.7, and 4.3.8.)

Perturbation theory corroborates the approximate equations in certain cases and further provides a method for incorporating the boundary layers which inevitably arise. If some stage of the approximation corresponds to simplified shell theory, one cannot expect to satisfy full-elasticity boundary conditions.

The basic small parameter ε of shell theory is the thickness over a characteristic length, say the sphere radius. The calculations here are based wholly on linear elasticity theory, that is, on small strains. Thus, the loads that are applied must be thought of as being sufficiently small so that the structure remains in the linear elastic range. Since the loads then occur linearly in the problem, they need not be considered in the perturbation scheme; all results are proportional to the loads. However, if large deformations or nonlinearities are to be considered, then the mutual dependence of load (made dimensionless with an elastic modulus) and ε is of vital importance.

In this Section we consider a simple special example of shell theory; namely, a segment of a spherical shell fastened rigidly around the edges and loaded by axi-symmetric pressure forces on the inner surface. (See Figure 4.3.5.) The problem is sufficiently general to illustrate all the essential features

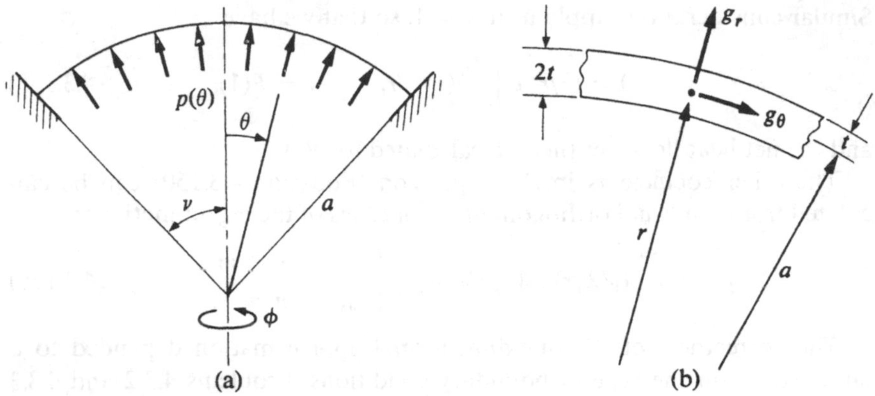

Figure 4.3.5 (a) Spherical Segment, (b) Detail of Thickness

of shell theory. First, the outer expansion valid away from the boundary is constructed and is shown to contain the membrane theory of thin shells. Then, the various boundary layers which must be added are discussed briefly.

The exact boundary-value problem demands a solution of the full equations for elasticity (written, for example, in terms of the displacements (q_r, q_θ, q_ϕ)) in polar coordinates (r, θ, ϕ) subject to the boundary conditions of prescribed stresses on the free surfaces and zero displacements on the fixed edges:

$$\begin{aligned} T_{rr}(a+t,\theta) &= T_{r\theta}(a+t,\theta) = 0, \qquad 0 \le \theta \le \nu; \\ T_{rr}(a-t,\theta) &= -p(\theta), \qquad T_{r\theta}(a-t,\theta) = 0, \qquad 0 \le \theta \le \nu; \\ q_r(r,\nu) &= q_\theta(r,\nu) = 0. \qquad a-t \le r \le a+t. \end{aligned} \tag{4.3.157}$$

Away from the edge ($\theta = \nu$), it can be expected that the solution to this problem should behave something like that of the full sphere under pressure. An exact solution of the latter problem (Reference 4.3.6, p. 142) is available for uniform pressure. A study of this exact solution shows that the deflections have an expansion starting with terms $O(1/\varepsilon)$, and that the hoop stresses $T_{\theta\theta}$, $T_{\phi\phi}$ are also $O(1/\varepsilon)$, as would be expected from an overall force balance. These facts can be used to start the expansion of our problem corresponding to membrane theory which is valid away from the edge $\theta = \nu$.

Our procedure is first to construct the membrane-theory expansion assumed valid away from the boundary, and then to construct the necessary boundary layer. A convenient starting point is the stress-equilibrium equations (Reference 4.3.6, p. 91) (div $T = 0$):

$$\frac{\partial T_{rr}}{\partial r} + \frac{1}{r}\frac{\partial T_{r\theta}}{\partial \theta} + \frac{1}{r}(2T_{rr} - T_{\theta\theta} - T_{\phi\phi} + T_{r\theta}\cot\theta) = 0, \tag{4.3.158}$$

$$\frac{\partial T_{r\theta}}{\partial r} + \frac{1}{r}\frac{\partial T_{\theta\theta}}{\partial \theta} + \frac{1}{r}[(T_{\theta\theta} - T_{\phi\phi})\cot\theta + 3T_{r\theta}] = 0. \tag{4.3.159}$$

The stress components are related to the strain components with the help of the elastic constants (λ, μ):

$$T_{rr} = \lambda\Delta + 2\mu\frac{\partial q_r}{\partial r}, \qquad T_{\theta\theta} = \lambda\Delta + 2\mu\left(\frac{1}{r}\frac{\partial q_\theta}{\partial\theta} + \frac{q_r}{r}\right). \quad (4.3.160)$$

$$T_{\phi\phi} = \lambda\Delta + 2\mu\left(\frac{\cot\theta}{r}q_\theta + \frac{q_r}{r}\right), \qquad \frac{1}{\mu}T_{r\theta} = \frac{\partial q_\theta}{\partial r} - \frac{q_\theta}{r} + \frac{1}{r}\frac{\partial q_r}{\partial\theta}.$$

$$T_{r\phi} = T_{\theta\phi} \equiv 0,$$

where the dilatation is

$$\Delta = \operatorname{div}\mathbf{q} = \frac{\partial q_r}{\partial r} + \frac{2}{r}q_r + \frac{1}{r}\frac{\partial q_\theta}{\partial\theta} + \frac{\cot\theta}{r}q_\theta. \quad (4.3.161)$$

Note also that

$$E = \text{modulus of elasticity} = \frac{\mu(3\lambda + 2\mu)}{\lambda + \mu},$$

$$\bar{\sigma} = \text{Poisson ratio} = \frac{\lambda}{2(\lambda + \mu)}.$$

Now consider a limit process $\varepsilon \to 0$ and coordinates fixed inside the shell (r^*, θ), where

$$r^* = \frac{r - a}{t} = \frac{r/a - 1}{\varepsilon}, \qquad -1 \le r^* \le 1, \qquad \frac{\partial}{\partial r} = \frac{1}{\varepsilon}\frac{\partial}{\partial r^*}. \quad (4.3.162)$$

The stresses in a thin shell thus have an asymptotic expansion of the form

$$T_{rr}(r, \theta; \varepsilon) = T(r^*, \theta) + \cdots, \qquad T_{r\theta}(r, \theta; \varepsilon) = S(r^*, \theta) + \cdots.$$
$$T_{\theta\theta}(r, \theta; \varepsilon) = \frac{H(\theta)}{\varepsilon} + \cdots, \qquad T_{\phi\phi}(r, \theta; \varepsilon) = \frac{J(\theta)}{\varepsilon} + \cdots. \quad (4.3.163)$$

If, in fact, stress $O(1/\varepsilon)$ were allowed in T_{rr}, $T_{r\theta}$, the equilibrium equations would immediately show that these stresses are functions of θ only, and the boundary conditions would show that these terms are zero.

The hoop stress terms $O(1/\varepsilon)$ here are assumed to depend only on θ; this mirrors their behavior for the full sphere where they are constant across the thickness. This can also be proved by a detailed consideration of the displacement equations and expansions. It is sufficient here to show consistency with the equations and boundary conditions. The equilibrium equations (Equations 4.3.158 and 4.3.159) have terms $O(1/\varepsilon)$, which are

$$\frac{\partial T}{\partial r^*} - (H + J) = 0, \quad (4.3.164)$$

$$\frac{\partial S}{\partial r^*} + \frac{dH}{d\theta} + (H - J)\cot\theta = 0. \quad (4.3.165)$$

Here we have used $r/a = 1 + \varepsilon r^*$, $a/r = 1 - \varepsilon r^* + \cdots$. These approximate stress-balance equations can be integrated at once to yield

$$T(r^*, \theta) = \tau(\theta) + \{H(\theta) + J(\theta)\}r^*, \tag{4.3.166}$$

$$S(r^*, \theta) = \sigma(\theta) - r^*(dH/d\theta) - r^*(H - J)\cot\theta, \tag{4.3.167}$$

where $\tau(\theta)$, $\sigma(\theta)$ are functions of integration. Now, applying the boundary conditions (Equations 4.3.157), we have four relations:

$$0 = T(1, \theta) = \tau(\theta) + H(\theta) + J(\theta),$$

$$-p(\theta) = T(-1, \theta) = \tau - (H + J),$$

$$0 = S(1, \theta) = \sigma - \frac{dH}{d\theta} - (H - J)\cot\theta,$$

$$0 = S(-1, \theta) = \sigma + \frac{dH}{d\theta} + (H - J)\cot\theta.$$

Elimination from this system provides the basic equations for H, J:

$$H + J = \frac{p}{2} = -\tau, \tag{4.3.168}$$

$$\frac{dH}{d\theta} + (H - J)\cot\theta = 0 = \sigma \tag{4.3.169}$$

or, in terms of H itself,

$$\frac{d}{d\theta}(H\sin^2\theta) = \frac{p}{2}\sin\theta\cos\theta. \tag{4.3.170}$$

The integral of (Equation 4.3.170), which has H bounded as $\theta \to 0$, is

$$H(\theta) = \frac{1}{\sin^2\theta}\int_0^\theta \frac{p(\alpha)}{2}\sin\alpha\cos\alpha\, d\alpha. \tag{4.3.171}$$

Here J follows from Equation (4.3.168):

$$J(\theta) = \frac{p(\theta)}{2} - \frac{1}{\sin^2\theta}\int_0^\theta \frac{p(\alpha)}{2}\sin\alpha\cos\alpha\, d\alpha. \tag{4.3.172}$$

For the special case where $p =$ const. the classical result is obtained:

$$H = J = p/4 = \text{const.} \tag{4.3.173}$$

Since τ, σ are given by Equations (4.3.168) and (4.3.169), we now have the $O(1)$ distribution of shear and normal stress across the section:

$$T(r^*, \theta) = -\frac{p(\theta)}{2}(1 - r^*), \tag{4.3.174}$$

$$S(r^*, \theta) = 0. \tag{4.3.175}$$

It is of interest now, and essential for matching later, to obtain the form of the deflection which corresponds to this distribution of stresses. As for the full sphere, the dominant terms of the deflection are $O(1/\varepsilon)$ and are functions of θ alone. The justification is similar to that given before. Thus, tentatively assume that

$$\frac{q_r}{a} = \frac{u(\theta)}{\varepsilon} + u_1(r^*, \theta) + \cdots, \tag{4.3.176}$$

$$\frac{q_\theta}{a} = \frac{v(\theta)}{\varepsilon} + v_1(r^*, \theta) + \cdots. \tag{4.3.177}$$

The stresses produced by this set of displacements are now studied. From Equation (4.3.161), we have

$$\Delta = \frac{1}{\varepsilon}\left\{\frac{\partial u_1}{\partial r^*} + 2u\frac{dv}{d\theta} + v\cot\theta\right\} + \cdots. \tag{4.3.178}$$

so that the expansion for T_{rr} starts out as

$$T_{rr} = \frac{1}{\varepsilon}\left\{\lambda\left(2u + \frac{dv}{d\theta} + v\cot\theta\right) + (\lambda + 2\mu)\frac{\partial u_1}{\partial r^*}\right\} + \cdots. \tag{4.3.179}$$

However, this term must be identically zero, since no T_{rr} of $O(1/\varepsilon)$ occurs in the problem. Thus, we have

$$\frac{\partial u_1}{\partial r^*} = -\frac{\lambda}{\lambda + 2\mu}\left(2u + \frac{dv}{d\theta} + v\cot\theta\right). \tag{4.3.180}$$

There is, in consequence, a dilatation of $O(1/\varepsilon)$ corresponding to the general stretching of the shell:

$$\Delta = \frac{1}{\varepsilon}\frac{2\mu}{\lambda + 2\mu}\left\{2u + \frac{dv}{d\theta} + v\cot\theta\right\} + \cdots. \tag{4.3.181}$$

Equations (4.3.168) and (4.3.169) can now be expressed as equations for the displacement of the shell:

$$\begin{aligned} T_{\theta\theta} + T_{\phi\phi} &= 2(\lambda + \mu)\Delta - 2\mu\left(\frac{\partial q_r}{\partial r} + \frac{q_r}{r}\right) \\ &= 2(\lambda + \mu)\frac{1}{\varepsilon}\left\{\frac{2\mu}{\lambda + 2\mu}\right\}\left\{2u + \frac{dv}{d\theta} + v\cot\theta\right\} - \frac{2\mu}{\varepsilon}\left(\frac{\partial u_1}{\partial r^*}\right) \end{aligned}$$

or

$$H + J = \frac{2\mu(3\lambda + 2\mu)}{\lambda + 2\mu}\left\{2u + \frac{dv}{d\theta} + v\cot\theta\right\} + \cdots. \tag{4.3.182}$$

Similarly, we obtain

$$T_{\phi\phi} - T_{\theta\theta} = \frac{2\mu}{\varepsilon}\left(v\cos\theta - \frac{dv}{d\theta}\right) + \cdots$$

or

$$J - H = 2\mu\left(v\cot\theta - \frac{dv}{d\theta}\right) + \cdots. \tag{4.3.183}$$

Thus, Equation (4.3.168) directly becomes

$$\frac{dv}{d\theta} + v\cot\theta + 2u = \frac{p(\theta)}{4\mu}\frac{\lambda + 2\mu}{3\lambda + 2\mu} \tag{4.3.184}$$

and, after a little elimination, Equation (4.3.169) becomes

$$\frac{du}{d\theta} - v = \frac{1}{2\mu}\frac{\lambda + \mu}{3\lambda + 2\mu}\frac{dp(\theta)}{d\theta}. \tag{4.3.185}$$

Equations (4.3.184) and (4.3.185) form the basic system of equations for the shape of the shell and are identical with the membrane equations mentioned in Reference 4.3.6, p. 584.

The equation for the tangential displacement alone, from Equations (4.3.184) and (4.3.185), is

$$\frac{d^2v}{d\theta^2} + \cot\theta\frac{dv}{d\theta} + (2 - \csc^2\theta)v = -\frac{1}{4\mu}\frac{dp}{d\theta}. \tag{4.3.186}$$

A particular solution corresponding to a rigid displacement is a solution of the homogeneous equations

$$v_p = A\sin\theta, \qquad u_p = -A\cos\theta. \tag{4.3.187}$$

The complete solution for the case $p =$ const. is, thus,

$$u = u_\infty - A\cos\theta, \qquad v = A\sin\theta, \tag{4.3.188}$$

where the constant u_∞ is the radial displacement of the full sphere under uniform pressure:

$$u_\infty = \frac{1}{8\mu}\frac{\lambda + 2\mu}{3\lambda + 2\mu}p. \tag{4.3.189}$$

It is clear that the solution represented by Equation (4.3.188) cannot satisfy the boundary condition of no displacement at $\theta = \nu$, even with a particular choice of the rigid displacement A. In fact, no such rigid displacement of $O(1/\varepsilon)$ is to be expected in this problem. Some kind of a boundary layer is needed near $\theta = \nu$.

If displacements q_r, q_θ of the same order occur in a thin layer near $O(\varepsilon)$ in thickness near $\theta = \nu$, plane-strain elasticity equations result. These are expressed in ($\theta^* = [\theta - \nu]/\varepsilon, r$). However, it is easy to show that no solution of these plane-strain equations in the "elasticity" boundary layer exists which matches to the membrane expansion. For matching, we would need

$$\frac{q_r}{a} = \frac{u^*(r, \theta^*)}{\varepsilon} + \cdots, \qquad \frac{q_\theta}{a} = \frac{v^*(r, \theta^*)}{\varepsilon} + \cdots, \tag{4.3.190}$$

and $u^* \to u_\infty$, $v^* \to 0$, $\theta^* \to -\infty (A = 0)$.

Thus, some intermediate boundary layer must be constructed, and its width must be greater than that of the elasticity layer. That is, a boundary-layer expansion is sought, in which

$$\bar{\theta} = \frac{\theta - v}{\delta(\varepsilon)}, \qquad r^* = \frac{(r/a) - 1}{\varepsilon}, \qquad (\delta(\varepsilon) \gg \varepsilon) \tag{4.3.191}$$

are fixed. Note that

$$\cot \theta = \cot v - \delta(1 + \cot^2 v)\bar{\theta}. \tag{4.3.192}$$

Returning to the stress equations, we assume an asymptotic expansion of the form

$$T_{rr}(r, \theta; \varepsilon) = \tau(r^*, \bar{\theta}) + \cdots, \tag{4.3.193}$$

$$T_{\theta\theta}(r, \theta; \varepsilon) = (1/\varepsilon)h(r^*, \bar{\theta}) + \cdots, \tag{4.3.194}$$

$$T_{\phi\phi}(r, \theta; \varepsilon) = (1/\varepsilon)g(r^*, \bar{\theta}) + \cdots, \tag{4.3.195}$$

$$T_{r\theta}(r, \theta; \varepsilon) = \beta(\varepsilon)\sigma(r^*, \bar{\theta}) + \cdots. \tag{4.3.196}$$

The orders of the hoop stresses are in accord with overall equilibrium ideas, the order of the normal stress with the boundary conditions, and the order of the shear is here undetermined. A large shear can be expected to be produced if substantial bending takes place near the boundary. Other possibilities should be investigated and ruled out. This assumption leads to an expansion capable of being matched. The dominant equations of stress equilibrium are, thus,

$$\frac{1}{\varepsilon}\frac{\partial \tau}{\partial r^*} + \frac{\beta(\varepsilon)}{\delta(\varepsilon)}\frac{\partial \sigma}{\partial \bar{\theta}} - \frac{1}{\varepsilon}(h + g) = 0 \tag{4.3.197}$$

$$\frac{\beta}{\varepsilon}\frac{\partial \sigma}{\partial r^*} + \frac{1}{\varepsilon\delta}\frac{\partial h}{\partial \bar{\theta}} = 0. \tag{4.3.198}$$

In order for Equation (4.3.198) to yield a nontrivial result, we need

$$\beta(\varepsilon) = \frac{1}{\delta(\varepsilon)}. \tag{4.3.199}$$

Then, the distinguished limit of Equation (4.3.197) occurs for $1/\varepsilon = \beta/\delta$ or

$$\delta = \sqrt{\varepsilon}, \qquad \beta(\varepsilon) = 1/\sqrt{\varepsilon}, \tag{4.3.200}$$

fixing the order of the boundary-layer thickness and shear stress. The resulting equations include all the terms of the "elasticity" boundary layer, and thus have at least the possibility of matching to an elasticity boundary layer.

Rewriting the basic equations (Equations 4.3.197, 4.3.198), we have

$$\frac{\partial \tau}{\partial r^*} + \frac{\partial \sigma}{\partial \bar{\theta}} - (h + g) = 0,$$

$$\frac{\partial \sigma}{\partial r^*} + \frac{\partial h}{\partial \bar{\theta}} = 0.$$

Next, consider the displacement field corresponding to the assumed orders of stress in Equations (4.3.194) through (4.3.196):

$$\frac{q_r}{a} = \frac{U(\bar{\theta})}{\varepsilon} + U_1(r^*, \bar{\theta}) + \cdots, \tag{4.3.201}$$

$$\frac{q_\theta}{a} = \frac{V(\bar{\theta}, r^*)}{\sqrt{\varepsilon}} + \sqrt{\varepsilon}\, V_1(r^*, \bar{\theta}) + \cdots. \tag{4.3.202}$$

It is necessary that $U = U(\bar{\theta})$ only, so that dilatation of $O(1/\varepsilon^2)$ does not occur. Then, for the dilatation Δ, we have

$$\Delta = \frac{1}{\varepsilon}\left\{\frac{\partial U_1}{\partial r^*} + 2U + \frac{\partial V}{\partial \bar{\theta}}\right\} + \cdots, \tag{4.3.203}$$

and the expressions for the $O(1/\varepsilon)$ components of hoop stress are

$$h = \lambda \frac{\partial U_1}{\partial r^*} + (\lambda + 2\mu)\frac{\partial V}{\partial \bar{\theta}} + 2(\lambda + \mu)U, \tag{4.3.204}$$

$$g = \lambda \frac{\partial U_1}{\partial r^*} + \lambda \frac{\partial V}{\partial \bar{\theta}} + 2(\lambda + \mu)U, \tag{4.3.205}$$

$$h + g = 2\lambda \frac{\partial U_1}{\partial r^*} + 2(\lambda + \mu)\left\{\frac{\partial V}{\partial \bar{\theta}} + 2U\right\}. \tag{4.3.206}$$

Considering next the normal stress T_{rr}, we have

$$T_{rr} = \frac{1}{\varepsilon}\left\{\lambda\left(\frac{\partial U_1}{\partial r^*} + 2U + \frac{\partial V}{\partial \bar{\theta}}\right) + 2\mu \frac{\partial U_1}{\partial r^*}\right\} + \tau(r^*, \bar{\theta}) + \cdots. \tag{4.3.207}$$

Again, the term $O(1/\varepsilon)$ in T_{rr} must vanish. This provides an expression for $\partial U_1/\partial r^*$ in terms of U, V and allows h, g to be expressed completely in terms of these quantities:

$$\frac{\partial U_1}{\partial r^*} = -\frac{\lambda}{\lambda + 2\mu}\left(\frac{\partial V}{\partial \bar{\theta}} + 2U\right). \tag{4.3.208}$$

We have, thus,

$$h = 4\mu \frac{\lambda + \mu}{\lambda + 2\mu} \frac{\partial V}{\partial \theta} + 2\mu \frac{3\lambda + 2\mu}{\lambda + 2\mu} U. \tag{4.3.209}$$

$$h + g = 2\mu \frac{3\lambda + 2\mu}{\lambda + 2\mu} \left(\frac{\partial V}{\partial \theta} + 2U \right). \tag{4.3.210}$$

A similar argument can be applied to the shear stress:

$$\frac{1}{\mu} T_{r\theta} = \frac{1}{\varepsilon^{3/2}} \frac{\partial V}{\partial r^*} + \frac{1}{\sqrt{\varepsilon}} \frac{\partial V_1}{\partial r^*} - \frac{V}{\sqrt{\varepsilon}}$$
$$+ (1 - \varepsilon r^* + \cdots) \left\{ \frac{1}{\varepsilon^{3/2}} \frac{dU}{d\theta} + \frac{1}{\sqrt{\varepsilon}} \frac{\partial U_1}{\partial \theta} \right\} + \cdots.$$

The term $O(1/\varepsilon^{3/2})$ must vanish:

$$\frac{\partial V}{\partial r^*} + \frac{\partial U}{\partial \theta} = 0, \tag{4.3.211}$$

which is one of the basic differential equations for the shell deflection. Also, the term $O(1/\sqrt{\varepsilon})$ is

$$\frac{\sigma(r^*, \bar{\theta})}{\mu} = \frac{\partial V_1}{\partial r^*} - V + \frac{\partial U_1}{\partial \theta} - r^* \frac{dU}{d\bar{\theta}}. \tag{4.3.212}$$

The consequence of Equation (4.3.211) is a linear variation of tangential displacement across the cross section

$$V(r^*, \bar{\theta}) = A(\bar{\theta}) - r^* \frac{dU}{d\bar{\theta}}, \tag{4.3.213}$$

and a corresponding linear variation of the hoop stresses from Equations (4.3.209) and (4.3.210). Introducing some special notation, we have

$$h = h^{(0)}(\bar{\theta}) + r^* h^{(1)}(\bar{\theta}), \tag{4.3.214}$$

$$g = g^{(0)}(\bar{\theta}) + r^* g^{(1)}(\bar{\theta}), \tag{4.3.215}$$

where

$$h^{(0)} = 4\mu \frac{\lambda + \mu}{\lambda + 2\mu} \frac{dA}{d\bar{\theta}} + 2\mu \frac{3\lambda + 2\mu}{\lambda + 2\mu} U,$$

$$h^{(1)} = -4\mu \frac{\lambda + \mu}{\lambda + 2\mu} \frac{d^2 U}{d\bar{\theta}^2},$$

$$h^{(0)} + g^{(0)} = 2\mu \frac{3\lambda + 2\mu}{\lambda + 2\mu} \left(\frac{dA}{d\bar{\theta}} + 2U \right).$$

$$h^{(1)} + g^{(1)} = -2\mu \frac{3\lambda + 2\mu}{\lambda + 2\mu} \frac{d^2 U}{d\bar{\theta}^2}.$$

The tangential equilibrium equation (Equation 4.3.198) can now be integrated in the form

$$\sigma(r^*, \bar{\theta}) = \sigma^{(0)}(\bar{\theta}) - r^* \frac{dh^{(0)}}{d\bar{\theta}} - \frac{r^{*2}}{2} \frac{dh^{(1)}}{d\bar{\theta}}. \qquad (4.3.216)$$

The boundary condition states that $\sigma(\pm 1, \theta) = 0$, so that we have

$$\frac{dh^{(0)}}{d\bar{\theta}} = 0 \qquad (4.2.217)$$

and

$$\sigma^{(0)} = \frac{1}{2} \frac{dh^{(1)}}{d\bar{\theta}}. \qquad (4.3.218)$$

The shear stress has a parabolic distribution across the thickness

$$\sigma(r^*, \bar{\theta}) = (1 - r^{*2})\tfrac{1}{2}(dh^{(1)}/d\bar{\theta}). \qquad (4.3.219)$$

A similar study of the radial equilibrium equation (Equation 4.3.197), using Equations (4.3.218), (4.3.214), and (4.3.215), allows the basic differential equation for the shell deflection to be found, and from its solution all the stresses can also be found. Integration of Equation (4.3.197) shows that

$$\tau(r^*, \bar{\theta}) = \tau^{(0)}(\bar{\theta}) + r^*\tau^{(1)}(\bar{\theta}) + r^{*2}\tau^{(2)}(\bar{\theta}) + r^{*3}\tau^{(3)}(\bar{\theta}) \qquad (4.3.220)$$

where

$$\tau^{(1)} = -\tfrac{1}{2}(d^2h^{(1)}/d\bar{\theta}^2) + h^{(0)} + g^{(0)},$$

$$\tau^{(2)} = \tfrac{1}{2}\{h^{(1)} + g^{(1)}\},$$

$$\tau^{(3)} = \tfrac{1}{6}(d^2h^{(1)}/d\bar{\theta}^2).$$

The boundary conditions at $r^* = \pm 1$ are

$$r^* = +1, \qquad 0 = \tau^{(0)} + \tau^{(1)} + \tau^{(2)} + \tau^{(3)} \qquad (4.3.221)$$

$$r^* = -1, \qquad -p(v) = \tau^{(0)} - \tau^{(1)} + \tau^{(2)} - \tau^{(3)} \qquad (4.3.222)$$

or

$$\tau^{(1)} + \tau^{(3)} = \frac{p(v)}{2} = -\frac{1}{3}\frac{d^2h^{(1)}}{d\bar{\theta}^2} + h^{(0)} + g^{(0)}. \qquad (4.3.223)$$

Equations (4.3.217) and (4.3.223) provide the basic systems of equations. Equations (4.3.217) states that $h^{(0)} =$ const. or

$$2(\lambda + \mu)(dA/d\bar{\theta}) + (3\lambda + 2\mu)U = (3\lambda + 2\mu)U_\infty, \qquad (4.3.224)$$

where $U \to U_\infty$ as $\bar{\theta} \to -\infty$, for matching the constant u of the membrane solution as $\theta \to v$; $dA/d\bar{\theta} \to 0$, $\bar{\theta} \to -\infty$. Equation (4.3.223) is, from the definitions of $h^{(0)}$, $h^{(1)}$, $q^{(0)}$,

$$\frac{4}{3}\mu\frac{\lambda + \mu}{\lambda + 2\mu}\frac{d^4U}{d\bar{\theta}^4} + 2\mu\frac{3\lambda + 2\mu}{\lambda + 2\mu}\left(\frac{dA}{d\bar{\theta}} + 2U\right) = \frac{p(v)}{2}. \qquad (4.3.225)$$

For the special case of $p(\theta) = \text{const.} = p$, which is all that will be considered further, the elimination of $dA/d\bar{\theta}$ from Equations (4.3.224) and (4.3.225) results in

$$\frac{d^4U}{d\bar{\theta}^4} + 4\kappa^4 U = \text{const.} = 4\kappa^4 U_\infty, \tag{4.3.226}$$

where

$$4\kappa^4 = \frac{3}{4}\frac{(3\lambda + 2\mu)(\lambda + 2\mu)}{(\lambda + \mu)^2}, \qquad U_\infty = \frac{1}{8\mu}\frac{\lambda + 2\mu}{3\lambda + 2\mu}\,p.$$

Equation (4.3.226) looks exactly like the equation of a beam on an elastic foundation and has oscillatory decaying solutions as $\bar{\theta} \to -\infty$. Discarding the solutions which grow as $\bar{\theta} \to -\infty$ and making $U(\bar{\theta}) = 0$ as $\bar{\theta} = 0$ to approach the fixed boundary condition at $\theta = v$, we have

$$U(\bar{\theta}) = U_\infty\{1 - e^{\kappa\bar{\theta}}\cos\kappa\bar{\theta}\} + be^{\kappa\bar{\theta}}\sin\kappa\bar{\theta}, \tag{4.3.227}$$

where the constant b is arbitrary. Next, from Equation (4.3.224), we calculate $A(\theta)$:

$$A(\bar{\theta}) = A_\infty + \frac{3\lambda + 2\mu}{4(\lambda + \mu)\kappa}\{(U_\infty + b)\cos\kappa\bar{\theta} + (U_\infty - b)\sin\kappa\bar{\theta}\}e^{\kappa\bar{\theta}} \tag{4.3.228}$$

and

$$\frac{dU}{d\bar{\theta}} = \kappa\{(U_\infty + b)\sin\kappa\bar{\theta} - (U_\infty - b)\cos\kappa\bar{\theta}\}e^{\kappa\bar{\theta}}. \tag{4.3.229}$$

This solution should match to the elasticity boundary layer as $\bar{\theta} \to 0$, $\theta^* \to -\infty$. The entire matching process can be expressed in terms of a suitable intermediate limit as usual, but here the details are omitted. The behavior of the "bending-layer" solution as $\bar{\theta} \to 0$ is

$$U(\bar{\theta}) \to (b - U_\infty)\kappa\bar{\theta} + b\kappa^2\bar{\theta}^2 + (U_\infty + b)\frac{\kappa^2\bar{\theta}^3}{3} + \cdots,$$

$$\frac{dU}{d\bar{\theta}} \to \kappa(b - U_\infty) + 2b\kappa^2\bar{\theta} + (U_\infty + b)\kappa^3\bar{\theta}^2 + \cdots,$$

$$A \to A_\infty + \frac{3\lambda + 2\mu}{4\kappa(\lambda + \mu)}\{U_\infty + b\} + 2U_\infty\kappa\bar{\theta} + \cdots, \tag{4.3.230}$$

and, from Equation (4.3.213)

$$V \to A_\infty + \frac{3\lambda + 2\mu}{4\kappa(\lambda + \mu)}\{V_\infty + b\} - r^*\kappa(b - U_\infty) + O(\bar{\theta}).$$

All attempts at matching this behavior to that of an "elasticity" boundary layer as $\theta^* \to -\infty$ fail, except if Equation (4.3.230) is made to satisfy the boundary condition at $\bar{\theta} = 0$ exactly; that is, the elasticity boundary layer is included in Equation (4.3.230). Thus, the arbitrary constants (A_∞, b) must be chosen so that $V = 0$ at $\bar{\theta} = 0$:

$$b - U_\infty = 0, \qquad A_\infty + \frac{3\lambda + 2\mu}{4\kappa(\lambda + \mu)}(U_\infty + b) = 0. \tag{4.3.231}$$

Thus, the bending boundary layer satisfies the fixed edge boundary condition exactly. As far as Equation (4.3.226) is concerned, this means that a fixed edge forces the boundary conditions. Thus, we have

$$U(\bar{\theta}) = \frac{dU}{d\bar{\theta}} = 0 \quad at\ \bar{\theta} = 0. \tag{4.3.232}$$

A consequence of this solution is that

$$V(r^*, \bar{\theta}) \to A_\infty = -\frac{3\lambda + 2\mu}{2(\lambda + \mu)\kappa} U_\infty \tag{4.3.233}$$

as $\bar{\theta} \to -\infty$. A term must be added to the membrane solution to match this deflection. This term can be a rigid displacement of $O(1/\sqrt{\varepsilon})$. In particular, we could have

$$\begin{aligned} \frac{q_r}{a} &= \frac{u(\theta)}{\varepsilon} - \frac{A_\infty \cos\theta}{\sqrt{\varepsilon}} + \cdots \\ \frac{q_\theta}{a} &= \frac{v(\theta)}{\varepsilon} + \frac{A_\infty \sin\theta}{\sqrt{\varepsilon}} + \cdots \end{aligned} \tag{4.3.234}$$

and this is the ultimate effect of the rigid boundary on the main part of the shell. The presence of an $O(1/\sqrt{\varepsilon})$ term in q_r implies a higher-order bending layer, etc.

It should be noted in conclusion that the considerations above are not valid for a shallow shell where ν is small. To study a shallow shell, $\nu(\varepsilon)$ must be assigned an order by studying the various limits.

4.3.5 Potential Induced by a Point Source of Current in the Interior of a Biological Cell

Certain boundary value problems become singular, in the perturbation sense, because the solution fails to exist for a limiting value of a parameter. For example, if a heat source is turned on and maintained inside a finite conducting body which is imperfectly insulated at its surface, a steady temperature will be reached. If the insulation is made more and more perfect, the body will heat up more quickly and in the limiting case of perfect insulation, a steady-state is never reached.

An analogous problem which occurs in electrophysiology is discussed in this section. In certain experiments, in order to measure passive electric properties, a micro-electrode was used to introduce a point source of current into a cell and the potential was measured at another point. The analysis of this experiment depends on the theoretical treatment sketched below.

The model for the cell is a finite body of characteristic dimension a enclosed by a membrane of thickness δ, surrounded by a perfectly conducting external medium (constant potential). The more general case of finite external conductivity can be worked out by similar methods and appears in (Reference 4.3.9). The geometry and coordinate system are shown in Figure 4.3.6.

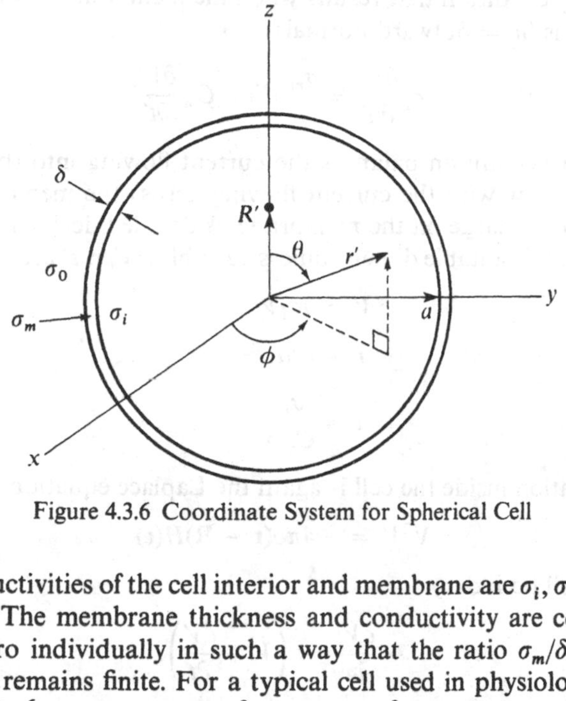

Figure 4.3.6 Coordinate System for Spherical Cell

The conductivities of the cell interior and membrane are σ_i, σ_m (mhos/cm) respectively. The membrane thickness and conductivity are considered to approach zero individually in such a way that the ratio σ_m/δ, the surface conductivity remains finite. For a typical cell used in physiological experiments $\delta = 10^{-6}$ cm and $a = 10^{-3}$ to 5×10^{-2} cm. The membrane is also assumed to have a surface capacitance $c_m \sim 1\mu$ farad/cm^2. Typical values are $\sigma_m = 3 \times 10^{-10}$ mhos/cm, $\sigma_i = 7 \times 10^{-3}$ mhos/cm and the basic dimensionless small parameter of the problem is

$$\varepsilon = \frac{\sigma_m a}{\delta \sigma_i} < 10^{-3}.$$

Let ( )′ denote temporarily quantities with physical dimensions and assume a point source of current (4π amps) at $\mathbf{r}' = \mathbf{R}'$ is turned on in a quiescent system. The current density $\mathbf{J}'$ amps/cm^2 is then given by

$$\nabla' \cdot \mathbf{J}' = 4\pi\, \delta(\mathbf{r}' - \mathbf{R}') H(t'). \tag{4.3.235a}$$

Ohm's Law is

$$\mathbf{J}' = -\sigma_i \nabla' V', \qquad V' = \text{potential (volts)},$$

and

$$H(t') = \begin{cases} 0 & t' < 0 \\ 1 & t' > 0 \end{cases} \quad \text{Heaviside step-function.}$$

Thus,

$$\nabla'^2 V' = -\frac{4\pi}{\sigma_i}\delta(\mathbf{r}' - \mathbf{R}')H(t'), \tag{4.3.235b}$$

The boundary condition that results when the membrane is considered as a discontinuity is (n' = outward normal)

$$-\sigma_i \frac{\partial V'}{\partial n'} = \frac{\sigma_m}{\delta} V' + C_m \frac{\partial V'}{\partial t'}. \tag{4.3.236}$$

This boundary condition balances the current flowing into the membrane under Ohm's Law with the current flowing across the membrane and the accumulation of charge on the membrane. A detailed derivation appears in Reference 4.3.9. If suitable dimensionless variables (V, r, t) are introduced by

$$V = a\sigma_i V'$$
$$r = r'/a \tag{4.3.237}$$
$$t = \frac{\sigma_m}{C_m \delta} t',$$

then the equation inside the cell is again the Laplace equation

$$\nabla^2 V = -4\pi\delta(\mathbf{r} - \mathbf{R})H(t) \tag{4.3.238}$$

and on the cell boundary

$$-\frac{\partial V}{\partial n} = \varepsilon\left(V + \frac{\partial V}{\partial t}\right). \tag{4.3.239}$$

This problem has the character discussed earlier. For $\varepsilon = 0$, no steady state exists.

We now wish to discuss the asymptotic behavior of $V(\mathbf{r}, t; \varepsilon)$ as $\varepsilon \to 0$.

The first expansion we consider is that associated with the limit $\varepsilon \to 0$ $\mathbf{r}$, t fixed in the coordinates as chosen. The limit problem with $\varepsilon = 0$ has no solution. This corresponds physically to the fact that a very large potential develops for small ε. Hence we try

$$V(\mathbf{r}, t; \varepsilon) = \alpha_0(\varepsilon)V_0(\mathbf{r}, t) + \alpha_1(\varepsilon)V_1(\mathbf{r}, t) + \cdots \tag{4.3.240}$$

where now

$$\alpha_0(\varepsilon) \to \infty$$

but still the $\alpha_j(\varepsilon)$ form an asymptotic sequence. If $\alpha_1(\varepsilon) = 1$, then we have

$$\nabla^2 V_0 = 0 \tag{4.3.241}$$

$$\nabla^2 V_1 = -4\pi\delta(\mathbf{r} - \mathbf{R})H(t) \tag{4.3.242}$$

$$\nabla^2 V_{2,3} = 0$$

with the following boundary condition

$$\frac{\partial V_0}{\partial n} = 0 \qquad \text{on the cell surface} \tag{4.3.243}$$

$$-\frac{\partial V_1}{\partial n} = \begin{cases} 0 & \text{if } \alpha_0(\varepsilon) \ll 1/\varepsilon, \\ V_0 + \partial V_0/\partial t & \text{if } \alpha_0(\varepsilon) = 1/\varepsilon. \end{cases}$$

It is clear that the latter case is the only case to have a solution so that

$$\alpha_0 = \frac{1}{\varepsilon}$$

$$-\frac{\partial V_1}{\partial n} = V_0 + \frac{\partial V_0}{\partial t} \quad \text{on the cell surface.} \tag{4.3.244}$$

Thus, $\alpha_2 = \varepsilon$, $\alpha_3 = \varepsilon^2$ etc., and $-\partial V_k/\partial n = V_{k-1} + \partial V_{k-1}/\partial t$, $k = 2, 3, \ldots$ on the cell surface.

The solution for V_0 is thus uniform inside the cell

$$V_0 = f_0(t). \tag{4.3.245}$$

As is typical in singular boundary value problems, this is all the information that can be obtained from a study of V_0. In order to find out more about $f_0(t)$, it is necessary to use a solvability condition derived from the equation and boundary condition for V_1. Integration of the equation (4.3.242) and the use of Gauss's theorem gives (d^3r = infinitesimal volume element)

$$\iiint_{\text{cell volume}} \nabla^2 V_1 d^3r = -4\pi = \iint_{\text{cell surface}} \frac{\partial V_1}{\partial n} dS. \tag{4.3.246}$$

Using the boundary condition (4.3.244), this gives

$$\iint_{\text{cell surface}} \left(\frac{df_0}{dt}(t) + f_0(t)\right) dS = 4\pi$$

$$\frac{df_0}{dt} + f_0 = 4\frac{\pi}{A}, \tag{4.3.247}$$

where A = surface area of the cell membrane. The solution for $f_0(t)$ is

$$V_0 = f_0 = \frac{4\pi}{A} + a_0 e^{-t}. \tag{4.3.248}$$

Thus, if we assume that the potential is initially zero, we have

$$V_0(\mathbf{r}, t) = \frac{4\pi}{A}(1 - e^{-t}). \qquad (4.3.249)$$

The result shows that the cell builds up to a large uniform potential independent of cell shape, but dependent on cell surface area A.

Next we consider the problem for V_1 using the result for V_0 to write the boundary condition (4.3.243) in the form

$$\frac{\partial V_1}{\partial n} = -\frac{4\pi}{A}. \qquad (4.3.250)$$

A corresponding solvability condition derived from the problem for V_2 exists for the potential V_1.

$$\iiint_{\text{cell volume}} \nabla^2 V_2 \, d^3 r = 0 = \iint_{\text{cell surface}} \frac{\partial V_2}{\partial n} dS.$$

Thus,

$$\iint_{\text{cell surface}} \left(V_1 + \frac{\partial V_1}{\partial t}\right) dS = 0. \qquad (4.3.251)$$

We can effectively split V_1 into a steady state part which is a characteristic function G_1 for the domain and a transient part. Let

$$V_1(\mathbf{r}, t) = G_1(\mathbf{r}) + f_1(\mathbf{r}, t), \qquad (4.3.252)$$

where

$$\nabla^2 G_1 = -4\pi\delta(\mathbf{r} - \mathbf{R})$$

$$-\frac{\partial G_1}{\partial n} = \frac{4\pi}{A} \quad \text{and} \quad \iint_{\text{cell surface}} G_1 dS = 0. \qquad (4.3.253)$$

The latter condition in (4.3.253) serves to define the arbitrary constant that would exist for G_1 otherwise. Correspondingly, we have the problem for f_1

$$\nabla^2 f_1 = 0$$

$$\frac{\partial f_1}{\partial n} = 0 \qquad (4.3.254)$$

$$\iint_{\text{cell surface}} \left(f_1 + \frac{\partial f_1}{\partial t}\right) dS = 0.$$

Again, we see that, because of the equations and boundary conditions, $f_1 = f_1(t)$ and it follows that

$$f_1 = a_1 e^{-t}. \qquad (4.3.255)$$

Now the representation for V_1 is

$$V_1 = G_1(\mathbf{r}, \mathbf{R}) + a_1 e^{-t} \tag{4.3.256}$$

and a new difficulty comes to light. The initial condition $V_1 = 0$ can not be satisfied so that this particular limit process expansion ($\varepsilon \to 0$, t fixed) is not initially valid.

In order to construct an initially valid expansion, we must take into account that there is another important short time scale in the problem, and that $\partial V/\partial t$ in the boundary condition (4.3.239) can be large. However, since the time is short, the potential has not yet had time to reach a large value. A consistent expansion which keeps the time derivative term in the boundary condition is:

$$V(\mathbf{r}, t; \varepsilon) = v_1(\mathbf{r}, t^*) + \varepsilon v_2(\mathbf{r}, t^*) + \cdots, \tag{4.3.257}$$

where $t^* = t/\varepsilon$. The limit process associated with this is, of course, $\varepsilon \to 0$, $\mathbf{r}$, t^* fixed. The following sequence of problems results

$$\begin{aligned} &\nabla^2 v_1 = -4\pi\delta(\mathbf{r} - \mathbf{R})H(t^*) \\ &-\frac{\partial v_1}{\partial n} = \frac{\partial v_1}{\partial t^*} \quad \text{on the cell surface} \\ &v_1 = 0 \quad \text{at } t = 0 \text{ on the cell surface} \end{aligned} \tag{4.3.258}$$

$$\begin{aligned} &\nabla^2 v_2 = 0 \\ &-\frac{\partial v_2}{\partial n} = \frac{\partial v_2}{\partial t^*} + v_1 \quad \text{on the cell surface} \\ &v_2 = 0 \quad \text{at } t = 0 \text{ on the cell surface.} \end{aligned} \tag{4.3.259}$$

Some indication of the general form the solution must have can be obtained by integration over the cell volume. From equation (4.3.258), we have for $t^* > 0$

$$\begin{aligned} \iiint_{\text{cell volume}} \nabla^2 v_1 d^3 r = -4\pi &= \iint_{\text{cell surface}} \frac{\partial v_1}{\partial n} dS \\ &= -\frac{\partial}{\partial t^*} \iint_{\text{cell surface}} v_1 dS. \end{aligned} \tag{4.3.260}$$

Thus, using the initial condition, we find

$$\iint_{\text{cell surface}} v_1 dS = 4\pi t^*. \tag{4.3.261}$$

Part of v_1 must increase linearly with t^*. The following decomposition of v_1 is suggested:

$$v_1(\mathbf{r}, t^*) = u_1(\mathbf{r}, t^*) + h(\mathbf{r})t^*, \tag{4.3.262}$$

where $u_1(\mathbf{r}, t^*)$ is a potential that does not grow with time. The boundary conditions of (4.3.258) become

$$\frac{\partial u_1}{\partial n} + t^* \frac{\partial h}{\partial n} = -\frac{\partial u_1}{\partial t^*} - h \quad \text{on cell surface.} \tag{4.3.263}$$

This implies that

$$\frac{\partial h}{\partial n} = 0 \quad \text{on cell surface} \tag{4.3.264}$$

since u_1 does not grow with time. Since $\nabla^2 h = 0$ inside the cell, it follows that

$$h(\mathbf{r}) = \text{const.} \tag{4.3.265}$$

This const. can be evaluated from the integral condition (4.3.261)

$$h(\mathbf{r}) = \frac{4\pi}{A}, \qquad A = \text{cell surface area.} \tag{4.3.266}$$

Thus we have

$$v_1(\mathbf{r}, t^*) = u_1(\mathbf{r}, t^*) + \frac{4\pi}{A} t^*. \tag{4.3.267}$$

Now, u_1 satisfies the problem

$$\begin{aligned} \nabla^2 u_1 &= -4\pi\delta(\mathbf{r} - \mathbf{R}) H(t^*) \\ \frac{\partial u_1}{\partial n} &= -\frac{\partial u_1}{\partial t^*} - \frac{4\pi}{A} \quad \text{on the cell surface} \\ u_1 &= 0 \quad \text{at } t^* = 0 \text{ on the cell surface.} \end{aligned} \tag{4.3.268}$$

It is clear that as u_1 approaches a steady state, it will tend to the characteristic function for the cell G_1, defined by (4.3.253). To get some idea how this approach might take place let us assume that the geometry is such the characteristic function can be represented by a separation of variables type of expansion:

$$G_1(\mathbf{r}, \mathbf{R}) = \sum_k c_k f_k(\rho_1) \psi_k(\rho_2, \rho_3). \tag{4.3.269}$$

Here $\rho_1 = \rho_c = \text{const.}$ is the cell surface and ρ_1 is a coordinate normal to the surface; ρ_2, ρ_3 are coordinates in the surface. $f_k(\rho_1)$ satisfies an equation of the form

$$\frac{d}{d\rho_1} \frac{K_2^2}{K_1} \frac{df_k}{d\rho_1} - \lambda_k^2 K_1 f_k = 0, \qquad K_{1,2}(\rho_1) > 0$$
$$\lambda_k^2 = \text{separation constant} \tag{4.3.270}$$

Typically, $\rho_1 = 0$ is a singular point inside the cell and depending on the type of expansion a delta function may appear on the right-hand side of

(4.3.270). In any case, the energy integral

$$\int_0^{\rho_c} f_k\left\{\frac{d}{d\rho_1}\left(\frac{K_2^2}{K_1}\frac{df_k}{d\rho_1}\right) - \lambda_k^2 K_1 f_k\right\}d\rho_1 = 0$$

implies that $f_k(df_k/d\rho_1) > 0$ at the cell surface, $((K_2^2/K_1)f_k(df_k/d\rho_1) \to 0$, $\rho_1 \to 0)$, or

$$\frac{df_k}{d\rho_1}(\rho_c) = \mu_k^2 f_k(\rho_c) \quad \text{on cell surface.} \tag{4.3.271}$$

Now we can try to expand u_1 in terms of the same functions

$$u_1(\mathbf{r}, t) = G_1(\mathbf{r}, \mathbf{R}) + \sum_k a_k(t^*) f_k(\rho_1)\psi_k(\rho_2, \rho_3), \tag{4.3.272}$$

where $a_k(0) = -c_k$ so that the singularity at $\mathbf{r} = \mathbf{R}$ is removed just at $t^* = 0$. But for all $t^* > 0$, it remains. The boundary condition of (4.3.268) then becomes for $t^* > 0$

$$\sum_k a_k(t^*)\frac{df_k}{d\rho_1}(\rho_c)\psi_k(\rho_2, \rho_3) = -\sum_k \frac{da_k}{dt^*} f_k(\rho_c)\psi_k(\rho_2, \rho_3) \tag{4.3.273}$$

or using (4.3.271)

$$\frac{da_k}{dt^*} + \mu_k^2 a_k = 0, \qquad \mu_k^2 = \frac{f_k'}{f_k} \quad \text{at cell surface} \tag{4.3.274}$$

or

$$a_k(t^*) = -c_k e^{-\mu_k^2 t^*}. \tag{4.3.275}$$

The calculations above can be regarded as symbolic, but are verified in detail for the explicit case of a sphere. In summary, for $t^* > 0$,

$$v_1 = G_1(\mathbf{r}, \mathbf{R}) - \sum_k c_k e^{-\mu_k^2 t^*} f_k(\rho_1)\psi_k(\rho_2, \rho_3) + \frac{4\pi}{A} t^*. \tag{4.3.276}$$

In a similar way the form of the short-time correction potential v_2 can also be found. In the calculation of v_1 only the term corresponding to membrane capacitance remains in the boundary condition. Now, for v_2, the effect of membrane resistance appears. Since (4.3.259) shows that

$$-\iint_{\text{cell surface}} \frac{\partial v_2}{\partial n} dS = 0 = \frac{\partial}{\partial t^*}\iint_{\text{cell surface}} v_2 dS - \iint_{\text{cell surface}} u_1 dS + 4\pi t^*. \tag{4.3.277}$$

A suitable decomposition for v_2 is

$$v_2(\mathbf{r}, t^*) = h_2(\mathbf{r})\frac{t^{*2}}{2} + u_2(\mathbf{r}, t^*) \tag{4.3.278}$$

where u_2 does not grow with t^*.

Then the boundary condition (4.3.259) is

$$-\frac{\partial h_2}{\partial n}\frac{t^{*2}}{2} - \frac{\partial u_2}{\partial n} = h_2(\mathbf{r})t^* + u_1 + \frac{4\pi}{A}t^* + \frac{\partial u_2}{\partial t^*}.$$

Again $\partial h_2/\partial n = 0$, $h_2 = \text{const.} = -4\pi/A$. The resulting problem for u_2 is

$$\nabla^2 u_2 = 0$$

$$-\frac{\partial u_2}{\partial n} = \frac{\partial u_2}{\partial t^*} + \sum_k c_k(1 - e^{-\mu_k^2 t^*}) f_k(\rho_k)\psi_m(\rho_2, \rho_3) \quad (4.3.279)$$

$$u_2 = 0 \quad \text{at } t^* = 0.$$

Thus, a representation for u_2 is sought

$$u_2 = \sum_k b_k(t^*) f_k(\rho_1)\psi_k(\rho_2, \rho_3). \quad (4.3.280)$$

The boundary condition (4.3.279) then gives

$$\frac{db_k}{dt^*} + \mu_k^2 b_k = -c_k(1 - e^{-\mu_k^2 t^*}), \qquad b_k(0) = 0 \quad (4.3.281)$$

so that

$$b_k(t^*) = -\frac{c_k}{\mu_k^2}(1 - e^{-\mu_k^2 t^*}) + c_k t^* e^{-\mu_k^2 t^*}. \quad (4.3.282)$$

Now we can discuss the match of the long-time and short-time expansion under the intermediate limit where $t_\eta = t/\eta(\varepsilon)$ is fixed $\varepsilon \ll \eta \ll 1$. Thus

$$t^* = \frac{\eta t_\eta}{\varepsilon} \to \infty, \qquad t = \eta t_\eta \to 0.$$

We have, for the long-time expansion

$$V_0 = \frac{4\pi}{A}(1 - e^{-\eta t_\eta}) = \frac{4\pi}{A}\left(\eta t_\eta - \frac{\eta^2 t_\eta^2}{2!} + \cdots\right)$$

$$V_1 = G_1(\mathbf{r}, \mathbf{R}) + a_1 e^{-\eta t_\eta} = G_1(\mathbf{r}, \mathbf{R}) + a_1(1 - \eta t_\eta + \cdots)$$

and for the short-time expansion, neglecting transcendentally small terms,

$$v_1 = G_1(\mathbf{r}, \mathbf{R}) + \frac{4\pi}{A}\frac{\eta t_\eta}{\cdot\,\varepsilon}$$

$$-v_2 = \sum_k \frac{c_k}{\mu_k^2} f_k(\rho_1)\psi_k(\rho_2, \rho_3) + \frac{4\pi}{A}\frac{\eta^2 t_\eta^2}{2\varepsilon^2}.$$

Now comparing $(1/\varepsilon)V_0 + V_1$ with $v_1 + \varepsilon v_2$, we see that the term $(4\pi/A)\eta t_\eta$ matches, that G_1 matches and that we must choose

$$a_1 = 0. \quad (4.3.283)$$

The term $(4\pi/A)(\eta^2 t_\eta^2/2)$ also matches and neglected terms are $O(\eta^3, \varepsilon^2)$.

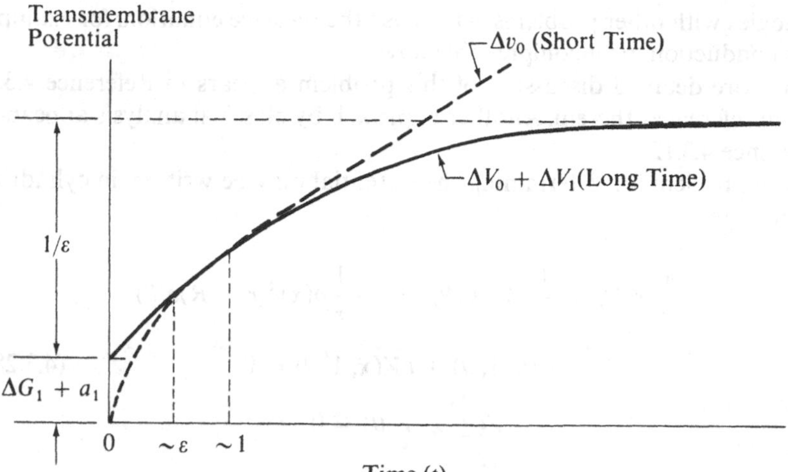

Figure 4.3.7 Matching of Short-Time and Long-Time Expansions

A figure indicating the voltage response at a typical point, as indicated by this theory, is given in Figure 4.3.7. For further discussion see also Reference 4.3.10.

4.3.6 Green's Function—Infinite Cylindrical Cell

A number of problems in biology require the solution of Laplace's equation with a boundary condition which describes the properties of the membrane surrounding a biological cell, separating the interior from the exterior, and buffering the internal environment from external disturbance. The membrane serves as an electrical buffer because its resistivity is much greater than the resistivity of the cell interior. The membrane boundary condition, therefore, contains a small parameter ε, the ratio of the internal resistance to the membrane resistance in appropriate units (cf. Section 4.3.5).

Here we consider a problem which arises when the electrical properties of very long cylindrical cells are investigated by the application of current to the interior of the cell from a microelectrode, a glass micropipette filled with conducting salt solution. The potential in the interior of the cell obeys Laplace's equation. The boundary condition is that the normal derivative of the potential at the inside surface of the membrane (proportional to the normal component of current) is proportional to the potential difference across the membrane. If the microelectrode is considered a point source of current, the solution to the problem is the Green's function for the electric potential in a cylinder with a membrane boundary condition.

The same method applies with other boundary conditions or source distributions in a part of the cell near the origin. There are also of course analogies with other problems, which use the Laplace equation, for example, heat conduction, or incompressible flow.

A more detailed discussion of this problem appears in Reference 4.3.11 and in references therein. Another approach by classical analysis appears in Reference 4.3.12.

The problem for determining the potential may be written, in cylindrical coordinates,

$$\frac{1}{r}(rV_r)_r + \frac{1}{r^2}V_{\theta\theta} + V_{xx} = -\frac{1}{r}\delta(x)\delta(r-R)\delta(\theta)$$
$$V_r(x, 1, \theta) + \varepsilon V(x, 1, \theta) = 0 \tag{4.3.284}$$
$$V(\pm\infty, r, \theta) = 0.$$

When ε is small, the boundary condition at $r = 1$ in (1) implies that the current flow will be predominantly in the axial direction, i.e., only a small fraction of the local current $O(\varepsilon)$, crosses the membrane in an axial distance of $O(1)$. We are tempted to try to find an expansion in the small parameter ε, in which the leading term is the potential for $\varepsilon = 0$. Denoting this term by $V_1(x, r, \theta)$, we see from (4.3.284) that V_1 satisfies

$$\frac{1}{r}(rV_{1_r})_r + \frac{1}{r^2}V_{1\theta\theta} + V_{1_{xx}} = -\frac{1}{r}\delta(x)\delta(r-R)\delta(\theta)$$
$$V_{1_r}(x, 1, \theta) = 0. \tag{4.3.285}$$

The boundary condition at $r = 1$ implies that no current crosses the membrane; all the current is confined to the interior of the cell. Consequently, V_1 must contain a part which is linearly decreasing with increasing $|x|$. This would lead to a potential of $V_1 \to -\infty$ as $|x| \to \infty$, making it impossible to satisfy the boundary condition at $|x| = \infty$, namely that $V = 0$ at $|x| = \infty$. To avoid this divergence, any expansion which contains V_1 can be valid only over a limited range of x, designated the near field, which contains the source point. At large distances from the source, we must look for another, far-field, expansion.

We expect that as $\varepsilon \to 0$, the region of validity of any near-field expansion of which V_1 is a part, becomes larger. If there is a linearly decaying potential over a large distance, and the potential approaches zero as $|x| \to \infty$, then the potential at $x = 0$ must be very large, i.e., $V(0, r, \theta) \to \infty$ as $\varepsilon \to 0$. Clearly, V_1 must be $O(1)$, and cannot be the leading term in the expansion. The leading term can be found by matching to the far-field solution, and therefore we first solve the far-field problem.

In the far field, a long distance from the source, current flow is predominantly in the axial direction. Since only a small fraction of the current within the cell, at any value of x, leaks out of the cylinder in an axial distance of $O(1)$, the variation in the x direction will be slow. We therefore, for convenience in ordering the far-field expansion, write the far-field potential in terms of a new slow variable $\tilde{x}$. Denoting the potential in the far field by W, we write the following expansion:

$$V = W(\tilde{x}, r, \theta; \varepsilon) = \zeta_0(\varepsilon) W_0(\tilde{x}, r, \theta) + \zeta_1(\varepsilon) W_1(\tilde{x}, r, \theta) + \cdots, \quad (4.3.286)$$

where the slow variable is defined by

$$\tilde{x} = \eta(\varepsilon) x, \qquad \eta \ll 1. \quad (4.3.287)$$

Thus,

$$\nabla^2 V = 0 = \zeta_0 \eta^2 W_{0_{\tilde{x}\tilde{x}}} + \zeta_1 \eta^2 W_{1_{\tilde{x}\tilde{x}}} + \cdots + \zeta_0 \nabla_t^2 W_0 + \zeta_1 \nabla_t^2 W_1 + \cdots,$$

where $\nabla_t^2 = (1/r)(\partial/\partial r)(r\partial/\partial r) + (1/r^2)(\partial^2/\partial\theta^2)$ is the transverse Laplacian, and on the boundary, $r = 1$, we have

$$W_r + \varepsilon W = 0 = \zeta_0 W_{0_r} + \zeta_1 W_{1_r} + \cdots + \varepsilon\zeta_0 W_0 + \varepsilon\zeta_1 W_1 + \cdots.$$

The corresponding approximating sequence of problems is

$$\begin{gathered} \nabla_t^2 W_0 = 0 \\ W_{0_r}(\tilde{x}, 1, \theta) = 0 \\ W_0(\pm\infty, r, \theta) = 0 \end{gathered} \quad (4.3.288)$$

$$\begin{gathered} \nabla_t^2 W_1 = -W_{0_{\tilde{x}\tilde{x}}} \\ W_{1_r}(\tilde{x}, 1, \theta) = -W_0(\tilde{x}, 1, \theta) \\ W_1(\pm\infty, r, \theta) = 0, \end{gathered} \quad (4.3.289)$$

where we have set

$$\varepsilon\zeta_0 = \zeta_1$$

to obtain the surface boundary condition in (4.3.289).

Writing an expansion for $\eta(\varepsilon)$ in the form

$$\eta(\varepsilon) = \eta_0(\varepsilon) + \eta_1(\varepsilon) + \eta_2(\varepsilon) + \cdots,$$

where the $\eta_i(\varepsilon)$ are an ordered sequence, we further set

$$\zeta_0 \eta_0^2 = \zeta_1$$

to obtain (4.3.289). Thus

$$\eta_0 = \sqrt{\varepsilon}.$$

We could take $\eta = \eta_0$ with $\eta_1 = \eta_2 = \cdots = 0$, and still obtain a sequence of problems of increasing order in ε. It will be seen below, however, that we would not be able to maintain uniform validity of the asymptotic expansion for W at large $\tilde{x}$. Assuming $\eta(\varepsilon)$ to have the more general form makes it possible to obtain a uniform expansion. Continuing this procedure with $\varepsilon\zeta_1 = \zeta_2$, $\eta_1 = \alpha_1 \varepsilon \eta_0$ etc., we find

$$\begin{aligned} \nabla_t^2 W_2 &= -(W_1 + 2\alpha_1 W_0)_{\tilde{x}\tilde{x}} \\ W_{2,r}(\tilde{x}, 1, \theta) &= -W_1(\tilde{x}, 1, \theta) \\ W_2(\pm\infty, r, \theta) &= 0 \end{aligned} \tag{4.3.290}$$

and

$$\begin{aligned} \nabla_t^2 W_3 &= -[W_2 + 2\alpha_1 W_1 + (\alpha_1^2 + 2\alpha_2) W_0]_{\tilde{x}\tilde{x}} \\ W_{3,r}(\tilde{x}, 1, \theta) &= -W_2(\tilde{x}, 1, \theta) \\ W_3(\pm\infty, r, \theta) &= 0, \end{aligned} \tag{4.3.291}$$

where the α_i are unknown constants. Thus, the far field expansion of the potential is taken in the form

$$W(\tilde{x}, r, \theta; \varepsilon) = \zeta_0(\varepsilon)[W_0(\tilde{x}, r, \theta) + \varepsilon W_1(\tilde{x}, r, \theta) + \varepsilon^2 W_2(\tilde{x}, r, \theta) + \cdots], \tag{4.3.292}$$

where the axial coordinate variable is

$$\tilde{x} = \sqrt{\varepsilon}(1 + \alpha_1 \varepsilon + \alpha_2 \varepsilon^2 + \cdots)x. \tag{4.3.293}$$

So far $\zeta_0(\varepsilon)$, the order of the leading term in the W expansion, is unknown. It will be determined by matching to the near field. The constants $\alpha_1, \alpha_2, \ldots$ in the expansion of $\tilde{x}$, which couple different orders of W in the sequence of problems, will be determined by requiring uniform validity of the W expansion for all values of $\tilde{x}$.

We now solve the sequence of problems. The solution to the first problem is independent of r and θ. Thus we have

$$W_0(\tilde{x}, r, \theta) = F(\tilde{x}), \tag{4.3.294}$$

where $F(\tilde{x})$ is an as yet arbitrary function of $\tilde{x}$. We must go to the second problem, to determine its functional form.

From (4.3.289) and (4.3.294) we obtain

$$\frac{1}{r}(rW_{1_r})_r + \frac{1}{r^2}W_{1_{\theta\theta}} = -F''(\tilde{x})$$

$$W_{1_r}(\tilde{x}, 1, \theta) = -F(\tilde{x}) \tag{4.3.295}$$

$$W_1(\pm\infty, r, \theta) = 0.$$

Since the inhomogeneous term in the equation, and the boundary condition at $r = 1$, are both independent of θ, clearly, W_1 is independent of θ. Examining (4.3.290, 1), the same reasoning then implies that $W_2, W_3, \ldots$ are all independent of θ.

Integrating (4.3.295), we obtain, for the solution which is bounded at $r = 0$

$$W_1(\tilde{x}, r) = -\frac{r^2}{4}F''(\tilde{x}) + G(\tilde{x}), \tag{4.3.296}$$

where $G(\tilde{x})$ is an arbitrary function of $\tilde{x}$ which cannot be determined until we go to the next problem for W_2.

Substituting the result (4.3.296) in the $r = 1$ boundary condition yields

$$F'' - 2F = 0 \tag{4.3.297}$$

$$W_0(\tilde{x}) = F(\tilde{x}) = Ae^{-\sqrt{2}|\tilde{x}|}, \tag{4.3.298}$$

where A is a constant to be determined by matching to the near field.

Continuing in the same way, we find

$$\frac{1}{r}(rW_{2_r})_r = (r^2 - 4\alpha_1)F - G'', \tag{4.3.299}$$

$$W_{2_r}(\tilde{x}, 1) = F/2 - G$$

$$W_2(\pm\infty, r) = 0$$

and then

$$W_2(\tilde{x}, r) = \frac{r^4}{16}F - r^2(\alpha_1 F + \tfrac{1}{4}G'') + H(\tilde{x}). \tag{4.3.300}$$

Substituting the expression for W_2 and F in the $r = 1$ boundary condition yields

$$G'' - 2G = -4A(\alpha_1 + \tfrac{1}{8})e^{-\sqrt{2}|\tilde{x}|}. \tag{4.3.301}$$

The right-hand side above is a homogeneous solution of the equation. Therefore the particular solution contains a term proportional to $\tilde{x}$ times

$\exp(-\sqrt{2}|\tilde{x}|)$. If such a term appears in G, then for sufficiently large $|\tilde{x}|$ the expansion will not be valid uniformly in $\tilde{x}$. To avoid this we require the right-hand side of (4.3.301) to vanish, which occurs if

$$\alpha_1 = -\frac{1}{8}. \tag{4.3.302}$$

It is now clear why we could not assume the simple relation $\tilde{x} = \sqrt{\varepsilon}x$ but required the more general form. The freedom to choose $\alpha_1, \alpha_2, \ldots$ allows us to force all of the $\tilde{x}$ dependence of W into $\exp(-\sqrt{2}\tilde{x})$, eliminating non-uniformities in the expansion.

The solution to (4.3.301) is thus

$$G(\tilde{x}) = Be^{-\sqrt{2}|\tilde{x}|}, \tag{4.3.303}$$

where

$$\tilde{x} = \sqrt{\varepsilon}\left(1 - \frac{\varepsilon}{8} + \cdots\right)x. \tag{4.3.304}$$

B will be determined by matching to the near field.

We obtain for the second term in the far-field expansion

$$W_1(\tilde{x}, r) = \left(-\frac{1}{2}Ar^2 + B\right)e^{-\sqrt{2}|\tilde{x}|}. \tag{4.3.305}$$

We now continue the same procedure and calculate $\alpha_2 = 5\varepsilon^2/384$.

The order of the expansion $\zeta_0(\varepsilon)$ and the three constants (A, B, C) are to be found by matching to the near field expansion with the help of a suitable intermediate limit ($\varepsilon \to 0$, x_σ fixed) where

$$x_\sigma = x\sigma(\varepsilon), \qquad \sqrt{\varepsilon} \ll \sigma(\varepsilon) \ll 1. \tag{4.3.306}$$

Under this limit the near field coordinate

$$x = x_\sigma/\sigma \to \infty$$

and the far field coordinate

$$\tilde{x} = \frac{\sqrt{\varepsilon}}{\sigma}x_\sigma\left(1 - \frac{\varepsilon}{8} + \frac{5}{384}\varepsilon^2 - \cdots\right) \to 0.$$

Because of the simple way $\tilde{x}$ enters the expansion it is appropriate to express the far field in terms of x as $\tilde{x} \to 0$ and to compare directly with the near field as $x \to \infty$.

The far field has the expansion

$$W\left[x\sqrt{\varepsilon}\left(1-\frac{\varepsilon}{8}+\frac{5\varepsilon^2}{384}-\cdots\right), r; \varepsilon\right]$$
$$=\zeta_0(\varepsilon)\left\{A-\varepsilon^{1/2}A\sqrt{2}|x|+\varepsilon\left[A\left(x^2-\frac{r^2}{2}\right)+B\right]\right.$$
$$+\varepsilon^{3/2}\sqrt{2}|x|\left[A\left(\frac{1}{8}-\frac{x^2}{3}+\frac{r^2}{2}\right)-B\right]$$
$$+\varepsilon^2\left[A\left(-\frac{x^2}{4}+\frac{x^4}{6}-\frac{r^2x^2}{2}+\frac{r^2}{8}+\frac{r^4}{16}\right)+B\left(x^2-\frac{r^2}{2}\right)+C\right]$$
$$+\varepsilon^{5/2}\sqrt{2}|x|\left[A\left(-\frac{5}{384}+\frac{x^2}{8}+\frac{x^2r^2}{6}-\frac{3r^2}{16}-\frac{r^4}{16}-\frac{x^4}{30}\right)\right.$$
$$\left.\left.+B\left(\frac{1}{8}-\frac{x^2}{3}+\frac{r^2}{2}\right)-C\right]+O(\varepsilon^3)\right\}. \tag{4.3.307}$$

Next we consider the sequence of near field problems implied by the far field behavior.

In the vicinity of the point source the potential is a rather complex function of position, and there is no simple mathematical representation in terms of elementary functions, as there is in the far field. The potential has a singularity at the source point; the current diverges from this point, half going toward $x = +\infty$, and half toward $x = -\infty$. Close to the source, the lines of current flow are diverging outward, equally in all directions. Those lines which are directed toward the membrane must curve to avoid the membrane as, again, only a small fraction of the local current leaves the cylinder. As the current flows down the cylinder, the lines become predominantly in the axial direction, and the potential joins smoothly onto the far-field potential.

In terms of the asymptotic expansions representing the near and far fields, this behavior requires that the near-field expansion increase in powers of $\sqrt{\varepsilon}$ so it can join the expansion (4.3.307) of the far field. Furthermore, in accordance with the earlier arguments which concluded that the $O(1)$ term in the near field has a linear dependence on $|x|$ as $|x| \to \infty$. we see that the second term in (4.3.307) must be $O(1)$ in order to match the near field. Consequently,

$$\zeta_0(\varepsilon)=\frac{1}{\sqrt{\varepsilon}}. \tag{4.3.308}$$

The near-field expansion must be of the form

$$V(x, r, \theta; \varepsilon) = \varepsilon^{-1/2}V_0(x, r, \theta) + V_1(x, r, \theta) + \varepsilon^{1/2}V_2(x, r, \theta) + \varepsilon V_3(x, r, \theta) + \cdots. \tag{4.3.309}$$

Thus, we obtain the following sequence of near-field problems:

$$\nabla^2 V_0 = 0$$

$$V_{0_r}(x, 1, \theta) = 0; \qquad V_0(x, r, \theta) \to A \quad \text{as } |x| \to \infty, \tag{4.3.310}$$

$$\nabla^2 V_1 = -\frac{1}{r}\delta(x)\delta(r-R)\delta(\theta)$$

$$V_{1_r}(x, 1, \theta) = 0$$

$$V_1(x, r, \theta) \to -A\sqrt{2}|x| \quad \text{as } |x| \to \infty, \tag{4.3.311}$$

$$\nabla^2 V_2 = 0$$

$$V_{2_r}(x, 1, \theta) = -V_0(x, 1, \theta)$$

$$V_2(x, r, \theta) \to A\left(x^2 - \frac{r^2}{2}\right) + B, \quad \text{as } |x| \to \infty, \tag{4.3.312}$$

$$\nabla^2 V_3 = 0$$

$$V_{3_r}(x, 1, \theta) = -V_1(x, 1, \theta)$$

$$V_3(x, r, \theta) \to \sqrt{2}|x|\left[A\left(\frac{1}{8} - \frac{x^2}{3} + \frac{r^2}{2}\right) - B\right] \quad \text{as } |x| \to \infty, \tag{4.3.313}$$

$$\nabla^2 V_4 = 0$$

$$V_{4_r}(x, 1, \theta) = -V_2(x, 1, \theta);$$

$$V_4(x, r, \theta) \to A\left(-\frac{x^2}{4} + \frac{x^4}{6} - \frac{r^2x^2}{2} + \frac{r^2}{8} + \frac{r^4}{16}\right)$$

$$+ B\left(x^2 - \frac{r^2}{2}\right) + C \quad \text{as } |x| \to \infty, \tag{4.3.314}$$

$$\nabla^2 V_5 = 0$$

$$V_{5_r}(x, 1, \theta) = -V_3(x, 1, \theta)$$

$$V_{5_r}(x, r, \theta) \to \sqrt{2}|x|\left[A\left(-\frac{5}{384} + \frac{x^2}{8} + \frac{x^2r^2}{6} - \frac{r^4}{16} - \frac{x^4}{30}\right)\right.$$

$$\left. + B\left(\frac{1}{8} - \frac{x^2}{3} + \frac{r^2}{2}\right) - C\right] \quad \text{as } |x| \to \infty. \tag{4.3.315}$$

The delta function source appears in the V_1 problem, consistent with the linear decrease with x as $|x| \to \infty$. All other orders of the potential are source free.

Each even(odd) order problem (except for the first two) is coupled to the preceding even(odd) order problem via the boundary condition on the $x = 1$ surface. The physical interpretation of this coupling is that the current crossing the membrane in the nth problem is proportional to the membrane potential in the $(n-2)$nd problem. The even order problems are coupled to the odd order problems by their asymptotic behavior as $|x| \to \infty$, i.e., the constants $A, B, C, \ldots$, appear in both even and odd order problems.

It should be noted that the $V_1, V_3, \ldots$ terms alone are sufficient to satisfy (4.3.284) at small x. It is only from considerations of the large $-|x|$ behavior required of the far-field potential, that we conclude that $V_0, V_2, \ldots$ are even necessary. These terms are thus known as "switchback" terms.

By direct substitution of the $|x| \to \infty$ asymptotic forms of V_0, V_2, and V_4 in the respective equations and boundary conditions (4.3.310, 12, 14), it is seen that the $|x| \to \infty$ forms are the solutions valid for all x. This part of the near field is thus completely contained in the far field

$$V_0 = A \tag{4.3.316}$$

$$V_2 = A\left(x^2 - \frac{r^2}{2}\right) + B \tag{4.3.317}$$

$$V_4 = A\left(-\frac{x^2}{4} + \frac{x^4}{6} - \frac{r^2x^2}{2} + \frac{r^2}{8} + \frac{r^4}{16}\right) + B\left(x^2 - \frac{r^2}{2}\right) + C. \tag{4.3.318}$$

Now we evaluate the constant A. Integrating (4.3.311) over the large volume of the cylinder between $-x$ and x, $|x| \to \infty$, and using the divergence theorem, we obtain

$$\begin{aligned} -1 &= \lim_{|x|\to\infty} \int_0^{2\pi} d\theta \int_0^1 rdr \int_{-x}^{x} dx \nabla^2 V_1 \\ &= \lim_{|x|\to\infty} \int_0^{2\pi} d\theta \int_0^1 rdr[V_{1_x}(x, r, \theta) - V_{1_x}(-x, r, \theta)] \\ &= -2\pi A\sqrt{2}, \end{aligned}$$

where in accordance with the $r = 1$ boundary condition, the integral over the surface of the cylinder is zero, leaving only the integral over the discs at $\pm x$. The last equality follows from substitution of the asymptotic behavior of V_1, as $|x| \to \infty$.

The problem (4.3.311) for V_1 is now definite. In order to solve the problem, it is convenient to decompose the near-field potential V_1 into two terms:

$$V_1(x, r, \theta) = \Phi_1(x, r, \theta) - \frac{|x|}{2\pi}. \tag{4.3.319}$$

We obtain the problem for Φ_1,

$$\begin{aligned} \nabla^2\Phi_1 &= -\delta(x)\left[\frac{1}{r}\delta(r-R)\delta(\theta) - \frac{1}{\pi}\right] \\ \Phi_{1_r}&(x, 1, \theta) = 0 \\ \Phi_1&(\pm\infty, r, \theta) = 0. \end{aligned} \tag{4.3.320}$$

The source term is a unit point source at $(0, R, 0)$ plus the uniform disc sink in the $x = 0$ plane. The net current source for Φ_1 is zero, i.e., all the current which enters the cylinder at the point $(0, R, 0)$ is removed uniformly in the cross section $(0, r, \theta)$. Unlike the problem for V_1; which contains unit current flowing outward as $|x| \to \infty$, the problem for Φ_1 contains no current flow as $|x| \to \infty$.

The boundary value problem may be solved by Fourier transformation in the θ- and x-coordinates. Defining the double Fourier transform of Φ_1 by

$$\psi_1^{(n)}(k, r) = \int_0^{2\pi} d\theta e^{-in\theta} \int_{-\infty}^{\infty} dx \cos(kx)\Phi_1(x, r, \theta)$$
$$\Phi_1(x, r, \theta) = \frac{1}{2\pi^2} \int_0^{\infty} dk \cos(kx) \sum_{n=-\infty}^{\infty} e^{in\theta}\psi_1^{(n)}(k, r) \tag{4.3.321}$$

noting that Φ_1 is even in x and θ, we see that the problem (4.3.320) becomes, in Fourier transform space,

$$\frac{1}{r}(r\psi_{1_r}^{(n)})_r - \left(k^2 + \frac{n^2}{r^2}\right)\psi_1^{(n)} = -\frac{1}{r}\delta(r - R) + 2\delta_{on}$$
$$\psi_{1_r}^{(n)}(k, 1, \theta) = 0,$$

where

$$\delta_{0n} = \frac{i}{2n\pi}\left[(-1)^n - 1\right].$$

The solution is

$$\psi_1^{(n)}(k, r) = -\frac{2\delta_{on}}{k^2} - I_n(kr)\frac{K_n'(k)}{I_n'(k)} I_n(kR) + \begin{cases} K_n(kR)I_n(kr), & 0 \le r \le R, \\ K_n(kr)I_n(kR), & R \le r \le 1. \end{cases} \tag{4.3.322}$$

Taking the inverse transform, we obtain

$$V_1(x, r, \theta) = -\frac{|x|}{2\pi} + \frac{1}{4\pi}(x^2 + r^2 + R^2 - 2rR\cos\theta)^{-1/2})$$
$$-\frac{1}{2\pi^2}\sum_{n=-\infty}^{\infty} e^{in\theta}\int_0^{\infty} dk \cos(kx)\left[\frac{K_n'(k)}{I_n'(k)} I_n(kR)I_n(kr) + 2\frac{\delta_{on}}{k^2}\right]. \tag{4.3.323}$$

The integral over k can be replaced by an equivalent sum by considering the integral in (4.3.323) as a portion of a contour integral, so that

$$V_1(x, r, \theta) = -\frac{|x|}{2\pi} - \frac{1}{2\pi}\sum_{n=-\infty}^{\infty} e^{in\theta}\sum_{s=1}^{\infty} e^{-\lambda_{ns}|x|} \times \frac{J_n(\lambda_{ns}R)J_n(\lambda_{ns}r)}{\lambda_{ns}\left(\frac{n^2}{\lambda_{ns}^2} - 1\right)J_n^2(\lambda_{ns})}, \tag{4.3.324}$$

where λ_{ns} is the sth zero of $J_n'(\lambda)$ excluding the one at $\lambda = 0$. We can see that as $|x| \to \infty$, $V_1 \to -|x|/2\pi$ plus terms which are exponentially small in Equation (4.3.324).

We now turn to the V_3 problem and evaluate the constant B. Integrating the Laplacian in (4.3.313) over the volume of a large cylinder extending from $-x$ to x, and using the divergence theorem, we have

$$0 = \lim_{|x|\to\infty} \int_{-x}^{x} dx \int_0^1 r dr \int_0^{2\pi} d\theta \nabla^2 V_3 \tag{4.3.325}$$

$$= \lim_{|x|\to\infty} \left[\int_{-x}^{x} dx \int_0^{2\pi} d\theta V_{3_r}(x, 1, \theta) + 2 \int_0^1 r dr \int_0^{2\pi} d\theta V_{3_x}(x, r, \theta) \right].$$

Using the boundary condition and the transform for V_1, we see that the first integral becomes

$$\begin{aligned} &- \lim_{|x|\to\infty} \int_{-x}^{x} dx \int_0^{2\pi} d\theta V_1(x, 1, \theta) \\ &= x^2 - \frac{1}{\pi} \int_{-\infty}^{\infty} dx \int_0^{\infty} dk \cos(kx) \psi_1^{(0)}(k, 1) \\ &= x^2 - \psi_1^{(0)}(0, 1). \end{aligned} \tag{4.3.326}$$

From (4.3.322), we obtain, using the Wronskian of I_n and K_n and the power series expansion of $I_n(k)$,

$$\psi_1^{(0)}(0, 1) = \lim_{k\to 0} \left(-\frac{2}{k^2} + \frac{I_0(kR)}{kI_1(k)} \right) = \frac{1}{2}\left(R^2 - \frac{1}{2} \right). \tag{4.3.327}$$

Using the asymptotic form for large $|x|$ for V_3 from (4.3.313), we see that the second integral in (4.3.325) becomes

$$\int_0^1 r dr \int_0^{2\pi} d\theta 2\sqrt{2} \left[A\left(\frac{1}{8} - x^2 + \frac{r^2}{2} \right) - B \right] = 2\pi\sqrt{2} \left[A\left(\frac{3}{8} - x^2 \right) - B \right]. \tag{4.3.328}$$

Thus, from (4.3.325)

$$B = \frac{\sqrt{2}}{4\pi} \left(\frac{5}{8} - \frac{R^2}{2} \right). \tag{4.3.329}$$

As a consequence of (4.3.329), W_1 and V_2 depend on R, the distance from the source to the axis of the cylinder, whereas lower order terms do not.

Having evaluated A and B, we have now obtained the near field and far field up to terms of $O(\varepsilon^{1/2})$, i.e., we have obtained V_0, V_1, V_2, W_0, and W_1. These terms represent that part of the potential which is numerically significant in a physiological experiment: all higher order terms are too small to detect anywhere in a cylindrical cell. In Reference 4.3.11, the calculations are carried out further to find the constant C.

The leading terms in the far-field expansion and in the near-field expansion are each of order $\varepsilon^{-1/2}$. In the near field, the leading term is a constant. Thus, near the point source, the interior of the cylinder is raised to a large, constant potential, relative to the zero potential at infinity. The physical basis for the large potential is that the membrane permits only a small fraction of the current to leave the cylinder per unit length. Consequently, most of the current flows a long distance before getting out, and a large potential drop is required to force this current down the cylinder. The existence of this large constant potential, and its magnitude of $O(\varepsilon^{-1/2})$, could only be deduced from considerations of the far field.

The leading term in the far field decays as $\exp(-\sqrt{2\varepsilon}|x|)$. Consequently, to lowest order, $1/e$ of the current leaves the cylinder in a distance of $1/\sqrt{2\varepsilon}$. The corresponding potential required to drive a current this distance is of $O(\varepsilon^{-1/2})$, which is the physical basis for the order of the large potential in the near field. The precise numerical values of the leading terms for V and W was determined by requiring in the limit $|x| \to \infty$, $\tilde{x} \to 0$, that the two terms be identical to the lowest order in ε. In the far far. field, i.e., $\tilde{x} = x\sqrt{\varepsilon}(1 - \varepsilon/8 + \cdots) \to \infty$, the potential is seen to approach zero exponentially.

The leading term in the far-field expansion is independent of r and θ. Thus, to the lowest order, the far-field current is distributed uniformly over the circular cross section of the cylinder. The leading term is the known result of one-dimensional cable theory Reference 4.3.13, Equation 14. The high order terms are all independent of the polar angle. They do, however, depend on the radial coordinate r. The dependence is in the form of a polynomial in r^2, the degree of the polynomial increasing by one in each successive term. We also see that the higher order terms also depend on R, the radial distance between the source and the axis of the cylinder. The potential is seen to be symmetric with respect to an interchange of r and R. This must be so because the potential is the Green's function (with source at $x = 0$, $\theta = 0$) for the cylindrical problem.

The solution obtained here is close to the solution derived by Barcilon, Cole and Eisenberg (Reference 4.3.14), using multiple scaling.

The result of the multiple scale analysis differs from our result only because Reference 4.3.14 contains a sign error and a secular term $V^{(3)}$ which has not been removed. If these errors are corrected, and the infinite sum over Bessel functions is written in closed form, the multiple scale result, the expansion of the exact solution, and the present results are identical.

Problems

1. Consider the self induced motion of a radially deforming slender-body of revolution in an incompressible fluid at rest. Assume that the body always remains neutrally buoyant by requiring the total displaced volume to be a constant. Thus, the velocity is always horizontal. Assume also, as in Section 4.3.1 that the body sheds no vortices and that the flow is everywhere irrotational.

If we fix our coordinate system in the body we have a problem similar to the one discussed in Section 4.3.1 except now the velocity at upstream-infinity (in the coordinate system fixed to the moving body) is an unknown $q(t, \varepsilon)$. It is easy to show that in this coordinate system the Bernoulli equation [cf. Equation (4.3.13)] for the case $\delta = 1$ is

$$C_p = -\phi_t + x\frac{dq}{dt} + \frac{1}{2}[q^2 - \phi_x^2 - \phi_r^2].$$

Consider the axial equation of motion in dimensionless form

$$m\frac{dq}{dt} = C_D(t),$$

where m is a dimensionless mass

$$m = 2\int_0^1 F^2(x, t)dx$$

and $C_D(t)$ is defined by Equation (4.3.14) with C_p as given above.

(a) By paralleling the discussion in Section 4.3.1 show that the inner expansion for the velocity potential is of the form

$$\phi = q(t, \varepsilon)x + \varepsilon^2 \log \varepsilon A_1(x, t) + \varepsilon^2[A_1(x, t)\log r^* + B_1(x, t)] + O(\varepsilon^4 \log \varepsilon),$$

where

$$A_1 = \frac{1}{2}G_t(x, t)$$

and

$$B_1 = -\frac{1}{4}\int_0^1 \frac{\partial^2 G}{\partial\xi\, \partial t}(\xi, t)\text{sgn}(x - \xi)\log 2|x - \xi|d\xi.$$

So far, the magnitude of q is assumed small ($q \ll 1$) but unknown.

(b) Using the expression for ϕ to calculate C_p gives the same result as Equation (4.3.44c) independently of q.

(c) Show also that the axial force coefficient now becomes

$$C_D(t) = \varepsilon^2 \log \varepsilon C_{D_0}(t) + \varepsilon^2 C_{D_1}(t) + \cdots,$$

where

$$C_{D_0}(t) = -\int_0^1 \frac{\partial}{\partial t}(G_t G_x)dx$$

$$C_{D_1}(t) = -\int_0^1 \frac{\partial}{\partial t}\left[\frac{1}{2}G_t G_x \log G + 2T_1 G_x\right]dx.$$

Thus, applying the axial equation of motion fixes the order of magnitude of q; in fact, we can expand

$$q(t, \varepsilon) = \varepsilon^2 \log \varepsilon q_0(t) + \varepsilon^2 q_1(t) + \cdots$$

and we have

$$m \frac{dq_0}{dt} = C_{D_0}(t)$$

$$m \frac{dq_1}{dt} = C_{D_1}(t).$$

(d) Assume again that G is periodic in t with period λ. It then follows that $C_{D_0}(t)$ is also periodic in t with the same period and *a zero average value*, i.e., the leading term in the axial equation of motion would be of the form

$$m \frac{dq_0}{dt} = \sum_{n=1}^{\infty} \alpha_n \sin\left(\frac{2n\pi t}{\lambda} + \beta_n\right),$$

where α_n and β_n are known constants once the body deformation is specified. Integrating the axial equation of motion with $q_0(0) = 0$ gives

$$q_0(t) = \frac{1}{m} \sum_{n=1}^{\infty} \left[-\frac{\lambda \alpha_n}{2n\pi} \cos\left(\frac{2n\pi t}{\lambda} + \beta_n\right) + \lambda \alpha_n \cos \beta_n / 2n\pi \right].$$

Thus, the body will acquire the constant average velocity

$$\langle q \rangle = (\varepsilon^2 \log \varepsilon) \frac{\lambda}{2m\pi} \sum_{n=1}^{\infty} \frac{\alpha_n}{n} \cos \beta_n + \cdots .$$

For example, if we assume $G(x, t)$ to have the simple form $G(x, t) = \sin \pi x + a \sin 2\pi \times \sin \omega t$ with $|a| < \frac{1}{2}$, it is easy to see that $G(x, t) > 0$ for all t and that $\int_0^1 G(x, t)dx = 2/\pi$. Hence, the volume of the body remains constant. Show that in this case

$$q_0(t) = \frac{2\pi}{3} a\omega(1 - \cos \omega t).$$

Therefore, the average velocity of motion is

$$\langle q \rangle = \frac{2\pi}{3} a\omega\varepsilon^2 \log \varepsilon + \cdots .$$

If we keep in mind that in a perfect fluid any initial velocity imparted to a rigid body is preserved, we conclude that radial deformations of the surface without vortex shedding do not provide a satisfactory mechanism for propulsion. This is because viscous forces would tend to reduce the already small value of $\langle q \rangle$ even further.

Fortunately for aquatic animals, they derive their propulsion by undulatory motions of their spines as well as vortex shedding from fins (when applicable). The interested reader on this subject can refer to the monograph in Reference 4.3.1.

2. Consider steady heat conduction in a cylindrical rod, $0 \le X \le L$, $0 \le R \le a$, with the following boundary conditions of temperature prescribed on all surfaces

$$\begin{aligned} &\text{at } X = 0: && T(0, R) = T^* F(R/a) \\ &\text{at } X = L: && T(L, R) = T^* G(R/a), \\ &\text{at } R = a: && T(X, a) = T^* H(X). \end{aligned}$$

Construct asymptotic expansions of the solution for $T/T^* = \theta(x, r; \varepsilon)$, when $x = X/L$, $r = R/a$, and $\varepsilon = a/L \to 0$.

Construct suitable boundary-layer solutions for the ends, and show how they match to the expansion valid away from the ends.

Does the solution constructed here represent one-dimensional heat conduction?

3. Consider a plane sound wave of frequency ω, wavelength λ incident on a sphere of radius a. Construct matched inner and outer expansions for the case $a/\lambda \ll L$. Compute the first two terms in the inner expansion.

The acoustic velocity potential satisfies the wave equation

$$\frac{\partial^2 \phi}{\partial x^2} + \frac{\partial^2 \phi}{\partial y^2} + \frac{\partial^2 \phi}{\partial z^2} - \frac{1}{c^2}\frac{\partial^2 \phi}{\partial t^2} = 0, \qquad c = \text{sound speed}.$$

The incoming plane wave is represented by (in complex notation)

$$\phi = A \exp\left[i\omega\left(t - \frac{x}{c}\right)\right].$$

The boundary condition at the surface of a rigid sphere is $\partial\phi/\partial r = 0$.

References

4.3.1 M. J. Lighthill, *Mathematical Biofluiddynamics*, Society for Industrial and Applied Mathematics, Philadelphia, Penn. 19103, 1975.

4.3.2 S. Kaplun, Low Reynolds number flow past a circular cylinder, *Journal of Math. and Mech.* **6**, 5 (1957), 595–603.

4.3.3 S. Kaplun and P. A. Lagerstrom, Asymptotic expansions of Navier–Stokes solutions for small Reynolds numbers, *Journal of Math. and Mech.* **6**, 5 (1957), 515–593.

4.3.4 P. A. Lagerstrom, Note on the Preceding Two Papers, *Journal of Math. and Mech.* **6**, 5 (1957), 605–606.

4.3.5 I. Proudman and J. R. A. Pearson, Expansions at small Reynolds number for the flow past a sphere and a circular cylinder, *Journal of Fluid Mechanics* **2**, Part 3 (1957).

4.3.6 A. E. H. Love, *A Treatise on the Mathematical Theory of Elasticity*. 4th ed., England: Cambridge University Press, 1927.

4.3.7 V. Z. Vlasov, *Allgemeine Schalentheorie und ihre Anwéndung in der Technik* (translated from Russian), Berlin: Akademie Verlag, 1958.

4.3.8 S. Timoshenko and S. Woinowsky-Krieger, *Theory of Plates and Shells*. 2nd ed., New York: McGraw-Hill Book Co., 1959.

4.3.9 A. Peskoff, R. S. Eisenberg, and J. D. Cole, *Potential Induced by a Point Source of Current in the Interior of a Spherical Cell*, University of California at Los Angeles, Rept. U.C.L.A.-ENG-7259, December 1972.

4.3.10 A. Peskoff and R. S. Eisenberg, The time-dependent potential in a spherical cell using matched asymptotic expansions, *J. Math. Bio.* **2**, 1975, pp. 277–300.

4.3.11 A. Peskoff, R. S. Eisenberg, and J. D. Cole. Matched asymptotic expansions of the Green's function for the electric potential in an infinite cylindrical cell, *S.I.A.M. Jnl. Appl. Math.* **30**, No. 2. March 1976, pp. 222–239.

4.3.12 A. Peskoff, Green's function for Laplace's equation in an infinite cylindrical cell, *J. Math. Phys.* **15** (1974), pp. 2112–2120.

4.3.13 R. D. Taylor, Cable theory, *Physical Techniques in Biological Research.* W. L. Nastuk ed. Vol. VIB. Academic Press, N.Y. 1963, pp. 219–262.

4.3.14 V. Barcilon, J. D. Cole, and R. S. Eisenberg. A singular perturbation analysis of induced electric fields in cells, *S.I.A.M. J. Appl. Math.* **21** (1971), pp. 339–353.

4.4 Multiple-Variable Expansions for Second-Order Partial Differential Equations

In Chapter 3 we saw a wide range of applications of multiple variable expansions to problems in ordinary differential equations. In contrast, applications to partial differential equations are more recent and have a more limited scope. In this section we will study some of the pertinent contributions and will point out the limitations of the various approaches with the aim of stimulating further research.

4.4.1 Initial Value Problems for Nonlinear Dispersive Waves

Here, we are concerned with some perturbation problems associated with nonlinear or weakly nonlinear dispersive waves. In all the examples we consider in this subsection, the solution will be a perturbation upon a uniform periodic solution, i.e., a solution consisting of a periodic wave-train propagating with constant speed in one direction. This assumption of nearly uniform solutions will lead to much simplification.

We point out that following the appearance of Reference 4.4.1 in 1967, there has been considerable interest in the "inverse scattering" method for calculating *exact* solutions for fairly general initial value problems for certain nonlinear dispersive problems. We do not discuss this remarkable technique as it would take us far beyond the scope of this book. At any rate, the technique is not a perturbation procedure and appears, at least so far, to be limited in applicability to only a few special problems. The reader can find a unified account of the inverse scattering method in Reference 4.4.2.

Linear Problem

To put the nonlinear theory into perspective we begin with a brief review of linear dispersive waves, using the simple model equation

$$u_{tt} - u_{xx} + u = 0. \tag{4.4.1}$$

A "uniform" solution of Eq. (4.6.1) is a function of the form

$$u = F(\theta); \qquad \theta = kx - \omega t, \tag{4.4.2}$$

where k and ω are constants. To see if Eq. (4.4.1) admits such solutions, we substitute the assumed form into Eq. (4.4.1) and find that F must satisfy

$$(\omega^2 - k^2)F'' + F = 0; \ ' = d/d\theta. \tag{4.4.3}$$

We discard unbounded solutions by requiring that $\omega^2 - k^2$ be positive. Normalizing the frequency of the resulting harmonic oscillator to be unity, we obtain the "dispersion relation"

$$\omega = \pm\sqrt{k^2 + 1} \tag{4.4.4}$$

and the uniform solutions, which in this case are periodic and sinusoidal:

$$F = A\sin(\theta + \phi), \tag{4.4.5}$$

where A and ϕ are the constant amplitude and phase.

The dispersion relation then specifies the allowable "frequency" ω corresponding to a given "wave number" k in order to have a uniform solution. Equation (4.4.5) is a uniform solution in the sense that an initial wave

$$u(x, 0) = A\sin(kx + \phi) \tag{4.4.6}$$

with A, k, and ϕ prescribed arbitrarily, propagates uniformly (without distortion) to the right (or left) with the "phase velocity"[5]

$$c_p = \omega/k = \pm\sqrt{k^2 + 1}/k. \tag{4.4.7}$$

Note also that Eq. (4.4.5) is the solution of the very special initial value problem

$$u(x, 0) = A\sin(kx + \phi) \tag{4.4.8a}$$

$$u_t(x, 0) = \mp A\sqrt{k^2 + 1}\cos(kx + \phi) \tag{4.4.8b}$$

for Eq. (4.4.1).

It turns out that for linear problems, these very special uniform solutions (corresponding to waves propagating to the right or left with the phase velocity $\pm\sqrt{k^2 + 1}/k$) play a fundamental role in the solution of problems with *arbitrary* initial conditions.

Consider, for example, the general initial value problem

$$u(x, 0) = f(x) \tag{4.4.9a}$$

$$u_t(x, 0) = g(x) \tag{4.4.9b}$$

for Eq. (4.4.1) on $-\infty < x < \infty$.

[5] The basic notions of linear wave propagation are exposed in detail in many texts, e.g., Reference 4.4.2, and will only be reviewed briefly here.

Using Fourier transforms, it is easily seen that the solution is

$$u(x, t) = \frac{1}{2\sqrt{2\pi}} \int_{-\infty}^{\infty} \left\{ \left[F(k) - \frac{iG(k)}{\sqrt{1 + k^2}} \right] e^{i(kx + \sqrt{1+k^2}\,t)} \right.$$

$$\left. + \left[F(k) + \frac{iG(k)}{\sqrt{1 + k^2}} \right] e^{i(kx - \sqrt{1+k^2}\,t)} \right\} dk, \tag{4.4.10a}$$

where F and G are the Fourier transforms of f and g, i.e.,

$$F(k) = \frac{1}{\sqrt{2\pi}} \int_{-\infty}^{\infty} e^{-ikx} f(x)dx; \qquad G(k) = \frac{1}{\sqrt{2\pi}} \int_{-\infty}^{\infty} e^{-ikx} g(x)dx. \tag{4.4.10b}$$

Thus, u is represented by a *continuous superposition of uniform waves traveling in both directions.*

To fix ideas, consider the fundamental solution of Equation (4.4.1), i.e.,

$$u_{tt} - u_{xx} + u = \delta(x)\delta(t); \qquad -\infty < x < \infty \tag{4.4.11a}$$

$$u(x, 0^-) = u_t(x, 0^-) = 0, \tag{4.4.11b}$$

where δ is the Dirac delta function. This represents the response to a concentrated unit "source" at $t = 0$, $x = 0$. Equations (4.4.11) are equivalent to the initial-value problem

$$u_{tt} - u_{xx} + u = 0 \tag{4.4.12}$$

$$u(x, 0^+) = 0 \tag{4.4.13a}$$

$$u_t(x, 0^+) = \delta(x) \tag{4.4.13b}$$

which, according to Eq. (4.4.10) has the solution

$$u(x, t) = \frac{1}{4\pi i} \int_{-\infty}^{\infty} \frac{e^{i(kx + \sqrt{1+k^2}\,t)} - e^{i(kx - \sqrt{1+k^2}\,t)}}{\sqrt{1 + k^2}} dk. \tag{4.4.14a}$$

Evaluating this, or proceeding directly from Equations (4.4.11) we can show that

$$u(x, t) = \begin{cases} (1/2)J_0(\sqrt{t^2 - x^2}), & t > |x|, \\ 0, & t < |x|. \end{cases} \tag{4.4.14b}$$

If we "discretize" the integrals appearing in Eq. (4.4.14a) we see that the solution is a composite of waves of all wave numbers $-\infty < k < \infty$, traveling in both the positive and negative directions, and obeying the dispersion relation (4.4.4). Moreover, the waves with higher wave numbers are of smaller amplitude because of the denominator $\sqrt{1 + k^2}$.

An interesting feature of the linear problem is the phenomenon of "dispersion" which is evidenced by the behavior for large times. For example, the solution (4.4.14) has the asymptotic form[6] in the limit $t \to \infty$, $\xi = x/t$ fixed.

$$u(x, t) = \frac{-1}{\sqrt{2\pi t}(1 - \xi^2)^{1/2}} \sin(k_0 x - \omega_0 t - \pi/4), \qquad 0 < \xi < 1, \quad (4.4.15)$$

where

$$k_0 = \xi/\sqrt{1 - \xi^2} \qquad (4.4.16a)$$

and

$$\omega_0 = \sqrt{1 + k_0^2} = 1/\sqrt{1 - \xi^2}. \qquad (4.4.16b)$$

This has the following geometrical interpretation. For an observer moving with constant velocity ξ, i.e., for points along a ray ξ = const., Equation (4.4.15) describes a sinusoidal wave with *constant* values of k_0 and ω_0. Conversely, for any fixed k_0, $\xi(k_0)$ is just the "group velocity" of waves with wave number k_0. The group velocity is defined by

$$c_g = d\omega_0/dk_0 = \frac{k_0}{\sqrt{1 + k_0^2}} = \xi(k_0) \qquad (4.4.16c)$$

and represents the velocity of the group of waves with wave numbers which are infinitesimally close to k_0. (See Reference 4.4.2 for further details.)

Thus, wave numbers and frequencies are propagated unchanged with the associated group velocity. Dispersion refers to the fact that for sufficiently large times different waves separate (disperse) as they travel with their group velocity. The amplitude of the waves decays like $1/\sqrt{t}$ along a given ray, and this result is consistent with energy conservation.

We also note that a stationary observer (at large times) sees progressively shorter waves, while the observer moving with speed $\xi(k_0)$ and watching a given phase sees this phase moving slowly[7] to the right because the local phase velocity is larger than ξ. In fact $c_p(k_0) > 1$ while $c_g(k_0) < 1$ for this example. Figure 4.4.1 is a schematic illustrating these features.

Nonlinear Problem, Uniform Periodic Waves

An interesting question is to what extent are the features described above for the linear problem still true in a nonlinear case. Consider as a model, the nonlinear equation

$$u_{tt} - u_{xx} + u + \varepsilon u^3 = 0, \qquad (4.4.17)$$

[6] Obtained, for example, by the method of stationary phase, from Eq. (4.4.14a) (see Reference 4.4.3) or by looking up the asymptotic behavior of J_0 in standard tables.

[7] We see that k_0 and hence ω_0 vary slowly because differentiating the expression for c_g with respect to x gives $\partial k_0/\partial x = O(t^{-1})$.

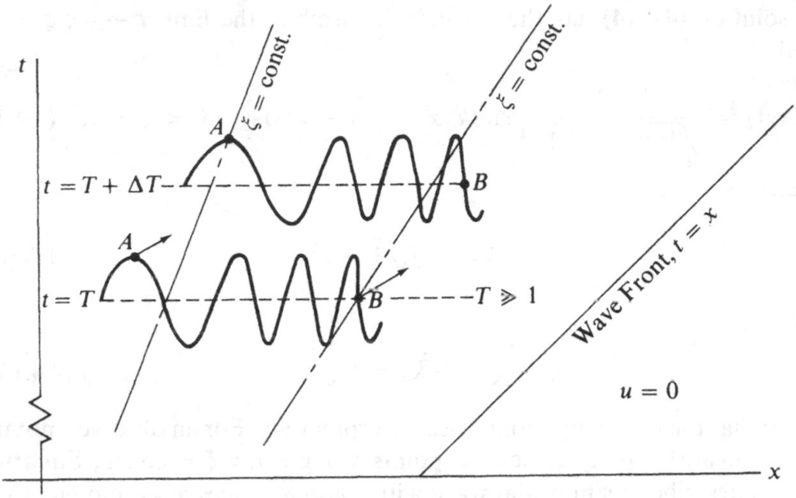

Figure 4.4.1 Dispersion at Large Times

where, for the time being, ε is not taken to be small. As in the linear case, if we seek a uniform solution of the form given by Eq. (4.4.2) we find that F obeys the nonlinear oscillator problem

$$(\omega^2 - k^2)F'' + F + \varepsilon F^3 = 0 \tag{4.4.18}$$

which also has periodic solutions given by elliptic functions [cf. Equation (3.6.44)]. These sn or cn waves are sometimes referred to as "cnoidal waves." The dispersion relation (4.4.4) was just a statement of the existence of bounded solution, which happened to be periodic (normalized to be 2π). The exact analog still holds for the nonlinear oscillator governed by (Equation 4.4.18). We first derive the energy integral

$$(\omega^2 - k^2)\frac{F'^2}{2} + \frac{F^2}{2} + \frac{\varepsilon F^4}{4} = E = \text{const.}; \qquad ' = \frac{d}{d\theta} \tag{4.4.19}$$

then note that for periodic solutions Equation (4.4.19) must describe a closed curve in the F, F' plane for any given E. Therefore, the period (normalized to be 2π) is given by

$$(\omega^2 - k^2)^{1/2} \oint \frac{dF}{\left[2E - F^2 - \dfrac{\varepsilon F^4}{2}\right]^{1/2}} = 2\pi. \tag{4.4.20}$$

For the linear case, $\varepsilon = 0$, this is simply Equation (4.4.4), a relation which is independent of the energy E. However, if $\varepsilon \neq 0$ the dispersion relation is energy (or amplitude) dependent and is an expression of the form

$$\omega = \omega(k, E; \varepsilon) \tag{4.4.21}$$

with a slightly more complicated structure.

Thus, the concept of a uniform periodic solution still holds for a nonlinear dispersive problem.[8] Unfortunately, this is as far as the analogy can be carried. The crucial result that *any* initial value problem can be expressed as a continuous superposition of uniform periodic waves is not true in general. Exact solutions are therefore not available except in those cases where the inverse scattering method applies.

There are various possible perturbation problems associated with the nonlinear Equation (4.4.17). The simplest, mentioned in Problem 3.1.1, consists of calculating the periodic uniform solution for the case $\varepsilon \ll 1$. Stokes calculated this result in 1847 and it is easily derived using a strained coordinate expansion. We summarize the results.

Assume a periodic solution of Equation (4.4.17) in the form

$$u = u_0(\theta^+) + \varepsilon u_1(\theta^+) + \cdots, \tag{4.4.22}$$

where, for waves traveling to the right, we set

$$\theta^+ = kx - \sqrt{1 + k^2}(1 + \varepsilon\omega_1 + \varepsilon^2\omega_2 + \cdots)t \tag{4.4.23}$$

for a given k and unknown ω_1. Applying the "method" of strained coordinates gives

$$u_0 = A \sin(\theta^+ + \phi), \qquad A \text{ and } \phi \text{ const.} \tag{4.4.24a}$$

$$u_1 = \frac{-A^3}{32} \sin 3(\theta^+ + \phi) \tag{4.4.24b}$$

$$\omega_1 = \frac{3}{8} A^2/(1 + k^2) \tag{4.4.24c}$$

etc. This is just the expansion of the exact solution one could obtain from Equation (4.4.19) in the limit as $\varepsilon \to 0$. In either case (ε arbitrary or ε small) the result is limited in the sense that it corresponds to the very special initial value problem

$$u(x, 0; \varepsilon) = A \sin(kx + \phi) - \varepsilon \frac{A^3}{32} \sin 3(kx + \phi) + \cdots \tag{4.4.25a}$$

$$u_t(x, 0; \varepsilon) = -A\sqrt{1 + k^2} \cos(kx + \phi) - \varepsilon A^3\left[\frac{3}{8\sqrt{1 + k^2}} \cos(kx + \phi) - \frac{3}{32}\sqrt{1 + k^2} \cos 3(kx + \phi)\right] + \cdots. \tag{4.4.25b}$$

Weakly Nonlinear Problem, Perturbed Initial Conditions

What happens if we alter the initial conditions of (4.4.25) slightly? Can one still express the solution in terms of one strained variable θ^+? Based on our experience so far we expect otherwise. In fact, suppose we set equal to

[8] The reader can verify that these results are true for an arbitrary nonlinear term in Equation (4.4.17) as long as the nonlinearity only involves u and the solutions are bounded.

zero the terms of order ε and higher in the initial conditions corresponding to Stokes' periodic solution, i.e., we wish to solve Equation (4.4.17) subject to (with $\phi = 0$)

$$u(x, 0; \varepsilon) = A \sin kx; \qquad A, k \text{ constant} \tag{4.4.26a}$$

$$u_t(x, 0; \varepsilon) = -A\sqrt{1 + k^2} \cos kx. \tag{4.4.26b}$$

We assume an expansion in the multiple variable form [cf. Equations (3.2.7)]

$$u(x, t; \varepsilon) = u_0(x, t^+, \tilde{t}) + \varepsilon u_1(x, t^+, \tilde{t}) + \cdots \tag{4.4.27}$$

$$t^+ = (1 + \varepsilon^2\omega_2 + \varepsilon^3\omega_3 + \cdots)t \tag{4.4.28a}$$

$$\tilde{t} = \varepsilon t \tag{4.4.28b}$$

and calculate the following equations for u_0, u_1, etc.

$$L(u_0) \equiv u_{0_{t^+t^+}} - u_{0_{xx}} + u_0 = 0 \tag{4.4.29a}$$

$$L(u_1) = -2u_{0_{t^+\tilde{t}}} - u_0^3 \tag{4.4.29b}$$

$$L(u_2) = -2u_{1_{t^+\tilde{t}}} + 2\omega_2 u_{0_{t^+t^+}} - u_{0_{\tilde{t}\tilde{t}}} - 3u_0^2 u_1. \tag{4.4.29c}$$

The solution of Equation (4.4.29a), for the given initial conditions, is

$$u_0 = a_0(\tilde{t})\sin[\psi_1^+(x, t^+) + \phi_0(\tilde{t})], \tag{4.4.30}$$

where we will use the notation

$$\psi_n^+ = nkx - \sqrt{1 + n^2k^2}\, t^+; \qquad n = 1, 2, \ldots \tag{4.4.31a}$$

$$\psi_n^- = nkx + \sqrt{1 + n^2k^2}\, t^+; \qquad n = 1, 2, \ldots \tag{4.4.31b}$$

to depict uniform periodic waves of wave number nk propagating to the right (ψ_n^+) or left (ψ_n^-). Thus a trigonometric function of ψ_n^+ or ψ_n^- represents a homogeneous solution of $L(u) = 0$. The slowly varying amplitude a_0 and phase ϕ_0 satisfy the conditions

$$a_0(0) = A \tag{4.4.32a}$$

$$\phi_0(0) = 0. \tag{4.4.32b}$$

To determine a_0 and ϕ_0 we study the solution of Equation (4.4.29b) for u_1. Substituting Equation (4.4.30) for u_0 gives

$$L(u_1) = 2\frac{da_0}{d\tilde{t}}(1 + k^2)^{1/2}\cos(\psi_1^+ + \phi_0) - \left[2a_0(1 + k^2)^{1/2}\frac{d\phi_0}{d\tilde{t}} + \frac{3}{4}a_0^3\right]\sin(\psi_1^+ + \phi_0) + \frac{a_0^3}{4}\sin 3(\psi_1^+ + \phi_0). \tag{4.4.33}$$

Since the first two terms on the right-hand side of Equation (4.4.33) are homogeneous solutions of $L(u_1) = 0$, they lead to unbounded contributions for large t in u_1. We eliminate these inconsistent terms by setting

$$\frac{da_0}{d\tilde{t}} = 0 \tag{4.4.34a}$$

$$\frac{d\phi_0}{d\tilde{t}} = -\frac{3a_0^2}{8\sqrt{1+k^2}}. \tag{4.4.34b}$$

Thus,

$$a_0(\tilde{t}) = A = \text{const.} \tag{4.4.35a}$$

$$\phi_0(\tilde{t}) = -3A^2\tilde{t}/8\sqrt{1+k^2}. \tag{4.4.35b}$$

We note that Equation (4.4.30) can also be written in the form

$$u_0 = A \sin\left\{kx - (1+k^2)^{1/2}\left[1 + \frac{3A^2\varepsilon}{8(1+k^2)} + \cdots\right]t\right\} \tag{4.4.36}$$

exhibiting the effect of the nonlinear term in perturbing the frequency. So far, the solution is identical to Stokes' periodic solution. However, since our initial conditions to $O(\varepsilon)$ are now different from those in Equations (4.4.25), we expect u_2, u_3 etc. to be more complicated.

It follows from the assumed form of the expansion for u that u_1 must satisfy the following initial conditions

$$u_1(x, 0, 0) = 0 \tag{4.4.37a}$$

$$u_{1_{t^+}}(x, 0, 0) = -u_{0_{\tilde{t}}}(x, 0, 0) = \frac{3A^3}{8\sqrt{1+k^2}} \cos kx \tag{4.4.37b}$$

The solution of what remains of Equation (4.4.33) for u_1 is assumed in series form

$$u_1(x, t^+, \tilde{t}) = \sum_{n=1}^{\infty} \{B_n(\tilde{t})\sin[\psi_n^+ + \alpha_n(\tilde{t})] + C_n(\tilde{t})\sin[\psi_n^- + \beta_n(\tilde{t})]\} - \frac{A^3}{32}\sin 3(\psi_1^+ + \phi_0). \tag{4.4.38}$$

Note that the last term in Equation (4.4.38) is the particular solution corresponding to the third harmonic term on the right-hand side of Equation (4.4.33) and is well behaved. This is because for Equation (4.4.17) (where in the limit $\varepsilon \to 0$ the problem is still dispersive) $\sin 3(\psi_1^+ + \phi_0)$ is not a homogeneous solution of $L(u) = 0$.

In contrast, for the problem

$$u_{tt} - u_{xx} + \varepsilon u^3 = 0 \tag{4.4.39}$$

which is non-dispersive in the limit $\varepsilon \to 0$, the third harmonic is a homogeneous solution and the associated particular solution will be a mixed secular term proportional to $t \cos 3k(x - t)$. Therefore, our approach does not apply to Equation (4.4.39). In fact, for this case, the third harmonic term must be eliminated at the outset, but this can only be accomplished if u_0 also contains a similar term. In that event, cubing u_0 produces, in addition, a fifth harmonic which must also be removed, etc. We see that, even for the simple initial value problem we are considering, u_0 must contain all the odd harmonics and be of the form

$$u_0 = \sum_{n=0}^{\infty} a_{2n+1}(\tilde{t})\sin[(2n+1)k(x - t^+) + \phi_{2n+1}(\tilde{t})]. \tag{4.4.40}$$

The calculations for the higher order terms are now much more complicated as we have to deal with products and powers of infinite series. We will return to the non-dispersive problem later on and consider an alternate approach which does not involve infinite series expansions.

For the case of Equation (4.4.17) the situation seems to be equally bleak in so far as u_1, u_2 etc., are concerned. Fortunately, all but a few of the coefficients B_n and C_n vanish. These coefficients are determined by requiring that u_2 be bounded and the details of the calculation are left as an exercise (Problem 4.4.1). We find that only B_1, B_3, C_1 and C_3 are needed and all the other coefficients must vanish. The resulting expression for u_1 subject to the initial conditions (4.4.37) is

$$\begin{aligned} u_1 &= \frac{3A^3}{16(1+k^2)}[-\sin(\psi_1^+ + \phi_0) + \sin(\psi_1^- - 2\phi_0)] \\ &\quad + \frac{A^3}{64}\left\{\left[1 + 3\sqrt{\frac{1+k^2}{1+9k^2}}\right]\sin\left(\psi_3^+ - \frac{3A^2\tilde{t}}{4\sqrt{1+9k^2}}\right)\right. \\ &\quad \left. + \left[1 - 3\sqrt{\frac{1+k^2}{1+9k^2}}\right]\sin\left(\psi_3^- + \frac{3A^2\tilde{t}}{4\sqrt{1+9k^2}}\right)\right\} - \frac{A^3}{32}\sin 3(\psi_1^+ + \phi_0). \end{aligned} \tag{4.4.41}$$

Consistency of u_1 on the $\tilde{t}$ scale also determines ω_2 to be

$$\omega_2 = A^4[-51 + 3k^2]/256(1 + k^2)^2. \tag{4.4.42}$$

This shows that u_1 contains waves of phase speed $(1 + k^2)^{1/2}/k$ associated with the initial condition, as well as waves with phase speed $(1 + 9k^2)^{1/2}/3k$. Moreover, both types of waves propagate to the right and the left. Thus, the effect of a slight perturbation on the initial conditions destroys the periodic nature of the Stokes solution.

Equation (4.4.17) also possesses a hierarchy of more complicated solutions corresponding to more general initial conditions. For example, instead of a single wave initially, we can study the solution corresponding to a discrete set of such waves (the case $N = 2$ is considered in Problem 4.4.2)

$$u(x, 0) = \sum_{i=1}^{N} A_i \sin k_i x \tag{4.4.43a}$$

$$u_t(x, 0) = -\sum_{i=1}^{N} A_i\sqrt{1 + k_i^2}\cos k_i x. \tag{4.4.43b}$$

As one might expect there will be resonance between these waves for certain ratios of the wave numbers, in exact analogy with the resonance between the normal modes of weakly nonlinear coupled oscillators [cf. Section 3.5.2]. Routine extension of the ideas in Sec. 3.5.2 can be used to study resonance cases. We will not go into the details here. The main results for $N = 2$ and $N = 3$ are outlined in parts (c) and (d) of Problem 4.4.2.

The most general initial value problem is, of course,

$$u(x, 0) = p(x) \tag{4.4.44a}$$

$$u_t(x, 0) = q(x) \tag{4.4.44b}$$

for arbitrary (well behaved) functions p and q. In this case u_0 is, in general, not expressible as a discrete set of uniform periodic waves, but must be of the form given in Equation (4.4.10a) where instead of F and G we must have functions of k and $\tilde{t}$ which have F and G as their initial values. The technical difficulties associated with the calculations to higher order are formidable and there is no general theory available (except for those cases which can be solved exactly by the inverse scattering method).

For the special case where the nonlinearities in the equation are in convolution form, the Fourier transform of the nonlinear partial differential equation is easily derived and reduces to a weakly nonlinear oscillator problem in Fourier space. This can be solved for arbitrary initial conditions then inverted. For further details we refer the reader to Problem 4.4.3 and Reference 4.4.4. This idea is not restricted to dispersive systems; in fact Chikwendu discusses dissipative and non-dispersive examples in Reference 4.4.4.

Slowly Modulated Waves (Weakly Nonlinear Case)

In view of the behaviour of the linear problem for large times, as depicted, for example, by Equation (4.4.15) it is natural to ask whether the nonlinear problem possesses solutions in the form of uniform periodic waves, but with slowly varying values of the amplitude, phase, wave number and frequency. The basic ideas for this class of solutions were put forward by Whitham in Reference 4.4.5. (The reader can also find a unified treatment in Reference 4.4.2) and are based on a variational approach using a Lagrangian formulation. In Reference 4.4.6 Luke gives a derivation based on multiple variable expansions, and we will adopt this point of view here.

To begin with, we consider the weakly nonlinear problem (4.4.17) with $\varepsilon \ll 1$, and seek solutions in the form

$$u(x, t; \varepsilon) = F(\theta, \tilde{x}, \tilde{t}; \varepsilon) = u_0(\theta, \tilde{x}, \tilde{t}) + \varepsilon u_1(\theta, \tilde{x}, \tilde{t}) + \cdots \tag{4.4.45}$$

which are 2π-periodic in the fast variable θ and which depend also on the slow variables

$$\tilde{x} = \varepsilon x \tag{4.4.46a}$$

$$\tilde{t} = \varepsilon t. \tag{4.4.46b}$$

The fast variable θ is a function of x, t and ε defined by the pair of equations

$$\theta_x = k(\tilde{x}, \tilde{t}) \tag{4.4.47a}$$

$$-\theta_t = \omega(\tilde{x}, \tilde{t}), \tag{4.4.47b}$$

where k and ω are as yet undetermined functions of $\tilde{x}$, $\tilde{t}$.

This is a direct generalization of Stokes' periodic solution (where k and ω were constants) to a class of solutions with slowly varying parameters.

We note that the definition of θ by Equations (4.4.47) is consistent only if $\theta_{xt} = \theta_{tx}$, i.e.,

$$k_{\tilde{t}}(\tilde{x}, \tilde{t}) = -\omega_{\tilde{x}}(\tilde{x}, \tilde{t}). \tag{4.4.48}$$

This represents conservation of waves and is one of the needed conditions linking the various unknowns occurring in the solution. The other conditions will be provided by requiring F to be periodic with respect to θ.

Once k and ω have been determined we can compute θ by quadrature in the form

$$\theta(x, t; \varepsilon) = \frac{1}{\varepsilon}\left[\int_0^{\tilde{x}} k(\xi, \tilde{t})d\xi - \int_0^{\tilde{t}} \omega(\tilde{x}, \tau)d\tau - \int_0^{\tilde{x}}\int_0^{\tilde{t}} k_{\tilde{t}}(\xi, \tau)d\xi\, d\tau\right]. \tag{4.4.49}$$

It is important to note that although θ can be conveniently expressed in terms of $\tilde{x}$ and $\tilde{t}$ it should be regarded as a function of x, t, and ε for the purposes of multiple variable expansion. We have assumed that $\theta(0, 0, \varepsilon) = 0$, and using Equation (4.4.48) shows that the integrand in the last term may also be set equal to $-\omega_{\tilde{x}}$.

Routine application of the multiple variable idea gives

$$u_t = -\omega F_\theta + \varepsilon F_{\tilde{t}} \tag{4.4.50a}$$

$$u_x = kF_\theta + \varepsilon F_{\tilde{x}} \tag{4.5.50b}$$

$$u_{tt} = \omega^2 F_{\theta\theta} - 2\varepsilon\omega F_{\theta\tilde{t}} - \varepsilon\omega_{\tilde{t}} F_\theta + \varepsilon^2 F_{\tilde{t}\tilde{t}} \tag{4.4.50c}$$

$$u_{xx} = k^2 F_{\theta\theta} + 2\varepsilon k F_{\theta\tilde{x}} + \varepsilon k_{\tilde{x}} F_\theta + \varepsilon^2 F_{\tilde{x}\tilde{x}}. \tag{4.4.50d}$$

If we now expand F, we obtain the following equations governing u_0, u_1, u_2.

$$L(u_0) \equiv (\omega^2 - k^2)u_{0_{\theta\theta}} + u_0 = 0 \tag{4.4.51a}$$

$$L(u_1) = 2\omega u_{0_{\theta\tilde{t}}} + 2ku_{0_{\theta\tilde{x}}} + (\omega_{\tilde{t}} + k_{\tilde{x}})u_{0_\theta} - u_0^3 \tag{4.4.51b}$$

$$L(u_2) = 2\omega u_{1_{\theta\tilde{t}}} + 2ku_{1_{\theta\tilde{x}}} + (\omega_{\tilde{t}} + k_{\tilde{x}})u_{1_\theta} - u_{0_{\tilde{t}\tilde{t}}} + u_{0_{\tilde{x}\tilde{x}}} - 3u_0^2 u_1. \tag{4.4.51c}$$

Thus, the assumption that the solution depends only on one fast variable (rather than on x and t individually) leads to an ordinary differential operator for each of the u_i.

Periodicity with respect to θ gives us the dispersion relation

$$\omega^2 - k^2 = \text{const.} = 1 \tag{4.4.52}$$

and, as before, we have normalized the period to be 2π by choosing the constant in Equation (4.4.52) to be unity.

The solution of Equation (4.4.51a) is

$$u_0 = a_0(\tilde{x}, \tilde{t})\sin\psi_0; \qquad \psi_0 = \theta + \eta_0(\tilde{x}, \tilde{t}), \tag{4.4.53}$$

where the amplitude a_0 and phase η_0 are unknowns to be determined by the periodicity requirement on u_1.

Using the above expression for u_0, we find that Equation (4.4.51b) reduces to

$$\begin{aligned} L(u_1) = {} & (2\omega a_{0_{\tilde{t}}} + 2ka_{0_{\tilde{x}}} + a_0\omega_{\tilde{t}} + a_0 k_{\tilde{x}})\cos\psi_0 \\ & -2\left[a_0\omega\eta_{0_{\tilde{t}}} + a_0 k\eta_{0_{\tilde{x}}} + \frac{3a_0^3}{8}\right]\sin\psi_0 \\ & + \frac{a_0{}^3}{4}\sin 3\psi_0. \end{aligned} \tag{4.4.54}$$

The terms proportional to $\cos\psi_0$ and $\sin\psi_0$ produce mixed secular contributions in u_1 and must be removed. This means that a_0 and η_0 obey the following first order partial differential equations

$$(\omega a_0^2)_{\tilde{t}} + (ka_0^2)_{\tilde{x}} = 0 \tag{4.4.55a}$$

$$\omega\eta_{0_{\tilde{t}}} + k\eta_{0_{\tilde{x}}} = -\tfrac{3}{8}a_0^2. \tag{4.4.55b}$$

The above, together with Equations (4.4.48) and (4.4.52) are a set of four equations for the four unknowns a_0, η_0, k, ω. We will now consider the solutions of these equations.

First, we use the dispersion relation to eliminate ω from the consistency condition (4.4.48). This reduces to the quasilinear equation

$$k_{\tilde{t}} + \frac{k}{\sqrt{1+k^2}}k_{\tilde{x}} = 0. \tag{4.4.56}$$

We solve this by integrating the characteristic equations

$$\frac{d\tilde{t}}{ds} = 1 \tag{4.4.57a}$$

$$\frac{d\tilde{x}}{ds} = \frac{k}{\sqrt{1+k^2}} \tag{4.4.57b}$$

$$\frac{dk}{ds} = 0 \tag{4.4.57c}$$

which pass through some initial curve

$$k(\tilde{x}, 0) = K(\tilde{x}). \tag{4.4.58}$$

The solution is conveniently expressed in parametric form

$$k(\tilde{x}, \tilde{t}) = K(\xi) \tag{4.4.59a}$$

$$\xi = \tilde{x} - \frac{K(\xi)\tilde{t}}{\sqrt{1 + K^2(\xi)}}. \tag{4.4.59b}$$

Thus, along the characteristic curve $\xi = \text{const.}$ (which is the straight line with slope $K/\sqrt{1 + K^2}$, given by Equation (4.4.59b)) k maintains its initial value. Notice that $K/\sqrt{1 + K^2}$ is the group velocity of waves with wave numbers near K according to the linear theory [cf. Equation (4.4.16c)]. Thus, the result we had for large t in the linear problem is now true *everywhere* for this special class of solutions of the weakly nonlinear problem. This is not surprising since at this stage the nonlinear term has not yet played a role, and Equations (4.4.59) just correspond to a necessary kinematic condition for slowly varying waves which obey the dispersion relation of Equation (4.4.52).

Observe also that the solution (4.4.59) becomes meaningless if the characteristics cross for $t > 0$. This means that the initial value K *must be a monotone non-decreasing function of* x, i.e., $K'(\xi) \geq 0$. For any initial value K which produces converging characteristics, i.e., $K' < 0$, Equation (4.4.17) cannot have a solution of the assumed form.[9]

Having found k, ω is now given by the dispersion relation and obviously along each of the characteristics (4.4.59b) $\omega(\tilde{x}, \tilde{t})$ also maintains its initial value

$$\omega(\tilde{x}, 0) = \sqrt{1 + K^2(\tilde{x})}. \tag{4.4.60}$$

To solve Equation (4.4.55a) for a_0 we use the known expressions for k and ω to find

$$a_{0_{\tilde{t}}} + \frac{k}{\sqrt{1 + k^2}} a_{0_{\tilde{x}}} = -\frac{k_{\tilde{x}}}{2(1 + k^2)^{3/2}} a_0. \tag{4.4.61}$$

This is linear and can easily be solved by integrating the characteristic equations. We also express the result in parametric form as follows

$$a_0 = \frac{A_0(\xi)}{\left\{1 + \dfrac{\tilde{t}K'(\xi)}{[1 + K^2(\xi)]^{3/2}}\right\}^{1/2}}, \tag{4.4.62}$$

where ξ is the solution of Equation (4.4.59b) and A_0 is the initial value of a_0, i.e.,

$$A_0(\tilde{x}) = a_0(\tilde{x}, 0). \tag{4.4.63}$$

[9] Since Equation (4.4.17) does not admit shocks, we cannot avoid the difficulty for $K' < 0$ by resorting to a weak solution of Equation (4.4.56).

Now, because of the inhomogeneous term on the right-hand side of Equation (4.4.61), we have the attenuation factor proportional to $\tilde{t}^{-1/2}$ for large $\tilde{t}$ in the amplitude. This result is also very reminiscent of the asymptotic expression for the linear problem [cf. Equation (4.4.15)] and is again to be expected since the nonlinear term has not been used yet; it only contributes to the solution for η_0.

With a_0 known, Equation (4.4.55b) for η_0 becomes

$$\eta_{0_{\tilde{t}}} + \frac{k}{\sqrt{1+k^2}}\eta_{0_{\tilde{x}}} = -\frac{3}{8}a_0^2. \tag{4.4.64}$$

This can also be easily solved to give

$$\eta_0(\tilde{x}, \tilde{t}) = \Psi_0(\xi) - \frac{3A_0^2(1+K^2)}{8K'}\log\frac{(1+K^2)^{3/2} + K'\tilde{t}}{(1+K^2)^{3/2}}, \tag{4.4.65}$$

where A_0 and K are the functions of ξ that we calculated earlier, and $\Psi_0(\tilde{x}) = \eta_0(\tilde{x}, 0)$.

Note that for $K' \neq 0$ we recover Stokes' result [cf. Equation (4.4.35)]

$$\eta_0 = \Psi_0(\xi) - 3A_0^2(\xi)\tilde{t}/8\sqrt{1+K^2} \tag{4.4.66}$$

along each characteristic. For $K' > 0$, η_0 decreases more slowly with $\tilde{t}$. Since $K' < 0$ is not meaningful we do not have to deal with an unpleasant logarithmic singularity in η_0 at finite $\tilde{t}$.

This completes the solution to $O(1)$ and the results to $O(\varepsilon)$ can be computed in a similar manner by considering the equation for u_2. (See Problem 4.4.4.)

The following limitations must be borne in mind regarding the above class of solutions.

(1) As in the case of Stokes' strictly periodic solution, our results correspond to very special initial conditions.

(2) These initial conditions are in the form of a given wave with spatially slowly varying amplitude, wave number (frequency) and phase, i.e. our results are a formal solution of Equation (4.4.17) in the limit $\varepsilon \to 0$ for the initial value problem

$$u(x, 0; \varepsilon) = A_0(\tilde{x})\sin\left[\frac{1}{\varepsilon}\int_0^{\tilde{x}} K(\xi)d\xi + \eta_0(\tilde{x}, 0)\right] + O(\varepsilon) \tag{4.4.67a}$$

$$u_t(x, 0; \varepsilon) = -A_0(\tilde{x})\sqrt{1+K^2(\tilde{x})}\cos\left[\frac{1}{\varepsilon}\int_0^{\tilde{x}} K(\xi)d\xi + \eta_0(\tilde{x}, 0)\right] + O(\varepsilon). \tag{4.4.67b}$$

(3) The slowly varying initial wave number K cannot be prescribed arbitrarily; we must have $K' \geq 0$ to avoid multiple valued solutions.

(4) Solutions to more general initial value problems cannot be calculated from the above by superposition.
(5) The asymptotic behavior for large times of an arbitrary initial value problem for the nonlinear equation (4.4.17) is not in the form of slowly modulated nearly periodic waves.

Slowly Modulated Waves (Strongly Nonlinear Case)

If we let $\varepsilon = 1$ in Equation (4.4.17), we can still derive solutions in the form of Equations (4.4.45) except now Equation (4.4.51a) is strictly nonlinear and we must proceed as discussed in Section 3.6. In fact, we can consider thc more general nonlinear problem

$$u_{tt} - u_{xx} + V'(u) = 0 \tag{4.4.68}$$

and using the expansion for u given by Equation (4.4.45) we obtain the following sequence of problems to solve for the u_i

$$\mu^2 u_{0_{\theta\theta}} + V'(u_0) = 0 \tag{4.4.69a}$$

$$\begin{aligned}\mu^2 u_{1_{\theta\theta}} + V''(u_0)u_1 &= 2\omega u_{0_{\theta\tilde{t}}} + 2k u_{0_{\theta\tilde{x}}} \\ &\quad + (\omega_{\tilde{t}} + k_{\tilde{x}})u_{0_\theta} \equiv p_1\end{aligned} \tag{4.4.69b}$$

$$\begin{aligned}\mu^2 u_{2_{\theta\theta}} + V''(u_0)u_2 &= 2\omega u_{1_{\theta\tilde{t}}} + 2k u_{1_{\theta\tilde{x}}} \\ &\quad + (\omega_{\tilde{t}} + k_{\tilde{x}})u_{1_\theta} - u_{0_{\tilde{t}\tilde{t}}} + u_{0_{\tilde{x}\tilde{x}}} - \tfrac{1}{2}V'''(u_0)u_1^2 \equiv p_2\end{aligned} \tag{4.4.69c}$$

etc., where

$$\mu^2 = \omega^2 - k^2. \tag{4.4.69d}$$

The small parameter ε appearing in the definitions of the slow variables $\tilde{x} = \varepsilon x$, and $\tilde{t} = \varepsilon t$ is now more difficult to pin down as it measures the departure of the initial data from that corresponding to the strictly periodic solution of Equation (4.4.68). More precisely, if Equation (4.6.68) possesses a *strictly periodic* solution, as discussed earlier in connection with Equation (4.4.17), this solution can be thought of as the *exact* solution corresponding to certain special initial conditions of the form

$$u(x, 0) = f(x, E, \omega, k) \tag{4.4.70a}$$

$$u_t(x, 0) = g(x, E, \omega, k). \tag{4.4.70b}$$

Here the energy E, the frequency ω and wave number k are *constants*, and the periodicity condition [cf. Equation (4.4.20)]

$$\mu \oint \frac{du}{[2E - V(u)]^{1/2}} = 2\pi \tag{4.4.71}$$

provides a relation of the form

$$\omega = \omega(k, E) \tag{4.4.72}$$

linking ω to k and E.

Now, suppose we wish to solve Equation (4.4.68) subject to initial conditions analogous to Equations (4.4.70) except that E, ω and k are allowed to vary slowly with x, the slowness of these variations being measured by ε. We have seen that it is not possible to *a priori* specify slowly varying initial conditions which result in solutions depending on θ only. In general one obtains waves traveling in both directions to higher orders. However, one can always calculate solutions of the type given in Equation (4.4.45) formally. Once such a solution is found it is easy to deduce, a posteriori, what initial conditions it corresponds to. In all cases, of course, these initial conditions reduce to Equations (4.4.70) as $\varepsilon \to 0$.

Aside from the above remarks regarding the role of the small parameter ε, there is nothing essentially different in the calculations now as compared with the ordinary differential equation problems of Section 3.6. Therefore, we just summarize the results of the solution to $O(1)$.

The periodicity condition, Equation (4.4.71), still holds for u_0

$$P = \mu \int \frac{du_0}{\sqrt{2E_0 - V(u_0)}} \equiv P(E_0, \mu). \tag{4.4.73}$$

If we require P to be a constant and normalize this constant to be 2π, we obtain from Equation (4.4.73) an expression of the form

$$\omega = \omega(k, E_0). \tag{4.4.74}$$

The consistency condition

$$k_{\tilde{t}} + \omega_{\tilde{x}} = 0 \tag{4.4.75}$$

gives a second relation between ω and k.

We write the solution of Equation (4.4.69a) in the form

$$u_0 = F(\psi, E_0, \mu); \qquad \psi = \theta + \eta_0, \tag{4.4.76}$$

where η_0 depends on $\tilde{x}$ and $\tilde{t}$ and represents a slowly varying phase. Thus, we need to determine ω, k, E_0 and η_0 to define the solution to $O(1)$. Equations (4.4.74) and (4.4.75) give two conditions, and two more conditions follow from requiring that u_1 be periodic with respect to θ. The details are identical to those discussed in Section 3.6 and we find

$$\int_0^{2\pi} p_1 F_\psi \, d\psi' = 0 \tag{4.4.77a}$$

$$\int_0^{2\pi} \left[\frac{p_1 l}{\mu^2 W} + \frac{P_{E_0} E_1}{P} + \frac{P_{E_0}}{P \mu^2 W} \int_0^{2\pi} p_1 F_\psi \, d\psi'' \right] d\psi' = 0, \tag{4.4.77b}$$

where, as in Equation (3.6.26), l is the periodic part of F_{E_0} defined by

$$l(\psi, E_0, \mu) = F_{E_0}(\psi, E_0, \mu) + \frac{\psi P_{E_0}}{P} F(\psi, E_0, \mu). \tag{4.4.78}$$

Also, W is the Wronskian of the linear operator for u_1, i.e.,

$$W(F_\psi, F_{E_0}) = F_\psi F_{E_0\psi} - F_{E_0} F_{\psi\psi}. \tag{4.4.79}$$

As in the oscillator problem, for the strictly nonlinear case one of the periodicity conditions Equation (4.4.77b), also involves E_1, which is one of the unknowns associated with the u_1 solution. The determination of E_1 requires considering the periodicity conditions (4.4.77) for u_2, but this would also involve η_1, the second unknown function appearing in the homogeneous solution of u_1, etc. This cascading effect would be disastrous were it not for the fact pointed out in Reference 4.4.6 that η_1 does not occur in the periodicity condition (4.4.77a) associated with u_2 and that this condition requires that $E_1 = 0$. We then use this result in Equations (4.4.77) for u_1 to calculate E_0 and η_0, and it is claimed in Reference 4.4.6 that η_0 is zero (or at worst a constant) also. The verification of these statements is left as an exercise for the special case $V'(u) = u^3$ in Problem 4.4.5.

The limitations of this class of solutions that we listed earlier apply even more stringently in the strongly nonlinear case.

4.4.2 Weakly Nonlinear Waves, Series Expansions

In the previous section we argued that in order to solve the problem

$$u_{tt} - u_{xx} + \varepsilon u^3 = 0; \qquad 0 < \varepsilon \ll 1 \tag{4.4.80a}$$

$$u(x, 0) = A \sin kx \tag{4.4.80b}$$

$$u_t(x, 0) = -Ak \cos kx \tag{4.4.80c}$$

which has the simple solution

$$u(x, t) = A \sin k(x - t) \tag{4.4.81}$$

if $\varepsilon = 0$, we must consider u_0 in the form

$$u_0(x, t^+, \tilde{t}) = \sum_{n=1}^{\infty} a_{2n-1}(\tilde{t}) \sin[(2n-1)k(x - t^+) + \phi_{2n-1}(\tilde{t})]. \tag{4.4.82}$$

All the odd harmonics must be retained in order to be able to derive a consistent expression for u_1. The equation governing u_1 is

$$u_{1_{t^+t^+}} - u_{1_{xx}} = -2u_{0_{t^+\tilde{t}}} - u_0^3. \tag{4.4.83}$$

We are now faced with the unpleasant task of cubing the infinite series given by Equation (4.4.82) and identifying the coefficients of all the homogeneous solutions. For the simple initial value problem (4.4.80) it is easily seen that u_0^3 will be a series in the form

$$u_0^3 = \sum_{n=1}^{\infty} A_{2n-1} \sin[(2n-1)k(x - t^+) + \Phi_{2n-1}], \tag{4.4.84}$$

where each A_{2n-1} depends on *all* the a_{2n-1}'s and each Φ_{2n-1} depends on all the ϕ_{2n-1}'s. Although it is possible to derive explicit expressions for the A_{2n-1} and Φ_{2n-1} we will not pursue this approach further as it is not clear how one would then solve the infinite set of coupled equations that result for the a_{2n-1} and ϕ_{2n-1} when we eliminate the mixed secular terms.

A Boundary-Value Problem

An example of a class of problems where this approach (of representing the solution in series form) has had limited success is given in Reference 4.4.7, and we consider this next.

We study the following initial/boundary value problem for the weakly nonlinear equation (dispersive when $\varepsilon = 0$)

$$u_{tt} - u_{xx} + u = \varepsilon f(u, u_t) \tag{4.4.85a}$$

$$u(0, t) = u(\pi, t) = 0 \tag{4.4.85b}$$

$$u(x, 0) = g(x) \tag{4.4.85c}$$

$$u_t(x, 0) = h(x) \tag{4.4.85d}$$

and assume a two variable expansion in the form

$$u(x, t; \varepsilon) = F(x, t^+, \tilde{t}; \varepsilon) = u_0(x, t^+, \tilde{t}) + \varepsilon u_1(x, t^+, \tilde{t}) + \cdots. \tag{4.4.86}$$

Then, u_0 and u_1 obey the following systems

$$u_{0_{t^+t^+}} - u_{0_{xx}} + u_0 = 0 \tag{4.4.87a}$$

$$u_0(0, t^+, \tilde{t}) = u_0(\pi, t^+, \tilde{t}) = 0 \tag{4.4.87b}$$

$$u_0(x, 0, 0) = g(x) \tag{4.4.87c}$$

$$u_{0_{t^+}}(x, 0, 0) = h(x) \tag{4.4.87d}$$

and

$$u_{1_{t^+t^+}} - u_{1_{xx}} + u_1 = -2u_{0_{t^+\tilde{t}}} + f(u_0, u_{0_{t^+}}) \tag{4.4.88a}$$

$$u_1(0, t^+, \tilde{t}) = u_1(\pi, t^+, \tilde{t}) = 0 \tag{4.4.88b}$$

$$u_1(x, 0, 0) = 0 \tag{4.4.88c}$$

$$u_{1_{t^+}}(x, 0, 0) = -u_{0_{\tilde{t}}}(x, 0, 0). \tag{4.4.88d}$$

We express the solution of Equations (4.4.87) using a series of eigenfunctions

$$u_0(x, t^+, \tilde{t}) = \sum_{n=1}^{\infty} [a_n \cos\sqrt{1 + n^2}\, t^+ + b_n \sin\sqrt{1 + n^2}\, t^+] \sin nx, \tag{4.4.89}$$

where the a_n and b_n are dependent on $\tilde{t}$. The initial conditions imply that

$$a_n(0) = \frac{2}{\pi}\int_0^\pi g(x)\sin nx\, dx \tag{4.4.90a}$$

$$b_n(0) = \frac{2}{\pi\sqrt{1+n^2}}\int_0^\pi h(x)\sin nx\, dx. \tag{4.4.90b}$$

We now determine the conditions governing the evolution of the a_n and b_n by considering u_1. In order that u_1 be bounded the right hand side of Equation (4.4.88a) must be orthogonal to all the homogeneous solutions

$$(\sin\sqrt{1+n^2}\,t^+)\sin nx, \qquad (\cos\sqrt{1+n^2}\,t^+)\sin nx.$$

This is a well known result for perturbed eigenvalue problems and can be proven directly [cf. Reference 4.4.7]. However, it is more instructive to derive the result by expanding u_1 in a series of eigenfunction.

Let

$$u_1 = \sum_{n=1}^{\infty} \alpha_n(t^+, \tilde{t})\sin nx. \tag{4.4.91}$$

Clearly, f must also have such an expansion, i.e.,

$$f(u_0, u_{0_{t^+}}) = \sum_{n=1}^{\infty} f_n(t^+, \tilde{t})\sin nx, \tag{4.4.92}$$

where

$$f_n(t^+, \tilde{t}) = \frac{2}{\pi}\int_0^\pi f[u_0(x, t^+, \tilde{t}), u_{0_{t^+}}(x, t^+, \tilde{t})]\sin nx\, dx. \tag{4.4.93}$$

Therefore, for each n, the α_n are governed by the equation

$$\frac{\partial^2\alpha_n}{\partial t^{+2}} + (1+n^2)\alpha_n = 2\frac{da_n}{d\tilde{t}}\sqrt{1+n^2}\sin\sqrt{1+n^2}\,t^+$$
$$-2\frac{db_n}{d\tilde{t}}\sqrt{1+n^2}\cos\sqrt{1+n^2}\,t^+ + f_n(t^+, \tilde{t}) \tag{4.4.94}$$

which is free of x and is very much analogous to what we had in Section 3.2 for weakly nonlinear oscillator problems.

We seek a solution for α_n by variation of parameters in the form [cf. the discussion in Section 3.2 following Equation (3.2.100)]

$$\alpha_n(t^+, \tilde{t}) = P_n(t^+, \tilde{t})\sin\psi + Q_n(t^+, \tilde{t})\cos\psi; \qquad \psi = \sqrt{1+n^2}\,t^+. \tag{4.4.95}$$

Substituting the above into Equation (4.4.94) gives

$$P_{n_{t^+}}\cos\psi - Q_{n_{t^+}}\sin\psi = \frac{da_n}{d\tilde{t}}\sin\psi - \frac{db_n}{d\tilde{t}}\cos\psi + \frac{f_n(t^+, \tilde{t})}{\sqrt{1+n^2}}, \tag{4.4.96a}$$

where we have set

$$P_{n_{t^+}} \sin \psi + Q_{n_{t^+}} \cos \psi = -\frac{da_n}{d\tilde{t}} \cos \psi - \frac{db_n}{d\tilde{t}} \sin \psi. \qquad (4.9.96b)$$

Therefore, solving Equations (4.4.96) we find

$$P_n = -\left[\frac{db_n}{d\tilde{t}} t^+ - \int_0^{t^+} \frac{f_n \cos \psi \, dt^+}{\sqrt{1 + n^2}}\right] \qquad (4.4.97a)$$

$$Q_n = -\left[\frac{da_n}{d\tilde{t}} t^+ + \int_0^{t^+} \frac{f_n \sin \psi \, dt^+}{\sqrt{1 + n^2}}\right]. \qquad (4.4.97b)$$

In order that u_1 be bounded[10], we must require that

$$\frac{db_n}{d\tilde{t}} - \lim_{t^+ \to \infty} \frac{1}{t^+} \int_0^{t^+} \frac{f_n \cos \psi \, dt^+}{\sqrt{1 + n^2}} = 0 \qquad (4.4.98a)$$

$$\frac{da_n}{d\tilde{t}} + \lim_{t^+ \to \infty} \frac{1}{t^+} \int_0^{t^+} \frac{f_n \sin \psi \, dt^+}{\sqrt{1 + n^2}} = 0. \qquad (4.4.98b)$$

Equations (4.4.98) are a system of infinitely many nonlinear equations for the a_n and b_n to be solved subject to the initial conditions in Equations (4.4.90).

We were able to derive compact formulas for the a_n and b_n because the boundary conditions (4.4.85b) allowed us to expand the solution in a series of eigenfunctions and to essentially eliminate the x-dependence from the results. This is not in general true for initial value problems on the infinite interval, and the corresponding expressions governing the slowly varying coefficients are more difficult to derive.

Having eliminated the mixed secular terms from u_1, the solution can easily be derived. Of course, it makes little sense to proceed further unless one can solve Equations (4.4.98). In general this is a formidable task and one would have to resort to some approximation scheme such as truncating the series after a relatively few terms. Given the notoriously slow convergence of trigonometric series, this does not appear to be a very useful approach either.

A rare example where Equations (4.4.98) were solved *exactly* appears in Reference 4.4.7 for the case where the perturbation function f is the Van der Pol function [cf. Equation (3.1.30)]

$$f = u_t - \tfrac{1}{3}u_t^3. \qquad (4.4.99)$$

The calculations are quite involved and will not be given here. In the next subsection we study a class of problems where we can avoid use of infinite series expansions and the associated difficulties.

[10] Note that since f_n depends on t^+ only through $\sin \psi$ and $\cos \psi$ the integrals in Equations (4.4.97) can, at worst, grow like t^+ as $t^+ \to \infty$.

4.4.3 Weakly Nonlinear Waves, Explicit Solutions

In view of the computational difficulties we encounter whenever we assume a series representation of the unperturbed solution, it is natural to investigate other alternatives.

For example, what would happen if in solving Equations (4.4.80) we took the solution for u_0 in the general (D'Alembert) form?

$$u_0(x, t^+, \tilde{t}) = f_0(\sigma, \tilde{t}) + g_0(\xi, \tilde{t}), \tag{4.4.100}$$

where σ and ξ are the characteristic variables

$$\sigma = x - t^+ \tag{4.4.101a}$$

$$\xi = x + t^+. \tag{4.4.101b}$$

As a consequence of the choice of initial conditions (4.4.80b, c) we set $g_0 = 0$, and substitute the result into Equation (4.4.83). This becomes

$$-4u_{1_{\sigma\xi}} = 2f_{0_{\sigma\tilde{t}}} - f_0^3. \tag{4.4.102}$$

Now, since the right-hand side is a function of σ only integration with respect to ξ will clearly produce an unbounded term proportional to ξ. To eliminate this we must set

$$2f_{0_{\sigma\tilde{t}}} - f_0^3 = 0 \tag{4.4.103}$$

which, unfortunately, is as difficult to solve as the original problem. In fact, Equation (4.4.80a) can be written in characteristic form as

$$4u_{\sigma\xi} - \varepsilon u^3 = 0 \tag{4.4.104}$$

and if we could solve Equation (4.4.103) we might as well have solved the exact equation (4.4.104).

Thus, we can make no progress, unless the nonlinear perturbation term *depends on a derivative of u*. In this event, Equation (4.4.103) is a first order equation in f_{0_σ} and can be integrated once. This idea was proposed and explored in Reference (4.4.8) on which much of this subsection is based. Fortunately, in most applications, the nonlinear perturbation terms indeed only involve derivatives of the dependent variable. Some examples are given in Chapter 5. We will discuss the essential features of the method next.

Cubic Damping (*Waves in One Direction*)

To begin with, let us consider the following simple problem

$$u_{tt} - u_{xx} + \varepsilon u_t^3 = 0, \qquad 0 < \varepsilon \ll 1 \tag{4.4.105}$$

of a wave equation with cubic damping and choose initial conditions which, for $\varepsilon = 0$, give only a wave travelling in the positive x direction. To simplify things even more we take a sinusoidal initial wave

$$u(x, 0) = A \sin kx; \qquad A, k, \text{const.} \tag{4.4.106a}$$

$$u_t(x, 0) = -Ak \cos kx. \tag{4.4.106b}$$

Assuming an expansion for u in the form

$$u(x, t; \varepsilon) = F(\sigma, \xi, \tilde{t}; \varepsilon) = u_0(\sigma, \xi, \tilde{t}) + \varepsilon u_1(\sigma, \xi, \tilde{t}) + \cdots \tag{4.4.107}$$

we have[11]

$$u_0(x, t^+, \tilde{t}) = f_0(\sigma, \tilde{t}) \tag{4.4.108}$$

with initial values

$$f_0(x, 0) = A \sin kx \tag{4.4.109a}$$

$$f_{0_\sigma}(x, 0) = Ak \cos kx. \tag{4.4.109b}$$

Writing the wave equation (4.4.105) in characteristic form and expanding, gives

$$-4u_{1_{\sigma\xi}} = 2f_{0_{\sigma\tilde{t}}} + f_{0_\sigma}^3. \tag{4.4.110}$$

To eliminate inconsistent terms proportional to ξ in u_1, we set the right-hand side equal to zero and can integrate the result with respect to $\tilde{t}$. We find

$$f_{0_\sigma} = \frac{1}{\sqrt{\tilde{t} + F_0(\sigma)}}, \tag{4.4.111a}$$

where $F_0(\sigma)$ is an "integration constant." The initial condition (4.4.109b) determines F_0, and f_{0_σ} is simply

$$f_{0_\sigma} = \frac{Ak \cos k\sigma}{[1 + A^2k^2\tilde{t} \cos^2 k\sigma]^{1/2}}. \tag{4.4.111b}$$

This can now be integrated with respect to σ and defines f_0 explicitly as[12]

$$f_0 = \frac{1}{k\tilde{t}^{1/2}} \sin^{-1}\left\{\sqrt{\frac{A^2k^2\tilde{t}}{1 + A^2k^2\tilde{t}}} \sin k\sigma\right\}. \tag{4.4.112}$$

Since f_{0_σ} is periodic and even in σ with zero average value over one period $(2\pi/k)$ in σ, f_0 is periodic and odd in σ. Therefore, in Equation (4.4.112) the absolute value of the arc sine function must be less than $\pi/2$.

The calculations for u_1 proceed in a similar manner and are given in Reference (4.4.8). We do not give the details here but simply point out that

[11] We are assuming here that u_0 only involves waves traveling to the right. In view of the initial conditions this is a plausible assumption. However, a rigorous justification requires retaining g_0 also in Equation (4.4.108) and showing that the initial conditions imply that $g_0(\xi, \tilde{t}) = 0$ (see Problems 4.4.6 and 4.4.10).

[12] The initial condition, Equation (4.4.109a) has been used, and it is easy to show that with $\sigma = x$ Equation (4.4.112) tends to Equation (4.4.109a) in the limit as $\tilde{t} \to 0$.

the solution for u_1 must include both $f_1(\sigma, \tilde{t})$ and $g_1(\xi, \tilde{t})$ functions. We will consider an example where u_0 involves both f_0 and g_0 presently.

To appreciate the difficulties that have been avoided in deriving this explicit result let us convert Equation (4.4.112) to series form. This is most directly accomplished by expressing f_{0_σ} in a Fourier cosine series then integrating the result term by term. We find an expression for f_0 in the form

$$f_0 = \sum_{n=1}^{\infty} B_{2n-1}(\tilde{t})\sin(2n-1)k\sigma, \tag{4.4.113}$$

where the B_{2n-1} are given by

$$B_{2n-1}(\tilde{t}) = \frac{2A}{(2n-1)\pi}\int_0^{\pi} \frac{\cos\xi\cos(2n-1)\xi\,d\xi}{\sqrt{1+A^2k^2\tilde{t}\cos^2\xi}}. \tag{4.4.114}$$

Each B_{2n-1} can be expressed as $1/\sqrt{1+A^2k^2\tilde{t}}$ times a power series in $J^2 = A^2k^2\tilde{t}/(1+A^2k^2\tilde{t}) < 1$ once the denominator in Equation (4.4.114) is expanded and the resulting expression is integrated term by term. For example,

$$B_1 = \frac{A}{\sqrt{1+A^2k^2\tilde{t}}}\left[1 + \frac{J^2}{8} + \frac{3}{64}J^4 + O(J^6)\right]. \tag{4.4.115}$$

The same results also follow by substituting the series (4.4.113) into the right-hand side of Equation (4.4.110) and eliminating terms independent of σ (which upon integration grow like ξ). At any rate the explicit result is certainly preferable to the series solution in Equation (4.4.113).

Cubic Damping (Waves in Both Directions)

Consider now the solution of Equation (4.4.105) with initial conditions

$$u(x,0) = 2\sin x \tag{4.4.116a}$$

$$u_t(x,0) = 0. \tag{4.4.116b}$$

Now the $\varepsilon = 0$ solution itself involves waves traveling in both directions. Therefore, we must take

$$u_0(\sigma,\xi,\tilde{t}) = f_0(\sigma,\tilde{t}) + g_0(\xi,\tilde{t}), \tag{4.4.117}$$

where f_0 and g_0 satisfy the initial conditions

$$f_0(x,0) = g_0(x,0) = \sin x. \tag{4.4.118}$$

The equation for u_1 is now

$$-4u_{1_{\sigma\xi}} = 2f_{0_{\sigma\tilde{t}}} - 2g_{0_{\xi\tilde{t}}} + f_{0_\sigma}^3 - 3f_{0_\sigma}^2 g_{0_\xi} + 3f_{0_\sigma}g_{0_\xi}^2 - g_{0_\xi}^3. \tag{4.4.119}$$

Deriving the boundedness conditions for u_1 now requires a little more care since some of the terms on the right-hand side of Equation (4.4.119) are harmless and should not be eliminated.

The analysis is considerably simplified by the fact that our initial conditions (4.4.118) are periodic. Thus, f_0 and g_0 are 2π-periodic functions of σ and ξ respectively. The terms which must be set equal to zero on the right-hand side of Equation (4.4.119) are then easily deduced to be

$$2f_{0_{\sigma\tilde{t}}} + 3b_0(\tilde{t})f_{0_\sigma} + f_{0_\sigma}^3 = 0 \tag{4.4.120a}$$

$$2g_{0_{\xi\tilde{t}}} + 3a_0(\tilde{t})g_{0_\xi} + g_{0_\xi}^3 = 0, \tag{4.4.120b}$$

where b_0 and a_0 are the average values of $g_{0_\xi}^2$ and $f_{0_\sigma}^2$ over one period, i.e.,

$$b_0(\tilde{t}) = \frac{1}{2\pi}\int_0^{2\pi} g_{0_\xi}^2(\xi, \tilde{t})d\xi \tag{4.4.121a}$$

$$a_0(\tilde{t}) = \frac{1}{2\pi}\int_0^{2\pi} f_{0_\sigma}^2(\sigma, \tilde{t})d\sigma. \tag{4.4.121b}$$

The two equations (4.4.120) are of the same form and, since for our initial value problem $f_0(x, 0) = g_0(x, 0)$, we need only consider one of these, say Equation (4.4.120a). Consider first the solution with $b_0(\tilde{t}) = c/3 = \text{const}$. We find

$$f_{0_\sigma} = \frac{c^{1/2}e^{-c\tilde{t}/2}}{[F_0(\sigma) - e^{-c\tilde{t}}]^{1/2}}, \tag{4.4.122a}$$

where $F_0(\sigma)$ is an arbitrary function of σ.

Having found the form of the solution for b_0 constant, assume that for $b_0(\tilde{t})$ the result has the same general structure and let

$$f_{0_\sigma} = \frac{\lambda(\tilde{t})}{[F_0(\sigma) + \phi(\tilde{t})]^{1/2}}, \tag{4.4.122b}$$

where $\lambda(\tilde{t})$ and $\phi(\tilde{t})$ are unknowns.

Substituting Equation (4.4.122b) into Equation (4.4.120a) shows that the assumed form is indeed a solution if λ and ϕ satisfy the pair of equations

$$2\lambda'(\tilde{t}) + 3b_0(\tilde{t})\lambda(\tilde{t}) = 0 \tag{4.4.123a}$$

$$\phi'(\tilde{t}) = \lambda^2(\tilde{t}), \tag{4.4.123b}$$

where primes denote differentiation with respect to $\tilde{t}$.

Similarly, if we assume that g_{0_ξ} is of the form

$$g_{0_\xi} = \frac{\theta(\tilde{t})}{[G_0(\xi) + \psi(\tilde{t})]^{1/2}} \tag{4.4.124}$$

we obtain by substitution into Equation (4.6.120b) the requirements that

$$2\theta'(\tilde{t}) + 3a_0(\tilde{t})\theta(\tilde{t}) = 0 \tag{4.4.125a}$$

$$\psi'(\tilde{t}) = \theta^2(\tilde{t}). \tag{4.4.125b}$$

In view of the symmetry of our initial condition it is clear that $a_0 = b_0$, hence $\theta = \lambda$ and $\phi = \psi$.

Equation (4.4.122b) will satisfy the initial condition $f_0(\sigma, 0) = \sin\sigma$ if $F_0(\sigma) = \sec^2\sigma$ and $\lambda(0) = 1$, $\phi(0) = 0$. Thus,

$$f_{0_\sigma} = \frac{\lambda(\tilde{t})\cos\sigma}{[1 + \phi(\tilde{t})\cos^2\sigma]^{1/2}}. \tag{4.4.126}$$

We now calculate a_0 by using the above in Equation (4.4.121b)

$$\begin{aligned} a_0 &= \frac{\lambda^2(\tilde{t})}{2\pi}\int_0^{2\pi}\frac{\cos^2\sigma\,d\sigma}{1 + \phi(\tilde{t})\cos^2\sigma} \\ &= \frac{\lambda^2(\tilde{t})}{\phi(\tilde{t})} - \frac{\lambda^2(\tilde{t})}{\phi(\tilde{t})\sqrt{1 + \phi(\tilde{t})}}. \end{aligned} \tag{4.4.127}$$

Using Equation (4.4.123a) with $a_0 = b_0$ and Equation (4.4.127) gives

$$-\frac{2}{3}\frac{\lambda'}{\lambda} = \frac{\phi'}{\phi} - \frac{\phi'}{\phi\sqrt{1 + \phi}} \tag{4.4.128}$$

which upon integration and use of Equation (4.4.123b) results in

$$\log\lambda^{-2/3} = \log(\phi')^{-1/3} = \log\frac{\phi[\sqrt{1 + \phi} + 1]}{\sqrt{1 + \phi} - 1} + \text{const.} \tag{4.4.129}$$

Therefore,

$$\phi' = 2^6[1 + \sqrt{1 + \phi}]^{-6}. \tag{4.4.130}$$

If we change variables from ϕ to m according to

$$m(\tilde{t}) = 1 + \sqrt{1 + \phi(\tilde{t})}; \qquad m(0) = 2 \tag{4.4.131}$$

we can integrate Equation (4.4.130) to find

$$m^8(\tilde{t}) - \tfrac{8}{7}m^7(\tilde{t}) = 2^8(\tilde{t} + \tfrac{3}{7}). \tag{4.4.132}$$

The solution of this algebraic equation defines $m(\tilde{t})$ and Equations (4.4.123b) and (4.4.130) give λ

$$\lambda = \sqrt{\phi'} = \frac{2^3}{m^3}. \tag{4.4.133}$$

Having determined ϕ and λ (hence θ and ψ) the solution for u_0 has the explicit form obtained by integrating Equation (4.4.126) with respect to σ

$$u_0 = \frac{\lambda(\tilde{t})}{\sqrt{\phi(\tilde{t})}}\left\{\sin^{-1}\left[\left(\frac{\phi(\tilde{t})}{1 + \phi(\tilde{t})}\right)^{1/2}\sin\sigma\right] + \sin^{-1}\left[\left(\frac{\phi(\tilde{t})}{1 + \phi(\tilde{t})}\right)^{1/2}\sin\xi\right]\right\}. \tag{4.4.134}$$

We emphasize that the simplicity and symmetry of the initial conditions were crucial in calculating this explicit result. Whether or not one can apply these ideas to more complicated perturbations and initial conditions depends on two criteria:

(i) The ability to isolate, as in Equations (4.4.120) the terms which must be eliminated;
(ii) The ability to solve the resulting system explicitly.

It is difficult to give necessary conditions under which these criteria are met. We know, for example, that for periodic initial conditions there is no difficulty in achieving the first goal. But periodic initial conditions do not exhaust the possibilities. Some examples of more general situations will be considered later. Let us now see how one can formally derive the conditions corresponding to Equations (4.4.120) for an arbitrary perturbation.

General Case

We study the wave equation

$$u_{tt} - u_{xx} + \varepsilon H(u_t, u_x) = 0; \qquad t \geq 0, -\infty < x < \infty \tag{4.4.135}$$

subject to initial conditions of the form

$$u(x, 0) = \rho(x) \tag{4.4.136a}$$

$$u_t(x, 0) = \mu(x). \tag{4.4.136b}$$

It is shown in Reference 4.4.8 that in order for the solution of Equation (4.4.135) to be bounded it is sufficient that ρ and μ be bounded and that $\varepsilon u_t H(u_t, u_x)$ be nonnegative for all E greater than some fixed E_0 in the u_t, u_x plane, where

$$E = \sqrt{u_t^2 + u_x^2}. \tag{4.4.137}$$

This is a condition analogous to the one derived in Section 3.2 [cf. Equation (3.2.78) and the discussion which follows] and is somewhat restrictive in the sense that there are many functions H which do not meet this requirement but which lead to bounded solutions.

At any rate, assuming that we are only dealing with bounded solutions, we expand u in multivariable form

$$u(x, t; \varepsilon) = u_0(x, t^+, \tilde{t}) + \varepsilon u_1(x, t^+, \tilde{t}) + \cdots \tag{4.4.138}$$

with

$$t^+ = (1 + \varepsilon^2\omega_2 + \varepsilon^3\omega_3 + \cdots)t \tag{4.4.139a}$$

$$\tilde{t} = \varepsilon t. \tag{4.4.139b}$$

As usual, it is easy to show that ω_1 need not be included in Equation (4.4.139a) as the dependence of the solution on εt is already accounted for through the $\tilde{t}$ variables. For a proof the reader may refer to Reference 4.4.8.

We then calculate the following equations for the first three terms in the expansion of u:

$$W(u_0) \equiv u_{0_{t^+t^+}} - u_{0_{xx}} = 0 \tag{4.4.140a}$$

$$W(u_1) = -2u_{0_{t^+\tilde{t}}} - H(u_{0_{t^+}}, u_{0_x}) \tag{4.4.140b}$$

$$W(u_2) = -2\omega_2 u_{0_{t^+t^+}} - 2u_{1_{t^+\tilde{t}}} - u_{0_{\tilde{t}\tilde{t}}}$$
$$- (u_{0_{\tilde{t}}} + u_{1_{t^+}}) \frac{\partial H}{\partial u_t}(u_{0_{t^+}}, u_{0_x}) - u_{1_x} \frac{\partial H}{\partial u_x}(u_{0_{t^+}}, u_{0_x}). \tag{4.4.140c}$$

The initial conditions imply that

$$u_0(x, 0, 0) = \rho(x); \qquad u_n(x, 0, 0) = 0, n \neq 0 \tag{4.4.141a}$$

$$u_{0_{t^+}}(x, 0, 0) = \mu(x); \qquad u_{1_{t^+}}(x, 0, 0) = -u_{0_{\tilde{t}}}(x, 0, 0)$$
$$u_{2_{t^+}}(x, 0, 0) = -u_{1_{\tilde{t}}}(x, 0, 0) - \omega_2 u_{0_{t^+}}(x, 0, 0), \quad \text{etc.} \tag{4.4.141b}$$

We write the general solution of Equation (4.4.140a) as

$$u_0(x, t^+, \tilde{t}) = f_0(\sigma, \tilde{t}) + g_0(\xi, \tilde{t}), \tag{4.4.142}$$

where σ and ξ are the characteristic variables

$$\sigma = x - t^+ \tag{4.4.143a}$$

$$\xi = x + t^+. \tag{4.4.143b}$$

It is more convenient to transform Equation (4.4.140b) to depend on σ and ξ. We find

$$-4u_{1_{\sigma\xi}} = 2f_{0_{\sigma\tilde{t}}} - 2g_{0_{\xi\tilde{t}}} - H(g_{0_\xi} - f_{0_\sigma}, g_{0_\xi} + f_{0_\sigma}). \tag{4.4.144}$$

In order that u_1 be bounded, it is first necessary that u_{1_σ} and u_{1_ξ} be bounded, we can solve for these derivatives directly as follows

$$-4u_{1_\sigma} = 2f_{0_{\sigma\tilde{t}}}\xi - 2g_{0_{\tilde{t}}} - \int^{\xi} H(g_{0_\xi} - f_{0_\sigma}, g_{0_\xi} + f_{0_\sigma})d\xi + \frac{\partial f_1}{\partial \sigma}(\sigma, \tilde{t}) \tag{4.4.145a}$$

$$-4u_{1_\xi} = -2g_{0_{\xi\tilde{t}}}\sigma + 2f_{0_{\tilde{t}}} - \int^{\sigma} H(g_{0_\xi} - f_{0_\sigma}, g_{0_\xi} + f_{0_\sigma})d\sigma + \frac{\partial g_1}{\partial \xi}(\xi, \tilde{t}) \tag{4.4.145b}$$

where f_1 and g_1 are arbitrary functions of their arguments to be determined by conditions on u_2.

Now, since u_0 is bounded, $g_{0_{\tilde{t}}}$ and $f_{0_{\tilde{t}}}$ are also bounded and we must have

$$\lim_{\xi\to\infty} \frac{1}{\xi} g_{0_{\tilde{t}}} = \lim_{\sigma\to\infty} \frac{1}{\sigma} f_{0_{\tilde{t}}} = 0. \tag{4.4.146}$$

Therefore, boundedness of u_{1_σ} and u_{1_ξ} requires (after dividing Equations (4.4.145) by ξ and σ respectively and taking the limits as ξ and σ tend to infinity) that

$$2f_{0_{\sigma\tilde{t}}} = \lim_{\xi \to \infty} \frac{1}{\xi} \int^{\xi} H(g_{0_\xi} - f_{0_\sigma}, g_{0_\xi} + f_{0_\sigma}) d\xi$$

$$\equiv y_0(f_{0_\sigma}, \tilde{t}) \tag{4.4.147a}$$

$$-2g_{0_{\xi\tilde{t}}} = \lim_{\sigma \to \infty} \frac{1}{\sigma} \int^{\rho} H(g_{0_\xi} - f_{0_\sigma}, g_{0_\xi} + f_{0_\sigma}) d\sigma$$

$$\equiv z_0(g_{0_\xi}, \tilde{t}). \tag{4.4.147b}$$

This is a system of two nonlinear equations for f_{0_σ} and g_{0_ξ}. Although it formally appears that the equations are uncoupled, they are in fact coupled through the $\tilde{t}$ dependence of y_0 and z_0. As we saw in the example of Equations (4.4.120), deriving the $\tilde{t}$ dependence of y_{0_ξ} and z_{0_σ} requires calculating certain averages involving g_{0_ξ} and f_{0_σ} respectively.

In Reference (4.4.9) a rigorous proof is given for the validity of Equations (4.4.147) in the case that f_{0_σ} and g_{0_ξ} are *periodic functions* of σ and ξ respectively with periods independent of $\tilde{t}$. This would be the case of periodic initial conditions as in the examples we have worked out so far for Equation (4.4.105).

The reader can verify that more general initial conditions can also be handled. (See for example Problem 4.4.11, and some of the applications discussed in Chapter 5.) The difficulty in applying Equations (4.4.147) indiscriminately is that y_0 and z_0 need not exist for certain perturbation functions H or initial conditions.

We point out that the ideas outlined above also apply without significant changes to cases where the perturbation function H is more general than in Equation (4.4.135); it may depend on higher derivatives of u or on the derivative of a nonlinear function of u, e.g., $(u^2)_x$ etc. Some examples are mentioned in Reference 4.4.8.

The theory as it applies to boundary value problems with fixed end points for Equation (4.4.135) is a straightforward extension of the foregoing results and is discussed in Reference (4.4.10). (See also Problem 4.4.12.)

Finally, we point out that the procedure discussed in this section carries over directly to elliptic equations if one expresses the solution in terms of complex characteristics. The idea is explored in Reference (4.4.11) which also contains some explicit examples.

PROBLEMS

1. Consider the solution to $O(\varepsilon)$ of Equation (4.4.17) with $0 < \varepsilon \ll 1$ and for the initial conditions in Equations (4.4.26).

   (a) Derive the initial conditions for the slowly varying parameters B_n, C_n, α_n, β_n, appearing in Equation (4.4.38).

(b) Show that boundedness of u_2 requires that all the B_n and C_n be constants. Hence the only nonvanishing terms correspond to $n = 1$ and 3.
(c) By requiring u_2 to be bounded with respect to $\tilde{t}$ show that ω_2 must obey Equation (4.4.42) and verify also the result in Equation (4.4.41).

2. Consider the initial value problem

$$u(x, 0) = A \sin kx + B \sin lx$$

$$u_t(t, 0) = -A\sqrt{1 + k^2} \cos kx - B\sqrt{1 + l^2} \cos lx$$

for Equation (4.4.17) with $0 < \varepsilon \ll 1$, and with A, B, k and l prescribed constants. Since the solution must involve at least two waves each with a different frequency shift, we canno longer use a single fast time in the form of Equation (4.4.28a). This is analogous to the situation discussed in Section 3.5.2 for the coupled weakly nonlinear oscillators. Guided by the approach used there assume a solution in the form

$$u = F(x, t_0, t_1, t_2, \ldots; \varepsilon) = u_0(x, t_0, t_1, \ldots) + \varepsilon u_1(x, t_0, t_1, \ldots) + \cdots,$$

where

$$t_0 = \varepsilon^0 t = t; \qquad t_1 = \varepsilon t; \qquad t_2 = \varepsilon^2 t, \qquad \text{etc.}$$

(a) Let u_0 be given by

$$u_0 = a_0 \sin(kx - \omega t + \phi_0) + b_0 \sin(lx - \mu t + \psi_0),$$

where

$$\omega = \sqrt{1 + k^2}, \qquad \mu = \sqrt{1 + l^2}$$

and a_0, b_0, ϕ_0 and ψ_0 depend on $t_1, t_2, \ldots$. Determine the dependence of a_0, b_0, ϕ_0 and ψ_0 on t_1 by requiring u_1 to be bounded with respect to t_0. Show that we encounter small divisors only when $|k/l| \approx 1$. (If $|k/l|$ is exactly equal to unity, the problem is really one of a single wave of constant amplitude and phase initially, and hence this limiting case is not interesting).
(b) Determine the dependence of a_0, b_0, ϕ_0 and ψ_0 on t_2 and the dependence of u_1 on t_1.
(c) Let $k = k_0$ and $l = k_0 + \varepsilon\kappa$, where $k_0(\neq 0)$ and κ are arbitrary finite constants independent of ε. Use trigonometric identities to reduce the initial conditions in this case to the form of a *single* slowly varying wave

$$u(x, 0) = A_0(\tilde{x})\sin[k_0 x + \Psi_0(\tilde{x})],$$

$$u_t(x, 0) = -A_0(\tilde{x})\sqrt{1 + k_0^2}\cos[k_0 x + \Psi_0(\tilde{x})].$$

where $\tilde{x} = \varepsilon x$, and A_0 and Ψ_0 are periodic functions of $\tilde{x}$ which reduce to constants as $\kappa \to 0$. Show that a multiple variable expansion for u as a function of x, t, $\tilde{x}$, and $\tilde{t} = \varepsilon t$ implies that, to order unity, u is a slowly modulated wave of the form given in Equation (4.4.53). Show also that k, a_0, and η_0 obey Equations (4.4.59), (4.4.62), and (4.4.65) respectively with $K' = 0$, i.e., $K = k_0 =$ const. Thus, the near resonant interaction between two waves for this problem merely corresponds to a periodic variation of the amplitude with

$$\xi = \tilde{x} - k_0\tilde{t}/\sqrt{1 + k_0^2}$$

as given by $A_0(\xi)$. The phase has a corresponding periodic variation, $\Psi_0(\xi)$, plus a secular contribution as given in Equation (4.4.66).

(d) For the case of three initial waves with wave numbers k, l, m, show that resonance occurs only if any two waves have nearly equal wave numbers. Set $k = k_0$, $l = l_0$ and $m = l_0 + \varepsilon\lambda$ with $k_0 \neq l_0$ and λ as three arbitrary constants independent of ε. Use an expansion procedure analogous to the one in part (c). to derive the solution to order unity. Show that in the special case $k_0 \approx l_0$ the problem reduces to that in part (c). Comment on the nature of the solution to order ε. In particular, does the unidirectional character of the wave motion persist to order ε?

3. Consider the weakly nonlinear problem

$$u_{tt} - u_{xx} + u + \frac{\varepsilon}{2\pi}\int_{-\infty}^{\infty}\int_{-\infty}^{\infty} u(x-\xi, t)u(\xi-\eta, t)u(\eta, t)d\xi\, d\eta = 0,$$

where the nonlinear term is in convolution form.

Fourier transformation of this equation gives

$$U_{tt} + (1 + k^2)U + \varepsilon U^3 = 0,$$

where

$$U(k, t; \varepsilon) = \frac{1}{\sqrt{2\pi}}\int_{-\infty}^{\infty} e^{-ikx}u(x, t; \varepsilon)dx$$

is the Fourier transform of u.

Consider the initial value problem

$$u(x, 0) = Ae^{-|x|}, \qquad A = \text{const.}$$

$$u_t(x, 0) = 0$$

and solve the nonlinear oscillator problem for U using the method of strained coordinates (Section 3.1) to $O(1)$.

Invert the result to derive u in the form

$$u(x, t) = \frac{1}{\sqrt{2\pi}}\int_{-\infty}^{\infty} e^{ikx}U(k, t; \varepsilon)dk$$

and study the behavior of the solution.

4. Having removed the secular terms from u_1, solve Equation (4.4.54) in the form

$$u_1 = a_1 \sin\psi_1 - \frac{a_0^3}{32}\sin 3\psi_0; \qquad \psi_1 = \theta + \eta_1,$$

where a_1 and η_1 depend on $\tilde{x}$ and $\tilde{t}$.

Consider Equation (4.4.51c) for u_2 and derive conditions on a_1 and η_1 by requiring u_2 to be periodic in θ. Solve these and study the behavior of the solution.

5. Carry out the details of the calculations for determining the solution to $O(1)$ of Equation (4.4.17) with $\varepsilon = 1$. In particular, show how one can calculate E_0 and η_0.

6. For the example of Equation (4.4.105) with initial conditions (4.4.106) show that the general form

$$u_0(x, t^+, \tilde{t}) = f_0(\sigma, \tilde{t}) + g_0(\xi, \tilde{t})$$

of the solution for u_0 leads to the conclusion that

$$g_0(\xi, \tilde{t}) = 0.$$

7. Study the signaling problem

$$u(0, t) = \sin pt, \qquad t \geq 0, p = \text{const.}$$

$$u(x, t) = 0, \qquad t < 0$$

for Equation (4.4.105). Assume an expansion for u in the form

$$u(x, t; \varepsilon) = u_0(x^+, \tilde{x}, t) + \varepsilon u_1(x^+, \tilde{x}, t) + \cdots,$$

where

$$x^+ = (1 + \varepsilon^2 \kappa_2 + \cdots)x, \qquad \tilde{x} = \varepsilon x.$$

8. Study the initial value problem defined by Equations (4.4.106) for the wave equation

$$u_{tt} - u_{xx} + \varepsilon[-\beta u_t + \alpha u_t^3] = 0,$$

where α and β are arbitrary positive constants.

In particular show that the initial sinusoidal wave tends to the "saw-tooth" wave with amplitude $(1/k)\sqrt{\beta/\alpha}$.

9. Consider the wave equation given in Problem 4.4.8 with the initial conditions given by Equations (4.4.116) and show that the first order solution u_0 is again given by Equation (4.4.134), where λ and m are now given by

$$\lambda(\tilde{t}) = 8e^{\beta\tilde{t}/2}/m^3(\tilde{t})$$

$$m^8 - \frac{8}{7}m^7 = 2^8\left[\frac{\alpha}{\beta}(e^{\beta\tilde{t}} - 1) + \frac{3}{7}\right]$$

and m and ϕ are related by Equation (4.4.131).

10. We wish to derive the conditions under which it is correct to assume that u_0 depends only on σ, t for the wave equation (4.4.135) when the initial conditions are in the special form

$$u(x, 0) = \rho(x)$$

$$u_t(x, 0) = -\rho'(x).$$

As pointed out in Reference (4.4.10), in order for $g_0(\xi, t)$ to vanish identically, f_0 must obey the differential equation

$$2f_{0_{\sigma\tilde{t}}} = H(-f_{0_\sigma}, f_{0_\sigma}).$$

Moreover, Equation (4.4.147b) reduces to

$$\lim_{\sigma\to\infty} \frac{1}{\sigma}\int^{\sigma} H(-f_{0_\sigma}, f_{0_\sigma})d\sigma = 0$$

which is a constraint on the allowable perturbation function H, and initial data ρ.

Verify that for H given by

$$H = u_t^3 \quad \text{or} \quad H = -\beta u_t + \alpha u_t^3$$

and the initial conditions

$$u(x, 0) = A \sin kx$$

$$u_t(x, 0) = -Ak \cos kx$$

the constraint condition is fulfilled. Find a simple perturbation function H and a function $\rho(x)$, for which the initial conditions $u(x, 0) = \rho(x)$, $u_t(x, 0) = -\rho'(x)$ *do not imply that* $g_0 = 0$.

11. Study the solution of Equation (4.4.105) to $O(1)$ for the *nonperiodic* initial conditions on $-\infty < x < \infty$.

$$u(x, 0; \varepsilon) = 2e^{-|x|}$$

$$u_t(x, 0; \varepsilon) = 0.$$

12. We wish to study the *initial/boundary* value problem for Equation (4.4.135) for the case of fixed ends over the interval $0 \leq x \leq l$, i.e.,

$$u(0, t; \varepsilon) = u(l, t; \varepsilon) = 0.$$

As pointed out in Reference (4.4.10), we can extend the definition of u over the entire x axis by requiring u to be odd and periodic with period $2l$, i.e.,

$$u(-x, t; \varepsilon) = -u(x, t; \varepsilon)$$

$$u(x + 2nl, t; \varepsilon) = u(x, t; \varepsilon), \qquad n = 0, \pm 1, \pm 2, \ldots .$$

The first of these conditions then implies that

$$u_t(-x, t; \varepsilon) = -u_t(x, t; \varepsilon)$$

$$u_x(-x, t; \varepsilon) = u_x(x, t; \varepsilon).$$

Therefore, the definition of H must be extended as follows

$$H(-u_t, u_x) = -H(u_t, u_x)$$

in order for the extended solution to satisfy Equation (4.4.135) for all values of x. Note that this requirement is automatically satisfied if H is an odd function of u_t. If it is even in u_t we must use the condition of odd extension and reverse the sign of H in the interval $(-l, 0)$.

Now, since u is odd and periodic we must have

$$g_0(\xi, \tilde{t}) = -f_0(-\xi, \tilde{t}).$$

Therefore, the solution for u_0 is

$$u_0 = f_0(\sigma, \tilde{t}) - f_0(-\xi, \tilde{t})$$

and the boundedness conditions, Equations (4.4.147), reduce to the simple condition

$$2f_{0_{\sigma\tilde{t}}}(\sigma, \tilde{t}) = \frac{1}{2l} \int_{-l}^{l} H[-f_{0_\sigma}(\sigma, \tilde{t}) + f_{0_\sigma}(-\xi, \tilde{t}), f_{0_\sigma}(\sigma, \tilde{t}) + f_{0_\sigma}(-\xi, \tilde{t})] d\xi.$$

Use the above results to study

$$u_{tt} - u_{xx} + \varepsilon\left(-u_t + \frac{u_t^3}{3}\right) = 0, \qquad 0 \le x \le \pi$$

$$u(x, 0) = 0$$

$$u_t(x, 0) = \sin x$$

$$u(0, t) = u(\pi, t) = 0.$$

In particular, derive an asymptotic form of the solution for large times (Reference 4.4.10).

References

4.4.1 C. S. Gardner, J. M. Greene, M. D. Kruskal, and R. M. Miura, Method for solving the Korteweg-de Vries equation, *Physics Review Letters* **19**, 1967, pp. 1095–1097.

4.4.2 G. B. Whitham, *Linear and Nonlinear Waves*, John Wiley and Sons, New York, 1974.

4.4.3 G. F. Carrier, M. Krook, and C. E. Pearson, *Functions of a Complex Variable, Theory and Technique*, McGraw-Hill Book Company, New York, 1966.

4.4.4 S. C. Chikwendu, Nonlinear wave propagation solution by Fourier transform perturbation, *International Journal of Nonlinear Mechanics*, to appear.

4.4.5 G. B. Whitham, A general approach to linear and nonlinear dispersive waves using a Lagrangian, *Journal of Fluid Mechanics* **22**, 1965, pp. 273–283.

4.4.6 J. C. Luke, A perturbation method for nonlinear dispersive wave problems. *Proceedings of the Royal Society, Series A*, **292**, 1966, pp. 403–412.

4.4.7 J. B. Keller and S. Kogelman, Asymptotic solutions of initial value problems for nonlinear partial differential equations, *S.I.A.M. Journal on Applied Mathematics* **18**, 1970, pp. 748–758.

4.4.8 S. C. Chikwendu and J. Kevorkian, A perturbation method for hyperbolic equations with small nonlinearities, *S.I.A.M. Journal on Applied Mathematics* **22**, 1972, pp. 235–258.

4.4.9 W. Eckhaus, New approach to the asymptotic theory of nonlinear oscillations and wave-propagation, *Journal of Mathematical Analysis and Applications* **49**, 1975, pp. 575–611.

4.4.10 R. W. Lardner, Asymptotic solutions of nonlinear wave equations using the methods of averaging and two-timing, *Quarterly of Applied Mathematics* **35**, 1977, pp. 225–238.

4.4.11 S. C. Chikwendu, Asymptotic solutions of some weakly nonlinear elliptic equations, *S.I.A.M. Journal on Applied Mathematics* **31**, 1976, pp. 286–303.

Chapter 5

Examples from Fluid Mechanics

In this chapter we consider some examples from fluid mechanics with the aim of illustrating the derivation of approximate equations, the role of several small parameters, and the calculation of uniformly valid results. Strictly speaking, there is no difference with the material discussed in Chapter 4. The emphasis here is on applications, so that we will not be dealing with model equations, but rather with nonlinear systems which reduce in the first approximation to some well known equation such as the transonic equation or the Korteweg-de Vries equation, etc. Typical to many problems is the occurrence of several small parameters. We will show how one must specify certain relations between these small parameters to account systematically for various competing perturbation terms and to derive meaningful approximations.

5.1 Weakly Nonlinear One-Dimensional Acoustics

Here, we study some problems concerning the propagation of weak disturbances in a quiescent gas. It is well known that in the first approximation (acoustics) the sound speed remains constant and disturbance quantities obey the one-dimensional wave equation. A regular perturbation based on this approximation fails in the far field because of the cumulative effects of the neglected nonlinear terms.

We will show that, for the case of outgoing waves (signaling problem), a correct description of the solution in the far field can be given by introducing a limit process expansion for large distances. In general, when both outgoing and incoming waves are present, as in the case for an initial value problem, we will derive a uniformly valid solution using a multiple variable expansion. This procedure is also applicable to signaling problems (see Problem 4.4.7) and is best suited for acoustics.

Problems in weakly nonlinear gas dynamics have also been worked out by different versions of the method of strained coordinates. These are surveyed in Reference 5.1.1 where it is shown that they fail in the higher orders and that uniformly valid results generally require multiple variable expansions. We will discuss this question in Sections 5.1.3 and 5.3.4.

5.1.1 Piston Problem, Near-Field Solution

Consider the classical problem of the one-dimensional propagation of waves produced by the motion of a piston at the end of a semi-infinite tube. The waves are described within the framework on an ideal (inviscid, non heat-conducting, etc.) gas. Various shock waves and regions of nonuniform flow are produced which propagate down the tube. The shock waves are treated as discontinuities. In actual practice, their structure and thickness depends on viscous and similar effects, but this thickness is assumed to be very small compared to the dimensions of interest. It is interesting to note, however, that a procedure similar to the one used here can also be used in the viscous case. When the piston motion is weak, that is, when velocities are produced which are small compared to the sound speed determined by the ambient state, then acoustics gives the usual first approximation. In acoustics, the special treatment of shock jumps can be forgotten, and they appear naturally in the linearized solution. The first step in our analysis of the problem is to work out first- and second-order acoustics. It is sufficient to consider isentropic flow, since only weak shock waves are to be produced, and the entropy changes are actually third-order in the shock jumps. This enables us to start the discussion with the following simple versions of the continuity and momentum equations:[1]

$$\frac{\partial c}{\partial T} + q\frac{\partial c}{\partial X} + \frac{\gamma - 1}{2} c \frac{\partial q}{\partial X} = 0, \qquad \text{continuity;} \tag{5.1.1}$$

$$\frac{\partial q}{\partial T} + q\frac{\partial q}{\partial X} + \frac{2}{\gamma - 1} c \frac{\partial c}{\partial X} = 0, \qquad \text{momentum.} \tag{5.1.2}$$

Here q and c are the flow velocity and local speed of sound, respectively, X is the distance, and T the time. In deriving these equations, the isentropic law for a perfect gas,

$$p/\rho^{\gamma} = \text{const.}, \tag{5.1.3}$$

[1] This of course is not necessary; continuity, momentum, and energy plus shock jumps could be used, but this method is simpler.

is assumed to apply throughout, where p = pressure, ρ = density, γ = ratio of specific heats and, by definition,

$$c^2 = \frac{\partial p}{\partial \rho} = \frac{\gamma p}{\rho} = (\text{const.})\gamma \rho^{\gamma-1},$$

$$2\frac{dc}{c} = (\gamma - 1)\frac{d\rho}{\rho}. \tag{5.1.4}$$

As we will see in Section 5.2, Equations (5.1.1) are precisely the same equations which govern long waves in shallow water of constant depth if we set $\gamma = 2$ and interpret q as the horizontal component of velocity and c^2 as the depth measured from the surface to a horizontal bottom.

An alternate form of Equations (5.1.1) and (5.1.2) expressing the rates of change of the Riemann invariants is derived by addition and subtraction of these equations:

$$\left\{\frac{\partial}{\partial T} + (q - c)\frac{\partial}{\partial X}\right\}\left\{\frac{c}{\gamma - 1} - \frac{q}{2}\right\} = 0, \tag{5.1.5}$$

$$\left\{\frac{\partial}{\partial T} + (q + c)\frac{\partial}{\partial X}\right\}\left\{\frac{c}{\gamma - 1} + \frac{q}{2}\right\} = 0. \tag{5.1.6}$$

The first of these equations states that the Riemann invariant $c/(\gamma - 1) - (q/2)$ does not change along the path of a left running sound wave propagating at the speed c relative to the fluid; the second corresponds to right running waves $(+X)$.

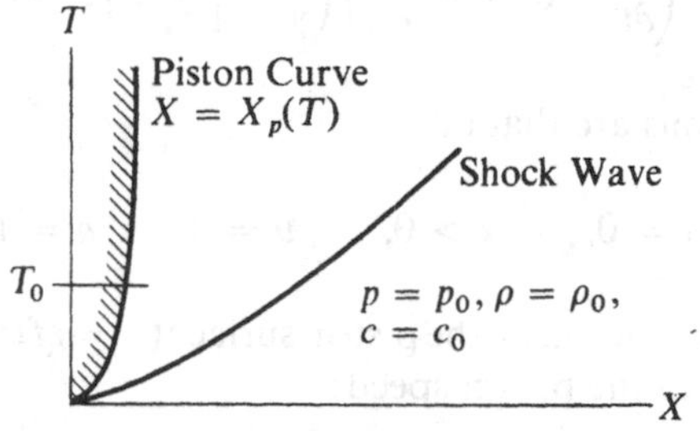

Figure 5.1.1 Piston Problem

Consider now the gas initially at rest (state p_0, ρ_0). The problem is specified by the motion of a piston with a characteristic time T_0, such that the distance traveled is small compared to $c_0 T_0$; c_0 = ambient speed of sound = $(\gamma p_0/\rho_0)^{1/2}$ (see Figure 5.1.1). This defines the small parameter of the problem, ε, and the piston curve $X_p(T)$ can be presented in terms of the dimensionless function f by

$$X_p(T) = \varepsilon c_0 T_0 f\left(\frac{T}{T_0}\right). \tag{5.1.7}$$

The corresponding piston speed q_p is

$$q_p = \frac{dX_p}{dT} = \varepsilon c_0 f'\left(\frac{T}{T_0}\right) \tag{5.1.8}$$

and is thus much less then c_0. Thus, the following dimensionless variables can be introduced:

$$u(x, t) = \frac{q(X, T)}{c_0}, \qquad a(x, t) = \frac{c(X, T)}{c_0}, \qquad x = \frac{X}{c_0 T_0}, \qquad t = \frac{T}{T_0}. \tag{5.1.9}$$

Equations (5.1.1), (5.1.2), and (5.1.5) and (5.1.6) are unchanged in form:

$$\frac{\partial a}{\partial t} + u\frac{\partial a}{\partial x} + \frac{\gamma - 1}{2} a \frac{\partial u}{\partial x} = 0, \tag{5.1.10}$$

$$\frac{\partial u}{\partial t} + u\frac{\partial u}{\partial x} + \frac{2}{\gamma - 1} a \frac{\partial a}{\partial x} = 0 \tag{5.1.11}$$

or

$$\left(\frac{\partial}{\partial t} + (u - a)\frac{\partial}{\partial x}\right)\left(\frac{a}{\gamma - 1} - \frac{u}{2}\right) = 0, \tag{5.1.12}$$

$$\left(\frac{\partial}{\partial t} + (u + a)\frac{\partial}{\partial x}\right)\left(\frac{a}{\gamma - 1} + \frac{u}{2}\right) = 0. \tag{5.1.13}$$

The initial conditions are that at

$$t = 0, \qquad x > 0, \qquad u = 0, \qquad a = 1 \tag{5.1.14}$$

and the boundary condition on the piston surface ($x = \varepsilon f(t)$) is that the velocity of the gas is equal to the piston speed:

$$u(\varepsilon f(t), t) = \varepsilon f'(t), \qquad t > 0. \tag{5.1.15}$$

The expansion procedure of acoustics has the limit process ($\varepsilon \to 0$, x, t fixed), and thus u and a have the following forms:

$$u(x, t) = \varepsilon u_1(x, t) + \varepsilon^2 u_2(x, t) + \cdots, \tag{5.1.16}$$

$$a(x, t) = 1 + \varepsilon a_1(x, t) + \varepsilon^2 a_2(x, t) + \cdots. \tag{5.1.17}$$

In the first few terms, only powers of ε appear, due to the simple occurrence of quadratic terms in the equations. From Equations (5.1.10) and (5.1.11), the first- and second-approximation equations are

$$\frac{\partial a_1}{\partial t} + \frac{\gamma - 1}{2}\frac{\partial u_1}{\partial x} = 0, \qquad (5.1.18)$$

$$\frac{\partial u_1}{\partial t} + \frac{2}{\gamma - 1}\frac{\partial a_1}{\partial x} = 0, \qquad (5.1.19)$$

$$\frac{\partial a_2}{\partial t} + \frac{\gamma - 1}{2}\frac{\partial u_2}{\partial x} = -u_1\frac{\partial a_1}{\partial x} - \frac{\gamma - 1}{2}a_1\frac{\partial u_1}{\partial x}, \qquad (5.1.20)$$

$$\frac{\partial u_2}{\partial t} + \frac{2}{\gamma - 1}\frac{\partial u_2}{\partial x} = -u_1\frac{\partial u_1}{\partial x} - \frac{2}{\gamma - 1}a_1\frac{\partial a_1}{\partial x}. \qquad (5.1.21)$$

For fixed t, the limit process $\varepsilon \to 0$ can be applied to the boundary condition (Equation 5.1.15) to produce

$$u(0, t) + \varepsilon f(t)u_x(0, t) + O(\varepsilon^2) = \varepsilon f'(t). \qquad (5.1.22)$$

The assumption that the expansion (Equation 5.1.16) is valid in $x \geq 0$ enables the terms in Equation (5.1.22) to be calculated:

$$\varepsilon u_1(0, t) + \varepsilon^2 u_2(0, t) + \varepsilon^2 f(t)u_{1_x}(0, t) + O(\varepsilon^3) = \varepsilon f'(t). \qquad (5.1.23)$$

The first- and second-order boundary conditions[2] for Equations (5.1.18), (5.1.19), (5.1.20) and (5.1.21) are, thus,

$$u_1(0, t) = f'(t), \qquad t \geq 0; \qquad (5.1.24)$$

$$u_2(0, t) = -u_{1_x}(0, t)f(t), \qquad t \geq 0. \qquad (5.1.25)$$

The first system is evidently equivalent to the simple wave equation with sound speed equal to unity in the units chosen:

$$\frac{\partial^2 u_1}{\partial t^2} - \frac{\partial^2 u_1}{\partial x^2} = 0. \qquad (5.1.26)$$

The general solution contains both outgoing $F(t - x)$ and incoming $G(t + x)$ waves, but it is clear from the physical problem that to the dominant order only outgoing waves should appear in the solution. This checks with the number of boundary conditions and enables a unique solution to be found. Thus, let

$$u_1(x, t) = F(t - x). \qquad (5.1.27)$$

It follows from the boundary condition (Equation 5.1.24) that u_1 is expressed in terms of the given piston speed by

$$u_1(x, t) = \begin{cases} f'(t - x), & t > x; \\ 0, & t < x; \end{cases} \qquad (5.1.28)$$

[2] Strictly speaking, an inner expansion in coordinates ($x^* = x/\varepsilon$, t) should be constructed to account for the boundary condition which is not applied at x fixed. However, the device in Equation (5.1.23) which transfers the boundary condition to the axis is equivalent to matching.

and Equation (5.1.18) or (5.1.19) gives

$$a_1(x, t) = \begin{cases} \dfrac{\gamma - 1}{2} f'(t - x), & t > x, \\ 0 & t < x. \end{cases} \tag{5.1.29}$$

Note that the Riemann invariant $a_1/(\gamma - 1) - u_1/2$ for incoming waves is const. $= 0$ in this flow. The flow is that of a simple wave of compression or expansion, the velocity and state being constant along one family ($t - x =$ const.) of characteristics of the ambient state (see Figure 5.1.2). If the velocity (and correspondingly the pressure) undergoes a positive jump at

$$t = 0, \qquad f'(0^+) > 0,$$

then this jump propagates with unit speed and is the acoustic approximation to the shock wave actually produced. Now, the second-order approximation can be considered by using Equations (5.1.28) and (5.1.29) in the right-hand sides of Equations (5.1.20) and (5.1.21) and by using the second-order boundary condition, Equation (5.1.25). The equations to be solved are

$$\frac{\partial a_2}{\partial t} + \frac{\gamma - 1}{2} \frac{\partial u_2}{\partial x} = \frac{\gamma^2 - 1}{4} f'(\tau) f''(\tau), \qquad \tau = t - x > 0, \tag{5.1.30}$$

$$\frac{\partial u_2}{\partial t} + \frac{2}{\gamma - 1} \frac{\partial a_2}{\partial x} = \frac{\gamma + 1}{2} f'(\tau) f''(\tau), \qquad \tau = t - x > 0, \tag{5.1.31}$$

and the boundary condition is

$$u_2(0, t) = f(t) f''(t), \qquad t > 0. \tag{5.1.32}$$

The solution can be split into a particular solution satisfying Equations (5.1.30) and (5.1.31) with the right-hand sides plus a solution of the homogeneous equations expressed as arbitrary functions of $(t + x, t - x)$. Just as before, the solution of the homogeneous system representing incoming waves is put equal to zero, and then the boundary condition (Equation 5.1.32) serves to

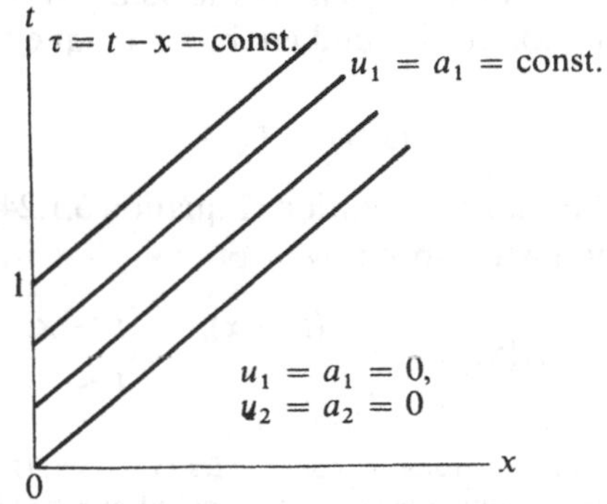

Figure 5.1.2 Wave Patterns

determine the solution uniquely. Note that even including the right-hand sides of Equations (5.1.30) and (5.1.31), the Riemann invariant for the incoming wave family is still equal to zero as it should according to Equation (5.1.12), that is,

$$a_2(x, t) = \frac{\gamma - 1}{2} u_2(x, t). \tag{5.1.33}$$

Thus, it is easy to verify that the solution of Equations (5.1.30) and (5.1.31) under the boundary condition (Equation 5.1.32) is

$$u_2(x, t) = \frac{\gamma + 1}{2} x f'(\tau) f''(\tau) + f(\tau) f''(\tau), \qquad \tau = t - x > 0, \tag{5.1.34}$$

$$a_2(x, t) = \frac{\gamma^2 - 1}{4} x f'(\tau) f''(\tau) + \frac{\gamma - 1}{2} f(\tau) f''(\tau), \qquad \tau = t - x > 0. \tag{5.1.35}$$

Here the first term is the particular solution, and in this term the coefficient x can be interpreted as a reflected-wave contribution [that is, $x \equiv \frac{1}{2}(t + x) - \frac{1}{2}(t - x))$]. In this representation, the sound waves of the second approximation are generated by an apparent distribution of sources and forces.

Thus, the final result of first- and second-order acoustics can be written

$$u(x, t; \varepsilon) = \varepsilon f'(\tau) + \varepsilon^2 \left\{ f(\tau) f''(\tau) + \frac{\gamma + 1}{2} x f'(\tau) f''(\tau) \right\} + O(\varepsilon^3), \tag{5.1.36}$$

$$a(x, t; \varepsilon) = 1 + \frac{\gamma - 1}{2} \varepsilon f'(\tau) + \frac{\gamma - 1}{2} \times \varepsilon^2 \left\{ f(\tau) f''(\tau) + \frac{\gamma + 1}{2} x f'(\tau) f''(\tau) \right\} + O(\varepsilon^3). \tag{5.1.37}$$

Various reasons for nonuniformity of the expansion are evident when the second term is compared to the first. The particular nonuniformity of interest to this Section is that of the far field. The second term of the expansion is comparable with the first when we have

$$\varepsilon x = O(1). \tag{5.1.38}$$

Near the wave front, this is equivalent to $\varepsilon t = O(1)$. In physical units, the distance away from the piston at which the acoustic formula becomes invalid is, thus roughly

$$\left(\frac{X}{X_p}\right) = O\left(\frac{c_0}{q_p}\right)^2. \tag{5.1.39}$$

If the piston speed is only $\frac{1}{10}$ of the sound speed, nonlinear effects begin to be important at a distance of the order of 100 times the piston displacement. Other nonuniformities appear, for example, when f'' is very large, but these are not discussed here. We only note here the second-order acoustic formula for the pressure on the piston. This follows from the fact that the Riemann invariant $[a/(\gamma - 1)] - (u/2) = 0$ to second-order, so that, in terms of the pressure on the piston, we have

$$\frac{p}{p_0} = \left(\frac{c}{c_0}\right)^{2\gamma/(\gamma-1)} = \left(1 + \frac{\gamma - 1}{2}\frac{q_p}{c_0}\right)^{2\gamma/(\gamma-1)}$$

$$= 1 + \gamma\frac{q_p}{c_0} + \frac{\gamma(\gamma+1)}{4}\left(\frac{q_p}{c_0}\right)^2 + \cdots. \tag{5.1.40}$$

5.1.2 Piston Problem, Uniformly Valid Solution

The results we derived in Section 5.1.1 are in terms of the two variables $\tau = t - x$ and x (or t) because we only have outgoing waves to order ε. Since the nonuniformity of the solution occurs when x becomes large along any characteristic $\tau = \text{const.}$, it is natural to seek a far field expansion where τ is retained and x is scaled as follows

$$\tilde{x} = \varepsilon x. \tag{5.1.41}$$

Moreover, as there were no singularities in either u or a, the requirement of matching the near and far field solutions dictates that the orders of the leading terms be the same. Therefore, we assume a far field expansion in the form[3]

$$u(x, t; \varepsilon) = \varepsilon\tilde{u}(\tau, \tilde{x}) + \varepsilon^2\tilde{u}_2(\tau, \tilde{x}) + \cdots \tag{5.1.42a}$$

$$a(x, t; \varepsilon) = 1 + \varepsilon\tilde{a}(\tau, \tilde{x}) + \varepsilon^2\tilde{a}_2(\tau, \tilde{x}) + \cdots. \tag{5.1.42b}$$

When we substitute these expansions into Equations (5.1.10) and (5.1.11) and note that

$$\frac{\partial}{\partial x} = -\frac{\partial}{\partial \tau} + \varepsilon\frac{\partial}{\partial \tilde{x}} \tag{5.1.43a}$$

$$\frac{\partial}{\partial t} = \frac{\partial}{\partial \tau} \tag{5.1.43b}$$

[3] It is easy to show that the assumed scales are correct by starting out with arbitrary choices and carrying out an order of magnitude analysis.

we obtain

$$\tilde{a}_\tau - \frac{\gamma - 1}{2}\tilde{u}_\tau = 0 \tag{5.1.44a}$$

$$\tilde{a}_{2_\tau} - \frac{\gamma - 1}{2}\tilde{u}_{2_\tau} = \tilde{u}\tilde{a}_\tau + \frac{\gamma - 1}{2}\tilde{a}\tilde{u}_\tau - \frac{\gamma - 1}{2}\tilde{u}_{\tilde{x}} \tag{5.1.44b}$$

$$\tilde{u}_\tau - \frac{2}{\gamma - 1}\tilde{a}_\tau = 0 \tag{5.1.44c}$$

$$\tilde{u}_{2_\tau} - \frac{2}{\gamma - 1}\tilde{a}_{2_\tau} = \tilde{u}\tilde{u}_\tau - \frac{2}{\gamma - 1}\tilde{a}_{\tilde{x}} + \frac{2}{\gamma - 1}\tilde{a}\tilde{a}_\tau. \tag{2.1.44d}$$

The terms of order ε, Equations (5.1.44a) and (5.1.44c), are identical, and give the Riemann invariant for incoming waves:

$$\frac{\tilde{a}}{\gamma - 1} - \frac{\tilde{u}}{2} = R(\tilde{x}). \tag{5.1.45}$$

This result also follows from Equation (5.1.12) for the exact Riemann invariant when the far-field expansions are used. We also note that $R(0) = 0$ in order to match with the Riemann invariant for the near field solution, i.e.,

$$\frac{a_1}{\gamma - 1} - \frac{u_1}{2} = 0 \tag{5.1.46}$$

and we will show that $R(\tilde{x}) = 0$ for all $\tilde{x} > 0$.

In order to derive the second relation linking $\tilde{a}$ and $\tilde{u}$ we eliminate the common homogeneous terms $\tilde{a}_{2_\tau} - (\gamma - 1)\tilde{u}_{2_\tau}/2$ from Equations (5.1.44b) and (5.1.44d). Next, we use Equation (5.1.45) to express $\tilde{a}$ in terms of $\tilde{u}$ and we obtain

$$\tilde{u}_{\tilde{x}} - \frac{\gamma + 1}{2}\tilde{u}\tilde{u}_\tau = R(\tilde{x})\tilde{u}_\tau - \frac{1}{\gamma - 1}\frac{dR}{d\tilde{x}}. \tag{5.1.47}$$

For a given $R(\tilde{x})$ this is a first order partial differential equation governing the solution in the far field. Since $R(\tilde{x}) = 0$ in the ambient region in front of the shock (because $\tilde{u}$ and $\tilde{a}$ are both equal to zero for all $\tilde{x}$ there), we can use the shock jump conditions to compute $R(\tilde{x})$ immediately behind the shock.

To work out the appropriate approximate shock jump conditions we start by deriving the exact shock relations using the ideas discussed in Section 4.1.3.

The generalization of Equation (4.1.123) to a system of n divergence relations

$$\frac{\partial P_i}{\partial T}(X, T, u_1, u_2, \ldots, u_n) + \frac{\partial Q_i}{\partial X}(X, T, u_1, \ldots, u_n) + R_i(X, T, u_1, \ldots, u_n) = 0; \qquad i = 1, \ldots, n \tag{5.1.48}$$

is the system of jump conditions

$$C \equiv \left(\frac{dX}{dT}\right)_{\text{shock}} = \frac{[Q_i]}{[P_i]}; \qquad i = 1, \ldots, n, \tag{5.1.49}$$

where [ ] denotes the jump of a quantity across the shock which is propagating with speed C.

For our problem, the appropriate conservation equations in divergence form are

$$\rho_T + (\rho q)_X = 0, \qquad \text{mass}; \tag{5.1.50a}$$

$$(\rho q)_T + (\rho q^2 + p)_X = 0, \qquad \text{momentum}; \tag{5.1.50b}$$

$$\left(\frac{\rho q^2}{2} + \frac{p}{\gamma - 1}\right)_T + \left\{q\left(\frac{\rho q^2}{2} + \frac{\gamma}{\gamma - 1} p\right)\right\}_X = 0, \qquad \text{energy}; \tag{5.1.50c}$$

and we calculate the following jump conditions (Rankine–Hugoniot relations).

$$p_s - p_0 = \rho_0 C q_s, \qquad \text{momentum}; \tag{5.1.51a}$$

$$\frac{\rho_0}{\rho_s} = 1 - \frac{q_s}{C}, \qquad \text{continuity}; \tag{5.1.51b}$$

$$C^2 - \frac{\gamma + 1}{2} C q_s - c_0^2 = 0, \qquad \text{wave speed}. \tag{5.1.51c}$$

Here $(\ )_s$ denotes conditions immediately behind the shock wave propagating into a uniform region of rest with the state $(\ )_0$. The wave-speed formula is derived from the energy equation with the use of continuity and momentum.

The shock speed C is also close to the acoustic speed c_0, since the disturbances are weak. Hence, let

$$\frac{C}{c_0} = 1 + \varepsilon \tilde{c}_s + \cdots,$$

so that the wave-speed formula gives

$$(1 + 2\varepsilon \tilde{c}_s + \cdots) - \frac{\gamma + 1}{2}(1 + \cdots)(\varepsilon \tilde{u}_s + \cdots) - 1 = 0$$

or

$$\tilde{c}_s = \frac{\gamma + 1}{4} \tilde{u}_s. \tag{5.1.52}$$

The local sound speed behind the shock is also easily calculated from Equations (5.1.51a) and (5.1.51b)

$$\frac{p_s}{p_0} = 1 + \gamma \varepsilon \tilde{u}_s + \cdots, \qquad \frac{\rho_0}{\rho_s} = 1 - \varepsilon \tilde{u}_s + \cdots$$

or

$$(1 + \varepsilon\tilde{a}_s + \cdots)^2 = \frac{p_s}{p_0}\frac{\rho_0}{\rho_s} = 1 + \varepsilon(\gamma - 1)\tilde{u}_s + \cdots. \tag{5.1.53}$$

Thus,

$$\frac{\tilde{a}_s}{\gamma - 1} - \frac{\tilde{u}_s}{2} = 0 \tag{5.1.54}$$

and this shows that we must set $R(\tilde{x}) = 0$ in Equation (5.1.45).

Therefore, the basic equation for the flow in the far field is

$$\tilde{u}_{\tilde{x}} - \frac{\gamma + 1}{2}\tilde{u}\tilde{u}_\tau = 0. \tag{5.1.55}$$

Equation (5.1.52) shows the difference in shock speed from the acoustic value, Equation (5.1.54), as a function of the shock strength $\tilde{u}_s$. Since $(\gamma + 1)/4 > (\gamma - 1)/2$ for $\gamma < 3$, the shock always moves faster than the corresponding sound wave, and this implies that a nonuniformity in a small region $x > t$ in the near field expansion, becomes a nonuniformity in a large region which must be accounted for in the far field.

The result in Equation (5.1.52) also follows from Equation (5.1.55) by regarding this to be a true conservation equation. Thus, if we apply the jump condition, Equation (5.1.49), directly to Equation (5.1.55) written in divergence form we obtain

$$\frac{d\tau}{d\tilde{x}} = -\frac{(\gamma + 1)}{4}\tilde{u}_s.$$

Let the shock path be given by $x_s(t)$ in physical coordinates. Then, using the definitions of τ and $\tilde{x}$ we calculate

$$\frac{d\tau}{dt} = 1 - \frac{dx_s}{dt}$$

$$\frac{d\tilde{x}}{dt} = \varepsilon\frac{dx_s}{dt}.$$

Therefore,

$$\frac{d\tau}{d\tilde{x}} = \frac{1 - dx_s/dt}{\varepsilon\, dx_s/dt}.$$

Equating the above two expressions for $d\tau/d\tilde{x}$ and solving the result for dx_s/dt to $O(1)$ we recover Equation (5.1.52) once we note that $dx_s/dt = C/c_0$.

The boundary condition needed to solve Equation (5.1.55) comes from matching the near and far field expansions and we consider this next.

Since τ is held fixed in both domains we match with respect to x and introduce the intermediate variable

$$x_\sigma = \frac{\tilde{x}}{\sigma(\varepsilon)}, \qquad \varepsilon \ll \sigma \ll 1$$

and matching to $O(\varepsilon)$ implies that

$$\lim_{\substack{\varepsilon \to 0 \\ \tau,\, x_\sigma \text{ fixed}}} \frac{\varepsilon u_1 - \varepsilon \tilde{u}}{\varepsilon} = 0. \tag{5.1.56}$$

It follows from Equation (5.1.28) for u_1 that we must set

$$\tilde{u}(\tau, 0) = f'(\tau). \tag{5.1.57}$$

The boundary condition for the far field is thus the same in this case as that of the near field. The near field generates disturbances which propagate unchanged along $\tau = \text{const.}$, and these appear near $\tilde{x} = 0$ as the boundary condition for the outer solution. The far field is thus uniquely determined, except that the shock wave has to be fitted in. The shock path can be represented in the coordinates of the far field as

$$x = t + \phi(\tilde{x}) \quad \text{or} \quad \tau = -\phi(\tilde{x}). \tag{5.1.58}$$

Note that the shock speed is

$$\frac{1}{1 + \varepsilon \tilde{c}_s} = \frac{c_0}{C} = \left(\frac{dt}{dx}\right)_s = 1 - \varepsilon\phi'(x),$$

so that

$$\tilde{c}_s = \phi'(\tilde{x}) = \frac{\gamma + 1}{4} \tilde{u}_s \tag{5.1.59}$$

on the shock $\tau = -\phi(\tilde{x})$. The solution to Equation (5.1.55) can be represented parametrically by the integrals of the characteristic differential equations with parameters λ, η (see Figure 5.1.3).

Thus, λ is the parameter which varies along a characteristic of Equation (5.1.55) while η is the parameter which is constant along this characteristic. Note that η identifies the characteristic which is along $\tau = t - x$ in the near field.

The characteristic equations

$$\frac{d\tilde{x}}{d\lambda} = 1, \qquad \frac{d\tau}{d\lambda} = -\frac{\gamma + 1}{2} \tilde{u}, \qquad \frac{d\tilde{u}}{d\lambda} = 0 \tag{5.1.60}$$

are easily solved in the form

$$\tilde{u}(\lambda, \eta) = \tilde{u}_0(\eta) \tag{5.1.61a}$$

$$\tilde{x}(\lambda, \eta) = \lambda \tag{5.1.61b}$$

$$\tau(\lambda, \eta) = -\frac{\gamma + 1}{2} \lambda u_0(\eta) + \tau_0(\eta). \tag{5.1.61c}$$

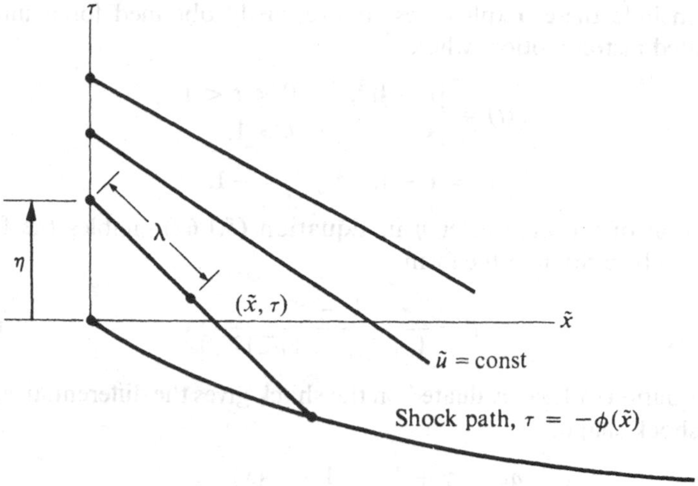

Figure 5.1.3 Far Field

Taking account of the boundary condition (5.1.57) defines $\tilde{u}_0$ and τ_0, and we have a one parameter representation of the far field solution:

$$\tilde{u}(\tilde{x}, \eta) = f'(\eta) \tag{5.1.62a}$$

$$\tau(\tilde{x}, \eta) = \eta - \frac{\gamma + 1}{2} \tilde{x} f'(\eta). \tag{5.1.62b}$$

For a given piston motion (f specified) we can eliminate η from these expressions to determine $\tilde{u}$ as a function of $\tilde{x}$ and τ. We can also use these relations together with Equation (5.1.59) for the shock speed to calculate the shock path. It can be noted from Equations (5.1.58) and (5.1.62) that the shock path bisects the characteristic directions of the outgoing waves since in the ambient state the characteristics are $\tau = 0$, while behind the shock on the lines $\eta =$ const., we have

$$\frac{d\tau}{d\tilde{x}} = -\frac{\gamma + 1}{2} f'(\eta). \tag{5.1.63a}$$

Now, according to Equation (5.1.58) $d\tau/d\tilde{x}$ at the shock is given by

$$\left(\frac{d\tau}{d\tilde{x}}\right)_{\text{shock}} = -\phi'(\tilde{x}) = -\frac{\gamma + 1}{4} f'(\eta) \tag{5.1.63b}$$

which is half the value of $d\tau/d\tilde{x}$ given by Equation (5.1.63a).

The picture in Figure 5.1.1 is drawn for the typical case of a decelerating piston in which the leading shock is followed by an expansion fan; acceleration or stoppage of the piston can produce following shocks, and the analysis has to include these. Explicit results are easily obtained for a uniformly decelerated piston motion, where

$$f(t) = \begin{cases} t - \frac{1}{2}t^2, & 0 < t < 1, \\ \frac{1}{2} & t > 1, \end{cases} \tag{5.1.64}$$

$$f' = 1 - t, \qquad f'' = -1.$$

Elimination of the parameter η in Equation (5.1.62) enables the far-field solution to be written in the form

$$\tilde{u} = \frac{1 - \tau}{1 + [(\gamma + 1)/2]\tilde{x}}. \tag{5.1.65}$$

Thus, Equation (5.1.59) evaluated on the shock gives the differential equation for the shock shape:

$$\frac{d\phi}{d\tilde{x}} = \frac{\gamma + 1}{4} \frac{1 + \phi(\tilde{x})}{1 + [(\gamma + 1)/2]\tilde{x}}. \tag{5.1.66}$$

With the initial condition $\phi(0) = 0$, the integral of Equation (5.1.66) gives a parabolic shock shape

$$\phi(\tilde{x}) = \sqrt{1 + [(\gamma + 1)/2]\tilde{x}} - 1. \tag{5.1.67}$$

In physical coordinates, this gives

$$x_s = t + \sqrt{1 + [(\gamma + 1)/2]\varepsilon x_s} - 1. \tag{5.1.68}$$

The asymptotic slope of the shock $(\tau, \tilde{x})$ is zero, so that all the characteristics up to that originating at $t = 1$ or $\eta = 1$ overtake the shock (see Figure 5.1.4). If the piston continues to decelerate after $t = 1$, these characteristics do not interact with the shock. The role of the shock in this problem is to establish the region in which the non-trivial solution of Equation (5.1.55) is to be used.

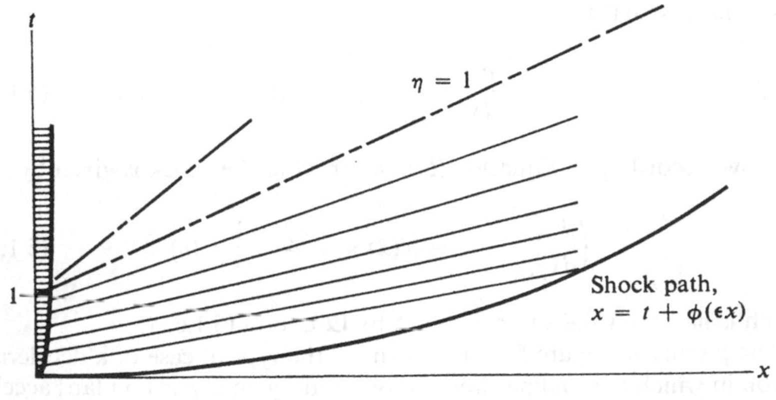

Figure 5.1.4 Asymptotic Field

In the example of this Section the "far" field equations contain the first approximation of the near field and, hence, provide the uniformly valid first approximation. However, if information on the piston surface only is required, the linear near-field theory can be used.

It is also interesting to derive the uniformly valid solution directly by a multiple variable exansion and shock conditions using the ideas discussed in Section 4.4.

In Problem 4.4.7 it was pointed out that for a signaling problem for a weakly nonlinear wave equations one can use a multiple variable expansion involving x^+, t and εx. Let us then expand u and a as follows

$$u(x, t; \varepsilon) = \varepsilon U_1(x^+, t, \tilde{x}) + \varepsilon^2 U_2(x^+, t, \tilde{x}) + \cdots \qquad (5.1.69a)$$

$$a(x, t; \varepsilon) = 1 + \varepsilon A_1(x^+, t, \tilde{x}) + \varepsilon^2 A_2(x, t, \tilde{x}) + \cdots, \qquad (5.1.69b)$$

where as usual

$$x^+ = (1 + \varepsilon^2 \kappa_2 + \cdots)x, \qquad \tilde{x} = \varepsilon x. \qquad (5.1.70)$$

The equations that result for U_1, U_2, A_1 and A_2 are

$$A_{1_t} + \frac{\gamma - 1}{2} U_{1_{x^+}} = 0 \qquad (5.1.71a)$$

$$U_{1_t} + \frac{2}{\gamma - 1} A_{1_{x^+}} = 0 \qquad (5.1.71b)$$

and

$$A_{2_t} + \frac{\gamma - 1}{2} U_{2_{x^+}} = -U_1 A_{1_{x^+}} - \frac{\gamma - 1}{2}(A_1 U_{1_{x^+}} + U_{1_{\tilde{x}}}) \qquad (5.1.72a)$$

$$U_{2_t} + \frac{2}{\gamma - 1} A_{2_{x^+}} = -U_1 U_{1_{x^+}} - \frac{2}{\gamma - 1}(A_1 A_{1_{x^+}} + A_{1_{\tilde{x}}}). \qquad (5.1.72b)$$

It is more convenient to use the characteristic coordinates, and since the shift κ_2 only affects the $O(\varepsilon^3)$ results we need not distinguish between x and x^+. Thus, let

$$\tau = t - x \qquad (5.1.73a)$$

$$\mu = t + x \qquad (5.1.73b)$$

and the equations for U_1 and A_1 become

$$\left[A_1 - \left(\frac{\gamma - 1}{2}\right)U_1\right]_\tau + \left[A_1 + \left(\frac{\gamma - 1}{2}\right)U_1\right]_\mu = 0 \qquad (5.1.74a)$$

$$\left[U_1 - \frac{2}{\gamma - 1} A_1\right]_\tau + \left[U_1 + \frac{2}{\gamma - 1} A_1\right]_\mu = 0. \qquad (5.1.74b)$$

By adding and subtracting we derive the general solution in terms of the Riemann invariants

$$\frac{A_1}{\gamma - 1} + \frac{U_1}{2} = F(\tau, \tilde{x}); \qquad \frac{A_1}{\gamma - 1} - \frac{U_1}{2} = G(\mu, \tilde{x}). \tag{5.1.75}$$

The same arguments that were used earlier to conclude that $R(\tilde{x}) = 0$ now give $G = 0$. Therefore,

$$U_1 = F(\tau, \tilde{x}) \tag{5.1.76a}$$

$$A_1 = \frac{\gamma - 1}{2} F(\tau, \tilde{x}). \tag{5.1.76b}$$

The boundary condition on the piston, Equation (5.1.15) becomes

$$U_1(0, t, 0) = f'(t), \qquad t > 0, \tag{5.1.77}$$

i.e.,

$$F(\tau, 0) = f'(\tau); \qquad \tau > 0. \tag{5.1.78}$$

If we now express U_2 and A_2 in terms of τ and μ and use Equations (5.1.76) for U_1 and $\dot{A}_1$, we transform Equations (5.1.72) to

$$\left[A_2 - \frac{(\gamma - 1)}{2} U_2\right]_\tau + \left[A_2 + \frac{(\gamma - 1)}{2} U_2\right]_\mu = \frac{(\gamma^2 - 1)}{4} FF_\tau - \frac{(\gamma - 1)}{2} F_{\tilde{x}} \tag{5.1.79a}$$

$$-\left[\frac{2A_2}{\gamma - 1} - U_2\right]_\tau + \left[U_2 + \frac{2A_2}{\gamma - 1}\right]_\mu = \frac{\gamma + 1}{2} FF_\tau - F_{\tilde{x}}. \tag{5.1.79b}$$

To solve for the Riemann invariant $(U_2/2) + A_2/(\gamma - 1)$, we multiply Equation (5.1.79a) by $1/(\gamma - 1)$ and Equation (5.1.79b) by $1/2$ and add. This gives

$$2\left[\frac{U_2}{2} + \frac{A_2}{\gamma - 1}\right]_\mu = \frac{\gamma + 1}{2} FF_\tau - F_{\tilde{x}}. \tag{5.1.80}$$

Now, the right hand side of Equation (5.1.80) depends only on τ and $\tilde{x}$. Therefore integration with respect to μ would introduce an inconsistent term proportional to μ in the solution for this Riemann invariant. Hence, we must set

$$\frac{\gamma + 1}{2} FF_\tau - F_{\tilde{x}} = 0 \tag{5.1.81}$$

and this is precisely Equation (5.1.55) when we identify $\tilde{u}$ with F. The boundary condition (5.1.78) is also the same as before and we duplicate our results for the uniformly valid solution when we use the shock condition, Equation (5.1.63b).

The multiple variable approach is seen to be significantly more efficient for problems of this type.

5.1.3 Initial Value Problem. Solution by Multiple Variable Expansions or Strained Coordinates

There are various physical situations which lead to an initial value problem on the infinite interval for Equations (5.1.1) and (5.1.2). For example, one can study the shock tube problem where a diaphragm at $x = 0$ separates two gases at rest with different pressures and densities in an infinite tube. At time $t = 0$, the diaphragm is removed instantaneously and a shock wave propagates into the low pressure gas while a rarefaction wave moves into the high pressure side. This problem can be solved exactly and if the initial pressures are not very different, we can also use perturbations where the small parameter ε measures the initial pressure difference. More complicated experiments can also be conceived.

A simpler physical model leading to a system of the form of Equations (5.1.1) and (5.1.2) is the case of long waves on shallow water which we will derive in Section 5.2. In this model, the appropriate equations are

$$\{u(1 + \eta)\}_t + \{u^2(1 + \eta) + (1 + \eta)^2/2\}_x = 0; \text{ conservation of momentum}$$

$$u_t + uu_x + \eta_x = 0 \tag{5.1.82a}$$

$$\{1 + \eta\}_t + \{u(1 + \eta)\}_x = 0; \text{ conservation of mass}$$

$$\eta_t + [u(1 + \eta)]_x = 0. \tag{5.1.82b}$$

The divergence forms (first lines) of these equations are derived in Reference 5.1.2, and it is easy to see that the simple forms (second lines) follow. Equations (5.1.82) are equivalent to Equations (5.1.10) and (5.1.11) if we set $\gamma = 2$ and $a^2 = 1 + \eta$. Here η is the height of the water surface above the undisturbed level, which is one unit of length above the horizontal bottom, and u is the horizontal velocity. (See Figure 5.2.1.)

Initial value problems for this system are also physically easy to realize. For example, the analog of the shock tube problem is the "dam breaking" problem where at time $t = 0$ a partition between two quiescent bodies of water at differing levels is broken.

In general, one can study the arbitrary initial value problem for Equations (5.1.82) where u and η are specified

$$\eta(x, 0) = \varepsilon F(x) \tag{5.1.83a}$$

$$u(x, 0) = \varepsilon G(x). \tag{5.1.83b}$$

Here ε measures the small departure of the initial state from equilibrium.

Equations (5.1.82) also admit shocks, and the jump conditions can be derived using Equation (5.1.49).

$$U = [u(1+\eta)]/[\eta], \qquad U = \left(\frac{dx}{dt}\right)_{\text{shock}}; \tag{5.1.84a}$$

$$U = [u^2(1+\eta) + (1+\eta)^2/2]/[u(1+\eta)]. \tag{5.1.84b}$$

As we will see in Section 5.2, the approximation leading to Equations (5.1.82) ceases to be valid when the gradients of u and η become large. Therefore the solution of these equations should become invalid *before* waves begin to break. However, as pointed out in Reference (5.1.2), hydraulic jumps are fairly well described by the jump conditions in Equations (5.1.84). For further discussion on the physics of the problem, as it pertains to water waves, we refer the reader to Reference (5.1.2). For our purposes, we will regard Equations (5.1.82) as a mathematical model for water waves when hydraulic jumps are present.

The material which follows is based on the work of Reference (5.1.3) where Equations (5.1.82) were analysed using multiple variable expansions for two types of initial conditions: (i) when F and G are periodic, and (ii) when F and G vanish outside some finite interval.

We assume a solution for u and η in the form

$$u(x, t, \varepsilon) = \varepsilon u_1(\sigma, \xi, \tilde{t}) + \varepsilon^2 u_2(\sigma, \xi, \tilde{t}) + \cdots \tag{5.1.85a}$$

$$\eta(x, t, \varepsilon) = \varepsilon \eta_1(\sigma, \xi, \tilde{t}) + \varepsilon^2 \eta_2(\sigma, \xi, \tilde{t}) + \cdots, \tag{5.1.85b}$$

where, as usual,

$$\sigma = x - t \tag{5.1.86a}$$

$$\xi = x + t \tag{5.1.86b}$$

$$\tilde{t} = \varepsilon t \tag{5.1.86c}$$

and we have not introduced a shift ω_2 in t because we will only consider the solution to $O(\varepsilon)$. We change variables using

$$\frac{\partial}{\partial t} = -\frac{\partial}{\partial \sigma} + \frac{\partial}{\partial \xi} + \varepsilon \frac{\partial}{\partial \tilde{t}} \tag{5.1.87a}$$

$$\frac{\partial}{\partial x} = \frac{\partial}{\partial \sigma} + \frac{\partial}{\partial \xi} \tag{5.1.87b}$$

and Equations (5.1.82) imply that the u_i and η_i satisfy the following system

$$2(u_1 + \eta_1)_\xi = 0 \tag{5.1.88a}$$

$$2(u_1 - \eta_1)_\sigma = 0 \tag{5.1.88b}$$

$$2(u_2 + \eta_2)_\xi = -(u_1 + \eta_1)_{\tilde{t}} - u_1(u_1 + \eta_1)_\sigma - u_1(u_1 + \eta_1)_\xi - \eta_1 u_{1_\sigma} - \eta_1 u_{1_\xi} \tag{5.1.89a}$$

$$2(u_2 - \eta_2)_\sigma = (u_1 - \eta_1)_{\tilde{t}} + u_1(u_1 - \eta_1)_\sigma + u_1(u_1 - \eta_1)_\xi - \eta_1 u_{1_\sigma} - \eta_1 u_{1_\xi}. \tag{5.1.89b}$$

The initial conditions imply that

$$\eta_1(x, x, 0) = F(x), \qquad \eta_2(x, x, 0) = 0; \tag{5.1.90a}$$

$$u_1(x, x, 0) = G(x), \qquad u_2(x, x, 0) = 0; \tag{5.1.90b}$$

etc.

The solution of Equations (5.1.88) is

$$\eta_1(\sigma, \xi, \bar{t}) = f_1(\sigma, \bar{t}) + g_1(\xi, \bar{t}) \tag{5.1.91a}$$

$$u_1(\sigma, \xi, \bar{t}) = f_1(\sigma, \bar{t}) - g_1(\xi, \bar{t}), \tag{5.1.91b}$$

where f_1 and g_1 satisfy the following initial conditions

$$f_1(x, 0) = \tfrac{1}{2}\{F(x) + G(x)\} \tag{5.1.92a}$$

$$g_1(x, 0) = \frac{1}{2}\{F(x) - G(x)\}. \tag{5.1.92b}$$

If we now use Equations (5.1.91) to simplify the right hand sides of Equations (5.1.89) we find

$$2(u_2 + \eta_2)_\xi = -2f_{1_{\bar{t}}} - 3f_1 f_{1_\sigma} + g_1 f_{1_\sigma} + f_1 g_{1_\xi} + g_1 g_{1_\xi} \tag{5.1.93a}$$

$$2(u_2 - \eta_2)_\sigma = -2g_{1_{\bar{t}}} + 3g_1 g_{1_\xi} - f_1 g_{1_\xi} - f_1 f_{1_\sigma} - g_1 f_{1_\sigma}. \tag{5.1.93b}$$

Integration gives

$$2(u_1 + \eta_2) = -\xi(2f_{1_{\bar{t}}} + 3f_1 f_{1_\sigma}) + \frac{1}{2} g_1^2 + f_1 g_1 + f_{1_\sigma} \int^\xi g_1 \, d\xi + p_2(\sigma, \bar{t}) \tag{5.1.94a}$$

$$2(u_2 - \eta_2) = -\sigma(2g_{1_{\bar{t}}} - 3g_1 g_{1_\xi}) - \frac{1}{2} f_1^2 - f_1 g_1 - g_{1_\xi} \int^\sigma f_1 \, d\sigma + q_2(\xi, \bar{t}). \tag{5.1.94b}$$

If the initial conditions are periodic, we choose the length scale for non-dimensionalizing distances to be the period of the initial disturbances and assume

$$\int_0^1 F(x) dx = 0 \tag{5.1.95a}$$

$$\int_0^1 G(x) dx = 0. \tag{5.1.95b}$$

This can be achieved by the appropriate choice of datum level and inertial frame of reference.

Since $F(x)$ and $G(x)$ are periodic in x with zero averages we also expect f_1 and g_1 to be periodic in σ and ξ respectively with zero averages. Thus, the integrals appearing in Equations (5.1.94) are bounded. The only unbounded terms are those exhibited with a ξ and σ factor, and we eliminate them by setting

$$f_{1_{\tilde{t}}} + \tfrac{3}{2} f_1 f_{1_\sigma} = 0 \qquad (5.1.96a)$$

$$g_{1_{\tilde{t}}} - \tfrac{3}{2} g_1 g_{1_\xi} = 0. \qquad (5.1.96b)$$

Note that for outgoing waves $g_1 = 0$, and if we set $\gamma = 2$, $\tau = \sigma$, $\tilde{t} = \tilde{x}$, $\tilde{u} = f_1$ to conform with the notation of Section 5.1.2, we recover Equation (5.1.96a) from Equation (5.1.47) for the piston problem.

If $F(x)$ and $G(x)$ vanish outside some bounded interval, we expect f_1 and g_1 to also vanish outside the domain bounded by the characteristics emanating from the end points of this interval. At any rate, the integrals in Equations (5.1.94) are bounded and Equations (5.1.96) still hold.

In Reference 5.1.3 the solution of the approximate equations (5.1.96) are compared with numerical integrations of the exact equations for the following two initial value problems $F(x) = \cos 2\pi x$, $G(x) = 0$, and

$$F(x) = \begin{cases} 0, & |x| > \tfrac{1}{2} \\ (1 - 4x^2)^2, & |x| < \tfrac{1}{2} \end{cases} \qquad (5.1.97)$$

$$G(x) = \tfrac{1}{2} F(x). \qquad (5.1.98)$$

In either case the numerical results are indistinguishable from the approximate solution for values of ε as large as 0.1.

Note that the procedure fails for initial data for which the integrals in Equations (5.1.94) are not bounded.

An interesting interpretation of the approximate results is derived in Reference 5.1.3 by regarding u and η to be defined by their first approximations εu_1 and $\varepsilon \eta_1$. Equation (5.1.96a) implies that f_1 maintains its constant initial value along the straight lines in the $\sigma, \tilde{t}$ plane with slope

$$\frac{d\sigma}{d\tilde{t}} = \frac{3}{2} f_1. \qquad (5.1.99a)$$

These lines have the slopes

$$\frac{dx}{dt} = 1 + \frac{3}{2} \varepsilon f_1 \qquad (5.1.99b)$$

in the x, t plane. Now if we identify εu_1 with u, and $\varepsilon \eta_1$ with η, we have

$$\eta + u = \varepsilon(\eta_1 + u_1) = 2\varepsilon f_1 \qquad (5.1.100)$$

so that $\eta + u$ *is constant along the straight lines with slopes*

$$\frac{dx}{dt} = 1 + \frac{3}{4}(\eta + u). \qquad (5.1.101)$$

This means that a uniformly valid approximation to $O(\varepsilon)$ for $\eta + u$ is given by the solution of the first order partial differential equation

$$(\eta + u)_t + \{1 + \tfrac{3}{4}(\eta + u)\}(\eta + u)_x. \qquad (5.1.102a)$$

The same reasoning applied to Equation (5.1.96b) leads to

$$(\eta - u)_t + \{-1 - \tfrac{3}{4}(\eta - u)\}(\eta - u)_x. \tag{5.1.102b}$$

There is an essential difference between Equations (5.1.102) and the result one obtains by regular perturbations (initially valid solution). To show this we proceed from Equations (5.1.82) written in characteristic form [cf. Equations (5.1.12) and (5.1.13)]

$$\left\{\frac{\partial}{\partial t} + (u - \sqrt{1+\eta})\frac{\partial}{\partial x}\right\}\left\{\sqrt{1+\eta} - \frac{u}{2}\right\} = 0 \tag{5.1.103a}$$

$$\left\{\frac{\partial}{\partial t} + (u + \sqrt{1+\eta})\frac{\partial}{\partial x}\right\}\left\{\sqrt{1+\eta} + \frac{u}{2}\right\} = 0. \tag{5.1.103b}$$

These equations imply that $u \pm 2\sqrt{1+\eta}$ remains constant along the characteristic curves with slope $(dx/dt) = u \pm \sqrt{1+\eta}$. Except for the case of a simple wave, the characteristics are not straight. If one now approximates these results for u and η small, i.e., by a regular perturbation, the conclusion is that

$$u \pm (2 + \eta) = \text{const.} \tag{5.1.104a}$$

along the curves (also not straight) with slope

$$\frac{dx}{dt} = u \pm \left(1 + \frac{\eta}{2}\right) = \pm\left\{1 + \left(\frac{\eta}{2} \pm u\right)\right\}. \tag{5.1.104b}$$

This initially valid result disagrees with what we found in Equations (5.1.102), i.e., that

$$u \pm \eta \quad \text{or} \quad u \pm (2 + \eta) = \text{const.} \tag{5.1.105a}$$

along the *straight lines* with slope

$$\frac{dx}{dt} = \pm\left\{1 + \frac{3}{4}(\eta \pm u)\right\}. \tag{5.1.105b}$$

In this example, the cumulative effect of the nonlinear terms is to "straighten" the curves along which the linear theory predicts that $u \pm \eta$ is constant to order ε.

It is also interesting to compare the results in Equations (5.1.102) with those corresponding to one version of the method of strained coordinates proposed in Reference 5.1.4 and applied to one-dimensional compressible flows in Reference 5.1.5.

As the procedure depends on the characteristic form of the equations, we introduce the *exact characteristic coordinates* $\alpha(x, t)$ and $\beta(x, t)$ and write

Equations (5.1.103) in the standard form

$$x_\alpha = (u + a)t_\alpha \tag{5.1.106a}$$

$$x_\beta = (u - a)t_\beta \tag{5.1.106b}$$

$$(u + 2a)_\alpha = 0 \tag{5.1.106c}$$

$$(u - 2a)_\beta = 0, \tag{5.1.106d}$$

where

$$a = \sqrt{1 + \eta}. \tag{5.1.106e}$$

We choose the origin of the α, β coordinate system such that $\alpha(x, 0) = \beta(x, 0) = x$ for all x. Thus, the x axis corresponds to the line $\alpha = \beta$. We wish to solve this system subject to the initial conditions given by Equations (5.1.83), so we must set $a(x, 0) = 1 + \varepsilon F(x)/2 + O(\varepsilon^2)$.

Now, we assume that *both dependent and independent variables* are expanded in a perturbation series with respect to ε holding α and β, the exact characteristics, fixed, i.e.,

$$x(\alpha, \beta; \varepsilon) = x_0(\alpha, \beta) + \varepsilon x_1(\alpha, \beta) + \cdots \tag{5.1.107a}$$

$$t(\alpha, \beta; \varepsilon) = t_0(\alpha, \beta) + \varepsilon t_1(\alpha, \beta) + \cdots \tag{5.1.107b}$$

$$u(\alpha, \beta; \varepsilon) = \varepsilon u_1^*(\alpha, \beta) + \varepsilon^2 u_2^*(\alpha, \beta) + \cdots \tag{5.1.107c}$$

$$a(\alpha, \beta; \varepsilon) = 1 + \varepsilon a_1^*(\alpha, \beta) + \varepsilon^2 a_2^*(\alpha, \beta) + \cdots. \tag{5.1.107d}$$

The initial conditions at $t = 0$ now translate to conditions when $\beta = \alpha$, and we have

$$x_0(\alpha, \alpha) = \alpha \tag{5.1.108a}$$

$$t_0(\alpha, \alpha) = 0 \tag{5.1.108b}$$

$$x_1(\alpha, \alpha) = 0 \tag{5.1.109a}$$

$$t_1(\alpha, \alpha) = 0 \tag{5.1.109b}$$

$$u_1^*(\alpha, \alpha) = G(\alpha) \tag{5.1.110a}$$

$$a_1^*(\alpha, \alpha) = F(\alpha)/2. \tag{5.1.110b}$$

When we substitute the expansions in Equations (5.1.107) into Equations (5.1.106) and collect terms with equal powers of ε we obtain

$$x_{0_\alpha} - t_{0_\alpha} = 0 \tag{5.1.111a}$$

$$x_{0_\beta} + t_{0_\beta} = 0 \tag{5.1.111b}$$

$$(u_1^* + 2a_1^*)_\alpha = 0 \tag{5.1.112a}$$

$$(u_1^* - 2a_1^*)_\beta = 0 \tag{5.1.112b}$$

$$x_{1_\alpha} - t_{1_\alpha} = t_{0_\alpha}(u_1^* + a_1^*) \tag{5.1.113a}$$

$$x_{1_\beta} + t_{1_\beta} = t_{0_\beta}(u_1^* + a_1^*). \tag{5.1.113b}$$

Integration of Equations (5.1.111) subject to the initial conditions in Equations (5.1.108) gives

$$x_0 = (\alpha + \beta)/2 \tag{5.1.114a}$$

$$t_0 = (\alpha - \beta)/2. \tag{5.1.114b}$$

We now integrate Equations (5.1.112) and use the initial conditions in Equations (5.1.110) to find

$$u_1^* = \tfrac{1}{2}[G(\beta) + G(\alpha)] + \tfrac{1}{2}[F(\beta) - F(\alpha)] \tag{5.1.115a}$$

$$a_1^* = \tfrac{1}{4}[G(\beta) - G(\alpha)] + \tfrac{1}{4}[F(\beta) + F(\alpha)]. \tag{5.1.115b}$$

When these results are used to compute the right hand sides of Equations (5.1.113) we obtain

$$x_{1_\alpha} - t_{1_\alpha} = \tfrac{3}{8}[G(\beta) + F(\beta)] + \tfrac{1}{8}[G(\alpha) - F(\alpha)] \tag{5.1.116a}$$

$$x_{1_\beta} + t_{1_\beta} = -\tfrac{1}{8}[G(\beta) + F(\beta)] - \tfrac{3}{8}[G(\alpha) - F(\alpha)]. \tag{5.1.116b}$$

Now these are easily integrated and the arbitrary functions of α and β which arise can be determined using Equations (5.1.109). The result is

$$x_1 = \frac{3}{16}(\alpha - \beta)[G(\alpha) + G(\beta) - F(\alpha) + F(\beta)] + \frac{1}{8}\int_\beta^\alpha G(s)ds \tag{5.1.117a}$$

$$t_1 = \frac{3}{16}(\alpha - \beta)[G(\alpha) - G(\beta) - F(\alpha) - F(\beta)] + \frac{1}{8}\int_\alpha^\beta F(s)ds. \tag{5.1.117b}$$

One could, at this point, invert the results obtained so far to compute u and a to order ε in terms of x and t. For this example it is easy to directly compare the prediction of this method with the results of Equations (5.1.102). If we restrict attention to the characteristics on which $\beta = \text{const.}$, i.e., those with slope $(dx/dt) = x_\alpha/t_\alpha$, we must have $u + 2a = \text{const.}$ According to the results in Equations (5.1.115) $u + 2a$ is indeed constant to order ε, as it should be, along the $\beta = \text{const.}$ characteristics. The next question is what is the approximation to $O(\varepsilon)$ of the $\beta = \text{const.}$ curves?

Using Equations (5.1.114) and (5.1.117), we calculate the following expression for the slope dx/dt along the $\beta = \text{const.}$ curves

$$\begin{aligned}\frac{dx}{dt} &= \frac{x_\alpha}{t_\alpha} = \frac{x_{0_\alpha} + \varepsilon x_{1_\alpha}}{t_{0_\alpha} + \varepsilon t_{1_\alpha}} + O(\varepsilon^2)\\ &= 1 + 2\varepsilon(x_{1_\alpha} - t_{1_\alpha}) + O(\varepsilon^2)\end{aligned}$$

and using Equation (5.1.116a) this becomes

$$= 1 + \frac{3}{4}\varepsilon[F(\beta) + G(\beta)] + \frac{\varepsilon}{4}[F(\alpha) - G(\alpha)] + O(\varepsilon^2). \quad (5.1.118)$$

Clearly these are not straight lines since

$$E = \frac{\varepsilon}{4}[F(\alpha) - G(\alpha)] \quad (5.1.119)$$

is, in general, not equal to zero.

In fact, Equation (5.1.118) is nothing more than Equation (5.1.104b) with the plus sign since $u + 1 + \eta/2 \approx 1 + \varepsilon(u_1^* + a_1^*)$, and using the expression we calculated for u_1^* and a_1^* gives Equation (5.1.118). A similar conclusion can be drawn for the other family (α = const.) of curves.

Thus, this method of strained coordinates is no better than a regular perturbation procedure (and certainly much more complicated) for this simple example. The correct uniformly valid result to $O(\varepsilon)$ is given by Equations (5.1.105) and this is confirmed by numerical integrations also.

In other versions of the method of strained coordinates a certain arbitrariness is introduced in the definition of the expansions. As a result, special rules are invoked, such as the statement that the second approximation should not have a stronger singularity than the first, in order to remove the arbitrariness. However, these methods, either in characteristic coordinates or by some other expansion of both dependent and independent variables, do not always give a uniformly valid solution for both near and far fields. What is worse, they sometimes superficially seem to work when in fact the results are incorrect. For this reason we do not devote much attention to the method as it has been used for studying partial differential equations.

Some further discussion of near and far field equations, comments on the method of strained coordinates, and application to other physically interesting examples of wave propagation including viscous effects, chemical reactions and radiation is presented in Reference 5.1.6.

References

5.1.1 L. Crocco, Coordinate perturbation and multiple scale in gas dynamics, *Philosophical Transactions, Royal Society of London, Series A* **A272**, 1972, pp. 272–301.

5.1.2 G. B. Whitham, *Linear and Nonlinear Waves*, John Wiley and Sons, New York, 1974.

5.1.3 C. E. Pearson, Dual time scales in a wave problem governed by coupled nonlinear equations, S.I.A.M. Review, **23**, 1981, pp. 425–433.

5.1.4 C. C. Lin, On a perturbation theory based on the method of characteristics, *Journal of Mathematics and Physics* **33**, 1954, 117–134.

5.1.5 P. A. Fox, Perturbation theory of wave propagation based on the method of characteristics, *Journal of Mathematics and Physics* **34**, 1955, 133–151.

5.1.6 W. Lick, Wave propagation in real gases, *Advances in Applied Mechanics* **10**, Fasc. 1, 1967, pp. 1–72.

5.2 Small Amplitude Waves on Shallow Water

The problem considered in this section is very similar to that in the last. Small amplitude waves are generated either by a wave-maker at $x = 0$ or by some initial disturbance of the surface. These waves then propagate to large distances and we must account for the small but cumulative nonlinear effects. In the case of water waves there is one added complication in that the problem involves two parameters. In addition to the ratio of wave height to wave length (which is analogous to the ε of the previous section), we also have the ratio of wave length to depth. Different relative orders of magnitudes for these two parameters lead to different expansions. We will first derive the equations for large amplitude waves on shallow water, and show that these are analogous to the equations of one dimensional compressible flow. We will then concentrate on small amplitude waves and derive the evolution equations corresponding to the most general value of the relative orders of magnitude of the two small parameters.

5.2.1 Derivation of Equations, Long Wave Expansion

Consider a channel of constant depth H with the origin of the X, Y coordinate system on the undisturbed free surface. The surface is defined by (see Figure 5.2.1)

$$Y = H\eta\left(\frac{X}{L}, \frac{T}{T_1}\right), \tag{5.2.1}$$

where T is the time, L is a characteristic wave length and T_1 is a characteristic time both of which will be defined presently.

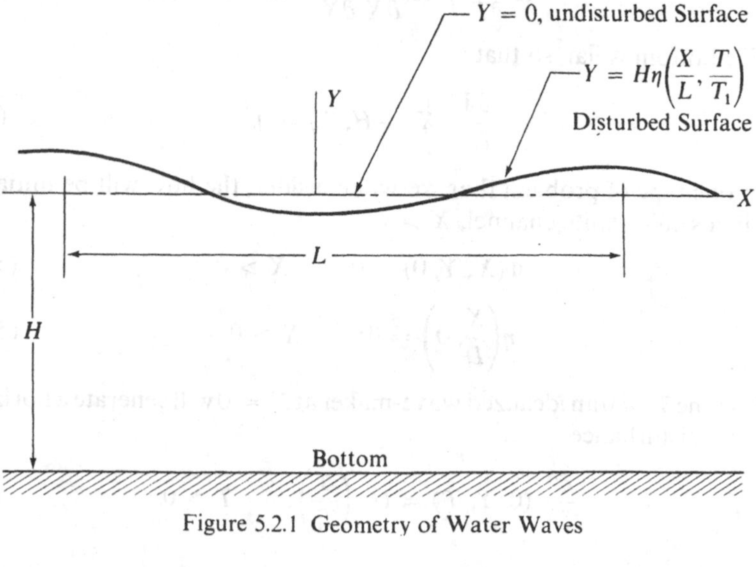

Figure 5.2.1 Geometry of Water Waves

The fluid is considered to be incompressible and of constant density and the flow irrotational so that the whole problem involves solving Laplace's equation for the velocity potential with boundary conditions on the bottom $Y = -H$ and on the free surface. In addition, we will consider as in Section 5.1, two types of motion. The first will be unidirectional, ($X > 0$) generated from rest by a wave-maker at $X = 0$. The second type of problem will involve waves traveling in both directions due to some initial surface disturbance.

The velocity vector $\mathbf{q}$ is given by

$$\mathbf{q} = \text{grad } \Phi(X, Y, T), \tag{5.2.2}$$

where Φ obeys

$$\frac{\partial^2 \Phi}{\partial X^2} + \frac{\partial^2 \Phi}{\partial Y^2} = 0, \qquad -H \leq Y \leq \eta H. \tag{5.2.3}$$

A Bernoulli equation, obtained by integrating the momentum equations, holds throughout the flow

$$\frac{\partial \Phi}{\partial T} + \frac{q^2}{2} + \frac{p}{\rho} + gY = \frac{p_\infty}{\rho}, \tag{5.2.4}$$

where $q = |\mathbf{q}|$, p is the pressure in the flow, ρ is the constant density, g is the constant gravitational acceleration, and p_∞ is the pressure at the free surface. This equation provides the following boundary condition on the free surface

$$\frac{\partial \Phi}{\partial T} + \frac{q^2}{2} + gH\eta = 0 \quad \text{on} \quad Y = H\eta. \tag{5.2.5}$$

The other boundary condition on the free surface is kinematic; a particle on the surface remains there

$$\frac{\partial \Phi}{\partial Y} = H \frac{\partial \eta}{\partial T} + H \frac{\partial \Phi}{\partial X} \frac{\partial \eta}{\partial X} \quad \text{on} \quad Y = H\eta. \tag{5.2.6}$$

The bottom is flat, so that

$$\frac{\partial \Phi}{\partial Y}(X, -H, T) = 0. \tag{5.2.7}$$

In one type of problem that we will consider, the flow will be initially at rest in a semi-infinite channel, $X > 0$

$$\Phi(X, Y, 0) = 0; \qquad X > 0 \tag{5.2.8a}$$

$$\eta\left(\frac{X}{L}, 0\right) = 0; \qquad X > 0 \tag{5.2.8b}$$

and at time $T = 0$ an idealized wave-maker at $X = 0$ will generate a horizontal velocity disturbance

$$\frac{\partial \Phi}{\partial X}(0, Y, T) = Uw\left(\frac{T}{T_0}\right); \qquad T > 0. \tag{5.2.9}$$

Here U and T_0 are characteristic velocity and time scales associated with the wave-maker.

A second category of problems for motion in the infinite channel is obtained by specifying an initial surface shape but zero velocity[4]

$$H\eta\left(\frac{X}{L}, 0\right) = Ah\left(\frac{X}{L_0}\right) \tag{5.2.10a}$$

$$\Phi(X, Y, 0) = 0, \tag{5.2.10b}$$

where A is a characteristic amplitude and L_0 is a characteristic horizontal scale for the initial surface shape.

In both problems, we have the characteristic velocity

$$V = \sqrt{gH} \tag{5.2.11}$$

defined by the parameters which occur in the Bernoulli equation. For the signaling problem, the characteristic horizontal length scale is then

$$L = VT_0 \tag{5.2.12}$$

associated with the wave-maker, and the characteristic time is

$$T_1 = L/V = T_0. \tag{5.2.13}$$

The two dimensionless parameters are

$$\delta = H/L = H/VT_0 = H^{1/2}/g^{1/2}T_0 \tag{5.2.14}$$

and

$$\varepsilon = U/V = U/\sqrt{gH}. \tag{5.2.15}$$

For the initial value problem, L is just L_0, the horizontal scale of the initial surface profile. Thus, the time scale is

$$T_1 = L_0/V = L/\sqrt{gH} \tag{5.2.16}$$

and δ and ε are given by

$$\delta = H/L_0 \tag{5.2.17}$$

$$\varepsilon = A/H. \tag{5.2.18}$$

Having defined L and T_1 in either case, we introduce the following dimensionless variables

$$x = \frac{X}{L}, \quad y = \frac{Y}{H}, \quad t = \frac{VT}{L}, \quad \phi = \frac{\Phi}{VL} \tag{5.2.19}$$

and obtain the dimensionless system

$$\delta^2\phi_{xx} + \phi_{yy} = 0 \tag{5.2.20}$$

[4] One could also specify an arbitrary initial velocity distribution or initial surface pressure distribution ($\partial\Phi/\partial T$), etc.

with boundary conditions

$$\phi_y(x, -1, t) = 0; \qquad \text{bottom} \tag{5.2.21}$$

$$\phi_t(x, \eta, t) + \frac{1}{2}\left[\phi_x^2(x, \eta, t) + \frac{1}{\delta^2}\phi_y^2(x, \eta, t)\right] + \eta = 0;$$

$$\text{Bernoulli, free surface} \tag{5.2.22}$$

$$\phi_y(x, \eta, t) = \delta^2[\eta_t + \phi_x(x, \eta, t)\eta_x]; \qquad \text{kinematic, free surface.} \tag{5.2.23}$$

For the signaling problem, we have the initial conditions

$$\phi(x, y, 0) = \eta(x, 0) = 0 \tag{5.2.24}$$

and the boundary condition

$$\phi_x(0, y, t) = \varepsilon w(t) \tag{5.2.25}$$

while for the initial value problem, we take

$$\phi(x, y, 0) = 0 \tag{5.2.26a}$$

$$\eta(x, 0) = \varepsilon h(x). \tag{5.2.26b}$$

The problem for ϕ is elliptic but depends on the parameter t, which enters the boundary conditions, in a nonlinear and differential way. An exact solution is certainly very difficult even numerically because of the complicated free surface conditions. However, a variety of approximate theories can be constructed. The fundamental approximation for this section is that of long waves, i.e., $\delta \ll 1$. Later on, this approximation will be combined with the assumption of small amplitudes, $\varepsilon \ll 1$.

The form of the long wave expansion is

$$\begin{aligned}\phi(x, y, t; \varepsilon, \delta) = \alpha(\delta)\{&\phi_0(x, y, t; \varepsilon) + \delta^2\phi_1(x, y, t; \varepsilon)\\ &+ \delta^4\phi_2(x, y, t; \varepsilon) + O(\delta^6)\}.\end{aligned} \tag{5.2.27}$$

In this limit x, y, t and ε are held fixed as $\delta \to 0$ and $\alpha(\delta)$ is to be determined later. The bottom boundary condition, Equation (5.2.21), dominates the resulting form of the expansion. The Laplace equation, Equation (5.2.20), is satisfied in a simple way equivalent to a power-series expansion in $(y + 1)$. The following successive approximations and solutions are found

$$\phi_{0_{yy}} = 0; \qquad \phi_0 = \varphi_0(x, t; \varepsilon) \tag{5.2.28}$$

$$\phi_{1_{yy}} = -\phi_{0_{xx}}; \qquad \phi_1 = \varphi_1(x, t; \varepsilon) - \frac{(y+1)^2}{2}\varphi_{0_{xx}}(x, t; \varepsilon) \tag{5.2.29}$$

$$\begin{aligned}\phi_{2_{yy}} = -\phi_{1_{xx}}; \qquad \phi_2 = \varphi_2(x, t; \varepsilon) &- \frac{(y+1)^2}{2}\varphi_{1_{xx}}(x, t; \varepsilon)\\ &+ \frac{(y+1)^4}{24}\varphi_{0_{xxxx}}(x, t; \varepsilon).\end{aligned} \tag{5.2.30}$$

The corresponding expansions of the velocity components in the x and y directions, respectively, are

$$\phi_x = \alpha\left\{\varphi_{0_x} + \delta^2\left[\varphi_{1_x} - \frac{(y+1)^2}{2}\varphi_{0_{xxx}}\right]\right.$$

$$\left. + \delta^4\left[\varphi_{2_x} - \frac{(y+1)^2}{2}\varphi_{1_{xxx}} + \frac{(y+1)^4}{24}\varphi_{0_{xxxxx}}\right] + O(\delta^6)\right\} \quad (5.2.31)$$

$$\phi_y = \alpha\left\{-\delta^2(y+1)\varphi_{0_{xx}} + \delta^4\left[-(y+1)\varphi_{1_{xx}} + \frac{(y+1)^3}{6}\varphi_{0_{xxxx}}\right] + O(\delta^6)\right\}. \quad (5.2.32)$$

We see that due to geometry of the domain, the main velocity component is parallel to the bottom, while the vertical component is an order of magnitude smaller.

5.2.2 Nonlinear Long-Wave Equations

The nonlinear long-wave equations are the first approximation equations resulting from the expansion in Equation (5.2.27) under the assumption that the dominant term in the shape of the free surface is independent of δ, and correspondingly that $\alpha = 1$. Thus, on the free surface, we have

$$y = \eta(x, t; \varepsilon, \delta) = \eta_0(x, t; \varepsilon) + \delta^2\eta_1(x, t; \varepsilon) + \cdots. \quad (5.2.33)$$

Now, if we substitute the expansion for ϕ and its derivatives into the free surface boundary conditions, Equations (5.2.22) and (5.2.23), we obtain the system

$$\varphi_{0_t} + \tfrac{1}{2}\varphi_{0_x}^2 + \eta_0 = 0 \quad (5.2.34a)$$

$$-(\eta_0 + 1)\varphi_{0_{xx}} = \eta_{0_t} + \varphi_{0_x}\eta_{0_x}. \quad (5.2.34b)$$

Rewriting these in terms of $u_0 = \varphi_{0_x}$, shows that they are exactly analogous to the equations of one-dimensional compressible flow [see Equations (5.1.10) and (5.1.11)] with $\gamma = 2$ and where we identify $(1 + \eta_0)^{1/2}$ with a, the dimensionless speed of sound.

$$\eta_{0_t} + [u_0(1 + \eta_0)]_x = 0 \quad (5.2.35a)$$

$$u_{0_t} + u_0 u_{0_x} + \eta_{0_x} = 0. \quad (5.2.35b)$$

Hydraulic jumps which are the analogues to shock waves in gas dynamics are possible for this system. In fact, the jump conditions were derived in Section 5.1.3 [see Equations (5.1.84)]. However, strictly speaking, the approximation $\delta \ll 1$ which leads to Equations (5.2.35) is not valid when waves steepen. Although the jump conditions, Equations (5.1.84), do sometimes accurately describe actual hydraulic jumps, the shallow water equations predict that all waves which carry a positive slope η_x will steepen and break. This can be easily seen either from the Riemann invariants or the small amplitude theory discussed in Section 5.1.3. However, observations indicate that some waves do not break because of the dispersive effects which have been ignored in this leading approximation.

For this reason, the Boussinesq equations, which supplement Equation (5.2.35b) with the leading dispersive term, are often used as a more accurate model for actual water waves. In our notation, the Boussinesq equations are[5]

$$\eta_t + [u(1+\eta)]_x = 0 \tag{5.2.36a}$$

$$u_t + uu_x + \eta_x = -\frac{\delta^2}{3}\eta_{xtt}. \tag{5.2.36b}$$

We will show later on that for small amplitude waves, the Boussinesq equations and the exact equations have the same uniformly valid solution at least to first order. Thus, even though Equations (5.2.36) are not obtained from the exact boundary conditions through a limit process, they do retain all the essential features of the exact equations. Unfortunately, the inclusion of the third derivative term prevents the occurrence of any hydraulic jumps and is therefore too drastic. For further discussion and proposed improved models, we refer the reader to Reference (5.2.1). We will not study the strictly nonlinear problem here. In the next two sections we will only consider the weakly nonlinear case of small amplitude waves.

5.2.3 Signaling Problem for Small Amplitude Waves

We will proceed as in Section 5.1 and develop our results first by constructing and matching two separate limit process expansions in the near and far fields. We will then show how a multiple variable expansion provides the uniformly valid solution directly. A crucial step in our calculations involves the choice of relative orders of magnitude for δ and ε since both are now assumed small. So we begin by regarding δ as an unknown function of ε.

We expand the free surface shape as follows

$$\eta(x, t; \delta(\varepsilon), \varepsilon) = \varepsilon\zeta_0(x, t) + \beta_1(\varepsilon)\zeta_1(x, t) + O(\beta_2), \tag{5.2.37}$$

where $\beta_1 = o(\varepsilon)$ is also unknown. Each term in the expansion for ϕ, Equation (5.2.27), must now be further expanded with respect to ε giving

$$\varphi_i(x, t, \varepsilon) = \varphi_{i0}(x, t) + \varepsilon\varphi_{i1}(x, t) + \cdots, \qquad i = 0, 1, 2, \ldots. \tag{5.2.38}$$

[5] In Reference 5.2.1, the reader can find a heuristic argument for inclusion of the term $-\delta^2\eta_{xtt}/3$ on the right-hand side of Equation (5.2.36b).

When we substitute all these expansions into the free surface boundary conditions, Equations (5.2.22) and (5.2.23), we find

$$\alpha\{\varphi_{00_t} + \varepsilon\varphi_{01_t} + \cdots + \delta^2[\varphi_{10_t} - \tfrac{1}{2}\varphi_{00_{xxt}} + \cdots]\} + \tfrac{1}{2}\alpha^2\varphi_{00_x}^2 + \varepsilon\zeta_0 + \beta_1(\varepsilon)\zeta_1 = O(\beta_2) + O(\alpha\varepsilon^2) + O(\alpha^2\delta^2) + O(\alpha^2\varepsilon) + O(\alpha\delta^4) + O(\alpha\delta^2\varepsilon) \tag{5.2.39a}$$

$$-\alpha\{\delta^2\varphi_{00_{xx}} + \varepsilon\delta^2\varphi_{01_{xx}} + \delta^4[\varphi_{10_{xx}} - \tfrac{1}{6}\varphi_{00_{xxxx}}] + \varepsilon\delta^2\zeta_0\varphi_{00_{xx}}\} = \delta^2\{\varepsilon\zeta_{0_t} + \beta_1\zeta_{1_t} + \alpha\varepsilon\varphi_{00_x}\zeta_{0_x}\} + O(\alpha\varepsilon^2\delta^2) + O(\delta^4\varepsilon\alpha) + O(\delta^2\beta_2) + O(\alpha\delta^6). \tag{5.2.39b}$$

The dominant terms in both Equations (5.2.39) balance only if we set

$$\alpha = \varepsilon \tag{5.2.40}$$

in which case we have

$$\varphi_{00_t} + \zeta_0 = 0 \tag{5.2.41a}$$

$$-\varphi_{00_{xx}} = \zeta_{0_t}. \tag{5.2.41b}$$

The next largest terms in Equation (5.2.39a) are of order $\alpha\varepsilon$, $\alpha\delta^2$, α^2, β_1, i.e., ε^2, $\varepsilon\delta^2$, β_1. Clearly, the richest equations for φ_{10} and ζ_1 result by choosing $\delta^2 = O(\varepsilon)$ and $\beta_1 = \varepsilon^2$. Thus, we introduce the similarity parameter κ, independent of δ and ε, such that

$$\delta = \kappa\sqrt{\varepsilon}. \tag{5.2.42}$$

Now, the remainder of the terms retained in Equation (5.2.39b) are $O(\varepsilon^3)$ and the neglected terms are $O(\varepsilon^4)$. Thus, in the second approximation we must satisfy

$$(\kappa^2\varphi_{10} + \varphi_{01})_t + \zeta_1 = \frac{\kappa^2}{2}\varphi_{00_{xxt}} - \frac{1}{2}\varphi_{00_x}^2 \tag{5.2.43a}$$

$$-(\kappa^2\varphi_{10} + \varphi_{01})_{xx} = \zeta_{1_t} + \varphi_{00_x}\zeta_{0_x} - \frac{\kappa^2}{6}\varphi_{00_{xxxx}} + \zeta_0\varphi_{00_{xx}}. \tag{5.2.43b}$$

We see that φ_{10} and φ_{01} always occur in the combination $\kappa^2\varphi_{10} + \varphi_{01}$, so we simplify the notation by setting

$$\varphi_{00}(x, t) \equiv \varphi_0^*(x, t) \tag{5.2.44a}$$

$$\kappa^2\varphi_{10}(x, t) + \varphi_{01}(x, t) \equiv \kappa^2\varphi_1^*(x, t) \tag{5.2.44b}$$

etc., and the system to be solved is

$$\varphi_{0t}^* + \zeta_0 = 0; \qquad -\varphi_{0_{xx}}^* = \zeta_{0_t} \tag{5.2.45}$$

$$\kappa^2\varphi_{1_t}^* + \zeta_1 = \frac{\kappa^2}{2}\varphi_{0_{xxt}}^* - \frac{1}{2}\varphi_{0_x}^{*2} \tag{5.2.46a}$$

$$-\kappa^2\varphi_{1_{xx}}^* = \zeta_{1_t} + \varphi_{0_x}^*\zeta_{0_x} - \frac{\kappa^2}{6}\varphi_{0_{xxxx}}^* + \zeta_0\varphi_{0_{xx}}^* \tag{5.2.46b}$$

corresponding to the expansion

$$\phi(x, y, t; \varepsilon, \delta) = \varepsilon\varphi_0^*(x, t) + \kappa^2\varepsilon^2\left[\varphi_1^*(x, t) - \frac{(y+1)^2}{2}\varphi_{0_{xx}}^*(x, t)\right] + O(\varepsilon^3). \tag{5.2.46c}$$

We expect this expansion to be valid for values of κ which are not large, because $\kappa^2\varepsilon^2$ will then no longer be $O(\varepsilon^2)$. To solve the system, we eliminate ζ_0 and ζ_1 and obtain

$$\varphi_{0_{tt}}^* - \varphi_{0_{xx}}^* = 0 \tag{5.2.47a}$$

$$\varphi_{1_{tt}}^* - \varphi_{1_{xx}}^* = \frac{1}{3}\varphi_{0_{xxtt}}^* - \frac{2}{\kappa^2}\varphi_{0_x}^*\varphi_{0_{xt}}^* - \frac{1}{\kappa^2}\varphi_{0_t}^*\varphi_{0_{tt}}^*. \tag{5.2.47b}$$

Once φ_0^* and φ_1^* have been determined ζ_0 and ζ_1 follow directly from Equations (5.2.45a) and (5.2.46a).

For the signaling problem defined by Equations (5.2.24) and (5.2.25), we have zero initial conditions for φ_0^* and φ_1^* and the following boundary conditions at $x = 0$:

$$\varphi_{0_x}^*(0, t) = w(t) = r'(t), \qquad \text{say,} \tag{5.2.48a}$$

$$\varphi_{1_x}^*(0, t) = 0. \tag{5.2.48b}$$

The solution for φ_0^* is

$$\varphi_0^*(x, t) = \begin{cases} -r(t-x); & t > x \\ 0; & t < x \end{cases} \tag{5.2.49}$$

and the equation for φ_1^* becomes

$$\varphi_{1_{tt}}^* - \varphi_{1_{xx}}^* = -\frac{1}{3}r^{(iv)}(t-x) - \frac{3}{\kappa^2}r'(t-x)r''(t-x)$$

$$\equiv s(t-x). \tag{5.2.50}$$

This is easily solved for the initial conditions $\varphi_1^*(x, 0) = \varphi_{1_t}^*(x, 0) = 0$ and the zero boundary condition, Equation (5.2.48b), at $x = 0$, but we do not write down this solution.

The essential point to notice is that the solution will contain terms proportional to x because, in terms of the characteristic variables $\tau = t - x$ and $\xi = t + x$, Equation (5.2.50) is

$$4\frac{\partial^2\varphi_1^*}{\partial\tau\,\partial\xi} = s(\tau). \tag{5.2.51}$$

Integration then produces terms proportional to ξ. Now, along any characteristic $\tau = \text{const.}$, $\xi \sim x$ as $x \to \infty$. Thus, the expansion in Equation (5.2.46c) is not uniformly valid when $x = O(1/\varepsilon)$.

The cumulative nonlinear effects demand a different far field expansion. Just as we showed in Section 5.1.2, we can construct a limit process expansion in the far field with respect to a coordinate system moving with the speed of the linear waves. Since in the first approximation, the solution for the near field depends only on $t - x$, this can be used as one of the coordinates. Then a study of the two principal possibilities (εx or εt) shows that εt is appropriate.[6]

Thus, we consider a limit process expansion with the fixed coordinates

$$\tau = t - x \tag{5.2.52a}$$

$$\tilde{t} = \varepsilon t. \tag{5.2.52b}$$

In these coordinates, the long wave expansion, Equation (5.2.27) is still valid since it only involves y. Thus, we assume a far-field expansion in the form

$$\begin{aligned}\phi(x, y, t; \varepsilon, \kappa\sqrt{\varepsilon}) &= \varepsilon\tilde{\varphi}_0(\tau, \tilde{t}) + \kappa^2\varepsilon^2\left[\tilde{\varphi}_1(\tau, \tilde{t}) - \frac{(y+1)^2}{2}\tilde{\varphi}_{0_{\tau\tau}}\right] \\ &+ \kappa^4\varepsilon^3\left[\tilde{\varphi}_2(\tau, \tilde{t}) - \frac{(y+1)^2}{2}\tilde{\varphi}_{1_{\tau\tau}} + \frac{(y+1)^4}{24}\tilde{\varphi}_{0_{\tau\tau\tau\tau}}\right] + O(\varepsilon^4).\end{aligned} \tag{5.2.53}$$

The derivatives of ϕ can now be calculated using

$$\frac{\partial}{\partial x} = -\frac{\partial}{\partial \tau} \tag{5.2.54a}$$

$$\frac{\partial}{\partial t} = \frac{\partial}{\partial \tau} + \varepsilon\frac{\partial}{\partial \tilde{t}}. \tag{5.2.54b}$$

We find

$$\phi_t = \varepsilon\tilde{\varphi}_{0_\tau} + \varepsilon^2\left[\tilde{\varphi}_{0_{\tilde{t}}} + \kappa^2\tilde{\varphi}_{1_\tau} - \frac{(y+1)^2}{2}\kappa^2\tilde{\varphi}_{0_{\tau\tau\tau}}\right] + O(\varepsilon^3) \tag{5.2.55a}$$

$$\phi_x = -\varepsilon\tilde{\varphi}_{0_\tau} + \kappa^2\varepsilon^2\left[-\tilde{\varphi}_{1_\tau} + \frac{(y+1)^2}{2}\tilde{\varphi}_{0_{\tau\tau\tau}}\right] + O(\varepsilon^3) \tag{5.2.55b}$$

$$\phi_y = \kappa^2\varepsilon^2[-(y+1)\tilde{\varphi}_{0_{\tau\tau}}] + \kappa^4\varepsilon^3\left[-(y+1)\tilde{\varphi}_{1_{\tau\tau}} + \frac{(y+1)^3}{6}\tilde{\varphi}_{0_{\tau\tau\tau\tau}}\right] + O(\varepsilon^4). \tag{5.2.55c}$$

The free surface has a corresponding expansion

$$\eta(x, t; \varepsilon, \kappa\sqrt{\varepsilon}) = \varepsilon\tilde{\eta}_0(\tau, \tilde{t}) + \varepsilon^2\tilde{\eta}_1(\tau, \tilde{t}) + O(\varepsilon^3) \tag{5.2.56a}$$

$$\eta_t = \varepsilon\tilde{\eta}_{0_\tau} + \varepsilon^2[\tilde{\eta}_{1_\tau} + \tilde{\eta}_{0_{\tilde{t}}}] + O(\varepsilon^3) \tag{5.2.56b}$$

$$\eta_x = -\varepsilon\tilde{\eta}_{0_\tau} - \varepsilon^2\tilde{\eta}_{1_\tau} + O(\varepsilon^3). \tag{5.2.56c}$$

[6] This is so because derivatives with respect to εt occur to order ε^2 while derivatives with respect to εx first occur to order ε^3.

Again, equations of motion result from the free surface boundary conditions. These take the form

$$\varepsilon\tilde{\varphi}_{0_\tau} + \varepsilon^2\left[\tilde{\varphi}_{0_{\tilde{t}}} + \kappa^2\tilde{\varphi}_{1_\tau} - \frac{\kappa^2}{2}\tilde{\varphi}_{0_{\tau\tau\tau}}\right] + \frac{1}{2}\varepsilon^2\tilde{\varphi}_{0_\tau}^2 + \varepsilon\tilde{\eta}_0 + \varepsilon^2\tilde{\eta}_1 = O(\varepsilon^3) \tag{5.2.57a}$$

$$\kappa^2\varepsilon\left[-(1+\varepsilon\tilde{\eta}_0)\varepsilon\tilde{\varphi}_{0_{\tau\tau}} + \kappa^2\varepsilon^2\left(-\tilde{\varphi}_{1_{\tau\tau}} + \frac{1}{6}\tilde{\varphi}_{0_{\tau\tau\tau\tau}}\right)\right]$$
$$= \kappa^2\varepsilon[\varepsilon\tilde{\eta}_{0_\tau} + \varepsilon^2(\tilde{\eta}_{1_\tau} + \tilde{\eta}_{0_{\tilde{t}}}) + \varepsilon^2\tilde{\eta}_{0_\tau}\tilde{\varphi}_{0_\tau}] + O(\varepsilon^4). \tag{5.2.57b}$$

The dominant terms in these equations produce, perhaps surprisingly, practically the same equations:

$$\tilde{\varphi}_{0_\tau} + \tilde{\eta}_0 = 0 \tag{5.2.58a}$$

$$\tilde{\varphi}_{0_{\tau\tau}} + \tilde{\eta}_{0_\tau} = 0. \tag{5.2.58b}$$

The integration function of $\tilde{t}$, which might distinguish the two forms of Equations (5.2.58), is zero from the conditions as $\tau \to \infty$. In these coordinates, the first approximation equation is not obtained until the second-order terms in Equations (5.2.57) are considered. Again, as far as $\tilde{\varphi}_1$ and $\tilde{\eta}_1$ are concerned, both of these relations are the same

$$\kappa^2\tilde{\varphi}_{1_\tau} + \tilde{\eta}_1 = -\tilde{\varphi}_{0_{\tilde{t}}} + \frac{\kappa^2}{2}\tilde{\varphi}_{0_{\tau\tau\tau}} - \frac{1}{2}\tilde{\varphi}_{0_\tau}^2 \tag{5.2.59a}$$

$$\kappa^2\tilde{\varphi}_{1_{\tau\tau}} + \tilde{\eta}_{1_\tau} = -\tilde{\eta}_{0_\tau}\tilde{\varphi}_{0_\tau} - \tilde{\eta}_0\tilde{\varphi}_{0_{\tau\tau}} + \frac{\kappa^2}{6}\tilde{\varphi}_{0_{\tau\tau\tau\tau}} - \tilde{\eta}_{0_{\tilde{t}}}. \tag{5.2.59b}$$

If we set $-\tilde{\eta}_{0_{\tilde{t}}} = \tilde{\varphi}_{0_{\tau\tilde{t}}}$ in Equation (5.2.59b) and integrate the result with respect to τ we find

$$\kappa^2\tilde{\varphi}_{1_\tau} + \tilde{\eta}_1 = -\tilde{\eta}_0\tilde{\varphi}_{0_\tau} + \frac{\kappa^2}{6}\tilde{\varphi}_{0_{\tau\tau\tau}} + \tilde{\varphi}_{0_{\tilde{t}}}. \tag{5.2.60}$$

Now the right hand sides of Equations (5.2.59a) and (5.2.60) must be the same, and this fact provides the second relation between $\tilde{\varphi}_0$ and $\tilde{\eta}_0$.

$$-\tilde{\varphi}_{0_{\tilde{t}}} + \frac{\kappa^2}{2}\tilde{\varphi}_{0_{\tau\tau\tau}} - \frac{1}{2}\tilde{\varphi}_{0_\tau}^2 = -\tilde{\eta}_0\tilde{\varphi}_{0_\tau} + \frac{\kappa^2}{6}\tilde{\varphi}_{0_{\tau\tau\tau}} + \tilde{\varphi}_{0_\tau}. \tag{5.2.61}$$

Using Equation (5.2.58a) to replace $\tilde{\eta}_0$ by $-\tilde{\varphi}_{0_\tau}$, we obtain the basic equation for the potential $\tilde{\varphi}_0$

$$2\tilde{\varphi}_{0_{\tilde{t}}} + \frac{3}{2}\tilde{\varphi}_{0_\tau}^2 = \frac{\kappa^2}{3}\tilde{\varphi}_{0_{\tau\tau\tau}}. \tag{5.2.62}$$

This is the well-known Korteweg-de Vries Equation, (see Reference 5.2.1) which is usually written in terms of $\sigma = x - t = -\tau$. We will see in Section

5.2.4 that it also follows from the Boussinesq equations for unidirectional waves.

There has been considerable recent work on the solution of this equation by the inverse scattering method. [See Reference 5.2.1 and the survey article in Reference 5.2.2]. A discussion of these results goes far beyond the scope of this book. Here we will merely derive the appropriate initial condition ($\tilde{t} = 0$) for our example.

It is convenient to change variables from $\tau, \tilde{t}$ to $\tau, \tilde{x} = \varepsilon x$.

Since

$$\tilde{t} = \varepsilon t = \varepsilon(t - x) + \varepsilon x = \varepsilon\tau + \tilde{x} \tag{5.2.63}$$

the solution of Equation (5.2.62)

$$\tilde{\varphi}_0 = F(\tau, \tilde{t}) \tag{5.2.64a}$$

can also be written in the form

$$\tilde{\varphi}_0 = F(\tau, \tilde{x}) + O(\varepsilon). \tag{5.2.64b}$$

Now, we introduce the intermediate variable

$$x_\eta = \frac{\tilde{x}}{\eta(\varepsilon)}, \qquad \varepsilon \ll \eta \ll 1 \tag{5.2.65}$$

and write both the near field and far field solutions in terms of τ and $\tilde{x}$. We find in the near field

$$\phi = \varepsilon\varphi_0^*(\tau) + O(\varepsilon\eta), \tag{5.2.66}$$

where the $O(\varepsilon\eta)$ remainder is the contribution from φ_1^*, and according to Equation (5.2.49),

$$\varphi_0^* = \begin{cases} -r(\tau), & \tau > 0 \\ 0, & \tau < 0. \end{cases} \tag{5.2.67}$$

In the far-field we have

$$\begin{aligned} \phi &= \varepsilon F(\tau, \eta x_\eta) + O(\varepsilon^2) \\ &= \varepsilon F(\tau, 0) + O(\varepsilon\eta). \end{aligned} \tag{5.2.68}$$

Therefore, matching requires that we set

$$F(\tau, 0) = \begin{cases} -r(\tau), & \tau > 0 \\ 0, & \tau < 0. \end{cases}$$

This means that the initial condition ($\tilde{t} = 0$) for the Korteweg-de Vries Equation is just $\tilde{\varphi}_0(\tau, 0) = F(\tau, 0)$ as defined, and this gives a well posed problem for solving Equation (5.2.62).

A simple class of solutions of Equation (5.2.62) are the uniform (translationally invariant) waves of the form

$$\tilde{\varphi}_0(\tau, \tilde{t}) = \psi(\tau - U\tilde{t}). \tag{5.2.69}$$

These are either solitary waves, or periodic (cn) waves and can be easily calculated from substitution of the assumed form into Equation (5.2.62). A remarkable result first observed numerically and later confirmed by use of the inverse scattering theory is that two solitary waves travelling at different speeds interact then separate essentially unchanged (except for phase shifts). The reader can find a discussion of this and other interesting results of the inverse scattering theory in Reference 5.2.1. In the next subsection we will show how the uniformly valid evolution equations for small amplitude waves on shallow water can be derived by a systematic two time expansion procedure.

5.2.4 Initial Value Problem, Multiple Variable Expansion

We now wish to approximate the solution of Equations (5.2.20)–(5.2.23) in the semi-infinite domain $-1 \leq y \leq \eta$, $-\infty < x < \infty$, with the initial conditions in Equations (5.2.26).

Again, we start with the long wave expansion, Equation (5.2.27), and regard the ϕ_i as functions of $x, y, t, \tilde{t}$. This implies that for $\varepsilon \ll 1$ we expand ϕ as follows

$$\phi(x, y, t; \varepsilon, \kappa\sqrt{\varepsilon}) = \varepsilon\theta_0(x, t, \tilde{t}) + \kappa^2\varepsilon^2\left[\theta_1(x, t, \tilde{t}) - \frac{(y+1)^2}{2}\theta_{0_{xx}}\right] + \kappa^4\varepsilon^3\left[\theta_2(x, t, \tilde{t}) - \frac{(y+1)^2}{2}\theta_{1_{xx}} + \frac{(y+1)^4}{24}\theta_{0_{xxxx}}\right] + O(\varepsilon^4). \tag{5.2.70}$$

The derivatives of ϕ have the expansions

$$\phi_t = \varepsilon\theta_{0_t} + \varepsilon^2\left\{\theta_{0_{\tilde{t}}} + \kappa^2\left[\theta_{1_t} - \frac{(y+1)^2}{2}\theta_{_{xxt}}\right] + O(\varepsilon^3)\right. \tag{5.2.71a}$$

$$\phi_x = \varepsilon\theta_{0_x} + \varepsilon^2\kappa^2\left[\theta_{1_x} - \frac{(y+1)^2}{2}\theta_{0_{xxx}}\right] + O(\varepsilon^3) \tag{5.2.71b}$$

$$\phi_y = -\kappa^2\varepsilon^2(y+1)\theta_{0_{xx}} + \kappa^4\varepsilon^3\left[-(y+1)\theta_{1_{xx}} + \frac{(y+1)^3}{6}\theta_{0_{xxxx}}\right] + O(\varepsilon^4). \tag{5.2.71c}$$

The surface shape η and its derivatives now have the expansions

$$\eta(x, t; \varepsilon, \kappa\sqrt{\varepsilon}) = \varepsilon\lambda_0(x, t, \tilde{t}) + \varepsilon^2\lambda_1(x, t, \tilde{t}) + O(\varepsilon^3) \tag{5.2.72a}$$

$$\eta_t = \varepsilon\lambda_{0_t} + \varepsilon^2(\lambda_{1_t} + \lambda_{0_{\tilde{t}}}) + O(\varepsilon^3) \tag{5.2.72b}$$

$$\eta_x = \varepsilon\lambda_{0_x} + \varepsilon^2\lambda_{1_x} + O(\varepsilon^3). \tag{5.2.72c}$$

Substitution of the assumed expansions into the free surface boundary conditions, Equations (5.2.22) and (5.2.23), now gives for the first approximation

$$\theta_{0_t} + \lambda_0 = 0 \tag{5.2.73a}$$

$$\theta_{0_{xx}} + \lambda_{0_t} = 0 \tag{5.2.73b}$$

and for the second approximation

$$\kappa^2\theta_{1_t} + \lambda_1 = -\theta_{0_{\tilde{t}}} + \frac{\kappa^2}{2}\theta_{0_{xxt}} - \frac{1}{2}\theta_{0_x}^2 \tag{5.2.74a}$$

$$\kappa^2\theta_{1_{xx}} + \lambda_{1_t} = -\lambda_{0_{\tilde{t}}} - \lambda_{0_x}\theta_{0_x} + \frac{\kappa^2}{6}\theta_{0_{xxxx}} - \lambda_0\theta_{0_{xx}}. \tag{5.2.74b}$$

We will solve this system by eliminating λ_0 and λ_1. Once θ_0 and θ_1 are known one can calculate λ_0 and λ_1 from Equations (5.2.73a) and (5.2.74a). We find that θ_0 and θ_1 obey wave equations

$$\theta_{0_{tt}} - \theta_{0_{xx}} = 0 \tag{5.2.75a}$$

$$\kappa^2(\theta_{1_{tt}} - \theta_{1_{xx}}) = -2\theta_{0_{t\tilde{t}}} + \frac{\kappa^2}{2}\theta_{0_{xxtt}} - 2\theta_{0_x}\theta_{0_{xt}} - \frac{\kappa^2}{6}\theta_{0_{xxxx}} - \theta_{0_t}\theta_{0_{xx}}. \tag{5.2.75b}$$

The initial conditions for θ_0 are

$$\theta_0(x, 0, 0) = 0 \tag{5.2.76a}$$

$$\theta_{0_t}(x, 0, 0) = -\lambda_0(x, 0, 0) = -h(x), \tag{5.2.76b}$$

and the initial conditions for θ_1 are

$$\theta_1(x, 0, 0) = 0 \tag{5.2.77a}$$

$$\theta_{1_t}(x, 0, 0) = -\frac{1}{\kappa^2}\left[\frac{\kappa^2}{2}h''(x) + \theta_{0_{\tilde{t}}}(x, 0, 0)\right]. \tag{5.2.77b}$$

We write the solution for θ_0 in the form

$$\theta_0 = f_0(\sigma, \tilde{t}) + g_0(\xi, \tilde{t}), \tag{5.2.78}$$

where

$$\sigma = x - t \tag{5.2.79a}$$

$$\xi = x + t. \tag{5.2.79b}$$

If we now express Equation (5.2.75b) in terms of the characteristic coordinates, we find

$$-4\kappa^2 \frac{\partial^2 \theta_1}{\partial \sigma\, \partial \xi} = \left(2f_{0_{\tilde{t}}} + \frac{\kappa^2}{3} f_{0_{\sigma\sigma\sigma}} + \frac{3}{2} f_{0_\sigma}^2\right)_\sigma + \left(-2g_{0_{\tilde{t}}} + \frac{\kappa^2}{3} g_{0_{\xi\xi\xi}} - \frac{3}{2} g_{0_\xi}^2\right)_\xi - f_{0_\sigma} g_{0_{\xi\xi}} + g_{0_\xi} f_{0_{\sigma\sigma}}. \tag{5.2.80}$$

Integrating gives

$$-4\kappa^2 \theta_1 = \xi\left(2f_{0_{\tilde{t}}} + \frac{\kappa^2}{3} f_{0_{\sigma\sigma\sigma}} + \frac{3}{2} f_{0_\sigma}^2\right) + \sigma\left(-2g_{0_{\tilde{t}}} + \frac{\kappa^2}{3} g_{0_{\xi\xi\xi}} - \frac{3}{2} g_{0_\xi}^2\right) - f_0 g_{0_\xi} + g_0 f_{0_\sigma} + f_1(\sigma, \tilde{t}) + g_1(\xi, t). \tag{5.2.81}$$

The first two terms (in parentheses) on the right hand side are unbounded for large times and must be removed, so f_0 and g_0 obey

$$2f_{0_{\tilde{t}}} + \frac{\kappa^2}{3} f_{0_{\sigma\sigma\sigma}} + \frac{3}{2} f_{0_\sigma}^2 = 0 \tag{5.2.82a}$$

$$-2g_{0_{\tilde{t}}} + \frac{\kappa^2}{3} g_{0_{\xi\xi\xi}} - \frac{3}{2} g_{0_\xi}^2 = 0. \tag{5.2.82b}$$

Each of these is a Korteweg-de Vries Equation (cf. Equation (5.2.62)]. For waves traveling to the right $\sigma = -\tau, g_0 = 0$ and we recover Equation (5.2.62). A similar result holds for waves traveling to the left. It is interesting that the f_0 and g_0 waves are uncoupled since for any given data $\theta_0(x, 0, 0)$ and $\lambda_0(x, 0, 0)$ we can calculate the corresponding initial values $f_0(\sigma, 0)$ and $g_0(\xi, 0)$.

We will now show that the Boussinesq equations, Equations (5.2.36), have precisely the same solutions as above, at least to $O(\varepsilon)$.

We expand u and η in terms of x, t and $\tilde{t}$

$$u = \varepsilon u_0(x, t, \tilde{t}) + \varepsilon^2 u_1(x, t, \tilde{t}) + O(\varepsilon^3) \tag{5.2.83a}$$

$$\eta = \varepsilon v_0(x, t, \tilde{t}) + \varepsilon^2 v_1(x, t, \tilde{t}) + O(\varepsilon^3). \tag{5.2.83b}$$

Substituting the above into Equations (5.2.36) and collecting terms gives the following system

$$v_{0_t} + u_{0_x} = 0 \tag{5.2.84a}$$

$$u_{0_t} + v_{0_x} = 0 \tag{5.2.84b}$$

which agrees with Equations (5.2.73) if we identify u_0 with θ_{0_x} and v_0 with λ_0 as we should since θ_0 is the first term in the expansion for the velocity potential and ν_0 and λ_0 are the leading terms in the respective expansions for η. In the second approximation for the Boussinesq equations we obtain

$$\nu_{1_t} + u_{1_x} = -\nu_{0_{\tilde{t}}} - u_0 \nu_{0_x} - \nu_0 u_{0_x} \tag{5.2.85a}$$

$$u_{1_t} + \nu_{1_x} = -u_{0_{\tilde{t}}} - u_0 u_{0_x} - \frac{\kappa^2}{3} \nu_{0_{xtt}}. \tag{5.2.85b}$$

These also agree with Equations (5.2.74) if we set

$$\nu_1 = \lambda_1 \tag{5.2.86a}$$

$$u_1 = \kappa^2\left(\theta_{1_x} - \frac{1}{6}\theta_{0_{xxx}}\right). \tag{5.2.86b}$$

The identification of ν_1 with λ_1 is obvious, and Equation (5.2.86b) implies that the expression to be used for u in the Boussinesq approximation must be the *average* value of the horizontal velocity over the depth, i.e.,

$$u = \frac{1}{1+\eta}\int_{-1}^{\eta} \phi_x(x, y, t; \varepsilon, \kappa\sqrt{\varepsilon})dy. \tag{5.2.87}$$

Equation (5.2.86b) now follows from Equation (5.2.87) if we use the expansion(5.2.71b) for ϕ_x.

References

5.2.1 G. B. Whitham, *Linear and Nonlinear Waves*, John Wiley and Sons, New York, 1974.

5.2.2 R. M. Miura, The Korteweg-de Vries equation: a survey of results, *S.I.A.M. Review* **18**, 1976, pp. 412–459.

5.3 Expansion Procedures of Thin Airfoil Theory at Various Mach Numbers

We consider now some of the expansion procedures used in airfoil theory. The usual assumption of the steady flow of a perfect, inviscid gas is made. The main question is the simplification of the rather complicated equations of motion when the perturbations are small (thin airfoils) and the free-stream Mach number can range from zero to infinity. The dependence on the two parameters, thickness ratio δ and Mach number at infinity M_∞, is an essential part of the considerations.

For simplicity, only families of shapes at zero angle of attack are considered. The airfoils are symmetric about the x-axis (Figure 5.3.1), and in the first problem that we consider the flow at upstream infinity is uniform with magnitude U. The unit of length is the airfoil chord.

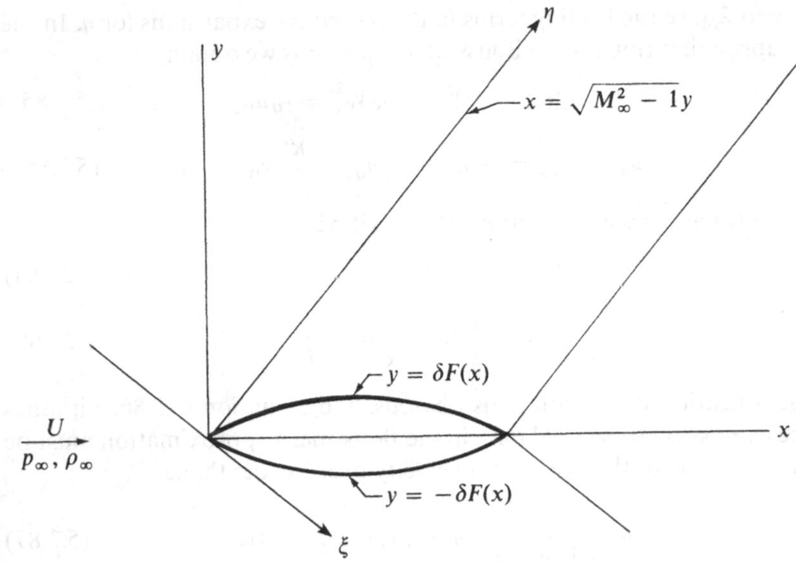

Figure 5.3.1 Thin Airfoil: Linearized Supersonic Flow Field

If $\mathbf{q} = (u, v)$ is the velocity, the boundary conditions are

$$q = U\mathbf{i} \text{ at upstream infinity,} \tag{5.3.1}$$

$$\frac{v(x, \delta F(x))}{u(x, \delta F(x))} = \delta F'(x) \qquad \text{tangent flow at body surface.} \tag{5.3.2}$$

By symmetry, it is sufficient to consider the domain $y \geq 0$. In Section 5.3.4 we will study the more general problem of a wind shear where U depends on y.

The equations to be solved come from conservation of mass, momentum, and energy, plus an equation of state. We must add shock-jump conditions to these equations since there is a possibility for shock waves to occur whenever the flow is locally supersonic ($M > 1$). The shock-jump conditions are derived from the conservation laws mentioned above using the ideas of a weak solution [see Sections 4.1 and 5.1]. A principal difficulty, associated with the intrinsic nonlinearity of the problem, is that the location of the shock waves is unknown in advance and must be found as part of the problem. Some of these difficulties are overcome when the shock waves are sufficiently weak, in which case jump conditions appear naturally in the solutions.

In all cases, for a perfect gas we can use the invariant corresponding to constant total enthalpy,

$$\frac{q^2}{2} + \frac{a^2}{\gamma - 1} = \frac{U^2}{2} + \frac{a_\infty^2}{\gamma - 1}, \tag{5.3.3}$$

where a = local speed of sound $= \sqrt{\partial p/\partial \rho} = \sqrt{\gamma R T}$, $\gamma = c_p/c_v$ = const. The only mechanism for entropy production and for the introduction of rotation into this model flow is a shock wave. The entropy is, thus, constant along a streamline but can jump across a shock wave. Isentropic flow, however, implies irrotational flow. As long as the Mach number is not too large, the flow can be assumed isentropic and irrotational .This assumption is not necessary but can be proved. Omission of the details will allow us to obtain the final result quickly. For details see Reference 5.3.1. Thus, the momentum equation can be written

$$(\mathbf{q}\cdot\nabla)\mathbf{q} = \nabla\left(\frac{q^2}{2}\right) = -\frac{1}{\rho}\nabla p = -\left(\frac{dp}{d\rho}\right)\frac{\nabla\rho}{\rho} = -a^2\frac{\nabla\rho}{\rho}. \tag{5.3.4}$$

The remaining information is contained in the continuity equation:

$$\operatorname{div}(\rho\mathbf{q}) = \mathbf{q}\cdot\nabla\rho + \rho\nabla\cdot\mathbf{q} = 0. \tag{5.3.5}$$

Now, from Equation (5.3.4), we have

$$\mathbf{q}\cdot\nabla\left(\frac{q^2}{2}\right) = -a^2\frac{\mathbf{q}\cdot\nabla\rho}{\rho} \tag{5.3.6}$$

and the use of the continuity equation enables ρ to be eliminated,

$$\mathbf{q}\cdot\nabla\frac{q^2}{2} = a^2\nabla\cdot\mathbf{q}. \tag{5.3.7}$$

Note that, as $a \to \infty$, the usual equation of incompressible flow results.

Introducing now the potential $\Phi(x, y; M_\infty, \delta)$ of the irrotational flow we have

$$u = \frac{\partial\Phi}{\partial x}, \qquad v = \frac{\partial\Phi}{\partial y}. \tag{5.3.8}$$

Equations (5.3.7) and (5.3.3) form the basic system of equations, and Equations (5.3.1) and (5.3.2) give the boundary conditions:

$$(a^2 - \Phi_x^2)\Phi_{xx} - 2\Phi_x\Phi_y\Phi_{xy} + (a^2 - \Phi_y^2)\Phi_{yy} = 0, \tag{5.3.9}$$

$$\frac{a^2}{\gamma - 1} + \frac{\Phi_x^2 + \Phi_y^2}{2} = \frac{a_\infty^2}{\gamma - 1} + \frac{U^2}{2} \tag{5.3.10}$$

$$\Phi_y(x, \delta F(x)) = \delta F'(x)\Phi_x(x, \delta F(x)), \tag{5.3.11}$$

$$\Phi \to Ux \quad \text{as } x \to -\infty. \tag{5.3.12}$$

Corresponding equations hold in (x, y, z). Equation (5.3.9) is quasi-linear; it is of elliptic type when the flow is locally subsonic, $\Phi_x^2 + \Phi_y^2 < a^2$, and hyperbolic when the flow is locally supersonic, $\Phi_x^2 + \Phi_y^2 > a^2$.

The problem is often simplified by various small disturbance approximations corresponding to thin airfoils. We will now discuss these expansions and the limit processes associated with them.

5.3.1 Linearized Theory

The content of this theory is equivalent to acoustics. The limit process considered is

$$\delta \to 0 \qquad (x, y; M_\infty \text{ fixed}). \tag{5.3.13}$$

The first term of the expansion represents a uniform free stream. Let

$$\Phi(x, y; M_\infty, \delta) = U\{x + \varepsilon_1(\delta)\phi_1(x, y; M_\infty) + \varepsilon_2(\delta)\phi_2(x, y; M_\infty) + \cdots\}. \tag{5.3.14}$$

The order of $\varepsilon_1(\delta)$ is fixed from the boundary condition, Equation (5.3.11), since we have

$$\Phi_y(x, y; M_\infty, \delta) = U\{\varepsilon_1(\delta)\phi_{1_y}(x, y; M_\infty) + \varepsilon_2(\delta)\phi_{2_y}(x, y; M_\infty) + \cdots\} \tag{5.3.15}$$

$$\Phi_x(x, y; M_\infty, \delta) = U\{1 + \varepsilon_1\phi_{1_x} + \varepsilon_2\phi_{2_x} + \cdots\}. \tag{5.3.16}$$

Thus, the boundary condition, Equation (5.3.11), has the expansion

$$\begin{aligned} &\varepsilon_1\phi_{1_y}(x, 0) + \varepsilon_1\delta F(x)\phi_{1_{yy}}(x, 0) + \cdots \varepsilon_2\phi_{2_y}(x, 0) + \cdots \\ &\quad = \delta F'(x)\{1 + \varepsilon_1\phi_{1_x}(x, 0) + \cdots\}, \qquad 0 < x < 1. \end{aligned} \tag{5.3.17}$$

It follows that $\varepsilon_1 = \delta$, and that reasonable boundary conditions for ϕ_2 appear if $\varepsilon_2 = \varepsilon_1\delta = \delta^2$. Here a power-series expansion of the solution near $y = 0$ has been assumed. In the plane-flow case, this turns out to be all right—an inner expansion in the manner of Section 4.3.1 would give this result trivially. For axial symmetry, of course, the situation is different, and a treatment along the lines of Section 4.3.1 is needed.

Thus Equation (5.3.14) is

$$\Phi = U\{x + \delta\phi_1 + \delta^2\phi_2 + \cdots\}. \tag{5.3.18}$$

Now, Equation (5.3.10) has the expansion

$$a^2 = a_\infty^2 - (\gamma - 1)U^2\delta\phi_{1_x} + O(\delta^2), \tag{5.3.19}$$

and the basic equation, Equation (5.3.9), is

$$\begin{aligned} &(a_\infty^2 - (\gamma - 1)U^2\delta\phi_{1_x} - U^2 - 2U^2\delta\phi_{1_x} + \cdots)(\delta\phi_{1_{xx}} + \delta^2\phi_{2_{xx}} + \cdots) \\ &\quad - 2U^2\delta^2\phi_{1_y}\phi_{1_{xy}} + \cdots + (a_\infty^2 - (\gamma - 1)U^2\delta\phi_{1_x} + \cdots) \\ &\quad \times (\delta\phi_{1_{yy}} + \delta^2\phi_{2_{yy}} + \cdots) = 0. \end{aligned} \tag{5.3.20}$$

First- and second-order approximate equations result:

$$O(\delta): \qquad (M_\infty^2 - 1)\phi_{1_{xx}} - \phi_{1_{yy}} = 0, \tag{5.3.21}$$

$$\begin{aligned} O(\delta^2): \qquad &(M_\infty^2 - 1)\phi_{2_{xx}} - \phi_{2_{yy}} \\ &= -M_\infty^2[(\gamma - 1)M_\infty^2 + 2]\phi_{1_x}\phi_{1_{xx}} - 2M_\infty^2\phi_{1_y}\phi_{1_{xy}}. \end{aligned} \tag{5.3.22}$$

Equation (5.3.21) has been used to simplify the right-hand side of Equation (5.3.22). Equation (5.3.21) is the steady version of the acoustic equation which results from a Galilean transformation with speed U. Equation (5.3.22) corresponds to a second-order version of acoustics. Equation (5.3.21) is of elliptic type for subsonic flight ($M_\infty < 1$), and of hyperbolic type for ($M_\infty > 1$), which in itself indicates that the expansion cannot be valid when M_∞ is close to unity. While it is true that the expansion becomes valid if δ is sufficiently small, the dependence on the parameter M_∞ may make the expansion not very useful. Some study of the properties of the solutions ϕ_1, ϕ_2, enables us to mark out the range of validity of the expansions with respect to both parameters (δ, M_∞).

The boundary conditions on the body surface which apply to Equations (5.3.21) and (5.3.22) are

$$\phi_{1_y}(x, 0) = F'(x), \tag{5.3.23}$$

$$\begin{aligned}\phi_{2_y}(x, 0) &= \phi_{1_x}(x, 0)F'(x) - F(x)\phi_{1_{yy}}(x, 0)\\ &= \phi_{1_x}(x, 0)F'(x) - (M_\infty^2 - 1)\phi_{1_{xx}}(x, 0)F(x).\end{aligned} \tag{5.3.24}$$

Consider now the supersonic case $M_\infty > 1$ which is the simpler to analyze. The solution of Equation (5.3.21), with the boundary conditions of tangent flow, Equation (5.3.23), and the further restrictions that only downstream running waves come from the body (due to the boundary conditions at upstream infinity), is

$$\phi_1(x, y) = -\frac{1}{\sqrt{M_\infty^2 - 1}} F(x - \sqrt{M_\infty^2 - 1}\,y),$$
$$0 < x - \sqrt{M_\infty^2 - 1}\,y < 1. \tag{5.3.25}$$

Note that

$$\phi_{1_y}(x, y) = F'(x - \sqrt{M_\infty^2 - 1}\,y)$$

so that Equation (5.3.23) is satisfied. From this potential, the first-order flow field and pressure can be found. For example, we have

$$\frac{p - p_\infty}{\rho_\infty U^2} = -\delta\phi_{1_x} = \frac{\delta}{\sqrt{M_\infty^2 - 1}} F'(x - \sqrt{M_\infty^2 - 1}\,y). \tag{5.3.26}$$

The potential is constant outside the wave zone $0 < x - \sqrt{M_\infty^2 - 1}\,y < 1$, and there is a jump of pressure, for example, across $x = \sqrt{M_\infty^2 - 1}\,y$ if $F'(0) \neq 0$. This jump is a linearized version of a shock wave. It is clear already that the expansion is not a good approximation unless

$$\frac{\delta}{\sqrt{M_\infty^2 - 1}} \ll 1,$$

but more precise information is obtained from the second approximation.

Introduce the characteristic coordinates of the first approximation

$$\xi = x - \sqrt{M_\infty^2 - 1}\,y, \qquad \eta = x + \sqrt{M_\infty^2 - 1}\,y, \tag{5.3.27}$$

and note that the right-hand side of Equation (5.3.22) is

$$-[(\gamma-1)M_\infty^2+2]\frac{M_\infty^2}{M_\infty^2-1}F'(\xi)F''(\xi)-2M_\infty^2F'(\xi)F''(\xi)$$
$$=-\frac{(\gamma+1)M_\infty^4}{M_\infty^2-1}F'(\xi)F''(\xi),$$

so that Equation (5.3.22) becomes

$$\frac{\partial^2\phi_2}{\partial\xi\,\partial\eta}=-\frac{\gamma+1}{4}\frac{M_\infty^4}{(M_\infty^2-1)^2}F'(\xi)F''(\xi). \tag{5.3.28}$$

Thus, the general solution for ϕ_2 can be written in $0<\xi<1$ as

$$\phi_2(\xi,\eta)=-\frac{\gamma+1}{8}\frac{M_\infty^4}{(M_\infty^2-1)^2}\eta F'^2(\xi)+G(\xi). \tag{5.3.29}$$

The arbitrary function $G(\xi)$ can be found from the boundary condition of Equation (5.3.24). Without working out the details, we see that the order of magnitude of the second term in now clear, so that the original expansion, Equation (5.3.14), now becomes

$$\Phi(x,y;M_\infty,\delta)=U\left\{x-\frac{\delta}{\sqrt{M_\infty^2-1}}F(x-\sqrt{M_\infty^2-1}\,y)\right.$$
$$\left.+O\left(\frac{\delta^2(\gamma+1)M_\infty^4}{(M_\infty^2-1)^2}(x+\sqrt{M_\infty^2-1}\,y)\right)\right\}. \tag{5.3.30}$$

We can say that the first two terms provide a valid approximation if $\delta\phi_2\ll\phi_1$ or if

$$\frac{(\gamma+1)\delta M_\infty^4}{(M_\infty^2-1)^{3/2}}(x+\sqrt{M_\infty^2-1}\,y)\ll 1. \tag{5.3.31}$$

The occurrence of nonuniformity with respect to the parameter M_∞ and the coordinates is a typical feature of expansion procedures with more than one parameter. Each nonuniformity indicated by Equation (5.3.31) corresponds to a different physical phenomenon. The type of nonuniformity and the conditions necessary for the linearized expansion to be a good approximation are indicated below, by eliminating unessential factors from Equation (5.3.31)

Transonic Regime

$$M_\infty\approx 1\qquad(\eta\text{ fixed}),\qquad\frac{\delta}{(M_\infty^2-1)^{3/2}}\ll 1,$$

Far Field

$$(x+\sqrt{M_\infty^2-1}y)\approx\infty\quad(M_\infty\text{ fixed}),\quad\delta(x+\sqrt{M_\infty^2-1}y)\ll 1,$$

Hypersonic Regime

$$M_\infty \approx \infty \qquad (\eta \text{ fixed}), \quad \delta M_\infty \ll 1.$$

The first two nonuniformities, transonic and far field, represent cumulative effects. For a body flying close to the speed of sound, small waves accumulate at the body and their nonlinear interaction eventually becomes important. The fact that the angle of a small wave is given incorrectly leads to a cumulative effect which means that even the location of the wave zone is given incorrectly in the far field. Lastly, the hypersonic difficulty is caused by an underestimation of the effects of the nonlinearity in producing pressure changes. In fact, the physical basis of linearized theory is invalid when $\delta M_\infty = O(1)$, since then the shock waves are strong, entropy effects must be included, and the basic equations must be reexamined.

Each of the regimes indicated above can be treated by introducing a suitable modified expansion procedure. The outline of the procedure for the transonic and hypersonic regimes is sketched in the following Sections. We will consider the uniformly valid solution over the near and far fields for the supersonic case in Section 5.3.4 in connection with the airfoil in a wind shear. The simpler case of uniform flow upstream is, in all respects, analogous to the piston problem in acoustics discussed in Section 5.1.2.

5.3.2 Transonic Theory

Linearized theory predicts catastrophic failure at $M_\infty = 1$ [cf. (5.3.26)] but the trouble is mathematical and not physical; the orders of magnitude of certain terms have been estimated incorrectly.

In order to arrive at the correct limit process and expansion, we go to the heart of the matter and set $M_\infty = 1$. If a suitable expansion can be found for $M_\infty = 1$, then it can certainly be extended to some neighborhood of $M_\infty = 1$ by considering that $M_\infty \to 1$ at a certain rate as $\delta \to 0$. It is clear, however, that the limit process $\delta \to 0$ with $M_\infty = 1$ cannot be carried out in the original system of coordinates, since this will merely produce the same result as in linearized theory. Instead, a coordinate $\tilde{y}$,

$$\tilde{y} = \beta(\delta) y, \tag{5.3.32}$$

must be kept fixed. Actually, we expect $\beta(\delta) \to 0$, since linearized theory indicates a larger extent of the perturbation field in the y-direction as $M_\infty \to 1$. We are striving for an expansion which has a possibility of satisfying the boundary conditions on the body[7] and, if possible, at infinity, so that the x-coordinate is fixed. Then $\beta(\delta)$ is to be found from these considerations, so that a distinguished limit process exists.

[7] In the case of axial symmetry, a suitable inner expansion can be used.

Expansion Procedure for $M_\infty = 1$

Instead of Equation (5.3.14), consider

$$\Phi(x, y; \delta) = U\{x + \varepsilon(\delta)\varphi(x, \tilde{y}) + \cdots\}. \tag{5.3.33}$$

The dominant terms which now appear correspond to those in the first and second approximations, Equations (5.3.21) and (5.3.22), with the orders adjusted appropriately $(\partial/\partial y) \to \beta(\partial/\partial\tilde{y})$.[8] Thus, from Equation (5.3.9), we find that

$$\varepsilon\beta^2\varphi_{\tilde{y}\tilde{y}} + \cdots = (\gamma + 1)\varepsilon^2\varphi_x\varphi_{xx} - 2\beta^2\varepsilon^2\varphi_{\tilde{y}}\varphi_{x\tilde{y}} + \cdots. \tag{5.3.34}$$

The distinguished limit occurs if $\beta \to 0$ as $\varepsilon \to 0$, in such a way that

$$\varepsilon\beta^2 = \varepsilon^2 \quad \text{or} \quad \beta = \sqrt{\varepsilon}. \tag{5.3.35}$$

All terms except that first two in Equation (5.3.34) are negligible, so that we obtain the transonic equation for $M_\infty = 1$:

$$\varphi_{\tilde{y}\tilde{y}} = (\gamma + 1)\varphi_x\varphi_{xx}, \qquad (M_\infty = 1). \tag{5.3.36}$$

This quasi-linear equation is of changing type:

$\varphi_x < 0$, locally subsonic, elliptic type;

$\varphi_x > 0$, locally supersonic, hyperbolic type.

The attempt to construct solutions to this equation leads to interesting and difficult problems (cf. Reference 5.3.2), but it is clear that this equation has the possibility of describing the physical phenomena correctly and, in fact, various experimental and recent numerical results indicate that it is very good for airfoils and nozzles of a practical shape. The problem is not complete until we discuss the boundary conditions which serve to fix the orders of magnitude.[9] On the body, Equation (5.3.11) now becomes

$$\varepsilon\beta\varphi_{\tilde{y}}(x, 0) + \cdots = \delta F'(x), \tag{5.3.37}$$

so that

$$\varepsilon\beta = \delta. \tag{5.3.38}$$

A comparison of Equations (5.3.38) and (5.3.35) shows us that

$$\varepsilon = \delta^{2/3}, \qquad \beta = \delta^{1/3} \tag{5.3.39}$$

for the distinguished limit. The boundary condition at upstream infinity is

$$\varphi \to 0, \qquad x \to -\infty. \tag{5.3.40}$$

[8] We can show that third- and higher-order-type terms cannot contribute to the dominant transonic approximations.

[9] The orders are different for axial symmetry: $\varphi_{\tilde{r}\tilde{r}} + (1/\tilde{r})\varphi_{\tilde{r}} = (\gamma + 1)\varphi_x\varphi_{xx}$, $\varepsilon = \delta^2$, $\beta = \delta$ (cf. Ref. 5.3.1).

This expansion procedure can also be carried to higher orders. The successive equations φ_2, φ_3 are *linear* but have variable coefficients depending on the lower approximations. This is typical for expansions where the first term satisfies a nonlinear equation (cf. van der Pol Oscillator, Section 2.6).

Expansion Procedure for $M_\infty \cong 1$

The concept of the limit process leading to the expansion must now be widened to consider $M_\infty \to 1$ as $\delta \to 0$, this can be expressed as

$$M_\infty^2 = 1 + K\nu(\delta), \qquad \nu(\delta) \to 0. \tag{5.3.41}$$

The quantity K is held fixed in the limit and may be $O(1)$. Here $K \equiv 0$ corresponds to $M_\infty = 1$. A significant order is to be found for $\nu(\delta)$. The expansion has the form

$$\Phi(x, y; M_\infty, \delta) = U\{x + \delta^{2/3}\varphi(x, \tilde{y}; K) + \cdots\}, \qquad \tilde{y} = \delta^{1/3}y, \tag{5.3.42}$$

and again the dominant terms come from the first and second approximations, Equations (5.3.21), and (5.3.22). Thus, we have

$$-\nu(\delta)\delta^{2/3}K\varphi_{xx} + \cdots + \delta^{4/3}\varphi_{\tilde{y}\tilde{y}} = \delta^{4/3}(\gamma + 1)\varphi_x\varphi_{xx} + \cdots. \tag{5.3.43}$$

The distinguished case is evidently $\nu(\delta) = \delta^{2/3}$; if $\nu(\delta) \gg \delta^{2/3}$, then a meaningless equation results, and if $\nu(\delta) \ll \delta^{2/3}$, the flow is that of $M_\infty = 1$, and the approximation is less general. The resulting transonic equation is

$$[K + (\gamma + 1)\varphi_x]\varphi_{xx} - \varphi_{\tilde{y}\tilde{y}} = 0, \tag{5.3.44}$$

with boundary conditions as before. The parameter K, which appears in the expansion, is often called the transonic similarity parameter:

$$K = \frac{M_\infty^2 - 1}{\delta^{2/3}} \tag{5.3.45}$$

and similarity rules or laws of similitude are expressed in terms of K. For example, the pressure at a given abscissa on a thin airfoil is expressed as

$$\frac{p - p_\infty}{\rho_\infty U^2\delta^{2/3}} = -\varphi_x(x, 0; K) = fn(K). \tag{5.3.46}$$

This sort of rule can be tested by experiments on a family of affine shapes varying (δ, M_∞) so that K is fixed. Note that the existence of the parameter K was implied in the discussion of the validity of linearized theory, when it was seen that linearized theory was valid for $K \gg 1$. The uniformity of the expansion, Equation (5.3.42), both with respect to the parameter K and the coordinates $(x, \tilde{y})$ can be studied as before.

The shock relations for the transonic equation follow from Equation (5.1.49) when we write Equation (5.3.44) in divergence form

$$\left\{K\varphi_x + \frac{\gamma + 1}{2}\varphi_x^2\right\}_x + \{-\varphi_{\tilde{y}}\}_{\tilde{y}} = 0 \tag{5.3.47a}$$

$$\{\varphi_{\tilde{y}}\}_x + \{-\varphi_x\}_{\tilde{y}} = 0. \tag{5.3.47b}$$

If we now identify T, X, C of Equation (5.1.48) with x, $\tilde{y}$ and $(d\tilde{y}/dx)_{\text{shock}}$ respectively, the shock jump conditions become

$$\left(\frac{d\tilde{y}}{dx}\right)_{\text{shock}} = -\frac{[\varphi_{\tilde{y}}]}{K[\varphi_x] + \dfrac{(\gamma + 1)}{2}[\varphi_x^2]} \tag{5.3.48a}$$

$$\left(\frac{d\tilde{y}}{dx}\right)_{\text{shock}} = -\frac{[\varphi_x]}{[\varphi_{\tilde{y}}]}, \tag{5.3.48b}$$

where [ ] denotes the jump in a quantity.

Equation (5.3.48b) states the well-known result that the velocity component tangent to the shock is conserved across the shock. The combination of Equations (5.3.48) and (5.3.47) expresses the jump conditions, giving the transonic form of a shock polar curve:

$$K[\varphi_x]^2 + \frac{\gamma + 1}{2}[\varphi_x^2][\varphi_x] = [\varphi_{\tilde{y}}]^2. \tag{5.3.49}$$

Finally, it is clear that there is some sort of matching between the transonic expansion and the linearized expansion. Near infinity where $\varphi_x \to 0$, the transonic equation, Equation (5.3.44), approaches the linear equation, (5.3.21), in form. The regions near infinity come closer to the airfoil for large K. Hence, in some sense, the linearized supersonic results should follow from transonic solutions as K gets large. The details of this matching were first worked out in Reference 5.3.3 for the problem of a supersonic thin airfoil in a wind shear and this will be reviewed in Section 5.3.4.

5.3.3 Hypersonic Theory

The discussion at the end of Section 5.3.1 shows that the linearized approximation is not valid when M_∞ is so large that $M_\infty \delta = O(1)$. In these circumstances, the Mach wave $x = \sqrt{M_\infty^2 - 1}\, y \cong M_\infty y$ is at a distance of the same order from the body as the body thickness δ; the body is in this sense no longer thin, and the shock waves are strong. However, small disturbances to the free stream are still produced, and an expansion procedure can be constructed. It is reasonable that a coordinate $y^* \sim M_\infty y$ be held fixed, so that details of the wave structure are preserved. The expansion procedure must now be applied to the exact equations of motion, since a potential does not exist if any shock waves appear. Some aspects of the procedure are sketched below. For the equations of motion in (u, v, p, ρ), we have the following.

(a) continuity: $$(\rho u)_x + (\rho v)_y = 0$$

(b) x-momentum:

$$uu_x + vu_y = -p_x/\rho \tag{5.3.50}$$

(c) y-momentum:

$$uv_x + vv_y = -p_y/\rho$$

(d) energy:

$$\rho\left(u\frac{\partial}{\partial x} + v\frac{\partial}{\partial y}\right)\left\{\frac{1}{\gamma - 1}\frac{p}{\rho} + \frac{u^2 + v^2}{2}\right\}$$
$$= -(pu)_x - (pv)_y.$$

The perfect-gas expression for internal energy per mass $(1/(\gamma - 1))(p/\rho)$ has been used in the energy equations. The energy equation can be replaced by the condition that the entropy is constant along a stream-line (but not across shock waves). Thus, we have

$$\left(u\frac{\partial}{\partial x} + v\frac{\partial}{\partial y}\right)\left(\frac{p}{\rho^\gamma}\right) = 0.$$

A study of the shock relations show that corresponding to a $v = O(\delta)$ we have perturbations in $p = O(\delta^2)$, in $\rho = O(1)$, and in $u = O(\delta^2)$. The basic idea used is that the nose shock angle to the free stream is of order δ. The perturbation is normal to the shock, so that the true velocity perturbations have different orders (cf. transonic case). For further details, consult Reference 5.3.4. Thus, with the limit process

$$\delta \to 0 \quad (x, y^*; H \text{ fixed}); \qquad y^* = \frac{y}{\delta} \qquad H = \frac{1}{M_\infty^2 \delta^2}$$

we associate the following expansion:

$$\frac{u(x, y; M_\infty, \delta)}{U} = 1 + \delta^2\bar{u}(x, y^*; H) + \cdots, \tag{5.3.51a}$$

$$\frac{v(x, y; M_\infty, \delta)}{U} = \delta\bar{v}(x, y^*; H) + \cdots, \tag{5.3.51b}$$

$$\frac{p(x, y; M_\infty, \delta)}{\rho_\infty U^2} = \delta^2\bar{p}(x, y^*; H) + \cdots, \tag{5.3.51c}$$

$$\frac{\rho(x, y; M_\infty, \delta)}{\rho_\infty} = \bar{\rho}(x, y^*; H) + \cdots. \tag{5.3.51d}$$

The derivative along a streamline becomes

$$\frac{1}{U}\left\{u\frac{\partial}{\partial x}+v\frac{\partial}{\partial y}\right\}\to\frac{\partial}{\partial x}+\bar{v}\frac{\partial}{\partial y^*}+O(\delta^2). \tag{5.3.52}$$

The dominant terms of Equation (5.3.50) are as follows.

continuity:

$$\bar{\rho}_x+(\bar{\rho}\bar{v})_{y^*}=0 \tag{5.3.53a}$$

x-momentum:

$$\bar{u}_x+\bar{v}\bar{u}_{y^*}=-\bar{p}_x/\bar{\rho} \tag{5.3.53b}$$

y-momentum:

$$\bar{v}_x+\bar{v}\bar{v}_{y^*}=-\bar{p}_{y^*}/\bar{\rho} \tag{5.3.53c}$$

energy:

$$\bar{\rho}\left(\frac{\partial}{\partial x}+\bar{v}\frac{\partial}{\partial y^*}\right)\left(\frac{1}{\gamma-1}\frac{\bar{p}}{\bar{\rho}}+\bar{u}+\frac{\bar{v}^2}{2}\right)=-\frac{\partial\bar{p}}{\partial x}-\frac{\partial(\bar{p}\bar{v})}{\partial y^*} \tag{5.3.53d}$$

entropy:

$$\left(\frac{\partial}{\partial x}+\bar{v}\frac{\partial}{\partial y^*}\right)\left(\frac{\bar{p}}{\bar{\rho}^\gamma}\right)=0. \tag{5.3.53e}$$

In the free stream we have

$$\bar{p}=\frac{p_\infty}{\rho_\infty U^2\delta^2}=\frac{1}{\gamma M_\infty^2\delta^2}=\frac{H}{\gamma}.$$

The simplification achieved in this expansion process is the uncoupling of the x-momentum equation so that, for example, the system (a), (c), and (e) of Equation (5.3.53) forms a complete system for $(\bar{p}, \bar{\rho}, \bar{v})$. The same remark applies to the shock wave relations that are not written out here and to the boundary condition of tangent flow on the airfoil surface (not applied at $y = 0$, now):

$$v(x, F(x)) = F'(x). \tag{5.3.54}$$

The equations (a), (c), and (e), shock relations, and boundary conditions are exactly analogous to those for unsteady one-dimensional motion ($x \to t$, $y^* \to x$) for the shock wave and flow produced by a piston moving with velocity v.

The similarity of hypersonic flows is thus expressed by the hypersonic similarity parameter $H = 1/M_\infty^2\delta^2$. The expansion is uniformly valid as $H \to 0$, corresponding to $M_\infty \to \infty$. For small H, the hypersonic solution joins on to linearized theory for $M_\infty \gg 1$, again in a sense which has not been made precise.

The nonlinear equations which appear for hypersonic theory are still difficult to solve, but some simplification and similarity has been discovered.

In all these expansions—linearized, transonic, hypersonic—various other local nonuniformities may appear due to a local violation of assumed orders of magnitude. The blunt nose region of a slender body in hypersonic flow is an example. Near the nose, a different expansion must be used to describe the flow, and the effect spreads downstream along streamlines forming a layer (entropy layer) near the body surface where some variables are estimated incorrectly. In constructing the local expansions and solutions in these regions and layers, the ideas of matching must once more be used (cf. Reference 5.3.4). The stagnation point of a body in transonic or linearized subsonic flow would demand a similar treatment if flow details were wanted. Very often, however, such details are unnecessary for overall results.

5.3.4 Supersonic Thin Airfoil Theory in a Wind Shear

Here we generalize the problem of Sections 5.3.1 and 5.3.2 so that the velocity at upstream infinity has a slow linear variation with altitude. If this free stream velocity decreases with altitude a point will be reached far below the airfoil where the freestream Mach number is nearly unity, and the velocity behind the weak shock originating from the airfoil leading edge becomes sonic. If we introduce a horizontal "ground" just above this point (to avoid dealing with a subsonic flow) we have a strictly hyperbolic problem in the domain above ground. The geometry is sketched in Figure 5.3.2 with dimensionless variables where lengths are normalized by the airfoil chord and velocities using the free stream value of U at the airfoil altitude. The pressure and density are normalized with respect to their constant values p_∞, ρ_∞ at upstream infinity.

The problem was studied in Reference 5.3.3 with three goals, (i) To show how a multiple variable expansion can be used to derive a uniformly valid approximation in the entire strictly supersonic region, (ii) To derive a generalization of the transonic equation, Equation (5.3.44), for the case of wind shear by a limit process from the exact equations, and (iii) To exhibit the matching between the solutions in (i) and (ii). We follow Reference (5.3.3) and consider in detail the first two goals; the matching between supersonic and transonic solutions is only sketched here.

We start with the equations of motion given in Equations (5.3.50) where all quantities are now dimensionless.

In the thin airfoil ($\delta \ll 1$) limit, the appropriate expansions are

$$u = u_0 + \delta u_1 + \delta^2 u_2 + O(\delta^3) \tag{5.3.55a}$$

$$v = \delta v_1 + \delta^2 v_2 + O(\delta^3) \tag{5.3.55b}$$

$$p = 1 + \delta p_1 + \delta^2 p_2 + O(\delta^3) \tag{5.3.55c}$$

$$\rho = 1 + \delta \rho_1 + \delta^2 \rho_2 + O(\delta^3). \tag{5.3.55d}$$

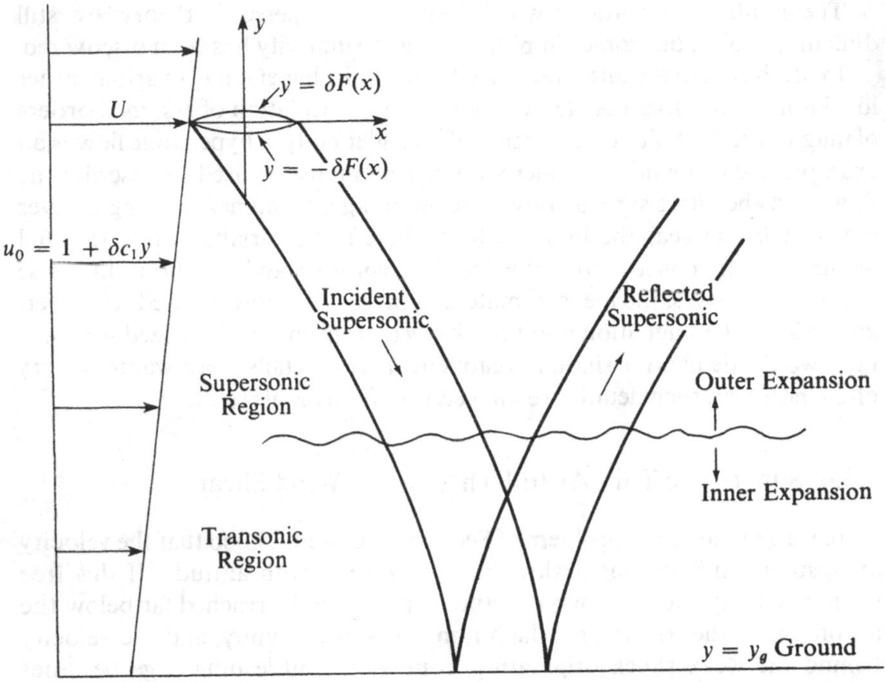

Figure 5.3.2 Airfoil in Supersonic Wind Gradient

It is easy to show that the entropy is constant up to order δ^2 and this result can be used to link the pressure to the density to $O(\delta^2)$. The details are given in Reference 5.3.5 on which Reference 5.3.3 is based. One finds

$$p_1 = \gamma \rho_1 \tag{5.3.56a}$$

$$p_2 = \gamma \rho_2 + \gamma(\gamma - 1)\rho_1^2/2 \tag{5.3.56b}$$

$$p_x = a^2 \rho_x + O(\delta^3) \tag{5.3.57a}$$

$$p_y = a^2 \rho_y + O(\delta^3), \tag{5.3.57b}$$

where a is the dimensionless speed of sound defined by

$$a^2 = \partial p/\partial \rho = \rho^{\gamma - 1}/M_\infty^2 + O(\delta^3) \tag{5.3.58a}$$

and M_∞ is the upstream Mach number at the airfoil altitude

$$M_\infty = \frac{U}{\gamma p_\infty/\rho_\infty}. \tag{5.3.58b}$$

Having eliminated p in favor of ρ from the problem of $O(\delta^2)$, we need only use the three equations of continuity, x and y momentum for the three dependent variables u, v, and ρ. The energy equation will then be automatically satisfied to order δ^2 when the above expressions linking p and its derivatives to ρ are used.

Expanding the continuity and momentum equations to second order and using the indicated expressions for p and its derivatives, we calculate

$$\delta[u_0 \rho_{1x} + u_{1x} + v_{1y}] + \delta^2[u_0 \rho_{2x} + u_{2x} + v_{2y} + (\rho_1 u_1)_x + (\rho_1 v_1)_y] = O(\delta^3), \tag{5.3.59a}$$

$$\delta[u_0 u_{1x} + \rho_{1x}/M_\infty^2 + v_1(u_0)_y] + \delta^2[u_0 u_{2x} + \rho_{2x}/M_\infty^2 + v_2(u_0)_y + u_1 u_{1x} + v_1 u_{1y} + (\gamma - 2)\rho_1 \rho_{1x}/M_\infty^2] = O(\delta^3), \tag{5.3.59b}$$

$$\delta[u_0 v_{1x} + \rho_{1y}/M_\infty^2] + \delta^2[u_0 v_{2x} + \rho_{2y}/M_\infty^2 + u_1 v_{1x} + v_1 v_{1y} + (\gamma - 2)\rho_1 \rho_{1y}/M_\infty^2] = O(\delta^3). \tag{5.3.59c}$$

In the above u_0 is a function of y and we have so far made no assumptions regarding this function. If $u_0(y)$ were arbitrary, the resulting flow would be rotational. Moreover, the presence of the functions $u_0(y)$ and $u_{0y}(y)$ in our equations to first order makes the solution very difficult analytically. From the standpoint of actual wind gradients encountered in the atmosphere,[10] it is more appropriate to assume that u_0 varies slowly with y over a length scale of the order of one chord. Thus, we will assume that $u_0 = u_0(\delta y)$, and for the purposes of this analysis we simplify this function further by taking it to be linear (see Figure 5.3.2) and set

$$u_0 = 1 + \delta c_1 y, \tag{5.3.60}$$

where c_1 is a nonnegative constant independent of δ. We note that the analysis presented in this section can be readily carried out with an arbitrary slowly varying function $u_0(\delta y)$. For any function $u_0(\delta y)$, $\partial u_0/\partial y = O(\delta)$, and the solution to first order can be carried out analytically.

Based on our previous studies in Sections 5.1 and 5.2, we know that even for u_0 = constant, the solution by a regular perturbation (i.e., assuming that the functions u_1, ρ_1, v_1 depend on x and y) will not lead to uniformly valid results in the far-field. This failure is particularly significant here as we intend to consider our flow sufficiently far away to have a transonic local Mach number. In the present problem, the fact that u_0 is a slowly varying function of y introduces a second requirement that the solution contain the two variables y and $\tilde{y} = \delta y$. It is therefore crucial to use multiple variable expansions.[11]

In order to motivate the choice of the fast y variable let us first consider a straightforward multi-variable expansion where we adopt x and y as fast variables and $\tilde{y} = \delta y$ as the slow variable. This means that in our formulas, partial derivatives with respect to y are replaced by $\partial/\partial y + \delta(\partial/\partial \tilde{y})$.

[10] It is shown in Ref. 5.3.3 that for the problem of an accelerating airfoil, if one considers the flow near a caustic, the appropriate model is one with a small wind shear. The problem of supersonic thin airfoil theory in an atmosphere with a slowly varying temperature is mathematically identical to the present problem and is sketched in Reference 5.3.5.

[11] Note that $\tilde{y}$ in this section is not the same as that used in Section 5.3.2 for transonic flow.

The equations to be solved to first order become

$$u_0\rho_{1x} + u_{1x} + v_{1y} = 0 \tag{5.3.61a}$$

$$M_\infty^2 u_0 u_{1x} + \rho_{1x} = 0 \tag{5.3.61b}$$

$$M_\infty^2 u_0 v_{1x} + \rho_{1y} = 0. \tag{5.3.61c}$$

We integrate Equation (5.3.61b) with respect to x and note that the resulting function of y and $\tilde{y}$ appearing from the integration must be identically equal to zero because both u_1 and ρ_1 vanish at $x = -\infty$ and all y. This follows from the fact that the boundary conditions of u and ρ at $x = -\infty$ are satisfied by the unperturbed solution. Using the result

$$M_\infty^2 u_0 u_1 + \rho_1 = 0 \tag{5.3.62}$$

to eliminate ρ_1 from Equation (5.3.61a) and (5.3.61b), then solving these for u_1 (a similar result is obtained for v_1) we find that

$$u_{1yy} - (M_\infty^2 u_0^2 - 1)u_{1xx} = 0. \tag{5.3.63}$$

Equation (5.3.63) is a wave equation with a slowly varying characteristic "speed",

$$\beta_0 = \sqrt{M_\infty^2 u_0^2 - 1} \equiv \beta_0(\tilde{y}) \tag{5.3.64}$$

as long as $u_0 > 1/M_\infty$, i.e., for the altitude range $0 \geq y \geq -[(M_\infty - 1)/(\delta M_\infty c_1)]$.

In analogy with the problem of an oscillator with slowly varying frequency [cf. Section 3.3.2] it is clear that y^+ (rather than y) defined by

$$\frac{dy^+}{dy} = \beta_0(\tilde{y}) \tag{5.3.65a}$$

or

$$y^+ = \frac{1}{\delta}\int_0^{\tilde{y}} \sqrt{M_\infty^2 u_0^2(s) - 1}\, ds \tag{5.3.65b}$$

is an appropriate fast variable.

In the above the quantity $M_\infty u_0(\tilde{y})$ is the local Mach number which varies slowly with altitude. Denoting

$$M_\infty u_0(\tilde{y}) = M_0(\tilde{y}) \tag{5.3.66}$$

we can integrate Equation (5.3.65b) to define y^+ as a function of y as follows:

$$y^+ = y_0^+ + \frac{1}{2M_\infty \delta c_1}\{M_0\sqrt{M_0^2 - 1} - \cosh^{-1}(M_0)\}, \tag{5.3.67}$$

where

$$y_0^+ = -\frac{1}{2M_\infty \delta c_1}\{M_\infty\sqrt{M_\infty^2 - 1} - \cosh^{-1}(M_\infty)\}. \tag{5.3.68}$$

Thus, $y^+ = 0$ when $y = 0$, and near $y = 0$ it is easily seen that

$$y^+ = \sqrt{M_\infty^2 - 1}\, y + O(\delta). \tag{5.3.69}$$

To summarize, we see that the presence of a wind gradient and the need for uniformity in the far-field, dictate the necessity of a $\tilde{y}$ dependence in the solution. In addition, the wind gradient introduces the slowly varying coefficient $\beta_0(\tilde{y})$ into the equations and this dictates the choice of the y^+ variable.

Supersonic Solution, Outer Expansion

Based on the above discussion, we now seek to express the solution in the form

$$u(x, y; \delta) = u_0(\tilde{y}) + \delta u_1(x, y^+, \tilde{y}) + \delta^2 u_2(x, y^+, \tilde{y}) + \cdots \quad (5.3.70a)$$

$$v(x, y; \delta) = \delta v_1(x, y^+, \tilde{y}) + \delta^2 v_2(x, y^+, \tilde{y}) + \cdots \quad (5.3.70b)$$

$$\rho(x, y; \delta) = 1 + \delta \rho_1(x, y^+, \tilde{y}) + \delta^2 \rho_2(x, y^+, \tilde{y}) + \cdots. \quad (5.3.70c)$$

In this study we specifically exclude the occurrence of subsonic regions by introducing a horizontal boundary (the ground) at some value of $y = y_g < 0$ to be determined later such that M_0 is strictly greater than unity in the flow field everywhere above y_g. We do anticipate however allowing M_0 to approach arbitrarily close to unity thus necessitating consideration of a transonic flow regime.

The boundary conditions on either side of the airfoil ($y = \pm\delta F(x)$) are given by

$$\frac{v(x, \pm\delta F(x); \delta)}{u(x, \pm\delta F(x), \delta)} = \pm\delta F'(x) \quad (5.3.71)$$

while on the ground, we have

$$v(x, y_g; \delta) = 0. \quad (5.3.72)$$

We use the fact that

$$\frac{\partial}{\partial y} = \beta_0(\tilde{y}) \frac{\partial}{\partial y^+} + \delta \frac{\partial}{\partial \tilde{y}} \quad (5.3.73)$$

in Equations (5.3.59) to calculate the following governing equations to first and second order:

$$u_0 \rho_{1x} + u_{1x} + \beta_0 v_{1y^+} = 0, \quad (5.3.74a)$$

$$M_\infty^2 u_0 u_{1x} + \rho_{1x} = 0, \quad (5.3.74b)$$

$$M_\infty^2 u_0 v_{1x} + \beta_0 \rho_{1y^+} = 0, \quad (5.3.74c)$$

$$u_0 \rho_{2x} + u_{2x} + \beta_0 v_{2y^+} = (\rho_1 u_1)_x + \beta_0(\rho_1 v_1)_{y^+} - v_{1\tilde{y}}, \quad (5.3.75a)$$

$$u_0 u_{2x} + \rho_{2x}/M_\infty^2 = -u_1 u_{1x} - \beta_0 v_1 u_{1y^+} - (\gamma - 2)\rho_1 \rho_{1x}/M_\infty^2 - v_1 \frac{du_0}{d\tilde{y}}, \quad (5.3.75b)$$

$$\begin{aligned} u_0 v_{2x} + \beta_0 \rho_{2y^+}/M_\infty^2 = &-u_1 v_{1x} - \beta_0 v_1 v_{1y^+} \\ &- (\gamma - 2)\beta_0 \rho_1 \rho_{1y^+}/M_\infty^2 - \rho_{1\tilde{y}}/M_\infty^2. \end{aligned} \quad (5.3.75c)$$

Consider first the solution to order δ. Equation (5.3.74b) can be integrated with respect to x, and, as discussed in connection with Equation (5.3.62), the resulting function of y^+ and $\tilde{y}$ on the right-hand side must be set equal to zero to find

$$M_\infty^2 u_0 u_1 + \rho_1 = 0. \tag{5.3.76}$$

This will define ρ_1, once u_1 is known.

Substituting $\rho_1 = -M_\infty^2 u_0 u_1$ into Equations (5.3.74a) and (5.3.74c) gives the hyperbolic system (as long as $\beta_0 \neq 0$)

$$\beta_0^2 u_{1x} - \beta_0 v_{1y^+} = 0 \tag{5.3.77a}$$

$$\beta_0 u_{1y^+} - v_{1x} = 0 \tag{5.3.77b}$$

which has the general solution

$$u_1 = [g(\sigma, \tilde{y}) - f(\xi, \tilde{y})]/\beta_0(\tilde{y}), \tag{5.3.78a}$$

$$v_1 = g(\sigma, \tilde{y}) + f(\xi, \tilde{y}). \tag{5.3.78b}$$

In the above, the variables σ and ξ are defined by

$$\sigma = x + y^+, \tag{5.3.79a}$$

$$\xi = x - y^+ \tag{5.3.79b}$$

and would be the characteristic coordinates of the system, Equations (5.3.77), if β_0 were constant. The fact that σ and ξ are characteristics of the $\beta_0 =$ const. case is actually of no importance, because as we shall see later on, whether $\beta_0 =$ cost. or $\beta_0 = \beta_0(\tilde{y})$, the solution is defined along a certain family of curves which reduces to the family $\sigma =$ const. only as $\tilde{y} \to 0$. The functions f and g are as yet unknown and will be determined by requiring the solution to $O(\delta^2)$ to be consistent.

To carry out the solution to order δ^2, we denote the right-hand sides of Equations (5.3.75a), (5.3.75b), and (5.3.75c) by E_1, E_2 and E_3 respectively. Next, we integrate Equation (5.3.75b) with respect to x in the form[12]

$$u_0 u_2 + \rho_2/M_\infty^2 = \int_{-\infty}^{x} E_2(s, y^+, \tilde{y})ds. \tag{5.3.80}$$

Using the above to eliminate ρ_2 from Equations (5.3.75a) and (5.3.75c) gives

$$v_{2_x} - \beta_0 u_{2_{y^+}} = \frac{1}{u_0}\left[E_3 - \beta_0 \int_{-\infty}^{x} \frac{\partial E_2(s, y^+, \tilde{y})}{\partial y^+} ds\right] \equiv H_3 \tag{5.3.81a}$$

$$v_{2_{y^+}} - \beta_0 u_{2_x} = \frac{1}{\beta_0}[E_1 - u_0 M_\infty^2 E_2] \equiv H_1. \tag{5.3.81b}$$

[12] As before, the function of y^+ and $\tilde{y}$ arising from the integration must be set equal to zero in order to satisfy the boundary condition at $x = -\infty$.

The solution of this system can also be expressed conveniently in terms of the σ and ξ variables in the form

$$v_2(\sigma, \xi, \tilde{y}) = k(\sigma, \tilde{y}) + h(\xi, \tilde{y}) + I(\sigma, \xi, \tilde{y}) + J(\sigma, \xi, \tilde{y}) \quad (5.3.82a)$$

$$u_2(\sigma, \xi, \tilde{y}) = \frac{k(\sigma, \tilde{y}) - h(\xi, \tilde{y})}{\beta_0} + \frac{J(\sigma, \xi, \tilde{y})}{\beta_0} - \frac{I(\sigma, \xi, \tilde{y})}{\beta_0}, \quad (5.3.82b)$$

where k and h are arbitrary functions, and I and J are the indefinite integrals defined by

$$\frac{\partial I}{\partial \sigma} = \frac{1}{4}[H_3(\sigma, \xi, \tilde{y}) + H_1(\sigma, \xi, \tilde{y})] \quad (5.3.83a)$$

$$\frac{\partial J}{\partial \xi} = \frac{1}{4}[H_3(\sigma, \xi, \tilde{y}) - H_1(\sigma, \xi, \tilde{y})]. \quad (5.3.83b)$$

Thus, once f and g are known one can successively calculate E_1, E_2, E_3, then H_1, H_3 and, finally, I and J as functions of σ, ξ and $\tilde{y}$.

Actually, the situation is considerably simpler. Consider the region marked "incident supersonic" in Figure 5.3.2. This region is bounded by the lower surface of the airfoil, the bow and trailing edge shocks and extends downward over all values of y for which the supersonic solution is valid. (This domain of validity will be established more precisely later on.)

It is proved in Reference (5.3.5) that a wave of order δ traveling along a σ = constant (ξ = constant) curve can only *reflect from a shock as a wave of order* δ^2 traveling along a ξ = constant (σ = constant) curve. This is a generalization of the result given in Reference (5.3.6) for a uniform flow upstream where the corresponding reflected wave is of order δ^3.

Now, in the incident supersonic region the disturbances u_1, v_1, ρ_1 generated by the airfoil propagate along the σ = constant curves. These disturbances vary slowly with $\tilde{y}$ due to the wind gradient and the effect of higher order terms. As in the case of no wind gradient, u_1, v_1, and ρ_1 cannot depend on the fast scale ξ since such a dependence could only arise as a reflection from the bow shock, and these reflections, now being of order δ^2, are accounted in the u_2, v_2 and ρ_2 terms.

Thus, if we confine our attention to the incident supersonic region (we can only consider the reflected supersonic region after the transonic flow problem has been solved) we must set $f(\xi, \tilde{y}) = 0$.

Since H_1 and H_3 now depend only on σ and $\tilde{y}$, the requirement that the solution to $O(\delta^2)$ be consistent is easily deduced. We first note that J is of the form

$$J = \tfrac{1}{4}\xi[H_3(\sigma, \tilde{y}) - H_1(\sigma, \tilde{y})] \quad (5.3.84)$$

and must be set equal to zero because it becomes unbounded for large ξ. On the other hand, since the flow variables vanish outside the incident supersonic region I is a bounded function of σ and must be retained in the solution.

The condition $J = 0$ or $H_1 = H_3$ means that

$$\frac{E_3}{u_0} - \frac{E_1}{\beta_0} + \frac{u_0 M_\infty^2}{\beta_0} E_2 - \frac{\beta_0}{u_0} \int_{-\infty}^{x} \frac{\partial E_2}{\partial y^+} dx = 0. \tag{5.3.85}$$

Since $E_2 = E_3(\sigma, \tilde{y})$,

$$\frac{\partial E_2}{\partial y^+} dx = \frac{\partial E_2}{\partial \sigma} \frac{\partial \sigma}{\partial y^+} \frac{dx}{d\sigma} d\sigma = \frac{\partial E_2(\sigma, \tilde{y})}{\partial \sigma} d\sigma.$$

Hence, Equation (5.3.85) becomes

$$E_3 + \frac{E_2}{\beta_0} - u_0 \frac{E_1}{\beta_0} = 0. \tag{5.3.86}$$

If we now use the fact that $f = 0$, Equations (5.3.78) give

$$u_1 = \frac{g(\sigma, \tilde{y})}{\beta_0(\tilde{y})} \tag{5.3.87a}$$

$$v_1 = g(\sigma, \tilde{y}) \tag{5.3.87b}$$

and we can calculate the E_i in the form

$$E_1 = \frac{\beta_0}{u_0}\left\{\frac{(1 + \beta_0^2)^2}{\beta_0^3} \frac{\partial g^2}{\partial \sigma} - \frac{u_0}{\beta_0} \frac{\partial g}{\partial \tilde{y}}\right\} \tag{5.3.88a}$$

$$E_2 = \beta_0\left\{-\frac{(\gamma - 1)(1 + \beta_0^2)}{2\beta_0^3} \frac{\partial g^2}{\partial \sigma} - \frac{g}{\beta_0} \frac{du_0}{d\tilde{y}}\right\} \tag{5.3.88b}$$

$$E_3 = -\frac{(\gamma - 1)(1 + \beta_0^2)}{2\beta_0} \frac{\partial g^2}{\partial \sigma} + \frac{\partial}{\partial \tilde{y}} \frac{(u_0 g)}{\beta_0}. \tag{5.3.88c}$$

Substituting the above into Equation (5.3.86) gives the following quasi-linear first order partial differential equation governing g:

$$-\frac{(\gamma + 1)(1 + \beta_0^2)^2}{\beta_0^3} g \frac{\partial g}{\partial \sigma} + \frac{2u_0}{\beta_0} \frac{\partial g}{\partial \tilde{y}} = \frac{g}{\beta_0} \frac{du_0}{d\tilde{y}} - g \frac{d}{d\tilde{y}}\left(\frac{u_0}{\beta_0}\right). \tag{5.3.89}$$

If θ is a parameter along the characteristics of Equation (5.3.89) [We distinguish between the characteristics of Equation (5.3.89) and the $\sigma =$ constant curves defined by Equations (5.3.79a) and (5.3.67). As we will see presently, the characteristics of the first order equation, Equation (5.3.89), have a much more important role in the solution.], then along these characteristics σ, $\tilde{y}$ and g, obey

$$\frac{d\sigma}{d\theta} = -\frac{(\gamma + 1)(1 + \beta_0^2)^2}{\beta_0^3} g \tag{5.3.90a}$$

$$\frac{d\tilde{y}}{d\theta} = \frac{2u_0}{\beta_0} \tag{5.3.90b}$$

$$\frac{dg}{d\theta} = \frac{g}{\beta_0} \frac{du_0}{d\tilde{y}} - g \frac{d}{d\tilde{y}}\left(\frac{u_0}{\beta_0}\right). \tag{5.3.90c}$$

The above can be solved to give the remarkably simple result that

$$g = g_0 \beta_0^{1/2} \tag{5.3.91a}$$

along the characteristics defined by

$$\sigma + \frac{\gamma + 1}{c_1} g_0 [\beta_0^{1/2} + \beta_0^{5/2}/5] = \sigma_0 = \text{const.} \tag{5.3.91b}$$

The value of the constant of integration σ_0 can be related to x_0, the value of x at $y = 0$, by setting $y = 0$ and $x = x_0$ above. Hence

$$\sigma_0 = x_0 + \frac{\gamma + 1}{c_1} g_0 [\beta_\infty^{1/2} + \beta_\infty^{5/2}/5], \tag{5.3.91c}$$

where

$$\beta_\infty^2 = M_\infty^2 - 1. \tag{5.3.91d}$$

If we use the fact that $\sigma = x + y^+$, and that y^+ is defined by Equation (5.3.67) in terms of $\tilde{y}$, the characteristic curve emanating from the point $y = 0$, $x = x_0$, has the explicit form

$$x_0 = x + \frac{1}{2M_\infty \delta c_1} \{M_0 \beta_0 - M_\infty \beta_\infty - \cosh^{-1}(M_0) + \cosh^{-1}(M_\infty)\}$$
$$+ \frac{\gamma + 1}{c_1} g_0 \left\{ \beta_0^{1/2} - \beta_\infty^{1/2} + \frac{1}{5} [\beta_0^{5/2} - \beta_\infty^{5/2}] \right\}. \tag{5.3.92}$$

It is easy to see that as $\tilde{y} \to 0$, i.e., near the airfoil, Equation (5.3.92) reduces to $x_0 = x + \beta_\infty y$. Thus, the characteristics of the reduced equation, Equation (5.3.89), are initially coincident with the $\sigma =$ const. curves.

It also follows from the above that in the limit $c_1 \to 0$, i.e., for a uniform freestream, the characteristics of Equation (5.3.89) tend to the straight lines

$$x + \beta_\infty y + \frac{(\gamma + 1)M_\infty^4}{2\beta_\infty^{3/2}} g_0 \tilde{y} = x_0 = \text{const.} \tag{5.3.93}$$

on which $g = g_0 =$ const. Had one used a coordinate straining technique, as first proposed in Reference (5.3.7) for this problem, Equation (5.3.93) would have appeared in the form of a "strained characteristic"

$$x + \left[\beta_\infty + \frac{\delta(\gamma + 1)M_\infty^4}{2\beta_\infty^{3/2}} g_0 + \cdots \right] y = \text{const.}, \tag{5.3.94}$$

where the term of order δ represents a correction (strain) to the characteristic $x + \beta_\infty y$ of the unperturbed problem. As pointed out in Reference (5.3.8) this idea is correct to order δ but cannot be carried out to higher orders; one must eventually use multiple scales.

We also note that for $c_1 = 0$, Equation (5.3.89) has a zero right-hand side. This equation was derived in Reference (5.3.9) using physical arguments and in Reference (5.3.10) by multiple variable expansions. Some examples of airfoil shapes using the homogeneous Equation (5.3.89) are given in Reference (5.3.9).

With $c_1 \neq 0$, the characteristic curves of Equation (5.3.89) are more complicated and one must use a multiple variable expansion right away. Moreover, g_0 now varies along these curves according to Equation (5.3.91a). The value of g_0 is given by the boundary condition on the airfoil, and we consider this next.

Expanding Equation (5.3.71) gives

$$v_1(x, 0, 0) = -F'(x) \tag{5.3.95a}$$

$$v_2(x, 0, 0) = F(x)\frac{\delta v_1}{\partial y}(x, 0, 0) - u_1(x, 0, 0)v_1(x, 0, 0), \tag{5.3.95b}$$

etc.

If we now use Equations (5.3.87b) and (5.3.91a) to evaluate v_1 at $y^+ = \tilde{y} = 0$ and set the result equal to the right-hand side of Equation (5.3.95a), we find that

$$g_0(x_0)\beta_\infty^{1/2} = -F'(x_0), \qquad -\tfrac{1}{2} \leq x_0 \leq \tfrac{1}{2}. \tag{5.3.96}$$

Thus,

$$g_0(x_0) = -F'(x_0)\beta_\infty^{-1/2} \tag{5.3.97}$$

and this defines the function g completely.

Using the above, the solution to order δ is given by

$$u_1 = -\frac{F'(x_0)}{(\beta_\infty \beta_0)^{1/2}} \tag{5.3.98a}$$

$$v_1 = -\left(\frac{\beta_0}{\beta_\infty}\right)^{1/2} F'(x_0) \tag{5.3.98b}$$

$$\rho_1 = M_\infty^2 u_0 \frac{F'(x_0)}{(\beta_\infty \beta_0)^{1/2}} \tag{5.3.98c}$$

$$p_1 = \frac{\gamma M_\infty^2 u_0 F'(x_0)}{(\beta_0 \beta_\infty)^{1/2}}. \tag{5.3.98d}$$

To map out the solution in the incident supersonic region, we note that through each point x_0 on the interval $-\frac{1}{2} \leq x_0 \leq \frac{1}{2}$, there emanates a curve $x_0 = x_0(x, y)$ defined by Equation (5.3.92). One each of the curves $x_0 =$ const. the flow quantities are defined by Equations (5.3.98) where $F'(x_0)$ is given by the airfoil shape.

The characteristics $x_0 =$ const. may envelope for certain choices of an airfoil shape. In this event one must introduce a shock in the flow field. For a normal convex airfoil, this occurrence is unlikely, and will not be considered here. Thus, we restrict our attention to airfoil shapes which only have a leading and trailing edge shock. These can be fitted in once the solution for u_1, v_1 are used with the appropriate shock jump conditions. It is shown in Reference (5.3.5) that the shock jump conditions imply that the shock slope is given by (this is the analogue of Equation (5.1.59))

$$\left(\frac{dy}{dx}\right)_{\text{shock}} = -\frac{1}{\beta_0} - \delta\frac{\gamma + 1}{4}\frac{M_0^4}{\beta_0^4}v_1 + O(\delta^2). \tag{5.3.99}$$

Therefore, knowing v_1, we can integrate Equation (5.3.99) with $y = 0$ at $x = \pm\frac{1}{2}$ to obtain the leading and trailing edge shocks.

We also note that as $\beta_0 \to 0$, i.e., as $y \to y_s = -[(M_\infty - 1)/\delta M_\infty c_1]$, u_1, ρ_1 and p_1 becomes singular. Thus, the outer expansion is only valid in the strictly supersonic region $(y + y_s)/y_s \gg 1$. If we chose some value of $y_g > y_s$ but such that $y_g/y_s \approx 1$, it is clear that our solution will fail near $y = y_g$. The solution in this transonic region will be considered next.

Transonic Solution, Inner Expansion

The underlying reasons for the failure of the supersonic solution, Equations (5.3.98), are very much the same as those discussed earlier in Section 5.3.2 for the case of uniform flow upstream ($c_1 = 0$). Since the supersonic solution now becomes singular near $y = y_s = -(M_\infty - 1)/\delta c_1 M_\infty$, we introduce an inner variable $\bar{y}$ which is a rescaled version of $y - y_s$

$$\bar{y} = \nu(\delta)\left(y + \frac{M_\infty - 1}{\delta M_\infty}\right), \qquad \nu \ll 1. \tag{5.3.100}$$

We have set $c_1 = 1$ for simplicity, and we anticipate that an order of magnitude analysis will give $\nu = \delta^{1/3}$, cf. Equations (5.3.32) and (5.3.39).

The situation in this problem is also analogous to that encountered in Sections 3.5.4 and 3.5.5 for passage through resonance in the sense that the outer expansion becomes invalid as a given slowly varying coefficient in the equations approaches a critical value. Moreover, as we shall see presently, here also, the inner region is large on the y scale but small on the $\bar{y}$ scale.

As in Section 5.3.2, we also rescale the dependent variables and assume expansions of the form

$$u(x, y; \delta) = u_0 + \alpha_1(\delta)\bar{u}_1(x, \bar{y}) + \cdots \tag{5.3.101a}$$

$$v(x, y; \delta) = \beta_1(\delta)\bar{v}_1(x, \bar{y}) + \cdots \tag{5.3.101b}$$

$$\rho(x, y; \delta) = \gamma_1(\delta)\bar{\rho}_1(x, \bar{y}) + \cdots, \tag{5.3.101c}$$

where

$$u_0 = 1 + \delta y = \frac{1}{M_\infty} + \frac{\delta}{\nu}\bar{y}. \tag{5.3.101d}$$

For the purposes of deriving the transonic equation it is more convenient to start with Equation (5.3.7) which follows from the continuity and momentum equations

$$(u^2 - a^2)u_x + (u_y + v_x)uv + (v^2 - a^2)v_y = 0. \tag{5.3.102}$$

The term $u^2 - a^2$ which is now small must be computed carefully. We use the uniformly valid result from Equation (5.3.58a) that $a^2 = \rho^{\gamma-1}/M_\infty^2 + O(\delta^3)$ and expand u and ρ using Equations (5.3.101a) and (5.3.101c) to calculate

$$a^2 = \frac{1}{M_\infty^2}[1 + (\gamma - 1)\bar{\rho}_1\gamma_1(\delta) + \cdots] \tag{5.3.103}$$

and

$$u^2 - a^2 = \frac{1}{M_\infty^2}\left[2M_\infty \bar{y}\frac{\delta}{v} + 2\bar{u}_1 M_\infty \alpha_1 - (\gamma - 1)\bar{\rho}_1\gamma_1 + \cdots\right]. \tag{5.3.104}$$

The most general result corresponds to the choice

$$\frac{\delta}{v} = \alpha_1(\delta) = \gamma_1(\delta). \tag{5.3.105}$$

With this choice Equation (5.3.102) has the following expansion

$$\frac{\alpha_1^2}{M_\infty^2}[2M_\infty \bar{y} + 2M_\infty \bar{u}_1 - (\gamma - 1)\bar{\rho}_1]\bar{u}_{1_x} - \frac{\beta_1 v}{M_\infty^2}\bar{v}_{1_{\bar{y}}}$$
$$+ \frac{\delta\beta_1}{M_\infty}\bar{v}_1(1 + \bar{u}_{1_{\bar{y}}}) + \beta_1^2 M_\infty \bar{v}_1\bar{v}_{1_x} + \cdots = 0. \tag{5.3.106}$$

Let us assume (this will be verified later) that $\delta\beta_1 \ll \alpha_1^2$ and $\beta_1 \ll \alpha_1$. Then, in order to have the richest equation in the limit we must set

$$\alpha_1^2 = \beta_1 v. \tag{5.3.107}$$

Using Equations (5.3.105) we conclude that

$$\beta_1 = \alpha_1^3/\delta \tag{5.3.108a}$$

$$v = \delta/\alpha_1 \tag{5.3.108b}$$

$$\gamma_1 = \alpha_1. \tag{5.3.108c}$$

Thus, once α_1 is determined, we can calculate all the other scales. To determine α_1 let us consider the expanded form of the y-momentum equation

$$\frac{\alpha_1^3}{\delta M_\infty}\bar{v}_{1_x} + \frac{\alpha_1^5}{\delta}\bar{v}_1\bar{v}_{1_{\bar{y}}} = -\frac{\delta}{M_\infty^2}\bar{\rho}_{1_{\bar{y}}} + \cdots. \tag{5.3.109}$$

Clearly, we must set $\alpha_1^3/\delta = \delta$, and this gives

$$\alpha_1 = \delta^{2/3} \tag{5.3.110a}$$

$$\beta_1 = \delta \tag{5.3.110b}$$

$$\nu = \delta^{1/3} \tag{5.3.110c}$$

$$\gamma_1 = \delta^{2/3}. \tag{5.3.110d}$$

We now verify that $\beta_1\delta \ll \alpha_1^2$ and $\beta_1 \ll \alpha_1$ as we assumed earlier.

To derive the governing equations, we note once again that the continuity equation to order $\delta^{2/3}$ may be integrated with respect to x and the resulting integration function of $\bar{y}$ set equal to zero, to obtain

$$\frac{\bar{\rho}_1}{M_\infty} + \bar{u}_1 = 0. \tag{5.3.111}$$

Equations (5.3.109) and (5.3.106) become

$$\bar{v}_{1_x} + \frac{1}{M_\infty}\bar{\rho}_{1_{\bar{y}}} = 0 \tag{5.3.112a}$$

$$[2M_\infty\bar{y} + 2M_\infty\bar{u}_1 + (\gamma - 1)\bar{\rho}_1]\bar{u}_{1_x} - \bar{v}_{1_{\bar{y}}} = 0. \tag{5.3.112b}$$

Eliminating $\bar{\rho}_1$ in favor of $\bar{u}_1$ using Equation (5.3.111) gives

$$\bar{v}_{1_x} - \bar{u}_{1_{\bar{y}}} = 0 \tag{5.3.113a}$$

$$M_\infty[2\bar{y} + (\gamma + 1)\bar{u}_1]\bar{u}_{1_x} - \bar{v}_{1_{\bar{y}}} = 0. \tag{5.3.113b}$$

and the terms which have been neglected are $O(\delta^{4/3})$. Equation (5.3.113a) implies that to this approximation the flow is irrotational so we introduce the potential $\bar{\phi}(x, \bar{y})$ defined by

$$\bar{\phi}_x = \bar{u}_1 \tag{5.3.114a}$$

$$\bar{\phi}_{\bar{y}} = \bar{v}_1 \tag{5.3.114b}$$

and can write Equations (5.3.113) in the form

$$M_\infty[2\bar{y} + (\gamma + 1)\bar{\phi}_x]\bar{\phi}_{xx} - \bar{\phi}_{\bar{y}\bar{y}} = 0 \tag{5.3.115}$$

which is the generalization of Equation (5.3.44) to the case of flow in a wind shear.

It is convenient to transform Equation (5.3.115) into the conventional form

$$(\bar{y} + \bar{\varphi}_{\bar{x}})\bar{\varphi}_{\bar{x}\bar{x}} - \bar{\varphi}_{\bar{y}\bar{y}} = 0 \tag{5.3.116}$$

by introducing the rescaled variables

$$\bar{x} = x/\sqrt{2M_\infty} \tag{5.3.117a}$$

$$\bar{\varphi} = (\gamma + 1)\bar{\phi}/2^{3/2}M_\infty^{1/2}. \tag{5.3.117b}$$

The details of the matching between the supersonic and transonic solutions can be found in References (5.3.3) and (5.3.5). Here we only sketch the results.

Since both the incident supersonic solution and the transonic solution in the matching region do not involve reflected waves to the orders considered, we can carry out the matching along the characteristics $x_0 = \text{const.}$ of Equation (5.3.92) and their counterparts for Equation (5.3.116). These latter are calculated in the form

$$\lambda = \bar{x} + \tfrac{2}{3}\bar{y}^{3/2} + A(\lambda)\bar{y}^{1/4} + \cdots, \tag{5.3.118a}$$

where we need only consider solutions of the form

$$\bar{\varphi} = \bar{\varphi}(\lambda) \tag{5.3.118b}$$

of Equation (5.3.116).

In Equation (5.3.118a) the first two terms correspond to the characteristics of the linear transonic equation (with the $\bar{\varphi}_{\bar{x}}$ term missing) and the $A(\lambda)\bar{y}^{1/4}$ term is the correction introduced by the nonlinearity.

The matching determines the function $A(\lambda)$ in terms of the airfoil shape, and the resulting far field solution for the transonic equation can be used as a boundary condition for solving Equation (5.3.116).

References

5.3.1 J. D. Cole and A. F. Messiter, Expansion procedures and similarity laws for transonic flow, *Zeitschrift für Angewandte Mathematik und Physik* **8**, 1957, pp. 1–25.

5.3.2 J. D. Cole, Modern developments in transonic flow, *S.I.A.M. Journal on Applied Mathematics* **29**, 1975, pp. 763–787.

5.3.3 G. Pechuzal and J. Kevorkian, Supersonic-transonic flow generated by a thin airfoil in a stratified atmosphere, *S.I.A.M. Journal on Applied Mathematics* **33**, 1977, pp. 8–33.

5.3.4 W. D. Hayes and R. F. Probstein, *Hypersonic Flow Theory*, 2nd Ed., Vol. 1, Academic Press, New York, 1966.

5.3.5 G. Pechuzal, Transition to transonic flow in the far field for a supersonic airfoil in a stratified atmosphere, PhD. Thesis, University of Washington, Seattle, Washington 98195, 1975.

5.3.6 M. J. Lighthill, *Higher Approximations. High Speed Aerodynamics and Jet Propulsion, Vol. 6, General Theory of High Speed Aerodynamics*, Ed. by W. R. Sears, Chapter E, pp. 345–489. Princeton University Press, 1954.

5.3.7 M. J. Lighthill, A technique for rendering approximate solutions to physical problems uniformly valid, *Philosophical Magazine* **40**, 1949, pp. 1179–1201.

5.3.8 L. Crocco, Coordinate perturbation and multiple scale in gas dynamics, *Philosophical Transactions of the Royal Society* **A272**, 1975, pp. 272–301.

5.3.9 W. D. Hayes, Pseudotransonic similitude and first-order wave structure, *Journal of the Aeronautical Sciences* **21**, 1954, pp. 721–730.

5.3.10 S. C. Chikwendu and J. Kevorkian, A perturbation method for hyperbolic equations with small nonlinearities, *S.I.A.M. Journal on Applied Mathematics* **22**, 1972, pp. 235–258.

Bibliography

C. M. Bender and S. A. Orszag, *Advanced Mathematical Methods for Scientists and Engineers*, McGraw-Hill, New York, 1978.

W. Eckhaus, *Matched Asymptotic Expansions and Singular Perturbations*. North-Holland, Amsterdam, 1973.

W. Eckhaus, Asymptotic Analysis of Singular Perturbations. North-Holland, Amsterdam, 1979.

S. Kaplun, *Fluid Mechanics and Singular Perturbations*. Ed. by P. A. Lagerstrom, L. N. Howard, and C. S. Liu, Academic, New York, 1967.

P. A. Lagerstrom and J. Boa, Singular Perturbations. To appear in *Encyclopedia of Mathematics and Its Applications*, Gian-Carlo Rota, Editor, Addison-Wesley, Reading, Massachusetts.

A. H. Nayfeh, *Perturbation Methods*, Wiley-Interscience, New York, 1973.

R. E. O'Malley, *Introduction to Singular Perturbations*, Academic Press, New York, 1974.

M. Van Dyke, *Perturbation Methods in Fluid Mechanics*, Parabolic Press, Palo Alto, California, 1975.

Author Index

Subject Index

Applied Mathematical Sciences

cont. from page ii

52. *Chipot:* Variational Inequalities and Flow in Porous Media.
53. *Majda:* Compressible Fluid Flow and System of Conservation Laws in Several Space Variables.
54. *Wasow:* Linear Turning Point Theory.
55. *Yosida:* Operational Calculus: A Theory of Hyperfunctions.
56. *Chang/Howes:* Nonlinear Singular Perturbation Phenomena: Theory and Applications.
57. *Reinhardt:* Analysis of Approximation Methods for Differential and Integral Equations.
58. *Dwoyer/Hussaini/Voigt (eds):* Theoretical Approaches to Turbulence.
59. *Sanders/Verhulst:* Averaging Methods in Nonlinear Dynamical Systems.
60. *Ghil/Childress:* Topics in Geophysical Dynamics: Atmospheric Dynamics, Dynamo Theory and Climate Dynamics.
61. *Sattinger/Weaver:* Lie Groups and Algebras with Applications to Physics, Geometry, and Mechanics.
62. *LaSalle:* The Stability and Control of Discrete Processes.
63. *Grasman:* Asymptotic Methods of Relaxation Oscillations and Applications.
64. *Hsu:* Cell-to-Cell Mapping: A Method of Global Analysis for Nonlinear Systems.
65. *Rand/Armbruster:* Perturbation Methods, Bifurcation Theory and Computer Algebra.
66. *Hlavácek/Haslinger/Necasl/Lovísek:* Solution of Variational Inequalities in Mechanics.
67. *Cercignani:* The Boltzmann Equation and Its Applications.
68. *Temam:* Infinite Dimensional Dynamical System in Mechanics and Physics.
69. *Golubitsky/Stewart/Schaeffer:* Singularities and Groups in Bifurcation Theory, Vol. II.
70. *Constantin/Foias/Nicolaenko/Temam:* Integral Manifolds and Inertial Manifolds for Dissipative Partial Differential Equations.
71. *Catlin:* Estimation, Control, and the Discrete Kalman Filter.
72. *Lochak/Meunier:* Multiphase Averaging for Classical Systems.
73. *Wiggins:* Global Bifurcations and Chaos.
74. *Mawhin/Willem:* Critical Point Theory and Hamiltonian Systems.
75. *Abraham/Marsden/Ratiu:* Manifolds, Tensor Analysis, and Applications, 2nd ed.
76. *Lagerstrom:* Matched Asymptotic Expansions: Ideas and Techniques.
77. *Aldous:* Probability Approximations via the Poisson Clumping Heuristic.
78. *Dacorogna:* Direct Methods in the Calculus of Variations.
79. *Hernández-Lerma:* Adaptive Markov Processes.
80. *Lawden:* Elliptic Functions and Applications.
81. *Bluman/Kumei:* Symmetries and Differential Equations.
82. *Kress:* Linear Integral Equations.
83. *Bebernes/Eberly:* Mathematical Problems from Combustion Theory.
84. *Joseph:* Fluid Dynamics of Viscoelastic Fluids.
85. *Yang:* Wave Packets and Their Bifurcations in Geophysical Fluid Dynamics.
86. *Dendrinos/Sonis:* Chaos and Socio-Spatial Dynamics.
87. *Weder:* Spectral and Scattering Theory for Wave Propagation in Perturbed Stratified Media.
88. *Bogaevski/Povzner:* Algebraic Methods in Nonlinear Perturbation Theory.
89. *O'Malley:* Singular Perturbation Methods for Ordinary Differential Equations.
90. *Meyer/Hall:* Introduction to Hamiltonian Dynamical Systems and the N-body Problem.
91. *Straughan:* The Energy Method, Stability, and Nonlinear Convection.